ROUTLEDGE HANDBOOK OF NATIONAL AND REGIONAL OCEAN POLICIES

This comprehensive handbook, prepared by leading ocean policy academics and practitioners from around the world, presents in-depth analyses of the experiences of fifteen developed and developing nations (Australia, Brazil, Canada, India, Jamaica, Japan, Mexico, New Zealand, Norway, Philippines, Portugal, Russian Federation, United Kingdom, United States, and Vietnam) and four regions of the world (East Asian Seas, European Union, Pacific Islands, and Sub-Saharan Africa) that have taken concrete steps toward cross-cutting and integrated national and regional ocean policy.

All chapters follow a common framework for policy analysis. While most coastal nations of the world already have a variety of sectorial policies in place to manage different uses of the ocean (such as shipping, fishing, and oil and gas development), in the last two decades the coastal nations and regions covered in this book have undertaken concerted efforts to articulate and implement an integrated, ecosystem-based vision for the governance of ocean areas under their jurisdiction. This includes goals and procedures to harmonize existing uses and laws, to foster sustainable development of ocean areas, to protect biodiversity and vulnerable resources and ecosystems, and to coordinate the actions of the many government agencies that are typically involved in oceans affairs.

The book highlights the serious conflicts of use in most national ocean zones and the varying attempts by nations to follow the prescriptions emanating from the 1982 UN Convention on the Law of the Sea (UNCLOS) and the outcomes of the 1992, 2002, and 2012 sustainable development summits. The case studies analyse the motivations underlying the development of ocean policy, the evolution of the policy over time, and the implementation and evaluation of the policy. An overview chapter provides a comparative analysis of ocean policy experiences, drawing lessons on factors contributing to ocean policy success.

Providing a definitive state-of-the-art review and analysis of national and regional ocean policies, the book should be of interest to governmental, non-governmental, and private sector practitioners involved in ocean and coastal management around the world, as well as to graduate and undergraduate students in marine and environmental policy.

Biliana Cicin-Sain is Professor of Marine Policy and Director of the Gerard J. Mangone Center for Marine Policy, School of Marine Science and Policy, University of Delaware, United States, and President of the Global Ocean Forum.

David L. VanderZwaag is Professor of Law and Canada Research Chair in Ocean Law and Governance with the Marine and Environmental Law Institute, Dalhousie University, Canada.

Miriam C. Balgos is Associate Scientist with the Gerard J. Mangone Center for Marine Policy, University of Delaware, United States, and Program Coordinator for the Global Ocean Forum.

ROUTLEDGE HANDBOOK OF NATIONAL AND REGIONAL OCEAN POLICIES

Edited by
Biliana Cicin-Sain, David L. VanderZwaag, and
Miriam C. Balgos

Routledge
Taylor & Francis Group

LONDON AND NEW YORK

First published 2015
by Routledge
2 Park Square, Milton Park, Abingdon, Oxon OX14 4RN

and by Routledge
711 Third Avenue, New York, NY 10017

Routledge is an imprint of the Taylor & Francis Group, an informa business

British Library Cataloguing-in-Publication Data
A catalogue record for this book is available from the British Library

Library of Congress Cataloging-in-Publication Data
Routledge handbook of national and regional ocean policies / edited by Billiana Cicin-Sain, David L. VanderZwaag and Miriam C. Balgos.
pages cm
Includes bibliographical references and index.
1. Ocean—Government policy—Case studies. 2. Oceanography and state—Case studies. 3. Marine resources—Government policy—Case studies. 4. Marine resources conservation—International cooperation—Case studies. I. Cicin-Sain, Biliana, editor of compilation. II. VanderZwaag, David L., editor of compilation. III. Balgos, M. C. (Miriam C.) editor of compilation. IV. Title: Handbook of national and regional ocean policies.
GC1018.5.R68 2015
333.91'64—dc23
2014037782

ISBN: 978-1-138-78829-9 (hbk)
ISBN: 978-1-315-76564-8 (ebk)

Cover images:
Clockwise from top left: tomwang/123RF; Johan Swanepoel/123RF; donald_gruener/ istock; Flanders Marine Institute (VLIZ) Maritime Boundaries Database (background map); Miriam Balgos; Federico Rostagno/123RF; Miriam Balgos; johnandersonphoto/ istock; Michal Tutko/123RF; jennybonner/istock; Bilian Cicin-Sain (central)

Typeset in Bembo
by Swales & Willis Ltd, Exeter, Devon, UK

Printed and bound in the United States of America by Publishers Graphics, LLC on sustainably sourced paper.

DEDICATION

The *Routledge Handbook on National and Regional Ocean Policies* is dedicated to the work and the enduring legacy of two giants in the field of ocean governance: **Robert W. Knecht** and **Jon M. Van Dyke**. These global leaders exemplified the vision of ocean stewardship and inspired and led others, through their writings and their policy actions, to exercise ocean stewardship at global and national levels, and to protect and enhance the lives of coastal and island peoples everywhere. Not only were the two men very good colleagues with each other, they were very good personal friends as well.

Robert W. Knecht, became known as the "Father of Coastal Management," through his tenure as the first director of the US Coastal Zone Management Program, the first-ever such program in the world, where he pioneered the application of new concepts and approaches to the management of the ocean and coast. He applied and extended these insights to the international level and to many other nations as well. He was active in both the UN Law of the Sea negotiations and in the UN Conference on Environment and Development and was the author or co-author of many articles and books, all woven around the themes of equitable sustainable development, benefits to people, global and national ocean stewardship, and the imperative of public participation. He served as a professor of marine policy at the College of Marine Studies at the University of Delaware where he co-directed the Center for Marine Policy, mentoring a generation of students in marine policy and the public service. Among many honors, he was the recipient of the Julius A. Stratton Award for Leadership, a US award given biennially to the person or group that has made the greatest difference in leading the cause for the coast. Prior to his ocean policy career, Professor Knecht, trained as an upper atmospheric physicist, served as Deputy Director of the NOAA Environmental Research Laboratories in Boulder, Colorado. In Boulder, he served as Mayor, pioneering green initiatives to safeguard the city's beauty and character.

Jon M. Van Dyke was the Carlsmith Ball Professor of Law at the William S. Richardson School of Law at the University of Hawaii at Manoa, where he taught Constitutional Law, International Law, International Ocean Law, and International Human Rights. Professor Van Dyke co-authored the casebook *International Law and Litigation in the U.S.* and other books on international ocean law and governance including the *Freedom for the Seas in the 21st Century: Ocean Governance and Environmental Harmony*, which received the Harold and Margaret Sprout

Award from the International Studies Association as the best book on international environmental policy. He was one of the best known interpreters of the UN Convention on the Law of the Sea, always reinforcing the rule of law and pushing the limits of ocean stewardship. He was internationally recognized for his scholarship and leadership on the festering ocean boundary delimitations in Asia. As well, he engaged in litigation on important constitutional and international issues before state, federal and foreign courts and international tribunals. He worked tirelessly for the rights of Native Hawaiians and authored a book on their land rights. He made significant contributions to the advancement of jurisprudence in the Pacific Islands, a region and peoples he so loved.

While we miss their presence dearly, their vision and dedication continues to inspire and spur us to action.

Biliana Cicin-Sain, David L. VanderZwaag,
and Miriam C. Balgos

CONTENTS

Contents

FIGURES

TABLES

List of Tables

CONTRIBUTORS

Ben Addison is Regional Manager, Illawarra, for the New South Wales Office of Environment and Heritage. He was a project officer for the Marine Division of the Australian Government's Department of the Environment and Heritage, involved in the East Marine Bioregional Planning process covering more than 2 million km² waters off the east coast of Australia under the Environment Protection and Biodiversity Conservation Act 1999 and Australia's Oceans Policy. Ben worked for the Australian government for more than ten years, during which time he focused primarily on coastal and marine planning, marine protected areas, and the development and implementation of Australia's Oceans Policy.

Juan Carlos Aguilar is a biologist of the Coordination for Adaptation to Climate Change, National Institute of Ecology and Climate Change, Secretariat of Environment and Natural Resources, Mexico.

Porfirio Alvarez-Torres is Executive Secretary of the Mexican Consortium of Research Institutions of the Gulf of Mexico, which aligns over one hundred and twenty academic and research institutions from Mexico and the United States to contribute to the best management practices and sustainable development of the Gulf of Mexico region. Prior to his current position, Dr Alvarez was Chief Technical Adviser for the Gulf of Mexico Large Marine Ecosystem for both the United Nations Industrial Development Organization (UNIDO) and the Mexican Ministry of Environment and Natural Resources (SEMARNAT). He received the Gulf Guardian award from the US Environmental Protection Agency (EPA) in 2013. Dr Alvarez received a PhD in Fisheries Science from the Tokyo University of Fisheries.

Milton L. Asmus is an oceanographer with a PhD in Marine Sciences from the University of South Carolina. He is Professor at the Federal University of Rio Grande (Brazil) and Collaborator Professor at the Republic University (Uruguay). Dr Asmus is a member of the Iberoamerican Network for Integrated Coastal Management and the Global Ocean Forum. Dr Asmus chairs the board of the Forum do Mar, Brazil. His research interests include oceanography and coastal ecology, and his work focuses mainly on integrated coastal management, systems ecology, estuarine ecology, and port environmental management.

Isaac Andres Azuz Adeath is Professor and Researcher at the Engineering School and Coordinator of the Master's Program in Environment and Sustainable Development of the Center for Higher and Technical Education (CETYS) University. He has been EU Erasmus Mundus Professor (2010) and an invited professor at several Mexican universities, and is currently a member of the National Research System (2014–16). He obtained a PhD in Marine Sciences from the Polytechnic University of Catalonia, Spain, and was a Mexican official delegate at the 2002 World Summit on Sustainable Development. He has also edited and authored several publications, including *Climate Change in Mexico: A Coastal/Marine Approach* (University of Campeche, CETYS University and Government of Campeche, 2010) and *Coastal Management in Mexico* (University of Campeche, CETYS University and SEMARNAT, 2004).

Miriam C. Balgos is Associate Scientist at the College of Earth, Ocean, and Environment, University of Delaware, and Program Coordinator for the Global Ocean Forum. Miriam leads the Gerard J. Mangone Center for Marine Policy team in the organization and conduct of multi-stakeholder dialogues and policy analyses in integrated ocean and coastal management. Her research has focused on integrated ocean and coastal management, marine protected areas, areas beyond national jurisdiction, and climate change adaptation. Miriam received a BSc in Fisheries and an MSc in Marine Biology from the University of the Philippines, and a PhD in Marine Studies from the University of Delaware.

Iwan Ball was formerly Research Associate in the School of Earth and Ocean Sciences at Cardiff University, working in the Marine and Coastal Environment Group. His research interests include marine policy and legislation, as well as environmental management of a broad range of marine sectors, particularly recreational boating, shipping, marine renewable energy, and integrated coastal zone management (ICZM). In 2006, he joined WWF Cymru as Marine Policy Officer, soon becoming Senior Policy Officer in WWF. For the past two years, he has been Marine Programme Manager (Oceans Governance) at WWF UK. He has an MSc and PhD in Marine Resource Management from Cardiff University.

Rhoda C. Ballinger, Senior Lecturer at Cardiff University's School of Earth and Ocean Sciences, has engaged in research on model institutional and policy frameworks for integrated coastal management (ICM) and on science-based coastal policymaking for more than twenty years. She has been involved in pan-European projects including Corepoint (Coastal Research and Policy Integration), IMCORE (Innovative Management for Europe's Changing Coastal Resource), and DeltaNet (Network of European Deltas), and in international, national and local coastal and estuary management projects, such as CoastNet (Coastal Partnerships Network), Arfordir Coastal Heritage, the Wales Coastal and Maritime Partnership, and the local Severn Estuary Partnership. She has a BSc in Geography, a Postgraduate Certificate in Education (PGCE), and a PhD from the University College of Wales, Aberystwyth.

Jay L. Batongbacal is Professor at the University of the Philippines College of Law, and Director of the Institute for Maritime Affairs and Law of the Sea. Jay holds a Master of Marine Management and PhD in Jurisprudential Science from Dalhousie University, Canada. Since 1997, he has undertaken research in and engaged with national and international maritime issues. He was the legal adviser to the Philippine team that successfully secured recognition by the Commission on the Limits of the Continental Shelf of the Philippines' continental shelf beyond 200 nautical miles in the Benham Rise region.

Stella Regina Bernad was formerly Legal Officer for Marine Affairs at the Partnerships in Environmental Management for the Seas of East Asia (PEMSEA). She was a member of the legal staff of PEMSEA responsible for the preparation of the Sustainable Development Strategy for the Seas of East Asia (SDS–SEA).

Gaëlle Brachet was a staff member of the Directorate for Research and Institutional Analysis and Public Policy, General Directorate of Environmental Policy and Regional and Sectoral Integration, Secretariat of Environment and Natural Resources, Mexico.

Sherry P. Broder is a civil litigation attorney who also teaches public international law and international ocean law at the University of Hawai'i Law School. She is also Adjunct Research Fellow at the East–West Center and Executive Director of the Jon Van Dyke Institute of International Law and Justice. Sherry was the first woman president of the Hawai'i State Bar Association, a finalist 'Trial Lawyer of the Year' (Trial Lawyers for Public Justice), and recognized as 'Solo Practitioner of the Year' by the American Bar Association (ABA). She co-edited the book *Governing Ocean Resources: A Tribute to Judge Choon-Ho Park* (Brill, 2013) and has published on international ocean law.

Fausto Efrén Burgoa is an economist of the Regional Integration Directorate, General Directorate of Environmental Policy and Regional and Sectoral Integration, Secretariat of Environment and Natural Resources, Mexico.

Chua Thia-Eng is Chair Emeritus of the East Asian Sea Partnership Council, PEMSEA, and board member of the International Ocean Institute and the Global Ocean Forum. He was formerly Executive Director of the PEMSEA Resource Facility and Regional Programme Director of PEMSEA. He has worked with the Food and Agriculture Organization of the United Nations (FAO), International Maritime Organization (IMO), United Nations Development Programme (UNDP), and ICLARM (now the WorldFish Center).

Biliana Cicin-Sain is Director of the Gerard J. Mangone Center for Marine Policy and Professor of Marine Policy at the University of Delaware. She is an expert in integrated coastal and ocean governance, with more than 100 publications in the field, and has forged international collaboration to advance the global oceans agenda as founder and President of the Global Ocean Forum. Dr Cicin-Sain received the 2010 Elizabeth Haub Award for Environmental Diplomacy, an honorary doctorate in maritime law from Korea Maritime University (2010), the 2007 Coastal Zone Foundation Award, the 2007 Elizabeth Mann Borgese Meerepreis, and a 2002 Ocean and Coastal Stewardship award. She received a PhD in political science from UCLA and postdoctoral training at Harvard University.

Sofia Cortina was the Director for Research and Institutional Analysis and Public Policy, General Directorate of Environmental Policy and Regional and Sectoral Integration, Secretariat of Environment and Natural Resources, Mexico.

Laleta Davis Mattis is currently University Counsel at the University of the West Indies Regional Headquarters, also serving as Adjunct Lecturer in International Environmental Law. Prior to her current position, Mrs Davis Mattis was Executive Director of the Jamaica National Heritage Trust and Director, Legal, Enforcement and Regulations Branch, with the National Environment and Planning Agency. Mrs Davis Mattis has represented Jamaica to the

UN Conference on the Law of the Sea, the Conference of the Parties to the Convention on Biological Diversity (CBD), and the Inter-American Network of Experts and Officials in Environmental Law, Enforcement, and Compliance. She also published on fisheries management in Jamaica and on international drug trafficking and the Law of the Sea.

Antonio Díaz-de-León-Corral is an independent consultant for Innovative Cutting Edge Solutions in Mexico City. Prior to his current position, he was Director General of the Mexican Ministry of Environment and Natural Resources (SEMARNAT), where he led thirty-one environmental state authorities in policy design and implementation of the Territorial Environmental Policy (Land and Ocean Use Planning), the Environment Fund, and Environment Capacity Building Programme. He received the National Oceanography Award from the Oceanographers Association of Mexico in 2010. Dr Díaz-de-León-Corral received a PhD in Natural Resources Management and Policy from Imperial College London.

Peter Edwards is a contractor with the National Oceanic and Atmospheric Administration (NOAA) based in the National Ocean Service, serving as Social Science Coordinator and Natural Resource Economist for the Coral Reef Conservation Program, responsible for incorporating social science and economic approaches to the Program's activities. Previously, he worked in NOAA's National Marine Fisheries Service as Natural Resource Economist for the Office of Habitat Conservation, responsible for leading the effort to monitor the economic impacts of federal expenditures on fifty habitat restoration projects under the American Recovery and Reinvestment Act of 2009. Dr Edwards has also led a number of ecosystem valuation studies in Jamaica.

Susan Farlinger is Director General of Canada's Department of Fisheries and Oceans (DFO), Pacific Region. Working primarily with the DFO, Susan has directed key regulatory portfolios in fisheries and aquaculture management, and in oceans, habitat and enhancement. She also served as Director General of the Pacific Forestry Centre with Natural Resources Canada (2005–08). She led the federal mountain pine beetle response, worked with placer miners and government agencies to develop a sustainable fish habitat regime, and managed aboriginal and fishery policy initiatives. Susan received the Queen's Jubilee Medal in 2012 in recognition of her significant contribution to compatriots, community, and Canada.

Sylvain Gambert is Policy Officer at the European Commission in Brussels, where he has worked on the development of the EU Integrated Maritime Policy since 2010. From 2012, he worked on the legislative proposal for Maritime Spatial Planning, which entered into force in 2014. Prior to joining the European Commission, he researched and published on the implementation of the European Union's environmental legislation in coastal and marine areas. In 2010, he received a PhD in Political Science from the European University Institute, Florence, where his research focused on the efficiency of collaborative mechanisms to support integrated environmental management.

Moritaka Hayashi is Professor Emeritus of Waseda University School of Law, Tokyo, where he taught from 1999 to 2008. He is currently Special Fellow at the Ocean Policy Research Foundation in Tokyo. Before joining Waseda University, he was Assistant Director-General (Head of the Fisheries Department) of FAO in Rome, Italy. Previously, he was a UN official in New York, largely serving its Office of Legal Affairs, most recently as Director of its Division for Ocean Affairs and the Law of the Sea. He also served the Japanese Mission to the United Nations as a counsellor and, subsequently, minister.

Indumathie Hewawasam is a marine policy specialist/consultant. As Marine Adviser for The Nature Conservancy, she worked on the financial architecture for the Coral Triangle Initiative. As Marine Adviser for the Government of Kenya, she helped to design the Kenya Coastal Development Project. She worked at the World Bank for twenty-two years, as Senior Environmental Specialist, mainstreaming environment within poverty reduction programmes and managing marine projects. She led the Tanzania Marine and Coastal Environmental Management Project, which brought Mainland Tanzania and Zanzibar together to promote sound governance in the exclusive economic zone and to reverse overexploitation of migratory fisheries.

Hoang Ngoc Giao is Associate Professor and Senior Lecturer in the School of Law at Vietnam National University and Director of the Center of Legislation Consultation. He was formerly Director of the Center of Data and Information at the Ministry of Foreign Affairs' Vietnam National Committee of Boundaries. Dr Giao is active in non-governmental organization (NGO) work on legislation and policy, including marine policy.

Alf Håkon Hoel is Research Director at the Institute of Marine Research in Norway. He was Chair of the Institute of Political Science, University of Tromsø, Norway. Hoel's research interests and publications include international relations in marine affairs, with an emphasis on the Arctic. He has authored numerous publications on Arctic governance, ecosystem approaches to marine management in the Arctic, and international fisheries management.

Ken Huffman has worked on oceans issues since the Canadian Oceans Act came into force in 1997. From 2007 to 2013, he was Manager of Regional Oceans Operations, Atlantic, at the Canadian Department of Fisheries and Oceans. In this position, Mr Huffman focused on domestic ocean and coastal integrated management, and the development and management of marine protected areas. Before his twenty-five years of public service, he was a consultant, researcher, and writer focusing on public service management and the transportation sector. Mr Huffman has an MPA from the University of Victoria, Canada.

Mariela Ibáñez was a staff member of the Directorate for Research and Institutional Analysis and Public Policy, General Directorate of Environmental Policy and Regional and Sectoral Integration, Secretariat of Environment and Natural Resources, Mexico.

Gerhard F. Kuska is Executive Director of the Mid-Atlantic Regional Association Coastal Ocean Observing System (MARACOOS), President of Ocean Strategies LLC, and Senior Fellow in Integrated Marine Policy at the University of Delaware. Previously, he was Associate Director of the White House Council on Environmental Quality, Director of Ocean and Coastal Policy in the Executive Office of the President, a member of the US Commission on Ocean Policy staff, and Senior Maritime Adviser to the Abu Dhabi government. He has worked for NOAA, the US Senate, and in global cargo industry management. His time with the US government included the establishment of four marine national monuments, totalling 867,834 km^2.

Terje Lobach, a lawyer specializing in the law of the sea, is Specialist Director in the Directorate of Fisheries in Norway. He has been employed by the Norwegian fisheries authorities and the Norwegian Foreign Service, has extensive experience in bilateral and multilateral negotiations, and has contributed to several publications on management of living marine resources. He served as Norway's representative to the Commission for the Conservation of Antarctic Marine

Living Resources (CCAMLR), FAO, the International Commission for the Conservation of Atlantic Tunas (ICCAT), Intergovernmental Oceanographic Commission Advisory Body of Experts of the Law of the Sea (IOC/ABE-LOS), the Northwest Atlantic Fisheries Organization (NAFO), the North East Atlantic Fisheries Commission (NEAFC), the South East Atlantic Fishery Organization (SEAFO), and the United Nations, was President of NAFO for four years and Chair of CCAMLR for two years, and has completed contractual work for the UN Division for Ocean Affairs and the Law of the Sea (DOALOS) and FAO.

Camille Mageau is Secretary of the Canadian Section of the Canada–United States International Joint Commission. Prior to taking on this position, she was Director of the Oceans Policy and Planning Branch of the Department of Fisheries and Oceans, during which time she led the development of Canada's vision for ocean management in 1995, and oversaw drafting and implementation of Canada's Oceans Act 1996. Dr Mageau has also served as a visiting scientist at NOAA and has worked with UNESCO's Intergovernmental Oceanographic Commission (IOC) to advance global ocean management. She holds MSc degrees in Marine Geology, Geophysics, and Oceanography.

Leopoldo Maraboli is a mining engineer with over three decades of experience at the World Bank, working on mineral sector development in Africa, Latin America, Asia, and Europe. He currently works in the private sector in Chile, advising on the development of the copper industry, including on social issues and mine safety.

Márcia Marques is Fellow of the Foundation of Science and Technology, working in the Intersectoral Oceanographic Commission of the Ministry of Education and Science. She has an MSc in Marine Biodiversity and Conservation, and a postgraduate course in Ocean Administrative Law. She has experience in controlling fisheries on board fishing vessels as a biologist.

Etiene Villela Marroni is Associate Professor at the Federal University of Pelotas (Brazil) and Coordinator/Researcher at the Laboratory for International Policy and Management of Oceanic Spaces (GIS Technology). Dr Marroni is currently dedicated to the study of diplomacy, involving multilateral relations regarding ocean space. She also studies how the world's coastal states embrace UNCLOS-related policies, coining in Brazil the expression 'coastal powers' and co-authoring (with Milton L. Asmus) two books on coastal management in Brazil. She holds a PhD in Political Science (International Politics).

Bernice McLean is an environmental specialist, with an interest in marine and coastal governance. She is working as an independent environmental consultant and as Associate to EcoAfrica Environmental Consultants. Bernice completed a PhD in Marine Policy at the University of Delaware, focusing on coastal policy development in Sub-Saharan Africa. Bernice worked in the Office of Global Programs of NOAA as a Sea Grant Fellow. Thereafter, she coordinated the marine and coastal work for the Endangered Wildlife Trust, and collaborated with other NGOs on strengthening fisheries governance in Southern Africa and in the Indian Ocean region.

Yuriy G. Mikhaylichenko is Adviser at the Ministry of Economic Development of the Russian Federation. He previously worked at the Ministry of Science and Technologies in the fields of planning and management of the Russian federal marine research programmes, and bilateral and international cooperation in ocean studies. He is a member of the editorial board of *Ocean & Coastal Management*, and has published two books and some 100 other publications in physical oceanography, integrated coastal management, and marine affairs. He received an MSc from MV Lomonosov Moscow State University and a PhD in Physical Oceanography from PP Shirshov Institute of Oceanology.

Norma Patricia Muñoz Sevilla is Secretary of Research and Graduate Studies at the Instituto Politécnico Nacional, Mexico. She holds a PhD in Biological Oceanography from the University of Aix-Marseille II (France), and a post-doctorate in Marine Biochemistry from the Practical School of Higher Studies, Paris. Dr Munoz has held other key positions in government and has served on various committees for the protection of the environment. She is a member of the President's Board of Climate Change and has co-authored a chapter on developing human resources for climate change in Climate *Change in Mexico, A Coastal and Marine Approach* (University of Campeche, CETYS University and Government of Campeche, 2011).

Magnus Ngoile was recently the Policy and Governance Coordinator of the UNDP/GEF Agulhas Somali Currents Large Marine Ecosystems (ASCLME) Project. He was founding member and first President/Chairman of the Western Indian Ocean Marine Science Association (WIOMSA) and has served on its board of trustees for several years. In 1999, he won the Pew Marine Conservation Award for his work to promote community-based marine and coastal management in Kilwa District, Tanzania. He holds a PhD in Fisheries Science from the University of Aberdeen (1987), an MSc in Fisheries Science from Humboldt State University, California (1977), and a BSc in Zoology and Botany from the University of Dar es Salaam, Tanzania (1974).

Nguyen Chu Hoi is Associate Professor and Senior Lecturer in the School of Science at Vietnam National University. He is also chairman of the Vietnam Association of Marine Environment and Nature and the Vietnam Association of Southeast Asian Nations (ASEAN) Working Group on Coastal and Marine Environment. Dr Chu Hoi was Deputy Administrator of the Vietnam Administration of Seas and Islands (2008–12), Director of the Vietnam Institute of Fisheries Economics and Planning (2004–07), and Director of the Haiphong Institute of Oceanology at the Vietnamese Academy of Science and Technology (1989–2001). Dr Chu Hoi has authored more than 180 publications on coastal and marine governance.

Gustavo Pérez-Chirinos is a biologist with the Regional Integration Directorate, General Directorate of Environmental Policy and Regional and Sectoral Integration, Secretariat of Environment and Natural Resources, Mexico.

Donna Petrachenko is Chief Adviser on International Biodiversity and Sustainability and Australia's Commissioner to the International Whaling Commission. She has more than twenty-five years of experience in Australia and Canada in maritime and oceans affairs, including internationally recognized skills in international initiatives and leading numerous delegations to the United Nations. Ms Petrachenko has led Australia's efforts on oceans policy and marine protection and management, including Australia's approach to marine bioregional planning, resulting in the completion of a network of marine protected areas around Australia's entire exclusive economic zone. She has spearheaded Australia's approach to high seas biodiversity conservation since 2003.

Tiago Pitta e Cunha is a consultant to the Portuguese president on science, environment and maritime affairs. He was a cabinet member of the European Commissioner for Maritime Affairs, and Coordinator of the EU Integrated Marine Policy and of the Portuguese Strategic

Commission for the Oceans. He represented the European Union at the United Nations in the field of maritime affairs (2000), and Portugal at the Conference of the Law of the Sea, the International Seabed Authority, and the UN Open-Ended Informal Consultative Process on Oceans and the Law of the Sea. Tiago holds a law degree and an LLM in European/International Law from the London School of Economics and Political Science.

Mary Power is a marine scientist with expertise in coastal management, coastal fisheries, and coastal geo-processes. She is currently Director, Strategic Partnerships, at the World Meteorological Organization (WMO) in Geneva, Switzerland. Her previous experience includes: Programme Director/Manager, Ocean and Islands Programme at the Pacific Islands Applied Geoscience Commission, Fiji; Coordinator, Reef Fisheries Observatory of the Secretariat of the Pacific Community (SPC), New Caledonia; Adviser, Coastal Management Programme South Pacific Regional Environment Programme, Samoa; and Regional Manager, State Government Great Barrier Reef Marine National Parks, Australia.

Tony George Puthucherril is a Research Associate at the Marine and Environment Law Institute, Dalhousie University. He holds a PhD in "Sea Level Rise, Coastal Climate Change Adaptation and Integrated Coastal Zone Management Law" from Dalhousie University, Canada, and an M. Phil in Water Resources Law from the National University of Juridical Sciences, India. He is the author of "*Towards Sustainable Coastal Development: Institutionalizing Integrated Coastal Zone Management and Coastal Climate Change Adaptation in South Asia*" (Martinus Nijhoff, 2014) and "*From Shipbreaking to Sustainable Ship Recycling: Evolution of a Legal Regime*" (Martinus Nijihoff, 2010). Puthucherril's research interests include international environmental law, climate change law, comparative environmental law, water rights, nuclear law, and coastal and marine law.

Raquel Ribeiro is Fellow of the Foundation of Science and Technology, working in the Intersectoral Oceanographic Commission of the Ministry of Education and Science. She has a degree in biology and spatial planning, and has co-published peer-reviewed articles on governance of transboundary marine protected areas.

Evelia Rivera-Arriaga is Professor at the EPOMEX Institute at the University of Campeche, Mexico. She studied biology at the National University of Mexico, obtained a PhD in Marine Studies at the University of Delaware, and has twenty-four years of experience in natural resources management and environmental policy. She currently serves as State Minister of Environment of Campeche, coordinates the Mexican Network for Integrated Coastal Management, is a member of the National Coordination for Environmental States Authorities, and oversees the Climate Change for the Yucatan Peninsula Regional Agreement. She has collaborated with the Inter-American Development Bank and World Bank on various initiatives.

Roberto Rosado is a biologist with the Land Use Planning Directorate, General Directorate of Environmental Policy and Regional and Sectoral Integration, Secretariat of Environment and Natural Resources, Mexico.

Mário Ruivo (University of Lisbon, 1950; University of Paris Sorbonne, Laboratoire Arago, 1951–54) is a biologist and was Professor of the Instituto de Ciências Biomédicas Abel Salazar (ICBAS)/University of Oporto in marine policy and management. Professor Ruivo currently chairs the Intersectoral Oceanographic Commission, the National Council for Environment and Sustainable Development, and the Portuguese Committee for the IOC. He was Director of the Division of Aquatic Resources and the Environment of the Department of

Fisheries of FAO, Head of the Portuguese Delegation to UNCLOS, Executive Secretary and Vice-Chairman of the IOC, a member and Coordinator of the Independent World Commission on the Oceans, and Scientific Adviser to the 1998 Lisbon World Exposition (EXPO'98).

Valentin P. Sinetsky was the author of more than ninety publications devoted to maritime affairs. He led the preparation of a wide variety of important governmental draft documents concerning the maritime activity of the Russian Federation. He was Deputy Head of the Scientific-Expert Council of the Maritime Board of the Government of the Russian Federation and Head of Division at the Council for Studies of Productive Forces. He was a first-rank captain of the Pacific Ocean Navy. Along with a PhD in Military Sciences, the late Dr Sinetsky held degrees from the Supreme Naval School in Leningrad and the Naval Academy.

Hance D. Smith was Reader and Coordinator of the Marine and Coastal Environment Group, Cardiff University, until 2011. His research interests span marine geography, the development of marine resources and maritime communities, and ocean use management and policy. He is editor of *Marine Policy*, general editor of the Routledge *Advances in Maritime Research* series, and joint editor of *The Routledge Handbook of Ocean Resources and Management* (Routledge, 2015). He was successively founder member, Secretary, and Chair of the International Geographical Union Commission on Marine Geography (1986–2000), and Fellow of the Royal Institution of Chartered Surveyors (1988–2008). Dr Smith holds a BSc and PhD in Geography from the University of Aberdeen.

Caitlin Snyder is a biologist in the Branch of Listing, Ecological Services Program of the US Fish and Wildlife Service (FWS). She provides expert policy advice related to listing endangered and threatened species and designating critical habitat under the Endangered Species Act of 1973. She holds a Master's in Environmental Management from Duke University. She has worked for the Global Ocean Forum (Delaware) and the Centre for Maritime Studies (Singapore). She served as Knauss Fellow in the FWS Office of Congressional and Legislative Affairs, and as Fulbright Fellow at the National University of Singapore's Tropical Marine Science Institute.

Anama Solofa began her career with the government of Samoa, in the field of coastal fisheries management, working with villages in the Community-based Fisheries Management Program (CFMP) to manage their fisheries resources. She undertook further studies and career development, branching into ocean governance, and was UN–Nippon Foundation of Japan Fellow (2009–10). Her research fellowship centred on ocean governance in the Pacific. She continues to work in the Pacific region as a consultant.

Tim Stojanovic is Lecturer in Sustainable Development and Geography at the University of St Andrews, where he also has membership in the Scottish Oceans Institute and Sustainability Institute. His research interests include the governance, planning, and management of coasts and oceans, and multidisciplinary approaches to environmental management, including coastal vulnerability and climate adaptation. Previously, Tim held an internship at the University of Delaware and NOAA, and was an investigator based in the Marine and Coastal Environment Group, Cardiff University. He holds a BSc and PhD in Marine Geography from Cardiff University.

Prue Taylor teaches environmental and planning law at the School of Architecture and Planning, University of Auckland. She is Deputy Director of the New Zealand Centre for

Environmental Law, and a member of the International Union for Conservation of Nature (IUCN) Commission of Environmental Law and its Ethics Specialist Group. She has authored numerous books and articles. Her book, *An Ecological Approach to International Law: Responding to the Challenges of Climate Change* (Routledge, 1998), won a New Zealand Legal Research Foundation Prize. In 2007, she received an outstanding achievement award from the IUCN in recognition of her contribution to law, ethics, and climate change.

Hiroshi Terashima is Executive Director of the Ocean Policy Research Foundation, Japan. His interests include integrated coastal and ocean management, and maritime safety and security. Terashima retired from the Japanese Ministry of Transport as Assistant Vice-Minister in 1994. He served as Executive Director of the Nippon Foundation (1994–2002). He has engaged in developing various ocean policy proposals, including the Ocean Policy Research Foundation (OPRF) *A Proposal for a 21st Century Ocean Policy* (2005), which led to the establishment of the 2007 Japanese Basic Act on Ocean Policy. He has written numerous papers and given lectures on these issues. He is also a member of the board of governors of the World Maritime University.

Tara Thrupp is currently working towards a PhD in Aquatic Toxicology at the Institute for the Environment, Brunel University. Her research explores the effect of human pharmaceutical chemicals in the aquatic environment, with particular focus on the steroidal pharmaceuticals and endocrine function in fish. She has experience in a wide range of research areas, including ecotoxicology, pathobiology, biodiversity, conservation, and ocean coastal policy. Her specialities and interests include pathobiology, microbiology, benthic ecology, animal behaviour, conservation, and marine and fisheries policy. She holds a BSc in Marine Geography from Cardiff University and an MSc (Research) in Aquatic Biology from Swansea University.

Jon M. Van Dyke was Professor of Law at the William S. Richardson School of Law at the University of Hawai'i at Manoa (1976–2011), where he taught constitutional law, international law, international ocean law, and international human rights. Professor Van Dyke co-authored the casebook *International Law and Litigation in the US* (Thomson West, 3rd edn, 2009), and other books on international ocean law and governance, including the *Freedom for the Seas in the 21st Century: Ocean Governance and Environmental Harmony* (Island Press, 1993), which received the 1994 Harold and Margaret Sprout Award from the International Studies Association as the best book on international environmental policy. He had also engaged in litigation on important constitutional and international issues before state and federal courts and international tribunals.

David L. VanderZwaag is Professor of Law and the Canada Research Chair in Ocean Law and Governance at the Marine and Environmental Law Institute, Dalhousie University, Canada. He is a member of IUCN's World Commission on Environmental Law (WCEL), Co-Chair of the WCEL's Specialist Group on Oceans, Coasts and Coral Reefs, and an elected member of the International Council of Environmental Law (ICEL). Dr VanderZwaag has published widely on marine and environmental law, recently including the books *Polar Oceans Governance in an Era of Environmental Change* (Edward Elgar, 2014) and *Recasting Transboundary Fisheries Management Arrangements in Light of Sustainability Principles: Canadian and International Perspectives* (Brill, 2010).

Guilherme G. Vieira is Associate Researcher at the Research Centre for Geography and Regional Planning at the Universidade Nova de Lisboa (UNL), Portugal, and at the Laboratory of Coastal Management at the Universidade Federal do Rio Grande (FURG), Brazil. Since 2004, he has conducted research on coastal management, governance and public policy, socio-environmental planning, and participatory management of natural resources. Guilherme is concurrently studying for an MSc at UNL, conducting comparative analyses of coastal management policies. He has a BSc in Geography and Oceanography from FURG, and received a Specialist Diploma on Territorial Management from UNL.

Kateryna M. Wowk is an expert in multidisciplinary approaches to problem-solving in the sustainable use of ocean and coastal spaces and resources. She is currently a federal consultant to NOAA, serving as Senior Social Scientist to NOAA's Chief Economist. She has written extensively on the impacts of climate change in the ocean, as well as on the valuation of marine ecosystem services. Dr Wowk holds a PhD in Marine Policy and International Relations from the University of Delaware, and an MSC in Environmental Science and Policy from Columbia University.

PREFACE

> We therefore commit to protect, and restore, the health, productivity and resilience of oceans and marine ecosystems, to maintain their biodiversity, enabling their conservation and sustainable use for present and future generations, and to effectively apply an ecosystem approach and the precautionary approach in the management, in accordance with international law, of activities having an impact on the marine environment, to deliver on all three dimensions of sustainable development.
>
> *(United Nations Conference on Sustainable Development (UNCSD) (2012)*
> *(The Future We Want, para. 158)*

We have come a long way since 1982, when the United Nations Convention on the Law of the Sea (UNCLOS) was adopted, providing the legal framework for the conservation and sustainable use of the oceans and their resources, and since 1992, when governments and the international community embraced a set of goals and targets to guide all actions on environment and development, including the oceans, at the United Nations Conference on Environment and Development (UNCED). More than twenty years later, much knowledge has been generated, various issues have been addressed, promising governance and management frameworks and paradigms have been put forward, and a plethora of best practices has been identified and disseminated. Yet it is acknowledged that the pace at which management and governance of the oceans is proceeding does not match the pace of degradation of the marine environment and its resources.

The 2012 United Nations Conference on Sustainable Development (UNCSD) (also known as 'Rio+20') provided an opportunity to review past management and governance performance, and to fast-track management and conservation efforts in a more targeted way. Rio+20 called for the formulation of the post-2015 development agenda, based on the provisions of the Rio+20 outcome document *The Future We Want*, incorporating a set of sustainable development goals and targets in which the oceans, among other priority sectors, are centrally addressed. This volume of case studies on national and regional ocean policies shares experiences and lessons learned that can inform the fulfilment of international goals and targets on oceans, as adopted in the 1992 UNCED, the 2002 World Summit on Sustainable Development (WSSD), the 2012 UNCSD, and the post-2015 development agenda.

Following the adoption of UNCLOS in 1982, states began to develop the enabling environment for its implementation by establishing maritime boundaries and national ocean policies, in view of the impending entry into force of the Convention. These policies, although sometimes lacking in the modern international principles that came later (as recommended, among other things, in the Rio Declaration on Environment and Development of 1992), served the purpose of codifying the countries' main responses to the requirements of UNCLOS. During this time period, as well, most coastal nations continued to put a variety of sectorial policies into place to manage different uses of the ocean (including shipping, fishing, oil and gas development, among others).

Responding to the reality of serious conflicts of use in most national ocean zones and to the prescriptions articulated in both UNCLOS and in the 1992 UNCED (given the inter-relationship among uses and processes in the coast and ocean, ocean and coastal governance must be 'integrated in content and precautionary and anticipatory in ambit'), in the past twenty years coastal nations have undertaken extensive efforts to articulate and implement an integrated vision for the governance of ocean areas under their jurisdiction. They have sought to harmonize existing uses and laws, to foster sustainable development of ocean areas, to protect biodiversity and vulnerable resources and ecosystems, and to coordinate the actions of the many government agencies that are typically involved in oceans affairs. The 2002 WSSD's Johannesburg Plan of Implementation specifically called for the promotion of integrated coastal and ocean management at the national level, and encouragement and assistance for countries in developing ocean policies and mechanisms on integrated coastal management. States thus received a precise directive regarding ocean management and governance, which prescribed the use of the integrated approach, along with the ecosystem-based approach, the precautionary approach, and other principles, as adopted at UNCED. This handbook attests to the outcomes of various countries' initiatives towards this target.

This book represents the main output of the project entitled 'The Nippon Foundation Research Task Force on National and Regional Ocean Policies', which was funded primarily by the Nippon Foundation of Japan. The project, led by Drs Biliana Cicin-Sain, David VanderZwaag, and Miriam C. Balgos, was carried out by the University of Delaware between 2004 and 2007, in collaboration with Dalhousie University and other partners, and aimed to:

1   develop a framework for cross-national analysis of national and regional ocean policies, and for drawing lessons useful to other cases in other countries/regions;
2   carry out systematic comparative analyses of national and regional ocean policies in a selected number of countries/regions on principles embodied, institutional arrangements, and other governance variables;
3   draw lessons from the comparative analyses and develop suggested guidance for other nations/regions contemplating national/regional ocean policy formulation and implementation; and
4   disseminate the results of the research work broadly.

A task force on national and regional policies was organized in 2004 to carry out these project objectives. The task force was convened over the course of two workshops (April 2004, Tokyo; June 2004, New York) to conceptualize the framework for the case studies, with technical input from an advisory committee composed of international ocean leaders. The outcomes of these analyses, and of initial analyses of national and regional case studies, were presented at an international conference (organized by the editors and by Professor Mario Ruivo), 'The Ocean Policy Summit: International Conference on Integrated Ocean Policy: National and Regional

Experiences, Prospects, and Emerging Practices' (TOPS 2005), held in Lisbon, Portugal, on 11–13 October 2005. Supported by the Global Environment Facility (GEF), the Portuguese government, and other partners, the conference brought together 218 ocean leaders from governments, regional organizations, United Nations agencies, academia, non-governmental organizations (NGOs), and industry representatives from fifty-three countries, focused on the institutional aspects of ocean management and governance at various levels.

Although, for some, it seemed premature to be looking at national and regional ocean policies at that time, given the relatively short time that had lapsed since the international goal on integrated national ocean policies was set at UNCED and at WSSD, with hindsight we can see that it was a key moment for a number of countries, such as Japan and Mexico, and for the European Union, which were motivated to develop their ocean policies, drawing, in part, on existing experiences discussed at TOPS 2005. While knowledge and experience of integrated ocean policies were still sparse at that time, countries and regions on the cusp of national/regional ocean policymaking nevertheless persevered and established ocean policies. As more countries and regions have initiated national and regional ocean policies, lessons may now be drawn for other countries to consider applying in future ocean policy development initiatives.

The book presents in-depth analyses of the experiences of fifteen nations (Australia, Brazil, Canada, India, Jamaica, Japan, Mexico, New Zealand, Norway, Philippines, Portugal, Russian Federation, United Kingdom, United States, and Vietnam) and four regions of the world (East Asian Seas, European Union, Pacific Islands, and Sub-Saharan Africa) that have taken concrete steps toward cross-cutting and integrated national and regional ocean policy. The case studies have been prepared by leading ocean policy academics and practitioners from around the world, each following a similar prescribed framework and format (see Appendix A and Appendix B for guides to the structure of the national and regional case studies, respectively). Following the guides, each case study took into consideration the motivations underlying the development of ocean policy, its scope, the evolution of the policy over time, and the implementation and evaluation of the policy. It is important to note that the case studies followed the framework of analysis to the extent possible, subject to the availability of data and applicability, since the policies analysed in the volume are at varying stages of policy development and implementation. A chapter providing guidance on domestic implementation of international agreements and a chapter providing a comparative overview of ocean policies are also included in the handbook.

The book should be of interest to a wide audience. The volume provides succinct lessons learned and emerging best practices, which are directly relevant to the growing number of nations and regions that are also beginning to pursue integrated ocean policies. The book should be useful to governmental, non-governmental, and private sector practitioners involved in ocean and coastal management around the world, as well as to graduate and undergraduate students in marine and environmental policy.

Many institutions and individuals contributed to this work, and we would like to thank them very sincerely, including:

- the authors of the book chapters, most of whom have been involved since the beginning of the project, in the conceptualization of the case studies, in the organization of TOPS 2005, and in the preparation of chapters, with multiple updates and revisions;
- the project advisory committee, composed of eminent ocean leaders,[1] including:

    o Dr Patricio Bernal, executive secretary, Intergovernmental Oceanographic Commission of the United Nations Educational, Scientific and Cultural Organization (IOC/UNESCO);

- o Dr Alfred Duda, senior adviser, International Waters, GEF;
- o Dr Serge Garcia, director, Fishery Resources Division, Food and Agriculture Organization of the United Nations (FAO);
- o Dr Jean-François Pulvenis de Séligny, director, Fishery Policy and Planning Division, FAO;
- o Dr Chua Thia-Eng, regional programme director, Partnerships in Environmental Management for the Seas of East Asia (PEMSEA);
- o Dr Isao Koike, professor, Ocean Research Institute, University of Tokyo;
- o Dr Gunnar Kullenberg, independent consultant, and former secretary, IOC/ UNESCO;
- o Dr Andrew Hudson, principal technical adviser, International Waters, United Nations Development Programme (UNDP);
- o Dr Veerle Vandeweerd, coordinator, United Nations Environment Programme (UNEP) Global Programme of Action for the Protection of the Marine Environment from Land-based Activities (GPA);
- o Mr Efthimios Mitropoulos, secretary-general, UN International Maritime Organization (IMO);
- o Mrs Annick de Marffy, former director, UN Office of Legal Affairs, Division for Ocean Affairs and the Law of the Sea;
- o Mr Philip Burgess, former co-chair, UN Open-ended Informal Consultative Process on Oceans and the Law of the Sea (UNICPOLOS); and
- o HE Ambassador Dr Felipe H. Paolillo, Uruguay, former co-chair, UNICPOLOS (Consultative Process);

- sponsors of the project, and of the TOPS 2005, including:

  - o the Nippon Foundation, Japan;
  - o the GEF;
  - o the Gerard J. Mangone Center for Marine Policy, University of Delaware;
  - o the Global Ocean Forum;
  - o the Portuguese Committee for the IOC;
  - o the Ministry of Foreign Affairs, Portugal;
  - o the Secretary of State for Maritime Affairs, Portugal;
  - o the Port Authority of Lisbon, Portugal;
  - o the Luso-American Development Foundation (FLAD);
  - o Oceanário de Lisboa;
  - o IOC/UNESCO;
  - o UNEP GPA;
  - o Canada's Department of Fisheries and Oceans (DFO); and
  - o the National Oceanic and Atmospheric Administration Coastal Services Center (NOAA CSC)

- ocean leaders who contributed to the organization of TOPS 2005, including Professor Mario Ruivo, Dr Tiago Pitta e Cunha, Dr Veerle Vandeweerd, Mr Charles Ehler, Mr John Roberts, Ms Diane James, Dr Anamarija Frankic, Dr Vladimir Golitsyn, Dr Gerard J. Mangone, Dr Alfred Duda, Ms Cristelle Pratt, Dr Indumathie Hewawasam, Mr Tom Laughlin, Dr Awni Behnam, Dr Magnus Ngoile, and Mr Franklin McDonald;

- the director of the Nippon Foundation and its staff, for their support throughout the duration of the project – that is, Mr Takashi Ito, Mr Kiyoshi Sasaki, Ms Yoko Yokouchi, Mr David Karashima, Ms Eiko Mizuno, Ms Sasako Hiraiwa, and Mr Mitsuyuki Unno;
- the GEF, especially Dr Alfred Duda, for support of TOPS 2005 and encouragement to promote integrated national and ocean policy around the world;
- the director of the Ship and Ocean Foundation (now the Ocean Policy Research Foundation) and its staff – that is, Mr Hiroshi Terashima, Mr John Dolan, and Ms Yumiko Tanaka;
- staff of the Gerard J. Mangone Center for Marine Policy, University of Delaware, who contributed to the organization of TOPS 2005 and to the production of the book – that is, Lindsay Williams, Amanda Wenczel, Kateryna Wowk, Shelby Hockenberry, Isabel Torres de Noronha, Bernice McLean, Nicole Ricci, Laverne Walker, Jason Didden, Kate Semmens, Caitlin Snyder, Gwenaelle Hamon, Alexis Martin, Erica Wales, Taylor Daley, Deanna Sigai, and Edward Carr;
- staff at Dalhousie University, especially Lauri MacDougall for her secretarial support and Molly Ross for her word-processing assistance;
- Susan Rolston, Seawinds Consulting Services, to whom we owe a debt of gratitude for copy-editing and assistance in the production of the book;
- the Schulich Academic Excellence Fund, Schulich School of Law, Dalhousie University, for supporting the production of this volume; and
- Tim Hardwick, senior commissioning editor, and Ashley Wright, senior editorial assistant, Routledge, for believing in the usefulness and relevance of the book, and for their patience in the final stages of book production, and Vanessa Plaister for her thorough copy-editing.

To all of the above, and to those whom we may have inadvertently forgotten to mention by name, we thank you very much for your part in the production of this handbook.

Biliana Cicin-Sain
David L. VanderZwaag
Miriam C. Balgos

Note

1 Please note that positions and affiliations indicated are as of 2004, the year in which the project was initiated and the advisory committee formed.

ABBREVIATIONS

ABNJ	areas beyond national jurisdiction
ACCOBAMS	Agreement on the Conservation of Cetaceans of the Black Sea, Mediterranean Sea, and Contiguous Atlantic Area
ADB	Asian Development Bank
AMA	aquaculture management area
AMAP	Arctic Monitoring and Assessment Programme
AMSA	Australian Maritime Safety Authority
ANZECC	Australian and New Zealand Environment Conservation Council
APEC	Asia–Pacific Economic Cooperation
APEI	area of particular environment interest
AQUIPESCA	Action for Aquaculture and Fisheries (Brazil)
ArcDev	Sustainable Philippine Archipelagic Development Framework
ASCLME	Agulhas and Somali Current Large Marine Ecosystem
ASEAN	Association of Southeast Asian Nations
AZRF	Arctic Zone of the Russian Federation
BCC	Benguela Current Commission
BCLME	Benguela Current Large Marine Ecosystem
BFAR	Bureau of Fisheries and Aquatic Resources (Philippines)
BIOMAR	Survey and Evaluation of Biotechnological Potential of Marine Biodiversity (Brazil)
BSIMPI	Beaufort Sea Integrated Management Planning Initiative (Canada)
CBCRM	community-based coastal resource management
CBD	Convention on Biological Diversity
CCA	Climate Change Adaptation
CCAMLR	Convention for the Conservation of Antarctic Marine Living Resources
CCFAM	Canadian Council of Fisheries and Aquaculture Ministers
CCMOA	Cabinet Committee on Maritime and Ocean Affairs (Philippines)
CCMS	Centre for Coastal and Marine Sciences (UK)
CCS	carbon capture and storage

CCSBT	Commission for the Conservation of Southern Bluefin Tuna
CCW	Countryside Council for Wales
CEC	Commission for Environmental Cooperation
CECAF	Fishery Committee for the Eastern Central Atlantic
CEO	Strategic Oceans Commission (*Comissão Estratégica dos Oceanos*)
CEPA	Canadian Environmental Protection Act 1999
CEQ	Council on Environmental Quality (US)
CERD	Committee on Elimination of Racial Discrimination (NZ)
CFE	Federal Electricity Commission (Mexico)
CFP	Common Fisheries Policy (EU)
CHONe	Canadian Healthy Oceans Network
CIAM	Interministerial Commission for Ocean Affairs (*Comissão Interministerial para os Assuntos do Mar*)
CIDPC	Interministerial Commission for the Delimitation of the Continental Shelf (*Comissão Interministerial para a Delimitação da Plataforma Continental*)
CIMARES	Mexican Intergovernmental Commission for Oceans and Coastal Affairs
CIRM	Interministerial Commission for Sea Resources (Brazil)
CITES	Convention on International Trade in Endangered Species of Wild Fauna and Flora
CLC	International Convention on Civil Liability for Oil Pollution
CLCS	Commission on the Limits of the Continental Shelf
CMA	Crown Minerals Act 1991 (NZ)
CMAs	coastal management areas (Canada)
CMC	Cays Management Committee (Jamaica)
CMM	Commission on Maritime Meteorology
CMOA	Commission on Maritime and Ocean Affairs (Philippines)
CMS	coastal marine strategies (NZ)
CMS	Convention on the Conservation of Migratory Species of Wild Animals
CMSP	coastal and marine spatial planning
CNA	National Water Commission (Mexico)
COAG	Council of Australian Governments
COARE	Coupled Ocean and Atmospheric Response Experiment
COGERCO	Coordination Group of Coastal Management (Brazil)
CONABIO	National Commission for Knowledge and Use of Biodiversity (Mexico)
CONAFOR	National Forestry Commission (Mexico)
CONAMA	National Environmental Council (Brazil)
CONANP	National Commission of Natural Protected Areas (Mexico)
CONAPESCA	National Fisheries Commission (Mexico)
CONAQUA	National Water Commission (Mexico)
COP	Committee on Ocean Policy (US)
COP	Conference of the Parties
COREP	Regional Fisheries Committee for the Gulf of Guinea
COSEWIC	Committee on the Status of Endangered Wildlife in Canada
COSMAR	Coastal and Marine Secretariat (Africa)

CPRM	Geological Survey of Brazil
CPV	Communist Party of Vietnam
CRC	Coastal Resources Center (Rhode Island)
CROP	Council of Regional Organisations in the Pacific
CROP MSWG	Council of Regional Organisations in the Pacific Marine Sector Working Group
CRZ	coastal regulation zone
CSAS	Canadian Science Advisory Secretariat
CSIRO	Commonwealth Scientific and Industrial Research Organisation (Australia)
DA	Department of Agriculture (Philippines)
DARD	Department of Agriculture and Rural Development (Northern Ireland)
DCMS	Department for Culture, Media, and Sports (UK)
DECC	Department of Energy and Climate Change (UK)
DEFRA	Department for Environment, Food, and Rural Affairs (UK)
DENR	Department of Environment and Natural Resources (Philippines)
DEWHA	Department of the Environment, Water, Heritage and the Arts (Australia)
DFA	Department of Foreign Affairs (Philippines)
DFID	Department for International Development (UK)
DFO	Department of Fisheries and Oceans (Canada)
DfT	Department of Transport (UK)
DGPAIRS	Environmental Policy, Sectoral and Regional Integration Division (Mexico)
DGPM	Directorate General for Maritime Policy (*Direção-Geral de Política do Mar*)
DILG	Department of Interior and Local Government (Philippines)
DMB	Department of Budget and Management (Philippines)
DND	Department of National Defense (Philippines)
DoC	Department of Conservation (NZ)
DOD	Department of Ocean Development (India)
DOE	Department of Energy (Philippines)
DOF	Department of Finance (Philippines)
DOI	Department of Interior (US)
DOJ	Department of Justice (Philippines)
DOLE	Department of Labor and Employment (Philippines)
DOST	Department of Science and Technology (Philippines)
DOT	Department of Tourism (Philippines)
DOTC	Department of Transportation and Communication (Philippines)
DPWH	Department of Public Works and Highways (Philippines)
DRR	disaster risk reduction
DSFAA	Deep Sea Fishing Authority Act, 1998 (Tanzania)
DTI	Department of Trade and Industry (Philippines)
DWFNs	distant water fishing nations
EA	Environment Agency (UK)
EACF	Brazilian Antarctic Station 'Commandante Ferraz'
EAF	ecosystem approach to fisheries

EAS	East Asian Seas
EBM	ecosystem-based management
EBSA	ecologically and biologically significant area
EC	European Commission
ECA	emission control area
ECA	excess crude amount
EcoGov	Environmental Governance (Philippines)
EcoQOs	Ecological Quality Objectives
EEA	European Economic Area
EEC	European Economic Community
EEZ	exclusive economic zone
EIA	environmental impact assessment
EMAM	Portuguese Task Force for Maritime Affairs (*Estrutura de Missão para os Assuntos do Mar*)
EMEPC	Portuguese Task Force for the Extension of the Continental Shelf (*Estrutura de Missão para a Extensão da Plataforma Continental*)
EMODNET	European Marine Observation and Data Network
EMSA	European Maritime Safety Agency
EN	English Nature
ENCOGERCO	Brazilian Summit on Coastal Management
ENDS	National Strategy for Sustainable Development (Portugal)
EO	ecosystem objective
EPA	Environmental Protection Agency (US)
EPBC	Environment Protection and Biodiversity Conservation Act 1999 (Australia)
EPJ	Environmental Policy of Jamaica
ERMA	Environment Risk Management Authority (NZ)
ESEA	environmental and socio-economic assessment
ESSIM	Eastern Scotian Shelf (Canada)
EU	European Union
EurOcean	European Centre for Information on Marine Science and Technology
FAO	Food and Agriculture Organization of the United Nations
FCT	Foundation for Science and Technology
FEEM	Business Forum for Ocean Economy (*Fórum Empresarial para a Economia do Mar*)
FISH	Fisheries for Improved Sustainable Harvest Project (Philippines)
FLAD	Luso-American Development Foundation (Portugal)
FMP	fishery management plan
FONATUR	National Fund for Tourism (Mexico)
FPAM	Standing Forum for Maritime Affairs (*Fórum Permanente Para os Assuntos do Mar*)
FSDEA	Sovereign Wealth Fund of Angola
FSI	Foreign Service Institute (Philippines)
FTP	federal target programme
FUND	International Convention on the Establishment of an International Fund for Compensation for Oil Pollution Damage
GBR	Great Barrier Reef

GBRMPA	Great Barrier Reef Marine Park Authority
GCC	Governance Coordinating Committee (US)
GDP	gross domestic product
GEF	Global Environment Facility
GERCO	National Program of Coastal Management (Brazil)
GI-GERCO	Integration Group of Coastal Management (Brazil)
GIWA	Global International Waters Assessment
GLOBEC	Global Ocean Ecosystem Dynamics
GNP	gross national product
GOF	Global Ocean Forum
GOOS	Global Ocean Observing System
GOOS/Brazil	Brazilian System of Ocean Observing and Climate
GOSLIM	Gulf of St Lawrence (Canada)
GPA	Global Programme of Action for the Protection of the Marine Environment from Land-based Activities
GSMMP	Global Studies and Monitoring of Pollution
HELCOM	Baltic Marine Environment Protection Commission
HLURB	Housing and Land Use Regulatory Board (Philippines)
HOTO	Health of the Oceans (Canada)
HSF	Hans Seidel Foundation
HTL	high tide line
IAEA	International Atomic Energy Agency
IBAMA	Brazilian Institute of Environment and Renewable Natural Resources
ICCAT	International Commission for the Conservation of Atlantic Tunas
ICES	International Council for the Exploration of the Sea
ICJ	International Court of Justice
ICM	integrated coastal [and ocean] management
ICOs	Interdepartmental Committees on Oceans (Canada)
ICOSRMI	Interagency Committee on Ocean Science and Resources Management Integration (US)
ICZM	integrated coastal zone management
IDG	interdepartmental group
IFCA	Inshore Fisheries Conservation Authorities (UK)
ILC	International Law Commission
IMCRA	Interim Marine and Coastal Regionalisation of Australia
IMF	International Monetary Fund
IMO	International Maritime Organization
IMP	Integrated Marine Policy
IMP	Integrated Maritime Policy
IMR	Institute of Marine Research
INE	National Ecology Institute (Mexico)
INP	National Fisheries Institute (Mexico)
IOC	Intergovernmental Oceanographic Commission
IOOS	Integrated Ocean Observing System
IOSEA	Indian Ocean and Southeast Asia
IOTC	Indian Ocean Tuna Commission
IPC	Interagency Policy Committee (US)
ISAP	Integrated Strategic Action Plan

ISBA	International Seabed Authority
ISPS	International Ship and Port Facility Security (Philippines)
ITLOS	International Tribunal for the Law of the Sea
ITSU	International Tsunami Information Centre
IUCN	International Union for the Conservation of Nature
IUU	illegal, unreported and unregulated (fishing)
IWC	International Whaling Commission
IWRM	integrated water resources management
JGOFS	Joint Global Ocean Flux Study
JOCI	Joint Ocean Commission Initiative (US)
JSOT	Joint Subcommittee on Ocean Science and Technology (US)
KCDP	Kenya Coastal Development Project
LBSMP	land-based source of marine pollution
LC/LP	London Convention and Protocol
LDP	Liberal Democratic Party (Japan)
LEPLAC	Plan for the Survey of the Brazilian Continental Shelf
LGU	local government unit
LME	large marine ecosystems
LNG	liquified natural gas
LOICZ	Land–Ocean Interaction in the Coastal Zone
LOMAs	large ocean management area (Canada)
LRTAP	long-range transport of air pollution
LTL	low tide line
MACEMP	Marine and Coastal Environmental Management Project (Tanzania)
MACOP	Ministerial Advisory Committee on Oceans Policy (NZ)
MAF	Ministry of Agriculture and Forestry (NZ)
MAFF	Ministry of Agriculture, Fisheries and Food (UK)
MAM	Ministry of Agriculture and Sea (*Ministério da Agricultura e Mar*)
MAMAOT	Ministry of Agriculture, Sea, Environment and Spatial Planning (*Ministério da Agricultura, do Mar, do Ambiente e do Ordenamento do Território*)
MAOTE	Ministry of Environment, Spatial Planning and Energy (*Ministério do Ambiente, Ordenamento do Território e Energia*)
MaPP	Marine Planning Partnership (Canada)
MARCO	Mid-Atlantic Regional Council on the Ocean (US)
MARD	Ministry of Agriculture and Rural Development (Vietnam)
MARINA	Maritime Industry Authority (Philippines)
MARPOL	International Convention for the Prevention of Pollution from Ships
MCEU	Marine Consents and Environment Unit (UK)
MCS	Marine Conservation Society
MCS	monitoring, control and surveillance
MCZ	marine conservation zone
MDGs	Millennium Development Goals
MEA	multilateral environment agreement
MED	Ministry of Economic Development (NZ)
MEPC	Marine Environment Protection Committee
MEXT	Ministry of Education, Culture, Sports, Science and Technology (Japan)

MFF	Mangroves for the Future
MFF	Multiannual Financial Framework
MFish	Ministry of Fisheries (NZ)
MHWS	mean high water spring
MLIT	Ministry of Land, Infrastructure, Transport and Tourism (Japan)
MLRA	Marine Living Resources Act, No. 18 of 1998 (South Africa)
MLWS	mean low water spring
MMA	Ministry of Environment (Brazil)
MME	Ministry of Mines and Energy (Brazil)
MMO	Marine Management Organisation (UK)
MMS	Minerals Management Service (US)
MOAC	Maritime and Ocean Affairs Center (Philippines)
MoD	Ministry of Defence
MoEF	Ministry of Environment and Forests (India)
MONRE	Ministry of Natural Resources and Environment (Vietnam)
MOST	Ministry of Science and Technology (Vietnam)
MoT	Ministry of Transport (NZ)
MoU	memorandum of understanding
MPA	marine protected area
MPI	Ministry of Planning and Investment (Vietnam)
MSFD	Marine Strategy Framework Directive (EU)
MSP	marine spatial planning
MSR	marine scientific research
MSRA	Magnuson–Stevens Fishery Conservation and Management Reauthorization Act of 2006 (US)
MSWG	Marine Sector Working Group
MSY	maximum sustainable yield
MW	megawatt
NA	National Assembly
NABST	National Advisory Board on Science and Technology
NACOMA	Namibian Coast Conservation and Management
NAFO	North Atlantic Fisheries Organization
NAMMCO	North Atlantic Marine Mammal Commission
NAMRIA	National Mapping and Resource Information Authority (Philippines)
NAO	North Atlantic Oscillation
NAPA	national adaptation programme of action
NASAPI	National Aquaculture Strategic Action Plan Initiative (Canada)
NASCO	North Atlantic Salmon Conservation Organization
NATO	North Atlantic Treaty Organization
NBSAP	National Biodiversity Strategy and Action Plans
NCWS	National Coast Watch System (Philippines)
NDPB	non–departmental public body
NEAFC	Northeast Atlantic Fisheries Commission
NEDA	National Economic Development Authority (Philippines)
NEMC	National Environmental Management Council (Tanzania)
NEMO	National Environmental Management of the Ocean (South Africa)
NEPA	National Environment and Planning Agency (Jamaica)
NEPAD	New Partnership for Africa's Development

NEPSDOC	National Environmental Policy for the Sustainable Development of Oceans and Coasts (Mexico)
NERC	Natural Environment Research Council (UK)
NES	national environmental significance
NFFO	National Federation of Fishermen's Organizations (UK)
NGO	non-governmental organization
NIPAS	National Integrated Protected Areas System (Philippines)
NIWA	New Zealand Institute of Water and Atmosphere
NM	nautical miles
NMP	National Marine Policy
NOAA	National Oceanic and Atmospheric Administration (US)
NOAA CSC	National Oceanic and Atmospheric Administration Coastal Services Center (US)
NOAG	National Oceans Advisory Group (Australia)
NOC	National Ocean Council (US); National Oceanography Centre (UK)
NOO	National Oceans Office (Australia)
NOP	National Ocean Policy
NPA	National Plan of Action
NPA	National Programme of Action (Canada)
NPC	National People's Congress (China)
NRCA	Natural Resources Conservation Authority (Jamaica)
NRCD	Natural Resources Conservation Department (Jamaica)
NRSMPA	National Representative System of Marine Protected Areas (Australia)
NSC	National Security Council (Philippines)
NSF	National Science Foundation (US)
NZ	New Zealand
NZCPS	New Zealand Coastal Policy Statement
OAP	Oceans Action Plan (Canada)
OBOM	Oceans Board of Management (Australia)
OCEANS 21	Ocean Conservation, Education, and National Strategy for the 21st Century Act of 2009 (US)
OCP	ocean and coastal policy
OCS	Offshore Constitutional Settlement (Australia)
OGD	other government departments (Canada)
OMRN	Ocean Management Research Network (Canada)
OPRC	International Convention on Oil Pollution Preparedness, Response, and Co-operation
OPRF	Ocean Policy Research Foundation (Japan)
OPSAG	Oceans Policy Science Advisory Group (Australia)
ORA	Oceans Resources Act (*Havressursloven*) of 2009 (Norway)
ORRAP	Ocean Research and Resources Advisory Panel (US)
OSPAR	Convention for the Protection of the Marine Environment of the North-East Atlantic
OSTP	Office of Science and Technology Policy (US)
OTEC	ocean thermal energy conversion
PACON	Pacific Conference on Science and Technology

PACCSAP	Pacific–Australia Climate Change Science and Adaptation Planning program
PACSICOM	Pan African Conference on Sustainable Integrated Coastal Management
PAF	Federal Action Plan
PAME	Protection of Arctic Marine Environment
PAWB	Protected Areas and Wildlife Bureau (Philippines)
PBGB	Placentia Bay – Grand Banks (Canada)
PCE	Parliamentary Commissioner for the Environment (NZ)
PCIMA	Pacific North Coast of British Columbia (Canada)
PDCTM	Program to Promote Ocean Sciences and Technologies (*Programa Dinamizador das Ciências e Tecnologias do Mar*)
PEGC	State Plan of Coastal Management (Brazil)
PEMEX	Mexican Oil Corporation (*Petróleos Mexicanos*)
PEMSEA	Partnerships for Environmental Management of the Seas of East Asia
PICES	North Pacific Marine Science Organization
PICs	Pacific Island countries
PICTs	Pacific Island countries and territories
PIF	Pacific Islands Forum
PI-GCOS	Pacific Islands Global Climate Observing System
PI-GOOS	Pacific Islands Global Ocean Observing System
PIL	public interest litigation
PIROF	Pacific Islands Regional Ocean Forum
PIROF-ISA	Pacific Islands Regional Ocean Policy and the Framework for Integrated Strategic Action
PIROP	Pacific Islands Regional Ocean Policy
PNCIMA	Pacific North Coast Integrated Management Area (Canada)
PNG	Papua New Guinea
PNGC	National Coastal Management Plan (Brazil)
PNMA	Brazilian National Policy of Environment
PNP	Philippine National Police
PNRM	National Policy for Sea Resources (Brazil)
PO	People's Organization
POP	persistent organic pollutant
PPA	Philippine Ports Authority (Philippines)
PPG-Mar	Training Resources in Marine Sciences (Brazil)
PRF	Partnerships for Environmental Management of the Seas of East Asia Resource Facility
PROANTAR	Brazilian Program for the Antarctic
PROAREA	Prospecting and Exploration of the Mineral Resources from International Areas of South and Equatorial Atlantic
PROARQUIPELAGO	Archipelago Program of São Pedro and São Paulo
PROFEPA	Federal Attorney for Environmental Protection (Mexico)
PROMAR	Maritime Mentality Program (Brazil)
PSRM	Sectorial Plan for Sea Resources (Brazil)
PSSA	particularly sensitive sea area
QMS	Quota Management System (NZ)
RCEP	Royal Commission on Environmental Pollution (UK)

REMPLAC	Program for Assessment of Mineral Potential of the Legal Brazilian Continental Shelf
REVIMAR	Evaluation, Monitoring and Conservation of Marine Biodiversity (Brazil)
REVIZEE	Program Evaluation of the Sustainable Potential of Living Resources in the Exclusive Economic Zone (Brazil)
REZ	renewable energy zone
RFB	regional fishery body
RFMO	Regional Fisheries Management Organization
RMA	Resource Management Act 1991 (NZ)
RMNC	Review of Marine Nature Conservation (UK)
ROP	Regional Ocean Partnerships (US)
RPA	Regional Programme of Action
RPB	Regional Planning Body (US)
RPF	Regional Partnership Fund
RSE	Royal Society of Edinburgh
RSP	Regional Seas Programme (UNEP)
RSPB	Royal Society for the Protection of Birds (UK)
SAC	special area of conservation
SADC	South African Development Community
SAGARPA	Ministry of Agriculture, Livestock, Rural Development, Fisheries and Food (Mexico)
SAP	strategic action plan
SAR	search and rescue
SCOR	Scientific Committee on Oceanic Research
SCT	Ministry of Communications and Transport (Mexico)
SD	sustainable development
SDS-SEA	Sustainable Development Strategy for the Seas of East Asia
SE	Ministry of Economy (Mexico)
SEA	strategic environmental assessment
SEAFO	South East Atlantic Fisheries Organization
SEC	Sustainable Environment Committee of Cabinet (Australia)
SECIRM	Secretariat of the Inter-Ministry Commission for Sea Resources (Brazil)
SECTUR	Ministry of Tourism (Mexico)
SEDESOL	Ministry of Social Development (Mexico)
SEGOB	Ministry of Government (Mexico)
SEMAM	Special Secretary of Environment (Brazil)
SEMAR	Ministry of the Sea–Navy (Mexico)
SEMARNAT	Ministry of Environment and Natural Resources (Mexico)
SENER	Ministry of Energy (Mexico)
SEZ	special economic zone
SFA	Seychelles Fishing Authority
SFC	Sea Fisheries Committee (UK)
SFF	Scottish Fishermen's Federation
SFI	Sea Fisheries Inspectorate (UK)
SI	statutory instrument
SIDA	Swedish International Development Agency

SIDS	small island developing States
SIMOR	Subcommittee on Integrated Management of Ocean Resources (US)
SIOFA	Southern Indian Ocean Fisheries Agreement
SMICZMP	Southern Mindanao Integrated Coastal Zone Management Project
SMRU	Sea Mammal Research Unit (UK)
SMST	Strategy for Marine Science and Technology (UK)
SNH	Scottish National Heritage
SoE	State of the Environment
SOPAC	Pacific Islands Applied Geoscience Commission
SPA	special protected area
SPAW	Specially Protected Areas and Wildlife
SPC	Secretariat of the Pacific Community
SPREP	South Pacific Regional Environment Programme
SPSLCMP	South Pacific Sea Level and Climate Monitoring Project
SRA	Ministry for Agrarian Reform (Mexico)
SRCF	Sub-regional Commission on Fisheries
SRE	Ministry of Foreign Affairs (Mexico)
SRFC	Sub-Regional Fisheries Commission
SSA	Health Ministry (Mexico)
SSA	Sub-Saharan Africa
SSHRC	Social Sciences and Humanities Research Council of Canada
SWIO	South West Indian Ocean
SWIOFC	South West Indian Ocean Fisheries Commission
SWIOFP	South West Indian Ocean Fisheries Project
TAC	total allowable catch
TACC	total allowable commercial catch
TCMP	Tanzania Coastal Management Partnership
TEEP	Temporal Environmental Employment Program (Mexico)
TOGA	Tropical Ocean and Global Atmospheric Project
TPDC	Tanzania Petroleum Development Corporation
TSEEZA	Territorial Sea and Exclusive Economic Zone Act, 1989 (Tanzania)
UK	United Kingdom
UKCS	United Kingdom Continental Shelf
UN	United Nations
UNCED	United Nations Conference on Environment and Development
UNCLOS	United Nations Convention on the Law of the Sea
UNCSD	United Nations Conference on Sustainable Development
UNCSR	United Nations Convention about Sea Rights
UNDP	United Nations Development Programme
UNEP	United Nations Environment Programme
UNESCO	United Nations Educational, Scientific, and Cultural Organization
UNFCCC	United Nations Framework Convention on Climate Change
UNFSA	United Nations Fish Stocks Agreement
UNGA	United Nations General Assembly
UNIDO	United Nations Industrial Development Organization
US	United States
USAID	US Agency for International Development

USOAP	US Ocean Action Plan
USSIWOC	Unified State System of Information on the World Ocean Conditions (Russian Federation)
USSR	Union of Soviet Socialist Republics
VASI	Vietnam Administration of Seas and Islands
VEPA	Vietnam Environmental Protection Agency
VMS	vessel monitoring system
WCPA	World Commission on Protected Areas (IUCN)
WG	Welsh Government
WHC	World Heritage Commission
WIO	Western Indian Ocean
WIOMSA	Western Indian Ocean Marine Science Association
WIOTO	Western Indian Ocean Tuna Organization
WMO	World Meteorological Organization
WOCE	World Ocean Circulation Experiment
WRI	World Resources Institute
WSSD	World Summit on Sustainable Development
WWF	World Wildlife Fund
ZOFEMATAC	Maritime and Coastal Federal Zone Administration (Mexico) (*Zona Federal Marítimo Terrestre y Ambientes Costeros*)

PART I

Introduction

1

A COMPARATIVE ANALYSIS OF OCEAN POLICIES IN FIFTEEN NATIONS AND FOUR REGIONS

Miriam C. Balgos, Biliana Cicin-Sain, and David L. VanderZwaag

> For too long, the world acted as if the oceans were somehow a realm apart – as areas owned by no-one, free for all, with little need for care or management. The Law of the Sea Convention and other landmark legal instruments have brought important progress over the past two decades in protecting fisheries and marine ecosystems. But this common heritage of all humankind continues to face profound pressures.
>
> *UN Secretary General Kofi Annan, at the 2004 meeting of the Seychelles and the United Kingdom, 'Reefs, Island Communities and Protected Areas: Committing to the Future', (Greenpeace, 2008: 2)*

Introduction

The management of oceans has always presented a challenge because of their fluidity, magnitude, geographic scope, and variety of uses. Hampered by powerful sectoral and traditional modes of governance, ocean management has become increasingly difficult as the oceans continue to absorb the unabated impacts of human uses, both downstream and upstream.

Beginning in 1945, the enclosure by coastal nations of ocean space adjacent to their coastlines as a means of asserting jurisdiction over the resources found in those areas – such as fisheries, and oil and gas – triggered the development of a new paradigm for international ocean governance: the worldwide acceptance of national jurisdiction over 200 nautical mile ocean zones. A series of international conferences, beginning in 1958, aimed to restore order and coherence at the international level after coastal nations put forward claims of extended ocean jurisdiction of various widths and powers. Three Law of the Sea Conferences, held under the auspices of the United Nations, were conducted to sort out the rights and duties of nations regarding oceans. The United Nations Convention on the Law of the Sea (UNCLOS) was concluded after nine years of negotiation in 1982 and came into force in 1994. The Convention represented a constitution for the world's oceans, and it detailed the rights, duties, and obligations of nations related to the ocean and its resources (Cicin-Sain and Knecht, 1998).

While the 320 Articles contained in UNCLOS address virtually all ocean issues and established international norms for future ocean governance, it generally provided little guidance to nations on how to govern ocean resources in an integrated manner, how to deal with the effects of one use on other uses, or how to bring ocean and coastal management together (Cicin-Sain and Knecht, 1998).

The United Nations Conference on Environment and Development (UNCED) convened in Rio de Janeiro in 1992 and produced five major outputs: the Rio Declaration on Environment and Development; the UN Framework Convention on Climate Change (UNFCCC); the Convention on Biological Diversity (CBD); Agenda 21; and a set of forest principles (Johnson, 1993). The Rio Declaration provided a set of twenty-seven non-binding principles to guide national and international actions on the environment, development, and social issues. Chapter 17 of Agenda 21, which was intended to act as a roadmap to sustainable development, stressed the importance of oceans and coasts in the global life-support system, as well as the positive opportunities that ocean and coastal areas and resources offer for sustainable development. Chapter 17 called for integrated policy and decision-making processes and institutions for the integrated management and sustainable development of coastal and marine areas, at both the local and national levels. It also suggested actions that coordinating institutions should undertake, and called for cooperation among states in the preparation of national guidelines for integrated management processes and in carrying out biodiversity conservation measures within their national jurisdiction. Integrated coastal and ocean management was also the recommended framework for dealing with coastal and ocean issues under the UNFCCC and the CBD (Cicin-Sain and Knecht, 1998).

These prescriptions were further reinforced at the 2002 World Summit on Sustainable Development (WSSD) held in Johannesburg, South Africa, at which the world's political leaders agreed to include among the major targets on oceans the promotion of integrated coastal and ocean management at the national level, and the encouragement and assistance to countries in developing ocean policies and mechanisms on integrated coastal management (United Nations, 2002). The Global Programme of Action for the Protection of the Marine Environment from Land-based Activities, the Barbados Programme of Action for the Sustainable Development of Small Island States, and the Mauritius Implementation Strategy also prescribe the use of integrated coastal [and ocean] management (ICM) as a framework for addressing ocean and coastal issues (Cicin-Sain et al., 2006).

Moving toward integrated oceans governance at national and regional levels

Until recently, most coastal nations of the world have relied on a variety of sectorial policies to manage different uses of the ocean, such as shipping, fishing, energy extraction, and recreation. This was sufficient in a time of resource abundance and few conflicts among ocean uses, but this time has long passed. In most nations and world regions, conflicts among the many uses of the oceans abound, and nations and regions face issues of resource depletion, pollution, and biodiversity loss, as well as a range of harmful effects from climate change. Countries are also facing important challenges from new uses of offshore waters, such as offshore aquaculture and wind farming, which necessitate the development of more elaborate and cross-cutting frameworks for national ocean policies.

Over the last two decades, a number of coastal nations have undertaken concerted efforts to articulate and implement an integrated vision for the governance of the entire ocean areas under their jurisdiction including their internal waters, territorial seas, and exclusive economic zones (EEZs). This is a very encouraging development, responding to the reality of serious conflicts of use in most national ocean zones, and to the prescriptions articulated in both UNCLOS ('the problems of ocean space are interrelated and must be treated as a whole') and in UNCED (given the interrelationship among uses and processes in the coast and ocean, ocean and coastal governance must be 'integrated in content and precautionary in ambit') (Cicin-Sain et al., 2006).

The move to create comprehensive national ocean policies to harmonize existing uses and laws, to foster sustainable development of ocean areas, to protect marine biodiversity and vulnerable resources and ecosystems, and to coordinate the actions of the many government agencies that are typically involved in oceans affairs is thus a growing practice, with more and more nations embarking on the development of a national ocean policy. Initial research indicates that there at least twenty-three countries and four regions of the world that have, or are taking concrete steps toward, cross-cutting and integrated national and regional ocean policies (not only separate sectoral policies). Examples include nations already in the phase of *implementing* national ocean policies (Australia, Brazil, Canada, China, Jamaica, Japan, Mexico, Norway, Portugal, Russian Federation, United Kingdom, and United States), nations in the process of *formulating* national ocean policies (New Zealand, the Philippines, South Africa, South Korea, and Vietnam), and nations in the *preparatory phase* of planning for national ocean policies (Colombia, France, India, Indonesia, Malaysia, and Thailand).

At the regional level, three regions have undertaken systematic efforts to formulate regional ocean policies: the Pacific Islands region, the East Asian Seas region, and the European Union. A fourth region, Africa, is beginning such a process in the context of the New Partnership for Africa's Development (NEPAD).

In all of the cases noted, nations and regional entities are facing the challenge of developing new concepts, procedures, and structures, and as such have much to gain from working with one another to share information and to draw lessons and best practices. As they embark on national- and regional-level ocean policy formulation, many nations – in particular, small island developing states – will need assistance in mapping and delimitation of their EEZs, and in developing new institutions and procedures.

The purpose of this book

The focus of this book is on the development and implementation of integrated ocean policies at both the national and regional levels. At the national level, by 'integrated ocean policy' we mean the cross–cutting policy framework for the integrated management of a nation's marine jurisdiction (territorial sea and EEZ). Integrated ocean policy is established for the purpose of providing a coordinated approach to the management of the whole range of marine uses and activities across all maritime sectors for the protection of the social, economic, and environmental values within a nation's marine jurisdiction (Petrachenko and Addison, Chapter 4). The development of an integrated ocean policy is an evolutionary process commonly initiated by the national government and involves consultation with various stakeholders.[1] It is guided by principles prescribed by international ocean law, and the policy itself will typically define guiding principles for the management of oceans and coasts under national jurisdiction. Policy might encourage integrated planning at various scales, for example the establishment of different tools for the planning of large ocean management areas (LOMAs) and coastal management areas (CMAs) such as in the case of Canada (Mageau et al., Chapter 3), or ecosystem-based planning in the case of Australia (Petrachenko and Addison, Chapter 4). The integrated ocean policy can take the form of comprehensive legislation, a range of policy initiatives that are marine-related and legislatively based, or an executive directive for the adoption of an oceans or marine strategy (Mageau et al., Chapter 3; Smith et al, Chapter 6; Petrachenko and Addison, Chapter 4). Integrated ocean policy does not replace sectoral policies related to oceans, but seeks to harmonize existing laws and policies, to reconcile conflicts, and to minimize duplication, inconsistencies, and overlaps in existing legislation or policy directives.

Regional entities have a useful role to play in assisting states both in developing national ocean policies for their ocean zones, as well as in articulating common approaches to the governance of particular ocean regions shared by various nations. Significant work along these lines is already taking place in the Pacific Islands region, the East Asia region (through the Partnerships in Environmental Management for the Seas of East Asia, or PEMSEA), the Asia Pacific region (through Asia-Pacific Economic Cooperation, or APEC), and the European Union. Significant work to bring nations together to address common transboundary regional issues related to the coasts and oceans is also being carried out in many regions of the world, through the Regional Seas Programme and the large marine ecosystem (LME) projects or programmes (Freestone et al., 2010). While systematic comparative analyses of these experiences are needed, this lies outside the scope of this book.

This book provides policy analyses of the experiences of fifteen countries (Australia, Brazil, Canada, India, Jamaica, Japan, Mexico, New Zealand, Norway, Philippines, Portugal, Russian Federation, United Kingdom, United States, and Vietnam) and four regions of the world (East Asian Seas, European Union, Pacific Islands, and Sub-Saharan Africa) in the development and implementation of integrated national and regional ocean policy (see Table 1.1 and Figures 1.1 and 1.2). The case studies have been commissioned from leading ocean policy academics and practitioners from around the world and follow a similar prescribed format.

The combined EEZs of the countries included in the study comprise 49.42 per cent of the world's EEZs (the countries total), while the combined EEZs of the regions covered comprise 54.65 per cent of the world's EEZs (the regions total), bringing these ocean areas under ongoing and evolving efforts of integrated management (see Table 1.2). This is an important and encouraging finding in and of itself, because it shows that a large portion of the world's EEZs are already the subject of deliberate efforts by national and regional authorities to improve oceans governance.

The aim of the book is to learn from this wide range of national and regional experiences, to understand the dynamics of integrated ocean policy development and implementation, to identify factors that contribute to successful policy development and implementation, and to draw lessons that may be useful to other nations and regions as they move to develop and implement national and regional ocean policies.

This book was prepared as part of the work of the Research Task Force on National Ocean Policies, initially organized in 2004 under the sponsorship of the Nippon Foundation of Japan and the University of Delaware Gerard J. Mangone Center for Marine Policy, to bring together academic experts and government officials working on national ocean policies from around the globe. The Research Task Force aimed to identify commonalities and differences among cases, to identify factors that promote or hinder cross-sectorial oceans policy, to draw lessons that

Table 1.1 List of countries and regions included in the study

	Country/ies	*Region*
Asia	India, Japan, Philippines, Vietnam	East Asian Seas
Oceania	Australia, New Zealand	Pacific Islands
Americas	Brazil, Canada, Mexico, United States	
Europe	Norway, Portugal, Russian Federation, United Kingdom	European Union
Africa		Sub-Saharan Africa
Caribbean	Jamaica	

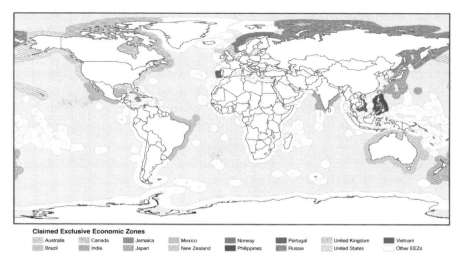

Claimed Exclusive Economic Zones

Australia	Canada	Jamaica	Mexico	Norway	Portugal	United Kingdom	Vietnam
Brazil	India	Japan	New Zealand	Philippines	Russia	United States	Other EEZs

Figure 1.1 The claimed exclusive economic zones (EEZs) in the fifteen national case studies

Note: The figure shows the claimed EEZs adjacent to the countries' mainland (i.e. it does not include non-adjacent areas of a country unless they are formal states of the country, such as the US state Hawaii); it depicts the EEZs as provided by the Flanders Marine Institute (VLIZ) database; and it does not include disputed areas.

Source: Prepared by Edward Carr, School of Marine Science and Policy, College of Earth, Ocean, and Environment, University of Delaware, based on VLIZ *Maritime Boundaries Geodatabase*, V.8, online at http://www.marineregions.org/ [accessed 23 August 2014].[2]

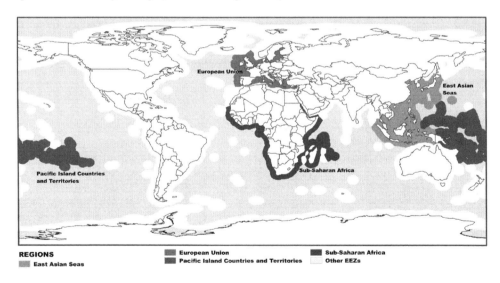

REGIONS

East Asian Seas	European Union	Sub-Saharan Africa
	Pacific Island Countries and Territories	Other EEZs

Figure 1.2 The four regions covered in the case studies

Note: East Asian Seas covers six large marine ecosystems (the South China Sea, the Gulf of Thailand, the East China Sea, the Yellow Sea, the Sulu-Celebes Sea, and the Indonesian Seas); *European Union* shows the claimed EEZs adjacent to the countries' mainland and does not include overseas EEZ claims; *Sub-Saharan Africa* includes the EEZs of African countries south of the Sahara (except Sudan), and of the island countries of Comoros, Madagascar, Mauritius, and Seychelles; and the figure depicts the EEZs as provided by the VLIZ database and does not include disputed areas.

Source: Prepared by Edward Carr, School of Marine Science and Policy, College of Earth, Ocean, and Environment, University of Delaware, based on VLIZ *Maritime Boundaries Geodatabase*, V.8, online at http://www.marineregions.org/ [accessed 23 August 2014]

Table 1.2 The Exclusive Economic Zones of countries and regions covered in the book

Country	km²	World (%)★
Australia	8,974,857	6.30
Brazil	3,648,308	2.56
Canada	6,006,154	4.22
India	2,290,268	1.61
Jamaica	263,283	0.18
Japan	4,050,000	2.84
Mexico	3,269,386	2.30
New Zealand	6,697,428	4.70
Norway	2,555,602	1.79
Philippines	2,265,684	1.59
Portugal	1,832,848	1.29
Russian Federation	8,095,881	5.68
United Kingdom	6,797,578	4.77
United States of America	12,234,403	8.59
Vietnam	1,396,299	0.98
TOTAL (COUNTRIES)	**70,377,979**	**49.42**
East Asia	17,215,373	12.09
European Union	24,511,751	17.21
Oceania	28,241,259	19.83
Sub-Saharan Africa	7,866,074	5.52
TOTAL (REGIONS)	**77,834,457**	**54.65**

★*EEZ world total*: 142,412,107 km²

Source: Sea Around Us Project (undated)

might be learned, to capture emerging good practices, to encourage other nations to formulate and implement national ocean policies, and to encourage networking among ocean policy entrepreneurs around the world.

For each selected case, the editors sought to invite at least one academic scholar and one government official working on national or regional ocean policy to develop a case study on their respective national or regional ocean policy. The project benefited from the expert guidance of an advisory committee composed of eminent ocean leaders, as noted in the Preface section. The research effort has greatly benefited mainly from the discussions held at a number of conferences, including: the Ocean Policy Summit, 'Integrated Ocean Policy: National and Regional Experiences, Prospects and Emerging Practices', held on 10–14 October 2005 in Lisbon, Portugal (TOPS 2005), at which initial versions of some of the case studies included in the book were presented; the Third Global Oceans Conference 2006; the Fourth Global Oceans Conference 2008; and the Fifth Global Oceans Conference 2010.[3]

Since some of the chapters in this volume were first written in 2006, substantial updating was required for publication. Authors were given the options of either adding a postscript providing updates to at least 2013 or incorporating necessary updates within the main body of their chapters. Readers of this volume will therefore find variations in approaches to updating.

The editors formulated a preliminary common framework for cross-national study of national and regional ocean policies, to categorize countries undergoing national ocean policy development into various categories, and to develop an initial sample and research framework for analysis. The national and regional efforts at creating cross-cutting and integrated national ocean policies presented are at different phases of policy formulation and implementation, defined as:

1   the *preparatory stage* – informal processes are ongoing to prepare the nation (or region) in the formal development of a cross-cutting and integrated national or regional ocean policy;

2   the *formulation stage* – a well-defined formal process is under way to develop a cross-cutting and integrated national or regional ocean policy; or

3   the *implementation stage* – cross-cutting and integrated national policy has already been enacted and is already being implemented (with funding).

The cases included in the study, grouped according to the stage at which they are in the policy development process, are as follows.

- *Implementation stage*: Countries – Australia (1998), Brazil (1980 and 2005), Canada (1996), Japan (2007), Jamaica (2002), Mexico (2006), Norway (2008), Portugal (2006 and 2013), Russian Federation (2001), United Kingdom (2009, 2010, and 2013), United States (2010); Regions – East Asian Seas (2003), European Union (2007), and Pacific Islands (2002)
- *Formulation stage*: Countries – New Zealand, Philippines, Vietnam
- *Preparatory stage*: Country – India; Region – Sub-Saharan Africa

The cases presented in the book were selected on the basis of the following criteria, which are designed to maximize the presentation of a wide variety of experiences:

1   they are found in *different regions of the world*;

2   they represent cases from both *developed and developing countries*;

3   they are found in *different phases of the ocean policy formulation and development process*; and

4   they include *examples of regional ocean policy formulation* that are likely to have important effects in shaping national ocean policies.

The framework of analysis and major questions posed

The study examined all of the national and regional cases using a common framework of analysis, which took into consideration the motivation underlying the development of ocean policy, its scope, the evolution of the policy over time, and the implementation and evaluation of the policy. It is important to note that the case studies followed the framework of analysis to the extent possible subject to the availability of data and applicability, since the case studies are in varying stages of policy development and implementation.

Appendix A presents a guide to comparative case studies on national ocean policies prepared by the editors to assist in the preparation of the national case studies. Regarding the regional case studies, a more simplified guide, adapted from the national case studies guide, provided guidance to the authors of the regional case studies and may be found in Appendix B. In the following section, we elaborate some of the major questions asked, first, of the national case studies, and second, of the regional case studies.

The significance of oceans and coasts in different nations

Case study authors were asked first to provide basic information about the country (such as geographic and demographic information – for example total land area, length of coastline, maritime zones, and human population – and maritime boundaries), as well as to depict the status of and utilization of marine and maritime resources, including economic data and conflicts among uses, thus providing the context for why oceans and coasts are an important policy area in the particular country.

A brief overview of the nature and evolution of the national ocean policy is given for each case study, including the factors that gave rise to it, such as use conflicts, environmental problems, increasing public demand for nature conservation and sustainable exploitation of natural resources, the need to expand coastal management efforts further offshore, opportunities for economic development, and results from large-scale research programmes, for example large marine ecosystems (LMEs), global international waters assessment, and the Global Ocean Observing System (GOOS). Descriptions of the stakeholders and key players involved in the process, including which groups were in favour and which were against policy development, along with other obstacles that were faced in national ocean policy formulation and implementation, are part of the information provided on the conflicts in the development of a national ocean policy.

The dynamics of national ocean policy development

Among the more important questions posed in each case study is how the policy development process began – that is, how it was initiated and by whom (for example through a legislative mandate, as a prime minister initiative, by a commission), the approach followed in the establishment of a national ocean policy, its objectives (for example to foster sustainable development, to protect biodiversity, to harmonize existing uses and laws, to coordinate government agencies), the principles that guided the process, and the steps that were followed during the policy preparatory or formulation phases.

Major questions asked

- How was the policy initiated and by whom?
- What objectives were addressed in the policy?

Guiding principles

The major principles that nations adopted or followed in national ocean policy development were often based, at least in part, on those prescribed by a range of international agreements, such as Chapter 17 of Agenda 21, the CBD, and UNCLOS, and include, most prominently: sustainable development/sustainability; integrated management; ecosystem-based management; good governance; adaptive management/best available science; the precautionary approach; the preservation of marine biodiversity; stewardship; multiple use management; and economic/social development and poverty alleviation. As discussed in the concluding section of this chapter, the different national ocean policies are remarkably congruent in terms of overall principles applied, and most confirm the importance of transparency, public and stakeholder involvement, incentives for cooperative action, and public accountability.

Major question asked

What major principles were adopted or followed in the national ocean policy?

Institutional considerations

Case studies address the institutional arrangements and processes followed in the formulation of the national ocean policy, including assessments of how well these worked. Some of the important questions posed relate to: the office responsible for formulating the policy; the processes that were followed; the political process being managed; the ways in which the stake-holders were involved; and the approaches that were adopted to harmonize sectorial issues and to resolve conflicts. Authors were also asked to analyse the specific strategies or interventions that were followed in the decision-making process, such as: consultation at ministerial and inter-ministerial levels; the organization of formal/informal advisory groups; definition by the leading agency of a draft vision for the nation's ocean, and guiding values and principles; the main goals and objectives of the policy; its scope and geographical application; the establishment of thematic working groups; an assessment of existing ocean-related policies and gap analysis; the formulation of policy options, including legislative and non-legislative; various iterations of consultation with relevant constituencies; and the issuance or adoption of a formal ocean policy.

Major questions asked

- What institutional arrangements and processes were followed to formulate the national or regional ocean policy? How well did they work?
- Who was in charge of formulating the policy?
- How were the stakeholders involved in the process?
- What approaches were adopted to harmonize sectorial issues and to resolve conflicts?
- What specific strategies or interventions were followed in the decision-making process?

The nature of the policy and legislation established

As important as the process itself are the results of the process, including whether it is an admin-istratively based or legislatively based policy. Each case study addresses the nature of the ocean policy established, for example the legal framework of the policy, what principles for ocean governance it provides, and whether it integrates sectorial legislation. The policy is assessed in terms of how well the policy applies the principles of integration, the precautionary principle, ecosystem-based management, public participation, community-based management, and the set of principles that the country has decided to adopt in the development and implementation of its national ocean policy.

Major questions asked

- Is the resulting policy administratively based or legislatively based, or both? Has a frame-work ocean law been adopted or is it planned? Does it incorporate constitutional or norma-tive principles for ocean governance?
- What is the legal framework of the national ocean policy? If there is a law, does it integrate sectorial legislation?
- What have been the major results of the programme? Measured against a set of evaluative criteria, how well did the programme achieve national ocean policy objectives?

Authority at national and subnational levels of government

The agency that took charge of implementing the policy, for example a lead agency, interim body, or a unit in the prime minister's office that serves as a focal point for implementation, is also described for each national ocean policy. In addition, how the implementing structure is organ-ized, including whether there is an advisory group, or a secretariat or staff in charge of the policy or programme, is included. Furthermore, the division of authority over ocean issues among national and subnational levels of government, and how the authority over land and water jurisdictions is divided, are among the general conflicts and competition between levels of government that are reviewed, along with the existing policies that address intergovernmental conflicts, the processes present in addressing these conflicts, and whether the processes are working well.

Major questions asked

- Who is in charge of implementing the policy? Is there a lead agency, an interim body, or a unit in the prime minister's executive office that serves as a focal point for implementation?
- How is the implementing structure organized? Is there an advisory group, or a secretariat or staff in charge of the policy or programme?
- What is the division of authority over ocean issues among national and subnational levels of government?
- How is the authority over land and over water jurisdictions divided?
- Are there general conflicts and competition among levels of government? Are there existing policies that address intergovernmental conflicts? What processes are present in addressing these conflicts? Are these processes working well?

The domestic implementation of international agreements

The case studies also examine the domestic implementation of selected international agreements, including UNCLOS, the CBD, the International Convention for the Prevention of Pollution from Ships (MARPOL 73/78), the 1972 London Convention on the Prevention of Marine Pollution by Dumping Wastes and Other Matter and its 1996 Protocol, and the Global Programme of Action for the Protection of the Marine Environment from Land-based Activities (GPA).

Major questions asked

- How are international agreements implemented at the national level?
- What organizational structures, methods, and resources are available to enforce the provisions of national ocean policy?
- Are there scientific research and technological development programmes to support the national ocean policy? Is there an ocean chapter in the national State of the Environment (SoE) report? Are there educational, training, and awareness programmes that build an 'ocean ethic'?

Implementing structure, financing, monitoring and evaluation, and outlook

For the countries that have reached the implementation stage in the process, information was provided on the organizational structure, methods, and resources used in the enforcement of national ocean policies, including violations and fines, and reports on compliance. Information is also provided on research, education, training and public awareness, and other programme components that are intended to support the implementation of the national policy, such as research programmes and an ocean chapter in the national SoE report.

The question of financing was dealt with especially by the nations that have gone further along in the implementation of their national ocean policies, including Australia, Brazil, Canada, Jamaica, Japan, the Russian Federation, and the United Kingdom. All provided information on how their national ocean programme is supported, whether the funds are adequate to administer the national ocean programme, and whether funds are available to develop related operational projects. For these countries, the case studies review the problems, issues, or obstacles addressed by the policy or programme of implementation, and whether the policy or programme is able to address these problems adequately, and if not, to determine the extent of impacts on the management and conservation of oceans, coasts, and islands. In this section of the case studies, information on the monitoring, evaluation, and adjustment processes carried out by the programme – including how lessons are identified and applied in programme implementation, and how factors are leading to successful or unsuccessful interventions – are also described. In the evaluation aspect, assessment as to whether the goals of the national ocean policy are attained – and, specifically, how economic security, maritime security, and social stability are achieved – through implementation of the policy is reviewed, along with indicators for success.

Major questions asked

- What organizational structures, methods, and resources does the country have to enforce provisions of national ocean policy?
- How is the national ocean programme financed – that is, what are the sources of its funding?
- How is the national ocean programme financed – that is, what are the sources of its funding? Are there adequate funds with which to administer the national ocean programme? Are there funds available to develop related operational projects?
- What were the major problems or obstacles encountered in the implementation of the national ocean policy? How does the policy address these? Do these problems continue to affect the management and conservation of oceans, coasts, and islands?
- What systems of monitoring and assessment or evaluation are used? What feedback process, learning, and adjustments are being carried out by the programme? How are lessons identified and applied in programme implementation and how are factors leading to successful or unsuccessful interventions recognized?
- How well are the goals of the national ocean policy attained? Specifically, how well are economic security, maritime security, and social stability attained through implementation of the policy?
- What indicators are used to measure success of the national ocean policy initiative? How were they developed?
- In your view, what is the short- and long-term outlook for the national ocean programme? Is this programme sufficiently rooted that it will be sustained for the next ten years?

Recommendations on the way forward

The authors' perspectives are added in the last section of the case studies, including their short- and long-term outlooks for the national or regional ocean policy or programme, and their viewpoints on ways of improving the national or regional ocean policy, the process by which it is being, or was, developed, and its implementation at the regional, national, or subnational levels.

Major question asked

What are your own ideas for improving the national ocean programme in your country, in terms of promoting national and subnational levels of action?

Regional policies

Similar analyses were undertaken of the four regions covered by the study. The same questions used in the analyses of national ocean policies were asked at the regional level, with the main difference being the additional focus on transboundary oceans issues, and a corresponding international integrated approach to management and governance of issues at the regional level, including questions about cohesion, consistency, and synergy in the principles, methods, and approaches employed among countries within the region.

Emerging findings

The cases present a wide array of experiences found at both national and regional levels, in developing as well as in developed countries, and in different regions of the world. This chapter provides a synopsis of current practice in integrated ocean policymaking and implementation that could be of use to nations and regions intending to develop their ocean policies in terms of expanding the options for carrying out the steps in the policy process. For a detailed description of the practices, it is imperative that readers review the specific case studies cited in order to ascertain the conditions under which the practices were (or were not) deemed effective in the context of the nation or region (Rose, 2004).

Appendix C summarizes each of the fifteen national ocean policies and four regional ocean policies covered by the study, providing a synopsis of the nature and content of existing policies, their status, and some highlights and recent developments.

The following sections examine the case studies according to the following key areas of national and regional ocean policy development:

- the nature and content of integrated national ocean policies;
- common catalysts;
- modes of getting started;
- the principles adopted;
- institutional aspects;
- national/subnational relations;
- funding;
- monitoring/adaptation; and
- supporting elements.

The nature of integrated national ocean policy

Form

Public policy can be defined generally as a system of laws, regulatory measures, plans or frameworks of action (e.g., projects, programs, and strategies) and funding priorities concering a particular topic, and implemented by an entity that has a mandate over the topic. Governmental policies are reflected in multiple ways. National ocean policies are commonly expressed either as a legislative policy (formulated in policy or appropriation bills by a legislature) or an executive policy (through executive orders and instructions to department or ministries). Examples of legislative policies are Canada's Oceans Act of 1996, Japan's 2007 Basic Act on Ocean Policy, Norway's Oceans Resources Act (*Havressursloven*) of 2009, and the UK's Marine and Coastal Access Act 2009 and Marine (Scotland) Act 2010. On the other hand, Australia's Oceans Policy, Mexico's National Environmental Policy for the Sustainable Development of

Oceans and Coasts, and the US National Ocean Policy are examples of executive policies. The differences in the form of policies are based on the institutional and political structure current at the time of the policy development process and how that structure supports or obstructs the type of policy adopted. For example, the United States, which has been trying to develop a legislatively based ocean policy for decades, recently adopted a national ocean policy by executive order instead, by means of a supportive presidential administration, following the recommendation of a Presidential Interagency Ocean Policy Task Force.

It is well recognized that legislatively based policies are more robust and resilient in the face of political changes, in contrast to executive-based policies (such as executive orders), which can be changed 'with the stroke of a pen' when a new administration comes to power.

The national ocean policy can be in an 'all in one' policy that contains all of the essential elements, such as Australia's Oceans Policy, or it could comprise multiple interrelated parts, such as that of Canada, which has been composed of:

1   the Oceans Act (providing the guiding principles on integrated management, sustainable development, and the precautionary approach; mandating the development and implementation of programmes to implement these principles; placing the existing authorities of the Department of Fisheries and Oceans within the context of oceans management; and providing guidance on how other mandated authorities should deliver their mandates);
2   the Oceans Action Plan (providing priority areas for action under four major themes – international leadership, sovereignty, and security; integrated oceans management for sustainable development; health of the oceans; and science and technology); and
3   the Oceans Strategy (expanding on the policy framework).

Content

Ideally, a national ocean policy identifies the priority issues that it aims to address, sets out the objectives that address the issues, provides the guiding principles on how the objectives are to be met, lays out the strategic approaches, plans, and programmes to be carried out in order to achieve the objectives, provides the institutional mechanism that will carry out the plans and programmes, and provides the requisite resources for implementation, monitoring, and evaluation.

The national ocean policies covered in this volume represented responses to a variety of issues that include: use conflicts; environmental problems; increasing public demand for nature conservation and sustainable exploitation of natural resources; the need to expand coastal management efforts further offshore; opportunities for economic development; and other issues identified from large-scale research programmes, such as the LME projects funded by the Global Environment Facility (GEF).

The objectives of the national ocean policies commonly include: achieving sustainable development (addressing all uses of the ocean); protecting ocean ecosystems, biodiversity, and vulnerable areas; promoting social and economic advancement; and ensuring maritime security. There are other objectives that may be unique to one or common to only a few countries, which include protecting sovereign rights and ensuring freedom of the high seas (Russian Federation), restoring ocean health (United States and Canada), promoting the scientific knowledge of the ocean (India, Japan, and Portugal), and integrating, coordinating, or harmonizing existing sectorial arrangements for ocean and coastal governance (Japan, Mexico, and Norway).

The principles found in the national ocean policies examined generally converge around principles that were adopted at the 1992 UNCED and the 2002 WSSD. These principles, along with the approaches, strategies, plans, and programmes formulated to achieve the policy objectives, are discussed in detail later in this chapter.

At the regional level, broad objectives are identified that address common issue areas, as well as issues that are transboundary in nature which are pertinent to the region, such as the management of shared or migratory fish stocks. For the European Union, East Asian Seas, and the Pacific Islands, the establishment of a framework for the development of national ocean policies was a primary objective.

Overall, the objectives identified in the national and regional case studies:

- recognize the interrelationships among marine ecosystems and multiple uses of the oceans and coasts;
- collectively do not aim to replace sectorial policies (such as fisheries), but strive to harmonize sectorial policies and, in many cases, also involve recommendations on improving particular sectorial policies; and
- recognize the common property nature of marine areas, duties to future generations, and duties of accountability and transparency.

The principles and other characteristics of integrated national ocean policies covered in the book are discussed in the following sections. Table 1.3 summarizes the form and content of the national and regional ocean policies covered in the book.

The policy development process

Common catalysts

There are various factors that can lead to policy change related to oceans, including catastrophic events, the personal experiences of decision-makers, high-visibility issues, social, economic, and technological changes, information dissemination about an issue, federal or national government influences on state and local policy, and commissions and work groups – all are examples of catalysts and sources of ideas for policy change (Tableman, 2005). The development of national ocean policies, in general, has been triggered by multiple catalysts that include, but are not limited to:

- multiple-use conflicts – conflicts among uses, users, and agencies;
- the decline in or degradation of resources and marine/coastal areas;
- recognition of the value of coastal and ocean resources in terms of ecological/ecosystem, social, and economic services;
- encouragement from the international level (in the form of international agreements, regional entities, and donor action); and
- inequities in accrued benefits for foreigners vs locals in ocean areas under national jurisdiction.

Unfortunately, for most nations, the need for integrated management of ocean and coastal resources has become apparent with the degradation of the marine environment and resources, and the resulting loss of a variety of associated economic opportunities. Upon recognition of the significant economic, as well as social and ecological potential of ocean and coastal resources to various users and stakeholders, countries have initiated the establishment of a policy for multiple uses through integrated and other modern approaches.

The evolution of national ocean policies often involves an initial foray into integrated coastal management (ICM) and the eventual realization of a need to expand from the coast out, to

Table 1.3 The characteristics of national and regional ocean policies

Country	Form	Content
Countries		
Australia	*Administrative framework*: Oceans Policy (1998)	Nine broad goals that were developed to meet the general expectations of the Australian community and address the range of issues facing the marine environment Eight major planning principles
Brazil	*Executive*: National Policy for Sea Resources (PNRM), instituted in 1980 and updated in 2005	Connects the various marine and coastal sectorial policies and plan initiatives and programmes for implementation through the National Plan of Coastal Management (PNGC) Revised PNRM (2005) included focus on marine biodiversity and exploration of non-living marine resources Provides for a non-centralized execution and involves states and civil organizations
Canada	*Legislative*: Oceans Act of 1996	Provides guiding principles on integrated management, sustainable development, and the precautionary approach; mandates the development of programmes to implement these principles; places the existing regulatory and management authorities of the Department of Fisheries and Oceans Canada (DFO) within the context of oceans management; offers guidance on how other mandated authorities should deliver their mandates Oceans Action Plan provided priority areas for action under four major themes (international leadership, sovereignty, and security; integrated oceans management for sustainable development; health of the oceans; science and technology) Oceans Strategy provides the policy framework
Jamaica	*Other*: Green Paper on Coastal and Oceans Management (2002)	Provides for enhancement of institutional capacities, integrated planning and management, prevention of environmental degradation, and community-based participatory approaches Specific action recommendations for each identified sustainable development issue, with institutional responsibility assigned to relevant agencies
Japan	*Legislative*: Basic Act on Ocean Policy of 2007	Six basic principles aimed at harmonization of development and use of the oceans with conservation of the marine environment, integrated management of the oceans, and international coordination and cooperation on ocean issues Stipulates the formulation and decision of a Basic Plan on Ocean Policy by the Cabinet Twelve basic measures to be taken in a comprehensive and systematic manner, including conservation of the marine environment, promotion of development, use and conservation in the EEZ and on the continental shelves, and integrated management of the coastal zones, among others Stipulates establishment of Headquarters for Ocean Policy in the Cabinet to promote measures with regard to the oceans intensively and comprehensively

(continued)

Table 1.3 (continued)

Country	Form	Content
Mexico	*Executive*: National Environmental Policy for the Sustainable Development of Oceans and Coasts (2006)	Contains main strategies and guidelines that define the environmental policy regarding the oceans and coasts of Mexico Main objectives are: developing a strategy for the integrated management of oceans and coasts; strengthening coordinated actions between and within coastal and marine-related institutions; promoting social and economic welfare through environmental and biodiversity conservation, and the sustainable use of coastal and marine natural resources; strengthening the legal, normative, and administrative framework for management of oceans and coasts; and developing an information system specific to oceans and coastal issues Principles: ecosystem-based management, multiple-use, sustainability, participatory governance, precautionary approach, adaptive management, and biodiversity protection
Norway	*Legislative*: Oceans Resources Act (Havressursloven) of 2009; Nature Diversity Act of 2009	Consolidates all relevant provisions for the management of living marine resources into a single piece of legislation, and lists a series of principles and concerns that are to be taken into consideration in the management of resources and genetic material, including: the precautionary approach; the ecosystem-based approach; optimum utilization and allocation of resources; the effective control of harvesting; the implementation of international law; transparency in decision-making; and regard for the Saami culture
Portugal	*Executive*: National Ocean Strategy 2013–20 (2014)	Reaffirms 'the national maritime identity in a modern, proactive and entrepreneurial framework', inserting 'Portugal, on a worldwide level, [as] a leading maritime nation and an undisputed partner of the IMP [Integrated Maritime Policy] and of the EU maritime strategy [sic], in particular for the Atlantic area' Calls for the adoption of an action plan (Plano Mar-Portugal), which sets out concrete attainable objectives, provides a timeline for meeting them and the resources needed to carry them out, and allows assessment of progress made
Russian Federation	*Executive*: Marine Doctrine (2001); Federal Target Programme (FTP) 'World Ocean'; Strategy for Development of Marine Activities (2010)	The Marine Doctrine aims to protect state sovereignty, sovereign rights, and high seas freedom, declares its national interests in the world ocean, and recognizes priorities and principles for a functional national marine policy The FTP 'World Ocean' aims to change the existing narrow, sectorial approach to a more integrated, effective nationwide system of regulation and management of marine activities, and addresses a number of issues, including the use of marine resource potential, the development of transport communications, the maintenance of sovereign rights and jurisdiction in Russia's coastal waters and international marine waters, control over the marine environment, and emergencies of a natural and human-caused character, as well as other specific problems related to providing safe and sustainable livelihoods both in coastal regions and throughout Russia as a whole The Strategy recognizes such goals as economic development and improvement of living standards in the country as priorities of marine activities, and defines as one of the major challenges the need for an integrated approach to complement the existing sectorial approach to marine activities planning

United Kingdom	Legislative: Marine and Coastal Access Act 2009; Marine (Scotland) Act 2010; Marine (Northern Ireland) Act 2013	Introduce new tools for marine and coastal conservation and management in the UK, providing a means for improved management and protection of marine species and habitats Marine conservation, one of the principal issues addressed, aims for expansion of the current network of marine protected areas (MPAs) in UK waters Include provisions for management of fisheries within the territorial sea, marine spatial planning (MSP), licensing marine activities, improving marine nature conservation through measures such as the development of marine conservation zones (MCZs), and the establishment of the Marine Management Organization
United States	Executive: National Ocean Policy (2010)	Provides a mechanism and methodology for ocean management already under increasing environmental and human demands One of its main purposes is to ensure the protection, maintenance, and restoration of the health of ocean, coastal, and Great Lakes ecosystems and resources Established a Cabinet-level National Ocean Council (NOC), representing an interagency assembly led by the Council on Environmental Quality (CEQ) and the White House Office of Science and Technology Policy (OSTP) to oversee implementation of the policy The NOC will develop strategic action plans for each of the priority objectives of the National Ocean Policy to help to coordinate actions, since existing legal authorities will carry out the implementation Nine priorities of the Interagency Ocean Policy Task Force (established in 2009 by President Obama to develop recommendations for a national ocean policy and for effective coastal and marine spatial planning) include promoting a healthy and productive ocean zone, and cover a wide array of issues, and foundational tools listed among the priorities include ecosystem-based management and coastal and marine spatial planning (CMSP) Environmental priorities include the reduction of harmful land-based impacts on the ocean, addressing the changing conditions of the Arctic, and climate change and ocean acidification adaptation Strengthening the observation, mapping, and infrastructure of the oceans, coasts, and Great Lakes for domestic and international observation efforts is also a priority, because the implementation of the policy is based on best available science In order to implement the policy effectively, support and improvement of inter-jurisdictional coordination is prioritized, along with the utilization of regional plans for ecosystem protection and restoration, and public education

(continued)

Table 1.3 (continued)

Country	Form	Content
Regions		
East Asian Seas (PEMSEA)		Provides for development of a policy framework for building partnerships, a regional marine environment resource facility, and a regional mechanism to support existing efforts
		Key components include the identification and incorporation of international instruments and conventions that pertain to the region, and the need for a more integrated approach toward these, with sub-strategies to sustain, preserve, protect, develop, implement, and communicate
European Union		Objectives include improving the status of marine resources and the ocean, and stimulating the growth and economic development of the European Union
		Ecosystem-based management and improvements in scientific knowledge are considered critical in achieving such objectives
		The Green Paper did not offer detailed policy solutions to marine-related issues, but rather posed a number of questions on integrated maritime policy and related marine issues
		The action plan cites several important projects, including the elimination of illegal fishing and high seas bottom trawling, the development of a strategy to address the impacts of climate change on coastal areas, and the formulation and implementation of integrated maritime policies by EU Member States
		Principles include stakeholder participation, subsidiarity and competitiveness, and the ecosystem approach as the Commission develops responses to the newly published action plan
Pacific Islands		Details the principles and actions that regional organizations should adopt to achieve sound ocean governance, and which nations should consider in the development of national policies and sustainable development plans
		Five key principles of PIROP: improving understanding of the ocean; encouraging sustainable development; maintaining ocean health; using the ocean for peaceful uses; and promoting cooperation and partnerships

include the entire EEZ and to address multiple-use conflicts and issues through integrated management frameworks. The United States is such an example, having established the Coastal Zone Management Act in 1972, and a national ocean policy encompassing the EEZ in 2010.

On a somewhat different trajectory, a number of countries (for example Brazil, India, and the Philippines) established national marine policies in the early 1990s as a means of preparation for carrying out their obligations in view of the impending entry into force of UNCLOS. Subsequently, with the continued degradation of coastal and marine resources and ecosystems, the need for more holistic and integrative approaches triggered the development of integrated national ocean policies. In the Philippines, the recognition that its marine ecosystems were suffering from the impacts of human use, which needed to be addressed, was bolstered by the impending entry into force of UNCLOS in 1994, which served as a motivation for then President Fidel Ramos to prepare the country to carry out its obligations under the Convention (Batongbacal, Chapter 15).

At the regional level, in the case of the European Union, the development of an integrated maritime policy was triggered by the recognition that European policies on maritime transport, industry, coastal regions, offshore energy, fisheries, the marine environment, and in other relevant areas had been developed separately, with no systematic linkages among them, highlighting the need for an integrated maritime policy at the European level that would both change the way in which policy is made and decisions taken, as well as develop a programme of work which would allow actions under different sectorial policies to develop in a coherent policy framework (Gambert, Chapter 18).

Getting started

The policy development process commonly begins with the identification and analysis of pressing ocean and coastal issues led by or carried out jointly by various government agencies and individuals, or ad hoc groups formed through executive order, or other official designations, such as Portugal's Strategic Commission for the Oceans, Japan's Basic Ocean Law Study Group, and the US Ocean Commission and Interagency Ocean Policy Task Force. These groups are normally composed of top-level technical and policy experts in the field. Their findings, including their recommendations, become the basis for drafting an ocean policy, which is subsequently adopted by executive order, legislative process, or endorsement by a government authority. In the case of Australia and Vietnam, experts were called on to prepare analytical reports. In the case of Norway, the constant input of new knowledge is noted to be a major impetus for the development of its Ocean Resources Act (Hoel and Lobach, Chapter 14).

In the United States, the National Ocean Policy was based on the recommendations of the Interagency Ocean Policy Task Force established by President Barack Obama in 2009. Composed of senior-level officials from executive departments, agencies, and offices of the federal government, the Task Force was charged with developing recommendations to enhance the country's ability to maintain healthy, resilient, and sustainable ocean, coastal, and Great Lakes resources (Cicin-Sain et al., Chapter 11).

In the European Union, the development of an integrated maritime policy started with the formation of a maritime policy task force in 2005 tasked with carrying out a broad consultation on a future maritime policy for the Union. Given the wide variety of interlinked maritime activities and the need to consider all sectors, a consultation document (Green Paper) was prepared and adopted by the European Commission in 2006, which raised a set of questions used for public consultation – the first step towards the establishment of a comprehensive EU Integrated Maritime Policy (Gambert, Chapter 18).

A regional marine strategy for the sustainable development of oceans and coasts in the East Asian Seas was facilitated by PEMSEA, a partnership arrangement involving various stakeholders of the Seas of East Asia, including national and local governments, civil society, the private sector, research and academic institutions, communities, international agencies, regional programmes, and financial institutions and donors, which started as a regional project on marine pollution prevention in the East Asian Seas sponsored by the Global Environment Facility (GEF) and implemented by the United Nations Development Programme (UNDP) and the International Maritime Organization (IMO) (PEMSEA, 2012). The strategy is largely based on relevant international conventions, and other international and regional instruments, as well as the lessons learned from the coastal and ocean governance experiences in the region – particularly from efforts and activities undertaken by PEMSEA. The development of the Sustainable Development Strategy for the Seas of East Asia (SDS-SEA) followed the process of consensus-building and consultation, including the formulation of the Shared Vision of the concerned stakeholders and their mission for achieving sustainable coastal and ocean development, followed by the identification of the values that the people of East Asia attach to the seas and the threats to them (Bernad and Chua, Chapter 20).

At their 1999 annual meeting, the leaders of the Pacific Islands Forum endorsed a set of recommendations emerging from the Pacific Regional Follow-up Workshop on the implementation of UNCLOS. One of the key recommendations – that a regional ocean policy be produced – was adopted by the leaders as a regional-level initiative. The Marine Sector Working Group of the Council of Regional Organizations in the Pacific (CROP MSWG) was tasked with developing a regional ocean policy, which was subsequently endorsed by the Thirty-third Pacific Islands Forum in 2002. The Pacific Islands Regional Ocean Policy (PIROP) was then launched at the WSSD in 2002 to serve as a framework for the region, along with an Integrated Strategic Action Plan (ISAP) that provides initiatives to implement the policy (Power and Solofa, Chapter 19).

In Africa, regional collaboration on ocean policy has taken place mainly through existing sub-regional agreements under the United Nations Environment Programme (UNEP) Regional Seas Programme (Abidjan and Nairobi Conventions), and through the development and implementation of the four African LME projects, and the development of an African LME Caucus. If momentum is generated towards the development of a regional ocean policy in the region, the case study authors argue that a regional or sub-regional ocean policy agenda should be driven by Africans themselves if it is to be acceptable to the various constituencies within the continent (Hewawasam and McLean, Chapter 21). The African Union (AU) is well placed to advance sub-regional ocean and coastal policies, building on the policy agendas already initiated with the Abidjan and Nairobi Conventions and in collaboration with partners such as the LME Caucus.

Table 1.4 summarizes the common catalysts, modes of getting started, and nature of policy processes taken by each country and region.

Stakeholder consultation

To ensure the successful implementation of a policy, support is needed from a range of stakeholders who may have diverse values and interests that need to be harmonized (UNEP, 2009). It is important that the role each of the stakeholders might play be identified, and that different methods and approaches be used to ensure their effective participation throughout the policy development process and implementation. The need for the policy to be based on strong political commitment and support at all levels of society and government, and by diverse stakeholders, highlights the need for effective public education and outreach and capacity development.

Table 1.4 Common catalysts, modes of getting started, and nature of policy development for national ocean policies

Country	Catalysts	Mode of getting started	Policy development process
Countries			
Australia	Expert reports called for an integrated approach to management of Australia's oceans and its resources. The 1992 Australian National Strategy for Ecologically Sustainable Development required government use of sustainable development principles in all natural resources management policies. A 1994 claim to a 200 nautical mile EEZ under UNCLOS Stakeholder recognition for a more strategic approach to oceans management	The Australian government commissioned technical papers on relevant topics, then consulted with stakeholders on what issues were important to Australians and how the oceans should be managed.	Development of Oceans Policy was led by the Department of the Environment and Heritage.
Brazil	The 7,300 km of Brazilian Atlantic coasts, along which 25 per cent of the national population live and which generate 70 per cent of the gross domestic product (GDP), were the main drivers for the government to establish a policy concerning the coastal zone and its population, the territorial sea, and the living, as well as non-living, resources of the continental shelf.	The National Policy for Sea Resources (PNRM) was instituted in 1980 by a presidential directive and is reviewed biannually by the Interministerial Commission for Sea Resources (CIRM), which aims to connect the various marine and coastal sectorial policies and plan initiatives and programmes for implementation through the National Plan of Coastal Management (PNGC).	
Canada	Environmental issues in oceanic areas and recognition of the need to maximize economic benefits from ocean resources in a sustainable manner	Extensive public consultation and subsequent recommendation to the Canadian prime minister from the National Advisory Board of Science and Technology for the development of a national policy and legislation	Formal federal, provincial, territorial, and public consultation processes

(continued)

Table 1.4 (continued)

Country	Catalysts	Mode of getting started	Policy development process
Countries			
India	Science and technology have ceased to be the government's central concerns, with the recognition of the need to restructure current administrative framework for ocean governance in order to address environmental protection, the allocation of fishing rights, the protection of marine species and the conservation of marine biodiversity, marine pollution, and offshore oil and gas exploration.	As far as the management of the coastal areas are concerned, the Coastal Regulation Zone 2011 was developed to provide for livelihood security to local coastal communities, to promote conservation and protection of coastal stretches, its unique environment, and its marine area, and to promote sustainability based on scientific principles taking into account the dangers posed by natural hazards and sea-level rise. However, apart from spatial integration, there is limited application of the concept of integration in this law.	
Jamaica	Recognition of competing and complementary issues and players in Jamaica's coastal and ocean resources	In 1996, the Jamaica Cabinet established the Interagency Advisory Committee to address coastal and oceans issues, which later evolved into the National Council for Oceans and Coastal Zone Management.	The National Council for Oceans and Coastal Zone Management led the policy formulation process, which resulted in the endorsement of the 2002 Green Paper on Coastal and Oceans Management by the Jamaican Cabinet.
Japan	Coastal and ocean developments since the 1960s have led to a number of conflicts among key stakeholders in various sectors, such as industrial and land development, fisheries, shipping, nature conservation, and recreation, as well as among government authorities.	The government initiated the development of the basic concepts and policy to advance ocean development with a long-term perspective in 1998. The *Diet* (Japanese parliament) facilitated coastal and ocean management by revising the Comprehensive National Land Development Act of 1950 in the form of the National Land Sustainability Plan Act of 2005, and adding to its scope the use and preservation of marine areas including the EEZ and the continental shelf. The Ministry of Land, Infrastructure, Transport and Tourism (MLIT) issued the MLIT Policy Outline on Ocean and Coastal Areas in 2006, which set out eight basic policy directions, including the promotion of integrated management of the ocean and coastal areas, followed by ninety-five specific measures that the government intends to implement.	In 2006, the Basic Ocean Law Study Group produced the Guidelines for Ocean Policy and an outline of the draft basic ocean law, followed up by the political parties, which drafted the Basic Act on Ocean Policy, passed by the Diet and entered into force in 2007.

Mexico	Several coastal issues arose that had to be addressed by more than one department within the Ministry, leading to the recognition of the need for coordination among federal agencies working sectorially and in isolation.	The 2001 diagnosis was reconsidered, and several tasks led by the Ministry's Undersecretariat for Planning and Environmental Policy through its General Directorate for Environmental Policy and Sectorial and Regional Integration (2004).
	Ministry of Environment and Natural Resources led identification of the main activities and difficulties faced by the Ministry regarding the coastal zone by its nineteen branches, although diagnosis did not lead to action (2001).	Development of an integrated ocean policy, the National Environmental Policy for the Sustainable Development of Oceans and Coasts (NEPSDOC) followed in 2006.
New Zealand	Recognition that since a significant amount of New Zealand's total biodiversity exists in the marine environment, it is critical that policy and law exists to manage the pressure exerted by users	In July 2000, the Cabinet agreed to the development of a national ocean policy as the basis for an overhaul of the system, which will involve all key government departments, working together.
	New Zealand has an extensive and integrated regime in place for managing the land, but not for managing the marine environment.	Guided by a Ministerial Advisory Committee on Oceans Policy (MACOP), ministers oversaw the development of the policy, envisioned to happen in three stages:
		1 definition of the goals, values and principles to guide the process;
		2 definition of the strategic and operation framework for the policy; and
		3 implementation of the policy.
		Stage 1 was completed in 2001 and included extensive stakeholder meetings; Stage 2 was in the process of being implemented when the government decided to suspend the process, pending resolution of some politically sensitive issues, including ownership of the public foreshore and seabed, and public access to the coast. The policy was reactivated in November 2005, but there have been no significant developments to date.

(continued)

Table 1.4 (continued)

Country	Catalysts	Mode of getting started	Policy development process
Countries			
Norway	The evolution of scientific knowledge, first-hand experience with problems related to pollution and over-fishing, the requirements of international agreements, and increasing conflicts between different uses of the oceans were the important driving forces in the development of an integrated national ocean policy.	A report to the Storting (the Norwegian parliament) in 2002 outlined the principles for a more integrated and ecosystem-oriented marine policy, followed by the development of an integrated management plan for the Norwegian part of the Barents Sea and the sea areas to the south and to the Lofoten Islands by 2006, and the initiation by the government of the development of a more modern legislative framework for the oceans.	The Ocean Resources Act of 2009 was developed by a ten-person, government-appointed committee representing the fishing industry, government prosecution, public administration, academia, and the Saami parliament. The draft Act consolidated all relevant provisions for the management of living marine resources into a single piece of legislation, and listed a series of principles and concerns that are to be taken into consideration in the management of resources and genetic material.
Philippines	Recognition that marine ecosystems are suffering from the stress of human use and of the potential contribution of oceans to national economic development The impending entry into force of UNCLOS in 1994 provided motivation for the country to prepare to carry out its obligations thereunder.	The research arm of the Department of Foreign Affairs commissioned several working papers and inter-agency consultations provided the basis for the development of an overall national marine policy. A national marine policy document was adopted, with the Sustainable Archipelagic Development Framework (1994), to express the guiding principles for management and development of the country's ocean space.	A new a broader Archipelagic Development Policy has been proposed.
Portugal	Triggered by major projects and ventures in which Portugal was directly involved, such as the hosting of EXPO'98, 'The Oceans: A Heritage for the Future', the adoption by the United Nations General Assembly (UNGA) of Portugal's proposal through the Intergovernmental Oceanographic Commission (IOC) of the United Nations Educational, Scientific, and Cultural Organization (UNESCO) to declare 1998 the 'International Year of the Ocean', and the presentation by the Independent World Commission on the Oceans, chaired by Mário Soares, former president of Portugal, of the report, The Ocean, Our Future	The national ocean policy development process has been marked by actions and follow-up efforts of all Portuguese governments in power since 1998, including legal instruments.	The first National Ocean Strategy (2006–16) was adopted by the government and resulted from the effort of working groups such as the Strategic Oceans Commission and the Task Force for Maritime Affairs. The National Ocean Strategy 2006–16 was updated as the current National Ocean Strategy (2013–20), under the responsibility of the Directorate General for Maritime Policy, taking into account, inter alia, the European Union's strategies, policies, and financial cycles. The Interministerial Commission for Ocean Affairs facilitates its short-term implementation.

Russian Federation	Government recognition that the lack of an integrated approach to the development of a marine policy did not allow for maintaining the unity and integrity of the government's sectoral activities and for monitoring of such activities for long-term planning and development After the collapse of the USSR, the transition to a new bureaucracy did not prove beneficial to the country's economic activities in its marine and coastal areas.	The FTP 'World Ocean' was developed to improve management, including state regulation, together with purposeful scientific and technical development of marine activity in the country. The concept of the FTP was approved by Decree of the President of the Russian Federation in January 1997. Forty-five federal and provincial executive agencies and thirteen research organizations took part in the development of this programme. The FTP 'World Ocean' has become the basis of a nationwide system of regulation and management of Russia's marine activity aimed at integration and increased effectiveness.	The Strategy for Development of Marine Activities of the Russian Federation adopted by the government of the Russian Federation in 2010 determined major challenges and perspectives for the development of marine activities of the country, outlined strategic goals and targets, and established milestones in the development of marine activities. The Strategy is the first national document of its kind that recognizes such goals as economic development and improvement of living standards in the country as priorities of marine activities. The Strategy defines as one of the major challenges the need for an integrated approach to complement the existing sectoral approach to marine activities planning. One of the strategic goals is the realization of the integrated approach to the development of specific coastal land and marine areas by recognizing them as special objects of governance.
United Kingdom	The need for coordination of government agencies involved in ocean affairs was recognized in the 1970s.	Consideration of integrated coastal and marine policies began in 2002, with the publication of the Marine Stewardship Report, a Marine Bill in the 2005–06 parliamentary sessions, and the White Paper in March 2007.	Development of the Marine and Coastal Access Act 2009, the Marine (Scotland) Act 2010, and the Marine (Northern Ireland) Act 2013 represent the recent legislative policy.

(continued)

Table 1.4 (continued)

Country	Catalysts	Mode of getting started	Policy development process
Countries			
United States	The need to develop a coherent system of ocean governance in the United States has long been recognized, and was formally endorsed by the National Ocean Commission Report (2004) and the Pew Oceans Commission Report (2003).	President Obama established the Interagency Ocean Policy Task Force (Task Force) on 12 June 2009, composed of twenty-four senior-level officials from executive departments, agencies, and offices across the federal government and led by the Chair of the Council on Environmental Quality (CEQ), charged with developing recommendations to enhance the country's ability to maintain healthy, resilient, and sustainable ocean, coasts, and Great Lakes resources for the benefit of present and future generations.	The Task Force prepared an Interim Report and an Interim Framework for Effective Coastal and Marine Spatial Planning in 2009, made available online for public comment. Based on about 5,000 written comments from Congress, stakeholders, and the public, the Task Force finalized its recommendations on 19 July 2010, which became the basis for the establishment of a National Policy for the Stewardship of the Ocean, Coasts, and Great Lakes (National Policy) and the creation of a National Ocean Council (NOC) to strengthen ocean governance and coordination.
Vietnam	The government became aware of the importance of coordinating and integrating all maritime and coastal activities into a national marine policy. Greater understanding of the importance of sustainable development, scientific research, environmental protection, biodiversity conservation, MPA management, and the conservation and management of marine habitats and coastal wetlands.	The government initiated the establishment of a National Strategy of Marine Economic Development towards 2020 and established the Vietnam Administration for Seas and Oceans, which is in charge of the process.	The preparation process involves a number of government ministries, marine sectors, and national consultants, and has already generated more than sixty reports on various themes related to the marine environment.

Regions		
Africa (NEPAD)	There is no concerted movement towards the development of an overall regional ocean policy so far. Regional collaboration is through existing sub-regional agreements under the UNEP Regional Seas Programme (RSP) (the Abidjan and Nairobi Conventions), and through the development and implementation of the four African LME projects and the development of an African LME Caucus. Both the Caucus and NEPAD seem to be in a position to initiate and lead the development of a regional ocean policy, if desired.	In 2001, the New Partnership for Africa's Development (NEPAD) was established as an integrated development plan that provides a vision and programme of action for the sustainable development of the African continent and the eradication of poverty. NEPAD has developed an environmental action plan to address priority environmental issues with four strategic directions: 1 capacity-building for environmental management; 2 securing political will to address environmental issues; 3 mobilizing and harmonizing international, regional and national resources, conventions and protocols; and 4 supporting best practice/pilot programmes.
East Asian Seas (PEMSEA)	The Partnership in Environmental Management for the Seas of East Asia (PEMSEA) initiated the process, as part of the activities that the programme planned to carry out to address ocean and coastal management issues in East Asia.	A regional marine strategy for the sustainable development of oceans and coasts in the East Asian Seas was facilitated by PEMSEA, largely based on relevant international conventions and other international and regional instruments, as well as the lessons learned from the coastal and ocean governance experiences in the region – particularly from efforts and activities undertaken by PEMSEA. Included in the strategy are the development of a policy framework for building partnerships, a regional marine environment resource facility, and a regional mechanism to support existing efforts. The development of the SDS-SEA followed the process of consensus-building and consultation, including the formulation of the Shared Vision of the concerned stakeholders and their mission for achieving sustainable coastal and ocean development, followed by the identification of the values that the people of East Asia attach to the seas and the threats to them; the SDS-SEA was a product of more than three years (2000–03) of extensive national and regional consultations among the countries of the region and other stakeholders. In 2003, the ministers of the countries of East Asia adopted the SDS-SEA.

(continued)

Table 1.4 (continued)

Country	Catalysts	Mode of getting started	Policy development process
Countries			
European Union	At the EU level, there was recognition that European policies on maritime transport, industry, coastal regions, offshore energy, fisheries, marine environment, and in other relevant areas had been developed separately, with no systematic links drawn between them, and the need for an integrated maritime policy at the European level that would change both the way in which policy was made and decisions taken, and a programme of work that would allow actions under different sectorial policies to develop in a coherent policy framework.	In 2005, the President of the European Commission asked the Commissioner for Maritime Affairs and Fisheries 'to steer a new Maritime Policy Task Force with the aim of launching a wide consultation on a future Maritime Policy for the Union'.	The Maritime Policy Task Force prepared the Green Paper on a Future Maritime Policy for the European Union, which the European Commission adopted in June 2006, providing an agenda for debate and questions for public consultation, as well as identifying how a policy framework would promote sustainable development and innovation. Based on 400 stakeholder comments received and 200 stakeholder events, the Commission published a report on the consultation, and its vision and action plans for an EU Integrated Maritime Policy, in 2007.
Pacific Islands	At their 1999 annual meeting, the leaders of the Pacific Islands Forum endorsed a list of recommendations emerging from the Pacific Regional Follow-up Workshop on the implementation of UNCLOS. One of the key recommendations – that a regional ocean policy be produced – was adopted as a regional-level initiative by the leaders.		The Marine Sector Working Group of the Council of Regional Organizations in the Pacific (CROP MSWG) was tasked with developing a regional ocean policy, which was subsequently endorsed by the Thirty-third Pacific Islands Forum in 2002. The Pacific Islands Regional Ocean Policy (PIROP) was then launched at the WSSD in 2002 to serve as a framework for the region, along with an Integrated Strategic Action Plan (ISAP) that provides initiatives to implement the policy.

Stakeholders include those who have a direct interest in ocean and coastal resources, such as fishers, coastal tourism operators, shipping industries, government ministries responsible for ocean and coastal uses and resources and ecosystem management, coastal communities, indigenous groups, academic institutions, environmental non-governmental organizations (NGOs), civil society organizations, and the private sector. Stakeholders also include those that are not traditionally included in the planning process on oceans and coasts – for example stakeholders in key economic sectors such as agriculture, energy, public works, and transport – but who are now recognized as having a potential role in integrated policy development on oceans and coasts.

The cases included in this volume defined stakeholders in different ways and have undertaken stakeholder consultation using a range of approaches. Australia, Canada, Jamaica, Mexico, New Zealand, Philippines, Portugal, the United Kingdom, the United States, the Pacific Islands, the East Asian Seas, and the European Union engaged in extensive consultation processes, which were a key feature in the ocean policy development process. Brazil, India, Japan, Norway, Russian Federation, Vietnam, and Africa have all carried out some form of government and public consultation, but to a lesser extent.

Stakeholder consultation at the regional level shows the extent and complexity in which varied interests have to be integrated in policymaking. In the European Union, the integrated approach to ocean policy development was effectively applied in a robust stakeholder process that brought together the various interests in the region to ensure better coverage of all relevant stakeholders and better results in taking into account their various interests. Given the interrelated maritime activities and the need to examine all sectors collectively, a consultation document (Green Paper) was adopted by the European Commission in June 2006, which contained a set of questions for public consultation, based on an accumulation of the ideas and input received from stakeholders over the preceding year. The Green Paper represented the first step towards the establishment of an all-embracing EU maritime policy:

> The consultation process demonstrated that the success of maritime policy would depend on the support and sense of ownership of stakeholders, including regional actors already very active in developing integrated maritime actions. Furthermore, the maritime regions of Europe are so diverse and region-based that action had to be different in focus according to each region.
>
> *(Gambert, Chapter 18)*

The Commission also adopted a roadmap on marine spatial planning, 'a process that provides a stable, reliable, and oriented planning framework for public authorities and stakeholders to coordinate their action, and to optimize the use of marine space to benefit economic development and the marine environment' (Gambert, Chapter 18).

In East Asia, the East Asian Seas Congress takes place every three years, and brings together stakeholders, experts, regional partners, and other actors from around the world to evaluate progress in the implementation of the regional strategy (SDS-SEA), to share their experiences, and to exchange information or ideas in different areas of concern on the sustainable development of coasts and oceans (Bernad and Chua, Chapter 20).

Principles in national and regional ocean policies

Principles adopted at the national level

National ocean policies commonly include a vision for the 200-mile zone or sea area, a set of principles to guide action in this area, and a set of processes and institutions for making decisions about uses, protection, and development of the area.

As evident in Table 1.5, which depicts the principles followed in the national ocean policies, there is considerable congruence in national ocean policy principles being adopted by nations, which are derived from the Rio Declaration and Chapter 17 of Agenda 21 (for example ecosystem-based, integrated, participatory, precautionary, equitable, sustainable), with sustainable development as an overarching goal for which political commitments were made in the case of ecosystem-based management (EBM) and integrated coastal management (ICM) at the 2002 WSSD (Freestone et al., 2010; Murawski et al., 2008).

In addition, the following additional principles were identified from the national ocean policies covered by this study:

- inter-generational equity (Australia, Jamaica, New Zealand, Philippines);
- intra-generational equity (New Zealand);
- fairness and equity (Portugal);
- national interests (Portugal, Vietnam);
- international responsibility (Jamaica, United States);
- ocean–land–atmosphere connection (Philippines, United States);
- subsidiarity/devolution (New Zealand, United Kingdom);
- capacity-building (India, Portugal);
- indigenous/traditional rights (Australia, Canada, Jamaica, New Zealand);
- the 'polluter pays' principle (Jamaica);
- understandable laws and clear decisions (United States); and
- spatial specificity (Portugal).

Principles adopted at the regional level

The following summarizes the principles found in the four regional ocean policies covered in this study.

- *Africa (NEPAD)* – economic/social development or poverty alleviation; equitable use; governance and capacity-building, and communication
- *European Union* – sustainable development/sustainability; integrated management; ecosystem-based management; good governance/transparency and public participation; subsidiarity; economic/social development; data dissemination; stewardship; international responsibility
- *Pacific Islands* – good governance/transparency and public participation; capacity-building and dissemination of information; sustainability; preservation of marine biodiversity; monitoring of progress/enhancement of enforceability; international responsibility
- *East Asia* – sustainability; integration; stewardship; science-based approach; ecosystem-based approach; multi-sectorial; participatory; balance between conservation and socioeconomic development; global responsibility; respect for indigenous rights; poverty alleviation

Although statements of principles in national ocean policies are useful, principles must be operationalized in ocean decision-making, and in decision-making at various levels and in particular sectors, in order to make a difference (Rothwell and VanderZwaag, 2006). National ocean policies influence the development of subnational (state- or province-level and local) policies. The availability of funding and the specifications and requirements of the national ocean policy impact subnational decision-making and the use of local resources in the application of the prescribed principles. Federal activities that impact state and local policy include federal and state-level programmes and initiatives, incentives and sanctions for state performance, and competitive grants.

Table 1.5 Principles adopted in national ocean policies

Country	National ocean policy (Date) Executive (E) / Legislative (L)	Sustainable development / Sustainability	Integrated management	Ecosystem-based management	Good governance	Adaptive management/Best available science	Precautionary approach	Preservation of marine biodiversity	Stewardship	Multiple-use management	Economic/social development / Poverty alleviation
Australia	Oceans Policy (1998) (E)	×	×	×		×	×	×	×	×	×
Brazil	National Marine Resources Policy (1980/2005) (E)	×	×	×	×	×	×		×	×	
Canada	Oceans Act (1996) (L)	×	×	×	×	×	×	×		×	
India	Ocean Policy Statement (1982) (E)	×	×	×	×		×				
Jamaica	Natural Resource Conservation Authority Act (1991) (L)	×	×		×		×		×		×
Japan	Basic Plan on Ocean Policy (2008) (E)	×	×	×	×	×		×	×	×	×
Mexico	Oceans Agenda (2001) (E)	×	×	×	×			×		×	
New Zealand	National Oceans Policy (2000, initiated process) (L)	×	×	×	×	×	×	×	×	×	
Norway	Ocean Resource Act (2008); Biodiversity Act (2009) (L)	×	×	×		×	×	×			×
Philippines	Sustainable Archipelagic Development Framework (proposed, 2004) (E)	×	×		×				×		
Portugal	National Ocean Strategy (2013–20) (E)	×	×	×	×	×	×		×		×
Russian Federation	Marine Doctrine (2001); Strategy for Development of Marine Activities (2010) (E)	×	×	×		×		×			×
United Kingdom	Marine and Coastal Access Act (2009); Marine (Scotland) Act (2010); Marine (Northern Ireland) Act (2013) (L)	×	×	×	×	×	×	×	×	×	×
United States	National Ocean Policy (2010) (E)	×	×	×	×	×	×	×	×	×	
Vietnam	Strategy of Marine Economy Development toward Year 2020 (E)	×	×	×	×	×		×			×

The institutional aspects of national ocean policy

Interagency mechanisms

With integration being one of the essential principles of national ocean policies, these policies typically involve an interagency/inter-sectorial coordinating mechanism and/or a lead implementing agency. This can take the form or some combination of:

- an interagency/inter-sectorial coordinating mechanism, such as a council or commission, for example the Interministerial Commission for Maritime Affairs (Portugal); and
- a lead implementing agency/agencies, such as the Department of Fisheries and Oceans (Canada), the Department of the Environment and Heritage (Australia), the Secretary of State for the Sea under the Ministry of Agriculture and Sea (Portugal), the Headquarters for Ocean Policy (Japan), the Russian Navy (Russian Federation), and the Ministry of the Environment (Norway).

Ideally, these lead ocean agencies or national coordinating bodies should: have clear terms of reference; involve coordination at the highest political levels; receive input from an external council of advisers; be transparent and allow some form of public involvement; and have incentives for joint action, such as joint budgets.

Especially important is the placing of the coordination entity at a level higher than the agencies that it is supposed to coordinate, for example in the Office of the Prime Minister.

Implementing bodies

It is important to have a national oceans office or similar to operationalize the national ocean policy and to oversee its implementation. The national oceans office or other variant should have its own separate budget and staff. The national oceans office is typically charged with a specific set of functions, such as the development of a national ocean plan, a 'state of the ocean' report, coordination of interagency activities, and collaboration with subnational authorities. A good example here is the Secretariat of Ocean Policy Headquarters in Japan, which serves as the overseer of the coordination of Japan's Basic Act of Ocean Policy, with a separate budget and more than thirty staff members.

In some cases, existing agencies are designated as lead agencies for the national ocean policy, with supporting or collaborating agencies also identified. The Department of Fisheries and Oceans Canada (DFO), for example, was given the responsibility for coordinating the implementation of the country's existing policy and statutory instruments following the prescriptions of the oceans policy.

Interagency and cross-sectorial collaboration

There are a host of challenges encountered in efforts to harmonize the management of ocean and coastal issues. These include how to achieve cross-sectorial agency collaboration, how to keep high-level participation from the various ministries, how to achieve multi-purpose outcomes, and how to achieve spatial integration (linking the management of freshwater, coasts, and oceans).

Interagency collaboration requires organizing across national agencies, and consulting and collaborating with a broad array of governments and stakeholders at various levels, as well as tribal or indigenous governments, and non-governmental, business, or industry sectors. Incentives that encourage inter-agency and cross-sectorial collaboration include: funding for interagency, cross-sectorial collaborative projects; information-sharing; technology transfer; development of complementary measures; and opportunities for synergies and trade-offs. Other incentives that could be used include perception of a common problem and the need for a joint solution, and fostering a common vision (soft incentives), as well as political and legislative mandates, joint tasking or product, and joint budgeting (hard incentives).

Achieving a multipurpose outcome

The expected outcomes of national ocean policies are the changes, benefits, and other consequences of their implementation, intended or unintended. Outcomes are commonly difficult to account for or to prove and normally require an understanding of the initial situation or problem for comparison (Community Matters, 2010), which underscores the necessity for benchmarking.

In Australia, marine bioregional planning focuses on the delivery of biodiversity conservation outcomes under the framework of ecologically sustainable development, primarily:

- an overarching, environmental outcomes-based framework; and
- the consolidation, through an integrated strategy, of conservation priorities and tools, per section 176 of the country's Environment Protection and Biodiversity Conservation Act 1999 (the EPBC Act).

Marine bioregional planning also provides the outcome-based framework for sectorial management to continue pursuing integrated oceans management in a strategic and consistent fashion. Moreover, continued investment in building the scientific knowledge of Australia's marine environment is considered a key input to ensuring the sustainable management of its natural resources. A greater focus on sustainability outcomes necessitates an implementation-driven planning process to ensure that the Oceans Policy has real outcomes for society (Petrachenko and Addison, Chapter 4).

In the European Union, the recent adoption of the 2014 Marine Spatial Planning Directive (MSPD) is expected to ensure better coordination, efficiency, and sustainability of multiple uses of the sea in EU countries (European Commission, 2014).

In the Pacific Islands, implementation of the Pacific Plan is expected to yield outcomes on the four goals of the Pacific Islands Forum – economic growth, sustainable development, good governance, and security – in the key regional priority areas of fisheries, shipping, tourism, maritime security and surveillance, and capacity development (Power and Solofa, Chapter 19).

In East Asia, the SDS-SEA identifies the desired institutional and operational changes and outcomes expected to occur in the region, which provide the rationale for pursuing the actions identified in the Strategy. It has been observed that national and local policy reforms in coastal and ocean governance are increasing throughout the region, which are expected to inspire and encourage national and local leaders to take more proactive roles in addressing cross-boundary environmental and resource use issues at the regional level, thereby building a momentum among the regional partners to continue achieving the goals and objectives of the SDS-SEA (Bernad and Chua, Chapter 20). The following four targets represent the desired outcomes of SDS-SEA implementation at the national and regional levels to 2015:

TARGET 1: a self-sustained regional partnership mechanism for the implementation of the SDS-SEA;
TARGET 2: national coastal and ocean policies and supporting institutional arrangements in place in at least 70% of Partner Countries;

TARGET 3: ICM programs for sustainable development of coastal and marine areas and climate change adaptation covering at least 20% of the region's coastline; and TARGET 4: a report on the progress of ICM programs every three years, including measures taken for climate change adaptation.

(PEMSEA, 2012: 11–12)

National/subnational relations

In many cases, national and subnational governments are both important and have different jurisdictions. Both have responsibilities for regulating and managing water and land resources and their uses, and both have critical technical expertise and information. However, inter-governmental conflicts arise because national and subnational governments have different responsibilities, legal authorities, approaches, and constituencies (Cicin-Sain and Knecht, 1998). Defining a collaborative relationship between national and state/provincial governments and a modus operandi that integrates the actions from both levels in very specific terms is a key function of national ocean policy. Integrated policymaking is required for at least three reasons:

1 A policy that addresses one issue can affect other issues;
2 Synergies among different issues exist and a policy intervention can be designed to achieve multiple benefits;
3 Successful implementation of a policy relies on support from a range of stakeholders who may have diverse values and interests that need to be harmonized.

(UNEP, 2009: 1)

Nested ocean governance

Nested systems of governance are essential because most environmental and societal issues both impact upon, and are affected by, conditions and actions at different levels in an ecosystem and governance hierarchy. Integrated policymaking requires full knowledge of how responsibility and decision-making authority is distributed within a layered governance system. This implies that a central feature of ecosystem-based governance is that all planning and decision-making must recognize and analyse conditions, issues, and goals at least at the next higher or lower level in the governance system. Thus ecosystem-based governance at the municipal scale must, at a minimum, be placed within the context of governance at the scale of the province, while governance at the scale of a province must, at a minimum, be analysed with an eye to governance at the scales of both municipalities and the nation (Olsen et al., 2009).

Defining collaborative national/subnational roles in regional marine planning and accompanying procedures is a key question that must be resolved. Defining how regional marine planning will be carried out, and through what procedures, methodologies, and processes, is also an important aspect. In Australia, for example, the primary tools for implementing the Oceans Policy are the bioregional plans, which set objectives, strategies, and actions for biodiversity conservation in the region. In the United States, regional planning bodies are being established in accordance with the National Ocean Policy to carry out regional marine planning. The scope, scale, and content of regional marine planning will be defined by the regions themselves, to solve problems that regions are concerned about, building on and complementing ongoing programmes, partnerships, and initiatives. The aim is to ensure that a region can develop an approach that works best, with a view to balancing regional and national interests, providing the most immediate regional benefits with the expectation that knowledge and experience accumulated over time will contribute to achieving national objectives.

Spatial integration

In the same manner, spatial integration – or the integration of management and governance across spatial boundaries, from freshwater/inland waters/watershed management, to coastal management, to EEZ management, to areas beyond national jurisdiction (ABNJs) – should ensure that planning and decision-making at one jurisdictional level should not adversely impact the same activities at another. This problem is especially challenging because freshwater/inland waters management and ocean and coastal management are commonly administered by different organizational units, and operate with different policy mandates and programme priorities (Jonch-Clausen, 2009). Linking the management of ABNJs to the management of freshwater, coastal, and EEZ areas further complicates the issue, since there is no single governing body that can carry out planning and decision-making in ABNJs. Integrated water resources management (IWRM) and integrated coastal [and ocean] management (ICM) are the preferred management approaches for inland waters, and for coastal and ocean waters, respectively.

More recently, within the EEZs, ecosystem-based marine spatial planning (MSP), a public process of analysing and allocating the spatial and temporal distribution of human activities in marine areas to achieve ecological, economic, and social objectives (Ehler and Douvere, 2009; Katsanevakis et al., 2011) has provided additional tools for strengthening spatial integration in different parts of the world (Jay et al., 2013; Secretariat of the CBD and the Scientific and Technical Advisory Panel of the GEF, 2012). The area for MSP is designated through a political process to be managed as a single unit and is more commonly applied at larger scales, such as an entire EEZ. Marine spatial planning incorporates all human uses in a single marine space into the planning process, which may be the key to truly integrated ocean and coastal planning and management (Ehler and Douvere, 2009).

Table 1.6 summarizes the interagency mechanisms and institutional arrangements developed by each country and region.

Domestic and regional implementation of international agreements

National governments and regional organizations have international obligations under UNCLOS and other international ocean and environmental agreements and principles of customary international law that constitute the international framework for governance and management of oceans and coasts (see Van Dyke and Broder, Chapter 2). The cases included in this volume examine how some of these agreements and principles, such as precautionary and ecosystem approaches, have been implemented domestically and regionally.

In many cases (such as Brazil, India, the Philippines), international prescriptions and responsibilities on oceans were identified among the catalytic factors leading to the initial development of national or regional coastal and ocean policies. The relationship between the national and international levels of action is complex, and is typically a two-way relationship – with nations influencing the establishment of new policies through their work at the United Nations and in multilateral negotiations, as well as in regional processes, and the implementation by national governments of internationally agreed prescriptions and obligations. As noted in the UK case study, the country's relationship with international law manifests two dimensions: the effects of international law on policymaking at the national level, and the contribution of the UK to policy and legal developments at the international level (Smith et al., Chapter 6).

Obligations and requirements of international agreements were identified among the driving forces behind greater coordination among various marine policy sectors and integrated oceans management in Norway (Hoel and Lobach, Chapter 14) whereas in other cases, such as in the Philippines, fully meeting these obligations remains part of ocean policy aspirations (Batongbacal, Chapter 15).

Table 1.6 Institutional aspects of national and regional ocean policies

Countries	Interagency mechanism	Administrative arrangement
Australia	National Oceans Ministerial Board (1999) Sustainable Environment Committee (2004) Oceans Board of Management National Oceans Advisory Group Oceans Policy Science Advisory Group	*Lead agencies:* National Oceans Office, Department of the Environment and Heritage (1999–2006) Marine and Biodiversity Division, Department of the Environment and Water Resources (2007) Department of the Environment (*formerly* Department of the Environment, Water, Heritage and the Arts) (from 24 November 2007) Department of the Environment and Water Resources (from 23 January 2007 to 24 November 2007) Department of the Environment and Heritage (prior to 23 January 2007)
Brazil	Inter-Ministry Commission for Sea Resources (CIRM) Integration Group of the Coastal Management	*Lead agency:* Brazilian Navy
Canada	Canadian Council of Fisheries and Aquaculture Ministers	*Lead:* Department of Fisheries and Oceans
India	Ocean Commission (proposed, Vision Perspective 2015)	*Lead:* Ministry of Earth Sciences
Jamaica	National Council on Ocean and Coastal Management	*Leads:* Ministry of Foreign Affairs (policymaking) National Environment and Planning Agency, and Maritime Authority (administrative)
Japan	Headquarters for Ocean Policy	*Lead:* Secretariat of Headquarters for Ocean Policy
Mexico	Intergovernmental Commission for Oceans and Coastal Affairs	*Lead:* Ministry of Environment and Natural Resources (SEMARNAT) (policymaking)
New Zealand	Ad hoc Ministerial Group of six Cabinet Ministers tasked to manage the development of national ocean policy Ministerial Advisory Committee on Oceans Policy	*Lead:* Oceans Policy Secretariat, Ministry of Environment

Norway	The 2008 Oceans Resources Act and the 2006 Integrated Management Plan provided for another layer of decision-making, rather than replaced existing structures.	*Leads:* Ministry of Industry, Trade and Fisheries Ministry of the Environment
Philippines	Commission on Maritime and Ocean Affairs	*Leads:* Department of Foreign Affairs (policymaking) Department of Environment and Natural Resources, Department of Agriculture, and Department of Transportation and Communications (administrative)
Portugal	Interministerial Commission for Maritime Affairs (CIAM), under the chairmanship of the Prime Minister	*Lead:* Secretary of State of the Sea/Ministry of Agriculture and Sea
Russian Federation	Marine Board of the Government of the Russian Federation	*Lead agency:* Russian Navy
United Kingdom	Initially, Inter-departmental Group on Coastal Policy Initially, Marine Consents and Environment Unit (cross-departmental unit of Department for Environment, Food and Rural Affairs, or Defra, and Department of Transport, or DoT); now Marine Policy Statement for all UK nations (England, Wales, Scotland, and Northern Ireland) Marine Management Organization (non-departmental public body independent of ministers to whom it is accountable)	*Lead:* Marine Management Organization for English waters
United States	White House/CEQ Committee on Ocean Policy White House/CEQ Interagency Ocean Policy Task Force National Ocean Council Regional Planning Bodies	*Lead:* Council on Environmental Quality (CEQ) (2004–09) Co-leads: CEQ and Office of Science and Technology Policy (OSTP) (2010–present)
Vietnam	None National maritime policy under development	*Lead:* Vietnam Administration for Seas and Islands, Ministry of Natural Resources and Environment

(continued)

Table 1.6 (continued)

Regions	Interagency mechanism	Administrative arrangement
Africa	None There is no overarching regional ocean policy	*Leads of various regional mechanisms:* Coastal and Marine Secretariat (CosMar), NEPAD Secretariats of the Abidjan and Nairobi Conventions LME projects Western Indian Ocean Marine Science Association (WIOMSA) Southwest Indian Ocean Fisheries Commission (SWIOFC)
East Asian Seas	The East Asian Seas (EAS) Partnership Council	*Lead:* PEMSEA Resource Facility (PRF)
European Union	Steering Group of Commissioners Interservice group of some 40 officials representing the different Directorates General in the European Commission Member States Expert Group	*Lead:* Directorate General for Maritime Affairs and Fisheries, European Commission
Pacific Islands	Marine Sector Working Group of the Council of Regional Organizations in the Pacific (CROP)	*Leads:* Pacific Islands Forum Secretariat Pacific Islands Applied Geoscience Commission Secretariat of the Pacific Regional Environment Program Forum Fisheries Agency University of the South Pacific

Moreover, although implementation of international agreements remains the prerogative and sovereign right of states parties, regional organizations and programmes increasingly provide funding, incentives, guidance, and other forms of support that facilitate national implementation of international agreements, and close collaboration and cooperation between adjacent nations and with other states (Bernad and Chua, Chapter 20; Gambert, Chapter 18; Hewawasam and McLean, Chapter 21; Mageau et al., Chapter 3). One particular example in which the roles and decision-making among national, regional, and global levels of governance are clearly delineated is in the achievement of the Aichi Biodiversity target of setting aside 10 per cent of marine and coastal areas for protection by 2020 (Target 11). Candidate ecologically or biologically significant areas (EBSAs) in open waters and deep-sea habitats are identified at regional workshops, but with the direct input of national agencies in the identification and description of these areas, using guidelines formulated at the global level, leaving it up to the governments to make decisions about management measures that should be put in place (Lee, 2013). Whether signatories to the CBD or not, countries use Target 11 as a reference for progress in national initiatives in marine biodiversity conservation, particularly in establishing marine protected areas (MPAs). Canada is very much behind the United States and Australia in the achievement of this target (Mageau et al., Chapter 3). Among the constraints identified are the long time frames required for stakeholder consultations, and the multiple departments and levels of government that need to reach agreement on MPA designations (Mageau et al., Chapter 3).

A broad assessment of the cases suggests that the linkage between international agreements and national ocean policymaking needs to be examined more closely than we have been able to do in this book. The linkage needs to be monitored and assessed through periodic reporting and public assessments, indicating the extent to which nations have actually implemented international obligations, and identifying successful factors, implementation gaps, and possible solutions for strengthening the linkages between international obligations and national actions.

Funding and other supporting elements

Funding

Dedicated oceans funding at national, regional, and global levels is essential for the success of integrated oceans governance. Unfortunately, in some cases, after the national ocean policy has been developed, no (or very few) funds are available for implementation. There have been efforts in different countries to develop special funds to support national and regional ocean policies, for example as dedicated funds derived from the leasing and exploitation of oil and gas resources on the outer continental shelf. This is a topic on which comparative analysis is difficult because the case studies generally do not provide sufficient information on funding considerations; it is therefore an important subject for further research. Part of the problem is related to the fact that, in many countries, it is very difficult to determine the national ocean budget – that is, which agencies are expending what level of funding for various ocean-related activities. This, too, is an important topic for further research and action.

However, some examples of the nature of funding support for the implementation of the national ocean policy may be gleaned from the case studies. Funding in support of Canada's Oceans Strategy is directed through the federal Department of Fisheries and Oceans Canada (DFO); partial funding came from the 'Health of the Oceans' initiative, which was carried out by five partner departments and agencies. The Russian Federation's Federal Target Programme (FTP) 'World Ocean' is financed from three sources: the federal budget, budgets of subjects of the Russian Federation, and off-budget sources. In the UK, by far the most important source of funding is the Consolidated Fund of HM Treasury. The UK does not impose the kind of taxation that might be used to fund marine policy projects directly, except for the Aggregate Sustainability Levy Fund, which is financing a substantial marine research programme under the Department for Environment, Food and Rural Affairs (Defra). Through taxation, hundreds of millions of pounds are devoted every year to central administration, agency operation, enforcement, scientific research, and higher education.

For countries in which national ocean policy is still under development, funding is provided either from the ministries involved in the process, or from external sources, as in the case of the Philippines, which received funding from PEMSEA, and in the case of Vietnam, which has received support from the World Bank, the Asian Development Bank (ADB), PEMSEA, the European Union, and various international NGOs.

Supporting elements

Supporting elements and conditions necessary to ensure the effectiveness of the ocean policy include research, science, and education support, and key actors, including a set of 'ocean policy entrepreneurs', inside and outside of government, who will be the watchdogs to ensure good implementation of the policy. Examples of the latter include Japan's Basic Ocean Law Study Group, a group of lawmakers and experts formed with the initiative and assistance of an NGO, which Group produced the Guidelines for Ocean Policy and an outline of the Basic Act on Ocean Policy. In the Philippines, the development of an ocean policy has received important support from NGOs and academic institutions. In Portugal, scientific contributions are provided by the Portuguese Foundation for Science and Technology of the Ministry of Education and Science.

Adaptive management

> From an oceans management programme perspective, monitoring, assessment, reporting, and re-evaluation of management measures applied to achieve the marine environmental quality objectives and social and economic objectives defined for integrated management and MPAs are an integral part of the operational frameworks of Oceans Act programmes.
>
> *(Mageau et al., Chapter 3)*

Over time, the management objectives also have to be reviewed to determine their relevance in light of evolving issues and their socio-economic context, and the identification of emerging competing issues. The importance of this component of national ocean policies cannot be understated given that the oceans sector will be competing with other sectorial issues on the national agenda for essential funds and resources, which need to be used in the most efficient and effective way. An effective monitoring and evaluation system would be able to identify successful activities that need to be replicated and expanded, weak or failed activities that need to be strengthened or dropped, or objectives that need to be prioritized, in terms of funding and other resources.

Australia's Oceans Policy required a comprehensive review of the policy every five years, which reviews were supposed to have been carried out in 2002, 2007, and 2012. The recommended changes from the 2002 review have been implemented and appear to have improved the efficiency in the implementation of the Oceans Policy (Petrachenko and Addison, Chapter 4). In a review article, however, it has been pointed out that, of the three original objectives of the policy (sectorial and jurisdictional integration, and ecosystem-based approaches), only the ecosystem-based approaches have been achieved, it having been too difficult for the federal government to achieve full integration across sectors and jurisdictions at the national level, suggesting the need for a legislative policy or a revision of the existing Oceans Policy (Vince, 2013).

In the case of Canada, the Oceans Act required a review of its implementation by Parliament within three years after its enactment. The 2001 Report on the Oceans Act by the Standing Committee on Fisheries and Oceans concluded that the Act was fundamentally sound. This report made twelve recommendations including: the establishment of a performance-based reporting system; reporting to Parliament on an annual basis; reporting on the state of the ocean on a periodic basis to track the health of the oceans, ocean communities, and related ocean industries. A 2005 review put forward additional recommendations that continued to be addressed, including through the Health of the Oceans Initiative involving five federal organizations and the ongoing convening of interdepartmental oceans committees.

Indicators are quantitative or qualitative statements or parameters that can be used to describe existing situations and to measure changes over time. Different sets of indicators, including indicators on input, process, outputs, and outcomes, can be analysed vis-à-vis the specific management issues that triggered the initiation of a management process (Belfiore et al., 2006). In Canada, the DFO has developed a Results-based Management and Accountability Framework with which to monitor the progress and implementation of the national ocean policy, which sets out performance measurement goals and indicators to assess departmental progress. The Framework was designed to track how the Department uses resources to undertake activities in order to affect the desired results and achieve stated outcomes. At the regional level, the SDS-SEA Implementation Strategy provides specific targets, actions, and indicators of progress (PEMSEA, 2012).

Moving forward on integrated ocean management

As we consider the way forward, and recalling the prescriptions of UNCLOS and Agenda 21, there is now broad consensus, in both principle and practice, that we can no longer manage the oceans as we did traditionally – sector by sector, use by use. Instead, we must adopt, as Chapter 17 of Agenda 21 put it, approaches that are 'integrated in content, and precautionary and anticipatory in ambit'. In fact, since 1992, the infrastructure for integrated, ecosystem-based ocean governance has been built in many nations and regions. Building on this foundation, in the next phase, the scope and reach of national and regional ocean policies must be expanded and enhanced. The national and regional experiences depicted in this volume, from fifteen nations and four regions – encompassing over half of the world's EEZs – clearly demonstrate solid forward movement toward achieving integrated ocean policy. Almost invariably, however, as is evident from the detailed descriptions of the national and regional cases, evolution of integrated ocean policy generally involves a slow and difficult process, typically beset by periodic setbacks and then (hopefully) periodic spurts forward.

A major reason for the difficulties that national and regional ocean policies encounter is the simple fact that sectorial marine policies (regarding shipping, fishing, aquaculture, tourism, among other things) and the interest groups associated with the sectorial policies are so entrenched in government. It is often difficult for countries to give sufficient power, authority, and financial resources to interagency mechanisms, such as a national ocean council, because of political fear on the part of sectorial interests that their situation and predictable management structure might undergo change. But it should be recalled here that integrated ocean policy is predicated on the basis of building on and enhancing sectorial programmes, rather than replacing them.

Reflecting on the factors that promote successful development and implementation in national and regional ocean policies, and considering avenues for encouraging the further implementation of ocean policies, we offer the following observations.

- *Embracing and implementing common ocean principles* It is clear from the studies in this volume that much of the world has already embraced major principles of integrated ecosystem-based ocean governance in their national or regional policies. The next step, in many cases, is further work in applying these principles to on-the-ground decisions about the ocean and coast by means of both integrated and sectorial policies, plans, and programmes. This work can be nurtured and assisted by global communities of practice on national/regional policies to make comparisons among various approaches and to draw lessons that are potentially useful to other nations.

- *Achieving an integrated outcome through formal coordination institutions* As we will see in many of the cases in this volume, having formal coordinating institutions to guide integrated ecosystem-based management of oceans and coasts, with independent input from external stakeholders, is an essential factor for success. While having informal mechanisms may work for a while, in the long run, given the typical ups and downs of the political process in each country and region, having a formal and legally established mechanism vested with appropriate authority generally provides a more enduring chance for success.

- *Ensuring and maintaining political support* As we will see in the evolution of many cases in this volume, political support of administrative and legislative leaders at the highest level is essential both for the development and continued implementation of national and regional ocean policies. Such policies may be established, but without continued support and legal basis, they may well flounder and become ineffective. Fostering of high-level political support must be carried out by ocean policy entrepreneurs in and out of government on a continuous basis.

- *Promoting binding policies* The cases in this volume clearly suggest that those policies that are formally embedded in law though legislative action tend to be more successful in the longer term than those that are solely based on executive action. Executive action can all too often be reversed with changing administrations.

- *Enabling stakeholders* The importance of external stakeholders motivating and providing input to ocean policy development processes and outcomes will be underscored in the majority of cases covered in this volume. Stakeholder engagement and support is essential, as well, for achieving continuing political support for the national/regional ocean policies.

- *Ensuring adequate funding and other supporting elements* Funding and other supporting elements, including research, science, and public education, is crucial for the initiation, adoption, and implementation of ocean policies. Adequate and consistent provision of financial and other forms of support has been instrumental in the steady and continued implementation over time of the national and regional ocean policies covered in the volume.

Rio +20 provides added impetus to integrated, ecosystem-based ocean governance

Thinking about the future, the work of ocean policy entrepreneurs around the world has been given additional impetus through the outcomes of the United Nations Conference on Sustainable Development (Rio+20), held at Rio de Janeiro, 20–22 June 2012. Oceans took centre stage in the Rio+20 process and at the Rio+20 summit itself. In the Rio+20 outcome document, *The Future We Want*, oceans are treated in twenty paragraphs, receiving central attention. A major accomplishment of the Rio+20 process is that oceans and their role in planetary survival and human well-being are now firmly established on the global agenda. As so eloquently said by Ms Elizabeth Thompson, Co-Executive Coordinator for the Rio+20 Conference at the Oceans Day at Rio+20:

> Oceans are the point at which planet, people, and prosperity come together. And that is what sustainable development is about. It is about all of us as shareholders of Earth, incorporated, acknowledging and acting on our responsibility to the planet, to the people, and to its bloodstream, the oceans.
>
> *(Quoted in Cicin-Sain, 2013)*

Paragraph 158 of the Rio+20 outcome documents highlights the importance of integrated, ecosystem-based ocean governance:

> We recognize that oceans, seas and coastal areas form an integrated and essential component of the Earth's ecosystem and are critical to sustaining it and that international law, as reflected in United Nations Convention on the Law of the Sea (UNCLOS), provides the legal framework for the conservation and the sustainable use of the oceans and their resources. We stress the importance of the conservation and sustainable use of the oceans and seas and of their resources for sustainable development, including through the contributions to poverty eradication, sustained economic growth, food security, creation of sustainable livelihoods and decent work, while at the same time protecting biodiversity and the marine environment and addressing the impacts of climate change. We therefore commit to protect, and restore, the health, productivity and resilience of oceans and marine ecosystems, and to maintain their biodiversity, enabling their conservation and sustainable use for present and future generations, and to effectively apply an ecosystem approach and the precautionary approach in the management, in accordance with international law, of activities impacting on the marine environment, to deliver on all three dimensions of sustainable development.
>
> *(Rio+20, para. 158)*

Future prospects and necessary actions

A major challenge in the next phase is to further enhance the implementation of integrated oceans policy, including its institutional aspects, at both national and regional levels, to consider appropriate applications in areas beyond national jurisdiction (ABNJs), and to consider how integrated governance could also be applied to the UN system to achieve greater effectiveness and coherence. Appendix D presents, in adapted form, the recommendations made by authors Cicin-Sain and Balgos, and other colleagues from the Global Ocean Forum, and presented to decision-makers during and after the Rio+20 process, charting out future directions to enhance integrated, ecosystem-based national and regional ocean policy, and to further disseminate the practices to other nations and regions of the world (Cicin-Sain, 2013).

An essential ingredient in continually promoting the continued enhancement and further development of integrated national and regional ocean policies is the steadfast engagement of ocean policy entrepreneurs around the world. These are the entrepreneurs, in and out of government, who, using whatever resources they can deploy, highlight the multi-use nature of ocean and coastal space, address and strive to harmonize multiple-use conflicts, call attention to the environmental issues of overfishing, pollution, and habitat degradation, are deeply concerned about the declines in marine biodiversity, and understand and prepare for the transformative force of climate change, which will especially affect coastal and island peoples in more than 200 coastal and island countries and territories around the world.

These national and regional ocean policy leaders need special help and nurturing in an era of change. We are in a new era in which climate change effects ineradicably pose a situation of higher risk and of possible tipping points. Changes to oceans, the effects on coastal communities, and the widespread displacement of coastal communities all pose prominent avenues for disaster. At the same time, as we chart the way to the new low-carbon economy and society, great opportunities for ambitious innovation are also prominent on the horizon (Cicin-Sain, 2011). With the threat of climate change, the importance of capacity development of national and regional ocean leaders, current and future professionals in the field, and the public becomes even more important and urgent. Moreover, the strengthening of national institutions dealing with oceans and coasts to respond to the challenges of climate change adaptation and mitigation represents an essential imperative (Cicin-Sain et al., 2011). Facing these challenges will entail strong commitment, active engagement, and resource mobilization by all parts of the international ocean community.

Notes

1 The term 'stakeholders' is used in this book in the broadest sense to include 'those with a direct interest in the resource either because they depend on it for their livelihoods or they are directly involved in its exploitation in some way', and 'those with a more indirect interest, such as those involved in institutions or agencies concerned with managing the resource or those who depend at least partially on wealth or business generated by the resource' (Townsley and FAO, 1998).
2 The VLIZ dataset is a series of geographic information system (GIS) shapefiles containing spatial information on the proposed extent of countries' EEZs. The dataset is compiled from existing geodatabases for Europe, the United States, and Australia. Further data on negotiated boundaries were added from a database of negotiated treaties according to UNCLOS. Remaining boundaries were calculated using GIS in accordance with UNCLOS by drawing a buffer 200 nautical miles from a country's baseline. In the absence of a reliable baseline for a given country, the shoreline was used as a proxy. If two countries are less than 400 nautical miles apart, the median line between the two was used to set the boundary, in accordance with UNCLOS.
3 The Global Oceans Conferences, organized by the Global Ocean Forum in collaboration with a host country, as well as a wide range of other international and national entities, mobilize high-level policy attention to review progress achieved (or lack thereof) in advancing the achievement of the global oceans agenda as expressed, inter alia, in the 2002 WSSD Johannesburg Plan of Implementation and in the Millennium Development Goals (MDGs), and to respond to new challenges such as climate change and improving governance of marine areas beyond national jurisdiction. See online at http://globaloceanforum.com/global-ocean-conferences/

References

Belfiore, S., Barbière, J., Bowen, R., Cicin-Sain, B., Ehler, C., Mageau, C., McDougall, D., and Siron, R. (2006) *A Handbook for Measuring the Progress and Outcomes of Integrated Coastal and Ocean Management*, ICAM Dossier No. 2, IOC Manuals and Guides No. 46, Paris: UNESCO, online at http://unesdoc.unesco.org/images/0014/001473/147313e.pdf [accessed 28 August 2014].

Cicin-Sain, B. (2011) 'Oceans and the Rio+20 Process: Achieving Significant Outcomes', Presentation to the Twelfth Meeting of the United Nations Open-Ended Informal Consultative Process on Oceans and the Law of the Sea, 22 June, New York.

Cicin-Sain, B. (2013) 'Rio+20 Implementation and Oceans: A Perspective Paper for Discussion at the 3rd International Symposium of Laureates of the Elizabeth Haub Prizes', 14–16 November, Murmau, Germany.

Cicin-Sain, B., and Knecht, R. (1998) *Integrated Coastal and Ocean Management: Concepts and Practices*, Washington, DC: Island Press.

Cicin-Sain, B., Balgos, M., Appiott, J., Wowk, K., and Hamon, G. (2011) *Oceans at Rio+20: How Well Are We Doing in Meeting the Commitments from the 1992 Earth Summit and the 2002 World Summit on Sustainable Development? Full Report and Summary for Decision Makers*, online at http://www.globaloceans.org/sites/udel.edu.globaloceans/files/Rio20SummaryReport.pdf [accessed 28 August 2014].

Cicin-Sain, B., Vandeweerd, V., Bernal, P., Williams, L. C., and Balgos, M. C. (2006) *Meeting the Commitments on Oceans, Coasts, and Small Island Developing States Made at the 2002 World Summit on Sustainable Development: How Well Are We Doing?*, Co-Chairs' Report – Volume 1, presented at the Third Global Conference on Oceans, Coasts and Islands, 23–28 January, UNESCO, Paris, online at http://globaloceanforumdotcom.files.wordpress.com/2013/06/wssdreport_100406d_0-2.pdf [accessed 28 August 2014].

Community Matters (2010) 'Question: What Are Outcomes, Outputs, Aims, and Objectives? How Do They Fit Together?', online at http://www.communitymatters.org.uk/content/532/What-are-outcomes-and-outputs [accessed 28 August 2014].

Ehler, C., and Douvere, F. (2009) *Marine Spatial Planning: A Step-by-Step Approach toward Ecosystem-Based Management*, IOC Manual and Guides No. 53, ICAM Dossier No. 6, Paris: UNESCO, online at http://www.unesco-ioc-marinesp.be/goto.php?id=ac1dd209cbcc5e5d1c6e28598e8cbbe8&type=docs [accessed 28 August 2014].

European Commission (2014) 'Commission Welcomes Parliament's Adoption of Maritime Spatial Planning Legislation', Press release, 17 April, online at http://europa.eu/rapid/press-release_IP-14-459_en.htm [accessed 19 August 2014].

Freestone, D., Cicin-Sain, B., Hewawasam, I., and Hamon, G. (2010) *Improving Governance: Achieving Integrated, Ecosystem-Based Ocean and Coastal Management*, online at http://globaloceans.org/sites/udel.edu.globaloceans/files/PolicyBrief-EBM-ICM.pdf [accessed 28 August 2014].

Greenpeace (2008) *Black Holes in Deep Ocean Space Closing the Legal Voids in High Seas Biodiversity Protection*, online at http://www.greenpeace.org/usa/Global/usa/report/2008/4/Black-Holes.pdf [accessed 28 August 2014].

Jay, S., Flannery, W., Vince, J., Liu, W.-H., Xue, J. G., Matczak, M., Zaucha, J., Janssen, H., van Tatenhove, J., Toonen, H., Morf, A., Olsen, E., Suaréz de Vivero, J. L., Mateos, J. C. R., Calado, H., Duff, J., and Dean, H. (2013) 'International Progress in Marine Spatial Planning', *Ocean Yearbook*, 27: 171–212.

Johnson, S. P. (1993) *The Earth Summit: The United Nations Conference on Environment and Development (UNCED) – Introduction and Commentary*, London: Graham & Trotman.

Jonch-Clausen, T. (2009) 'Linking the Management of Freshwater and Oceans', Presentation to the Roundtable Discussion on Meeting Human and Environmental Needs through Linking Integrated Management of Freshwater Basins with Downstream Coastal Areas and their Ecosystems at the Fifth World Water Forum, 20 March, Istanbul.

Katsanevakis, S., Stelzenmüller, V., South, A., Sørensen, T. K., Jones, P. J. S., Kerr, S., Badalamenti, F., Anagnostou, C., Breen, P., Chust, E., D'Anna, G., Duijn, M., Filatova, T., Fiorentino, F., Hulsman, H., Johnson, K., Karageorgis, A. P., Kröncke, I., Mirto, S., Pipitone, C., Portelli, S., Qiu, W., Reiss, H., Sakellariou, D., Salomidi, M., van Hoof, L., Vassilopoulou, V., Fernández, T. V., Vöge, S., Weber, A., Zenetos, A., and ter Hofstede, R. (2011) 'Ecosystem-Based Marine Spatial Management: Review of Concepts, Policies, Tools and Critical Issues', *Ocean & Coastal Management*, 54: 807–20.

Lee, J. (2013) 'Impacts and Challenges to Marine Biodiversity beyond Areas of National Jurisdiction', Intersessional Workshop on Conservation and Management Tools, 6–7 May, online at http://www.un.org/Depts/los/biodiversityworkinggroup/workshop2_lee.pdf [accessed 8 September 2014].

Murawski, S., Cyr, N., Davidson, M., Hart, Z., Balgos, M., Wowk, K., and Cicin-Sain, B. (2008) 'Ecosystem-Based Management and Integrated Coastal and Ocean Management and Indicators for Progress', Presentation to the Fourth Global Conference on Oceans, Coasts, and Islands, April, Hanoi, Vietnam, online at http://www.globaloceans.org/sites/udel.edu.globaloceans/files/EBM-ICM-PB-April4.pdf [accessed 28 August 2014].

Olsen, S. B., Page, G. G., and Ochoa, E. (2009) *The Analysis of Governance Responses to Ecosystem Change: A Handbook for Assembling a Baseline*, LOICZ Reports & Studies No. 34, Geesthacht: GKSS Research Center.

PEMSEA (2012) *Sustainable Development Strategy for the Seas of East Asia (SDS-SEA) Implementation Plan 2012–16*, online at http://www.pemsea.org/sites/default/files/sdssea-implementation-plan.pdf [accessed 28 August 2014].

Rose, R. (2004) *Learning from Comparative Public Policy: A Practical Guide*, London: Routledge.

Rothwell, D. R., and VanderZwaag, D. L. (2006) 'The Sea Change: Towards Principled Oceans Governance', in D. R. Rothwell and D. L. VanderZwaag (eds) *Towards Principled Oceans Governance: Australian and Canadian Approaches and Challenges*, London: Routledge.

Sea Around Us Project (undated) 'Exclusive Economic Zones (EEZ)', online at http://www.seaaroundus.org/eez/ [accessed 26 August 2014].

Secretariat of the Convention on Biological Diversity and the Scientific and Technical Advisory Panel of the Global Environment Facility (2012) *Marine Spatial Planning in the Context of the Convention on Biological Diversity: A Study Carried out in Response to CBD COP Decision X/29*, Technical Series No. 68, Montreal: GEF.

Tableman, B. (2005) *How Governmental Policy is Made*, Michigan State University Best Practice Briefs No. 34, online at http://outreach.msu.edu/bpbriefs/issues/brief34.pdf [accessed 28 August 2014].

Townsley, P., and Food and Agriculture Organization of the United Nations (FAO) (1998) *Social Issues in Fisheries*, online at http://www.fao.org/docrep/003/w8623e/w8623e00.htm#Contents [accessed 28 August 2014].

United Nations (2002) *Report of the World Summit on Sustainable Development*, Johannesburg, South Africa, 26 August–4 September 2002, UN Doc. A/CONF.199/20, online at http://daccess-ods.un.org/access.nsf/Get?Open&DS=A/CONF.199/20&Lang=E [accessed 28 August 2014].

United Nations Environment Programme (UNEP) (2009) *Integrated Policymaking for Sustainable Development: A Reference Manual*, online at http://www.unep.ch/etb/publications/IPSD%20manual/UNEP%20IPSD%20final.pdf [accessed 28 August 2014].

Vince, J. (2013) 'Marine bioregional plans and implementation issues: Australia's oceans policy process', *Marine Policy*, 38: 325–9.

2

INTERNATIONAL AGREEMENTS AND CUSTOMARY INTERNATIONAL PRINCIPLES PROVIDING GUIDANCE FOR NATIONAL AND REGIONAL OCEAN POLICIES

Jon M. Van Dyke and Sherry P. Broder

Introduction

The acceptance by the negotiators at the Third United Nations Conference on the Law of the Sea[1] of the simple direct and elegant language of Article 192 of the 1982 Convention on the Law of the Sea (UNCLOS) marked a turning point in the human stewardship of the ocean: 'States have the obligation to protect and preserve the marine environment.'[2] Each word has importance and power. The operative word 'obligation' makes it clear that countries have positive duties and responsibilities, and must take action. The verbs 'protect' and 'preserve' reinforce each other, to emphasize that countries must respect the natural processes of the ocean and must act in a manner that demonstrates understanding of these processes and ensures that they continue for future generations. The 'marine environment' is a purposively comprehensive concept covering all aspects of the ocean world, including the water itself, its resources, the air above, and the seabed below, and all jurisdictional zones: internal waters, territorial seas, contiguous zones, exclusive economic zones (EEZs); continental shelves; archipelagic waters; and the high seas. Article 192 thus recognizes the profound responsibility that all countries have to govern the oceans in a manner that respects the marine creatures that inhabit them. The marine environment must be preserved for the benefit of those who will come later to exploit its resources, to study its mysteries, and to enjoy the many pleasures that it offers.

This chapter provides guidance to national governments and regional organizations regarding what their obligations are under UNCLOS and the many other treaties, agreements, and pronouncements and principles of customary international law related to the management of coasts and coastal waters. The 1982 UNCLOS and the 1995 Agreement for the Implementation of the Provisions of the United Nations Convention on the Law of the Sea of 10 December 1982 Relating to the Conservation and Management of Straddling Fish Stocks and Highly Migratory Fish Stocks (the 1995 UN Fish Stocks Agreement, or UNFSA 1995) provide details regarding management of coastal ecosystems and shared fisheries. Other principles of customary international law and relevant treaties provide further guidance, as do the initiatives undertaken by UN bodies and regional organizations. The governance and management of coastal resources is still a work in progress, but much has been accomplished in recent years, and countries now have a set of principles that explain their responsibilities and guide their activities.

The 1982 UN Convention on the Law of the Sea

In addition to the overarching obligations identified in Article 192, UNCLOS contains numerous specific duties that nations must fulfil. Article 56 recognizes coastal state sovereignty over the living resources in the 200 nautical mile EEZs, but Articles 61, 62, 69, and 70 require the coastal state (a) to cooperate with international organizations to ensure that species are not endangered by overexploitation, (b) to manage species in a manner that protects 'associated or dependent species' from overexploitation, (c) to exchange data with international organizations and other nations that fish in its EEZ, and (d) to allow other states (particularly developing, landlocked, and geographically disadvantaged states) to harvest the surplus stocks in its EEZ. Article 63 addresses stocks (or stocks of associated species) that 'straddle' adjacent EEZs, or an EEZ and an adjacent high seas area, and requires the states concerned to agree (either directly or through an organization) on the measures necessary to ensure the conservation of such stocks. Article 64 requires coastal states and distant-water fishing states that harvest highly migratory stocks such as tuna to cooperate (either directly or through an organization) to ensure the conservation and optimum utilization of such stocks. Article 65 contains strong language requiring nations to 'work through the appropriate international organization' to conserve, manage, and study cetaceans such as whales and dolphins. Article 66 gives the states of origin primary responsibility for anadromous stocks (such as salmon and sturgeon), but requires the states of origin to cooperate with other states whose nationals have traditionally harvested such stocks and those states through whose waters these fish migrate.

On the high seas, Articles 118 and 119 require states to cooperate with other states whose nationals exploit identical or associated species. Article 118 mandates that nations '*shall enter into negotiations* with a view to taking the measures necessary for the conservation of the living resources concerned' (emphasis added), and suggests creating regional fisheries organizations, as appropriate. Article 120 states that the provisions of Article 65 on marine mammals also apply on the high seas.

Article 209(2) requires states parties to 'adopt laws and regulations to prevent, reduce and control pollution of the marine environment from activities in the Area undertaken by vessels, installations, structures and other devices flying their flag or of their registry or operating under their authority', and further requires that these laws and regulations 'shall be no less effective than the international rules' applicable to activities in the area. The requirement that a legal system be available to provide prompt and adequate compensation is confirmed in Article 235(2) of the Convention, which requires states parties to 'ensure that recourse is available in accordance with their legal systems for prompt and adequate compensation or other relief in respect of damage caused by pollution of the marine environment by natural or juridical persons under their jurisdiction'. Article 235(3) further suggests that state parties should establish 'compulsory insurance or compensation funds', where appropriate. This approach is also supported by Principle 6(2) of the International Law Commission (ILC) 2006 Principles on the Allocation of Loss in the Case of Transboundary Harm Arising out of Hazardous Activities, which requires that states ensure that '[v]ictims of transboundary damage should have access to remedies in the State of origin that are no less prompt, adequate and effective than those available to victims that suffer damage, from the same incident, within the territory of that State', and by Principle 6(4), which provides further that 'States may provide for recourse to international claims settlement procedures that are expeditious and involve minimal expenses' (ILC, 2006).

Article 199 of UNCLOS requires states parties to 'jointly develop and promote contingency plans for responding to pollution incidents in the marine environment', and this obligation is supported by Principle 16 of the ILC 2001 Principles of Transboundary Harm from Hazardous Activities, requiring the state of origin to 'develop contingency plans for responding to emergencies' (ILC, 2001).

The common, but differentiated, responsibilities imposed upon countries to implement UNCLOS

The notion that countries have common, but differentiated, responsibilities to act under international treaties has developed from the principle of equity (or justice) in international law. It has been recognized that the formal equality of states does not inevitably mean that all states have the same duties, because some have better means with which to protect the global environment and to assist other states. This idea was identified in Principle 23 of the 1972 Stockholm Declaration, which explained that 'it will be essential in all cases to consider . . . the extent of applicability of standards which are valid for the most advanced countries but which may be inappropriate and of unwarranted social cost for developing countries' (United Nations, 1973). Principle 7 of the 1992 Rio Declaration went on to say more directly that, '[i]n view of the different contributions to global environmental degradation, States have common but differentiated responsibilities'.

The 1982 Convention recognizes these different responsibilities in several Articles, including, for instance, Article 207 on land-based pollution, which refers to the economic capabilities of developing states when articulating the responsibility to deal with this problem.[3] Other provisions in UNCLOS providing special preferences for developing and otherwise disadvantaged countries include:

- Article 62(2) and (3), granting developing countries preferential rights to the surplus stocks in the EEZs of other coastal states in their regions;
- Articles 69 and 70, giving developing landlocked and geographically disadvantaged states preferential rights to the surplus stocks in EEZs of coastal states in their regions;
- Article 82, exempting developing states from making payments from continental shelf resources beyond 200 nautical miles and giving them preferential rights to payments made by other states;
- Article 119, giving developing countries some preferential rights to the living resources of the high seas;
- Article 194(1), stating that states must prevent, reduce, and control pollution of the marine environment, 'using for this purpose the best practicable means at their disposal *in accordance with their capabilities*' (emphasis added);
- Article 199, requiring states to develop contingency plans for responding to pollution incidents '*in accordance with their capabilities*' (emphasis added);
- Articles 202 and 203, stating that developing states are entitled to training, equipment, and financial assistance from developed states and international organizations with regard to the prevention, reduction, and control of marine pollution;

- Article 206, explaining that the duty to assess environmental impacts of planned activities extends '*as far as practicable*' (emphasis added); and
- Articles 266–269, stating that developing countries are entitled to receive marine science and marine technology on fair and reasonable terms and conditions.

The 1995 UN Fish Stocks Agreement also contains a number of provisions recognizing the special rights of developing countries, as follows.

- The Preamble recognizes the need for specific assistance, including financial, scientific, and technological assistance, in order that developing states can participate effectively in the conservation, management, and sustainable use of straddling fish stocks and highly migratory fish stocks.
- Article 3(3) provides that 'States shall give due consideration to *the respective capacities of developing States* to apply articles 5, 6 and 7 within areas under national jurisdiction and their need for assistance as provided for in this Agreement' (emphasis added).
- Article 11(f) gives developing states a preference to enter into a fishery and into a fishery organization as a new member.
- Article 24 addresses the financial needs of developing countries:

States shall give full recognition to *the special requirements of developing States* in relation to conservation and management of straddling fish stocks and highly migratory fish stocks and development of fisheries for such stocks. To this end, *States shall*, either directly or through the United Nations Development Programme, the Food and Agriculture Organization of the United Nations and other specialized agencies, the Global Environment Facility, the Commission on Sustainable Development and other appropriate international and regional organizations and bodies, *provide assistance to developing States*.

(Emphasis added)

- Article 25 provides some more specific language regarding these obligations:

States shall cooperate, either directly or through subregional, regional or global organizations:

    a to *enhance the ability of developing States, in particular the least-developed among them and small island developing States*, to conserve and manage straddling fish stocks and highly migratory fish stocks and to develop their own fisheries for such stocks;

    b to assist developing States, *in particular the least-developed among them and small island developing States*, to enable them to participate in high seas fisheries for such stocks, including *facilitating access to such fisheries* subject to articles 5 and 11; and

    c to facilitate the participation of developing States in subregional and regional fisheries management organizations and arrangements.

(Emphasis added)

- Funding is addressed in Article 26:

    1 States shall cooperate to *establish special funds to assist developing States* in the implementation of this Agreement, including assisting developing States to meet the costs involved in any proceedings for the settlement of disputes to which they may be parties.

2 States and international organizations *should assist developing States in establishing new subregional or regional fisheries management organizations or arrangements*, or in strengthening existing organizations or arrangements, for the conservation and management of straddling fish stocks and highly migratory fish stocks.

(Emphasis added)

Other treaties use contextual norms – that is, by requiring a certain conduct 'as far as possible' or 'according to [a state's] abilities'. Some agreements have adopted the idea of calling upon the developed countries to provide financial support for developing countries to comply with international agreements. These two elements, asymmetry of obligations and financial support for developing countries, are thus the prominent features of the principle of common, but differentiated, responsibility.

However, in the context of deep seabed mining, the Seabed Chambers of the International Tribunal for the Law of the Sea (ITLOS) set the highest standards of due diligence for all sponsoring states, irrespective of whether they are developed or developing states and of the financial capabilities of the state.[4] In the unanimous 2011 *Advisory Opinion on Responsibilities and Obligations of States Sponsoring Persons and Entities with Respect to Activities in the Area*, Judge Tullio Treves reasoned that '[e]quality of treatment between developing and developed sponsoring States is consistent with the need to prevent commercial enterprises based in developed States from setting up companies in developing States, acquiring their nationality and obtaining their sponsorship'.[5] Significantly, the Seabed Chambers was very concerned because '[t]he spread of sponsoring States "of convenience" would jeopardize uniform application of the highest standards of protection of the marine environment'.[6] This conclusion – not to apply the common, but differentiated, principle because it could create a sponsoring state 'of convenience' – is significant. Because of its inherent fairness in preventing sponsoring states of convenience, it may be applied in other contexts in the future. The Seabed Chambers was, however, careful to include an acknowledgement of the continuing need for the common, but differentiated, idea and also concluded that 'Developing States should receive necessary assistance including training'.[7]

The underlying principles of customary international law

Among the principles of international environmental law that are now obligatory under customary international law are the following.

The 'no-harm' rule

The 'no harm' rule, stating that countries must not allow their territories to be used in a manner that harms their neighbours or shared common areas (in Latin, *sic utere tuo ut alienum non laedas*), is reflected in Principle 21 of the 1972 Stockholm Declaration (United Nations, 1973) and Principle 2 of the Rio Declaration, and has been accepted as a central component of international law.[8] The International Court of Justice (ICJ) stated in its 2010 Opinion in the *Pulp Mills Case*[9] that this rule prohibits damage to shared spaces, as well as to national territory:

The existence of the general obligation of States to ensure that activities within their jurisdiction and control respect the environment of other States or of *areas beyond national control* is now part of the corpus of international law relating to the environment.[10]

This duty to avoid causing injury to others is reflected in Article 194(2) of UNCLOS, which requires states 'to take all measures necessary to ensure that activities under their jurisdiction or control are so conducted as not to cause damage by pollution to other States and their environment'. Another form of this principle can be found in Article 87(2) of the Convention, which says – after listing the freedoms of the high seas – that '[t]hese freedoms shall be exercised by all States with due regard for the interests of other States in their exercise of the freedoms of the high seas'. Further, Article 235(1) notes that states parties 'shall be liable in accordance with international law' if they fail to fulfil 'their international obligations concerning the protection and preservation of the marine environment'. This obligation is confirmed again in Article 263(3) with regard to damages caused by marine scientific research.

The duty to cooperate

The duty to cooperate is one of the central and most venerable principles of international law.[11] As Professor Alan Boyle (1990: 278) has explained in simple and direct terms: 'States are required to co-operate with each other in controlling transboundary pollution and environmental risks.' Principle 24 of the 1972 Stockholm Declaration states that:

> International matters concerning the protection and improvement of the environment *should be handled in a co-operative spirit by all countries*, big and small, on an equal footing. Cooperation through multilateral or bilateral arrangements or other appropriate means is essential *to effectively control, prevent, reduce and eliminate adverse environmental effects* resulting from activities conducted in all spheres, in such a way that due account is taken of the sovereignty and interests of all States.
>
> *(United Nations, 1973; emphasis added)*

This principle was utilized by the Arbitral Tribunal in the 1957 *Lake Lanoux Arbitration*,[12] the Tribunal holding that, as a matter of customary international law, a state engaging in behaviour likely to impact the environment of another state significantly is obliged to involve the affected state in discussions regarding these activities. Inherent in this process is the duty to listen to the concerns expressed by the affected nations, along with their ideas about how best to reduce potential risks. Suggestions that are helpful and constructive should, of course, be accepted and acted upon. If a country rejects a suggestion, it should explain its rejection. These consultations are designed to anticipate and reduce risks. Preparing contingency plans for emergencies can be done only with a full understanding of the dangers involved. A nation that is consulted about a project outside its borders does not have a veto power over that project, but it does have the right to understand the risks created by the project and to offer constructive advice about how best to reduce those risks.

The duty to cooperate includes the duty to notify other affected countries,[13] the duty to exchange information, the duty to listen to the concerns of affected countries, the duty to respond to these concerns, and the duty to negotiate in good faith. In some situations, countries also have the duty to reach an agreement and a duty to submit the dispute to third-party adjudication if they cannot resolve the matter.[14] That countries engaging in activities impacting shared spaces have procedural obligations to consult and cooperate with other affected countries was made clear in *Pulp Mills*, in which the ICJ ruled that 'Uruguay breached its procedural obligations to inform, notify and negotiate'.[15] The ICJ also recognized this duty to inform in the 1949 *Corfu Channel Case*,[16] in which Albania was held to have had the duty to disclose the presence of mines in the Channel. Likewise, France was required to consult in good faith with Spain over riparian rights in the 1957 *Lac Lanoux Arbitration*.[17] Article 198 of UNCLOS requires that when a state becomes aware that its activities are causing, or are likely to cause, damaging pollution to the marine environment, it shall immediately notify other states likely to be affected by such damage. Similarly, the 1986 Convention on Early Notification of a Nuclear Accident requires notification of nuclear accidents.[18] The 1989 Basel Convention on the Control of Transboundary Movements of Hazardous Wastes and Their Disposal and the International Atomic Energy Agency (IAEA) Code of Practice on the same (IAEA, 1990) both require a state to notify and obtain the consent of the sending, receiving, and transit states in accordance with their respective laws and regulations.

The duty to cooperate played a central role in the judgment of the ICJ in 1997 in *Gabcikovo Nagymaros Project*,[19] which, as described by Professors Birnie and Boyle (2002: 108), had '[t]he effect of . . . requir[ing] the parties to co-operate in the joint management of the project, and to institute a continuing process of environmental protection and monitoring'. The authors explained that:

> The Court's environmental jurisprudence is not extensive but its judgments affirm the existence of a legal obligation to prevent transboundary harm, to co-operate in the management of environmental risks, to utilize shared resources equitably and, albeit less certainly, to carry out environmental impact assessment and monitoring.
>
> *(Birnie and Boyle, 2002: 108)*

ITLOS confirmed the importance of the duty to cooperate in two of its cases. In the 2001 *MOX Plant Case*,[20] the Tribunal ruled that the duty to cooperate required the two countries to exchange information concerning the risks created by the plant, to monitor the effects of the plant on the marine environment, and to work together to reduce those risks. Similarly, in the 2003 *Case Concerning Land Reclamation by Singapore In and Around the Straits of Johor (No. 12)*,[21] the Tribunal issued a ruling stating that:

> [G]iven the possible implications of land reclamation on the marine environment, *prudence and caution* require that Malaysia and Singapore establish mechanisms for exchanging information and assessing the risks or effects of land reclamation works and devising ways to deal with them in the areas concerned.[22]

To give teeth to this duty to cooperate, the Tribunal went on to prescribe provisional measures with which the parties had to comply:

> Malaysia and Singapore *shall cooperate* and shall, for this purpose, enter into consultations forthwith in order to:
>
> a establish promptly a group of independent experts with the mandate
>
> 1 *to conduct a study*, on terms of reference to be agreed by Malaysia and Singapore, to determine, within a period not exceeding *one year* from the date of this Order, the effects of Singapore's land reclamation and to propose, as appropriate, measures to deal with any adverse effects of such land reclamation . . .
>
> b exchange, on a regular basis, information on, *and assess risks or effects of*, Singapore's land reclamation works . . . [23]

Finally, the Tribunal directed Singapore 'not to conduct its land reclamation in ways that might cause irreparable prejudice to the rights of Malaysia or serious harm to the marine environment, taking especially into account the reports of the group of independent experts'.[24]

The duty to prepare an environmental impact assessment

The duty to cooperate requires the preparation of an environmental impact assessment (EIA), as explained by the ICJ in the *Pulp Mills Case,* in which the Court said that the requirement to undertake an EIA 'may now be considered a requirement under general customary international law . . . where there is a risk that the proposed industrial activity may have a significant adverse impact in a transboundary context, in particular on a shared resource'.[25]

In the ITLOS Seabed Chamber's 2011 Advisory Opinion, the Tribunal very significantly found that the requirement for the preparation of an EIA is now clearly an obligation that is generally required. The Chamber also decided that the obligation to conduct an EIA is a direct and mandatory obligation under Article 206 of UNCLOS and a general obligation under international law, as recognized by the ICJ in *Pulp Mills.*

The obligation to prepare EIAs is also found, for instance, in Articles 204–206 of UNCLOS,[26] Regulation 31(6) of the International Seabed Authority (ISA) Regulations on Prospecting and Exploration of Polymetallic Nodules in the Area (ISA, 2000) and Regulation 33(6) of the ISA Regulations on Prospecting and Exploration for Polymetallic Sulphides in the Area (ISA, 2010), Principle 17 of the 1992 Rio Declaration, and Principle 7 of the 2001 ILC Principles of Transboundary Harm from Hazardous Activities. The Espoo Convention also requires an EIA for activities that are likely to cause a significant transboundary impact.[27]

The precautionary principle

The precautionary principle (sometimes called the 'precautionary approach'[28]) has evolved into a customary international law norm,[29] as confirmed in Principle 15 of the 1992 Rio Declaration, which states:

> In order to protect the environment, the precautionary approach shall be widely applied *by States* according to their capabilities. Where there are threats of serious or irreversible damage, *lack of full scientific certainty shall not be used as a reason for postponing cost-effective measures to prevent environmental degradation.*
>
> *(Emphasis added)*

This principle continues to develop and is presently recognized by governments and international organizations as an authoritative norm guiding activities affecting the environment. It flows directly from the responsibility of 'due diligence' that is a component of the no-harm rule, and it constitutes 'an obligation of diligent prevention and control' (Birnie and Boyle, 2002: 115).

The essential components of the precautionary principle are that:

- developments and initiatives affecting the environment should be thoroughly assessed before action is taken;
- the burden is on the developer or initiator to establish that the new programme is safe;
- alternative technologies should be explored;
- the absence of full scientific certainty should not limit cost-effective precautionary measures to protect the environment; and
- whenever serious or irreversible damage is anticipated, the action should be postponed or cancelled.

In its 2011 Advisory Opinion on seabed mining, the ITCLOS Seabed Chamber ruled that there are obligations that flow from UNCLOS and from Nodules and Sulphides Regulations, and which include assisting the authority, applying the precautionary approach and 'best environmental practices', ensuring that the contractor complies with its obligation to conduct an EIA, and providing effective methods for compensation in the case that harm results from the mining activity. The Chamber noted that the Regulations, by embodying the precautionary approach defined in Principle 15 of the 1992 Rio Declaration, had the effect of transforming a non-binding concept into a binding obligation. The Chamber noted too that the precautionary approach has been incorporated into a growing number of international treaties and other instruments, many of which reflect the formulation of Principle 15. In the view of the Chamber, this has initiated a 'trend' towards making the approach part of customary international law. Moreover, the Chamber concluded in a very significant finding that 'it is appropriate to point out that the precautionary approach is also an integral part of the general obligation of due diligence of sponsoring States, which is applicable even outside the scope of the Regulations'.[30] In addition, the Chamber was clear that riskier activities require a higher standard of due diligence.

How application of the precautionary principle impacts the burden of proof for parties in international disputes remains unclear. Different tribunals have taken varied stances on how the precautionary principle impacts the parties' burden of proof. The ICJ, in *Pulp Mills* (2010), held that the burden of proof, as articulated in its 1975 Statute of the River Uruguay (the treaty between Uruguay and Argentina), was not reversed by application of the precautionary principle, nor did it place the burden equally on both parties. However, the ICJ was relying specifically on a particular treaty. In another context, there has been an acknowledgement that the precautionary principle does require a reversal of the burden of proof by application of the precautionary principle. As Judge Wolfrum expressed in his separate Opinion in the 2001 *MOX Plant Case*:

> There is no general agreement as to the consequences which flow from the implementation of this principle *other than* the fact that the burden of proof concerning the possible impact of a given activity is reversed. A State interested in undertaking or continuing a particular activity has to prove that such activities will not result in any harm, rather than the other side having to prove that it will result in harm.[31]

The precautionary principle has been rather controversial, because some commentators view it as being too vague,[32] while others view it as unrealistic, but it is now a major presence at all international negotiations, and it appears regularly in treaties and documents because it reflects the view that it is necessary to be extra-vigilant in our stewardship of resources, especially in light of the many mistakes that we have made in recent years.[33] Although the exact application of the precautionary principle is still a subject of discussion, it clearly serves as a guiding principle of customary international law for a state seeking to initiate an environmentally sensitive activity that may cause transboundary harm or harm to shared resources. Certainly, the inclusion of the precautionary standard in the 1996 Protocol to the London Dumping Convention[34] and in the 1995 UN Fish Stocks Agreement[35] provides strong evidence that the approach is here to stay.[36]

Hawai'i's Supreme Court has also recognized the importance of the precautionary principle with regard to decisions affecting the allocation of water. In 2000, it explained that:

> [T]he precautionary principle simply restates the [Water] Commission's duties under the [Hawai'ian] constitution and [Hawai'i's Water] Code. Indeed, the lack of full scientific certainty does not extinguish the presumption in favor of public trust purposes or vitiate the Commission's affirmative duty to protect such purposes whenever feasible.[37]

The 1995 UN Fish Stocks Agreement

On 4 December 1995, the nations of the world settled on the text of an important document with the cumbersome title of 'Agreement for the Implementation of the Provisions of the United Nations Convention on the Law of the Sea of 10 December 1982 Relating to the Conservation and Management of Straddling Fish Stocks and Highly Migratory Fish Stocks', known as the 1995 UN Fish Stocks Agreement, or UNFSA 1995.[38] The goal of this agreement was to stop the dramatic overfishing that has decimated fish populations in many parts of the world (Pitt, 1993).[39] It builds on existing provisions in UNCLOS described above, but it also introduces a number of new strategies and obligations that have been requiring fishers to alter their operations in a number of significant ways. In addition to strengthening the role of regional organizations, as explained next, it also promotes peaceful dispute resolution by applying the dispute resolution procedures of the 1982 Convention to disputes involving straddling and migratory stocks.

The duty to cooperate

The guiding principle that governs UNFSA 1995 is the duty to cooperate. This core concept is given specific new meaning within the context of fisheries management, and the coastal nations and distant-water fishing nations of each region are now required to share data and to manage the straddling fisheries together. Article 7(2) requires that:

> Conservation and management measures established for the high seas and those adopted for areas under national jurisdiction *shall be compatible* in order to ensure conservation and management of the straddling fish stocks and highly migratory fish stocks in their entirety.
>
> *(Emphasis added)*

This duty gives the coastal state a leadership role in determining the allowable catch to be taken from a stock that is found both within and outside its exclusive economic zone (EEZ), as evidenced by the requirement in Article 7(2)(a) that contracting parties 'take into account' the conservation measures established by the coastal state under Article 61 of UNCLOS for its EEZ 'and ensure that measures established in respect of such stocks for the high seas do not undermine the effectiveness of such measures'. This polite, diplomatic language indicates clearly that catch rates outside a 200 nautical mile EEZ cannot differ significantly from those within the EEZ.

The duty to work through an existing or new fisheries organization

The 1995 Agreement requires coastal and island nations to work together with distant-water fishing nations in an organization or arrangement to manage shared fisheries. Article 8(3) addresses this issue and is quoted here in full because its somewhat ambiguous language requires close examination:

> Where a subregional or regional fisheries management organization or arrangement has the competence to establish conservation and management measures for particular straddling fish stocks or highly migratory fish stocks, *States fishing for the stocks on the high seas and relevant coastal States shall give effect to their duty to cooperate* by becoming

a member of such an organization or a participant in such an arrangement, or by agreeing to apply the conservation and management measures established by such an organization or arrangement. *States having a real interest in the fisheries concerned may become members of such organizations or participants in such arrangement. The terms of participation of such organizations or arrangements shall not preclude such States from membership or participation*; nor shall they be applied in a manner which discriminates against any State or group of States having a real interest in the fisheries concerned.

(Emphasis added)

It is hard to read this language without concluding that the coastal and island nations must cooperate with the distant-water fishing nations fishing in adjacent high seas areas either by allowing them into an existing fishery management organization or by creating a new one that all can join. All states 'having a real interest' in the shared fishery stock must be allowed into the organization. Only those states that join a regional organization or agree to observe its management regulations can fish in a regional fishery (Article 8(4), and see Article 17(1)). Article 13 requires existing fisheries management organizations to 'improve their effectiveness in establishing and implementing conservation and management measures'.

Article 11 addresses the difficult question of whether *new* distant-water fishing nations must be allowed into such an organization once established. Do the nations that have established fishing activities in the region have to allow new entrants? The language of Article 11 does not give a clear answer to this question, but it seems to indicate that some new entrants could be excluded if the current fishing nations had developed a dependency on the shared fish stock in question. Furthermore, developing nations from the region would appear to have a greater right to enter the fishery than would developed nations from outside the region.

The precautionary approach

Article 5(c) of the 1995 Agreement lists the precautionary approach among the principles that govern conservation and management of shared fish stocks, and Article 6 elaborates on this requirement in some detail, focusing on data collection and monitoring. Then, in Annex II, the Agreement identifies a specific procedure that must be used to control exploitation and monitor the effects of the management plan. For each harvested species, a 'conservation' or 'limit' reference point, as well as a 'management' or 'target' reference, must be determined. If stock populations go below the agreed-upon conservation/limit reference point, then 'conservation and management action should be initiated to facilitate stock recovery' (Annex II(5)). Overfished stocks must be managed to ensure that they can recover to the level at which they can produce the maximum sustainable yield (Annex II(7)). The continued use of the maximum sustainable yield approach indicates that the Agreement has not broken free from those approaches that have led to the rapid decline in the world's fisheries,[40] but the hope is that the conservation/limit reference points will lead to early warnings of trouble that will be taken more seriously.

The duty to assess, and to collect and share data

Article 5(d) reaffirms the duty to 'assess the impacts of fishing, other human activities and environmental factors' of stocks, and Articles 14 and 18(3)(e) explain the data collection requirements necessary to facilitate such assessments. Article 14 requires contracting parties to mandate fishing vessels flying their flags to collect data 'in sufficient detail to facilitate effective stock assessment' (Article 14(1)(b)). Annex I then explains the specific information that must be collected, which includes the amount of fish caught by species, the amount of fish discarded, the types of fishing method used, and the locations of the fishing vessels (Annex I, Article 3(1)). In order to permit stock assessment, each nation must also provide data to the regional fishery organization on the size, weight, length, age, and distribution of its catch, plus 'other relevant research, including surveys of abundance, biomass surveys, hydro-acoustic surveys, research on environmental factors affecting stock abundance, and oceanographic and ecological studies' (Annex I, Article 3(2)). These requirements, if taken seriously, will revolutionize the fishing industry, in which the competitive nature of the quest for fish has encouraged each nation to hide its activities from others to the greatest extent possible. The data collected 'must be shared with other flag States and relevant coastal States through appropriate subregional or regional fisheries management organizations or arrangements' in a 'timely manner', although the 'confidentiality of nonaggregated data' should be maintained (Annex I, Article 7). Decision-making at regional fishery organizations must now be 'transparent' under Article 12, and international and non-governmental organizations (NGOs) must be allowed to participate in meetings and to observe the basis for decisions.

The methods of enforcement

Article 18 of the 1995 Agreement further requires contracting parties to establish 'national inspection schemes', 'national observer programmes', and 'vessel monitoring systems, including, as appropriate, satellite transmitter systems', to manage their flag fishing vessels with some rigour. Article 21(1) gives these requirements teeth by authorizing the ships of a nation that is party to a regional fisheries agreement to board and inspect on the high seas any ship flying the flag of any other nation that is a party to the same agreement.[41] If the boarded vessel is found to have committed a 'serious violation', it can be brought into the 'nearest appropriate port' for further inspection (Article 21(8)). The term 'serious violation' is defined in Article 21(11) to include using prohibited fishing gear, having improper markings or identification, fishing without a licence or in violation of an established quota, and failing to maintain accurate records or tampering with evidence needed for an investigation.

Dispute resolution

Part VIII of the Agreement requires contracting parties to settle their disputes peacefully and extends the dispute-resolution mechanisms of UNCLOS to disputes arising under this new 1995 Agreement. These procedures are complicated and somewhat untested, but should provide flexible and sophisticated mechanisms to allow nations to resolve their differences in an orderly fashion.

New treaties governing non-highly migratory fish stocks

On 21 June 2012, the 2006 Southern Indian Ocean Fisheries Agreement (SIOFA) entered into force. This treaty seeks to ensure the long-term conservation and sustainable use of non-highly migratory fish stocks in the high seas of the southern Indian Ocean. SIOFA is a non-tuna regional fishing management organization (RFMO), and has the potential to perform an important role in regulating and controlling overexploitation for fish stocks in the high seas, which were previously unregulated.

The current parties to SIOFA are Australia, the Cook Islands, the European Union, Mauritius, and the Seychelles (FAO Fisheries and Aquaculture Department, undated*a*). The goal of the treaty is to promote the long-term conservation and sustainable use of fisheries resources. The treaty reflects a newer generation philosophy and incorporates as guiding principles the precautionary approach, ecosystem-based approaches to fisheries management and effective monitoring, and control and surveillance measures to ensure compliance with the objectives of the treaty (Articles 4(c), 4(a), and 6(h)). It is dedicated also to 'taking into account the needs of developing States that are Contracting Parties to the Agreement, and in particular the least-developed among them and Small Island developing States' (Article 4(g)). Unfortunately, SIOFA has no permanent secretariat and this may limit its effectiveness.

On 24 August 2012, the Convention on the Conservation and Management of High Seas Fishery Resources in the South Pacific Ocean entered into force, establishing the South Pacific Regional Fisheries Management Organisation. The members of the Convention are Australia, Belize, the Republic of Chile, the People's Republic of China, the Cook Islands, the Republic of Cuba, the European Union, the Kingdom of Denmark in respect of the Faroe Islands, the Republic of Korea, New Zealand, the Russian Federation, Chinese Taipei, and the Republic of Vanuatu (SPRFMO, undated*a*). The secretariat for this new RFMO is based in Wellington, New Zealand. The Convention covers a huge high seas areas beyond national jurisdiction, generally from Australia and the south coast of Australia to the entire coastline of South America to the Marshall Islands, and then generally south and around the outer limits of the national jurisdictions of Pacific States and territories (FAO Fisheries and Aquaculture Department, undated*b*). The non-highly migratory fisheries covered by the Convention are high seas stocks, although some stocks straddle the high seas and the EEZs of coastal states. These fisheries are both pelagic and demersal (SPRFMO, undated*b*). Although there are other RFMOs whose jurisdiction includes parts of this area, this is the only RFMO that covers non-highly migratory species in its Highs Seas.

This new RFMO has a difficult task ahead. Some of the species under its jurisdiction are seriously threatened commercial fish stocks. Among the most important threatened species are the orange roughy and jack mackerel (Schiffman, 2010). The orange roughy is a deep-sea species that was first harvested only in the 1970s, but rapidly became depleted. The orange roughy is a slow-maturing fish and starts breeding at about age thirty. It is harvested using highly destructive bottom-trawling, which results in serious by-catch and which also threatens the South Pacific seamounts (Deep Sea Conservation Coalition, 2011). The jack mackerel is another species fished using bottom-trawling. The stocks of the jack mackerel have dropped 63 per cent between 2006 and 2011 (Rosenblum and Cabra, 2012). This new 2012 Convention is similarly required to apply the precautionary approach and an ecosystem-based approach to fisheries management, and to recognize the needs of the developing states and small island developing states, and the needs of developing state coastal communities (Articles 2(a), 2(b), and 1(a)(viii)).

These recent developments demonstrate the continuing international efforts to close jurisdictional gaps in international conservation and management of fish stocks, and the ecosystems in which they occur, particularly on the high seas with non-migratory stocks.

The United Nations Conferences on the Environment

The 1992 Rio Conference

In 1992, the UN Conference on Environment and Development took place in Rio de Janeiro (known as the 'Earth Summit'). This Summit resulted in several non-binding legal instruments relevant to ocean policy, including the 1992 Rio Declaration on Environment and Development, Agenda 21, the Convention on Biological Diversity (CBD), and the United

Nations Framework Convention on Climate Change (UNFCCC). Each of these instruments is briefly discussed next.

The 1992 Rio Declaration

The 1992 Rio Declaration consists of twenty-seven principles intended to guide future sustainable development around the world. The Rio Declaration espouses generally accepted principles for global governance of the environment, including the precautionary principle, sustainable development, common, but differentiated, responsibilities, and the duty to conduct an environmental impact assessment (EIA). The Rio Declaration has been signed by almost all nations and remains a guiding instrument to this day.

Agenda 21

The international community promulgated Agenda 21, a non-binding, voluntarily implemented action plan. A total of 178 governments voted to adopt the programme. It focused on a plan of action for sustainable development and contains many important innovations, including an emphasis on EIAs, calling for the '[f]urther development and promotion of the widest possible use' of environmental assessments. Chapter 17 of Agenda 21 focuses on 'Protection of the Oceans, All Kinds of Seas, Including Enclosed and Semi-Enclosed Seas, and Coastal Areas and the Protection, Rational Use and Development of Their Living Resources', and lays down a broad agenda related to integrated management and sustainable development of coastal areas, protection of the marine environment, sustainable use and conservation of marine resources of the high seas and those under national jurisdiction, climate change, regional coordination, and sustainable development of small islands (UNGA, 1992). It has been said that 'Agenda 21 goes beyond reactive remedies and emphasizes the precautionary approach as an essential requirement for integrated management and sustainable development of coastal areas and the marine environment under national jurisdiction' (Yankov, 1999: 271).

Agenda 21 noted that '[l]and-based sources contribute 70 percent of marine pollution', but that '[t]here is currently no global scheme to address marine pollution from land-based sources'. Land-based pollution has been difficult to control, because it comes from so many sources. Indeed, it is frequently difficult even to identify the source of the pollution, particularly for non-point sources such as surface run-off from development and clearing activities, mining, forestry, animal husbandry, and agriculture.

With regard to sea-based pollution, Chapter 17 of Agenda 21 identifies the need for additional measures in four priority areas to combat the growth of marine pollution from shipping, dumping, operation of offshore oil and gas platforms, and from ports. Controlling these sources of pollution is a challenging assignment for many countries, as the other chapters in this volume will demonstrate. Chapter 17 has been useful in establishing a framework for evaluating national achievements, but it has also been criticized for failing to include any enforcement capability or any reference to dispute resolution, and for failing to address the special problems of polar regions.[42]

The United Nations Framework Convention on Climate Change

The UNFCCC is an international environmental treaty negotiated at the Earth Summit in 1992. The objective of the treaty is to 'stabilize greenhouse gas [GHG] concentrations in the atmosphere at a level that would prevent dangerous anthropogenic interference with the climate system'. The treaty sets no binding limits on GHG emissions for individual countries and contains no enforcement mechanisms. It entered into force in 1994 and, as of May 2014, it had 196 parties (UNFCCC, undated). Article 3(1) of the Convention states that parties should act to protect the climate system on the basis of common, but differentiated, responsibilities, and that developed country parties should take the lead in addressing climate change.

The parties to the Convention have met annually from 1995 in Conferences of the Parties (CoPs) to assess progress in dealing with climate change. In 1997, the Kyoto Protocol was concluded and set forth legally binding obligations for developed countries to reduce their GHG emissions. The Kyoto Protocol's compliance period ended in 2012 and it has been impossible for the nations of the world to reach agreement on a new treaty to combat climate change, despite the dire predictions of scientists and the ongoing experiences of sea-level rise, severe weather events, accelerating melting of ice in the Arctic and Antarctic, and ocean acidification.

The Fifteenth CoP (COP-15) was held in Copenhagen in 2012 and attended by 193 countries. A document was produced – the Copenhagen Accord[43] – but because of objections from Bolivia, Cuba, Nicaragua, Sudan, Tuvalu, and Venezuela, it was agreed to 'take note' of the Copenhagen Accord rather than formally to adopt it and make it legally binding (UNFCCC, 2009; see also Broder, 2010), meaning that nations can choose to make a political commitment. The Accord does endorse the continuation of the commitments under the Kyoto Protocol. It acknowledges that 'the scientific view that the increase in global temperature should be below 2 degrees Celsius' and calls for an assessment to be concluded by 2015 that would include consideration of reducing this long-term goal to 1.5C (UNFCCC, 2009). The developed countries listed in Annex A agreed to 'commit to implement' economy-wide GHG emissions reductions by 2020, while the commitment of developing countries was to implement '[n]ationally appropriate mitigation actions' (UNFCCC, 2009).

Since the 2012 Copenhagen COP-15, very limited progress has been made on the drafting and acceptance of a new binding successor or complement to the Kyoto Protocol. In the meantime, global emissions of GHGs continue to rise. Coastal and marine ecosystems are deteriorating and threatened by the many effects of climate change, such as sea-level rise, loss of ice sheets, acidification, ocean warming, and changes in weather patterns. The health of marine species and the oceans is severely affected. It is clear that immediate action is needed from everyone to reduce GHG emissions significantly.

The Global Programme of Action for the Protection of the Marine Environment from Land-Based Activities

The Global Programme of Action for the Protection of the Marine Environment from Land-Based Activities was adopted on 3 November 1995 at an intergovernmental conference (ICG) that met in Washington, DC, in part owing to the need for such a document, as identified in Agenda 21, Chapter 17 (UNEP, 1995). This document assigned to the United Nations Environment Programme (UNEP) the responsibility of coordinating efforts to control land-based pollution sources. Coastal countries are required to apply integrated coastal area management approaches, to recognize linkages between freshwater bodies and coastal environments, and to apply environmental assessment techniques when evaluating options (UNEP GPA, undated). They are also obliged to work with and through regional and international bodies, and to focus in particular on sewage, persistent organic pollutants, radioactive substances, heavy metals, hydrocarbons, nutrients, sediments, and litter. Although the Global Programme of Action has 'no legally binding character it is nevertheless expected that [its] impact on the development of environmental law and practice will be significant', because it is linked to other treaties and contains 'the first comprehensive global framework with pragmatic approach and focused on sustainable and integrated environmental management with appropriate institutional arrangements and mechanisms for monitoring [its] implementation' (Yankov, 1999: 284).

The 2002 Johannesburg Conference

In 2002, the World Summit on Sustainable Development (WSSD) convened in Johannesburg, South Africa, and promulgated a 'Plan of Implementation' (WSSD, undated). This document reviewed the themes outlined in Chapter 17 of Agenda 21 and offered enthusiastic endorsement of these goals, but in many cases it failed to set clear targets and timetables for implementation of its commitments, and it created no enforcement mechanism at the international level. In the area of fisheries, it continued to rely upon the 'maximum sustainable yield' approach – a single-species and high-risk management strategy that reduces natural abundance of fish stocks by a third to a half, or more, and often leads to overfishing (WSSD, undated). This approach departs dramatically from the precautionary approach utilized in the 1995 UN Fish Stocks Agreement (the precautionary approach not even being mentioned in the Plan). The Plan does, however, require countries to '[e]liminate subsidies that contribute to illegal, unreported and unregulated fishing and to overcapacity', and to establish a network of marine protected areas (MPAs) by 2012 (WSSD, undated). As discussed later in this chapter, the international community has since made significant progress in the implementation of MPAs. However, the prevalence of illegal, unreported, and unregulated fishing continues to be a pressing issue in marine resource management.

The Rio+20 Conference

The United Nations Conference on Sustainable Development, also known as 'Rio 2012', 'Rio+20', or 'Earth Summit 2012', was the third international conference on sustainable development aimed at reconciling the economic and environmental goals of the global community. Rio+20 represented a twenty-year follow-up to the 1992 Earth Summit held in the same city, and marked the tenth anniversary of the 2002 WSSD in Johannesburg. The primary result of the Conference was the non-binding document entitled 'The Future We Want', subsequently endorsed by the UN General Assembly (UNGA) as Resolution 66/288.[44] World leaders of the 192 governments in attendance renewed their political commitment to sustainable development and declared their commitment to the promotion of a sustainable future, setting forth a common vision for sustainable development in the future. The parties supported increased funding for sustainable activities and projects, and pledged US$513 billion from governments, the private sector, and others (United Nations News Center, 2012). However, Rio+20 was widely criticized as a failure, largely because of the discord between the goals of developing and developed nations, and the strong focus on economic development, with environmental protection a secondary goal.[45]

Resolution 66/288 reaffirms previous action plans such as Agenda 21. The Resolution includes several commitments that focus on the oceans. There was agreement to take action to reduce marine pollution from land-based sources, especially plastics, as well as persistent organic pollutants, heavy metals, and nitrogen-based compounds. It was recognized that it was essential to work on reducing ocean acidification and sea-level rise, and to enhance efforts to address sea-level rise and coastal erosion and to end fishing subsidies and overfishing. Rio+20 acknowledges the importance of area-based conservation measures, including MPAs, as a tool for conservation of biological diversity and sustainable use of its components.

The parties recommitted to eliminating illegal, unreported, and unregulated fishing, as advanced in the Johannesburg Plan of Implementation, by supporting national and regional action plans, implementing effective and coordinated measures to identify vessels engaged in such fishing, and depriving offenders of the financial benefits accruing therefrom.[46]

No decision was reached to negotiate a new agreement for the conservation and management of biodiversity beyond national jurisdictions. However, an Ad Hoc Open-Ended Informal Working Group to study issues relating to the conservation and sustainable use of marine biological diversity beyond areas of national jurisdiction has been assigned to work on this issue and is expected to produce a report to the UNGA for 2015.[47]

Agreements for the protection of biological diversity

While UNCLOS provides generally for the protection and preservation of the marine environment under Article 192, and the 1995 UN Fish Stocks Agreement addresses the impact of commercial fishing on certain stocks, neither provides conservation measures specifically targeted towards the protection of biological diversity. The following international agreements concentrate on the impact of human activities on biological diversity, particularly with regard to threatened species.

The 1992 Convention on Biological Diversity

The Convention on Biological Diversity (CBD), produced under the auspices of the UNEP and opened for signature at the 1992 Rio Conference, marked a turn in environmental consciousness away from a focus on endangered species toward a focus on the importance of protecting 'the overall richness of life on earth' (Nanda and Pring, 2003: 171). This focus on ecosystem diversity should logically lead to a protection of larger areas, including land, coastal, and marine environments (Nanda and Pring, 2003). The Convention established that the conservation of biological diversity is a 'common concern of humankind, although it also recognizes that countries have right to the sustainable use of their biological resources'. The Convention encourages countries to develop plans to manage their resources, including establishing MPAs and addressing threats to coral reefs (Hunter et al., 2002).

Article 8(c) of the CBD requires contracting parties to '[r]egulate or manage biological resources important for the conservation of biological diversity . . . with a view to ensuring their conservation and sustainable use', which should be done by means of a national plan. Article 10(b) reinforces this obligation by requiring contracting parties to '[a]dopt measures relating to the use of biological resources to avoid or minimize adverse impacts on biological diversity'. Other provisions state that countries should inventory and monitor their species, establish protected areas, maintain natural habitats, utilize environmental assessments, exchange information with neighbours, and prepare contingency plans for emergencies.

Agreement on all issues related to biodiversity has proved to be difficult, because the developing countries have 'perceived outside attempts to curtail internal development as a threat to sovereignty' and have sought compensation from the developed countries with regard to any restrictions on development designed to protect biodiversity (Guruswamy, 2003: 136). The treaty has been criticized as 'aspirational', 'because its most general obligations contain heavily qualified language', but nonetheless has provided a mechanism for discussing issues and maintaining momentum toward future solutions (Guruswamy, 2003: 136). A CoP meets every two years as a legislative body, and a secretariat functions year-round to coordinate with other international bodies and prepare for the meetings of the parties. The parties have also created a Subsidiary Body on Scientific, Technical and Technological Advice (SBSTTA) to provide advice on the status of biodiversity and the effects of national actions, and a Clearing-House Mechanism for technical and scientific cooperation to promote collaborative research and conservation activities.

The Nagoya Protocol

The 2010 Nagoya Protocol on Access to Genetic Resources and the Fair and Equitable Sharing of Benefits Arising from the Utilization to the Convention on Biological Diversity is a supplementary agreement to the CBD. It was adopted on 29 October 2010 in Nagoya, Japan, and has not yet entered into force.[48] It provides a legal framework for the implementation of the CBD's mandate for the fair and equitable sharing of benefits arising out of the utilization of genetic resources, the goal of which is supposed to be the conservation and sustainable use of biodiversity. Parties are supposed to encourage the directions of the 'benefits towards the conservation of biological diversity and sustainable use' (Article 9).

The Nagoya Protocol also includes very specific protections for local communities, indigenous peoples, and users of traditional knowledge associated with genetic resources. It provides that these communities are entitled to protection of their interests and that they should receive benefits through a legal framework that includes their rights (Articles 6 and 12). The principle of prior informed consent of these communities is required and is a key component of the process (Article 6).

The 1973 Convention on International Trade in Endangered Species of Wild Fauna and Flora

The 1973 Convention on International Trade in Endangered Species of Wild Fauna and Flora (CITES) provides protection for marine species in addition to any protections afforded by domestic laws and other international agreements. This Convention plays an important role in setting guidelines and standards for commercial trade of wildlife. It is both a trade and conservation treaty, which prohibits commercial trade across national boundaries of some 830 species listed in its Appendix I. This list includes all sea turtles and all of the great whales covered by the moratorium established by the 1946 International Convention for the Regulation of Whaling. The trade of another 25,000 species (including, for instance, polar bears and narwhals) listed in Appendix II are regulated to ensure species' survival.[49]

CITES prohibits not only trade across national land boundaries, but also specifically the taking of listed marine species on the high seas; accordingly, it includes a broad definition of trade, which covers the 'export, re-export, import and introduction from the sea' of a species. 'Introduction from the sea' is further defined as 'transportation into a State of specimens of any species which were taken in the marine environment not under the jurisdiction of any State'. The Convention thus governs trade of marine species in areas that are largely beyond the reach of state regulation. It recognizes the vulnerability of the high seas and other maritime regions beyond national jurisdictions.

The 1946 International Convention for the Regulation of Whaling

Article 65 of UNCLOS is explicit in requiring states to 'work through the appropriate international organizations for [the] conservation, management and study' of cetaceans (whales and dolphins). The International Whaling Commission (IWC), established under the

1946 International Convention for the Regulation of Whaling, is presently the appropriate international organization and it has maintained a moratorium on all harvesting of whales since 1986, except for limited kills allocated to indigenous people, mostly in the Arctic region.[50]

Since 1986, the IWC has established a moratorium prohibiting whaling for commercial purposes pursuant to paragraph 10(e) of the Schedule adopted that year, which stated: 'Catch limits for the killing for commercial purposes of whales from all stock for the 1986 coastal and 1985/86 pelagic seasons and thereafter shall be zero.' This moratorium was adopted because of the gross overharvesting of whales that occurred during the nineteenth century and was still occurring, and because of a new appreciation of the importance of cetaceans within the marine environment. The relentless slaughter of whales resulted in the severe depletion of all species from which virtually all of the species, including the blue whale, the right whale, and the West Pacific Gray whale, have never recovered, remaining critically endangered.[51] The status of the populations of many species is largely unknown. Even with a moratorium, whales face grave threats from other sources, such as global and seawater warming (which can reduce the abundance of krill and change the polar environment necessary for certain whales), marine pollution, noise pollution in the ocean from military and seismic sonar, overfishing of the whales as food sources, entanglement in fish gear, and ship strikes.[52]

The smaller cetaceans have also been dramatically overfished and are in a difficult situation in many parts of the world. To give just one example, between 1870 and 1983, a fishery existed in the Black Sea for the bottlenose dolphin, the harbour porpoise, and the common dolphin, using harpoons and later purse seining, for food and oil. Turkey, Romania, Bulgaria, and Russia/USSR, the four countries involved in this fishery, harvested some 200,000 dolphins each year – an extraordinary 20 per cent of the population. Between 1931 and 1941, the USSR took between 110,000 and 130,000 dolphins per year, but in later years its harvests dropped to 75,000 per year because of the population losses, and in 1966 the fishery operations in all countries except Turkey stopped because of a collapse of the population. The present population of dolphins in the Black Sea is 500,000 – half its previous level – and they still suffer from accidental killings, gill net fishing, destruction of coastal ecosystems, and pollution. The World Wide Fund for Nature (WWF) has estimated that, worldwide, some 300,000 cetaceans are killed inadvertently each year, including those trapped in nets or caught on lines intended for other species (Pohl, 2003). The 1996 Agreement on the Conservation of Cetaceans of the Black Sea, Mediterranean Sea and Contiguous Atlantic Area (ACCOBAMS) prohibits the deliberate taking of any cetaceans and requires contracting parties to 'co-operate to create and maintain a network of specially protected areas to conserve cetaceans',[53] but two of the major Black Sea countries, Russia and Turkey, have not yet ratified this treaty.

The precarious status of the cetaceans has been further threatened by the scientific whaling that has been expanding, and by efforts to lift the IWC moratorium and resume commercial whaling. These efforts appear to violate the obligation to preserve and protect the marine environment. Scientific whaling can be undertaken only if it is truly scientific – that is, if it has passed neutral peer review and leads to scientific publications made available to all. The impact of proposed scientific whaling programmes on endangered species calls out for a full environmental impact assessment (EIA) under Articles 204–206 of UNCLOS. A lifting of the moratorium on commercial whaling at the present time, when the number of whales remains at a small fraction of the previous population, would be irresponsible.

The IWC has designated two whale sanctuaries in which commercial whaling is prohibited: the Southern Ocean Whale Sanctuary and the Indian Ocean Whale Sanctuary. The IWC's 1986 moratorium contains a scientific permit exception under which a permit may be issued to 'kill, take and treat whales for purposes of scientific research'.[54] Japan interpreted the scientific research exception broadly and designed a very generous 'scientific' whaling programme in the Antarctic allowing the harvesting of a large number of whales. This so-called scientific whaling programme has been the subject of much international criticism.

The IWC met in Panama in 2012. Once again, Japan submitted a proposal to make its limited coastal whaling programme acceptable. It was rejected, as it has consistently been in the past; a proposal for a Southern Atlantic Ocean sanctuary, which has been supported by the Latin American member nations, was also rejected – again – as was Greenland's application for a renewal of its aboriginal subsistence quota (IWC, 2012).

In 2010, Australia filed a case against Japan in the International Court of Justice (ICJ) claiming that Japan's so-called scientific whaling programme in the Southern Antarctic Ocean was a violation of the IWC moratorium and in fact commercial whaling; New Zealand intervened.[55] The Southern Ocean Whale Sanctuary was established by the IWC in 1994, and Japan was the only country opposing it (International Fund for Animal Welfare, undated). The ICJ decided twelve-to-four in favour of Australia, finding that Japan was actually engaged in commercial whaling in the Southern Ocean, which did not conform to the scientific research exception of the ICW moratorium. The judgment was narrowly tailored to emphasize that it was not the scientific whaling programme, but rather the way in which Japan had chosen to implement it, that violated the ICW moratorium. Japan publicly stated that it agreed to abide by the Court's judgment – but it remains to be seen how and to what extent it will reduce its current whaling programme – given that, less than a month after the ICJ decision, Japan stubbornly proceeded with a scaled-down version of its annual Northwest Pacific whaling campaign (Lies, 2014).

The conservation of marine areas

Another method by which states can protect the marine environment is to enact conservation measures protecting or preserving entire areas or regions of the ocean. Ecosystem-based environmental management protects biodiversity, as well as preserves ecosystem functions and services, and guarantees the ongoing sustainability of marine resources. States and international organizations have undertaken such ecosystem-based management by means of a variety of international organizations and agreements, with varying degrees of success.

Marine protected areas

Article 194(5) of UNCLOS requires countries to take measures to protect and preserve rare or fragile ecosystems, as well as the habitat of depleted, threatened, or endangered species and other forms of marine life. This key responsibility requires countries to establish marine protected areas (MPAs) and to establish management regimes in these areas that honour and protect the marine life within. These marine sanctuaries are taking many forms, including in some circumstances 'no take' zones, which are similar to wilderness areas on land.

One recent dramatic example of an MPA is the Papahānaumokuākea Marine National Monument established jointly by the United States and the State of Hawai'i around the northwestern Hawai'ian Islands. On 29 September 2005, Hawai'i's Governor Linda Lingle signed new state rules creating the Northwestern Hawai'ian Islands State Marine Refuge, banning fishing and limiting public access in state waters extending 3 miles from the shores of the islands, which are home to delicate coral reefs, scores of fish species, and endangered Hawai'ian monk seals and sea turtles. Then, on 15 June 2006, then US President George W. Bush signed Proclamation 8031 creating the world's largest marine conservation area, covering nearly 140,000 square miles of ocean territory. Papahānaumokuākea Marine National Monument is the single largest conservation area under the US flag and one of the largest marine conservation areas in the world. It encompasses 139,797 square miles (362,073 km²) of the Pacific Ocean (Papāhanaumokuākea Marine National Monument, undated). On 3 April 2008, Papahānaumokuākea was officially internationally recognized as a particularly sensitive sea area (PSSA) by the International Maritime Organization (IMO). Special zones, known as 'areas to be avoided', will appear on international nautical charts to direct ships away from coral reefs, shipwrecks, and other ecologically or culturally sensitive areas, and there will be a mandatory ship reporting system. On 30 July 2010, Papahānaumokuākea was inscribed as a mixed (natural and cultural) UNESCO World Heritage Site.

Particularly sensitive sea areas

Since the ratification of UNCLOS, international shipping has increased dramatically, while the global marine environment has degraded rapidly.[56] With the expansion of maritime trade and the growth in the size of fleets, risks for the marine environment have increased. The balance that UNCLOS struck in favour of the freedom of navigation is no longer fair. Coastal states have difficulty in protecting their marine environment from the dangers and hazards presented by global shipping. The Convention protects navigational freedom by placing heavy constraints on coastal states' jurisdiction in their EEZs. In so doing, UNCLOS curtails the ability of coastal states to implement and enforce measures protecting marine resources.

The Convention limits coastal state environmental efforts in order to protect the freedom of navigation by providing coastal states with few options for imposing protective measures even in navigationally challenging or ecologically sensitive areas. Article 211(6)(a) provides that, where an area in an EEZ is particularly navigationally challenging or ecologically sensitive, a coastal state may 'direct a communication' to 'a competent international organization' (which has been generally interpreted to refer to the IMO), to permit the adoption of coastal state regulations in that area that are more stringent than international ones. This Article provides coastal states with few effective options, however, because Article 211(6)(c) mandates that requested restrictions cannot include 'design, construction, manning or equipment standards other than generally accepted international rules and standards'.

In 1990, after determining that the Great Barrier Reef (GBR) was an area of ecological, social, cultural, economic, and scientific importance, the IMO's Marine Environment Protection Committee (MEPC) recognized the GBR as the first PSSA.[57] The IMO approved compulsory pilotage, backed by criminal penalties, which is not permitted under other international conventions. Australia was eager to protect a particularly vulnerable part of the GBR, between Mackay Island and the Tropic of Capricorn. The area of the Reef covered by the PSSA 'extends 2,300 kilometres along the east coast of Queensland and covers an area of 346,000 square kilometres', passing through both Australia's territorial sea and its EEZ. The Torres Strait was not part of the Great Barrier Reef Region PSSA initially, but the IMO extended the Reef PSSA to include the Torres Strait in 2005. The area is also an MPA under Australian domestic law and a 'special area' under the 1972 International Convention for the Prevention of Pollution from Ships (MARPOL), as modified by the Protocol of 1978 (AMSA, undated).

In 1991, the IMO passed Assembly Resolution 720(17), establishing 'Guidelines for Designating and Identifying Particularly Sensitive Sea Areas (PSSAs)',[58] under which PSSAs are defined as 'areas with "ecological, socio-economic, or scientific" importance'. The IMO established that it can designate areas as PSSAs not only in states' territorial seas, but also in their EEZs.

The IMO language reflects the language of Article 211(6)(a) of UNCLOS, but the PSSA designation goes one step further: it allows the IMO to impose measures to be taken in all maritime zones of a coastal state, including measures that affect design, construction, manning, or equipment standards. The creation of the PSSA mechanism was therefore an important and significant step forward in expanding coastal states' ability to protect marine resources, because it allows the imposition of new restrictive measures in sensitive areas of EEZs.

By 2014, the IMO had approved fourteen PSSAs, including: the Great Barrier Reef, Australia (1990); the Sabana-Camagüey Archipelago in Cuba (1997); the sea area around Malpelo Island, Colombia (2002); the marine area around the Florida Keys, United States (2002); the Wadden Sea, Denmark, Germany, and The Netherlands (2002); the Paracas National Reserve, Peru (2003); the Western European Waters, Belgium, France, Ireland, Portugal, Spain, and the United Kingdom (2004); the Torres Strait, as an extension to the Great Barrier Reef, Australia and Papua New Guinea (2005); Canary Islands, Spain (2005); the Galapagos Archipelago, Equador (2005); the Baltic Sea area, Denmark, Estonia, Finland, Germany, Latvia, Lithuania, Poland, and Sweden (2005); the North-West Hawai'ian Islands, Papahānaumokuākea Marine National Monument, United States (2008); the Strait of Bonifacio, France and Italy (2011); and Saba Bank (the Caribbean island of Saba), The Netherlands (2012) (IMO, undated*b*). Current guidelines on designating a PSSA are contained in Revised Guidelines for the Identification and Designation of Particularly Sensitive Sea Areas (PSSAs).[59]

The number of PSSAs is growing, reflecting the strong interest of coastal states in protecting the marine environment from pollution and other forms of environmental degradation. The expanse of ocean space that can be included in a PSSA is large. Coastal states may adopt national legislation and regulations to implement the PSSAs, which constitute a recognition and acceptance of local priorities by international interests. The definition and coverage of PSSAs has been expanded to better protect the marine environment and to readjust the imbalance between the freedom of navigation and the interests of the coastal state.

International Seabed Authority areas of particular environmental interest

The International Seabed Authority (ISA) is the international organization that has control and management over the sea floor and the seabed beyond national jurisdiction, which is part of the common heritage of humanity and is described in UNCLOS as the 'Area' (Article 1(1)). Within the Area, the Convention defines resources only as 'all solid, liquid or gaseous mineral resources in situ in the Area at or beneath the seabed, including polymetallic nodules', making no mention of how non-mineral living marine resources should be managed or regulated by the ISA (Article 133). Modern exploration of the Area has led to increasing awareness of the vulnerable and unique marine ecosystems therein, and the ISA has taken steps to protect certain portions of the Area from exploitation by designating them 'areas of particular environmental interest' (APEIs) (ISA, 1999: ch. 9). The ISA Guidelines require that mining be prohibited within the APEIs. The ISA has created the APEIs to be large enough to reflect a variety of environmental conditions on the sea floor and to have significant species biodiversity. Furthermore, APEIs are designated in locations in which mining activity outside the APEI will not negatively impact the marine environment through drifting plumes of sediment (ISA, 1999: ch. 9, §5.6).

UNESCO World Heritage Sites

The 1972 Convention Concerning the Protection of the World Cultural and Natural Heritage (the UNESCO World Heritage Convention) is another avenue for conservation of the marine environment. This Convention recognizes the importance of intergenerational equity and, as such, affirms that '[t]he properties on the World Heritage List are assets held in trust to pass on to generations of the future as their rightful inheritance'. Under the Convention, countries have a duty to protect and manage World Heritage Sites, and this duty includes legislative and regulatory measures that ensure the protection of sites against activities that might negatively impact their integrity or outstanding universal value (Article 5). Countries are also requested to ensure effective implementation of these measures; failure to protect and manage a site adequately may result in its listing on the 'World Heritage Sites in Danger' list and then ultimately, if justified, its deletion from the World Heritage List.

Listings can encourage states to implement stricter conservation measures to protect a location. But, at the same time, the Convention explicitly recognizes that listing a place as a World Heritage Site gives rise to a duty to cooperate:

> Whilst fully respecting the sovereignty of the States on whose territory the cultural and natural heritage mentioned in Articles 1 and 2 is situated, and without prejudice to property right provided by national legislation, the States Parties to this Convention recognize that such heritage constitutes a world heritage for whose protection it is the duty of the international community as a whole to co-operate.
>
> *(Article 6(1))*

The World Heritage Committee comprises representatives from twenty-one states parties to the World Heritage Convention (Article 11). The Convention requires that states appoint to the Committee 'persons qualified in the field of the cultural or natural heritage' (Article 9(3)), which is supposed to exclude professional diplomats or 'mere' political appointees (Scovazzi, 2008: 147, 155). The Committee has several advisory bodies, the International Centre for the Study of the Preservation and Restoration of Cultural Property, the International Council on Monuments and Sites, and the International Union for Conservation of Nature to assist it in making impartial and objective decisions.

The World Heritage Committee has several important powers under the 1972 Convention: it establishes the criteria for inclusion of sites on the World Heritage List; it approves sites to be added to the list; it places sites on the World Heritage Sites in Danger list or removes them from the World Heritage List altogether;[60] and it considers and approves requests by member states for assistance in protecting their World Heritage Sites. However, there is no enforcement power other than the pressure that can be applied at the international level.

There are forty-six marine World Heritage Sites at time of writing. The Great Barrier Reef was inscribed as the first such site in 1981. The most recent, the Phoenix Islands, an oceanic archipelago located in the Republic of Kiribati in the central Pacific Ocean, was designated in 2010 and became the largest World Heritage Site. Protected from most human interactions by their physical remoteness and difficulty of access, their lack of adequate freshwater sources for habitation, and their distance from the major oceanic trading routes, the marine environment of the Phoenix Islands is pristine (UNESCO World Heritage Convention, undated).

The 1972 Convention on the Prevention of Marine Pollution by Dumping of Wastes and Other Matter

The transformation of the 1972 Convention on the Prevention of Marine Pollution by Dumping of Wastes and Other Matter (known initially as the 'London Dumping Convention') is certainly one of the most intriguing environmental stories of the past three decades. This Convention was drafted shortly after the 1972 Stockholm meeting, which launched international environmental consciousness. As originally written, it contained a 'black list' of materials (such as high-level radioactive wastes) that could never be dumped into the ocean and a 'grey list' of items (such as low-level radioactive wastes) that could be dumped in appropriate locations if proper governmental permits were obtained. This treaty was a step forward, but it still permitted a substantial amount of dumping, and efforts were made at annual London meetings of its contracting parties to tighten its provisions, so that no radioactive materials whatsoever could be dumped and so that the dumping of other hazardous materials would similarly be prohibited.[61] Although some developed nations resisted restrictions on their ability to dump low-level radioactive wastes,[62] after many debates and many preliminary meetings, a new Protocol was adopted in 1996 that virtually rewrites the London Convention, known as the 'London Protocol' (Hunter et al., 2002: 733). In fact, the name of the treaty was even changed, because the contracting parties did not want the public to think that it authorized dumping, and it is now titled simply the '1972 London Convention'.

Under the 1996 London Protocol, the presumptions are reversed, and the dumping of all wastes is prohibited unless the item to be dumped is explicitly listed in Annex I. Even these Annex I materials, which include dredged material, sewage sludge, vessels, and ocean platforms, cannot be dumped without a permit (Annex 1 and Article 4(2)). Permits can be granted only after assessments are undertaken that evaluate options and describe the potential effects of the dumping (Annex II). Incineration at sea and the dumping of industrial wastes are completely prohibited (Article 5). This new Protocol is thus based on the precautionary approach,[63] as well as the 'polluter pays' principle.[64] The burden has thus shifted from dumping unless it were proven harmful toward no dumping unless it can be shown that there are no alternatives (Hunter et al., 2002: 735). The Protocol also contains a number of provisions to assist developing countries in dealing with their wastes and to encourage them to become parties. It establishes a Technical Cooperation and Assistance Programme (TCAP) to help countries relying upon the oceans for the dumping of wastes, and seven such programmes were established by the IMO in 1997–98 (Hunter et al., 2002: 735–6).

This makeover of the London Convention illustrates the progress of the IMO toward a more environmentally conscious focus on sustainable development and a sense of shared responsibility for the common areas of the planet. In a 16 June 1995 communication to the contracting parties to the London Convention, issued by the Secretary General of the IMO, it was clearly stated that, under Article 210(6) of UNCLOS, parties to the Law of the Sea Convention are bound by the requirements of the London even if they are not parties to the latter:

> Through that article [UNCLOS, Article 210] States parties to UNCLOS are legally bound to adopt laws and regulations, and take other measures to prevent, reduce and control pollution by dumping, which must no less effective than the global rules and standards. *The London Convention 1972*, which has been ratified or acceded to by 74 countries, *is the global instrument containing rules and standards on the prevention of marine pollution from dumping at sea.*[65]

As of May 2014, eighty-seven countries had become contracting parties to the London Convention and forty-four countries had ratified the 1996 Protocol (which came into force on 24 March 2006, when it received its twenty-sixth ratification) (IMO, undated*a*). This assessment of the London Convention 1972 as the global instrument is therefore even stronger, and the authoritative status of the London Protocol is also to be considered.

This 1972 Convention and its 1978 Protocol stipulate the standards that international shipping has to meet with regard to all forms of ship-generated marine pollution.[66] It builds on Article 211 of UNCLOS, which requires countries to regulate vessel-source pollution, and it establishes mandatory discharge standards, rules on construction, design, equipment and staffing of vessels, and navigation restrictions that limit navigation in ecologically sensitive areas. Its most important innovation was to ensure compliance by means of explicit documentation require-ments governing record books, oil discharge data, and operating certificates. Contracting parties must provide reception facilities for wastes listed in Annexes I, II, IV, and V, such as oil, sewage, and garbage, and they have the right to inspect vessels for illegal discharges.

Air pollution from ships is governed by the Protocol of 1997, which entered into force in 2005 and became Annex VI of MARPOL. It entered into force on 19 May 2005. As of May 2014, Annex VI has been ratified by seventy-eight countries, representing about 95 per cent of the gross tonnage of the world's merchant shipping fleet (IMO, undated*c*). It limits the main air pollutants produced by ship emissions, including sulphur oxides and nitrous oxides, and prohib-its the deliberate emissions of ozone-depleting substances.

The Kyoto Protocol assigns the responsibility of dealing with the shipping industry's con-tribution to greenhouse gases (GHGs) to the IMO.[67] Global shipping uses bunker fuel, which is the waste product of petroleum distillation and a tar-like sludge.[68] When burned as fuel, it produces emissions containing a high content of sulphur oxide (a sooty pollutant associated with production of acid rain), nitrogen oxide (a smog-causing pollutant), carbon dioxide, carbon monoxide, black carbon, particulate matter, and other pollutants.

Following the entry into force of MARPOL Annex VI on 19 May 2005, the IMO's Marine Environment Protection Committee (MEPC) began efforts to revise the Annex to reduce emis-sion limits. Starting in 2008, the MEPC implemented changes to Annex VI to require progres-sive reductions on the amount of harmful emission from nitrogen oxide, sulphur oxide, and particulate matter. Subsequently, the MEPC developed technical and operations measures, and set forth new energy-efficient ship design, management plans, and operational guidelines – first, on a voluntary basis, and more recently, as mandatory (IMO, 2011).

The MEPC struggled with the issue of technology transfer to developing nations for these new energy-efficient design requirements, but so far there is only a voluntary system of transfers in place (IMO, 2013a: paras 4.3–4.16). Even adopting a resolution on technology transfer was controversial and required several years of consideration because of the objections by developed nations, including the United States, to the use of the 'common, but differentiated' principle.[69] In addition, because of these same difficult-to-resolve differences relating to responsibilities, no method for assessing a carbon tax has yet been agreed upon at IMO despite many years of discussion (IMO, 2013b).

One of the most significant recent changes to MARPOL Annex VI in the effort to reduce air pollution from shipping was the creation of emission control areas (ECAs). In certain designated coastal regions, the emissions of sulphur oxide, nitrous oxide, and particulate matter can be more stringently controlled than in other ocean areas. An ECA was established and entered into force for each of the Baltic Sea in 2005 (sulphur oxide), the North Sea in 2006 (sulphur oxide), the North American seas in 2012 (sulphur oxide, nitrous oxide, and particulate matter), and the US Caribbean Sea in 2013 (sulphur oxide, nitrous oxide, and particulate matter) (IMO, undated*d*). The large North American ECA includes waters adjacent to the Pacific and Atlantic/Gulf coasts and the eight main Hawai'ian Islands, and extends up to 200 nautical miles from coasts of the United States, Canada, and the French territories (EPA, 2010). China objected to the proposals because 'the designation of an ECA of this size might set a precedent, leading to a situation that in the future all of the world's oceans could become ECAs', and because the it feared that because the area covered is huge and includes many shipping lanes, its requirements would become the rule rather than an exception (IMO, 2009: para. 4.14). Despite China's objections, this broad coastal ocean region became the third official ECA designated by the IMO.

The issue of the common, but differentiated, principle and how it would be applied has been the subject of debate for many years at the IMO with regard to implementation of carbon market-based reduction measures (IMO, 2009: paras 4.106–4.129). As a result of this ongoing controversy, no decisions have been made on this issue.

The IMO has undertaken to provide guidance and a permit system for geo-engineering projects proposed to take place in the ocean. One example is the concept of 'ocean iron fertilization', which is the intentional introduction of iron to the upper ocean to stimulate phytoplankton growth in iron-deficient areas, which will then capture carbon from the atmosphere (Broder and Haward, 2013b). This process is thought by some to be a very dangerous scientific experiment and a violation of the precautionary principle, while others look to it as a viable geo-engineering option with which to address climate change. Iron fertilization might increase the ability of the oceans' biological pump to draw carbon dioxide (CO_2) from the atmosphere and increase the food supply for marine life, but science has not yet established the effectiveness or safety of this procedure to the marine and human environment.[70] Iron-deficient waters in the Southern Ocean around Antarctica are currently the prime place of interest for experimentation.

The IMO has passed a Resolution affirming that ocean iron fertilization is within the 'scope' of the London Convention and Protocol, and that any activities outside its assessment framework are contrary to the London Convention and Protocol.[71] In the Resolution, the IMO clarified and applied the precautionary principle, stating that, given the insufficient knowledge on its effectiveness and potential environmental impacts, 'utmost caution' should be exercised and only 'legitimate scientific experiments' would be permitted, using an assessment framework to be developed by the London Convention/London Protocol (LC/LP) Scientific Groups.[72] This Resolution differs dramatically from Decision IX/16 of the Convention on Biological Diversity in that it rejects the complete moratorium preferred by the parties to the CBD (for all iron fertilization other than for small-scale projects in coastal waters), but agrees that the precautionary approach should apply to all iron fertilization activities.[73] There is no liability and compensation regime, and there is legal uncertainty about the binding mandate of resolutions of the London Convention and Protocol. Nonetheless, the parties adopted Resolution LC/LP.2, which included the Assessment Framework for Scientific Research Involving Ocean Fertilization: 'This Assessment Framework provides criteria for an initial assessment of OIF proposals and detailed steps for completion of an environmental assessment, including risk management and monitoring.'[74] The Resolution and Assessment Framework commit the joint treaty system to a 'global, transparent, and effective control and regulatory mechanism for ocean fertilization activities', and requires sharing of research results, the prohibition of carbon credits at this time, the imposition of the precautionary approach, and the establishment of a baseline as a condition precedent. Objections were raised about the proposed requirement of obtaining 'prior informed consent' from parties in the region of potential impact and, as a result, the final version merely required the seeking of consent.

In 2013, the IMO adopted an amendment to the 1996 Protocol to regulate marine geo-engineering, including ocean iron fertilization (IMO, 2013c: paras 4.2–4.17). A new Annex V includes an Assessment Framework.[75] This development demonstrates the awareness of the need by this international regulatory body to keep pace with scientific advancements in the fast-growing field of marine geo-engineering.

The IMO is in the process of developing a mandatory International Code of Safety for Ships Operating in Polar Waters (a 'Polar Code'), to cover the full range of design, construction, equipment, operational, training, search and rescue (SAR), and environmental protection matters (IMO, undated*c*). A new MARPOL Regulation, to protect the Antarctic from pollution by heavy-grade oils, was adopted by the MEPC at its Sixtieth Session in March 2010 and entered into force on 1 August 2011 (IMO, undated*c*).

Threats to the marine environment

It is difficult to list adequately all of the abuses of the oceans and their devastating impacts. Rising sea temperatures and sea-level rise are overwhelming threats. At the same time, there other looming examples, such as illegal, unreported, and unregulated fishing, unsustainable overexploitation of fish and marine mammals, trawling of high seas seamounts, the use of low- and mid-frequency sonar, floating nuclear power plants, geo-engineering, oil pollution, drilling for oil and gas in the Arctic Ocean, Gulf of Mexico, and other places, land pollutants that make their way to the ocean (releases from nuclear power plants, untreated sewage, garbage, fertilizers, pesticides, industrial chemicals), marine debris, alien species introduction, and shipment of ultrahazardous nuclear and other cargoes.[76] The world community has made significant progress in developing norms with which to govern the marine environment, but climate change, technological advances, and increasing world population continue to pose threats and challenge policymakers to determine how these norms should be applied to new problems and how to achieve truly sustainable development.

Conclusion

This chapter has shown the way in which the international community has come together to address coastal and ocean issues, has set goals and made a commitment to integrated management of coastal and offshore ocean areas, and has created timetables and deadlines, but has not yet created robust enforcement mechanisms. The challenge for the next phase of this effort is to move from talk to action, to make a more serious commitment to tackling the threats that prevent coastal communities from taking full advantage of coastal resources, to monitor closely progress toward these goals, and to penalize non-compliance.

Notes

1 This Conference, which continued from 1974 to 1982, produced the 1982 United Nations Convention on the Law of the Sea (UNCLOS).
2 See generally Van Dyke (1993a; 2000).
3 UNCLOS, Art. 207(4) (emphasis added): 'States, acting especially through competent international organizations or diplomatic conference, shall endeavour to establish global and regional rules, standards and recommended practices and procedures to prevent, reduce and control pollution of the marine environment from land-based sources, *taking into account* characteristic regional features, *the economic capacity of developing States and their need for economic development.*'
4 ITLOS, *Case No. 17: Responsibilities and Obligations of States sponsoring Persons and Entities with Respect to Activities in the Area (Request for Advisory Opinion submitted to the Seabed Disputes Chamber)* (2011), online at http://www.itlos.org/index.php?id=110 [accessed 29 October 2014] (the ITLOS 2011 Advisory Opinion). See generally Broder and Haward (2013a).
5 ITLOS 2011 Advisory Opinion, [159].
6 Ibid, [159].
7 Ibid, [163].
8 See generally Van Dyke (1993b).
9 *Case Concerning Pulp Mills on the River Uruguay (Argentina v. Uruguay)* [2010] ICJ 135.
10 Ibid, [29], citing *Legality of Nuclear Weapons Advisory Opinion* [1996] ICJ 226, 241–2 (emphasis added).

11 Some of the material that follows is adapted from Van Dyke (2006).

12 *Affaire du Lac Lanoux (France v. Spain)* (1957) 12 RIAA 281.

13 Rio Declaration, Principle 19: 'States shall provide prior and timely notification and relevant infor-mation to potentially affected States on activities that may have a significant adverse transboundary environmental effect and shall consult with those States at an early stage and in good faith.' As to the obligation to notify under customary international law as an aspect of the principle of good faith, see Lammers (1986).

14 In most cases, 'an obligation to negotiate does not imply an obligation to reach an agreement': *Pulp Mills*, [150], citing *Railway Traffic Between Lithuania & Poland Advisory Opinion* (1931) PCIJ Series A/B, No. 42, at 116.

15 *Pulp Mills*, [158].

16 *Corfu Channel Case (United Kingdom v. Albania)* [1949] ICJ 4.

17 *Affaire du Lac Lanoux (France v. Spain)* (1957) 24 ILR 101, 128.

18 Principle 4 of the 2001 International Law Commission (ILC) Principles of Transboundary Harm from Hazardous Activities requires states to 'cooperate in good faith', Principle 8 requires states to notify other states of anticipated risks, and Principle 12 requires the exchange of information (ILC, 2001).

19 *Gabcikovo Nagymaros Project (Hungary v. Slovakia)* [1997] ICJ 7.

20 *MOX Plant Case (ITCLOS Case No. 10) (Ireland v. United Kingdom)* (2001) 41 ILM 405.

21 *Case Concerning Land Reclamation by Singapore In and Around the Straits of Johor (No. 12) (Malaysia v. Singapore)* (2003) 126 ILR 487

22 Ibid, [99] (emphasis added).

23 Ibid, [106] (emphasis added).

24 Ibid, [106].

25 *Pulp Mills*, [204].

26 UNCLOS, Art. 206, requires states undertaking 'activities under their jurisdiction or control [that] may cause substantial pollution of or significant and harmful changes to the marine environment [to], as far as practicable, assess the potential effects of such activities on the marine environment and . . . communicate reports of the results of such assessments' to nations that may be affected by the project. See generally Van Dyke (1993b: 402–7).

27 1991 Convention on Environmental Impact Assessment in a Transboundary Context, Art. 2.1 (requir-ing contracting parties to take all appropriate measures to prevent, reduce, and control significant adverse transboundary environmental impacts from proposed activities). Article 31(6) of the Seabed Authority Regulations 2000 requires states parties sponsoring activities in the area to 'cooperate with the Authority in the establishment and implementation of programmes for monitoring and evaluating the impacts of deep seabed mining on the marine environment'.

28 The precautionary approach and precautionary principle are often considered to be interchangeable, but some generally consider the precautionary approach a softening of the principle. The United States has opposed the use of the term 'principle' on the grounds that this word has special legal implications, and that a principle can be considered a source of law and compulsory. The precautionary principle is considered to have a legal implication in the European Union. However, in some particular cases, an approach could be binding (Recuerda, 2008).

29 See, e.g., Van Dyke (2004: 357).

30 ITLOS 2011 Advisory Opinion, [131].

31 *MOX Plant*, 134.

32 See, e.g., Bodansky (1991b: 4–5): 'Although the precautionary principle provides a general approach to environmental issues, it is too vague to serve as a regulatory standard because it does not specify how much caution should be taken.' But cf. Bodansky (1991a: 413): 'Indeed, so frequent is its invocation that some commentators are even beginning to suggest that the precautionary principle is ripening into a norm of customary international law.' See generally Fullem (1995); Hickey and Walker (1995).

33 See generally Van Dyke (1996b).

34 1996 Protocol to the 1972 Convention on the Prevention of Marine Pollution by Dumping of Wastes and Other Matter, as amended in 2006, Art. 3 (reversing the presumptions established in the original Convention, so that the dumping of all wastes is prohibited unless the item to be dumped is explicitly listed in Annex I).

35 UNFSA 1995, Arts 5(c) and 6 (listing the 'precautionary approach' among the principles that govern the conservation and management of shared fish stocks and elaborating on this requirement in some detail, focusing on data collection and monitoring).

36 See, e.g., Western Pacific Regional Fishery Management Council (1998: 26), stating proudly that the Council has established 'a precautionary management approach to fishery conservation and management', as evidenced by its establishment of a moratorium and then a limited-entry programme 'in response to the rapid entry of longline vessels into the Hawai'i-based fleet'. Article 31(2) of the Seabed Authority Regulations 2000 requires states parties sponsoring activities in the area to 'apply a precautionary approach, as reflected in Principle 15 of the Rio Declaration to such activities', '[i]n order to ensure effective protection for the marine environment from harmful effects'.

37 *In the Matter of the Water Use Permit Applications, Petitions for Interim Instream Flow Standard Amendments, and Petitions for Water Reservations for the Waiahole Ditch Combined Contested Case Hearing*, 94 Hawai'i 97, 155, 9 P.3d 409 (2000) (emphasis added). The court continued: 'As with any general principle, its meaning must vary according to the situation and can only develop over time. In this case, we believe the [Water] Commission describes the [precautionary] principle in its quintessential form: at minimum, the absence of firm scientific proof should not tie the Commission's hands in adopting reasonable measures designed to further the public interest.'

38 See generally Colburn (1997); Kedziora (1997); Hayaski (1996a; 1996b); Balton (1996); Mack (1996); Christopherson (1996); Van Dyke (1996b).

39 See also generally Van Dyke (1993b). Among the stocks that are now seriously depleted are Atlantic halibut, New Zealand orange roughy, bluefin tuna, rockfish, herring, shrimp, sturgeon, oysters, shark, Atlantic and some Pacific Northwest salmon, American shad, Newfoundland cod, and haddock and yellowtail flounder off of New England (Associated Press, 1998: A-19, col. 2, quoting from a study led by Stanford biologist Harold Mooney and funded by the National Research Council, an arm of the National Academy of Sciences).

40 Fishing to attain the maximum sustainable yield inevitably means reducing the abundance of a stock, sometimes by half or two-thirds. This reduction can threaten the stock in unforeseeable ways and also will impact on other species in the ecosystem.

41 Nations already have the power to board, inspect, and arrest vessels violating laws established to 'control and manage the living resources in the exclusive economic zone': UNCLOS, Art. 73(1).

42 See, e.g., Basiron (1998); Swan (1995).

43 The text of the Accord is provided in UNFCCC (2010).

44 *Resolution 66/288 adopted by the General Assembly on 27 July 2012: The Future We Want*, UN Doc. A/RES/66/288, online at http://www.un.org/en/ga/search/view_doc.asp?symbol=%20A/RES/66/288 [accessed 29 October 2014].

45 See Montague (2012); Vidal (2012); Meyer (2012).

46 On 28 March 2013, the International Tribunal for the Law of the Sea (ITLOS) received a request from the Sub-Regional Fisheries Commission (SRFC) to render an Advisory Opinion on the responsibilities of flag states and others for illegal, unreported, and unregulated fishing in the EEZ and on the High Seas: ITLOS, *Case No. 21: Request for an Advisory Opinion Submitted by the Sub-Regional Fisheries Commission* (pending), online at http://www.itlos.org/index.php?id=252&L=0 [accessed 25 May 2014]. The SRFC is a regional fishery body (RFB) and includes Cape Verde, Gambia, Guinea, Guinea-Bissau, Mauritania, Senegal, and Sierra Leone. This case raises important questions about state responsibility and international liability, and the decision will play an important role in the implementation of international fisheries conservation and whether compensation might be required in the event of breach of duties and responsibilities.

47 Under *Resolution 68/70 Adopted by the General Assembly on 9 December 2013 on Oceans and the Law of the Sea*, UN Doc. A/RES/68/70, online at http://www.un.org/en/ga/search/view_doc.asp?symbol=A/RES/68/70 [accessed 25 May 2014].

48 The Nagoya Protocol was opened for signature by CBD parties between 2 February 2011 and 1 February 2012, and during that time was signed by ninety-two states. As of May 2014, there were thirty-seven ratifications to the Nagoya Protocol. See CBD (undated).

49 See Hunter et al. (2002).

50 Because they wished to continue harvesting whales, Norway, Iceland, the Faroe Islands (Denmark), and Greenland (Denmark) created the North Atlantic Marine Mammal Commission (NAMMCO) in 1992. Norway has consistently objected to the moratorium established by the IWC and has been harvesting minke whales in the North Atlantic under the blessing of NAMMCO. See generally Scheiber (1998; 2000); Caron (1995); Bjorndal and Conrad (1998). Canada, which is a member of neither the IWC nor NAMMCO, but sends observers to meetings of both organizations, has authorized Inuit to harvest limited numbers of bowhead whales. See generally McDorman (1998). McDorman (1998: 181)

notes that 'there is a degree of inconsistency in Canada's permitting whaling but remaining outside the IWC while at the same time complaining bitterly about the nations that fish outside Canada's East Coast EEZ without joining NAFO'.

51 Among the many documented abuses is the report that the Soviet Union harvested 48,477 humpback whales from 1948 to 1973, instead of the 2,710 that it officially reported to the IWC, and the discovery that whales recently harvested by Japan, ostensibly pursuant to its 'scientific' whaling for Antarctic minke whales, included humpback, fin, and Arctic minke whales (Caron, 1995: 171–3, citing Angier, 1994, and Szabo, 1994). Professor Caron has explained that this sad history supports the position 'that, regardless of population size, the notion that a common resource such as whales can be sustainably managed is illusory', not only because of the difficulties of attaining scientific certainty, but also because the historical record indicates that 'some of the users may act in bad faith and the capacity of the resource manager to police such users is insubstantial' (Caron, 1995: 171). See also Scheiber (2000: 155): 'Within the IWC, moreover, these nations collectively failed entirely to halt or ameliorate the slaughter even when it was obvious to all that some of the stocks were going to crash and species be endangered with possible extinction.'

52 See Van Dyke et al. (2004).

53 Agreement on the Conservation of Cetaceans of the Black Sea, Mediterranean Sea and Contiguous Atlantic Area (ACCOBAMS), Art. 2(1).

54 International Convention for the Regulation of Whaling, Art. VIII. Japan is the only ICW member to have issued scientific permits to take whales under the authority of this exception. Iceland had also issued scientific permits under the ICW, but it withdrew in 1992. When it rejoined in 2004, Iceland objected to the whaling moratorium set by the ICW and, based on that objection, has continued to hunt whales. Norway also objected to the moratorium and, on that basis, has continued its whaling programme. Both Iceland and Norway conduct whaling only in their own EEZs. Japan has been hunting in international waters.

55 *Whaling in the Antarctic (Australia v. Japan, New Zealand intervening)*, 31 March 2014, ICJ, online at http://www.icj-cij.org/docket/files/148/18136.pdf [accessed 25 May 2014].

56 Some of the material in this section has been adapted from Van Dyke and Broder (2012).

57 *MEPC Resolution 44(30) Adopted on 16 November 1990: Identification of the Great Barrier Reef Region as a Particularly Sensitive Area*, Doc. No. W/1633D/EWD, online at http://www.imo.org/blast/blastData-Helper.asp?data_id=17630&filename=44(30).pdf [accessed 29 October 2014].

58 *IMO Assembly Resolution A.720(17) Adopted on 6 November 1991: Guidelines for the Designation of Special Areas and the Identification of Particularly Sensitive Areas*, Doc. No. W/4580x/EWP, online at http://www.imo.org/blast/blastDataHelper.asp?data_id=22581&filename=A720(17).pdf [accessed 29 October 2014].

59 *IMO Assembly Resolution A.982(24) Adopted on 1 December 2005: Revised Guidelines for the Identification and Designation of Particularly Sensitive Sea Areas (PSSAs)*, Doc. No. A 24/Res.982, online at http://www.imo.org/blast/blastDataHelper.asp?data_id=25322&filename=A982(24).pdf [accessed 29 October 2014].

60 The WHC Operational Guidelines state that protection and management of UNESCO World Heritage Sites by countries should include legislative and regulatory measures that ensure the protection of sites against activities that might negatively impact their integrity or outstanding universal value. Failure to adequately protect and manage a site may ultimately result in its being placed on a World Heritage in Danger List and, ultimately, deletion from the World Heritage List.

61 See Van Dyke (1988); Davis and Van Dyke (1990).

62 During the Seventh Consultative Meeting in 1983, the contracting parties passed a resolution imposing a moratorium on the dumping of all low-level radioactive wastes, but the Soviet Union, China, Belgium, France, the United Kingdom, and the United States voted against the resolution, and several other industrialized nations abstained. The dissenting nations did not feel that they were bound by this resolution, and the British government sought to continue its dumping programme. But the British unions refused to load the low-level waste on British ships in 1985, and thus the British were forced to adhere to the moratorium by their own people (Van Dyke, 1988).

63 1996 London Protocol, Art 3(1): 'In implementing this Protocol, Contracting Parties shall apply a precautionary approach to environmental protection from dumping of wastes or other matter whereby appropriate preventative measures are taken when there is reason to believe that wastes or other matter introduced into the marine environment are likely to cause harm even when there is no conclusive evidence to prove a causal relation between inputs and their effects.'

64 1996 London Protocol, Art. 3(2): 'Taking into account the approach that the polluter should, in principle, bear the cost of pollution, each Contracting Party shall endeavor to promote practices whereby those it has authorized to engage in dumping or incineration at sea bear the cost of meeting the pollution prevention and control requirements for the authorized activities, having due regard to the public interest.'

65 Letter from the Secretary-General of the International Maritime Organization (IMO) addressed to the States Parties to the 1982 United Nations Convention on the Law of the Sea which are not yet States Parties to the 1972 London Convention on the Prevention of Marine Pollution by Dumping of Wastes and Other Matter London Convention issued by the Division for Ocean Affairs of the United Nations Office of Legal Affairs (UN Division for Ocean Affairs and the Law of the Sea/Office of Legal Affairs, 1999: 55).

66 See generally Hunter et al. (2002: 707–24).

67 The Kyoto Protocol, Art. II.II, provides the following regarding maritime emissions: 'The Parties included in Annex I shall pursue limitation or reduction of emissions of greenhouse gases not controlled by the Montreal Protocol from aviation and marine bunker fuels, working through the International Civil Aviation Organization and the IMO, respectively.'

68 When crude oil is refined, the lighter fractions with lower boiling points, such as gasoline, kerosene, propane, jet fuel, and naphtha, are removed; heavier petroleum products, such as diesel and lubricating oil, are distillated out more slowly. The heavier materials, which become bunker fuel, are at the 'bottom of the barrel' because their boiling points are much higher. The process of heating at high temperatures causes the rearrangement of molecules, creating larger molecules that require an even higher boiling temperature, and these larger molecules become part of the residue. Contaminants such as toxic heavy metals also sink to the bottom during distillation. Bunker fuel thus consists of all of the materials that could not be removed in the distilling process (Yujuico, 2007).

69 *IMO Resolution MEPC.229(65) Adopted on 17 May 2013 on Promotion of Technical Co-operation and Transfer of Technology Relating to the Improvement of Energy Efficiency of Ships*, Doc. No. MEPC 65/22, online at http://www.imo.org/KnowledgeCentre/IndexofIMOResolutions/Documents/MEPC%20-% 20Marine%20Environment%20Protection/229(65).pdf [accessed 18 May 2014].

70 In September, 2009, the Royal Society – Britain's premier scientific organization – released its first analysis of a host of controversial methods for intentionally altering Earth's climate, known broadly as 'geo-engineering', which could slow or halt climate change by either restricting the amount of sunlight heating Earth's surface or reducing levels of carbon dioxide in the atmosphere (Royal Society, 2009). This report is the first major national academy report solely devoted to this topic. The Royal Society (2009: 1) defines 'geo-engineering' as 'the deliberate large-scale manipulation of the planetary environment to counteract anthropogenic climate change'. It is sometimes also referred to as 'climate engineering', 'climate remediation', and 'climate intervention'. Geo-engineering encompasses a range of techniques to remove CO_2 from the atmosphere or to block incoming sunlight.

71 *IMO Resolution LC/LP.1 Adopted on 31 October 2008 on the Regulation of Ocean Fertilization*, Doc. No. IMO LC 30/16, online at http://www.imo.org/blast/blastData.asp?doc_id=14101&filename=1.doc [accessed 29 October 2014].

72 The parties '[agree] that the scope of the London Convention and Protocol includes ocean fertilization activities': ibid: para. 1.

73 *Decision IX/16 Adopted by the Conference of the Parties to the Convention on Biological Diversity at its Ninth Meeting: Biodiversity and Climate Change*, Doc. No. UNEP/CBD/COP/EC/IX/16, online at http:// www.cbd.int/doc/decisions/cop-09/cop-09-dec-16-en.pdf [accessed 9 October 2008].

74 *IMO Resolution LC-LP.2(2010) on the Assessment Framework for Scientific Research Involving Ocean Fertilization*, Annex 5 to IMO (2010).

75 *Resolution LP.4(8) Adopted on 18 October 2013 on the Amendment to the London Protocol To Regulate the Placement of Matter for Ocean Iron Fertilization and other Marine Geoengineering Activities*, Annex 4 in IMO (2013d).

76 See Van Dyke et al. (2004); Van Dyke (2002); Currie and Van Dyke (2005).

References

Angier, N. (1994) 'DNA Tests Find Meat of Endangered Whales for Sale in Japan', *New York Times*, 13 September, p. C4.

Associated Press (1998) 'Steps Must Be Taken to Counter Overfishing, US Panel Warns', *Honolulu Star-Bulletin*, 23 October, p. A-19, col. 2.

Australian Maritime Safety Authority (AMSA) (undated) 'Torres Strait Particularly Sensitive Sea Area', online at https://www.amsa.gov.au/environment/legislation-and-prevention/torres-strait-pssa/ [accessed 12 March 2012].

Balton, D. (1996) 'Strengthening the Law of the Sea: The New Agreement on Straddling Fish Stocks and Highly Migratory Fish Stocks', *Ocean Development & International Law*, 27: 125.

Basiron, M. (1998) 'The Implementation of Chapter 17 of Agenda 21 in Malaysia: Challenges and Opportunities', *Ocean & Coastal Management*, 41: 1–17.

Birnie, P., and Boyle, A. (2002) *International Law and the Environment*, 2nd edn, Oxford: Oxford University Press.

Bjorndal, T., and Conrad, J. (1998) 'A Report on the Norwegian Minke Whale Hunt', *Marine Policy*, 22: 161–74.

Bodansky, D. (1991a) 'Remarks: New Developments in International Environmental Law', *American Society of International Law Proceedings*, 85: 401–13.

Bodansky, D. (1991b) 'Scientific Uncertainty and the Precautionary Principle', *Environment*, 33: 4–8.

Boyle, A. (1990) 'Nuclear Energy and International Law: An Environmental Perspective', *British Yearbook of International Law*, 60: 257–78.

Broder, J. (2010) 'Climate Goal is Supported by China and India', *New York Times*, 10 March, p. A9, online at http://www.nytimes.com/2010/03/10/science/earth/10climate.html [accessed 27 November 2014].

Broder, S., and Haward, M. (2013a) 'Geoengineeering: Ocean Iron Fertilization and the Challenges for International Regulatory Action', in H. Scheiber and J. Paik (eds) *Regions, Institutions, and Law of the Sea: Studies in Ocean Governance*, Leiden: Brill/Martinus Nijhoff Publishers.

Broder, S., and Haward, M. (2013b) 'The International Legal Regimes Governing Ocean Iron Fertilization', in H. Scheiber and J. Paik (eds) *Regions, Institutions, and Law of the Sea: Studies in Ocean Governance*, Leiden: Brill/Martinus Nijhoff Publishers.

Caron, D. (1995) 'The International Whaling Commission and the North Atlantic Marine Mammal Commission: The Institutional Risks of Coercion in Consensual Structures', *American Journal of International Law*, 89: 154–74.

Christopherson, M. (1996) 'Toward a Rational Harvest: The United Nations Agreement on Straddling Fish Stocks and Highly Migratory Species', *Minnesota Journal of Global Trade*, 5: 357–80.

Colburn, J. (1997) 'Comment: Turbot Wars – Straddling Stocks, Regime Theory, and a New UN Agreement', *Journal of Transnational Law & Policy*, 6: 323.

Convention of Biological Diversity (CBD) (undated) 'Status of Signature, and Ratification, Acceptance, Approval or Accession', online at http://www.cbd.int/abs/nagoya-protocol/signatories [accessed 18 May 2014].

Currie, D., and Van Dyke, J. (2005) 'Recent Developments in the International Law Governing Shipments of Nuclear Materials and Wastes and Their Implications for SIDS', *Review of European Community & International Environmental Law (RECIEL)*, 14(2): 117–24.

Davis, W., and Van Dyke, J. (1990) 'Dumping of Decommissioned Nuclear Submarines at Sea: A Technical and Legal Analysis', *Marine Policy*, 14: 467–76.

Deep Sea Conservation Coalition (2011) 'Where Have All the Orange Roughy Gone?', online at http://savethedeepsea.blogspot.com/2011/07/where-have-all-orange-roughy-gone.html [accessed 25 May 2014].

Environmental Protection Agency (EPA) (2010) *Designation of North American Emission Control Area to Reduce Emissions from Ships*, EPA Regulatory Announcement No. EPA-420-F-10-015, online at http://www.epa.gov/otaq/regs/nonroad/marine/ci/420f10015.pdf [accessed 25 May 2104].

Food and Agriculture Organization of the United Nations (FAO) Fisheries and Aquaculture Department (undateda) 'Regional Fishery Bodies Summary Descriptions: South Indian Ocean Fisheries Agreement (SIOFA)' online at http://www.fao.org/fishery/rfb/siofa/en [accessed 25 May 2014].

Food and Agriculture Organization of the United Nations (FAO) Fisheries and Aquaculture Department (undatedb) 'FAO Regional Fishery Bodies Summary Descriptions: South Pacific Regional Fisheries Management Organisation (SPRFMO)', online at http://www.fao.org/fishery/rfb/sprfmo/en [accessed 23 May 2014].

Fullem, G. (1995) 'The Precautionary Principle: Environmental Protection in the Face of Scientific Uncertainty', *Willamette Law Review*, 31: 495–522.

Guruswamy, L. (2003) *International Environmental Law in a Nutshell*, 2nd edn, St. Paul, MN: Thomson West.

Hayashi, M. (1996a) 'Enforcement by Non-Flag States on the High Seas under the 1995 Agreement on Straddling and Highly Migratory Fish Stocks', *Georgetown International Environmental Law Review*, 9: 1–36.

Hayashi, M. (1996b) 'The 1995 Agreement on the Conservation and Management of Straddling and Highly Migratory Fish Stocks: Significance for the Law of the Sea Convention', *Ocean and Coastal Management*, 29: 51–69.

Hickey, Jr., J., and Walker, V. (1995) 'Refining the Precautionary Principle in International Environmental Law', *Virginia Environmental Law Journal*, 14: 423–53.

Hunter, D., Salzman, J., and Zaelke, D. (2002) *International Environmental Law and Policy,* 2nd edn, New York: Foundation Press.

International Atomic Energy Agency (IAEA) (1990) *Code of Practice on the International Transboundary Movement of Radioactive Waste (IAEA Code)*, 13 November, Doc. No. INFCIRC/386, online at https://www.iaea.org/Publications/Documents/Infcircs/Others/inf386.shtml [accessed 29 October 2014].

International Fund for Animal Welfare (undated) 'The Southern Ocean Sanctuary', online at http://www.ifaw.org/united-states/our-work/whales/southern-ocean-sanctuary [accessed 25 May 2014].

International Law Commission (ILC) (2001) 'Prevention of Transboundary Harm from Hazardous Activities', in *Report of the International Law Commission, Fifty-Third Session*, UN Doc. A/56/10, online at http://www.un.org/documents/ga/docs/56/a5610.pdf [accessed 29 October 2014].

International Law Commission (ILC) (2006) 'Draft Principles on the Allocation of Loss in the Case of Transboundary Harm Arising Out of Hazardous Activities', in *Report of the International Law Commission, Fifty-Eighth Session*, UN Doc. A/61/10, online at http://legal.un.org/ilc/reports/english/a_61_10.pdf [accessed 29 October 2014].

International Maritime Organization (IMO) (undated*a*) 'Particularly Sensitive Sea Areas (PSSA)', online at http://www.imo.org/ourwork/environment/pollutionprevention/pssas/Pages/Default.aspx [accessed 25 May 2014].

International Maritime Organization (IMO) (undated*b*) 'Special Areas under MARPOL', online at http://www.imo.org/OurWork/Environment/PollutionPrevention/SpecialAreasUnderMARPOL/Pages/Default.aspx [accessed 25May 2014].

International Maritime Organization (IMO) (undated*c*) 'Shipping in Polar Waters: Development of an International Code of Safety for Ships Operating in Polar Waters (Polar Code)', online at http://www.imo.org/MediaCentre/HotTopics/polar/Pages/default.aspx [accessed 19 May 2014].

International Maritime Organization (IMO) (2009) *Report of the Marine Environment Protection Committee on its Fifty-ninth Session*, Doc. No. MEPC 59/24, online at http://www.uscg.mil/imo/mepc/docs/mepc59-report.pdf [accessed 29 October 2014].

International Maritime Organization (IMO) (2011) 'Marine Environment Protection Committee (MEPC) – 62nd Session: 11 to 15 July 2011', 15 July, online at http://www.imo.org/MediaCentre/MeetingSummaries/MEPC/Pages/MEPC-62nd-session.aspx [accessed 1 May 2014].

International Maritime Organization (IMO) (2013a) *Report of the Marine Environment Protection Committee on its Sixty-Fifth Session*, 15 July, Doc. No. MEPC 65/22, online at http://www.mpa.gov.sg/sites/pdf/mepc65-add2.pdf [accessed 29 October 2014].

International Maritime Organization (IMO) (2013b) *Adoption of the Agenda*, 4 June, Doc. No. MEPC 66/1/1.

International Maritime Organization (IMO) (2013c) 'Proposal to Amend the London Protocol to Regulate Placement of Matter for Ocean Fertilization and Other Marine Geoengineering Activities', in *Report of the Thirty-Fifth Consultative Meeting of the Contracting Parties to the London Convention and Eighth Meeting of the Contracting Parties to the London Protocol*, 21 October, Doc. No. LC-LP 35/15, pp. 10–11.

International Maritime Organization (IMO) (2013d) *Report of the Thirty-Fifth Consultative Meeting of the Contracting Parties to the London Convention and Eighth Meeting of the Contracting Parties to the London Protocol*, 21 October, Doc. No. LC-LP 35/15.

International Maritime Organization (IMO) (2014a undated*a*) 'Convention on the Prevention of Marine Pollution by Dumping of Wastes and Other Matter', online at http://www.londonprotocol.imo.org [accessed 1 May 2014].

International Maritime Organization (IMO) (2014b undated*d*) 'Status of Conventions', online at http://www.imo.org/About/Conventions/StatusOfConventions/Pages/Default.aspx [accessed 1 May 2014].

International Seabed Authority (ISA) (1999) *Deep-Seabed Polymetallic Nodule Exploration: Development of Environmental Guidelines – Proceedings of the International Seabed Authority's Workshop held in Sanya, Hainan Island, China (People's Republic of China) (1–5 June 1998)*, Kingston: ISA.

International Seabed Authority (ISA) (2000) *Regulations on Prospecting and Exploration for Polymetallic Nodules in the Area*, Doc. No. ISBA/6/A/18, online at http://www.isa.org.jm/files/documents/EN/Regs/ MiningCode.pdf [accessed 25 May 2014].

International Seabed Authority (ISA) (2010) *Regulations on Prospecting and Exploration for Polymetallic Sulphides in the Area*, online at http://www.isa.org.jm/files/documents/EN/Regs/PolymetallicSulphides.pdf [accessed 30 November 2014].

International Whaling Commission (IWC) (2012) *Status of Agenda Items at IWC/64 as of Friday 6 July 2012 17.30*, 7 July, Doc. No. 64/15/Rev, online at *http://iwc.int/private/downloads/5c3m15872lwcos4088ww oss00/Status.pdf* [accessed 25 May 2014].

Kedziora D. (1997) 'Gunboat Diplomacy in the Northwest Atlantic: The 1995 Canada–EU Fishing Dispute and the United Nations Agreement on Straddling and Migratory Fish Stocks', *Northwestern Journal of International Law and Business*, 17: 1132–62.

Lammers, H. (1986) *Transfrontier Pollution and International Law*, The Hague: Centre for Studies and Research in International Law and International Relations, Hague Academy of International Law.

Lies, E. (2014) 'Japan Will Conduct Pacific Whale Hunt in Wake of Court Ruling', *Reuters*, 18 April, online at http://www.reuters.com/article/2014/04/18/us-japan-whaling-idUSBREA3G08C20140418 [accessed 25 May 2014].

Mack, J. (1996) 'Comment: International Fisheries Management – How the UN Conference on Straddling and Highly Migratory Fish Stocks Changes the Law of Fishing on the High Seas', *California Western International Law Journal*, 26(2): 313–33.

McDorman, T. (1998) 'Canada and Whaling: An Analysis of Article 65 of the Law of the Sea Convention', *Ocean Development & International Law*, 29: 179–94.

Meyer, A. (2012) 'Rio+20: Too Little, Too Late?', *Union of Concerned Scientists The Equation blog*, online at http://blog.ucsusa.org/rio20-too-little-too-late [accessed 18 June 2012].

Montague, B. (2012) 'Comment: Analysis – Rio+20: Epic Failure', *The Bureau of Investigative Journalism*, 22 June, online at http://www.thebureauinvestigates.com/2012/06/22/analysis-rio-20-epic-fail [accessed 25 May 2014].

Nanda, V., and Pring, G. (2003) *International Environmental Law for the 21st Century*, 2nd edn, Leiden/ Boston, MA: Martinus Nijhoff.

Papāhanaumokuākea Marine National Monument (undated), 'About', online at http://www.hawaiireef. noaa.gov [accessed 15 May 2014].

Pitt, D. (1993) 'Despite Gaps, Data Leave Little Doubt that Fish Are in Peril', *The New York Times*, 3 August, p. C4, col. 1.

Pohl, O. (2003) 'World Panel Will Now Act to Conserve the Whale Population', *The New York Times*, 17 June, p. A11.

Recuerda, M. (2008) 'Dangerous Interpretations of the Precautionary Principle and the Foundational Values of the European Union Food Law: Risk Versus Risk', *Journal of Food Law & Policy*, 4: 1–43.

Rosenblum, M., and Cabra, M. (2012) 'In Mackerel's Plunder, Hints of Epic Fish Collapse', *The New York Times*, 25 January, online at http://www.nytimes.com/2012/01/25/science/earth/in-mackerels-plunder-hints-of-epic-fish-collapse.html?pagewanted=all [accessed 28 November 2014].

Royal Society, The (2009) *Geoengineering the Climate: Science, Governance, and Uncertainty*, 1 September, RS Policy Document 10/09, online at http://royalsociety.org/Geoengineering-the-climate/ [accessed 25 May 2014].

Scheiber, H. (1998) 'Historical Memory, Cultural Claims, and Environmental Ethics in the Jurisprudence of Whaling Regulation', *Ocean & Coastal Management*, 38: 5–40.

Scheiber, H. (ed.) (2000) *Law of the Sea: The Common Heritage and Emerging Challenges*, Leiden/Boston, MA: Brill.

Schiffman, H. (2010) 'The Evolution of Fisheries Conservation and Management: A Look at the New South Pacific Regional Fisheries Management Organizationi Law and Policy', *Thomas M. Cooley Law Review*, 28(2): 181–88.

Scovazzi, T. (2008) 'Articles 8–11 World Heritage Committee and World Heritage List', in F. Francioni (ed.) *The 1972 World Heritage Convention: A Commentary*, Oxford/New York: Oxford University Press.

South Pacific Regional Fisheries Management Organisation (SPRFMO) (undateda), 'Status of the Convention', online at http://www.southpacificrfmo.org/status-of-the-convention/ [accessed 21 May 2014].

South Pacific Regional Fisheries Management Organisation (SPRFMO) (undatedb) 'Species and Ecosystems', online at http://www.southpacificrfmo.org/about-the-sprfmo/ [accessed 18 May 2014].

Swan, J. (1995) 'Protection of Oceans and Their Living Resources', in Projet de Société (ed.) *Canadian Responses to Agenda 21: An Assessment*, online at http://www.iisd.org/worldsd/canada/projet/a21toc. htm [accessed 1 May 2014].

Szabo, M. (1994) 'DNA Test Traps Whale Tenders', *New Scientist*, 28 May, p. 4.

UNESCO World Heritage Convention (undated) 'Phoenix Islands Protected Area', online at http://whc.unesco.org/en/list/1325 [accessed 28 November 2014].

United Nations (1973) *Report of the UN Conference on the Human Environment, 5–16 June 1972*, Doc. No. A/CONF/48/14/Rev.1, online at http://daccess-dds-ny.un.org/doc/UNDOC/GEN/NL7/300/05/IMG/NL730005.pdf?OpenElement [accessed 29 October 2014] (1972 Stockholm Declaration).

United Nations Environment Programme (UNEP) (1995) *Global Programme of Action for the Protection of the Marine Environment from Land-Based Activities*, 5 December, Doc. No. UNEP (OCA)/LBA/IG.2/7, online at http://daccess-dds-ny.un.org/doc/UNDOC/GEN/K96/000/14/PDF/K9600014.pdf?OpenElement [accessed 29 October 2014].

United Nations Division for Ocean Affairs and the Law of the Sea/Office of Legal Affairs (1995) *Law of the Sea Informational Circular, No. 2*, online at http://www.un.org/depts/los/LEGISLATIONANDTREATIES/losic/losic2e.pdf [accessed 28 November 2014].

United Nations Environment Programme Global Programme of Action for the Protection of the Marine Environment from Land-Based Activities (UNEP GPA) (undated) 'Aims of the Global Programme of Action', online at http://www.gpa.unep.org/index.php/about-gpa [accessed 1 May 2014].

United Nations Framework Convention on Climate Change (UNFCCC) (undated) 'Status of Ratification of the Convention', online at http://unfccc.int/essential_background/convention/status_of_ratification/items/2631.php [accessed 15 May 2014].

United Nations Framework Convention on Climate Change (UNFCCC) (2009) *Report of the Conference of the Parties on its Fifteenth Session, Held in Copenhagen, from 7 to 19 December 2009*, Doc. No. FCCC/CP/2009/11/Add.1, online at http://unfccc.int/resource/docs/2009/cop15/eng/11a01.pdf [accessed 30 March 2010] (Copenhagen Accord).

United Nations Framework Convention on Climate Change (UNFCC) (2010) *Report of the Conference of the Parties on its Fifteenth Session*, 30 March, Doc. No. FCCC/CP/2009/11/Add.1, online at http://unfccc.int/resource/docs/2009/cop15/eng/11a01.pdf [accessed 28 November 2014].

United Nations General Assembly (UNGA) (1992) 'Protection of the Oceans, All Kinds of Seas, Including Enclosed and Semi-Enclosed Seas, and Coastal Areas and the Protection, Rational Use and Development of Their Living Resources', in *Report of the United Nations Conference on Environment and Development, Vol. II*, 13 August, Doc. No. A/CONF.151/26, online at http://www.un.org/documents/ga/conf151/aconf15126-2.htm [accessed 29 October 2014].

United Nations News Centre (2012) 'Rio+20: $513 Billion Pledged towards Sustainable Development', 22 June, online at http://www.un.org/apps/news/story.asp?NewsID=42312#.U4O3qfldV9o [accessed 25 May 2014].

Van Dyke, J. (1988) 'Ocean Disposal of Nuclear Wastes', *Marine Policy*, 12: 82–95.

Van Dyke, J. (1993a) 'International Governance and Stewardship of the High Seas and its Resources', in J. Van Dyke, D. Zaelke, and G. Hewison (eds) *Freedom for the Seas in the 21st Century: Ocean Governance and Environmental Harmony*, Washington, DC: Island.

Van Dyke, J. (1993b) 'Sea Shipment of Japanese Plutonium under International Law', *Ocean Development and International Law*, 14: 399–430.

Van Dyke, J. (1996a) 'Applying the Precautionary Principle to Ocean Shipments of Radioactive Materials', *Ocean Development & International Law*, 27: 379–97.

Van Dyke, J. (1996b) 'The Straddling and Migratory Stocks Agreement and the Pacific', *International Journal of Marine and Coastal Law*, 11: 406–15.

Van Dyke, J. (2000) 'Sharing Ocean Resources in a Time of Scarcity and Selfishness', in H. Scheiber (ed.) *The Law of the Sea: Inherited Doctrine and a Regime for the Common Heritage*, Leiden: Brill.

Van Dyke, J. (2002) 'The Legal Regime Governing Sea Transport of Ultrahazardous Radioactive Materials', *Ocean Development and International Law*, 33: 77–108.

Van Dyke, J. (2004) 'The Evolution and International Acceptance of the Precautionary Principle', in D. Caron and H. Scheiber (eds) *Bringing New Law to Ocean Waters*, Leiden: Brill/Martinus Nijhoff.

Van Dyke, J. (2006) 'Liability and Compensation for Harm Caused by Nuclear Activities', *Denver Journal of International Law & Policy*, 35: 13–46.

Van Dyke, J., and Broder, S. P. (2012) 'Particularly Sensitive Sea Areas; Protecting the Marine Environment in the Territorial Seas and Exclusive Economic Zones', *Denver Journal of International Law & Policy*, 40: 472–81.

Van Dyke, J., Gardner, E., and Morgan, J. (2004) 'Whales, Submarines, and Active Sonar', *Ocean Yearbook*, 18: 330–63.

Vidal, J. (2012) 'Rio+20: Earth Summit Dawns with Stormier Clouds than in 1992', *The Guardian*, 19 June, online at http://www.theguardian.com/environment/2012/jun/19/rio-20-earth-summit-1992-2012 [accessed 18 June 2012].

Western Pacific Regional Fishery Management Council (1998) *A 20-Year Report*, Honolulu: Western Pacific Regional Fishery Management Council.

World Summit on Sustainable Development (WSSD) (undated) *Plan of Implementation*, online at http://www.un.org/jsummit/html/documents/summit_docs/2309_planfinal.htm [accessed 1 May 2014].

Yankov, A. (1999) 'The Law of the Sea Convention and Agenda 21: Marine Environmental Implications', in A. Boyle and D. Freestone (eds) *International Law and Sustainable Development*, Oxford/New York: Oxford University Press.

Yujuico, E. (2007) 'Powering the Shipping Industry w/ Drrrty Fuel', *International Political Economy Zone blog*, 2 December, online at http://ipezone.blogspot.com/2007/12/powering-shipping-industry-w-drrrty.html [accessed 24 May 2014].

Conventions and treaties

1946 International Convention for the Regulation of Whaling, 2 December 1946, (1946) 161 UNTS 72.

1972 Convention Concerning the Protection of the World Cultural and Natural Heritage (the 'World Heritage Convention'), 16 November 1972, in force 15 December 1975, 1037 UNTS 151, online at http://whc.unesco.org/en/conventiontext [accessed 28 November 2014].

1972 Convention on the Prevention of Marine Pollution by Dumping of Wastes and Other Matter ('London Dumping Convention') (1973) 11 ILM 129.

1973 Convention on International Trade in Endangered Species of Wild Flora and Fauna (CITES) (1973) 27 UST 1087, TIAS No. 8249.

1982 United Nations Convention on the Law of the Sea (UNCLOS), 10 December 1982, UN Doc. A/CONF.62/122, (1982) 21 ILM 1261.

1986 Convention on Early Notification of a Nuclear Accident, 26 September 1986, in force 27 October 1986, IAEA Doc. INFCIRC/335, (1986) 25 ILM 1370.

1989 Basel Convention on the Control of Transboundary Movements of Hazardous Wastes and Their Disposal (1989) 28 ILM 649.

1992 Convention on Biological Diversity (CBD), UN Doc. DPI/130/7, (1992) 31 ILM 818.

1992 Report of the United Nations Conference on Environment and Development, Vols I–III ('Agenda 21'), 13 August, UN Doc. No. A/CONF.151/26.

1992 Rio Declaration on Environment and Development ('Rio Declaration'), 14 June 1992, UN Doc. A/CONF.151/5/Rev.1, (1992) 31 ILM 874.

1995 Agreement for the Implementation of the Provisions of the United Nations Convention on the Law of the Sea of 10 December 1982 Relating to the Conservation and Management of Straddling Fish Stocks and Highly Migratory Fish Stocks (the 'UN Fish Stocks Agreement', or 'UNFSA 1995') (1995) 34 ILM 1542, 8 September, UN Doc. A/CONF.164/37.

1996 Protocol to the 1972 Convention on the Prevention of Marine Pollution by Dumping of Wastes and Other Matter (the 'London Protocol') (1996) 36 ILM 1.

1997 Kyoto Protocol to the United Nations Framework Convention on Climate Change ('Kyoto Protocol'), 10 December 1997, UN Doc. FCCC/CP/1997/7/Add.1, (1998) 37 ILM 22, online at http://unfccc.int/resource/docs/convkp/kpeng.pdf [accessed 25 May 2014].

2006 Southern Indian Ocean Fisheries Agreement (SIOFA), 12 June 2006, entered into force 21 June 2012, UN No. I-49547, online at http://treaties.un.org/doc/Publication/UNTS/No%20Volume/49647/Part/I-49647-08000002803296a6.pdf [accessed 25 May 2014].

2009 Convention on the Conservation and Management of High Seas Fishery Resources in the South Pacific Ocean, 14 November 2009, online at http://www.southpacificrfmo.org/assets/Convention-and-Final-Act/2353205-v2-SPRFMOConvention-textascorrectedApril2010aftersignatureinFebruary-2010forcertificationApril2010.pdf [accessed 18 May 2014].

2010 Nagoya Protocol on Access to Genetic Resources and the Fair and Equitable Sharing of Benefits Arising from Their Utilization to the Convention on Biological Diversity ('Nagoya Protocol'), 29 October 2010, online at http://www.cbd.int/abs/text/ [accessed 25 May 2014].

PART II

National Ocean Policies

3

OCEANS POLICY
A Canadian case study*

*Camille Mageau,** David L. VanderZwaag, Ken Huffman,** and Susan Farlinger***

Introduction

Over the years, Canada, like most other coastal nations, has developed an intricate set of policies and regulatory instruments focused on the management of traditional sectoral uses of the oceans. Nearly two decades ago, the necessary steps were taken to modernize the way in which Canadian authorities manage ocean-based activities. Canada did not set out to design 'one' comprehensive, all-inclusive oceans policy. The primary approach taken was to identify, through Canada's Oceans Act enacted in 1996, one federal lead authority responsible for the coordination and harmonization of existing policy and statutory instruments, and to formulate a national vision and guiding principles for oceans management within which existing and emerging policies and laws would be interpreted and implemented.

This chapter outlines Canada's statutory and policy instruments and implementation approach to oceans management. The political and environmental context within which a new management approach was developed will be described, as well as the processes that led to the development of the Oceans Act, its policy framework, Canada's Oceans Strategy (Fisheries and Oceans Canada, 2002a), the Canadian government's blueprint for action, Canada's Oceans Action Plan (Fisheries and Oceans Canada, 2005a), and the evolution of the approach in the face of changing government priorities. The relationship between key ocean-related agreements and Canadian domestic law and practice is summarized. In closing, lessons learned during the past two decades will be examined, as will the challenges that lie ahead.

Ocean policy context, processes, and institutional arrangements

Basic information

Canada is a maritime nation that borders the North Pacific, the Arctic, and the North Atlantic oceans, with marine areas covering a broad range of ocean climactic and oceanographic environments. Canada's current ocean regions total almost 6 million km² (Fisheries and Oceans Canada, 2014), and this will likely increase significantly once the extended continental shelf is delimited under the 1982 United Nations Convention on the Law of the Sea (UNCLOS).

Eight out of the ten provinces and all three territories border oceans, and approximately 24 per cent of Canada's population inhabits the coastal zone along a coastline that is one of the longest in the world, at about 245,000 km (Fisheries and Oceans Canada, 2014). The oceans provide the recreational, environmental, employment, income, and cultural staples to more than 7 million Canadians who live in coastal communities (Fisheries and Oceans Canada, 2014).

Canada has, in the past, defined itself as a fishing and shipping nation, with a long history and culture based on the rich productivity and diversity of its ocean resources. With the emergence of several other ocean-related industries, many of which vie for access to the same ocean space, the footprint of each industry and the cumulative impact of these activities have taken their toll on the environment resulting in:

- failing oceans health, including declining fish stocks, increasing numbers of marine species at risk and invasive species, declining biodiversity, and marine habitat loss;
- growing conflicts between oceans users and administrative, jurisdictional, and regulatory complexities; and,
- an oceans industry sector that is significantly weaker than its potential.

In addition to these domestic challenges, new challenges are emerging, particularly in the Arctic, as ice cover diminishes, and global demand for resources and efficient movement of goods are pushing for new shipping routes and development in the Arctic. The legal status of the Northwest Passage – whether internal waters, as claimed by Canada, or a strait used for international navigation, as argued by the United States – is an ongoing debate (Byers, 2013), and Canada is still facing many infrastructure deficits, such as adequate port facilities in the North (VanderZwaag, 2014a).

The marine areas that border Canada are vastly different from one another. The Pacific coast of Canada is characterized by a relatively narrow continental shelf about 50 km in width and a very indented coastal area of bays, fjords with inlets, wetlands, and estuaries. In addition to shipping, and aboriginal, recreational, and commercial fishing activities, the dominant industries include ecotourism, with an increasing focus on aquaculture in some areas of the coast. In the past five years, shipping of crude oil out of Vancouver has increased, and public attention and debate has hit the national scene, initially focussed on the controversial Northern Gateway pipeline with its export terminal at Kitimat on the British Columbia North Coast. More recently, the government of British Columbia has made liquefied natural gas (LNG) development the centrepiece of its economic plan for the province.

The Atlantic coast has a much wider continental shelf. Offshore areas have traditionally supported extensive and varied fishing, marine transportation activities, and, increasingly, initiatives related to oil and gas, ecotourism, and aquaculture. New hope for economic development is emerging, with tidal energy development as a focus in the Bay of Fundy and new proposals for oil and LNG terminals in Quebec and the Maritimes.

The Arctic marine area along the northern coast of Canada and its archipelago is characterized by a broad shallow shelf and land-fast ice. Transportation activities in the Arctic are largely seasonal and predominantly oriented around community resupply. Land mining, oil and gas exploration, ecotourism, and subsistence harvesting all contribute to the marine-based northern economy. In September 2013, the *Nordic Orion*, carrying a load of coking coal from Vancouver to Finland, became the first commercial bulk carrier to transit the Northwest Passage and further commercial transits are expected (Weber, 2014). The pace and scale of transit shipping in the Arctic, however, continues to be subject to considerable uncertainty (Lasserre, 2014). Canada still has unresolved ocean boundaries (VanderZwaag, 1995). In the Arctic, the offshore boundary in the Beaufort Sea between Alaska and the Yukon remains in dispute (McDorman, 2009), while Canada and Denmark (Greenland) reached a tentative agreement in November 2012 on the Lincoln Sea maritime boundary, with a treaty text to be prepared based on the negotiations (Foreign Affairs, Trade and Development Canada, 2012). On the Pacific coast, Canada has maritime boundary issues with the United States in the Dixon Entrance region (British Columbia–Alaska) and seaward of the Juan de Fuca Strait (British Columbia–Washington). In the Gulf of Maine, on the Atlantic coast, Canada and the United States continue to dispute the ownership of Machias Seal Island in the Bay of Fundy and jurisdiction over adjacent waters (McDorman et al., 1985; McDorman, 2009).

Over the last fifteen years, the oceans have been a dynamic growth sector for the Canadian economy and (based on an estimate using 2006 data) generated about CAN$28 billion, in direct and indirect gross domestic product (GDP) through ocean-related industries. More recent unpublished estimates using 2008 data put the value at about CAN$39 billion (Fisheries and Oceans Canada, 2009a). While there is considerable variation in estimates and debate on methodology, oceans activity makes a modest (about 2 per cent) contribution to the Canadian economy, but is a significant economic driver regionally. For example, in British Columbia, the oceans sector represents about 5 per cent of gross provincial product (GPP) and almost 7 per cent of jobs. In Atlantic Canada (including Quebec), the comparative figures are 5 per cent of GPP and almost 4 per cent of jobs. Commercial fishing and aquaculture continue to make an annual contribution to Canada's oceans economy totalling CAN$2 billion, supplemented by a further CAN$2 billion from the fish-processing industry. Employment in the fisheries and aquaculture and spin-off jobs is 65,000, providing more than CAN$2 billion in labour income, often in coastal communities with limited other opportunities. Offshore oil and gas production has increased in annual investment value over the past decade from CAN$250 million to almost CAN$10 billion. In GDP value, oil and gas production on the east coast represents almost a third of the total oceans sector – and is growing. Considerable renewable energy resources, such as offshore wind, wave, and tidal energy, have been identified on all three of Canada's coasts, and a number of initial project proposals have been developed over the past decade, but no commercial projects are in place currently except for a small tidal power station on the Annapolis River in Nova Scotia.

Recreation and tourism have continued to grow, with coastal and cruise ship tourism being regionally important, although a relatively small contributor to the oceans economy. As a maritime nation, Canada has a significant and vibrant shipping industry. The marine transportation sector contributes CAN$5.5 billion to GDP, 75,000 jobs, and CAN$3.7 billion in labour income to the Canadian economy. International trade worth CAN$205.3 billion moved through Canada's national marine transportation system in 2011 (Transport Canada, 2013).

Aboriginal communities have the longest history of coastal occupancy. Coastal aboriginal cultures are tied to ocean resources for food, social, and ceremonial reasons. Commerce between First Nations, and after contact between aboriginal communities and Europeans, was often based on oceans activities or resources.

Canada is a confederation of ten provinces and three northern territories. Federal jurisdiction extends to marine navigation and shipping, international affairs, defence, and environmental protection, as well as the protection of living resources within offshore areas (Fisheries and Oceans Canada, 1997a). Provinces, the subnational authorities within Canada, may also exert jurisdiction over some offshore waters. In general, provinces own and manage the seabed within the coastal inter-tidal area. Provinces have constitutional authority over property and civil rights within the province, pursuant to section 92(13) of the Constitution Act, 1867. Canadian case law has recognized two legal foundations for provincial offshore jurisdiction, marine areas considered *inter fauces terrae* ('between the jaws of land') and marine areas considered to be part of a province at the time of confederation.[1]

Management of activities within Canadian marine waters has developed on a sector or regional basis, and is therefore diverse and lacks a cohesive approach. For example, there are ten major and thirteen minor federal agencies that have mandates that impact on oceans. There are roughly fifty federal statutes directly impacting ocean-related activities and more than eighty provincial laws that affect coastal and marine planning (Fisheries and Oceans Canada, 1997b; 2009b).

In addition to this legislated division of power, Canada sets as a high priority its constitutional obligations to aboriginal peoples. The Constitution Act, 1982 (section 35) recognizes and affirms existing aboriginal and treaty rights. Where land claim agreements have been settled, and include specific resource management responsibilities and commitments by the federal government to cooperate and collaborate with the signatories, the situation is clear. In some cases, however, claims that may impact on ocean areas have not yet been settled, and interim arrangements that do not prejudice the outcomes of land claims discussions are in place, being developed, or needed (Brown and Reynolds, 2004; Moodie, 2004; Ginn, 2006; Jones, 2006).

The Oceans Act, section 2(1), contains an explicit provision to provide certainty that it does not abrogate or derogate from existing aboriginal and treaty rights. This provision sets out the framework for the relationship of Oceans Act programmes and activities with aboriginal peoples. While integrated planning and the development of marine protected areas (MPAs) are without prejudice to rights and title, the involvement and support of aboriginal peoples is clearly required where their interests are potentially affected. Many coastal communities, of and by themselves, have large aboriginal populations, and in some areas specific arrangements respecting harvesting and co-management have been made with aboriginal authorities.

The importance of the oceans to the federal, provincial, First Nations, municipal, and local communities, stakeholders, and interest groups requires engagement of these interests in setting priorities and planning oceans activities. It is this context that informed the development of an Oceans Act. The federal Department of Fisheries and Oceans Canada (DFO) is the lead federal agency responsible for the coordination of both domestic and international oceans policy. This mandate is in addition to more traditional marine responsibilities related to the management of aboriginal, commercial, and recreational fisheries, marine safety and communication, environmental response, and the provision of marine scientific advice and research.

A brief overview of the nature and evolution of national oceans policy

Although the development of a national oceans policy and legislation was first proposed in 1987 (Fisheries and Oceans Canada, 1987), the first steps towards the elaboration of a national oceans policy for Canada were taken when the government of Canada enacted the Oceans Act, in 1996. This statute formalizes, in a comprehensive way, how Canada's oceans are to be defined and managed.

The Oceans Act lays the foundation for the oceans policy by committing to a number of principles and is structured to delineate the geographic area over which Canada intends to apply its ocean management approach. The Act defines the guiding principles of integrated management, sustainable development, and the precautionary approach, provides the mandate to develop and implement programmes to implement these principles, and situates the DFO's existing regulatory and management authorities within the context of oceans management. The Act also recognizes other mandated authorities and provides guidance on how their mandates should be delivered within the marine environment.

The development and review of the Oceans Act, by means of the public and parliamentary processes, was complemented by a broad public consultation process, which led to Canada's Oceans Strategy, the overarching oceans policy framework for the integrated management of Canada's oceans (Fisheries and Oceans Canada, 1999a). During the five years immediately following the proclamation of the Oceans Act, the ocean management programmes outlined in the statute were piloted in the field to better define the policy guidance required and to inform the development of the federal Oceans Action Plan (OAP). Unlike most legislation, there was no centrally approved funding channelled through the budget to implement the Oceans Act, leaving the DFO to reallocate funds internally. For the first few years of the programme, DFO internal funding was 'B-based' – or contingent on an annual exercise to find savings in other programme areas or to use a portion of the Department's approved carry-forward from the previous year. In the early 2000s, after the Oceans Act components became a key component of the DFO's Strategic Plan, the programme was given stable funding by DFO management and oceans programming included in the DFO core ('A-base') budget.

Flowing from the call to advance modern ocean management and political commitment in the October 2004 Speech from the Throne (Government of Canada, 2004a) and the 2005 Budget Speech (Department of Finance Canada, 2005), Canada's OAP outlined and funded priority areas for action under four major themes: international leadership, sovereignty, and security; integrated oceans management for sustainable development; health of the oceans; and science and technology (Fisheries and Oceans Canada, 2005a). The priority was reinforced by the designation by the prime minister of a parliamentary secretary to support implementation of the OAP. The profile and political will secured the first funding for a government-wide initiative. The OAP Phase I (2005–07) consisted of CAN\$28.5 million for eighteen initiatives across seven federal departments and agencies. Following Phase I – and a change of government – the 2007 federal budget introduced the National Water Strategy, which proposed CAN\$19 million over two years for the Health of the Oceans (HOTO) programme, to further support sustainable development, management and protection of ocean resources, and water quality (Department of Finance, 2007). The two-year HOTO funding was subsequently increased to five years and CAN\$61.5 million when the details were approved and announced by Cabinet. The HOTO programme became Phase II of the OAP, but crucially the focus was narrowed to only one of the four pillars identified in the 2005 OAP, with twenty-two initiatives spread across five departments. In 2012–13 and 2013–14, one-year funding extensions were granted for HOTO, although in the latter case the scope was narrowed as some initiatives were completed and others, such as Transport Canada's National Aerial Surveillance Programme of pollution monitoring and prevention, were rolled under other funding umbrellas. In the February 2014 budget, five-year funding was included under what is now the National Conservation Plan. Funding of CAN\$37 million was announced by the prime minister on 15 May 2014 to 'strengthen marine and coastal conservation' (Prime Minister of Canada, 2014a).

Policy development processes

In Canada, the development of an oceans policy has been, and continues to be, an evolutionary process. In 1994, the National Advisory Board on Science and Technology (NABST), following extensive public consultations, recommended to the prime minister that Canada move decisively to address environmental issues confronting oceanic areas and to maximize the economic benefits that could be derived by managing ocean resources more sustainably (NABST, 1994). Specific recommendations focused on the need to develop a national policy, as well as legislation focused on the management of ocean and coastal spaces and resources.

Although similar calls had been made in the past, there was, at this time, a convergence of domestic and international fishing and pollution issues, primarily in the North Atlantic, that served to focus public, as well as political, interest (Commissioner of the Environment and Sustainable Development, 2005). As a result of this heightened profile, drafting of a comprehensive Oceans Act was initiated and the resulting 1996 Act came into force on 31 January 1997.

The Oceans Act

The Oceans Act comprises three parts. Part One of the Act recognizes Canada's maritime zones and commits the government of Canada to meeting its conservation and management responsibilities within these marine areas. Consistent with the terms of the 1982 United Nations Convention on the Law of the Sea (UNCLOS), Canada has defined its territorial sea, contiguous zone, exclusive economic zone (EEZ), and continental shelf excluding the outermost extent. Canada has worked diligently to delimit the outer extent of the continental shelf and submitted its initial submission to the Commission for the Limits of the Continental Shelf by the required deadline in December 2013. A full submission was delivered for the Atlantic Ocean, but only preliminary information for the Arctic (Foreign Affairs, Trade and Development Canada, 2013).

Part Two of the Act provides the Minister of Fisheries and Oceans with specific policy and programme authorities to implement Canada's approach to oceans management in estuarine, coastal, and marine ecosystems. Section 29 of the Oceans Act provides for the development of a national strategy, Canada's Oceans Strategy, which constitutes the policy framework for modern oceans management and serves as guidance for the development and updating of sector-based policies and processes. The Act calls upon the Minister to develop this strategy in collaboration with federal colleagues, provincial and territorial governments, affected aboriginal organizations, coastal communities, and other persons and bodies, including those bodies established under land claims agreements (section 29). Finally, the Act includes provisions for the development of specific programme areas: marine protected areas (MPAs), and integrated management plans. In addition, for purposes of implementing integrated management plans, the Minister may establish marine environmental quality guidelines, objectives, and standards (section 32), and there is provision to prescribe marine environmental quality requirements and standards through regulation (section 52(1)). These programmes are the key tools used to implement the national ocean policy objectives: understanding and protecting the marine environment, supporting sustainable economic opportunities, and providing international leadership.

Part Three of the Oceans Act sets out the accountabilities for the Act. It identifies the Minister of Fisheries and Oceans as the lead federal authority responsible for oceans management within Canada and situates the existing resource management, scientific, hydrographic, coastguard, and other responsibilities of the department within an oceans management context.

Following adoption of the Oceans Act, the DFO reallocated modest funds to support its implementation through a series of pilot projects and the development of Canada's Oceans Strategy (Fisheries and Oceans Canada, 2002a) in consultation with Canadians. Pilot projects were selected based on feasibility criteria, including the complexity of the ocean issues involved, the receptivity of potential partners, the level of scientific information available, and the conservation imperatives of the area. Projects included the identification of areas of interest for MPAs and the announcement of several pilot MPAs, such as the Sable Gully and Endeavour Hot Vents in 1998 (Fisheries and Oceans Canada, 1998). Pilot integrated management initiatives were also established in the areas of the Eastern Scotian Shelf (ESSIM) in 1998, the Beaufort Sea in 2000, and the Pacific North Coast of British Columbia (PCIMA) in 2001 (Commissioner of the Environment and Sustainable Development, 2005). The final two pilot projects, Placentia

Bay – Grand Banks (PBGB) and Gulf of St Lawrence (GOSLIM), were identified by 2005. The pilot projects provided lessons with respect to policy integration, the building of relationships, the development of governance structures, and related arrangements.

The policy development process continued its course with two public engagement and consultation processes. The first was focused on the vision for the Oceans Act (Fisheries and Oceans Canada, 1994); the other focused on a structured consultation on Canada's Oceans Strategy and was designed to solicit federal, provincial, First Nations, and public input. Over a period of five years, the DFO engaged the views and perspectives of Canadians by supporting a wide range of discussions, workshops, and consultation activities across the country. Two Oceans Ambassadors and a Minister's Advisory Council on Oceans played key roles in the consultations.

Canada's Oceans Strategy

Canada's Oceans Strategy (Fisheries and Oceans Canada, 2002a) and its companion Integrated Management and Operational Framework (Fisheries and Oceans Canada, 2002b) were released in 2002 following formal federal, provincial, territorial, aboriginal, and public consultations. Presented to Cabinet, the Oceans Strategy received government endorsement and became the basis upon which federal activities were to be conducted in marine waters.

The release of the Canada's Oceans Strategy as a policy of the government of Canada set out the achievement of its objectives as a shared responsibility for all federal departments with an oceans mandate (Fisheries and Oceans Canada, 2002a). The following fundamental principles are set out in the Oceans Act and Canada's Oceans Strategy:

- *integrated management* – to plan and manage human activities impacting on oceans in a comprehensive fashion, while considering all factors necessary for the conservation and sustainable use of marine resources and the shared use of ocean space;
- *sustainable development* – to integrate social, economic, and environmental aspects of decision-making; and
- *the precautionary approach* – to err on the side of caution in making management decisions.

Integrated management is a spatially based planning process that results in common understanding of ecosystem and human activity objectives on the part of regulators, stakeholders, and interested parties, and the production of an 'integrated management plan' for a geographic area (Cicin-Sain and Knight, 1998). The plan provides a framework within which to conduct activities, and to develop and implement integrated and adaptive management strategies and actions. The plans are based on the recognition that integrated management planning must occur in an ecosystem context for the decisions to be environmentally sound and ocean activities sustainable.

Canada's Oceans Strategy commits the government to work collaboratively within the federal government and among levels of government, to share responsibility for achieving common objectives, and to engage Canadians in ocean-related decisions in which they have a stake (Fisheries and Oceans Canada, 2002a). Integrated management planning includes the establishment of institutional governance mechanisms as a cornerstone of the national oceans approach. This integration is not limited to policies and legislative authorities that oversee the management of oceans activities; its primary focus is planning and managing activities on a geographic, but also on an ecological, basis. The ecological link is key, since it ties back to the principles of the Oceans Act, including the adoption of an ecosystem-based approach.

Integration is required to achieve *sustainable development*, which in itself requires that conservation issues be addressed and that economic diversification and multiple uses be recognized as legitimate objectives. The ability to adapt management decisions to reflect new scientific and technical developments, to reflect changing economic and social objectives, and to respond to positive or negative environmental responses is key to achieving the principles of integrated management and sustainable development.

The *precautionary approach* should be followed as part of the decision-making process for integrated management. When there is a risk of serious or irreversible harm and there is significant scientific uncertainty, then decisions and management options should err on the side of caution. Within the context of oceans management, application of the precautionary principle is inextricably linked to two other concepts: an ecosystem-based and a science-based approach to decision-making (Cobb et al., 2005).

The ecosystem-based approach relies on the identification of ecosystem objectives that, together with social and economic objectives, form the basis for integrated management planning and related decision-making. These ecosystem objectives are based on an assessment of ecological information and an evaluation of the risk posed to ecosystem structure and function based on both available information and uncertainties. In this way, the risks of uncertainty are incorporated into decisions and are managed into the future through adaptive management.

Institutional arrangements and processes used

Following prime ministerial acceptance of the recommendation by NABST's Committee on Oceans and Coasts (NABST, 1994) that Canada formulate an overall oceans policy framework and develop ocean-focused legislation, a ministerial Vision Paper on oceans management was released (Fisheries and Oceans Canada, 1994). Public comments on the Vision Paper served to form the basis of the draft legislation. While parliamentary procedures do not allow for public review of draft legislation, information sessions outlining the intent of the legislation were held. The normal parliamentary consultation procedures, which involve formal publication of draft legislation by the House of Commons, as well as targeted consultations with affected parties, were conducted. Witnesses to the parliamentary review process, including potentially affected stakeholders, environmental non-government organizations (NGOs), aboriginal peoples, coastal communities, and academics, broadened the scope of the Act.

The DFO also led the development of Canada's Oceans Strategy (Fisheries and Oceans Canada, 2002a), incorporating the lessons learned from the pilot application of the Oceans Act programme and the views expressed during public engagement processes led by Oceans Ambassadors appointed by the Minister. Policy development entailed consulting a range of governmental and non-governmental stakeholders, and using different mechanisms to connect with subnational and aboriginal authorities, and the academic community. Between 1997 and 2002, the DFO engaged the views and perspectives of Canadians by supporting a wide range of discussions, workshops, and consultation activities across the country. These activities included a discussion document, *Toward Canada's Oceans Strategy* (Fisheries and Oceans Canada, 1997c), interactive websites, public opinion polls and research, an international Oceans Stewardship Conference (Fisheries and Oceans Canada, 2001a), international workshops on integrated management, cross-country consultation sessions, a Minister's Advisory Council on Oceans (Fisheries and Oceans Canada, 2002c), and a national oceans discussion series in cooperation with the Canadian Broadcasting Corporation (CBC) and the International Oceans Institute of Canada (CBC, 2001). Bilateral meetings were conducted with key national stakeholders, including environmental NGOs, industry organizations, and the main aboriginal organizations.

The development of a national oceans policy therefore involved a mix of legislation, policy development, pilot projects, and relationship building. While legislation and policy development take place at the national level in federal departments such as the DFO, coordination and collaboration are required at many levels to create the environment in and tools with which to implement such a horizontal collaborative initiative. Governance arrangements and processes are described next, and Table 3.1 offers an indication of the complexity of these relationships.

A Minister's Advisory Council on Oceans was established in September 2000 for a three-year term to provide advice on ocean management policy issues and to help to engage the public and private sectors in issues related to oceans management (Fisheries and Oceans Canada, 2002c). The Council consisted of nine individuals from diverse backgrounds representing a range of interest groups, including aboriginal peoples, industry members, and academics (Fisheries and Oceans Canada, 2000). As such, the Council was instrumental in increasing public understanding and awareness of the nature and intent of Canada's ocean management approach.

In 2001, federal, provincial, and territorial ministers agreed that an Oceans Task Group would be established under the aegis of the Canadian Council of Fisheries and Aquaculture Ministers to help to develop and implement Canada's Oceans Strategy (Canadian Council of Fisheries and Aquaculture Ministers, 2001). For a decade, this Task Group provided a forum for federal–provincial issues on oceans management, with its work guided by an annual work plan approved by ministerial council.

To aid federal government coordination and input to ocean policy development, a system of interdepartmental committees on oceans (ICOs) was established at the deputy minister, assistant deputy minister, and programme levels. In 2004, interdepartmental working groups were formed to focus on the four 'pillars' set out in Canada's Oceans Action Plan – that is, international leadership, sovereignty and security, integrated oceans management for sustainable development, health of the oceans, and oceans science and technology (Fisheries and Oceans Canada, 2004a).

Overall, the various governance mechanisms and agreements were effective in developing a policy framework and action plan that reflected a range of stakeholder interests. These front-end initiatives were endorsed at the highest levels of government.

The nature of the policy and legislation established

The nature of the resulting policy

The Oceans Act is enabling legislation, designed to provide the Minister of Fisheries and Oceans with the responsibility of focusing current federal legislative and policy tools to increase the linkages among and overall effectiveness of federal government efforts in specific geographic areas. This collaborative aspect of the legislation is the most challenging to implement, in that willing partners are needed to advance ocean management. Intergovernmental agreements have been required, as well as negotiations with industry and aboriginal authorities. Implementation of Oceans Act programmes have moved at different paces in different areas, with more rapid progress achieved in ocean management areas in which collaborative mechanisms were already in place. As lead and facilitator, the DFO has had to concentrate on building the relationships while at the same time developing the science-based tools and fostering the governance arrangements needed to incorporate the values and interests of others.

The Oceans Act and the oceans policy framework neither supersede nor fetter other policies or statutes, but provide the context within which other ocean-related mandates should operate. On this basis, both the Act and Canada's Oceans Strategy (Fisheries and Oceans Canada, 2002a) provide the broad framework to guide further federal policy development to work with other levels of government and provide a new context within which to interpret older policies.

Table 3.1 National, subnational, and local oceans governance structures and agreements

	Examples of governance structures			Examples of agreements
	National	Subnational	Local	
International	Membership in international committees, councils, and science organizations, including regional fisheries management organizations, the Arctic Council, Asia-Pacific Economic Cooperation (APEC), the International Maritime Organization (IMO), and the Intergovernmental Oceanographic Commission (IOC)			
Other government departments (OGDs)	Deputy ministers' committee Support committees (assistant deputy and director general level)	Subnational implementation committees	OGD planning or regulatory processes	National Marine Protected Area Strategy
Provinces and territories	Canadian Council of Fisheries and Aquaculture Ministers' Oceans Task Group	Regional oceans management committees in most large ocean management areas (LOMAs)	Lead on coastal planning	National Framework for Canada's Network of Marine Protected Areas Canada/Quebec St Lawrence Action Plan
Aboriginal organizations		Co-management bodies established pursuant to the Inuvialuit Final Land Claims Agreement directly involved in Beaufort Sea ocean management planning bodies	Planning process/ traditional ecological knowledge consultation	Turning Point Agreement (a British Columbia–First Nations agreement relating to the Pacific North Coast LOMA)
Stakeholders				
Local communities	Subnational implementation committees	Advisory/planning process		
Industry stakeholders	Subnational implementation committees	Advisory/planning process		Ocean Management Research Network Canadian Association of Petroleum Producers/Statement of Canadian Practice with respect to the Mitigation of Seismic Sound in the Marine Environment
Oceans interest groups	Subnational implementation committees	Advisory/planning process		Membership on Canadian delegations

Together, they provide the principles and key tools with which to implement modern oceans governance, within which existing policies and statutes, and traditional relationships between regulators and their traditional 'clients', may operate. As the guiding principles such as precaution and adaptive management are interpreted and utilized in integrated management planning, they will be integrated into new sectoral policies. Since the building blocks of Canada's oceans policy framework, and the associated implementation programmes, are solidly anchored on precaution, ecosystem-based management, and sustainable development, these principles are, by definition, embedded in decisions that will be taken within the integrated management planning areas.

The implementation of principles

In Canada, an ecologically based framework to guide the development of integrated management plans has been developed. The integrated management planning framework extends from the large- to the small-scale – that is, from large ocean management areas (LOMAs) to smaller scale (for example in Newfoundland and Labrador) coastal management areas (CMAs) (VanderZwaag et al., 2012: 332). In other ocean regions, scale has evolved beyond the initial LOMA, for example since the Eastern Scotian Shelf Integrated Management (ESSIM) plan was developed, the planning area has expanded conceptually to the bio-region scale, including the entire Scotian Shelf and the Bay of Fundy. The Canadian approach to integrated management recognizes that management objectives and planning practices should reflect the fact that ecosystems nest within other ecosystems. Governance structures, practices, and decisions respecting resource and activities management should be made with explicit consideration of ecosystem impact. The logic is that the precautionary approach is built into integrated management through the identification of ecosystem objectives that activities must respect within specified planning areas. The practice has proven to be a significant – and ongoing – challenge. A brief review of Canada's incorporation of the principles of ecosystem-based management, integrated management, the precautionary approach, and public participation and community-based management follows.

Ecosystem-based management

The Preamble to the Oceans Act states that 'conservation, based on an ecosystem approach, is of fundamental importance to maintaining biological diversity and productivity in the marine environment'. An ecosystem-based approach to management recognizes that human activities must be managed in consideration of the interrelationships between organisms, their habitats, and the physical environment, based on the best science available. The Act further holds that human activities should be managed such that marine ecosystems – their structure (for example biological diversity), function (for example productivity), and overall environmental quality (for example water and habitat quality) – are not compromised, and are maintained at appropriate temporal and spatial scales. It is in these key areas that ecosystem objectives will be set for each of the integrated management areas (Cicin-Sain, 2003).

Significant domestic and international efforts have been invested in making this principle operational (IOC, 2003). In 2001, Canada held a scientific workshop to develop a preliminary framework that had conservation of species and habitats and the sustainability of human use as its two overarching objectives (Jamieson et al., 2001). Work has continued in Canada, and internationally, to further refine the initial objectives identified at this meeting. Three overarching ecosystem objectives have been identified: to maintain populations, species, and communities within bounds of natural variability; to conserve the function of each component of the ecosystem so that it can play its natural role in the food web; and to conserve the physical and chemical properties of the ecosystem (Jamieson et al., 2001: 16–20). This work has resulted in the development of a process and tools with which to apply ecosystem-based management to decision-making within Canada's LOMAs.

Implementation of ecosystem-based management begins with the identification of marine eco-regions, which are based on ecological features and functions (Powles et al., 2004). Existing scientific and traditional information on the state and condition of the ecosystem bound within the planning area is then collected, and a science-based review of that information and an evaluation of the risks posed to ecosystem structure and function conducted. As a result of the review and evaluation of known scientific information, ecologically and biologically significant areas, ecologically and biologically significant species, and community properties, as well as degraded areas and depleted species of special concern, are identified (Fisheries and Oceans Canada, 2005b). Priority ecosystem-based conservation objectives and limits are defined within these eco-regions. Management decisions and the choice of management measures adopted are informed by the conservation objectives (Fisheries and Oceans Canada, 2004b). As the approach has evolved and in an effort to provide a better focus for effort with limited resources, a risk management approach has been layered throughout the integrated oceans management process. Thus, as Figure 3.1 reveals, in the current version of the six-step process, risk management assessment tools have been refined to guide management action.

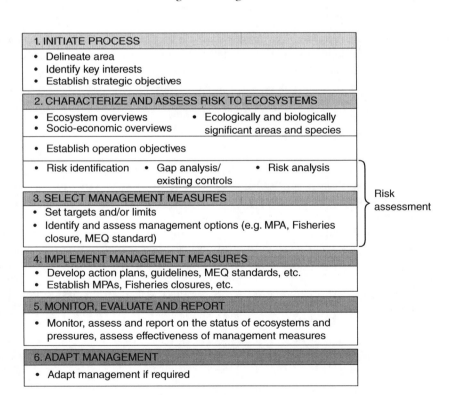

Figure 3.1 The six-step integrated oceans management process

Source: Fisheries and Oceans Canada (undated*a*)

It is important to reiterate that integrated management is a means to achieve an end: the sustainable management of ocean resources and spaces. For this reason, Canada's integrated management processes are designed to initially identify conservation objectives that must be respected by any activity to be operated in the planning area if the ecosystem is to continue to function and sustain the pressures of resource extraction and other ocean uses. Once the 'conservation limits' are defined, the Canadian integrated management process focuses on the identification of social, cultural, and economic objectives or desirable targets that subnational and local governments, stakeholders, and the public wish to achieve in the planning area.

Ecosystem considerations are being incorporated into fisheries management policies, plans, and practices. For example, in Canadian waters in which relatively unique and highly sensitive marine ecosystems are known to exist, and in which there is scientific evidence that fishing practices are having a long-term adverse effect on the ecosystem, action has been taken to mitigate these effects through the application of management measures. These measures include:

- fishing gear modifications, mesh and hook size considerations, and other measures to ensure that fishing practices conform to specific habitat conservation requirements;
- the application of seasonal and area fishing closures or gear-type closures if impacts cannot be mitigated;
- the establishment of marine protected areas (MPAs) in which long-term protection measures cannot be adequately addressed through fishing closures and other measures; and
- monitoring of the area for compliance and management effectiveness.

However, ecosystems do not respect political or administrative boundaries. As a result, it has been important to give effect to the concept of collaborative planning and management systems (O'Boyle and Jamieson, 2006). Domestic decision-making across ecosystems needs to be connected by the participation of federal, provincial, territorial, aboriginal, and local authorities and programmes. The minister has the option to use bilateral agreements with provinces or territories, and co-management arrangements with aboriginal groups, to implement and achieve ecosystem objectives. For example, in 2004, the governments of Canada and British Columbia signed the Memorandum of Understanding Respecting the Implementation of Canada's Oceans Strategy on the Pacific Coast of Canada (Fisheries and Oceans Canada, 2004c), with a commitment to develop sub-agreements focused on integrated management, MPAs, and information-sharing, although political differences have stymied the actual conclusion of further subsidiary memoranda of understanding (McCrimmon and Fanning, 2010). In the Arctic, the Beaufort Sea Integrated Management Planning Initiative (BSIMPI), guided by the Senior Management Committee, a collaborative body composed of representatives from government, aboriginal, and industry stakeholder groups (Elliott and Spek, 2004; Siron et al., 2009), developed the 2009 Integrated Management Plan. Ongoing leadership and direction is provided by a regional coordination committee under what is now called the Beaufort Sea Partnership. Regionally, agreements have been signed at the senior management level, for example the 2011 Canada–Nova Scotia Memorandum of Understanding, which focuses on the development of a bioregional network plan for conservation.

Ecosystem-based management objectives for LOMAs are set at the scale of ecosystem or broad eco-region. Integrated management planning units, and sectoral management plans nested within these areas, do not necessarily correspond to an entire eco-region. Consequently, the Oceans Act, section 32(d), provides the authority to set marine environmental quality guidelines, requirements, and standards, which can be specific to one particular planning area, but can also apply to activities across bioregions and complement the broader-scale ecosystem objectives.

Monitoring programmes tied to the eco-region-level ecosystem objectives and the marine environmental quality targets linked to specific management plans provide a mechanism for tracking change over time and triggering management action.

Canada's Fisheries Act, the main legislative tool to protect ecosystems on which fish depend, was amended in 2012 and the habitat protection provisions were substantially curtailed.[2] The previous habitat protection provision (section 35) was broad and quite strong. It prohibited persons from carrying on any work or undertaking that resulted in the harmful alteration, disruption, or destruction of fish habitat except under conditions authorized by the Minister of Fisheries and Oceans or under regulations pursuant to the Fisheries Act. The new provisions prohibit persons from carrying on any work, undertaking, or activity that results in serious harm to fish that are part of a commercial, recreational, or aboriginal fishery, or to fish that support such a fishery (that is, food chains). Therefore various habitat protections were potentially removed, for example water bodies not subject to a fishery such as salt marshes (Hutchings and Post, 2013).

Implementation of the ecosystem approach at the bilateral and regional levels has been varied. For example, while Canada has supported formal adoption of the ecosystem approach within the Northwest Atlantic Fisheries Organization (NAFO), along with various closures to bottom trawling to protect vulnerable marine ecosystems, Canada's transboundary management arrangements with the United States in the Gulf of Maine, and with St Pierre and Miquelon in the North Atlantic, have largely focused on the allocation and conservation of a limited number of commercially important fish stocks (Russell and VanderZwaag, 2010a).

Integrated management

Recognizing that integration must carry over to the planning of conservation areas as well, the Oceans Act, section 35(2), calls for the Minister of Fisheries and Oceans to lead and coordinate the development and implementation of a national system of MPAs on behalf of the government of Canada (Fisheries and Oceans Canada, 1999b). Three federal agencies – the DFO, Parks Canada, and Environment Canada – are mandated to establish federal MPAs, and provincial authorities also are active in protecting areas within their areas of jurisdiction (Government of Canada, 1998). To maximize the effectiveness of federal intervention and to ensure that the appropriate tools are being used, the DFO, along with other federal departments, developed a Federal Marine Protected Areas Strategy to achieve a national network of MPAs (Government of Canada, 2005a). Efforts to broaden the national network to include provincial and territorial MPAs resulted in the National Framework for Canada's Network of Marine Protected Areas (Government of Canada, 2011) and development of an inventory of existing federal, provincial, and territorial MPAs to inform the gap analysis phase of the National Network's development (Government of Canada, 2010). The involvement of federal, provincial, and territorial authorities in the five integrated oceans management priority areas within which ecologically and biologically significant areas, significant species, and significant community properties have been identified has facilitated network design. Under the National Conservation Plan, efforts are now under way to advance this work.

As part of the Oceans Action Plan (OAP), implementation of integrated management focused on five priority geographic areas in which mandated federal, provincial, territorial, and aboriginal authorities were working cooperatively to develop integrated ocean management plans: Placentia Bay/Grand Banks off Newfoundland; the Scotian Shelf off Nova Scotia; the Beaufort Sea in the western Arctic; the Gulf of St Lawrence; and the Pacific North Coast, or Queen Charlotte Basin, off British Columbia (see Figure 3.2). As they move beyond the pilot

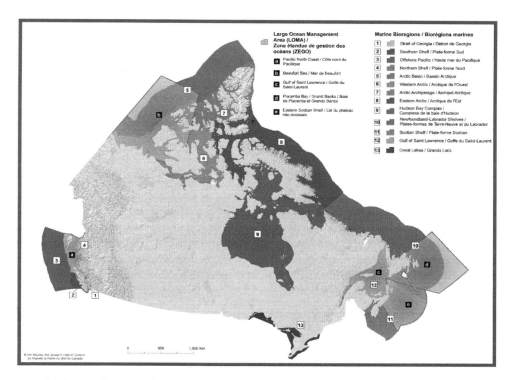

Figure 3.2 Priority bioregional integrated management planning areas

Source: Department of Fisheries and Oceans Canada, July 2014.

stage and as the DFO's knowledge of ecological properties has improved, these areas have been modified to become bioregional planning areas. In some regions, the boundaries have largely remained the same (such as the Pacific), while in others areas the boundaries have changed significantly (for example the Scotian Shelf).

Activities undertaken within each of the planning areas include the assessment and overview of the state of health of marine ecosystems, which provides mandated authorities and stakeholders with information on marine and coastal ecosystems, and recommendations to support planning and management decisions. In collaboration with the Geological Survey of Canada, the DFO undertook seabed mapping to better characterize benthic habitats, to define bottom communities, and to support identification of the most appropriate management actions (Fisheries and Oceans Canada, 2005a). Areas, species, and community properties in need of special management and/or conservation measures have also been identified, as have degraded areas and depleted species. Governance arrangements to foster federal, provincial, territorial, and aboriginal collaboration were established, as were forums to engage citizens and stakeholders.

While some of these activities were already well advanced in some of the priority LOMAs as a result of previous federal investments and efforts, the additional funds and the more visible accountability attached to the special budget allocations supported implementation of OAP initiatives (Fisheries and Oceans Canada, 2005a) within a prescribed period of time. The final draft of the ESSIM Plan was released in July 2006 (ESSIM Planning Office, 2006), while the Integrated Oceans Management Plan for the Beaufort Sea Large Ocean Management Area was released in 2010 (Fisheries and Oceans Canada, 2010a). In the other priority LOMAs of the Pacific North Coast, the Gulf of St Lawrence, and Placentia Bay/Grand Banks, ecosystem over-view reports and assessments, as well as other key pieces such as identification of ecologically and biologically significant areas, species, and properties, conservation objectives' prioritization, and human use overviews, were completed. By 2014, the three remaining plans had been released (Fisheries and Oceans Canada, undated*b*), although significant challenges have meant that the plan for the Pacific North Coast (PNCIMA, 2013) is still in draft form.

Integrated management is more than the development of spatially based management plans. Effective management requires integration at a variety of levels (Chircop and Hildebrand, 2006). There are numerous examples of spatial integration in which efforts between provincial authori-ties, responsible for land-based issues and inter-tidal seabed, and federal authorities, responsible for overlying waters and resources, are being coordinated to establish the necessary protection measures on land and in coastal waters to achieve the objectives of coastal MPAs. For exam-ple, coastal sand dunes adjacent to the Basin Head Marine Protected Area, off Prince Edward Island, have been protected under the authority of the provincial Natural Areas Protection Act (Government of Canada, 2005b). In June 2010, a ground-breaking agreement was signed to establish the Gwaii Haanas Marine Area, a National Marine Conservation Area and Haida Heritage Site, to integrate with the existing land area for a unique mountain top to deep sea conservation area (Parks Canada, undated*a*). In March 2011, Canada and Nova Scotia signed a memorandum of understanding pledging to strengthen cooperation in coastal and oceans man-agement, with subsidiary agreements and/or working groups to be developed (Government of Nova Scotia, 2011).

In British Columbia, integrated ocean and coastal planning has taken place at both the federal and provincial levels, and has varied over time in terms of leadership focus, resources, and level of collaboration. In the 1990s, the provincial government ventured first into marine planning, seeking to apply principles of land-use planning to marine areas in the Central Coast, with the goal of addressing use conflicts and First Nations concerns. The federal government followed, with the Pacific North Coast Integrated Management Area (PNCIMA) planning exercise for the North Coast, working well with First Nations and the province until 2011, when issues arose over US Moore Foundation funding through Tides Canada of some federal, provincial, and aboriginal components of the planning exercise. Lobbying by shipping and energy interests, linked to Northern Gateway pipeline and other North Coast energy proposals, moved the fed-eral government to withdraw from the funding agreement, which offered significant resources (up to CAN$8.3 million) to PNCIMA marine planning (Living Oceans Society, 2011). The federal government indicated that a streamlined plan to meet timelines no longer required funds for additional science and consultation (O'Neil, 2012; BC Shipping News, 2011). With the federal retreat to a high-level strategic plan, the Province of British Columbia and northern coastal First Nations forged ahead with the Marine Planning Partnership for the North Pacific Coast (MaPP), and in the spring of 2014 released draft plans for public input for four coastal area sub-regions, including a Haida Gwaii Draft Marine Plan (MaPP, undated). Federal participation in the MaPP process to date has been limited and issues remain over how to address federal jurisdiction (such as fishing) issues in the MaPP plans.

There are numerous opportunities for science and spatial co-location of federal and pro-vincial science programmes in the five geographic areas. A primary example is the targeted use of seabed mapping using side scan sonar to support integrated management within the prior-ity areas while still addressing the primary agency's geological mandate. A further example is provided by the development of the Federal Marine Protected Areas Strategy by the DFO, Parks Canada, and Environment Canada. The strategy requires the three federal agencies with

MPA mandates to establish a network of MPAs, to integrate information, to engage public interests, and to determine the best means with which to achieve the objectives of the MPA (Government of Canada, 2005a).

Integration among sectors is multifaceted. One example was the establishment of ONE OCEAN in 2002. This stakeholder-driven information and public education group was established in Newfoundland by leaders in the oil and gas industry and the fishing industry to resolve issues of common concern by means of informal interventions and information exchanges (ONE OCEAN Secretariat, 2003). A further example is the ongoing coordination of activities of the submarine cable industry and the fishing industry by the DFO, Maritimes Region, to address concerns and conflicts around the location of fibre optic cables.

At the international level, Canada worked with the United States and the Intergovernmental Oceanographic Commission (IOC) to develop a handbook on the identification and use of governance, socioeconomic, and ecological objectives, and related indicators (IOC, 2006). These objectives and indicators measure the effectiveness of integrated coastal and oceans management.

The precautionary approach

Canada has recognized the importance of the precautionary approach in key legislation and policy documents. The Preamble to the Oceans Act calls for a precautionary approach to marine resources management. Section 30 of the Act mandates that Canada's national oceans strategy be founded on the principles of sustainable development, integrated management, and the precautionary approach.

Other Canadian legislation also incorporates the precautionary approach. The Canadian Environmental Protection Act, 1999 (CEPA), for example, requires that administrative decisions under the Act, such as whether to allow new chemical substances into Canada, follows the precautionary principle. The 1999 Act also encourages pollution prevention approaches. The Canadian Environmental Assessment Act, 2012, section 4(2), requires officials administering the Act to exercise their powers in a manner that protects the environment and human health, and which applies the precautionary principle. However, the 2012 Act leaves considerable discretion as to which projects will be subject to environmental assessment other than projects under the authority of the Canadian Nuclear Safety Commission (CNSC) or the National Energy Board (Doelle, 2012).

Through an interdepartmental consultation process, Canada has developed guiding principles to be followed by departments or agencies in applying precaution. The Framework for the Application of Precaution in Science-based Decision Making about Risk is broad and applicable to all federal mandates (Government of Canada, 2003).

The Framework is, however, only one element that guides implementation of the precautionary approach. In oceans management, the primary guidance for the precautionary approach remains Canada's Oceans Strategy (Fisheries and Oceans Canada, 2002a) and, in more detail, the Policy and Operational Framework for Integrated Management (Fisheries and Oceans Canada, 2002b: 9). The latter specifies that priority will be given to maintaining ecosystem health and integrity, especially in the case of uncertainty. The DFO's Aquaculture Policy Framework also notes the need for aquaculture development to occur in the context of a precautionary approach (Fisheries and Oceans Canada, 2002d: Principle 2). Other DFO policies, such as the Wild Pacific Salmon Policy (Fisheries and Oceans Canada, 2005c), the New Emerging Fisheries Policy (Fisheries and Oceans Canada, 2001b), and the development of an ecosystem-based model for recovery strategy development for endangered and threatened species, all require reference to ecosystem considerations and uncertainty (Sheppard et al., 2005).

Application of the precautionary approach to fisheries has been receiving increased attention. A Policy on New Fisheries for Forage Species explicitly urges the precautionary approach to be followed, whereby biomasses of forage species used as limit reference points in management should ensure both the non-impairment of future recruitment of target species and the non-depletion of food supply for predators (Fisheries and Oceans Canada, 2009c). A Policy for Managing the Impacts of Fishing on Sensitive Benthic Areas recognizes the special need for precaution in managing fishing activities in areas of the Arctic and waters deeper than 2,000 metres, for which little or no information is available regarding the benthic features (Fisheries and Oceans Canada, 2009d). A Fishery Decision-Making Framework Incorporating the Precautionary Approach has been adopted that encourages the categorization of fish stocks into three status zones ('healthy', 'cautious', and 'critical') by means of the use of precautionary reference points (Fisheries and Oceans Canada, 2009e).

Much work remains for all levels of government in working out the application of precaution, with laws varying between strong and weak versions. Canada has adopted a strong precautionary approach to ocean dumping in a 'reverse listing' approach, under which only wastes on an acceptable list may be disposed of at sea (CEPA 1999, Schedule 5). Although the Fisheries Act, section 36(3), prohibits the deposit of deleterious substances into waters frequented by fish, discharge standards for six major industries, including pulp and paper mills and petroleum refineries, are set in regulations that do not explicitly emphasize pollution prevention and precaution. New regulations, adopted in April 2014, will allow ministerial regulations to be passed, authorizing various other deposits including from aquacultural operations.[3]

Tensions have arisen in Canada over how the precautionary principle or approach should be applied (VanderZwaag, 1998; VanderZwaag et al., 2002–03). For example, concerns have been raised with respect to the potential risks associated with escapees and the possible spread of parasites from finfish aquaculture operations. There have been calls for the removal of existing open-pen salmon farms and the prohibition of new farms (Leggatt, 2001). Instead of a prohibitory approach to precaution, governments have largely responded with various regulatory and licensing controls to mitigate the impact of fish farms, including mandatory monitoring programmes with specific intervention measures (British Columbia Ministry of Agriculture and Lands, 2005). In May 2013, the government of Nova Scotia appointed a two-person independent panel to review the adequacy of the province's existing aquaculture regulatory framework, and the panel is expected to finalize its recommendations by the end of 2014 (Government of Nova Scotia, 2013).

The Supreme Court of Canada has opened the legal door for Canadian courts to review administrative decisions in light of adherence to the precautionary principle. In the 2001 *Spraytech* case,[4] Justice L'Heureux-Dubé referred to the wide acceptance of the precautionary principle in international law and policy, and relied on the principle to help to justify a broad interpretation of provincial legislation as authorizing municipalities to regulate pesticides. She recognized that the values and principles reflected in international law may help to inform the contextual approach to statutory interpretation and judicial review (Chapman, 2002).

Canada is also party to various international bodies, working groups, regional fisheries management organizations, and international scientific organizations within which the precautionary approach continues to evolve, and implementation tools have been developed for fisheries.[5] A study of Canada's transboundary fisheries management arrangements identifies two continuing challenges in putting precaution into practice: the common overriding of precautionary scientific advice, and limited scientific information and understanding (Russell and VanderZwaag, 2010b).

At a meeting of officials from the five Arctic Council states in Nuuk, Greenland, 24–26 February 2014, Canada supported moving forward with precautionary interim measures to prevent unregulated fishing in the central Arctic Ocean beyond national jurisdiction. The meeting agreed to develop a Ministerial Declaration for signature or adoption by the five states, and the meeting looked forward to a broader process of engaging additional states in taking interim measures.[6]

Public participation and community-based management

Canadian ocean management policy supports a commitment to citizen engagement, although the strength of that commitment and the capacity to undertake it has varied with the stage of development of policy and over time. The overall objective is to create governance mechanisms that foster a greater involvement of the people most affected by decisions. The LOMAs primarily address large-scale ecosystem and economic development issues; they also provide the context for nesting a network of smaller areas or other ocean management tools, such as MPAs.

Participants in ocean and coastal management are clearly identified, including the federal government, provincial/territorial/local authorities, aboriginal organizations and communities, industry, NGOs, community groups, and the academic/science/research community. At least in the formative stage, and in keeping with the enabling (rather than directive) and collaborative nature of the Oceans Act, oceans management programmes in Canada clearly directed and enabled community involvement in the design and management of integrated management plans and MPAs (Fisheries and Oceans Canada, 2003). As the programme evolved and focused, and (arguably) resources and capacity declined, public participation became much more targeted on specific initiatives or issues,

Coastal management areas (CMAs) enable communities to play a stronger role in issues affecting their future by matching local capabilities and development priorities to the opportunities and carrying capacity of the local ecosystem. Local economic issues, such as inshore fisheries, conventional tourism and ecotourism, aquaculture sites, ports, and other transportation facilities, may all be matters considered. Following initial pilots, particularly in Quebec and the Newfoundland and Labrador regions, resource limitations, capacity to deliver on plans, and other issues led to a focus at the LOMA level. In some cases, provinces have taken on a more expansive coastal and marine role, for example in British Columbia, Newfoundland and Labrador, New Brunswick, and Nova Scotia (Ricketts and Hildebrand, 2011: 15).

Authority at the national level

In addition to leading and facilitating the development and implementation of an oceans management strategy, the Minister of Fisheries and Oceans is authorized to:

- coordinate the activities of ocean stakeholders to develop a strategy;
- develop tools and coordinate with stakeholders the development of specific plans to implement the strategy;
- develop integrated management plans for all Canadian marine waters;
- establish, as required, subnational and local bodies to assist with the implementation plans;
- establish and enforce measures or regulations associated with MPAs; and
- develop marine environmental quality guidelines.

In the October 2004 Speech from the Throne, the government of Canada made the better management of its ocean spaces and resources a government-wide priority and called for the development of:

an Oceans Action Plan by maximizing the use and development of oceans technology, establishing a network of marine protected areas, implementing integrated management plans and enhancing the enforcement of rules governing oceans and fisheries, including rules governing straddling fish stocks.

(Government of Canada, 2004a: 13)

The government also made a significant investment in strengthening initiatives related to international fisheries and oceans governance. These efforts were focused on improving compliance within the Northwest Atlantic Fisheries Organization (NAFO), creating conditions for change, and strengthening global fisheries and oceans governance.

With the endorsement of the government-wide Oceans Action Plan (OAP), seven federal departments were responsible for the delivery of specific elements of this national work plan. Their tasks ranged from international coordination, completion of ecosystem overview reports, and developing governance arrangements to seabed mapping. Table 3.2 identifies the key activities in Phase 1 of the OAP.

An Oceans Action Plan Secretariat coordinated integration of the interdepartmental efforts to deliver the OAP. In addition to housing the Secretariat, the DFO is responsible for the implementation of ocean programmes key to plan implementation (integrated management, MPAs, and marine environmental quality). While the coordination role continued in terms of tracking and evaluation of results, the original thematic subgroups did not meet once the funding was in place and the renewal efforts took a more centralized approach. In the end, funding was approved only for the Health of the Oceans (HOTO) theme in the 2007 budget. This created challenges for the implementation of the OAP, but the DFO, and other departments, reacted to the changed environment by practising adaptive management and continuing work on the Plan – particularly the legislatively mandated integrated management component through focus on the LOMAs and on advancing the science necessary for decision-making.

The national and subnational division of authority

While there is a clear federal responsibility for the protection of the marine environment and the sustainable use of marine resources, effective environmental protection and conservation require broad-based partnerships. Provincial, territorial, and local governments have roles and responsibilities with regards to oceans activities. Provinces and territories have primary responsibility for their lands, the shoreline, and specific seabed areas, and municipalities have responsibility for many of the land-based activities affecting the marine environment, such as sewage and waste disposal. Aboriginal authorities also have a key governance role to play where settled land claims include marine resource management responsibilities.

There is a strong provincial/territorial desire and a practical need for subnational engagement. In the formative years, the federal, provincial, and territorial governments collaborated under the auspices of the Canadian Council of Fisheries and Aquaculture Ministers (CCFAM), through the Oceans Task Group (Canadian Council of Fisheries and Aquaculture Ministers, 2001) and, more recently, through existing and developing regional governance mechanisms, to develop joint work plans and approaches. The development of a national framework for a network of MPAs is a good recent example of working through a federal–provincial/territorial mechanism to achieve results in a challenging and potentially conflictual environment (Government of Canada, 2011).

An Aquaculture Task Group under the CCFAM, composed of both federal and provincial representatives, facilitated discussions on clarifying and coordinating federal–provincial

Table 3.2 Key activities of Phase 1 of Canada's Oceans Action Plan 2005–07

OAP Phase 1 initiative	Key activities
International leadership, sovereignty, and security	
1 Gulf of Maine Canada–United States collaboration	Joint ecosystem overview and objectives setting for integrated management planning
2 Arctic Marine Strategic Plan	Eight countries address key issues in the circumpolar Arctic via the Working Group for the Protection of the Arctic Marine Environment (PAME) of the Arctic Council
3 International fisheries and oceans governance	Ecosystemic research, with a focus on the Grand Banks Appointment of an ambassador for fisheries conservation Strengthening global governance
Integrated management in large ocean management areas (LOMAs)	
4 Ecosystem overview and assessment reports	Review and assessment of scientific knowledge in five LOMAs
5 Ecologically and biologically significant areas (EBSAs)	Identification of areas and species requiring special management measures in LOMAs
6 Seabed mapping	Characterization of habitat in LOMAs
7 Ecosystem objectives (EOs)/Smart regulations	Ecosystem-specific EOs and possible regulatory options
8 Economic assessment and analysis	Documentation of value of activities in support of integrated management planning
9 Targeted subnational consultations	Engagement of affected and responsible parties in LOMAs and MPAs
10 Agreements with provinces, territories, and aboriginal authorities	Development of agreements on roles and responsibilities
11 Subnational management and advisory bodies	Intergovernmental and stakeholder for LOMA planning and management
Health of the oceans	
12 Oceans Act MPAs	Key MPAs designated by 2007
13 Canadian Wildlife Service marine wildlife areas	Key marine wildlife areas designated
14 National Marine Protected Area Strategy to establish a network	Implementation of federal MPA strategy to establish a network
15 Science research and advice for MPAs	Development of tools, including selection criteria for EBSAs
16 Ballast water and marine pollution regulations	Science support and completion of the regulatory process
17 Pollution prevention surveillance for sea-based sources	Increased surveillance
Oceans science and technology	
18 Oceans technology network	Support of technology networks and research priorities
19 Placentia Bay Technology Demonstration Project	Integration of real-time data to support oceans management decisions

responsibilities in relation to aquaculture (Canadian Council of Fisheries and Aquaculture Ministers, 2006). In November 2010, the CCFAM endorsed the National Aquaculture Strategic Action Plan Initiative 2011–15 (NASAPI), which sets out a strategic vision for ensuring environmental, social, and economic sustainability in the aquaculture sector (Fisheries and Oceans Canada, 2010b). The NASAPI indicates that the CCFAM Strategic Management Committee, composed of senior managers for each province/territory and the federal government, will oversee implementation and preparation of annual progress reports. The NASAPI is supplemented by five specific action plans, covering East Coast shellfish, East Coast finfish, freshwater aquaculture, West Coast shellfish, and West Coast finfish (Fisheries and Oceans Canada, 2010b: 14).

The substantial delegation of licensing authority over aquaculture by the federal government to provinces (VanderZwaag et al., 2006) has been the subject of litigation, and a shift towards federal licensing in British Columbia has occurred in the wake of litigation. On 9 February 2009, British Columbia's Supreme Court, concluding that marine finfish aquaculture is a fishery and therefore subject to exclusive federal jurisdiction, declared that much of the provincial regulatory regime was beyond the province's constitutional jurisdiction.[7] The federal government subsequently issued the Pacific Aquaculture Regulations in 2010. The Regulations establish a federal aquaculture licensing regime applicable to internal waters and the territorial sea, and allow various licensing conditions to be imposed, such as measures to minimize the escape of fish and measures to minimize the impact on fish habitat.

Management and advisory bodies are currently in place to support specific integrated management plans and MPA management plans. The bodies generally include federal and provincial/territorial representation, and they provide a forum for stakeholders, including industry, academia, NGOs, aboriginal peoples, and citizens. Their goals are to provide ongoing communication, information-sharing, and input, and to effectively inform oceans management planning processes. As integrated ocean management plans are completed, the role of these bodies shifts to targeted measures to implement the plans.

Various other federal–provincial coordination mechanisms also exist. For example, councils of federal–provincial/territorial ministers address environment, wildlife, and energy issues. Joint federal–provincial offshore petroleum boards have been established for Nova Scotia, and Newfoundland and Labrador, through accords and mirror federal–provincial legislation (Penick, 2001; Taylor and Dickey, 2001). The boards are responsible for reviewing the environmental impacts of proposed offshore hydrocarbon activities and for imposing operational conditions.[8] With the hope and expectation of a vibrant offshore marine renewable energy industry, a joint process is evolving to streamline the review of project proposals in this new field, with an initial focus on tidal energy in the Bay of Fundy.

The domestic implementation of international agreements

The effectiveness of Canada's management efforts in the Arctic, Pacific, and Atlantic oceans requires close collaboration and cooperation with adjacent nations and with other states. Canada has worked with the United States and Mexico through the Commission for Environmental Cooperation (CEC) since 1994 (CEC, 2003). Canada also participates in the Arctic Council,[9] which provides a mechanism for eight circumpolar nations to collaborate with respect to addressing Arctic marine environmental issues (Kiovurova and VanderZwaag, 2007; VanderZwaag, 2014b).

While a broad array of international environmental agreements have relevance to the oceans, this chapter briefly discusses Canada's implementation efforts and challenges under five key documents:

- the 1982 United Nations Convention on the Law of the Sea (UNCLOS);
- the 1992 Convention on Biological Diversity (CBD);
- the 1973 International Convention for the Prevention of Pollution from Ships, as amended by the Protocol of 1978 (MARPOL);
- the 1996 Protocol to the London Convention; and
- the Global Programme of Action for the Protection of the Marine Environment from Land-based Activities.

The 1982 UN Convention on the Law of the Sea

Although Canada was a leading country in negotiations for UNCLOS and signed the Convention in 1982 (McDorman, 1988: 536–8), it did not ratify the Convention until 7 November 2003, with the Convention entering into force for Canada on 7 December 2003 (McDorman, 2004). Delays in ratification were, in part, the result of deep concerns relating to high seas and straddling stock fisheries issues. Canada had already, under the Oceans Act, incorporated into domestic law its maritime zones and the jurisdictional entitlements set out in UNCLOS – namely, a 12 nautical mile territorial sea, a contiguous zone out to 24 nautical miles from the territorial sea baselines, a 200 nautical mile exclusive economic zone (EEZ), and a continental margin extending beyond the EEZ, in accordance with Article 76 of UNCLOS.

Federal funding enabled Canada to undertake the process of delimiting the outer extent of its continental shelf. Canada made a submission for areas of the Atlantic shelf to the Commission for the Limits of the Continental Shelf on 6 December 2013. It reserved the right to conduct further investigations with respect to delineation of its claim in the Arctic Ocean (Government of Canada, 2013).

A number of challenges related to implementation of UNCLOS face Canada. They include issues related to revenue-sharing responsibilities of federal and provincial authorities for oil and gas production beyond 200 nautical miles, and the scope of Canada's powers to regulate shipping as new areas become accessible in the Arctic as a result of climactic variations (VanderZwaag, 2014a).

By ratifying the 1995 UN Agreement on Straddling and Highly Migratory Fish Stocks (the 1995 UN Fish Stocks Agreement, or UNFSA 1995) in August 1999 (Government of Canada, 2004b), Canada made international fisheries reform and modernization a major priority. In May 2005, Canada hosted a major international conference on high seas fisheries governance (Government of Canada, 2005c), and Canada continued to push for more effective addressing of illegal, unreported, and unregulated (IUU) fishing (Government of Canada, 2005d). Various high seas biodiversity and fishing issues remain to be worked out, not only in Canadian ocean policy, but also globally (Russell and VanderZwaag, 2010c). For example, how might discrete high seas fish stocks be better managed, and how should the sustainable use and conservation of marine biodiversity beyond national jurisdiction be addressed? An Ad Hoc Open-ended Informal Working Group to study issues relating to the conservation and sustainable use of marine biological diversity beyond areas of national jurisdiction has met since 2006, under the auspices of the UN General Assembly (UNGA), to discuss such issues as access to marine genetic resources and a possible legal instrument to address the conservation of marine biodiversity in areas beyond national jurisdiction, but agreement has not been reached on future governance arrangements (Broggiato et al., 2014).

The 1992 Convention on Biological Diversity

Under the CBD, governments undertake to conserve and sustainably use biodiversity by means of various commitments (VanderZwaag et al., 2012: 314). Parties are required to develop national biodiversity strategies and action plans, and to integrate these into broader national plans for environment and development. The Convention also calls upon parties to develop networks of MPAs, to prevent and control the introduction of alien species, and to develop legislation and/or other regulatory provisions for the protection of threatened species and populations. Following the adoption of a Canadian Biodiversity Strategy in 1995 (Government of Canada, 1995), Canada's progress has varied in implementing the key commitments.

Canada is far from completing the establishment of an MPA network. Under the authority of the Oceans Act, section 35, only eight offshore MPAs have been established (see Figure 3.3):

- the Endeavour Hydrothermal Vents (2003) and the Bowie Seamount (2008) off British Columbia;
- the Gully (2004) off Nova Scotia;
- Basin Head (2005) off Prince Edward Island;
- Gilbert Bay (2005) off Labrador;
- Eastport (2005) off Newfoundland;
- the Musquash Estuary Marine Protected Area (2006) off New Brunswick; and
- Tarium Niryutait Marine Protected Area in the Beaufort Sea (2010).

The Aichi Biodiversity Target, adopted in 2010 under the CBD, urges the protection of 10 per cent of coastal and marine areas by 2020, but Canada appears unlikely to meet that target.

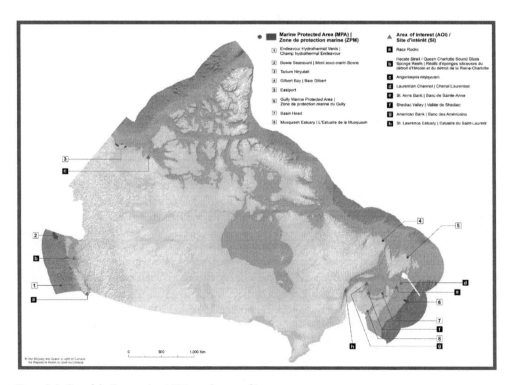

Figure 3.3 Canada's Oceans Act MPAs and areas of interest

Source: Department of Fisheries and Oceans Canada, July 2014.

It may take decades for Canada to establish a fully functioning MPA network (Commissioner of the Environment and Sustainable Development, 2012). A further eight areas of interest for MPA establishment have been identified. No national wildlife areas have been established in the EEZ, as authorized under the Canada Wildlife Act.

Only four national marine conservation areas have been established pursuant to the Canada National Marine Conservation Areas Act: the Saguenay–St Lawrence Marine Park, Quebec; Fathom Five National Marine Park, Georgian Bay, Ontario; Lake Superior National Marine Conversation Park of Canada; and Gwaii Haanas National Marine Conservation Reserve, British Columbia (Parks Canada, undated*b*). With the addition of more than 700 provincial and territorial MPAs (Government of Canada, 2010), Canada has only formally protected about 1 per cent of its oceans estate (Commissioner of the Environment and Sustainable Development, 2012). A 2014 analysis showed Canada lagging behind all major marine estate nations, with 1.3 per cent formal protection versus 11.6 per cent for Russia, 16 per cent for France and Britain, and more than 30 per cent for the United States and Australia (CPAWS, 2014). While these figures are well below targets, other area-based measures, such as fisheries closures, do provide a level of protection within an integrated management context. The current state of the art – and argument – is lacking in criteria and agreement on what should count as effective protection for these 'informal' measures. Non-regulatory measures have the distinct advantage of being relatively easy, cheap, and quick to implement, but many argue that their ease of implementation makes them equally vulnerable to change or removal based on economic or other criteria.

Progress on regulations is slow, and has become more complex and time-consuming in an era in which Canadian governments are viewing much regulation as a burden. Marine protection regulation-making is caught up in this environment at the federal level. In 2011, a Red Tape Reduction Commission consulted across the country and, in 2012, released an action plan to reduce red tape. Among the measures affecting all new regulations are a 'one for one' rule requiring the removal of one regulation for every new one to offset the new administrative burden imposed, a small business lens, and a number of reporting requirements. While the goal is logical and laudable, the impact in the implementation phase has been delays, while process and tools are established, along with time and funding resources to carry out new procedures (such as economic analyses of burden) (President of the Treasury Board, 2012).

With respect to the introduction of new alien aquatic species via ballast water in ships, Canada initially relied upon voluntary measures set out in the Guidelines for the Control of Ballast Water Discharge from Ships in Waters under Canadian Jurisdiction (Transport Canada, 2001). However, in light of the 2004 International Convention for the Control and Management of Ships' Ballast Water and Sediments, Canada has issued binding Ballast Water Control and Management Regulations. The Regulations prescribe management measures for ballast water, requiring exchange at least 200 nautical miles from shore and in water depths greater than 2,000 metres before entering Canadian waters. Emergency ballast exchange within Canadian waters is also restricted to specific zones. These zones are identified based on lowest ecological risk.

In December 2002, Canada enacted the Species at Risk Act (SARA). The Act is part of the government of Canada's three-pronged strategy for the protection of wildlife species at risk, which also includes commitments under the 1996 national Accord for the Protection of Species at Risk and activities under the Habitat Stewardship Programme for Species at Risk. The 2002 Act implements key elements of the Canadian Biodiversity Strategy. It requires recovery strategies and action plans to be prepared for listed endangered and threatened species, and management plans, for species of special concern. The Act formally recognizes the role of the independent advisory Committee on the Status of Endangered Wildlife in Canada (COSEWIC) in assessing species at risk. The Act applies to all federal lands in Canada, all wildlife species listed as being at risk, and their critical habitat. It also puts in place various prohibitions, such as prohibiting persons from killing, harming, harassing, or taking an individual of a listed endangered or threatened species, and from damaging or destroying the residence of one or more individuals of a listed endangered or threatened species. The need to better define, with scientific rigour, key provisions of the Act relating to critical marine habitat and residences, as well as the shared accountability between federal ministers, and between federal and provincial ministers, makes it difficult to fully assess the effectiveness of the statute.

Various implementation difficulties also stand out (VanderZwaag and Hutchings, 2005). Listing of some marine fish species has been a challenge, since listing under SARA involves a political decision rather than scientific determination (Findlay et al., 2009; Waples et al., 2013). For example, COSEWIC has listed as endangered Cultus Lake and Sakinaw Lake sockeye salmon populations, Interior Fraser River coho salmon, the Newfoundland and Labrador population of Atlantic cod, and the porbeagle shark, and has categorized as threatened the Laurentian North population of Atlantic cod (COSEWIC, 2007). Because of the potential social and economic impacts of SARA listing for these populations, the Canadian government chose against listing.[10] Delays in developing recovery strategies and action plans (VanderZwaag et al., 2011; Lemkow and VanderZwaag, 2014), and limited identification of critical habitats have also been problematic (Mooers et al., 2010; David Suzuki Foundation et al., 2009; Taylor and Pinkus, 2013).

The CBD's Programme of Work on Marine and Coastal Biological Diversity includes consideration of protected areas beyond national jurisdiction (CoP, 2004). High seas issues, particularly as they relate to ecosystem health, are of interest to Canada. Canada is working with existing governance bodies and their scientific advisers to integrate scientific knowledge and expertise to provide best available scientific advice to inform decisions. For example, in December 2005, Canada hosted an international scientific experts' workshop to review and assess ecologically based criteria for the identification of areas and/or resources that are ecologically and biologically significant and may require special management measures, including protected area status in the high seas (IUCN, 2006). The aim of the workshop was to provide integrated advice to authorities such as the CBD, the United Nations Food and Agriculture Organization (FAO), and the International Maritime Organization (IMO), for their consideration.

MARPOL

Canada has formally accepted the six annexes of MARPOL dealing with oil pollution (Annex I), noxious liquid substances carried in bulk (Annex II), harmful substances carried in packaged form (Annex III), sewage (Annex IV), garbage (Annex V), and air pollution (Annex VI). The Vessel Pollution and Dangerous Chemicals Regulations, issued under the Canada Shipping Act, 2001, bring Canada into line with MARPOL standards for most vessel-source pollutants. Prevention of pollution from harmful substances in packaged form continues to be addressed by the Transportation of Dangerous Goods Regulations.

Canada has chosen to apply stricter vessel-source pollution control standards for its Arctic waters. Pursuant to the Arctic Waters Pollution Prevention Act, passed in 1970, Canada has imposed special pollution discharge and other restrictions for vessels operating in the Arctic. For example, oil deposits from ships are generally prohibited with only a few exceptions, such as when resulting from stranding or collision and from engine exhaust (under the Arctic Shipping Pollution Prevention Regulations). Effective 1 August 2009, Canada extended the special shipping standards out to 200 nautical miles in the Arctic in light of Article 234 of UNCLOS.[11] Article 234 grants coastal states special legislative and enforcement jurisdiction over vessels navigating in ice-covered waters (McRae and Goundrey, 1982; Bartenstein, 2011).

The 1996 Protocol to the 1972 London Convention

Becoming the tenth country to accede to the 1996 Protocol (Government of Canada, 2001), which takes a precautionary approach to ocean disposal, Canada ensured its implementation through the provisions of the Canadian Environmental Protection Act, 1999. The Act adopts a 'safe list' approach by allowing ocean disposal of only a limited list of wastes listed in Schedule 5; any disposal must be in accordance with the conditions of a Canadian permit. Before issuing an ocean disposal permit, the Minister of Environment is required to subject the application to a waste assessment process, set out in Schedule 6 of the Act, which, among other things, requires refusal of a permit if reuse, recycling, or treatments of the waste are practical options.

The Global Programme of Action for the Protection of the Marine Environment from Land-based Activities

Canada became the first country to develop a National Programme of Action for the Protection of the Marine Environment from Land-based Activities (NPA) in 2000 (NPA Advisory Committee, 2000). The NPA sets national priorities for addressing land-based marine pollution and activities through a 'high', 'medium', and 'low' ranking approach. Listed as high-contaminant priorities are sewage and persistent organic pollutants. Responding to shoreline construction and alteration, and wetland and salt marsh alteration, are also listed as high priorities. In separate chapters for four main coastal regions (the Pacific, the Arctic, Southern Quebec/St Lawrence, and the Atlantic), the NPA also describes regional problems, priorities, and needed actions. A federal/provincial/territorial committee, established in 1996 soon after the Global Programme of Action for the Protection of the Marine Environment from Land-based Activities (GPA) Washington Conference and co-chaired by Environment Canada and the DFO, was responsible for the development and implementation of the NPA.

Tracking implementation activities is difficult because of the numerous sources of land-based marine pollution, the multiple jurisdictions and programmes involved along Canada's extensive coastlines (Wells, 2002), and the lack of a dedicated funding for GPA implementation. Canada's report to the 2001 Intergovernmental Review Meeting on Implementation of the GPA included an annex highlighting more than ninety key programmes, within government, NGOs, and communities that address the goals and priorities of the GPA (NPA Advisory Committee, 2001). For example, the collaborative development, by federal, provincial, and local authorities, of integrated management processes and plans at the scale of the coastal management area (CMA) contributed directly to the implementation of the NPA. In a national report on GPA implementation prepared for the Second Intergovernmental Review Meeting in October 2006, Canada described various other projects contributing to GPA implementation including a technology investigation for enhancing municipal wastewater treatment in Arctic climates and an inventory of land-based sources of pollution in the Hudson Bay watershed (Environment Canada, 2006).

Canada has contributed to the GPA by advancing GPA activities at the regional level. The Regional Programme of Action for the Protection of the Arctic Marine Environment from Land-based Activities (RPA), adopted by Arctic Council ministers in 1998, established two high priorities for regional action – namely, addressing persistent organic pollutants and heavy metals – and identified pollution hot spots in the Russian Federation (Arctic Council, 1999). The Arctic Council's Protection of the Arctic Marine Environment (PAME) Working Group updated the RPA, under the lead of Canada and Iceland (PAME, 2008), with revisions approved by Arctic Council ministers in April 2009 (Arctic Council, 2009).

Interest in the GPA appears to have waned over the years, and Canada has struggled to address sewage pollution in particular. With over 150 billion litres of raw or undertreated sewage discharged annually, wastewater comprises Canada's largest source of water pollution (Environment Canada, undated). In 2012, Canada passed the Wastewater Systems Effluent Regulations under the Fisheries Act in an effort to address this concern, but the Regulations fall short of comprising a regulatory framework that could facilitate adequate wastewater management for many reasons.

- The Regulations do not apply to some of the largest geographic regions in Canada: the Northwest Territories, Nunavut and northern Newfoundland and Labrador, and Quebec.
- The Regulations do not apply to commercial, industrial, or institutional wastewater systems that contain under 50 per cent blackwater by volume, or to pulp and paper mills.
- The Regulations apply only to wastewater systems that have the capacity to collect 100 cubic metres or more of wastewater daily.
- Finally, the timelines for implementation of the proscribed effluent quality standards may be delayed owing to temporary and transitional authorizations that allow the continued discharge of deleterious effluents. Transitional authorizations may be granted based on degree of assessed risk and can extend until 2040, while temporary authorizations, which allow the discharge of acutely lethal concentrations of un-ionized ammonia, may be granted for three-year periods. These timelines have been criticized for providing inadequate environmental and public protection (Bowman, 2010).

Enforcement

While each federal statute pertaining to oceans has its own set of regulations, enforcement procedures, penalties, and fines, section 35 of the Oceans Act provides the Minister of Fisheries and Oceans with the authority to develop specific regulations pertaining to the designation of MPAs and the prescription of measures needed to achieve the conservation objectives of the MPA. Section 37 of the Act provides for penalties if prescribed measures are contravened, with persons liable to a fine not exceeding CAN$100,000 on summary conviction or up to CAN$500,000 for an indictable offence. The Act also provides the authority to make regulations prescribing marine environmental quality requirements and standards. In practice, this is intended to give effect to those ecosystem objectives that require the force of regulation.

With respect to enforcement and surveillance, the approach adopted by the Canadian government is to multi-task pollution prevention among fishery officers and other federal and provincial enforcement officers active in the geographic area in which the oceans conservation or management measure is being applied. Notwithstanding this, enforcement is only one of many measures on the compliance continuum. Consequently, substantial effort is dedicated in both the integrated management and MPA processes to engaging stakeholders and involving them in advisory and management bodies. Better understanding and 'ownership' of the management plans and associated regulatory measures provides support and potentially reduces the more costly surveillance and enforcement efforts.

Regulations developed under the Oceans Act include those to designate eight current MPAs and, to date, no contraventions have been detected, although there have been activities and proposals that necessitated management actions (such as the Musquash Estuary proposal for a provincial aquaculture licence outside the boundary of the MPA, but with potential risk through current flows). Regulations focused on the mitigation of seismic sound in the marine environment were also under development. The DFO held targeted public consultations in 2005 and

2006, and revised its draft Statement of Canadian Practice Respecting the Mitigation of Seismic Sound in the Marine Environment (Fisheries and Oceans Canada, 2006). The Statement of Canadian Practice has now been given effect under the authority of the Newfoundland and Labrador and Nova Scotia Petroleum Boards for oil and gas applications, but proposed Oceans Act regulations for non–oil-and-gas seismic surveys did not advance.

Canada has been a leader in developing legislative provisions supportive of effective enforcement and creative sentencing options for those convicted of environmental and fisheries offences. Most federal and provincial statutes provide for strict liability offences where the Crown does not have to show fault (intentional, reckless, or negligent behaviour) by the offender, but only a guilty act, such as a deleterious deposit into waters frequented by fish. Many statutes allow judges to be innovative in issuing sentencing orders beyond the traditional sanctions of fines or imprisonment. For example, section 79(2) of the Fisheries Act allows courts to impose various requirements on offenders, including prohibiting activities that may continue or repeat the offence, directing remedial and avoidance measures, directing convicted persons to publish the facts relating to the offence, requiring persons to pay governmental costs of remedial or preventative actions, ordering persons to perform community service, directing persons to contribute funds for the purpose of promoting fish habitat conservation and fisheries management, and requiring persons to comply with any other conditions for securing the person's good conduct.

A more recent legislative effort to bolster enforcement in the oceans sector was aimed at more effectively countering ship-source pollution, especially in contravention of MARPOL standards, which has had damaging consequences to migratory seabirds. The 2005 amendments to the Migratory Birds Conventions Act, 1994 and the Canadian Environmental Protection Act, 1999 expanded the scope of persons who may be held responsible for offences, extended the jurisdiction of Canadian courts to cover infringements in the EEZ, and substantially increased penalties.[12]

Research and education

Canada's Oceans Strategy emphasizes the need to base decisions on sound science and to address uncertainties in our knowledge base so that management actions can be adjusted as new scientific information becomes available (Fisheries and Oceans Canada, 2002a: 12–13, 22). The importance given to improving our understanding of marine ecosystems, their properties and critical functions, as well as the impacts of single and multiple activities on these parameters, has resulted in a shift in the orientation and organizational structure of the research and scientific support services within the DFO and by other service providers. Increased partnerships with academia, international scientific organizations, and sister agencies in other governments have facilitated the development of tools for the application of ecosystem-based considerations of ocean issues and the building of a rigorous scientific advisory peer review process designed to support all ocean managers.

Further, to foster the scientific understanding necessary to support ocean management policy, and to bridge the gap between natural and social sciences, an Ocean Management Research Network (OMRN) was established as a joint initiative between the Social Science and Humanities Research Council (SSHRC) and the DFO in 2001. The OMRN created a national network of interdisciplinary and cross-sectoral research working groups to develop knowledge and best practices for sustainable oceans management (OMRN, 2004). The OMRN continued as a loose network of academics and professionals after the end of its government funding in 2008.

Another network with a science focus was established in 2008: the Canadian Healthy Oceans Network (CHONe) (CHONe, undated). As a university–government partnership, CHONe focuses on conservation and sustainable use of Canada's oceans under three themes: marine biodiversity, population connectivity, and ecosystem function. The creation of this network aligned with and supported the gradual shift of DFO Science's focus from single species to the broader ecosystem within which species live. Both of these networks took substantial progress in bridging the oft-noted gaps between academic enquiry and the more operational research needs in government.

The commitment to advance ocean science and technology was anchored in Canada's Oceans Action Plan (OAP), with the objectives to improve information-sharing through connecting information networks, to promote innovation and new technologies by supporting prototype development and targeted research and development, and to enhance commercialization through demonstration projects in the priority LOMAs (Fisheries and Oceans Canada, 2005a).

In the last five years, DFO Oceans became a significant client of the DFO Science Canadian Science Advisory Secretariat (CSAS), which coordinates the peer review of science issues for the DFO. As the focus of the oceans programme and funding narrowed, significant effort was invested to fill gaps in knowledge related to identification of ecologically and biologically significant areas and ecologically significant species and science-based conservation objectives, as well as specific studies on species of interest or features, such as corals and sponges. Oceans also was a significant DFO Science client in efforts to better understand the impacts of specific human use activities (such as marine renewable energy and marine transportation) on ecosystem components and functions. All of this effort was aimed at gaining a better understanding of risks, so as to better focus resources and choose appropriate conservation tools and management actions, and ultimately to implement integrated management and marine conservation under the Oceans Act.

Financing

As was noted previously, no new funds were provided to implement the Oceans Act or Canada's Oceans Strategy (2002). Until the federal government's approval of the Oceans Action Plan (OAP) in 2005, funding for implementation of the national ocean management approach had been achieved through reallocation of funds within the DFO. The programmes delivered in the six administrative regions of the DFO have been dependant on transfers of national funds on an annual basis. Since 1997, the Department has redirected approximately CAN$140 million to fund the activities in support of the Oceans Strategy.

The OAP, however, provided the first new central funding of CAN$28.5 million over two years across seven departments (Fisheries and Oceans Canada, 2005a). The 2007 federal budget proposed funding of CAN$19 million over two years to help to clean and protect Canada's oceans, and to support increased water pollution prevention, surveillance, and enforcement along its coasts (Department of Finance Canada, 2007). Once approved by Cabinet and Treasury Board, the Health of the Oceans (HOTO) commitment grew from CAN$19 million over two years to CAN$61.5 million over five years, projected through 2011–12 and allocated to five federal departments or agencies as follows:

- Transport Canada – CAN$23.85 million;
- Fisheries and Oceans Canada – CAN$23.173 million;
- Environment Canada – CAN$8 million;
- Parks Canada – CAN$6.25 million; and
- Indian and Northern Affairs Canada – CAN$0.175 million.

Budget 2012 provided a full one-year extension of HOTO, followed by a smaller conservation-focused extension in Budget 2013 and the recently announced CAN$37 million National Conservation Plan funding for five years to 2018–19 to be shared among several departments.

The broader financial picture for the government of Canada and DFO in the years following the establishment of HOTO has had a significant, and growing, impact on the implementation of the Oceans Act. In 2007, the federal government initiated strategic reviews on a four-year cycle of every federal organization's budget, with the goal of realizing savings to reinvest in other priorities and to address the federal budget deficit. As the programme evolved, and the economy suffered through the global financial crisis of 2008–09, savings that were initially, to some extent, reinvested in departments were increasingly used for fiscal stimulus and other priorities, and in the third year (the 2010 budget), the documents stopped identifying reinvestments of strategic review savings as the stimulus Economic Action Plan became *the* priority. The strategic review arrived at the DFO in 2009 (year three) and the results were announced in the February 2010 budget. The DFO was to implement cuts growing over three years to CAN$58.8 million (or less than 5 per cent of the global budget). Decisions were made to backload the cuts with more than CAN$40 million coming in the final year 2013–14. In 2010, strategic and operating reviews were added, so that by 2012–13 the DFO had significant cuts to absorb. The Department began rolling out major changes, including a long-anticipated rewrite of the Fisheries Act, new delivery mechanisms for fisheries licensing, significant changes to the habitat programme, and financial and human resource pressures in most areas of the Department.

For the oceans programme, there was some stability (for example the five-year HOTO funding to 2012), with heavy doses of uncertainty. Some of the uncertainty was generated by several senior management changes, with the programme shifted in three reorganizations in six years. Corporate reorganizations and budget cuts often trigger unanticipated consequences, for example in some cases cuts in other programmes through strategic review to 'focus on their core role' eliminated or reduced activities that were key to processes in other departments. In the case of MPA regulations, assessments required for the regulatory process were not available for a time, then became subject to cost recovery, creating budget issues and introducing delays. In addition, the Treasury Board Secretariat, as the overseer of the regulatory process, revised the regulatory process, for excellent reasons, to introduce more rigour and to ensure that regulations were indeed the right tool for the issue being addressed. For Oceans Act MPAs, this meant timing and budget issues as new processes were laid out and new requirements added. Finally, in 2011–12, the government moved aggressively on its red tape reduction initiative, which again, during the transition, created delays and uncertainty as departments adapted to the new tests (such as the requirement at the departmental level to remove one regulation for every new one adopted). From its peak in the mid-2000s, it appears that the oceans programme in 2014–15 has lost about a third of its authorized complement of people – dropping from roughly 115–120 full-time equivalents (FTEs) to somewhere in the high 70s – and the A-base budget is down at least 30 per cent (from CAN$15 million to CAN$9.2 million).

Implementation and the long-term outlook

There are many challenges in implementing an oceans policy that seeks integration of the planning and management of ocean activities among various levels of government. Reorienting single-species, single-activity decisions to decisions focused on the sustainability of the ecosystem, and therefore of the industries and traditions dependent upon ocean resources, is a major hurdle. Moving from the theoretical level to the application of concepts, such as ecosystem-based management and precaution, in day-to-day governance decisions is fraught with science challenges, as well as concerns about changes. Perhaps the greatest challenges are implementing the institutional changes and building the relationships and capacities essential to achieving integration.

Implementation review efforts

Various governmental and independent review efforts have occurred to assess Oceans Act implementation, and the long-term outlook for ocean law and policy development in Canada remains quite uncertain. The Oceans Act, section 52, required a review of its administration by Parliament within three years after its enactment. The October 2001 Report on the Oceans Act by the Standing Committee on Fisheries and Oceans concluded that the Act was fundamentally sound. It made twelve recommendations, including that a performance-based reporting system be established and reports provided to Parliament on an annual basis. A further recommendation called for the preparation of a 'state of the ocean' report on a periodic basis to track the health of the oceans, ocean communities, and related ocean industries (Standing Committee on Fisheries and Oceans, 2001).

On 29 September 2005, the Commissioner of the Environment and Sustainable Development reported to the House of Commons on Oceans Act implementation and issued key recommendations. Recommendations directed to the DFO included:

- recognizing and managing Canada's Oceans Action Plan (OAP) as a government *horizontal initiative*;
- finalizing and implementing operational guidance for integrated management planning, including MPAs, in the five priority ocean areas;
- planning and managing resources to ensure commitments and targets set out in departmental documents, such as the annual report on plans and priorities, are met, as well as the 2002 World Summit on Sustainable Development (WSSD) oceans commitments;
- finalizing and implementing an accountability framework for its management activities; and
- improving communications to the public, including periodic information on the state of the oceans (Commissioner of the Environment and Sustainable Development, 2005).

The recommendations were addressed by the government of Canada in Phase 1 of the OAP (Fisheries and Oceans Canada, 2005a), released in May 2005. The recommendations continued to be addressed, in part, through the horizontal management of the Health of the Oceans (HOTO) initiative involving five federal organizations, as well as the ongoing convening of interdepartmental oceans committees (ICOs) at the director general, assistant deputy minister, and deputy minister levels.

Federal departments are required to provide a performance report to Parliament as part of their annual report on plans and priorities. Information on programmes, their budgets, plans, and expected results for integrated management, MPAs, and other ocean management activities are provided for public scrutiny (Fisheries and Oceans Canada, 2004a).

The DFO has developed a Results-based Management and Accountability Framework to monitor the progress and implementation of the national ocean policy (Commissioner of the Environment and Sustainable Development, 2005). This framework sets out performance measurement goals and indicators to assess departmental progress. It was designed to track how the DFO uses resources to undertake activities in order to affect the desired results and achieve stated outcomes.

The DFO has completed reviews of the Eastern Scotian Shelf Integrated Management (ESSIM) initiative and the interdepartmental HOTO initiative. The ESSIM evaluation highlighted the need for political support for integrated planning and the decline in enthusiasm when stakeholders do not perceive a link between high-level objectives and concrete management measures (McCuaig and Herbert, 2013). The HOTO evaluation emphasized the short-term nature of HOTO funding and the need for longer-term coordinated policy direction to address ocean management issues (Fisheries and Oceans Canada, 2012).

A 2012 Expert Panel Report by the Royal Society of Canada on Sustaining Canada's Marine Biodiversity has offered further critiques of Canada's ocean policy implementation (Hutchings et al., 2012). The report concluded that completed integrated management plans for LOMA pilot projects stand out for their aspirational generality and for setting overall goals, objectives, and management strategies, but *not* for marine spatial planning. Major offshore areas have remained outside the integrated planning ambit, including the Bay of Fundy, the Gulf of Maine, and the central and eastern Arctic. Coastal area management planning initiatives have been relatively few and hard to track. The primary focus of the DFO has been on LOMAs, as exemplified by the ESSIM integrated planning process, which chose to address management issues beyond the 12 nautical mile territorial sea.

Outlook

Funding for Phase 1 of the OAP, renewed funding in the 2007 federal budget, and interest shown by other levels of governments to develop collaborative governance arrangements and processes augured well for short-term implementation of Canada's Oceans Strategy. Integrated management processes were undertaken in five LOMAs, with plans completed for four regions, and eight MPAs have been designated. Work is progressing towards the designation of eight additional candidate MPAs. Health of the oceans funding and, more recently, National Conservation Plan funds secured support for the integrated management process with respect to the advancement of federal and national MPA networks.

The Canadian integrated management approach is clearly in a process of transition. With the substantial completion of LOMA pilot projects, Canada is moving towards a bioregional planning agenda. The National Framework for Canada's Network of Marine Protected Areas proposes the formulation of bioregional planning teams in twelve oceanic bioregions and the Great Lakes, with planning linked to LOMA processes where they exist (Government of Canada, 2011). The Framework promises to facilitate the development of bioregional MPA network action plans.

However, the future of integrated coastal/ocean planning appears quite uncertain. Exactly how bioregional planning will proceed remains unclear. Marine spatial planning is not referenced within Canadian ocean-related policy or legislation (Hall et al., 2011). Prime Minister Harper announced a new National Conservation Plan on 15 May 2014, but the announcement was quite thin on details (Prime Minister of Canada, 2014b). The Plan promises to commit CAN$252 million over the next five years to support a series of conservation initiatives that facilitate meeting the 2020 CBD biodiversity targets of protecting 17 per cent of land and inland waters, and 10 per cent of marine and coastal areas. Included in the funding areas are CAN$100 million for the Nature Conservancy of Canada to protect sensitive lands, and CAN$37 million for marine and coastal conservation.

Lessons learned

While Canada, like other countries, is still learning in the complex field of ocean policy and governance, nine major lessons do stand out.

1   *Enabling ocean management legislation provides a useful guide.*

Canada's Oceans Act has provided an important framework for directing how human uses of Canada's oceans may be better managed. The Act has defined Canada's maritime zones, and recognized the attendant rights and responsibilities within those zones, in conformity with UNCLOS. The Act has clearly designated the DFO as the lead federal authority for developing integrated management plans for marine areas, for setting the environmental quality standards that must be met, and for designating/establishing MPAs. The Act facilitated the development of a broad policy framework and a government-wide plan of action. Internationally, Canada enjoyed traction as a model for oceans law and policy development. It is important to note that the enabling nature of the legislation (more 'may' than 'shall') fails to create accountabilities for effective delivery beyond those granted to the Minister of Fisheries and Oceans. This can work in a collaborative environment, but not so well when resources and political focus are tight.

2   *Passing an Oceans Act should not detract from the need for other legislative and regulatory reforms.*

While Canada's Oceans Act has substantially advanced ocean governance initiatives and arrangements, there remain several sectoral laws that do not yet reflect the modern ocean governance commitment of the government of Canada. For example, Canada's Fisheries Act, dating back to 1868, has yet to be 'modernized' to reflect modern ocean governance principles, although the policies guiding its application have evolved over time.

In response to this problem, the Minister of Fisheries and Oceans introduced two proposed revisions of the Fisheries Act: Bill C-45, in December 2006, and Bill C-32, in November 2007. The proposed revisions explicitly supported the application of the principles of sustainable development, including the ecosystem-based approach, precautionary approach, and increased stakeholder participation in decision-making. However, both Bills died on the Order Paper when the parliamentary sessions were prorogued (that is, formally ended). At time of writing, a new Fisheries Bill had not been reintroduced.

3   *Including sustainable development principles in national oceans-related legislation is very important.*

While principles, by their nature, tend to be general and open to various interpretations, principles such as integration, precaution, and the ecosystem-based approach do serve useful functions (Rothwell and VanderZwaag, 2006). At the very least, principles invite decision-makers and others to rethink traditional management approaches. Principles may be considered part of the search for 'good governance'. They facilitate discussions and debate within government bureaucracies, but also among the broader public. The principles and aspirations expressed in the Preamble to the 1996 Act were useful in starting to define an oceans strategy.

4   *Developing integrated management plans and establishing MPAs takes time.*

Building the relationships and capacity required to bring participants at all levels to the table takes time and requires skilled negotiation. The special relationship of the government with aboriginal peoples must be considered and managed in the development of MPAs and integrated management planning processes. Both of these processes involve multiple steps, all of them requiring, to a greater or lesser extent, the involvement of other government authorities and meaningful consultation with affected parties.

In going forward, one of the major tests will be the management of public expectations for timely and focused intervention to address issues of immediate concern to them. User conflicts and environmental degradation have evolved over years. To change human relationships and behaviours and to detect positive responses in the marine environment will likely require decades.

5   *Federated states face particular challenges in achieving integrated coastal/ocean management.*

Being a country with eight provinces and three territories fronting ocean areas, Canada faces special challenges in achieving integrated coastal/ocean management. Canada's Oceans Act, section 28, recognizes the constitutional limitations of the federal government by limiting integrated management planning to marine waters and not directly encompassing provincial coastal lands and rivers. The requirement under the Act for the federal government to collaborate with other levels of government seeks to draw in other government authorities as partners in the integrated management process, while respecting the current division of powers. The extent to which integrated management planning initiatives will influence provincial laws, policies, and interests remains to be seen.

The complexity of shared federal–provincial responsibilities may also affect the pace of legislative and regulatory developments. For example, development and enactment of Canada's Species at Risk Act was prolonged in part as a result of the jurisdictional complexities and sensitivities surrounding species at risk. Several other ocean-related activities, such as aquaculture management, involve both federal and provincial authorities, and therefore present significant challenges because of federal–provincial jurisdictional issues.

The relationship of the federal government with provinces and territories continues to develop, and much of the success of integrated planning will depend on continuing progress. It is through these inter-jurisdictional relationships, and relationships between regulators, that an existing fragmented set of laws and policies will be coordinated in the domestic management of oceans activities.

6   *Limited marine ecosystem understanding continues to be a major challenge.*

While Canada is firmly committed to implementing an ecosystem-based approach to management, including fisheries management, the limited scientific data and understanding of complex marine ecosystems remains a challenge. Canada's Oceans Action Plan (OAP) recognized that ecosystem-based science needed to be strengthened, and one of the pillars of the plan was to enhance ocean science and technology (Fisheries and Oceans Canada, 2005a). Further, the relationship between human uses, including cultural uses, and the value of marine ecosystem goods services remains a huge challenge that will continue to limit good policy and operational decisions.

7   *Incentives are critical for changes in governance and accountability.*

Ecosystem-based integrated management of oceans requires changes in governance both within the federal agencies and between levels of government. Until implementation of the OAP was initiated, neither the necessary inter-agency structures, nor other departmental accountabilities, were in place. During the first years of implementation of Canada's Oceans Act and oceans policy, both accountability and financing (internal reallocation) were located with only one department (the DFO). This situation did not support a coordinated federal approach.

As recommended in the 2005 Report of the Commissioner of the Environment and Sustainable Development, a horizontal, all-of-government approach is a fundamental requirement for success in bringing all federal regulators to the table. Subnational authorities (provincial, territorial, aboriginal) and stakeholders may require capacity-building and incentives to participate in a national programme. Financial investment is required to build integrated management and may be an important incentive both at the federal and subnational levels.

8 *Demonstrating progress is crucial, particularly in long-term planning.*

For integrated oceans management planning, there is a clear need to demonstrate progress (value for money), especially to political masters. This is a challenge when time frames are long, when progress depends on others, and when resources are limited. Auditors and external reviewers can be significant allies. For the Oceans Act, statutory, evaluation, and auditor (such as the Commissioner of Environment and Sustainable Development) reviews have provided an important platform from which to tell the oceans 'story' and, in a number of cases, have provided recommendations that supported receiving new resources to achieve mandated tasks.

9 *Enacting bare bones integrated ocean planning legislation may carry implementation risks.*

Leaving planning timelines, area choices, and procedures to bureaucratic discretion may contribute to considerable uncertainty and variability in integrated planning practice. Lack of a specific marine spatial planning mandate and clear provisions on how plans will be given legal effect may encourage general and aspirational planning results.

Conclusion

Experience gained since the promulgation of the Oceans Act and adoption of Canada's Oceans Strategy (Fisheries and Oceans Canada, 2002a) as the federal policy framework, has highlighted the need for clear implementation strategies. Efforts will need to continue on advancing:

- intersectoral and inter–departmental buy–in;
- intergovernmental (federal–provincial-territorial) relationships (Canadian Council of Fisheries and Aquaculture Ministers and federal–provincial-territorial agreements);
- increased collaboration internationally to address issues of common concern; and
- clear guidelines for the interpretation and implementation of ecosystem–based management.

Implementing a results–based system of monitoring and reporting for government-wide initiatives is daunting, with ministerial accountabilities continuing to be linked to single activities, as opposed to the horizontal target of integrated oceans management. Generating the political will, profile, and resources with which to support a robust policy and effective implementation of the integrated approach continue to be long–term goals.

Various strategic next steps are possible. They include introduction of a new Oceans Action Plan, updating the existing national Oceans Strategy and issuance of a detailed bioregional planning document. One thing is clear: Canada's quest for integrated coastal/ocean management is far from over.

Notes

★ This chapter, intended to be accurate as of May 2014, is an updated and revised version of a paper published in Koivurova et al. (2009: 35–87). The research assistance of Emily Mason, JD Candidate Schulich School of Law, is gratefully acknowledged, as is the research support of the Social Sciences and Humanities Research Council of Canada (SSHRC). The authors also wish to thank Fisheries and Oceans staff at National Headquarters and in Regions who provided comments on drafts and factual information.

★★ The views expressed in this paper do not necessarily reflect the views of the International Joint Commission or the Government of Canada.

1 *Reference Re. Ownership of the Bed of the Strait of Georgia* [1984] 1 SCR 388.

2 By the Jobs, Growth and Long-term Prosperity Act, SC 2012, c. 19, s. 142.

3 Regulations Establishing Conditions for Making Regulations under Section 36 (s. 2) of the Fisheries Act, SOR/2014-91.
4 Case No. 114957 *Canada* Ltée *(Spraytech, Société d'arrosage) v. Hudson (Ville)* [2001] 2 SCR 241.
5 See, e.g., FAO (1996).
6 Chairman's Statement, Meeting on Arctic Fisheries, Nuuk, Greenland, 24–26 February 2014 (copy on file with the authors).
7 *Morton v. British Columbia (Minister of Agriculture & Lands* (2009) 92 BCLR 4th 314, 42 CELR (3d) 79 (BCSC).
8 More information on the Canada–Nova Scotia Offshore Petroleum Board is available online at http://www.cnsopb.ns.ca/ and for the Canada–Newfoundland and Labrador Offshore Petroleum Board at http://www.cnlopb.ca/
9 Information on the activities of the Arctic Council is available online at http://www.arctic-council.org
10 See Orders Giving Notice of Decisions not to add Certain Species to the List of Endangered Species, SI/2005-2 (decision not to list Cultus and Sakinaw salmon), SI/2006-61 (decision not to list Newfoundland and Labrador population of Atlantic cod and Interior Fraser population of coho salmon), and SI/2006-110 (decision not to list the porbeagle shark). Other tools, such as government programmes and initiatives by non-governmental organizations (NGOs) and industry, are expected to protect and assist recovery of these non-listed species.
11 An Act to amend the Arctic Waters Pollution Prevention Act, SC 2009, c. 11.
12 Persons responsible for depositing a substance harmful to migratory birds not authorized under the Canada Shipping Act, 2001, may include masters, chief engineers, owners, and operators of a vessel, and directors or officers of a corporation that is the owner or operator of a vessel (An Act to amend the Migratory Birds Convention Act, 1994 and the Canadian Environmental Protection Act, 1999, SC 2005, c. 23, s. 5(4)). Persons or vessels contravening provisions of the Migratory Birds Convention Act, 1994, or its regulations are subject to a fine of up to CAN$1 million or to imprisonment for a term up to three years or both upon conviction by indictment, and to a fine of not more than CAN$300,000 or to imprisonment for a time of up to six months or to both upon summary conviction. Persons and vessels may be convicted for a separate offence for each day on which the offence is committed or continued.

References

Books, articles, and other documents

Arctic Council (1999) *Arctic Council Regional Programme of Action for the Protection of the Arctic Marine Environment from Land-based Activities*, Ottowa, ON: Minister of Public Works and Government Services Canada.
Arctic Council (2009) *Arctic Council Regional Programme of Action for the Protection of the Arctic Marine Environment from Land-based Activities*, Akureyri: PAME International Secretariat.
Bartenstein, K. (2011) 'The Arctic Exception in the Law of the Sea Convention: A Contribution to Safer Navigation in the Northwest Passage?', *Ocean Development and International Law*, 42: 22–52.
BC Shipping News (2011) 'Vancouver Sun Report: Harper Withdraws Support of PNCIMA', online at http://www.bcshippingnews.com/home/industry-news/vancouver-sun-report-harper-withdraws-support-pncima-process [accessed 26 June 2014].
Bowman, L. (2010) 'Canadian Government Decides that Your Sewage Doesn't Stink for up to 30 Years from Now', *Environmental Law Centre (Alberta)*, 21 June, online at http://www.environmentallawcentre.wordpress.com/2010/06/21/canadian-government-decides-that-your-sewage-doesn%E2%80%99t-stink-for-up-to-30-years-from-now/ [accessed 24 June 2014].
British Columbia Ministry of Agriculture and Lands (2005) *Sea Lice Management 2005*, Victoria, BC: Ministry of Agriculture, Food and Fisheries.
Broggiato, A., Arnaud-Haond, S., Chiarolla, C., and Greiber, T. (2014) 'Fair and Equitable Sharing of Benefits from Utilization of Marine Genetic Resources in Areas beyond National Jurisdiction: Bridging the Gaps between Science and Policy', *Marine Policy*, 49: 176–85.
Brown, C. R., and Reynolds, J. I. (2004) 'Aboriginal Title to Sea Spaces: A Comparative Study', *University of British Columbia Law Review*, 37(2): 449–93.
Byers, M. (2013) *International Law and the Arctic*, New York: Cambridge University Press.
Canadian Broadcasting Corporation (CBC) (2001) 'Oceans Explorations 2001: Learning from Our Oceans', *Ideas with Paul Kennedy*, CBC Radio One, February–March.

Canadian Council of Fisheries and Aquaculture Ministers (2001) 'Fisheries and Aquaculture Ministers Make Progress in Key Areas', News release No. 830–729/04, 20 September.

Canadian Council of Fisheries and Aquaculture Ministers (2006) 'Ministers Renew Approach to Fisheries and Aquaculture Challenges', News release No. 830–891/004, 10–11 October.

Canadian Health Oceans Network (CHONe) (undated) 'About', online at http://chone.marinebiodiversity.ca/about [accessed 12 November 2014].

Canadian Parks and Wilderness Society (CPAWS) (2014) *Dare to be Deep: Charting Canada's Course to 2020*, online at http://cpaws.org/uploads/CPAWS_DareDeep2020_final.pdf [accessed 24 June 2014].

Chapman, K. (2002) '114957 *Canada Ltée (Spraytech, Société d'arrosage) v. Hudson (Ville)*: Application of the Precautionary Principle in Domestic Law', *Canadian Journal of Administrative Law & Practice*, 15(1): 123–36.

Chircop, A., and Hildebrand, L. (2006) 'Beyond the Buzzwords: A Perspective on Integrated Coastal and Ocean Management in Canada', in D. R. Rothwell and D. L. VanderZwaag (eds) *Towards Principled Oceans Governance: Australian and Canadian Approaches and Challenges*, London: Routledge.

Cicin-Sain, B. (ed.) (2003) 'The Role of Indicators in Integrated Coastal Management', *Oceans & Coastal Management*, 46(3–4) (special issue).

Cicin-Sain, B., and Knight, R. W. (1998) *Integrated Coastal and Ocean Management: Concepts and Practices*, Washington, DC: Island Press.

Cobb, D., Berkes, M. K., and Berkes, F. (2005) 'Ecosystem-Based Management and Marine Environmental Quality Indicators in Northern Canada', in F. Berkes, R. Huebert, H. Fast, M. Manseau, and A. Diduck (eds) *Breaking Ice: Renewable Resource and Ocean Management in the Canadian North*, Calgary, AB: University of Calgary Press.

Commissioner of the Environment and Sustainable Development (2005) 'Fisheries and Oceans Canada: Canada's Oceans Management Strategy', Chapter 1 in *Report of the Commissioner of the Environment and Sustainable Development to the House of Commons*, online at http://publications.gc.ca/site/eng/446551/publication.html [accessed 22 May 2014].

Commissioner of the Environment and Sustainable Development (2012) 'Marine Protected Areas', Chapter 3 in *Fall Report of the Commissioner of the Environment and Sustainable Development*, Ottawa, ON: Office of the Auditor General of Canada.

Commission for Environmental Cooperation (CEC) (2003) *Strategic Plan for North American Cooperation in the Conservation of Biodiversity*, online at http://www.cec.org/files/pdf/BIODIVERSITY/Biodiversitystrategy.pdf [accessed 20 May 2014].

Committee on the Status of Endangered Wildlife in Canada (COSEWIC) (2007) *Canadian Species at Risk*, Ottawa, ON: COSEWIC.

Conference of the Parties (CoP) (2004) Decision VII/5 on Marine and Coastal Biological Diversity: Review of the Programme of Work on Marine and Coastal Biodiversity, Seventh Conference of the Parties to the Convention on Biological Diversity (CBD COP-7), paras 9–20 February, Kuala Lumpur, Malaysia

David Suzuki Foundation, Ecojustice, Environmental Defence, and Nature Canada (2009) *Canada's Species at Risk Act: Implementation at a Snail's Pace*, online at http://www.ecojustice.ca/publications/reports/canadas-species-at-risk-act-implementation-at-a-snails-pace/attachment [accessed 22 May 2014].

Department of Finance Canada (2005) *The Budget Speech 2005*, online at http://www.fin.gc.ca/budget05/pdf/speeche.pdf [accessed 21 May 2014].

Department of Finance Canada (2007) *The Budget Speech 2007*, online at http://www.budget.gc.ca/2007/pdf/speeche.pdf [accessed 21 May 2014].

Doelle, M. (2012) 'CEAA 2012: The End of Federal EA as We Know It?', *Journal of Environmental Law & Practice*, 24(1): 1–17.

Eastern Scotian Shelf Integrated Management (ESSIM) Planning Office (2006) *Eastern Scotian Shelf Integrated Ocean Management Plan (2006–2011): Final Draft*, Dartmouth, NS: Fisheries and Oceans Canada.

Elliott, G. M., and Spek, B. (2004) 'Integrated Management Planning in the Beaufort Sea: Blending Natural and Social Science in a Settled Land Claim Area', in N. W. P. Munro, T. B. Herman, K. Beazley, and P. Dearden (eds) *Making Ecosystem-based Management Work: Proceedings of the Fifth International Conference on Science and Management of Protected Areas*, 11–16 May 2003, Victoria, BC.

Environment Canada (undated) 'Wastewater', online at http://www.ec.gc.ca/eu-ww/ [accessed 24 June 2014].

Environment Canada (2006) *National Report on GPA Implementation*, Ottawa, ON: Environment Canada.

Federal/Provincial/Territorial Advisory Committee on Canada's National Programme of Action for the Protection of the Marine Environment from Land-based Activities (NPA Advisory Committee) (2000)

Canada's National Programme of Action for the Protection of the Marine Environment from Land-based Activities (NPA), Ottawa, ON: Government of Canada.

Federal/Provincial/Territorial Advisory Committee on Canada's National Programme of Action for the Protection of the Marine Environment from Land-based Activities (NPA Advisory Committee) (2001) *Implementing Canada's National Programme of Action for the Protection of the Marine Environment from Land-based Activities: National Report to the 2001 Intergovernmental Review Meeting on Implementation of the Global Programme of Action*, Ottawa, ON: Environment Canada.

Findlay, C. S., Elgie, S., Giles, B., and Burr, L. (2009) 'Species Listing under Canada's Species at Risk Act', *Conservation Biology*, 23(6): 1609–17.

Fisheries and Oceans Canada (undated*a*) 'Integrated Management of Oceans Resources', online at http://www.dfo-mpo.gc.ca/oceans/management-gestion/integratedmanagement-gestionintegree/index-eng.htm [accessed 30 October 2014].

Fisheries and Oceans Canada (undated*b*) 'Large Oceans Management Areas', online at http://www.dfo-mpo.gc.ca/oceans/marineareas-zonesmarines/loma-zego/index-eng.htm [accessed 26 June 2014].

Fisheries and Oceans Canada (1987) *Oceans Policy for Canada: A Strategy to Meet the Challenges and Opportunities on the Oceans Frontier*, Ottawa, ON: Fisheries and Oceans Canada, Information and Publications Branch.

Fisheries and Oceans Canada (1994) *A Vision for Ocean Management*, Ministerial Vision Paper, Ottawa, ON: Fisheries and Oceans Canada.

Fisheries and Oceans Canada (1997a) *The Role of the Federal Government in the Oceans Sector*, Ottawa, ON: Fisheries and Oceans Canada, Oceans Directorate.

Fisheries and Oceans Canada (1997b) *The Role of the Provincial and Territorial Governments in the Oceans Sector*, Ottawa, ON: Fisheries and Oceans Canada, Oceans Directorate.

Fisheries and Oceans Canada (1997c) *Toward Canada's Oceans Strategy: Discussion Paper*, Ottawa, ON: Fisheries and Oceans Canada.

Fisheries and Oceans Canada (1998) 'Statement by David Anderson, Minister of Fisheries and Oceans Canada: Announcement on Offshore Marine Protected Areas', 8 December, Fisheries and Oceans Canada, Ottawa, ON.

Fisheries and Oceans Canada (1999a) *Canada's Oceans: Experience and Practices*, Monograph No. 7, *Sustainable Development in Canada* Series, Ottawa, ON: Minister of Supply and Services.

Fisheries and Oceans Canada (1999b) *National Framework for Establishing and Managing Marine Protected Areas*, Work document, Ottawa, ON: Fisheries and Oceans Canada.

Fisheries and Oceans Canada (2000) *Members of the Minister's Advisory Council on Oceans (MACO): Backgrounder*, September, Ottawa, ON: Fisheries and Oceans Canada.

Fisheries and Oceans Canada (2001a) *Partnerships for Living Oceans*, Canadian Oceans Stewardship Conference Report, 6–8 June, Fisheries and Oceans Canada, Oceans Directorate, Vancouver, BC/Ottawa, ON.

Fisheries and Oceans Canada (2001b) *New Emerging Fisheries Policy*, online at http://www.dfo-mpo.gc.ca/fm-gp/policies-politiques/efp-pnp-eng.htm [accessed 22 May 2014].

Fisheries and Oceans Canada (2002a) *Canada's Oceans Strategy: Our Oceans, Our Future*, online at http://www.dfo-mpo.gc.ca/oceans/publications/cos-soc/pdf/cos-soc-eng.pdf [accessed 22 May 2014].

Fisheries and Oceans Canada (2002b) *Policy and Operational Framework for Integrated Management of Estuarine, Coastal and Marine Environments in Canada*, Ottawa, ON: Fisheries and Oceans Canada, Oceans Directorate.

Fisheries and Oceans Canada (2002c) 'Thibault Appoints Two New Members to Minister's Advisory Council on Oceans', News release, 10 December.

Fisheries and Oceans Canada (2002d) *DFO's Aquaculture Policy Framework*, online at http://publications.gc.ca/site/eng/415165/publication.html [accessed 22 May 2014].

Fisheries and Oceans Canada (2003) 'National Engagement Summary on Canada's Oceans Strategy' (unpublished).

Fisheries and Oceans Canada (2004a) *Performance Report for the Period Ending March 31, 2004*, online at http://www.dfo-mpo.gc.ca/Library/350930_P1.pdf [accessed 22 May 2014].

Fisheries and Oceans Canada (2004b) *Habitat Status Report on Ecosystem Objectives*, Canadian Science Advisory Secretariat (CSAS) Habitat Status Report No. 2004/001, online at http://www.dfo-mpo.gc.ca/csas/Csas/status/2004/HSR2004_001_e.pdf [accessed 22 May 2014].

Fisheries and Oceans Canada (2004c) *Memorandum of Understanding Respecting the Implementation of Canada's Oceans Strategy on the Pacific Coast of Canada*, online at http://www.dfo-mpo.gc.ca/oceans/publications/bc-cb/index-eng.asp [accessed 22 May 2014].

Fisheries and Oceans Canada (2005a) *Canada's Oceans Action Plan for Present and Future Generations*, online at http://www.dfo-mpo.gc.ca/oceans/publications/oap-pao/pdf/oap-eng.pdf [accessed 22 May 2014].

Fisheries and Oceans Canada (2005b) *Identification of Ecologically and Biologically Significant Areas*, Canadian Science Advisory Secretariat (CSAS) Ecosystem Status Report No. 2004/006, online at http://www.dfo-mpo.gc.ca/csas/Csas/status/2004/ESR2004_006_e.pdf [accessed 22 May 2014].

Fisheries and Oceans Canada (2005c) *Canada's Policy for Conservation of Wild Pacific Salmon*, online at http://www.pac.dfo-mpo.gc.ca/publications/pdfs/wsp-eng.pdf [accessed 22 May 2014].

Fisheries and Oceans Canada (2006) *Performance Report for the Period ending March 31, 2006*, online at http://www.tbs-sct.gc.ca/dpr-rmr/0506/FO-PO/fo-po_e.pdf [accessed 22 May 2014].

Fisheries and Oceans Canada (2009a) *Gardner Pinfold Consulting Economists Economic Impact of Marine Related Activities in Canada*, online at http://www.dfo-mpo.gc.ca/ea-ae/cat1/no1-1/no1-1-eng.htm [accessed 30 October 2014]. (An unpublished update to this report was produced for Fisheries and Oceans Canada in 2012 by Acton White Associates Inc. in association with Innova Quest Inc. The former report used 2006 data; the latter, 2008.)

Fisheries and Oceans Canada (2009b) *The Role of the Provincial and Territorial Governments in the Oceans Sector*, Ottawa, ON: Fisheries and Oceans Canada, Oceans Directorate.

Fisheries and Oceans Canada (2009c) *Policy on New Fisheries for Forage Species*, online at http://www.dfo-mpo.gc.ca/fm-gp/peches-fisheries/fish-ren-peche/sff-cpd/forage-eng.htm [accessed 20 May 2014].

Fisheries and Oceans Canada (2009d) *Policy for Managing the Impacts of Fisheries on Sensitive Benthic Areas*, online at http://www.dfo-mpo.gc.ca/fm-gp/peches-fisheries/fish-ren-peche/sff-cpd/benthi-back-fiche-eng.htm [accessed 20 May 2014].

Fisheries and Oceans Canada (2009e) *A Fishery Decision-making Framework Incorporating the Precautionary Approach*, online at http://www.dfo-mpo.gc.ca/fm-gp/peches-fisheries/fish-ren-peche/sff-cpd/pre-caution-eng.htm [accessed 20 May 2014].

Fisheries and Oceans Canada (2010a) 'Announcement of an Integrated Oceans Management Plan for the Beaufort Sea', News release, 27 August, online at http://www.dfo-mpo.gc.ca/media/npress-communique/2010/hq-ac43-eng.htm [accessed 22 May 2014].

Fisheries and Oceans Canada (2010b) *National Aquaculture Strategic Action Plan Initiative (NASAPI) 2011–2015*, online at http://www.dfo-mpo.gc.ca/aquaculture/lib-bib/nasapi-inpasa/index-eng.htm [accessed 21 May 2014].

Fisheries and Oceans Canada (2012) *Evaluation of the Health of the Oceans (HOTO) Initiative*, online at http://www.dfo-mpo.gc.ca/ae-ve/evaluations/12-13/6b135-eng.htm [accessed 24 May 2014].

Fisheries and Oceans Canada (2014) 'Foreword', in *Canada's Oceans Action Plan*, online at http://www.dfo-mpo.gc.ca/oceans/publications/oap-pao/pdf/oap-eng.pdf [accessed 26 June 2014].

Food and Agriculture Organization of the United Nations (FAO) (1996) *Precautionary Approach to Capture Fisheries and Species Introductions*, FAO Technical Guidelines for Responsible Fisheries No. 2, Rome: FAO.

Foreign Affairs, Trade and Development Canada (2012) 'Canada and Kingdom of Denmark Reach Tentative Agreement on Lincoln Sea Boundary', News release, 28 November, online at http://www.international.gc.ca/media/aff/news-communiques/2012/11/28a.aspx?lang=eng [accessed 21 May 2014].

Foreign Affairs, Trade and Development Canada (2013) 'Canada Marks Major Milestone in Defining Its Continental Shelf', Press release, 9 December, online at http://www.international.gc.ca/media/aff/news-communiques/2013/12/09a.aspx?laag=eng [accessed 21 May 2014].

Ginn, D. (2006) 'Aboriginal Title and Oceans Policy in Canada', in D. R. Rothwell and D. L. VanderZwaag (eds) *Towards Principled Oceans Governance: Australian and Canadian Approaches and Challenges*, London: Routledge.

Government of Canada (1995) *Canadian Biodiversity Strategy: Canada's Response to the Convention on Biological Diversity*, Ottawa, ON: Canadian Museum of Nature.

Government of Canada (1998) *Working Together for Marine Protected Areas: A National Approach*, Ottawa, ON: Fisheries and Oceans Canada.

Government of Canada (2001) 'Disposal at Sea Regulations, SOR/2001-275: Regulatory Impact Analysis Statement', *Canada Gazette II*, 135(17).

Government of Canada (2003) *Framework for the Application of Precaution in Science-based Decision Making about Risk*, online at http://www.pco-bcp.gc.ca/index.asp?lang=eng&page=information&sub=publications&doc=precaution\precaution-eng.htm [accessed 22 May 2014].

Government of Canada (2004a) 'Speech from the Throne to Open the First Session of the Thirty-Eighth Parliament of Canada', 5 October 2004, Prime Minister's Office, Ottawa, ON, online at http://www2.parl.gc.ca/Parlinfo/Documents/ThroneSpeech/38-1-e.html [accessed 21 May 2014].

Government of Canada (2004b) 'Regulations Amending the Coastal Fisheries Protection Regulations, SOR/2004–110: Regulatory Impact Analysis Statement', *Canada Gazette II*, Extra, 138(6).

Government of Canada (2005a) *Canada's Federal Marine Protected Areas Strategy*, online at http://www.dfo-mpo.gc.ca/oceans/publications/fedmpa-zpmfed/pdf/mpa-eng.pdf [accessed 22 May 2014].

Government of Canada (2005b) 'Basin Head Marine Protected Area Regulations: Regulatory Impact Analysis Statement', *Canada Gazette I*, 139(25).

Government of Canada (2005c) *Conference Report*, Conference on the Governance of High Seas Fisheries and the UN Fish Agreement, 'Moving from Words to Action', 1–5 May, St. John's, Newfoundland and Labrador, online at http://www.dfo-mpo.gc.ca/fgc-cgp/conf_report_e.pdf [accessed 21 May 2014].

Government of Canada (2005d) *Canada's National Plan of Action to Prevent, Deter and Eliminate Illegal, Unreported and Unregulated Fishing*, online at http://www.dfo-mpo.gc.ca/npoa-pan/npoa-iuu/npoa-iuu_e.pdf [accessed 21 May 2014].

Government of Canada (2010) *Spotlight on Marine Protected Areas in Canada*, Ottawa, ON: Fisheries and Oceans Canada, Communications Branch.

Government of Canada (2011) *National Framework for Canada's Network of Marine Protected Areas*, Ottawa, ON: Fisheries and Oceans Canada.

Government of Canada (2013) *Partial Submission of Canada to the Commission on the Limits of the Continental Shelf regarding its continental shelf in the Atlantic Ocean*, online at http://www.un.org/depts/los/clcs_new/submissions_files/submission_can_70_2013.htm [accessed 30 October 2014].

Government of Nova Scotia (2011) *Memorandum of Understanding between Canada and Nova Scotia Respecting Coastal and Oceans Management in Nova Scotia*, online at http://www.novascotia.ca/coast/consultation-mou.asp [accessed 26 June 2014].

Government of Nova Scotia (2013) 'New Aquaculture Regulations to Better Protect Coastal Communities', News release, 1 May, online at http://novascotia.ca/news/release/?id=20130501003 [accessed 29 May 2014].

Hall, T., MacLean, M., and Coffen-Smout, S. (2011) 'Advancing Objectives-Based, Integrated Ocean Management through Marine Spatial Planning: Current and Future Directions on the Scotian Shelf off Nova Scotia, Canada', *Journal of Coastal Conservation*, 15(2): 247–55.

Hutchings, J. A., and Post, J. R. (2013) 'Gutting Canada's Fisheries Act: No Fishery, No Fish Habitat Protection', *Fisheries*, 38(11): 497–501.

Hutchings, J. A., Côté, I. M., Dodson, J. J., Fleming, I. A., Jennings, S., Mantua, N. J., Peterman, R. M., Riddell, B. E., Weaver, A. J., and VanderZwaag, D. L. (2012) *Sustaining Canada's Marine Biodiversity: Responding to the Challenges Posed by Climate Change, Fisheries, and Aquaculture*, Ottawa, ON: Royal Society of Canada.

Intergovernmental Oceanographic Commission (IOC) (2003) *A Reference Guide on the Use of Indicators for Integrated Coastal Management*, ICAM Dossier 1, IOC Manuals and Guides No. 45, Paris: UNESCO.

Intergovernmental Oceanographic Commission (IOC) (2006) *A Handbook for Measuring the Progress and Outcomes of Integrated Coastal and Ocean Management*, ICAM Dossier 2, IOC Manuals and Guides No. 46, Paris: UNESCO.

International Union for Conservation of Nature (IUCN) (2006) *Update on MPAs beyond National Jurisdiction, February 2004–February 2006*, online at http://cmsdata.iucn.org/downloads/unga_wg_hsmpa_13_17_02_06___1_pager_update____a4_1.pdf [accessed 30 October 2014].

Jamieson, G., O'Boyle, R., Arbour, J., Cobb, D., Courtenay, S., Gregory, R., Levings, C., Munro, J., Perry, I., and Vandermeulen, H. (2001) *Proceedings of the National Workshop on Objectives and Indicators for Ecosystem-based Management*, Sidney, BC, 27 February–2 March, Canadian Science Advisory Secretariat Proceedings Series 2001/09, Ottawa, ON: Fisheries and Oceans Canada.

Jones, R. (2006) 'Canada's Seas and her First Nations', in D. R. Rothwell and D. L. VanderZwaag (eds) *Towards Principled Oceans Governance: Australian and Canadian Approaches and Challenges*, London: Routledge.

Koivurova, T., and VanderZwaag, D. L. (2007) 'The Arctic Council at 10 Years: Retrospect and Prospects', *University of British Columbia Law Review*, 40(1): 121–94.

Koivurova, T., Chircop A., Franckx, E., Molenaar, E. J., and VanderZwaag, D. L. (2009) *Understanding and Strengthening European Union–Canada Relations in Law of the Sea and Ocean Governance*, Rovaniemi: Northern Institute for Environmental and Minority Law.

Lasserre, F. (2014) 'Simulations of Shipping along Arctic Routes: Comparison, Analysis and Economic Perspectives' *Polar Record*, online at http://dx.doi.org/10.1017/S0032247413000958 [accessed 30 October 2014].

Leggatt, S. (2001) *Clear Choices, Clean Waters: The Leggatt Inquiry into Salmon Farming in British Columbia, Report and Recommendations*, online at http://davidsuzuki.org/publications/reports/2001/clear-choices-clean-waters/ [accessed 21 May 2014].

Lemkow, A., and VanderZwaag, D. L. (2014) 'Recovery Planning under Canada's Species at Risk Act in a Changing Ocean: Gauging the Tides, Charting Future Coordinates', *Journal of Environmental Law & Practice*, 26: 122–56.

Living Oceans Society (2011) 'Federal Government Breaks Commitment to Comprehensive Management Plan for Pacific North Coast', Media release, 9 September, online at http://www.livingoceans.org/media/releases/federal-government-breaks-commitment-comprehensive-management-plan-pacific-north [accessed 26 June 2014].

Marine Planning Partnership for the North Pacific Coast (MaPP) (undated) 'Marine Planning Partnership for the North Pacific Coast (MaPP)', online at http://mappocean.org [accessed 26 June 2014].

McCrimmon, D., and Fanning, L. (2010) 'Using Memoranda of Understanding to Facilitate Marine Management in Canada', *Marine Policy*, 34(6): 1335–40.

McCuaig, J., and Herbert, G. (eds) (2013) *Canadian Technical Report of Fisheries and Aquatic Sciences 3025: Review and Evaluation of the Eastern Scotian Shelf Integrated Management (ESSIM) Initiative*, Dartmouth, AB: Fisheries and Oceans Canada, Oceans and Coastal Management Division, Maritimes Region.

McDorman, T. L. (1988) 'Will Canada Ratify the Law of the Sea Convention?', *San Diego Law Review*, 25(3): 535–79.

McDorman, T. L. (2004) 'Editorial: Canada Ratifies the 1982 United Nations Convention on the Law of the Sea', *Ocean Development & International Law*, 35(2): 103–14.

McDorman, T. L. (2009) *Salt Water Neighbors: International Ocean Law Relations between the United States and Canada*, New York: Oxford University Press.

McDorman, T. L., Saunders, P. M., and VanderZwaag, D. L. (1985) 'The Gulf of Maine Boundary: Dropping Anchor or Setting a Course?', *Marine Policy*, 9(2): 90–107.

McRae, D. M., and Goundrey, D. J. (1982) 'Environmental Jurisdiction in Arctic Waters: The Extent of Article 234', *University of British Columbia Law Review*, 16(2): 197–228.

Moodie, D. J. R. (2004) 'Aboriginal Maritime Title in Nova Scotia: An "Extravagant and Absurd Idea"?', *University of British Columbia Law Review*, 37(2): 495–540.

Mooers, A. O., Doak, D. F., Findlay, C. S., Green, D. M., Grouios, C., Mana, L. L., Rashvand, A., Rudd, M. A., and Whitton, J. (2010) 'Science, Policy, and Species at Risk in Canada', *BioScience*, 60(10): 843–9.

National Advisory Board on Science and Technology (NABST) (1994) *Opportunities from our Oceans: Report of the Committee on Oceans and Coasts*, Ottawa, ON: NABST.

NPA Advisory Committee *See* Federal/Provincial/Territorial Advisory Committee on Canada's National Programme of Action for the Protection of the Marine Environment from Land-based Activities.

O'Boyle, R., and Jamieson, G. (2006) 'Observations on the Implementation of Ecosystem-Based Management: Experiences on Canada's East and West Coasts', *Fisheries Research*, 79: 1–12.

O'Neil, P. (2012) 'Documents Show Enbridge Concerns about Tides Canada', *National Post*, 23 April, online at http://business.financialpost.com/2012/04/23/documents-show-enbridge-concerns-about-tides-Canada [accessed 26 June 2014].

Ocean Management Research Network (OMRN) (2004) Review and Update from the National Secretariat: Network News, June, Ottawa, ON: OMRN.

ONE OCEAN Secretariat (2003) *ONE OCEAN: Identifying Industry Workshop Priorities and Future Direction, A Report Submitted to the ONE OCEAN Advisory Board by the ONE OCEAN Secretariat*, online at http://www.oneocean.ca/pdf/ONE%20OCEAN%20REPORT.pdf [accessed 21 May 2014].

Pacific North Coast Integrated Management Area (PNCIMA) (2013) *Draft Pacific North Coast Integrated Management Area Plan*, online at http://www.pncima.org/media/documents/pdf/draft-pncima-plan-may-27--2013.pdf [accessed 28 June 2014].

Parks Canada (undated*a*) 'Gwaii Haanas Management', online at http://www.pc.gc.ca/eng/pn-np/bc/gwaiihaanas/plan/plan1.aspx [accessed 26 June 2014].

Parks Canada (undated*b*) 'National Marine Conservation Areas of Canada: National Marine Conservation Area System', online at http://www.pc.gc.ca/progs/amnc-nmca/system/index_e.asp [accessed 22 May 2014].

Penick, V. (2001) 'Legal Framework in the Canadian Offshore', *Dalhousie Law Journal*, 24(1): 1–22.

Powles, H., Vendette, V., Siron, R., and O'Boyle, B. (2004) *Proceedings of the Canadian Marine Ecoregions Workshop*, 23–25 March, Ottawa, DFO Canadian Science Advisory Secretariat Proceedings Series 2004/016, online at http://www.dfo-mpo.gc.ca/csas/Csas/Proceedings/2004/PRO2004_016_B.pdf [accessed 21 May 2014].

President of the Treasury Board of Canada (2012) *Red Tape Reduction Action Plan*, online at http://www. tbs-sct.gc.ca/rtrap-parfa/rtrapr-rparfa-eng.pdf [accessed 21 May 2014].

Prime Minister of Canada (2014a) 'National Conservation Plan', online at http://pm.gc.ca/eng/ news/2014/05/15/national-conservation-plan [accessed 12 November 2014].

Prime Minister of Canada (2014b) 'PM Launches National Conservation Plan', Press release, 15 May, online at http://pm.gc.ca/eng/news/2014/05/15/national-conservation-plan [accessed 30 October 2014].

Protection of the Arctic Marine Environment Working Group (PAME) (2008) *PAME Progress Report to Senior Arctic Officials*, 19–20 November, Kautokeino, Norway, online at http://arcticportal.org/ uploads/BW/p8/BWp8LeLvRW6AKx9HfLECHQ/Dec-08.PAME-Report-to-SAOs-Nov-2008. pdf [accessed 21 May 2014].

Ricketts, P. J., and Hildebrand, L. (2011) 'Coastal and Ocean Management in Canada: Progress or Paralysis?' *Coastal Management*, 39(1): 4–19.

Rothwell, D. R., and VanderZwaag, D. L. (2006) 'The Sea Change towards Principled Oceans Governance', in D. R. Rothwell and D. L. VanderZwaag (eds) *Towards Principled Oceans Governance: Australian and Canadian Approaches and Challenges*, London: Routledge.

Russell, D. A., and VanderZwaag, D. L. (eds) (2010a) *Recasting Transboundary Fisheries Management Arrangements in Light of Sustainability Principles: Canadian and International Perspectives*, Leiden: Martinus Nijhoff.

Russell, D. A., and VanderZwaag, D. L. (2010b) 'Challenges and Future Directions in Transboundary Fisheries Management: Concluding Reflections', in D. A. Russell and D. L. VanderZwaag (eds) *Recasting Transboundary Fisheries Management Arrangements in Light of Sustainability Principles*, Leiden: Martinus Nijhoff.

Russell, D. A., and VanderZwaag, D. L. (2010c) 'The International Law and Policy Seascape Governing Transboundary Fisheries', in D. A. Russell and D. L. VanderZwaag (eds) *Recasting Transboundary Fisheries Management Arrangements in Light of Sustainability Principles*, Leiden: Martinus Nijhoff.

Sheppard, V., Rangeley, R., and Laughren, J. (2005) *An Assessment of Multi-Species Recovery Strategies and Ecosystem-Based Approaches for Management of Marine Species at Risk in Canada: WWF Canada Report for Fisheries and Oceans Canada*, Ottawa, ON: Fisheries and Oceans Canada.

Siron, R., VanderZwaag, D. L., and Fast, H. (2009) 'Ecosystem-Based Ocean Management in the Canadian Arctic', in A. H. Hoel (ed.) *Best Practices in Ecosystem-Based Oceans Management in the Arctic*, Tromsø: Norwegian Polar Institute.

Standing Committee on Fisheries and Oceans (FOPO) (2001) *Report on the Oceans Act*, online at http:// www.parl.gc.ca/HousePublications/Publication.aspx?DocId=1032010 [accessed 21 May 2014].

Taylor, A., and Dickey, J. (2001) 'Regulatory Regime: Canada–Newfoundland/Nova Scotia Offshore Petroleum Board Issues', *Dalhousie Law Journal*, 24(1): 51–86.

Taylor, E. B., and Pinkus, S. (2013) 'The Effects of Lead Agency, Nongovernmental Organizations, and Recovery Team Membership on the Identification of Critical Habitat for Species at Risk: Insights from the Canadian Experience', *Environmental Reviews*, 21(2): 93–102.

Transport Canada (2001) *Guidelines for the Control of Ballast Water Discharge from Ships in Waters under Canadian Jurisdiction*, Transport Canada Publication No. 13617, Ottawa, ON: Canadian Marine Advisory Council, Transport Canada.

Transport Canada (2013) *Transportation in Canada 2012: Annual Report*, online at http://www.tc.gc.ca/ eng/policy/anre-menu.htm [accessed 21 May 2014].

VanderZwaag, D. L. (1995) *Canada and Marine Environment Protection: Charting a Legal Course towards Sustainable Development*, London: Kluwer Law International.

VanderZwaag, D. L. (1998) 'The Precautionary Principle in Environmental Law and Policy: Elusive Rhetoric and First Embraces', *Journal of Environmental Law & Practice*, 8(3): 362–3.

VanderZwaag, D. L. (2014a) 'Canada and the Governance of the Northwest Passage: Rough Waters, Cooperative Currents, Sea of Challenges', in D. D. Caron and O. Nilufer (eds) *Navigating Straits: Challenges for International Law, A Law of the Sea Institute Publication*, Leiden/Boston, MA: Brill/Nijhoff.

VanderZwaag, D. L. (2014b) 'The Arctic Council and the Future of Arctic Ocean Governance: Edging Forward in a Sea of Governance Challenges', in T. Stephens and D. L. VanderZwaag (eds) *Polar Oceans Governance in an Era of Environmental Change*, Cheltenham: Edward Elgar.

VanderZwaag, D. L., and Hutchings, J. A. (2005) 'Canada's Marine Species at Risk: Science and Law at the Helm, but a Sea of Uncertainties', *Ocean Development & International Law*, 36(3): 219–59.

VanderZwaag, D. L., Chao, G., and Covan, M. (2006) 'Canadian Aquaculture and the Principles of Sustainable Development', in D. L. VanderZwaag and G. Chao (eds) *Aquaculture Law and Policy: Towards Principled Access and Operations*, London: Routledge.

VanderZwaag, D. L., Engler-Palma, M. C., and Hutchings, J. A. (2011) 'Canada's Species at Risk Act and Atlantic Salmon: Cascade of Promises, Trickles of Protection, Sea of Challenges', *Journal of Environmental Law & Practice*, 22(3): 267–307.

VanderZwaag, D. L., Fuller, S. D., and Myers, R. A. (2002–03) 'Canada and the Precautionary Principle/ Approach in Ocean and Coastal Management: Wading and Wandering in Tricky Currents', *Ottawa Law Review*, 34(1): 119–23.

VanderZwaag, D. L., Hutchings, J. A., Jennings, S., and Peterman, R. M. (2012) 'Canada's International and National Commitments to Sustain Marine Biodiversity', *Environmental Reviews*, 20: 312–52.

Waples, R. S., Nammack, M., Cochrane, J. F., and Hutchings, J. A. (2013) 'A Tale of Two Acts: Endangered Species Listing Practices in Canada and the United States', *BioScience*, 63(9): 723–34.

Weber, B. (2014) 'More Northwest Passage Travel Planned by Danish Shipper', *CBC News*, 3 January, online at http://www.cbc.ca/news/Canada/north/more-northwest-passage-travel-planned-by-dan-ish-shipper-1.2482731 [accessed 28 May 2014].

Wells, P. G. (2002) 'Invigorating the United Nations Global Programme of Action (GPA) for the Protection of the Marine Environment from Land-based Activities: Utilizing Both Bottom-up and Top-down Approaches', *Marine Pollution Bulletin*, 44(8): 719–21.

Statutes and statutory instruments

An Act to amend the Migratory Birds Convention Act, 1994 and the Canadian Environmental Protection Act, 1999, SC 2005, c. 23

Arctic Shipping Pollution Prevention Regulations, CRC, c. 353

Arctic Waters Pollution Prevention Act, RSC 1985, c. A-12

Ballast Water Control and Management Regulations, SOR/2011-237

Basin Head Marine Protected Area Regulations, SOR/2005-293

Bill C-45, An Act Respecting the Sustainable Development of Canada's Seacoast and Inland Fisheries, 2006, First Session, Thirty-ninth Parliament

Bowie Seamount Marine Protected Area Regulations, SOR/2008-124

Canada Shipping Act, 2001, SC 2001, c. 26

Canadian Environmental Assessment Act, SC 2012, c. 19

Canadian Environmental Protection Act, 1999, SC 1999, c. 33

Constitution Act, 1867 (UK), 30 & 31 Vict. c. 3, reprinted in RSC 1985, Appendix II, No. 5

Constitution Act, 1982, being Schedule B to the Canada Act, 1982 (UK), 1982, c. 11

Eastport Marine Protected Areas Regulations, SOR/2005-294

Endeavour Hydrothermal Vents Marine Protected Area Regulations, SOR/2003-87

Fisheries Act, RSC 1985, c. F-14

Gilbert Bay Marine Protected Regulations, SOR/2005-295

Gully Marine Protected Area Regulations, SOR/2004-112

Manicouagan Marine Protected Area Regulations, 30 September 2006, *Canada Gazette I*, 140(39)

Migratory Birds Convention Act, 1994, SC 1994, c. 22

Musquash Estuary Marine Protected Area Regulations, SOR/2006-354

Natural Areas Protection Act, RSPEI 1988, c. N-2

Oceans Act, SC 1996, c. 31

Pacific Aquaculture Regulations, SOR/2010-270

Species at Risk Act, SC 2002, c. 29

Tarium Niryutait Marine Protected Areas Regulations, SOR/2010-190

Transportation of Dangerous Goods Regulations, SOR/2001-286

Vessel Pollution and Dangerous Chemicals Regulations, SOR/2012-69

Wastewater Systems Effluent Regulations, SOR/2012-139

Conventions and treaties

1973 International Convention for the Prevention of Pollution from Ships, as amended by the Protocol of 1978 (MARPOL), published as International Maritime Organization (IMO) (2011) *MARPOL Consolidated Edition 2011*, London: IMO

1982 United Nations Convention on the Law of the Sea (1982)1833 UNTS 397 (UNCLOS)

1992 Convention on Biological Diversity (1992) 1760 UNTS 79 (CBD)

1995 Agreement for the Implementation of the Provisions of the United Nations Convention of the Law of the Sea of 10 December 1982 Relating to the Conservation and Management of Straddling Fish Stocks and Highly Migratory Fish Stocks (1995) 2167 UNTS 88 (1995 UN Fish Stocks Agreement, or UNFSA 1995)

1996 Protocol to the Convention on the Prevention of Marine Pollution by Dumping of Wastes and Other Matter, 1972 and Resolutions Adopted by the Special Meeting (1996) 36 ILM 7 (London Protocol)

2004 International Convention for the Control and Management of Ships' Ballast Water and Sediments, IMO Doc. No. BWM/CONF/36 (Ballast Water Management Convention, or BWM Convention)

4

AUSTRALIA'S OCEANS POLICY[1]

Donna Petrachenko and Ben Addison

Introduction

In 1998, the Australian government set Australia's Oceans Policy in place to provide a framework for the integrated and ecosystem-based planning and management of Australia's marine jurisdiction. The aim of the policy was to bring the whole range of marine uses and activities across all maritime sectors in Australia under a coordinated approach to management. As a nation with strong ties to the ocean, the Australian people consider the marine environment to be a major asset for the country. The Oceans Policy was built with a grand and all-encompassing scope in mind. It is an ambitious and challenging policy designed to protect the social, economic, and environmental values of Australia's marine jurisdiction.

This chapter will explore the development of Australia's Oceans Policy, and discuss the challenges and obstacles facing its implementation since its inception, along with what lessons have been learnt in that time. The first section will provide some contextual information about Australia, describing the nation's governance arrangements, and offering a brief description of the physical, social, and economic characteristics relevant to the Oceans Policy.

The second section will describe how Australia's Oceans Policy was developed, including what drivers and processes shaped the final policy, and what objectives and principles lay at its heart. This section will also describe past and present institutional arrangements that have been put in place to support the ongoing development of the Oceans Policy.

The third section discusses the specific nature and content of Australia's Oceans Policy and the mechanisms by which it is implemented. Authority, reporting, and budgetary structures surrounding the policy are described, as are supporting enforcement and research arrangements. This section also describes in some detail the range of Australia's international obligations that are addressed by the Oceans Policy.

The fourth section reviews some of the key issues that have faced the implementation of Australia's Oceans Policy, and describes the monitoring and evaluation mechanisms associated with it. A brief forecast for the future of Australia's Oceans Policy has also been included.

The fifth section includes a few concluding remarks on areas of Australia's Oceans Policy in which improvements are being sought to secure substantial outcomes from the policy process, while the final section provides an update on developments to 2014.

Basic information and an overview of Australia's Oceans Policy

Basic information on Australia

Government structure

Australia was founded as a British colony in Sydney in 1788. Six colonies were formed in the eighteenth and nineteenth centuries, eventually federating in 1901 to form the Commonwealth of Australia. The six colonies became the Australian states. Two territories were subsequently formed: the Northern Territory and the Australian Capital Territory.

Although Australia is an independent nation, the formal head of state is the monarch of Great Britain, Queen Elizabeth II, who is represented in Australia by the governor general. The governor general is appointed by the queen on the advice of the prime minister as the Australian head of government. The role of the governor general is, by convention, ceremonial, with decision-making being the responsibility of the elected government through the federal Parliament.

In addition to the federal jurisdiction, each of the states and territories has its own parliament. State parliaments are subject to the national Constitution, as well as their own state constitutions. Federal law overrides state law when there is inconsistency between the two levels of legislation.

Local government acts as a third sphere of governance in Australia. Local governments are formed under state or territory legislation (Department of Foreign Affairs and Trade, undated).

Land area

In terms of area, Australia is the world's sixth largest country, with a total of 7,692,024 km² of land on continental Australia (Table 4.1). External territories, including the Australian Antarctic Territory, add a further 5,897,500 km² to Australia's territory.

Table 4.1 Australian state and territory land areas

State or territory	Land area (km²)
Australian states and territories	
Western Australia	2,529,875
Queensland	1,730,648
South Australia	1,349,129
Northern Territory	983,482
New South Wales	800,642
Victoria	227,416
Tasmania	68,401
Australian Capital Territory	2,431
Total	**7,692,024**
Australian external territories	
Australian Antarctic Territory	5,896,500
Territory of Heard and McDonald Islands	370
Territory of Christmas Island	135
Coral Sea Islands Territory	81
Norfolk Island	35
Territory of Cocos (Keeling) Island	14
Territory of Ashmore and Cartier Islands	2
Total	**5,897,137**

Source: Geoscience Australia (undatedc).

Length of coastline

Australia has a coastline 59,736 km in length (Geoscience Australia, undated*a*). The Australian Antarctic Territory has a coastline length of 7,272 km, but that figure can increase to as much as 7,657 km if allowances are made for the coastal ice shelf (Australian Antarctic Division, undated).

Marine jurisdiction

Australia has one of the largest ocean jurisdictions of any country in the world. Australia has responsibilities for an area encompassing 14 million km^2 (Department of the Environment and Heritage, 2004). Australia's oceans are broken up into a number of maritime zones, consistent with international law, as follows.

- Coastal waters extend from the territorial sea baseline out to 3 nautical miles from the coastline. They fall within the jurisdiction of Australian state and territory governments.
- The Australian territorial sea extends 12 nautical miles from the baseline.
- The contiguous zone includes the area between 12 and 24 nautical miles from the baseline.
- The exclusive economic zone (EEZ) extends 200 nautical miles from the baseline (National Oceans Office, 2001).
- Australia has rights to sedentary organisms, and the mineral and non-living resources of the seabed and subsoil of the continuous continental shelf that extends beyond the 200 nautical mile limit of the EEZ under the Seas and Submerged Lands (Limits of the Continental Shelf) Proclamation 2012.

Biogeography

The Australian continent is a large land mass that extends from 10° 41'21"S at Cape York Peninsula in northern Queensland to 43° 38'40"S at South East Cape in southern Tasmania – a distance of about 3,700 km (Geoscience Australia, undated*b*). Consequently, continental Australia includes waters in both the temperate and tropical climatic zones. Australia also has responsibility for Macquarie Island, Heard and MacDonald Islands, and a large portion of Antarctica, all occurring within the polar climatic zone. The question of how the marine environment in these zones is broken down into distinct bio-geographical regions has been the subject of considerable debate over the years (GBRMPA, 1995).

In response to the need for a serviceable bioregional model for marine planning efforts, the governments of Australia cooperated on the development of the Interim Marine and Coastal Regionalisation of Australia (IMCRA v3.3). This version of IMCRA focused on continental shelf waters of less than 200 metres in depth, and identified bioregions based on species distribution and physical characteristics. These bioregions were typically hundreds to thousands of kilometres in size (IMCRA, 1998).

On 13 October 2005, the Australian government announced the launch of the National Marine Bioregionalisation. The National Marine Bioregionalisation was to be an extension of the IMCRA v3.3 project and characterize the deeper waters of Australia into distinct bioregions to assist in integrated oceans planning projects (National Oceans Office, 2005).

In April 2006, a new Integrated Marine and Coastal Regionalisation of Australia (IMCRA v4.0) was endorsed by the governments of Australia. IMCRA v4.0 was the product of the combination of IMCRA v3.3 and the National Marine Bioregionalisation projects. In combining the two national-scale marine regionalizations, IMCRA v4.0 covered Australia's waters from the coast to the edge of the EEZ, excluding Antarctica and Heard and MacDonald Islands (Department of the Environment and Heritage, 2006).

Australia's marine environment is influenced by three major ocean currents.

- The East Australian Current carries warm and salty water from the central Pacific Ocean, south along the east side of the Australian continent.
- The Leeuwin Current carries warm tropical water from the Indonesian through-flow south along the west Australian coastline, extending around the southern edge of the continent into the Great Australian Bight.
- The Antarctic Circumpolar Current is the largest single ocean current in the world, circling the Antarctic continent in an easterly direction. This current directly impacts on Australia's polar region and feeds nutrient-rich cold waters into the temperate areas of southern Australia (National Oceans Office, 2002).

Australia's marine environment is diverse and has an extraordinarily high level of endemism in some areas. For example, of the estimated 3,400 species of fish in Australia, 600 are found in southern Australian waters. Of those 600 species, 85 per cent are believed endemic to Australia and a further 11 per cent are shared only with neighbouring New Zealand (Poore, 1995).

Population

The population of Australia is relatively low, with an estimated 23.4 million as of 12 March 2014 (Australian Bureau of Statistics, undated*b*). Table 4.2 sets out Australia's population by state or territory. Australia's population is massively urbanized and concentrated in the southeast. More than 80 per cent of Australians lived within the coastal zone in 1994 (Australian Academy of Science, 1994), and this continues to be the case some twenty years later.

Economic data

Australia's oceans are an important economic resource. In 2002–03, Australia's marine industries directly contributed approximately AUS$26.7 billion to the Australian economy and, through this activity, contributed indirectly a further $46 billion in value to other sectors of the economy. This economic activity equated to around 14 per cent of Australia's gross domestic product (GDP) in that year.

Australia's marine industries are important employers within the Australian economy. In 2002–03, some 253,000 persons were directly employed within the marine sector, with a

Table 4.2 Australia's 2014 population by state or territory

State or territory (Capital city)	Population
Australian Capital Territory (Canberra)	385,600
New South Wales (Sydney)	7,500,600
Northern Territory (Darwin)	243,700
Queensland (Brisbane)	4,708,500
South Australia (Adelaide)	1,682,600
Tasmania (Hobart)	514,700
Victoria (Melbourne)	5,821,300
Western Australia (Perth)	2,565,600
Total	**23,425,700**

Source: Australian Bureau of Statistics (undated*a*)

further 690,000 jobs associated with supporting the sector. As a result, in 2002–03, marine industries were directly and indirectly responsible for employing over 9 per cent of the Australian workforce.

In 2002–03, the largest marine industry in Australia, in terms of value added to the Australian economy, was marine tourism, which contributed around 42 per cent of the total marine industries economic activity. Second largest was Australia's oil and gas sector, which contributed around 41.8 per cent of marine industries economic activity. In terms of growth within Australia's marine sector, oil and gas was the fastest growing component over the period 1995–96 to 2002–03, recording an average annual growth rate of 6 per cent (Allen Consulting Group, 2004). In 2012, it was estimated that the ocean contributed $44 billion annually to the current economy and it is projected to reach $100 billion by 2015 (Australian Government OPSAG, 2013: 5).

The importance of ecological sustainable development to Australia's Oceans Policy

The Australian National Strategy for Ecologically Sustainable Development (Commonwealth of Australia, 1992) requires that the Australian government use the principles of ecological sustainable development in all natural resource management policies. The concept of 'ecologically sustainable development' was a fundamental principle in the design of the Oceans Policy. The principles of ecologically sustainable development have also been embedded in the Environment Protection and Biodiversity Conservation Act 1999 (the EPBC Act), the Australian government's primary environment legislation.

With such a vast ocean territory, Australia commands large reserves of natural resources in a relatively pristine marine environment. To maximize Australia's ability to use these resources now and for future generations, the Oceans Policy requires that Australia's marine jurisdiction be managed sustainably.

A brief overview of the nature of and evolution of Australia's Oceans Policy

On 1 August 1994, Australia claimed a 200 nautical mile EEZ under the 1982 United Nations Convention on the Law of the Sea (UNCLOS), assuming responsibility for a massive area of ocean and natural resources (Commonwealth of Australia, 1994). By this time, it was obvious that Australia knew very little about its oceans and the wealth that they potentially contained. The nations of the world were turning to the oceans for consumable natural resources, and it was becoming increasingly apparent that the marine environment could be a source of future growth and prosperity (McKinnon et al., 1989).

The management of Australia's marine jurisdiction is shared between seven state and territory governments, in addition to the federal Australian government. Within each government's sphere of influence, management is divided along sectoral lines that operate independently of one another, with varying degrees of coordination.

The significance of the oceans to Australia's interests and the complexity of management arrangements in Australian waters led many stakeholders to call for a more strategic approach to oceans management. Given that Australia's marine environment is still in relatively good condition, marine experts from many fields saw an opportunity to take a proactive role in oceans management and to protect the sustainability of the marine environment before extensive damage occurred (Australian State of the Environment Committee, 2001).

As a result, Australia's Oceans Policy was developed and launched in 1998, with the goal of coordinating marine activities in Australia to create an effective and efficient oceans management regime. By following the principles of the Oceans Policy, the development of ocean resources could continue in an organized and environmentally sustainable fashion. There were, however, several challenges associated with the early life of policy. Not only did the sheer diversity of interests prove to be difficult to integrate, but also the state and territory governments were not signatories to the Oceans Policy. Any internal cross-jurisdictional arrangements therefore had to be negotiated in the political arena.

The policy development process

Initiation of the policy

By 1997, the Australian government had recognized that Australia needed to implement a policy for the management of its oceans. Numerous reports from a range of experts had been calling for an integrated approach to oceans management from as early as 1989 (McKinnon et al., 1989). The policy development process was led by the Department of the Environment and Heritage. However, given that Australia's oceans fall within the jurisdiction of both state and federal governments, an extensive consultation process was planned that would include all Australian governments, the community, key interest groups, and other major stakeholders.

The Australian government commissioned several technical papers on topics such as best practice oceans management planning methods, Indigenous Australian ocean policy interests, cultural considerations, and international obligations. At the same time, the Australian government consulted with the community to identify what issues were important to Australians and how the public thought the oceans should be managed. This consultation period culminated in the Australian Oceans Forum, held in Canberra on 2 December 1997 (Department of the Environment, Sports and Territories, 1998). The Forum hosted representatives from all of the coastal governments, scientists, policymakers, and representatives of a broad range of industry, conservation, and other non-government interests.

Based on the information gathered by the Forum and the consultation process, Australia's Oceans Policy was drafted for release in 1998, the International Year of the Ocean. The policy was completed and approved by the prime minister in late 1998. The state and territory governments chose not to sign up to the policy, however, and as a result it was binding only on the federal Australian government.

Although the Oceans Policy is not itself legislative, it must take into consideration all of the existing legislative instruments that govern Australia's oceans. The piece of legislation that is most relevant to the Oceans Policy is the Environment Protection and Biodiversity Conservation Act 1999 (the EPBC Act). This Act is the federal Australian government's key environmental statute, and is designed to protect the biodiversity and sustainability of Australia's marine and terrestrial environment. Marine bioregional planning – the planning model first designed for the ongoing implementation of the Oceans Policy – sits within the framework of section 176 of the EPBC Act.

Objectives

Australia's Oceans Policy begins with a vision statement:

> Healthy oceans: cared for; understood and used wisely for the benefit of all, now and in the future.

Under this vision, Oceans Policy lists nine goals:

1   To exercise and protect Australia's rights and jurisdiction over offshore areas, including offshore resources.
2   To meet Australia's international obligations under the United Nations Convention on the Law of the Sea and other international treaties.
3   To understand and protect Australia's marine biological diversity, the ocean environment and its resources, and ensure ocean uses are ecologically sustainable.
4   To promote ecologically sustainable economic development and job creation.
5   To establish integrated oceans planning and management arrangements.
6   To accommodate community needs and aspirations.
7   To improve our expertise and capabilities in ocean-related management, science, technology and engineering.
8   To identify and protect our natural and cultural marine heritage.
9   To promote public awareness and understanding.

These nine goals were developed to meet the general expectations of the Australian community and to provide scope to address the range of potential issues that the Oceans Policy may face. As a result, the goals are broad and generalized statements of intent rather than focused and quantifiable objectives (Commonwealth of Australia, 1998).

The wide geographical range of Australia's oceans has implications for setting management objectives. Priorities vary from region to region, reflecting the different needs and issues of communities and the environment. In order to focus the Oceans Policy in a regional context, a series of 'regional marine plans' were to be developed, one for each area of reasonably ubiquitous ocean territory. These plans were intended to be the primary implementation tool for the Oceans Policy. They were to include an objective-setting exercise for the purpose of creating a set of regionally specific objectives against which the effectiveness of management options could be measured. Regional objectives would meet the specific needs of ocean users in that area, but still fall under the broader goals of the policy. They were to be drafted on the basis of extensive consultation with stakeholders from within the region.

Although the framework used in regional marine planning was essentially correct, the wide-ranging interests and issues in each marine region made it difficult to bring regional marine planning to bear on specific problems that required attention. As a result, a new model for regional specific planning was created in the form of marine bioregional planning. As already noted, marine bioregional plans were to be created within the framework of the EPBC Act.

Under section 176 of the EPBC Act, marine bioregional plans were to focus on the delivery of biodiversity conservation outcomes using the principles of ecologically sustainable development. The plans were to offer a consolidated view of the Australian government's environmental roles, responsibilities, and priorities pertaining to each specific marine planning region. As a result, the plans were to give greater guidance to industry about their legal obligations.

Major principles

Australia's Oceans Policy incorporates several major planning principles.

Ecological sustainable development

All Australian governments are committed to ecologically sustainable development, as a way of 'using, conserving, and enhancing the community's resources so that ecological processes, on which life depends, are maintained and the total quality of life, now and in the future, can be increased' (Commonwealth of Australia, 1992: para. 1). A key component of the concept of 'ecological sustainable development' is intergenerational equity: ecologically sustainable development can be achieved only if the ecosystem can be maintained to such a level that the resources it provides will remain available for future generations.

The ability to continue to use natural resources in the long term is dependent on maintaining the health and integrity of the ecosystem that provides them. The Oceans Policy requires that resources be used sustainably and that the security of ecosystem health is paramount. The policy must take into account existing legislation that protects biodiversity.

The principle of ecological sustainable development is now a fundamental part of environmental and resource management legislation at all levels of Australian government, and is part of the 1992 Intergovernmental Agreement on the Environment (IGAE). The Agreement facilitates a nationally cooperative approach on the environment in Australia and clarifies the roles of each sphere of government (Department of the Environment and Heritage, undated).

Ecosystem-based management and integration

Australia's ocean ecosystems and their marine biological diversity are core national assets. If Australia's use of them is well managed, it will be able to meet a broad range of economic, social, and cultural aspirations. These ecosystems also effect a range of essential environmental 'services' that would be extremely costly or impossible to restore or replace if ecosystem functioning were to be impaired.

Traditional management of natural resources occurs on a sector-by-sector basis and is often based on political boundaries. Management agencies exist within each jurisdiction to manage the same resource. Regional marine plans used entire ecosystems as the basic planning unit, and sought to integrate across jurisdictions and sectors to ensure that all impacts on the ecosystem were considered concurrently. Bioregional plans were also to be based on marine ecosystems and seek to maximize integration.

Multiple-use management

Managing the oceans for multiple uses is a major platform of the Oceans Policy. In 1997, a report on multiple-use management was produced for the purpose of policy development (Sainsbury et al., 1997). Entitled *Multiple Use Management in the Australian Marine Environment: Principles, Definitions and Elements*, the report stated that implementation of multiple-use management requires scientific support for decision-making and in the operational use of performance measures.

Multiple-use management requires all ocean resource use (extractive and non-extractive) to be considered jointly. This allows the cumulative impacts of ocean resource use and the interactions between different uses to be understood. Measures can then be put into place to manage conflicting uses. Multiple-use management seeks to maintain ecosystem health, while providing opportunities for a variety of uses that offer the greatest long-term community benefits (taking economic, environmental, social, and cultural values into account).

Precautionary approach

In explaining the precautionary approach, Australia's Oceans Policy states that incomplete information on possible impacts should not postpone action intended to reduce or avoid unacceptable levels of change, or to prevent serious or irreversible environmental degradation of the oceans. Also, if the potential impact of an activity is uncertain, priority should be given to maintaining ecosystem health and productivity.

Best available science

The Oceans Policy acknowledges that our knowledge of the marine environment is far from complete. The vast area, depth, isolation, and inaccessibility of the oceans have made it difficult to explore Australian waters fully. Recognizing this fact, the Oceans Policy requires that ongoing study and research of Australia's oceans continue, to ensure that the implementation of the goals of the policy is supported by the best available knowledge.

Australia already has a strong and effective marine science research community across a wide range of research institutions and agencies within government, academic, and industry sectors. The challenge is to improve the coordination and integration of the research effort, to maximize its effectiveness and efficiency – particularly for government-funded research. A key priority for the Australian government is to develop a strategic approach to marine science research and data management to support its oceans policy by means of investigating measures for improved coordination, efficiency, communication, and strategic priority-setting.

Adaptive management

Oceans are naturally dynamic systems. Environmental changes, whether natural or human-induced disturbances, can have far reaching effects on the resources and use of the marine environment. Our oceans are poorly understood: vast areas of the open sea remain unexplored, and our knowledge of key processes is limited and usually focused on coastal and inshore areas. Managing a little-known and variable system requires an adaptive approach, and this is achieved in Australia through integrated marine planning. Regional marine plans and the successive bioregional planning model aim to identify useful indicators of sustainability (ecological, economic, social, and cultural) and to link them to clearly articulated and agreed long-term objectives.

Indigenous rights

The Indigenous people of Australia were living on this land for many thousands of years before the European colonial settlement in 1788. The Indigenous people are extraordinarily culturally diverse. Many Indigenous communities have strong cultural and spiritual ties to the ocean, or 'sea country'. The history of the Indigenous people in Australia extends back to long before the last ice age, and has seen the natural rise and fall of sea level over the millennia. Some sacred sites that existed on land have since disappeared below the surface, but the significance of those sites to the Indigenous people remains. Australia's Oceans Policy recognizes the importance of the oceans to the Indigenous people and the role that they can play in shaping how Australia's oceans are managed. The policy therefore provides for the participation and representation of Indigenous people in all processes related to its implementation.

Stewardship and transparency

One of the founding concepts of Oceans Policy is that responsibility for managing the oceans rests with all of the users of the marine environment. Although the policy is binding only on the federal Australian government, many of the outcomes sought under the Oceans Policy require the participation of all maritime sectors and jurisdictions if they are to be effective. Given that there is no direct legislation to enforce the policy, it is essential that all government and non-government interest groups work cooperatively to negotiate effective outcomes. The institutional arrangements that support the Oceans Policy include several stakeholder groups whose role is to provide leadership and, ultimately, stewardship on behalf of the interest groups whom they represent.

Institutional arrangements

The wide scope of Australia's Oceans Policy required that several administrative bodies be created in support of its implementation. Numerous technical working groups, representative groups, and advisory panels were established to work on specific projects under the policy. The following are the key administrative bodies that have been involved in its development and implementation.

The National Oceans Ministerial Board

The National Oceans Ministerial Board (NOMB) was originally formed in 1999 as the lead decision-making body for matters related to Australia's Oceans Policy. The Board was established to oversee the implementation of the policy and the development of regional marine plans. Its focus was on high-level strategic issues, such as government policy-setting and the overall direction of the Oceans Policy. The Board was composed of the ministers responsible for the following portfolios:

- Environment and Heritage (chair);
- Agriculture, Fisheries and Forestry;
- Industry, Tourism and Resources;
- Education, Science and Training; and
- Transport and Regional Services.

Following the October 2004 federal election, then Prime Minister John Howard made a number of changes to the administrative arrangements for the Oceans Policy. He decided to abolish the NOMB and directed that major ocean issues should be given to the Sustainable Environment Committee of the Cabinet of the federal Australian government for deliberation.

The Oceans Board of Management

The Oceans Board of Management (OBOM) comprised principally the heads of the federal Australian government departments with significant interest in oceans policy specifically and marine matters generally. The formal membership included the heads of the following departments and agencies:

- the Department of the Environment and Water Resources (chair);
- the Department of Agriculture, Fisheries and Forestry;
- the Department of Industry, Tourism and Resources;
- the Department of Education, Science and Training;
- the Department of Transport and Regional Services;
- the Department of Defence;
- the Department of Finance and Administration;
- the Department of the Prime Minister and Cabinet; and
- the Australian Fisheries Management Authority.

The OBOM was a forum for constructive discussion between senior government department officials on oceans management issues. As the most senior group to consider oceans issues at the departmental or operational level, it provided advice directly to the Minister for the Environment, Heritage and the Arts and to the broader government as required. Although the Board no longer meets in a formal capacity, discussion and cooperation between Australian government agencies continues to occur on marine policy issues as they occur as part of the routine business of government.

The National Oceans Advisory Group

The National Oceans Advisory Group (NOAG) is a high-level non-government stakeholder group, consisting of representatives from the following key marine sectors and interest groups:

- tourism;
- Indigenous Australians;
- conservation;
- ports;
- commercial shipping;
- recreational and light commercial industries;
- commercial fishing;
- recreational fishing;
- oil and gas;
- minerals;
- community;
- science;
- maritime policy and law; and
- the Oceans Policy Science Advisory Group.

The Group is an independent body that reports directly to the Minister for the Environment, Heritage and the Arts. Its purpose is to provide government ministers with advice on the overall scope and effectiveness of the implementation of the Oceans Policy.

As a non-government representative, NOAG is uniquely placed to provide insight on the response of the wider community to oceans policy issues. The Group has also proven to be an excellent forum for the exchange of ideas between all sectors without the government acting as an intermediary.

The Marine and Biodiversity Division and the National Oceans Office

Formed as an executive agency of the Commonwealth in 1999, the National Oceans Office (NOO) was created to support the NOMB in the implementation of the Oceans Policy. Primarily, NOO was responsible for the development of regional marine plans, intended to be the primary implementation tool for the Oceans Policy, and for providing support and advice on a number of technical issues. The Office coordinated the day-to-day delivery of Australia's Oceans Policy, and provided advice and technical support to the NOMB, OBOM, NOAG, and the other administrative bodies that support the policy.

Following the October 2004 federal election, the NOO was integrated into the wider organization of the Department of the Environment and Heritage as part of a broader restructure of the public service. Following the November 2007 federal election and the resulting change of government, then Prime Minister Kevin Rudd restructured the ministry of the Australian government. Responsibility for the implementation and development of oceans policy came to rest with the Marine and Biodiversity Division of the Department of the Environment, Water, Heritage and the Arts, which provided central coordination and policy advice to the Australian government on the marine environment.

In addition to the implementation of the Oceans Policy by means of marine bioregional planning, the Marine and Biodiversity Division was responsible for the delivery of a national representative system of marine protected areas (MPAs), programmes for the protection of marine and migratory biodiversity, Australia's cetacean recovery policy, the assessment of sustainable fisheries, and the participation of the Australian government in international marine forums. The Marine and Biodiversity Division also provided policy and technical advice to other government agencies on a variety of marine-related matters, and produced and distributed information to the community.

The nature and implementation of Australia's Oceans Policy

The nature of Australia's Oceans Policy

The existing management arrangements for Australia's oceans are extraordinarily complex. There are scores of existing instruments that regulate or control how affairs are conducted in the marine environment. Australia's Oceans Policy is an 'umbrella' policy in the sense that it does not replace existing policies and legislation; rather, it is designed to integrate and coordinate existing mechanisms without adding to existing oceans management arrangements.

By the virtue of its function, the Oceans Policy is high-level and general in nature. It provides a broad direction on how Australia's oceans need to be managed rather than specific details on operational matters. The policy provides the scope within which marine plans can be produced, and it is these plans that provide the specifics on oceans management. The Oceans Policy cannot replace or remove existing instruments.

Australia's Oceans Policy is administrative only, with no supporting legislation. At this stage, there have been no plans to institute an Oceans Act, although, as already noted, the Environment Protection and Biodiversity Conservation Act 1999 plays a significant role in implementing the principles of the Oceans Policy. Given that the marine bioregional planning model uses the EPBC Act as a framework, the role of this piece of legislation in the implementation of the policy will increase with time.

Implementing the principles: Detailed assessment

The primary tool for implementing Australia's Oceans Policy was originally intended to be regional marine planning. Regional marine planning was to be integrated – that is, it was to bring together all relevant interests, fields of discipline, and jurisdictions. It was also to be participatory: key stages of the process required the direct involvement and participation of ocean users and other stakeholders. Regional marine planning was additionally to be strategic and was to focus on delivering priorities over a long period of time.

Regional marine planning used an 'ecosystem-based approach' to management, a concept that is central to the Oceans Policy. The entire Australian exclusive economic zone (EEZ) has been divided into large marine domains – or large marine ecosystems – which are based on broad biophysical patterns and represent the units for strategic integrated planning. The large marine domains were originally defined by the Commonwealth Scientific and Industrial Research Organisation (CSIRO) using bathymetry, bathymetry variance, water column properties, and seafloor plate age. The exact boundaries of the domains were subject to some variation as new information came to light about their characteristics. More information on this process can be found in Australia's Marine Science and Technology Plan (Jensen and Marine Science and Technology Working Group, 1999: 45). Figure 4.1 shows the boundaries for marine planning in Australia in 2006, which were based on large marine domains.

Regional marine planning has considerably increased our knowledge of the oceans and its resources, and our ability to manage for sustainability and biodiversity protection. The consolidation of existing and new biophysical data has resulted in a deep-water marine bioregionalization (see 'Research and education' later in this chapter). This considerable achievement has enabled Australia to embark on the establishment of a network of representative marine protected areas (MPAs).

The first regional marine plan for South-east Australia was finalized in 2004 (National Oceans Office, 2004). It was the first of what were to be six regional marine plans, encompassing thirteen large marine domains, to be developed around Australia. Regional marine planning was under way in the Northern Marine Region and about to commence in the South-west Marine Region when the Australian government approved the development of the marine bioregional planning model for the implementation of the Oceans Policy.

As a reflection of Oceans Policy, regional marine planning had been an extremely broad-level planning construct. It had become apparent that it would be very difficult to achieve effective, on-the-ground outcomes from such a mechanism with so wide a scope. In response, bioregional plans were an evolution of regional marine planning, designed to provide an increased focus on key sustainability objectives. These plans were be implemented under the existing framework

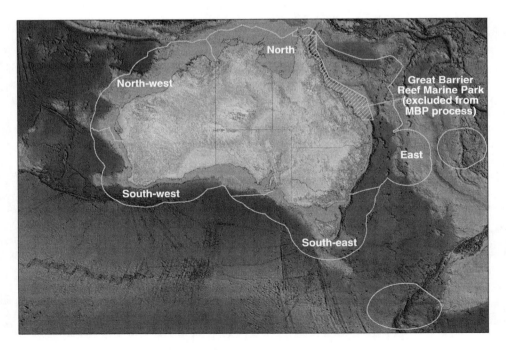

Figure 4.1 Marine bioregional planning areas

Source: Department of the Environment and Water Resources (2007)

of section 176 of the EPBC Act. With the support of a legislated framework, bioregional plans represented the next logical step in strengthening the effective implementation of the Oceans Policy in Australia.

Under a section 176 marine bioregional plan, marine planning is narrowed to focus on the delivery of biodiversity conservation outcomes under the framework of ecologically sustainable development. It delivers two primary things: first, an overarching, environmental-outcomes-based framework; and second, the consolidation, through an integrated strategy, of conservation priorities and tools.

Bioregional plans represent a comprehensive ecological profile of the region, identifying key conservation values or features and threatening processes. The plans set objectives, strategies, and actions for biodiversity conservation in the region that will be delivered over the life of the plans. The plans also include a comprehensive overview of the Australian government's current suite of biodiversity conservation powers and initiatives, such as threatened species recovery planning, threat abatement plans, strategic assessments, MPA development, and fisheries assessments. The plans operate as a key reference document to guide industry engagement with the government in its administration of the EPBC Act.

In addition to marine planning, there are several other government processes that furthered the goals of Oceans Policy. Integration of activities across the state and federal governments was pursued through the Natural Resource Management Ministerial Council. The Council included representatives of all state and federal governments in Australia, and was empowered to make decisions that cross jurisdictional boundaries. An example of such a cross-jurisdictional process was the development of the Framework for a National Cooperative Approach to Integrated Coastal Zone Management. The Framework recognized that the coastal environment of Australia is subject to a number of environmental pressures resulting from the high concentration of population in the coastal zone. The purpose of the Framework was to promote a cooperative approach to managing coastal zone issues that would be relevant to all Australian governments. The Framework had the strategic goal of providing a basis for a cooperatively developed implementation plan to address coastal issues in Australia (Natural Resource Management Ministerial Council, 2003).

Authority at national level

Prior to the November 2007 federal election, ultimate responsibility for the implementation of the Oceans Policy rested with the Sustainable Environment Committee of the Cabinet. The Committee made all of the high-level strategic decisions, and provided final approval for the release of key documents and processes. Following the change of government in the 2007 election, however, authority for the implementation of the Oceans Policy came to rest with the Minister for the Environment, Heritage and the Arts. Formal mechanisms for broader government decision-making on environmental issues were to be developed by the new government.

Prior to the election, the Oceans Board of Management (OBOM) reported, first, to the Minister for the Environment and Water Resources, and then to other ministers as required. The Board's membership comprised the departmental and agency heads of portfolios, plus government organizations with interests in the marine environment. As a result, OBOM had a significant influence on the management of marine issues. Although it no longer meets in a formal capacity, the heads of government departments and agencies still provide direction and leadership on the management of operational issues relating to oceans policy and provide advice to government on high-level policy matters.

Under the direction of the Minister for the Environment, Heritage and the Arts, the Marine and Biodiversity Division of the Department was responsible for coordinating the implementation of the Oceans Policy. The Marine and Biodiversity Division acted as a coordinator and facilitated the exchange of information between the government and stakeholders. It maintained the administrative functions of the Oceans Policy and provided the Minister with technical advice on oceans issues. The Marine and Biodiversity Division was also responsible for developing bioregional plans for Australia's oceans, and represented the Australian government in many technical fora both domestically and internationally.

As noted earlier, the National Oceans Advisory Group (NOAG) also provided independent advice to government on policy matters relating to oceans issues and on the work of the Marine and Biodiversity Division. As an independent non-government advisory body, NOAG also ensures transparency of process for the community.

The national/subnational division of authority and interaction

In the early 1970s, the state governments of Australia challenged the Commonwealth's assertion of sovereignty over the then 3 nautical mile territorial sea. The Commonwealth and the states subsequently came to a series of arrangements collectively known as the Offshore Constitutional Settlement (OCS). The purpose of the OCS was to give the states a greater legal and administrative role in offshore areas. The principle legislation implementing the OCS – the Coastal Water States, Power and Title Act 1982 – entered into force in February 1983.

The 1982 Act made it clear that, should the territorial sea subsequently be extended from 3 nautical miles to 12 nautical miles, the OCS arrangements would continue to apply only to the 3 nautical mile limit. In 1990, the territorial sea was extended to the 12 nautical mile limit, but the relevant limit for the purposes of the OCS remained at 3 nautical miles.

In effect, through the OCS, the Commonwealth agreed to give the states primary responsibility over coastal waters from the territorial baseline out to 3 nautical miles (except in the case of external territories, which are directly administered by the federal government). Beyond that boundary, the federal Australian government retains primary responsibility. The OCS also included a number of cooperative arrangements for the management of resources offshore, such as fisheries and petroleum. These cooperative arrangements are reflected in relevant federal, state, and territory legislation (Commonwealth of Australia, 1998).

The territorial division of responsibility for Australia's oceans means that management options for issues that cross the 3 nautical mile jurisdictional boundary need to be negotiated between the state, territory, and Australian governments. There are several forums through which this can be achieved, although the primary institution for such negotiations is the Council of Australian Governments (COAG).

The Council is the peak intergovernmental forum in Australia, comprising the prime minister, state premiers, territory chief ministers, and the president of the Australian Local Government Association. Under the leadership of COAG, a number of ministerial councils with representatives from each jurisdiction meet to discuss specific policy matters. The council most commonly associated with the Oceans Policy is the Natural Resource Management Ministerial Council.

Domestic implementation of international agreements

The 1982 United Nations Convention on the Law of the Sea
Australia has established zones of offshore jurisdiction in accord with the UNCLOS. On 1 November 1979, Australia proclaimed an exclusive fishing zone (the Australian Fishing Zone) extending 200 nautical miles from the territorial sea baseline (Commonwealth of

Australia, 1979). Australia then extended its territorial sea from 3 nautical miles to 12 nautical miles, with effect from 20 November 1990 (Commonwealth of Australia, 1990), and proclaimed a 200 nautical mile EEZ with effect from 1 August 1994 (Commonwealth of Australia, 1994) and a 24-mile contiguous zone with effect from 7 April 1999 (Commonwealth of Australia, 1999).

On 21 April 2008, the United Nations Commission on the Limits of the Continental Shelf (CLCS) confirmed Australia's right to claim 2.5 million km² of extended continental shelf beyond 200 nautical miles from the territorial sea baseline (Minister for Resources and Energy, 2008). Australia has the exclusive right to the resources of the seabed of this area, but not to the water column above, which remains international waters.

The Convention on Biological Diversity

The 1992 Convention on Biological Diversity (CBD) is one of the more significant multilateral environment agreements, both in terms of the broad issues covered and the high number of countries that have either ratified or acceded to it (194, including Australia). Australia maintains close relations in CBD negotiations with a number of likeminded agricultural exporter countries. Australia is working closely and collaboratively with New Zealand on a broad range of CBD issues.

The ever-expanding breadth and reach of the Convention's agenda has necessitated taking a strategic whole-of-government approach to Australia's participation. This will assist in better focusing available resources and effort. It will also guide Australian input into the programme of work, which often encroaches upon – and sometimes conflicts with – the substantial business of other international instruments. This approach should enable government to identify the most appropriate forum at which these key interests should be addressed.

The obligations of parties to the Convention are, perhaps, best summarized by Article 6 ('General Measures for Conservation and Sustainable Use'), which states that:

> Each contracting Party, shall, in accordance with its particular conditions and capabilities:
>
> a Develop national strategies, plans or programmes for the conservation and sustainable use of biological diversity, or adapt for this purpose existing strategies, plans or programmes which shall reflect, inter alia, the measures set out in this Convention relevant to the Contracting Party concerned; and
> b Integrate, as far as possible and as appropriate, the conservation and sustainable use of biological diversity into relevant sectoral or cross-sectoral plans, programmes and policies.

National representative system of marine protected areas

Both the CBD and UNCLOS commit parties to protection of the marine environment. In Australia, this international commitment has translated into one of the major initiatives under Australia's Oceans Policy, which provides for the accelerated development of a National Representative System of Marine Protected Areas (NRSMPA). The establishment of the NRSMPA also supports the programme of the World Conservation Union (IUCN) World Commission on Protected Areas (WCPA) to promote the establishment of a global representative system of MPAs. The implementation of the NRSMPA is guided by the Guidelines for Establishing the NRSMPA (ANZECC, 1998).

Excluding the Great Barrier Reef Marine Park (which is governed under its own legislation, the Great Barrier Reef Marine Park Act 1975), twenty-seven MPAs have been declared in Commonwealth waters to-date, protecting a total of 498,377 km². However, despite these achievements, a number of gaps remain in Australia's distribution of MPAs. For this reason, the government has made it a priority to establish MPAs in large-scale bioregions that are not already represented within the NRSMPA.

The development of the NRSMPA has been linked to the regional marine planning initiative, providing opportunities to consider other conservation measures when designing representative MPAs and to ensure that MPAs are not identified in isolation from the management of sustainable resource use. This relationship continued under the bioregional marine planning model. Indeed, it was expected that the links between MPAs and other conservation measures would be much stronger in this approach than they had been in the past.

The total area of Australia's MPA estate, including the Great Barrier Reef, exceeds 842,620 km², an area equal to about a third of the entire global MPA estate (Department of the Environment and Heritage, 2006b; Sea Around Us Project, 2005).

Protection of marine species

Although Australia supports the marine biodiversity conservation objectives of the CBD, it is also party to a number of other multilateral environment agreements (MEAs) and non-legally binding memoranda of understanding (MoUs), which aim to protect, conserve, and manage marine species at the national, regional, and global levels. The principal national environment legislation through which Australia implements its obligations and actions from MEAs and MoUs is the Environment Protection and Biodiversity Conservation Act 1999. Examples of MEAs and MoUs to which Australia is a party and which aim to protect, conserve, and manage marine species include:

- the 1982 Convention on the Conservation of Migratory Species of Wild Animals (CMS);
- the 1973 Convention on International Trade in Endangered Species of Fauna and Flora (CITES); and
- the 2001 Indian Ocean and South-east Asia Turtle Memorandum of Understanding (IOSEA Turtle MoU)

Species listed in the CMS appendices for which Australia is a range state are automatically listed as migratory species under the EPBC Act. Under the Act it is an offence to kill, injure, take, trade, keep, or move a member of a listed migratory species in Commonwealth waters without a permit. Breach of this prohibition may result in a conviction or a fine.

Listed migratory species are considered matters of national environmental significance (NES) under the EPBC Act. Any action that has, will have, or is likely to have a significant impact on a matter of NES, including listed migratory species, needs to be subjected to environmental impact assessment (EIA) prior to approval. Conditions may be attached to approval to minimize impacts.

Species listed in the CITES appendices are automatically covered by the export and import regulations under the EPBC Act. It is an offence to export or import CITES-listed species from or into Australia without the appropriate CITES permits.

Because some of the listed migratory species are also listed as threatened under the EPBC Act, the Australian government is required to develop and implement recovery plans for those species. For example, Australia implements the IOSEA Turtle MoU by means of the Recovery

Plan for Marine Turtles in Australia. Actions under the IOSEA Turtle MoU and the Recovery Plan are implemented through Australian government project funding, and are undertaken by state agencies, research institutions, industry, and non-government organizations (NGOs).

The Australian government ensures that its obligations and actions under MEAs and MoUs complement national legislation, and that national legislation is consistent with its international obligations. Australian government agencies, state and territory governments, and stakeholders are consulted on species listing, and the Australian federal Parliament assesses treaty actions to ensure that they are in the national interest prior to accepting legally binding obligations.

The 1973 International Convention for the Prevention of Pollution from Ships, as modified by the Protocol of 1978 (MARPOL)

Regulations to prevent pollution from ships in Australian waters are implemented by both federal and state and territory governments. They are based on the 1973 International Convention for the Prevention of Pollution from Ships, as modified by the Protocol of 1978, known as MARPOL. This Convention is in force in 152 countries (as of May 2014) and is the main international convention covering prevention of ship-sourced pollution in the marine environment.

MARPOL deals with pollution that might result from accidents such as collisions or groundings, as well as all types of waste generated during the normal operation of ships, known as 'operational waste'. Ships are permitted to discharge small quantities of this operational waste, subject to very strict controls. A 'discharge' is any release from a ship, whatever the cause. The Convention has separate technical annexes dealing with preventing pollution by oil, chemicals, harmful substances in packaged forms, sewage, garbage, and air emissions.

There are special protection measures for Australia's Great Barrier Reef. The Great Barrier Reef Marine Park Authority (GBRMPA) introduced new offences and increased penalties under the Great Barrier Reef Marine Park Act 1975, which apply to ships when within the boundary of the marine park. These included the introduction of new provisions that make it an offence if:

- a person uses or enters a zone for a purpose other than a purpose permitted under the relevant zoning plan (which provision imposes strict liability);
- a ship operates in a zone in which a ship is not permitted to be operated under the relevant zoning plan (a two-tiered offence whereby the first tier provides for a maximum penalty of AUS$1.1 million and the second tier imposes strict liability);
- a ship operates in a zone without permission of the GBRMPA where such permission is required under the relevant zoning plan (a two-tiered offence whereby the first tier provides for a maximum penalty of $1.1 million and the second tier imposes strict liability); and
- a person operates a ship contrary to the conditions of a permission (the maximum penalty being $1.1 million);
- a person negligently operates a vessel in the marine park in circumstances in which that operation results in, or is likely to result in, damage to the park (a two-tiered offence whereby the first tier provides for a maximum penalty of $1.1 million and the second tier imposes strict liability) (AMSA, 2001).

MARPOL places an obligation on governments to ensure that ports provide adequate facilities for the disposal of the various waste products generated on board ship. The authority responsible for managing the port is required to ensure that the waste reception facilities are adequate for the type of shipping traffic that passes through the port. These facilities can be fixed, such as those normally found at oil terminals to receive tank washing from oil tankers, or mobile, such as road tankers operated by private waste removal contractors (ANZECC, 1997).

The Australian Maritime Safety Authority (AMSA) conducts an extensive programme of inspecting ships visiting Australian ports to ensure compliance with the relevant International Maritime Organization (IMO) conventions, a programme known as 'port state control'. AMSA is responsible for the application and enforcement of MARPOL in areas of Commonwealth jurisdiction, which is to the limit of the 200 nautical mile EEZ. State governments are responsible for coastal waters up to 3 nautical miles (5.5 km) offshore, and the GBRMPA is responsible for enforcement activity in respect of illegal discharges in the Great Barrier Reef Marine Park.

Penalties under MARPOL legislation administered by AMSA are up to $1.1 million for the shipowner and $220 000 for the master of a ship discharging in contravention of the MARPOL regulations. The legislation provides wide powers for AMSA marine surveyors to board ships and obtain evidence, such as oil samples, and enables ships to be detained while investigations are carried out.

While the focus of AMSA, the GBRMPA, and the various state government agencies involved in enforcing MARPOL will always be on preventing pollution incidents, enforcement action is becoming increasingly effective. From 1991 to 2003, there were ninety-four successful prosecutions in Australian courts. Further successful prosecutions were also conducted in foreign courts following pollution incidents in Australian waters.

The 1972 Convention on the Prevention of Marine Pollution by Dumping Wastes and Other Matter

The Environment Protection (Sea Dumping) Act 1981 (the Sea Dumping Act) was enacted to fulfil Australia's international responsibilities under the 1972 Convention on the Prevention of Marine Pollution by Dumping of Wastes and Other Matter (London Convention). Australia ratified the 1996 Protocol to the London Convention in 2001 and implemented it domestically even before it came into force.

The Sea Dumping Act is administered by the Department of the Environment, Water, Heritage and the Arts (DEWHA), or the GBRMPA if dumping is to take place within the Great Barrier Reef Marine Park. The Act applies in respect of all Australian waters (other than waters within the limits of a state or the Northern Territory), from the low-water mark out to the limits of the EEZ. The Act regulates the deliberate loading and dumping of wastes and other matter at sea. It applies to all vessels, aircraft, or platforms in Australian waters, and to all Australian vessels or aircraft in any part of the sea. The Act does not cover operational discharges from ships, such as sewage and galley scraps. Those are regulated by the protection of the sea legislation administered by AMSA.

Permits from the DEWHA (or the GBRMPA) are required for all sea dumping operations and permits will be considered only for those categories of waste that are specified in Annex 1 to the 1996 Protocol. In deciding whether to grant a permit, consideration is given to the type of material proposed to be dumped, the dump site, and the potential impacts on the marine environment. Dredge spoil is assessed in accordance with Australia's National Ocean Disposal Guidelines for Dredged Material (Commonwealth of Australia, 2002). Currently, about thirty permits are issued in Australia per year, mainly for the dumping of uncontaminated dredge spoil, disposal of vessels, and for burials at sea.

The DEWHA also administers the Sea Installations Act 1987 and the Sea Installations Levy Act 1987. The Sea Installations Act provides the legislative basis for the Australian government to:

- ensure that sea installations are operated with regard to the safety of the people using them, and the people, vessels, and aircraft near to them;
- apply appropriate laws in relation to such sea installations; and
- ensure that such sea installations are operated in a manner that is consistent with the protection of the environment.

A 'sea installation' refers to any man-made structure that when in, or brought into physical contact with, the seabed or when floating can be used for an environment-related activity. An 'environment-related activity' is defined as any activity relating to tourism or recreation, the carrying on of a business, exploring, exploiting, or using the living resources of the sea, seabed, or subsoil of the seabed, marine archaeology, or any other prescribed activity. Examples of the sorts of structures that are defined as sea installations include floating hotels, tourism pontoons, artificial islands, and submarine power cables. There are also several exclusions set out under the Act. Basically, the Act applies from the 3 nautical mile state limit out from the coast to the outermost limits of Australian waters. It applies from the coast outwards in the case of external territories.

The Global Programme of Action for the Protection of the Marine Environment from Land-based Activities

Although Australia has not formally adopted a national action plan under the Global Programme of Action for the Protection of the Marine Environment from Land-based Activities (GPA) of the United Nations Environment Programme (UNEP), priorities for action in controlling pollution from land-based activities have been identified through the Framework for a National Cooperative Approach to Integrated Coastal Zone Management. The Framework was endorsed by the Natural Resource Management Ministerial Council in October 2003. It sets the scene for national cooperation in managing coastal issues and achieving ecologically sustainable development outcomes in the coastal zone over the next decade. The Framework addresses cross-border and sectoral issues, harmonizes joint actions towards management of common issues, and builds on existing, and encourages new, investments from all jurisdictions.

The Framework identifies land-based sources of pollution as a key issue that should be addressed cooperatively. It states that:

> Land-based sources of pollution, especially diffuse source pollution, whether derived from agricultural or urban sources, has been identified as one of the greatest threats to the health, productivity and biodiversity of Australia's coasts and oceans. The effects of land-based sources of pollution tend to be reasonably well understood where those effects are obvious and extreme. However, these effects tend to be very poorly understood where they are small, non-linear and/or spatially exclusive. A national cooperative approach to improve estuarine, coastal and marine water quality will enhance ecologically sustainable outcomes.
>
> *(Natural Resource Management Ministerial Council, 2006: 17)*

The Natural Resource Management Ministerial Council was considering an implementation plan that would contain nationally agreed actions to address land-based sources of pollution. Additionally, there are numerous smaller mitigation projects being carried out by the Australian government in cooperation with one or more state or territory governments, which are relevant to the reduction of land-based sources of marine pollution.

Enforcement

Australia's Oceans Policy has no supporting legislation and therefore there are no direct enforcement activities related to the policy. However, there are several existing pieces of legislation that relate to activities that it describes. Each of these instruments has its own enforcement and compliance programme, which will vary depending on the requirements of legislation. Enforcement operations under natural resource management legislation at the federal level are usually carried out by law enforcement agencies and the military, such as the Australian Federal Police, the Royal Australian Navy, Coastwatch, and the Australian Customs Service.

The marine bioregional planning model existed within the framework of the EPBC Act. Section 176(5) states that, '[s]ubject to this Act, the Minister [for the Environment, Heritage and the Arts] must have regard to a bioregional plan in making any decision under this Act to which the plan is relevant'. The exact enforcement and compliance implications of the bioregional plans are yet to be fully explored.

Research and education

To meet national science priorities approved by the Oceans Board of Management (OBOM), several research projects have been commissioned to further the implementation of the Oceans Policy.

The National Science Work Program provided strategic science support for the development and implementation of the Oceans Policy. Projects under the programme aimed to:

- build on existing knowledge, and to make data and information about oceans widely available;
- survey and map the biological, geological, and oceanographic characteristics of Australia's oceans, including the uses occurring within them; and
- incorporate the best available knowledge into the development of marine planning.

Work under the National Science Work Program focused on three main areas:

1 development of national marine bioregionalization, describing broad patterns of biodiversity, to support the planning and management framework;
2 providing the socio-economic information for marine planning to assist the planning process and advance national assessments of marine issues and uses in Australia's oceans; and
3 advancing the understanding and use of ecosystem-based and adaptive management within marine planning.

The Australian government contributes to collaborative research voyages that enhance understanding of the biophysical environment under Australia's marine jurisdiction. Given the expense of conducting marine research, several institutions contribute to voyage expenses to defray costs. Each voyage carries scientists from contributing organizations and will run several projects simultaneously to capitalize on the limited availability of ship time. Partnerships typically include museums, universities, and government agencies, often from other nations. For example, in early 2003, the National Oceans Office (NOO) coordinated two marine science voyages with the governments of France and New Zealand: one across the Great Australian Bight; the other around Lord Howe Island and Norfolk Ridge. During these research cruises, several hundred undescribed species were discovered and previously unknown underwater features were mapped, stimulating worldwide media interest. The data collected will be a crucial input to future marine planning efforts.

Further marine science voyages are planned, with the research effort to be delivered through the Australian government's research vessel, *RV Southern Surveyor*, which has been upgraded with modern ocean floor imaging equipment, funded by a multi-agency partnership. The investment has significantly enhanced Australia's marine science research capabilities.

The results of a major project to define Australia's deep-water marine bioregions were publicly announced on 13 October 2005. The National Marine Bioregionalisation was a major national scientific project that divided Australia's oceans into bioregions to assist in the development of oceans planning and management. This bioregionalization covered an area of approximately 10 million km^2. The project used information on the physical environment and the distribution of marine species to identify each unique bioregion. Each bioregion has a unique combination of seascape features (such as canyons and seamounts), and species of fish, sponges, plankton, and other marine organisms (National Oceans Office, 2005).

In April 2006, the Natural Resources Management Ministerial Council endorsed a new Integrated Marine and Coastal Regionalisation of Australia, known as 'IMCRA v4.0', resulting from the combination of the Interim Marine and Coastal Regionalisation of Australia, or IMCRA v3.3 (which provided a marine regionalization of continental shelf waters) with the National Marine Bioregionalisation (which regionalized off-shelf waters). In combining the two national-scale marine regionalizations, IMCRA v4.0 covers Australia's waters from the coast to the edge of the EEZ, excluding Antarctica and Heard and MacDonald Islands. IMCRA v4.0 is being used as a key tool for marine planning and MPA development in Australian waters (Department of the Environment and Heritage, 2006a).

The Marine and Biodiversity Division produced numerous communication and educational products related to the Oceans Policy that can be accessed online at http://www.environment.gov.au/coasts/discovery/index.html. The site contains detailed information on many oceans-related subjects, including reports, media releases, animated seafloor fly-throughs, and research voyage diaries.

Most government agencies and research organizations with marine expertise maintain websites that contain information relevant to their mandate. Many organizations also run an education programme as part of their communications policy.

Under Australia's Oceans Policy, the Australian government is committed to:

- developing a long-term marine education policy and programme for all school ages (from kindergarten to Year 12) to be incorporated in all Australian states and territories;
- continuing to develop relevant resource materials for use in schools, and technical and further education colleges, in cooperation with professional bodies; and
- continuing to support provision of quality practical educational material for teachers and students.

The Australian government is also committed under the policy to a variety of training measures including:

- continuing to develop training courses, open learning courses, and summer school programmes that focus on integrated marine management and enhancement of practical skills;
- establishing training and development programmes to meet the specific needs and requirements of Indigenous landholders and managers; and
- establishing workplace training to ensure the continued development and dissemination of best practice marine and coastal management skills.

Another important communication tool is national state of the environment (SoE) reporting. The national-level SoE reports provide information about environmental and heritage conditions, trends, and pressure points for the Australian continent, surrounding seas, and Australia's external territories. National SoE reports are released every five years, with the most recent released in 2011 (State of the Environment 2011 Committee, 2011). The SoE report includes themes specifically on coasts and the marine environment.

Financing

Funding for the implementation of Australia's Oceans Policy is drawn through the consolidated revenue appropriated by the Department of the Environment, Water, Heritage and the Arts by means of the standard government agency budget review process. The implementation of the Oceans Policy began in 2000, with the establishment of the National Oceans Office in Hobart, Tasmania. By the end of the 2005–06 financial year, AUS$25.95 million was spent on establishing administrative arrangements to support the Oceans Policy and on the development of regional marine planning. An additional $13 million was spent over the same period on science-based research to support the development and implementation of the Oceans Policy. As part of its 2006–07 budget, the Australian government committed a further $37 million over four years to fund the development of the marine bioregional plans.

Implementation, evaluation, and long-term outlook

Review of problems, issues, or obstacles addressed by Australia's Oceans Policy

The issues faced in delivering Australia's Oceans Policy are complicated and vary from region to region. None of these issues is insurmountable, but the complex administrative arrangements for Australia's oceans require good integration of management across sectors and jurisdictions if they are to produce effective outcomes.

One fundamental issue that must be addressed before the Oceans Policy can be considered to be fully effective is the question of integration. The division of responsibility across the federal, state, and territory governments, and across all maritime sectors and interest groups, creates a situation in which the simplest of issues can require a response from many authorities and organizations. This tendency for issues to become increasingly complex can be addressed only through strong communication and integration across sectors and jurisdictions.

This complexity, coupled with the inherently broad nature of the Oceans Policy, has proven to be a major obstacle in achieving substantial on-the-ground outcomes in its delivery. Regional marine planning was intended to be the vehicle with which to deliver the policy, but the goals of the policy were so diverse and the scale of the plans so large that there was a tendency for the plans to try to achieve too much – effectively, to be every possible solution to every problem.

It should be acknowledged that regional marine planning has moved government agencies and industry to the point at which they were able to conceptualize the management of ocean use in an integrated way. Significant advances were made under the Oceans Policy in improving understanding of the marine environment, and regional marine planning was a valuable tool for informing the management of Australia's marine resources. However, it is clearly apparent that regional marine planning lacked the focus to produce real outcomes in all policy areas.

Given that Australia's Oceans Policy was one of the first national marine policies, it is not surprising that there is a substantive element of 'learn by doing' associated with the regional marine planning process. The early lessons of the southeast and northern Australian planning processes led to a major adjustment of the planning model in 2004 to focus on three major outputs: a description of the marine region; ten-year ecological sustainability indicators; and an outline of strategies and actions. This framework is essentially correct, but some adjustment was needed to bring the focus more directly to biodiversity conservation priorities at the regional level, and to create the long-term security of access and certainty of process for existing and future marine-based industries that was one of the key objectives of the Oceans Policy.

The first regional marine plan for southeast Australia was completed and approved by government in April 2004. Given that the plan took more than four years to complete and is primarily a process document, which has not tackled the issue of multiple-use decision-making in a practical manner, it received some negative reaction. The plan did not include a mechanism for marine use zoning or resource access, but provided the basis for improved *coordination* of government processes (as opposed to the *integration* envisaged under the Oceans Policy). This is not surprising, and reflects the relatively mature and effective sectoral arrangements already in place in Australia for dealing with industry impacts on the marine environment.

It should be noted that the EPBC Act was only a Bill before Parliament at the time that the Oceans Policy was published. It appears that the objectives of integrated oceans management have been pursued without sufficient reference to or consideration of the opportunities presented by the EPBC Act. Australia's Oceans Policy included a commitment to examine whether there was a need to establish a statutory basis for the development and implementation of regional marine planning. Consideration of this issue represented a substantial piece of work and involved questioning whether existing sectoral legislation was capable of delivering on oceans policy objectives. It was found that the existing management system had the capacity to implement management on an ecosystem basis, whether through permits or management plans, although there was a degree of uncertainty as to the extent that this was within legislative mandates.

The bioregional planning model sits within the existing legislated framework of the EPBC Act. The advent of the EPBC Act and the strong emerging capacity for the whole of government working on key cross-sectoral issues in the marine environment indicates that those aspects of the regional marine planning model focused on the allocation of resource access and use across and within sectors are essentially superfluous. This will remain the case while the government manages sectoral activity according to the principles of ecologically sustainable development.

Monitoring, evaluation, and adjustment

The last review of the Oceans Policy took place in 2002 and resulted in a number of recommendations on how to improve the delivery of its goals. The recommended changes were implemented and appear to have improved the efficiency of the delivery of the policy.

Regional marine plans were subjected to review as they developed. As experience with these large-scale marine plans increased, the government was able to assess its progress and make adjustments where necessary. The latest evolution of regional marine plans into marine bioregional plans marked the most significant shift to date in how the Oceans Policy was to be implemented. This planning regime has been subject to regular scrutiny to ensure that it is delivering good outcomes.

Marine bioregional plans were also to include baseline measures for ecological sustainability, which would be reported in the Australian SoE report every five years (the 2011 report including a theme on the marine environment and coasts). The sustainability indicators would be agreed with Commonwealth agencies, and could then be used in sector-based planning and regulatory processes.

Outlook

Australia's Oceans Policy has had cross-party political support in Australia, and there is no reason to think that the basic principles underpinning it will not be maintained in the long term. It would be no surprise, however, if there were some review of the details of the Oceans Policy in light of the experience gained since the launch of the policy in 1998, with a view to updating it to better reflect the current state of affairs.

Continued investment in building the scientific knowledge of Australia's marine environment will be a key input to ensuring the sustainable management of its natural resources. A greater focus on sustainability outcomes will create an implementation-driven planning process and ensure that the Oceans Policy has real outcomes for the community. A planning framework enshrined in legislation will guide government decision-making and provide greater certainty for industry groups, guiding their interactions with the Australian government.

The new approach described by bioregional planning will bring a sharper focus to marine conservation outcomes. At the same time, it will provide the outcome-based framework for sectoral management to continue pursuing integrated oceans management in a strategic and consistent fashion. The plans will be developed within an adaptive management framework that allows for regular monitoring and periodic review.

Recommendations for improving Australia's Oceans Policy

It would not be proper for the authors, as public officials, to suggest improvements to Australia's Oceans Policy in the absence of explicit government consideration and endorsement. However, it has been very apparent that, for the Oceans Policy to be fully effective, there needs to be a greater level of engagement with the state governments of Australia. In the early days of the Oceans Policy, it proved to be difficult for the federal and state governments to develop meaningful partnerships in relation to its application. Over time, engagement has improved, but there is further scope for enhanced working relationships.

The integration of oceans management is a daunting and difficult task. Ongoing investment in scientific knowledge, and regular self-analysis and review, are vital for ensuring continued and meaningful growth in the development and implementation of the Oceans Policy. The policy faces many challenges, and its success can be achieved only by means of a persistent drive to achieve sound policy objectives. To ensure ongoing support for the Oceans Policy, tangible outcomes need to be delivered and recognized by the Australian community. Australia's Oceans Policy and its associated planning framework must be seen as a sound investment in securing the wealth of Australia's marine natural resources.

Developments to 2013

Since 2013, Australia has had two federal elections: the 2010 election, returning a Labour government; and the 2013 election, resulting in a Liberal/National coalition government. In neither of the elections did Australia's Oceans Policy feature per se, the focus instead being on implementation of specific initiatives that had their origins in the policy.

By 2012, the Australian government had completed the marine bioregional planning process, which resulted in bioregional plans for each of Australia's five marine bioregions (Department of the Environment, undated*b*) and a national network of marine protected areas (MPAs) (see Figure 4.2). The development of the national MPA network increased Commonwealth marine reserves from twenty-seven to sixty, expanding the national estate by 2.3 million km² to cover a total of 3.1 million km², which equates to 36 per cent of waters in Commonwealth jurisdiction.

By virtue of establishing the MPA network, the Australian government met the commitment made at the 2002 United Nations World Summit on Sustainable Development (WSSD) – that is, to establish a national representative system of MPAs – and exceeded Aichi Target 11 established under the 1992 Convention on Biological Diversity (CBD) – that, by 2020, at least 17 per cent of terrestrial and inland waters and 10 per cent of coastal and marine areas (especially areas of particular importance for biodiversity and ecosystem services) are to be conserved effectively and equitably managed, ecologically representative, and connected systems of protected areas and other effective area-based conservation measures, integrated into the wider landscape and seascape.

It is fair to say that, during the period from 2008 to 2012, the government was focused on efforts to complete the national network. This effort encompassed developing overarching goals and principles for the design, the utilization of science to guide the identification of potential areas of interest as part of the marine bioregional profiles for each of the marine regions, conducting social and economic assessments, and undertaking extensive public consultation. The Commonwealth marine reserves officially came into effect on 17 November 2012, adding to the southeast Commonwealth marine reserves network proclaimed in 2007 (Department of the Environment, undated*a*).

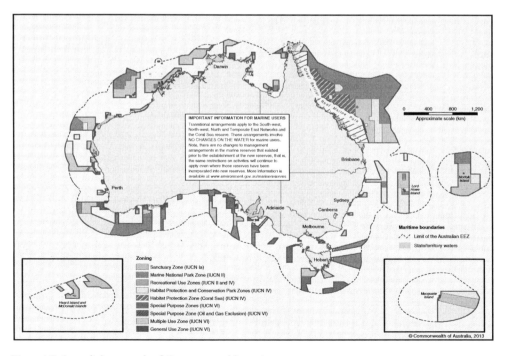

Figure 4.2 Australia's network of Commonwealth marine reserves

Source: Department of the Environment (undated*a*)

During the 2013 election campaign, the then Opposition party committed to suspend and review the management plans, owing to what it viewed as a flawed consultation process. As a result, upon winning the election, the new government moved to set aside the management plans put in place by its predecessor. The (again) renamed Department of the Environment is, at time of writing, working to establish an expert scientific panel to examine the science underpinning the reserves and to establish bioregional advisory panels to rectify perceived weaknesses in the previous consultation process.

The positive side is that the government remains committed to the overall network, leaving the boundaries in place and concentrating on internal zoning arrangements. It needs to be emphasized that the national network is a network of multiple-use MPAs with zoning appropriate to the conservation values that they are designed to protect. The next step that the new government intends to take is to reassess the reserves and identify appropriate types of zoning, and then to develop new management plans.

Since 2008, the Australian government has not solely been focused on domestic implementation, but has also made substantial progress in neighbouring waters. Australia has partnered with Indonesia, Papua New Guinea, Solomon Islands, Malaysia, and the Philippines, in the Coral Triangle Initiative on Coral Reefs, Fisheries and Food Security. The Coral Triangle partnership was formed in 2007, and Australia's participation represents recognition that, because oceans are shared resources, effective oceans management requires that no one country can rely on measures implemented solely within its own EEZ. This is particularly important in the northern regions of Australia, given the connectivity with neighbouring marine environments. In this same vein, Australia has been making efforts to explore potential transboundary arrangements with New Caledonia (France) in the Coral Sea. The government emphasizaed the importance of working across boundaries in November 2013 when Parliamentary Secretary for the Minister of the Environment, Senator the Honourable Simon Birmingham stated, at the 'Developments in Global Oceans Governance and Conservation' seminar held in Sydney, that:

> Unlike, of course, many other nations, we have no shared terrestrial borders, nor the transboundary issues associated with managing the natural systems that straddle those borders. Our transboundary management issues relate to that biggest shared resource in the world – our oceans.
>
> *(Birmingham, 2013)*

He also stated that:

> . . . [W]e know that it is in Australia's national interest to promote effective management of oceans beyond national jurisdiction; that, with marine jurisdiction such as we have bordered by high seas in parts of the Pacific, Indian and Southern Oceans, . . . we want to ensure that ecosystems, biodiversity and marine resources extend beyond not just our boundaries and that there is complementary management in those oceans bordering Australia's jurisdiction, and that our trans-boundary marine industries operate in a regional and global context and that they . . . [will] operate in the future hopefully along the types of high standards that Australia imposed in its terrestrial waters, but also that they will benefit hopefully from a clearer and level playing field regulatory environment in those open waters.
>
> *(Birmingham, 2013)*

From this statement, it is clear what the priorities may be for oceans management from the perspective of the new government. It could be argued that this direction is fully in line with the thinking behind the 1998 Oceans Policy. Perhaps the new government will follow the principles laid out in the policy even though the words 'Oceans Policy' have not been used in government in Australia for almost a decade.

Regarding institutional arrangements, in 2012 after completion of marine bioregional plans and the marine reserves, the then Marine Division of the Department of Environment, Sustainability, Water, Population and Communities was disbanded. Responsibility for implementing the national network was assigned to the Director of National Parks and the associated overarching marine policy to a newly created division entitled Wildlife, Heritage and Marine Division.

Given these changes, it is an open question where responsibility for broader oceans policy resides. Theoretically Australia's Oceans Policy is still on the books. It has not been rescinded or superseded, yet it can be found only in the Department's archive online. For all intents and purposes it is no longer valid as an operating policy. In terms of institutional arrangements put in place to implement the Oceans Policy, only the Oceans Policy Science Advisory Group (OPSAG) still exists, and is associated with the Australian Institute of Marine Science. It continues to coordinate scientific information–sharing between government bodies and the broader marine science community in Australia. In December 2012, it released the document *Marine Nation 2025: Marine Science to Support Australia's Blue Economy* (Australian Government OPSAG, 2013).

Even the Natural Resource Management Ministerial Council has been subsumed within another body, the Standing Council on Primary Industries (SCoPI). Under the new government, there is no standing intergovernmental body with a remit for natural resource management issues generally or oceans management issues specifically; instead, the current approach is one in which federal, state, and territorial ministers can meet on an ad hoc basis should the need arise. This is in recognition that Australia's states are 'sovereign in their own sphere' and 'should be able to get on with delivering on their responsibilities, with appropriate accountability and without unnecessary interference from the Commonwealth' (Council of Australian Governments, 2013).

Needless to say, then, the future for oceans policy in Australia is uncertain.

Is the current state of affairs a reflection of a view that, with marine bioregional planning complete, implementation is under way so that no further focus is required on oceans management? Stakeholders are still raising marine issues with government. These issues range from a shark cull on the western Australian coast, to dumping of dredge spoils in the Great Barrier Reef and the annual public outcry when the Japanese undertake so-called 'scientific whaling' in the Southern Ocean. Oceans issues are still top of mind for many coastal residents and summer beachgoers, but appear to be quietly moving from being a top government priority, as was the case in the late 1990s and beginning of the new century. It is still, however, early days for the new government and its view on oceans policy is yet to be tested.

Notes

1 This chapter was originally drafted in 2007, updated in 2009, and then again in April 2014, to reflect major developments in oceans policy implementation. On 24 November 2007, there was a federal election in Australia, resulting in a change of government. Under that government, primary responsibility for Australia's Ocean Policy fell to the Department of the Environment, Water, Heritage and the Arts. Previously, the department had been known as the Department of the Environment and Water Resources (between 23 January 2007 and 24 November 2008) and as the Department of the Environment and Heritage (before 23 January 2007). At time of writing (2014), the department is known as the Department of the Environment.

2 For the purposes of this chapter, the title of the department that was in use at the time of events described in the text has been used, but it should be recognized that all department names refer to the same body responsible for supporting the delivery of the environment portfolio priorities of the Australian government of the day.

References

Allen Consulting Group (2004) *The Economic Contribution of Australia's Marine Industries 1995–96 to 2002–03*, Hobart, Tas: National Oceans Office.

Australian Academy of Science (1994) *The Australian Coastal Zone and Global Change: Research Needs*, Canberra: Australian Academy of Science.

Australian and New Zealand Environment Conservation Council (ANZECC) (1997) *Best Practice Guidelines for Waste Reception Facilities at Ports, Marina and Boat Harbours in Australia and New Zealand*, online at http://www.environment.gov.au/archive/coasts/pollution/waste-reception/index.html [accessed 14 September 2005].

Australian and New Zealand Environment Conservation Council (ANZECC) (1998) *Guidelines for the Establishing the National System of Representative Marine Protected Areas*, online at http://www.environment.gov.au/resource/guidelines-establishing-national-representative-system-marine-protected-areas [accessed 1 November 2014].

Australian Antarctic Division (undated) 'Areas, Lengths, Heights and Distances', online at http://www.antarctica.gov.au/about-antarctica/environment/geography/areas-lengths-heights-and-distances [accessed 14 January 2005].

Australian Bureau of Statistics (undated*a*) 'Australian Demographic Statistics, June 2013', online at http://www.abs.gov.au/ausstats/abs@.nsf/mf/3101.0/ [accessed 1 November 2014].

Australian Bureau of Statistics (undated*b*) 'Population Clock', online at http://www.abs.gov.au/Ausstats/abs@.nsf/0/1647509ef7e25faaca2568a900154b63?OpenDocument [accessed 4 June 2008].

Australian Government Oceans Policy Science Advisory Group (OPSAG) (2013) *Marine Nation 2025: Marine Science to Support Australia's Blue Economy*, online at http://www.aims.gov.au/documents/30301/550211/Marine+Nation+2025_web.pdf/bd99cf13-84ae-4dbd-96ca-f1a330062cdf [accessed 12 March 2014

Australian Maritime Safety Authority (AMSA) (2001) *Review of Ship Safety and Pollution Prevention Measures in the Great Barrier Reef*, Canberra: AMSA.

Australian State of the Environment Committee (2001) *Coasts and Oceans, Australian State of the Environment Report 2001 (Theme Report)*, Canberra: Department of the Environment and Heritage.

Birmingham, Senator the Hon. S. (2013) 'Speech: Developments in Global Oceans Governance and Conservation Seminar', 7 November, Sydney, online at http://www.environment.gov.au/minister/birmingham/2013/sp20131107.html [accessed 1 November 2013].

Commonwealth of Australia (1979) *Commonwealth of Australia Gazette*, No. S189, 26 September, Canberra: Australian Government Publications Service.

Commonwealth of Australia (1990) *Commonwealth of Australia Gazette*, No. S297, 13 November, Canberra: Australian Government Publications Service.

Commonwealth of Australia (1992) *National Strategy of Ecological Sustainable Development*, Canberra: Australian Government Publications Service.

Commonwealth of Australia (1994) *Commonwealth of Australia Gazette*, No. S290, 29 July, Canberra: Australian Government Publications Service.

Commonwealth of Australia (1998) *Australia's Oceans Policy*, Canberra: Environment Australia.

Commonwealth of Australia (1999) *Commonwealth of Australia Gazette*, No. S148, 7 April, Canberra: Australian Government Publications Service.

Commonwealth of Australia (2002) *National Ocean Disposal Guidelines for Dredged Material*, Canberra: Environment Australia.

Council of Australian Governments (2013) 'COAG Comuniqué, COAG Meeting, 13 December 2013', online at http://www.coag.gov.au/node/516 [accessed 12 March 2014].

Department of Foreign Affairs and Trade (undated) 'Australian Now: Australia's System of Government', online at http://www.dfat.gov.au/facts/sys_gov.html [accessed 14 January 2005].

Department of the Environment (undated*a*) 'Commonwealth Marine Reserves', online at http://www.environment.gov.au/topics/marine/marine-reserves [accessed 1 November 2014].

Department of the Environment (undated*b*) 'Marine Bioregional Plans', online at http://www.environment.gov.au/marine/marine-bioregional-plans [accessed 1 November 2014].

Department of the Environment and Heritage (undated) '1992 Intergovernmental Agreement on the Environment', online at http://www.environment.gov.au/about-us/esd/publications/intergovernmental-agreement [accessed 14 January 2005].

Department of the Environment and Heritage (2004) 'The Imperatives for National Coastal Policy', Transcript of Opening Address by the Hon Dr David Kemp MP at Coast to Coast'04, 19 April,

Hobart, Tas, online at http://www.environment.gov.au/minister/archive/env/2004/tr19apr04.html [accessed 1 November 2014].

Department of the Environment and Heritage (2006) *A Guide to the Integrated Marine and Coastal Regionalisation of Australia, Version 4.0, June 2006 (IMCRA v4.0)*, online at http://www.environment.gov.au/resource/guide-integrated-marine-and-coastal-regionalisation-australia-version-40-june-2006-imcra [accessed 1 November 2014].

Department of the Environment and Water Resources (2007) *Marine Bioregional Planning in Commonwealth Waters: The Way Ahead for Australia's Northern Oceans*, online at http://www.environment.gov.au/system/files/resources/03398bf4-d25c-43cd-ad18-c5bdf0997f84/files/north-brochure.pdf [accessed 29 November 2014].

Department of the Environment, Sports and Territories (1998) *Australia's Oceans Policy: Report of the Forum Held in Canberra on 2–3 December 1997*, Canberra: Commonwealth of Australia.

Great Barrier Reef Marine Park Authority (GBRMPA) (1995) 'Towards a Marine Regionalisation for Australia', Proceedings of a Workshop Held in Sydney, New South Wales, 4–6 March 1994, GBRMPA, Townsville, Qld.

Geoscience Australia (undateda) 'Border Lengths', online at http://www.ga.gov.au/scientific-topics/geographic-information/dimensions/border-lengths [accessed 14 January 2005].

Geoscience Australia (undatedb) 'Continental Extremities', online at http://www.ga.gov.au/scientific-topics/geographic-information/dimensions/continental-extremities [accessed 14 January 2005].

Geoscience Australia (undatedc) 'Dimensions', http://www.ga.gov.au/scientific-topics/geographic-information/dimensions [accessed 31 January 2007].

Interim Marine and Coastal Regionalisation for Australia Technical Group (IMCRA) (1998) *Interim Marine and Coastal Regionalisation for Australia: An Ecosystem-Based Classification for Marine and Coastal environments, Version 3.3*, online at http://www.environment.gov.au/resource/interim-marine-and-coastal-regionalisation-australia-version-33 [accessed 1 November 2014].

Jensen, R., and Marine Science and Technology Working Group (1999) *Australia's Marine Science and Technology Plan*, Canberra: Department of Industry, Science and Resources.

McKinnon, K. R., Hammond, L. S., Hickman, B., Lamberton, D., Male, R., Taylor, A. R., Thomson, J. M., Williams, M., and Weir, A. D. (1989) *Oceans of Wealth*, Canberra: Australian Government Printing Service.

Minister for Resources and Energy (2008) 'UN Confirms Australia's Rights over Extra 2.5 Million Square Kilometres of Seabed', online at http://www.coml.org.au/UNConfirmsAustraliaRightsoverSeabed21Apr08.pdf [accessed 4 June 2008].

National Oceans Office (2001) *Snapshot of the South-East: A Description of the South-east Marine Region*, Hobart, Tas: Commonwealth of Australia.

National Oceans Office (2002) *Ecosystems: Nature's Diversity*, Hobart, Tas: Commonwealth of Australia.

National Oceans Office (2004) *South-East Regional Marine Plan: Implementing Australia's Oceans Policy in the South-east Marine Region*, Hobart, Tas: Commonwealth of Australia.

National Oceans Office (2005) *National Marine Bioregionalisation of Australia*, Hobart, Tas: Commonwealth of Australia.

Natural Resource Management Ministerial Council (2006) *Framework for a National Cooperative Approach to Integrated Coastal Zone Management*, online at http://www.environment.gov.au/resource/national-cooperative-approach-integrated-coastal-zone-management-framework-and [accessed 30 November 2014].

Poore, G. (1995) 'Biogeography and Diversity of Australia's Marine Biota', in L. Zann and P. Kailola (eds) *State of the Marine Environment Report for Australia Technical, Annex 1: The Marine Environment*, Townsville, Qld: Great Barrier Reef Marine Park Authority.

Sainsbury, K., Haward, M., Kriwoken, L., Tsamenyi, M., and Ward, T. (1997) *Multiple Use Management in the Australian Marine Environment: Principles, Definitions and Elements*, Canberra: Environment Australia.

Sea Around Us Project (2005) 'The Global Network of Marine Protected Areas in 2005', online at http://www.seaaroundus.org/ecosystemsmaps/images/mpaglobal_worldmap.pdf [accessed 13 July 2006].

State of the Environment 2011 Committee (2011) *Australia State of the Environment 2011L Independent Report to the Australian Government Minister for Sustainability, Environment, Water, Population and Communities*, Canberra: Department for Sustainability, Environment, Water, Population and Communities (DSEWPaC).

5

THE MARINE POLICY OF THE RUSSIAN FEDERATION

Its formation and realization

Yuriy G. Mikhaylichenko and Valentin P. Sinetsky[1]

Introduction: Basic information

History and reality

With coasts bordering three oceans, Russia, with a total land area of 17.1 million km² and a population size of 145.7 million, has one of the most extensive coastlines in the world (62,185 km, without taking into account small islands – see Figure 5.1). This area includes coasts on the Arctic Ocean (39,940 km), the Far East (17,740 km), the Azov and Black Seas (2,385 km), the Baltic Sea (660 km), and the Caspian Sea (1,460 km) (Table 5.1). The length of the country's maritime border is more than 39,000 km. The area of its exclusive economic zone (EEZ) totals more than 6 million km².

Situated in climatic zones ranging from the Arctic to subtropical zones, Russia's coastal zone is characterized by strong heterogeneity on all parameters. Approximately 18 million people

Figure 5.1 The coasts of the Russian Federation

Table 5.1 Federal districts bordering on sea, coastal subjects of the Russian Federation, and their coastline lengths

Federal districts (okrugs)		Coastal subjects of the Russian Federation	Coastline length (km)
North Western Federal District (District centre – Saint Petersburg)	1	Kaliningrad Province (*Oblast*) (administrative centre – Kaliningrad)	140
	2	Leningrad Province (*Oblast*) (Saint Petersburg)	410
	3	City of Saint Petersburg	110
	4	Murmansk Province (*Oblast*) (Murmansk)	1,730
	5	Republic of Karelia (Petrozavodsk)	600
	6	Arkhangelsk Province (*Oblast*) (Arkhangelsk)	8,850, including Novaya Zemlya Islands (4,320) and Franz Josef Land Islands (3,410)
	7	Nenetsky Autonomous District (AD) (*Okrug*) (Naryan–Mar)	3,180
Ural Federal District (Yekaterinburg)	8	Yamalo Nenetsky AD (*Okrug*) (Salekhard)	5,570
Siberian Federal District (Novosibirsk)	9	Krasnoyarsk Province (*Kray*) (Krasnoyarsk)	9,510, including Severnaya Zemlya Islands (2,510)
Far Eastern Federal District (Khabarovsk)	10	Republic of Sakha (Yakutia) (Yakutsk)	7,840, including Lyakhov Islands (560) and Novosibirsk Islands (1,860)
	11	Chukotsky AD (*Okrug*) (Anadyr)	5,080
	12	Kamchatka Province (*Kray*) (Petropavlovsk Kamchatskiy)	4,380
	13	Magadan Province (*Oblast*) (Magadan)	1,970
	14	Khabarovsk Province (*Kray*) (Khabarovsk)	3,390
	15	Sakhalin Province (*Oblast*) (Yuzhno Sakhalinsk)	4,300, including Kuril Islands (1,790)
	16	Primorsky Province (*Kray*) (Vladivostok)	1,280
Southern Federal District (Rostov on Don)	17	Astrakhan Province (*Oblast*) (Astrakhan)	660
	18	Republic of Kalmykia (Elista)	170
	19	Krasnodar Province (*Kray*) (Krasnodar)	1,010
	20	Rostov Province (*Oblast*) (Rostov on Don)	175
North Caucasian Federal District (Pyatigorsk)	21	Republic of Dagestan (Makhachkala)	630
Crimean Federal District (Simferopol)	22	Republic of Crimea (Simferopol)	~1,100
	23	City of Sevastopol	~100
TOTAL			**62,185**

Note: Map scale 1:5,000,000

Sources: Kalinina et al. (1992) and others

populate the country's coastal zone (about 12 per cent of the total population). Of this total, about a half live in the coastal zone of the Black, Azov, and Baltic Seas (which accounts for only 2 per cent of the entire coastal zone area), whereas only about 15 per cent live in the coastal zone of the Arctic regions (which accounts for 67 per cent of the entire coastal zone area). The natural/geographic and economic variety of coastal regions is reflected in the maritime activity of Russia.

A historical feature of Russia is its aspiration to expand to the sea – to operate, develop, and research its expanses and resources. Russians started to develop the coasts and islands of the White and Barents Seas in the twelfth century. Active Russian expansion towards the north and east began in the middle of the sixteenth century, when Tsar Ivan the Terrible ruled. Later in the sixteenth and in the first half of the seventeenth centuries, the future Russian cities in the Arctic region were established. By the middle of the seventeenth century, Russian estates had reached the Pacific coast. Russia finally achieved access to the Baltic Sea early in the eighteenth century, when Tsar Peter the Great ruled. In the second half of the eighteenth century, it achieved access to the Black Sea. Eventually, in the middle of the nineteenth century, the borders of the Russian state in the Far East were established.

The aspiration to expand to the sea was not without problems, arising from complicated and dynamic internal and international conditions. These problems in development of the world's oceans were exacerbated by the fact that the theories of implementing marine activity were not independently designed in Russia. Although the strategic necessity of a comprehensive approach to the development of marine activity was highlighted long ago, it was not fully understood even during the period of planned management of the national economy of the Union of Soviet Socialist Republics (USSR). At that time, the sectoral approach prevailed. This led to skewed sectoral developments and to their short-sighted accommodation, and to the unjustified dispersion of capital investments, especially towards development of coastal infrastructure. As a result, there was organizational and technological duplication of functions of similar maritime economic complexes between different government departments, leading to contradictions and even antagonism between departments. Nevertheless, the USSR consistently proved itself as a leading sea power throughout the world's oceans. This was especially true during the 1960s and 1970s, when the USSR's sea power became stronger owing to the upgrading of the civil and military fleet, active development of coastal infrastructure, and constant enhancement of scientific capacity.

The absence of an integrated approach to formation of marine policy complicated the realization of sea power for the USSR. For many reasons, the lack of development of such an approach is still the most vulnerable aspect of marine policy in modern Russia.

Significant economic activity on the coasts of the USSR, in the adjacent seas, and in the world's oceans was carried out under conditions of a strictly centralized management system based on the sectoral (ministerial) approach and exclusive state ownership of land, resources, and means of production. The sectoral ministries were responsible, first of all, for the fulfilment of their own plans and, as a result, did not pay attention to complex/integrated regional programmes. In practice, this meant that, in terms of a command system, there was no necessity for proven recommendations for complex/integrated use of coastal resources, and for objective mechanisms of coordination of interests and settlement of conflicts among marine resource users.

The fundamental changes that have taken place in Russia since 1991, connected with the transformation of the USSR, have also influenced its maritime activity. These elements of change include privatization and changed patterns of ownership, a significant revision of the governing role of national, provincial, and local authorities, and a noticeable weakening of the level of attention paid to nature protection under conditions of deterioration in the socio-economic situation. We should also note the problems and difficulties – typical of a transition period – with the organization of the new system of economic relations dealing with ownership and use of natural resources, linking nationwide, provincial, and local interests with the interests of private business. As a result, we can clearly see the insufficiency of an appropriate normative legal base and institutional structures for integrated management of marine resource use and the absence of sufficient skills in market regulation. Also, Russia lacks an investment policy at several levels, does not demonstrate a practical demand for the results of scientific studies, and has not developed the modern coastal zone profiles required for efficient management purposes.

The general results of the political transformation

According to the Constitution, adopted in December 1993, the Russian state (the Russian Federation or Russia) consists of eighty-five 'subjects of federation' (provinces and a few big cities). The subjects of federation are recognized as equal in rights in mutual relations with federal government agencies and independent in issues of their own competence. Twenty-three subjects of federation have direct access to the sea.

In 2000, a decree of the president of the Russian Federation divided the country into federal districts in which presidential plenipotentiaries provide constitutional powers of the head of the state in the territory of the appropriate district. There are nine federal districts at time of writing. Each federal district includes a group of subjects of federation. Out of nine federal districts, seven have access to sea, with four of these basically being coastal subjects of federation. The northern two federal districts have one coastal subject of federation in which maritime activity is seasonal.

Some of the current coastal subjects of federation (autonomous districts on the north and northeast coasts of the country) do not appear to be ready to assume the status and executive duties of state functions. Before the Constitution of 1993 was adopted, they had been components of larger administrative-territorial units. Clause 2 of article 11 of the Constitution of the Russian Federation gives subjects of federation the right to form bodies of state power. Article 77 of the Constitution says that the system of bodies of state power of subjects of federation is established independently, including the provision that federal executive authorities and executive authorities of subjects of federation form a unified system of executive power. However, while these changes to the state system provided for unified implementation, in the particular case of marine areas, the power remains within the purview of the federal authorities.

The new internal state structure of the Russian Federation is characterized by a high degree of independence of the subjects of federation. However, they enjoy little discretion as far as the management of the marine activities is concerned because all marine areas fall under federal jurisdiction.

In addition to the new federal structure, municipal governments were introduced throughout the Russian Federation. At the same time, article 12 of the Constitution of the Russian Federation says, 'institutions of municipal government are not included into the system of state power'. Article 8 of the Constitution and federal legislation have established a high level of independence for municipal government institutions, and have provided them with a large number of rights and duties. Formation of municipal governments on this new basis has changed relations between the authorities of coastal subjects of federation and authorities of the coastal municipalities. These relations have not been fully defined. Regarding finances, the dependence of local government on the institutions of subjects of federation, and through them, from federal executive authorities, has not only been maintained, but also frequently exacerbated. At the same time, there is no clear definition of the new order and rules of participation for coastal local government institutions in implementing maritime activity and national marine policy. As a result, the Russian Federation governing system is based on three authority levels. The relationship between these levels concerning marine activity in many respects remains uncertain.

Currently, according to Russian legislation, the legal status of sea expanses is established on the basis of a new approach. Clause 1 of article 67 of the Constitution of the Russian Federation states that the territory of the Russian Federation includes the territories of its subjects, the internal waters, and the territorial sea and air space above them. This implies that internal maritime waters and the territorial sea are not part of territories of subjects of the Russian Federation. Clause 2 of article 67 says that 'The Russian Federation has sovereign rights and carries out jurisdiction over the continental shelf and in the exclusive economic zone of the Russian Federation in the order determined by the federal law and international law regulations' (not by subjects of the Russian Federation law regulations). Simultaneously, article 71(m) of the Constitution, which defines the status of the territorial sea, exclusive economic zone (EEZ), and the continental shelf, refers to the exclusive competence of the Russian Federation. These provisions make it difficult to actively engage subjects of federation in implementing national marine policy and developing marine activity.

In the field of joint competence of the Russian Federation and its subjects, there is no direct mention of marine activities. Article 72(c) of the Constitution of the Russian Federation specifies that 'water resources' are a joint competence of the Russian Federation and its subjects. However, internal maritime waters and the territorial sea are declared in the Water Code of the Russian Federation to be federal property. At the same time, the operational regulation in article 36 of the Water Code specifies that management of the federal property on water objects is to be carried out by the government of the Russian Federation. Responsibilities for the management of federal property on water objects, according to the Constitution of the Russian Federation and the Water Code, can be transferred by the government of the Russian Federation to interested federal executive agencies and executive agencies of subjects of the federation. Under current legislation, subjects of the Russian Federation do not have regulatory power in the sphere of marine activity, although they can (in some cases) participate in water objects management.

As already noted, definition of the status of the internal maritime waters, the territorial sea, and the continental shelf, according to the Constitution of the Russian Federation, is the exclusive competence of the Russian Federation. The status of the specified sea expanses is determined in accordance with international law and regulated by federal laws, such as the 1995 Law on the Continental Shelf of the Russian Federation, the 1998 Law on the Internal Maritime Waters, Territorial Sea and Contiguous Zone of the Russian Federation, and the 1998 Law on the Exclusive Economic Zone of the Russian Federation.

The establishment of such a detailed legal regime by means of federal laws and other legislative acts does not, however, exclude participation by coastal subjects of the Russian Federation in the implementation of federal powers on the preservation, use, and management of marine resources and expanses. According to the Constitution, federal executive authorities have the right to establish such regimes for those sea expanses that provide an optimum level of competence of coastal subjects of the Russian Federation in the designated area. However, in the existing legislation, these opportunities have not been realized.

Thus, changes that happened in the internal administrative-territorial division at all levels and in the federal structure of the Russian Federation have an extremely deep and qualitative character. Previous concepts of state government were completely rejected because the Soviet-era legislation did not appear to be suitable for measuring the socio-political, financial, and economic realities of modern life in Russia. The system of state power and management has been created already. It comprises several regulatory acts, but the process of reform has not been completed. The changed state system, and system of authority and management, as well as a multitude of conceptual documents, and legislative and statutory acts, are not cohesive and require additional work. The discussion of next steps in the reform of the administrative-territorial system of the Russian Federation is ongoing.

The general results of the social and economic transformation

In the process of modernizing its social and economic system, Russia has – partly success-fully – undergone several stages of social and economic transformation. The first stage, during the 1990s, was directed at dismantling the old socialist system, which had, by that time, gone through economic, socio-political, and values collapse. By the end of 1991, Russia had practi-cally no institutions reliably functioning in all spheres of political, economic, social, and cultural life. Large-scale market imbalances, such as economic recession, empty shops, and threats of famine and cold, demonstrated that economic institutions were destroyed. An even greater dan-ger was the collapse of institutions of state power, with actual – and later, formal – disintegration of the USSR. This situation threatened the existence of Russia as a sovereign state. The first task was the restoration of primary institutions crucial for the functioning of the country – that is, state institutions, basic economic mechanisms, and core relations of ownership.

During this stage, a chain of crises, experienced by society with great difficulty, generated the basic institutions of a market economy and democracy, and restored both macroeconomic and political stability. By the end of the 1990s, the following tasks had been accomplished:

- basic political institutions had been created, the most significant being the adoption of the Constitution of the Russian Federation and the regulation of federative relations;
- macroeconomic stabilization had been carried out, providing the country with a steady currency and balanced budget; and
- mass privatization had been conducted, which laid the foundation for transition of the Russian economy to a market economy.

The creation and development of private property institutions became a key factor creating the base for the beginning of economic growth.

The next stage, the period 1999–2008, was a time of restoration and growth of the economy. The federal government had an opportunity to solve strategic problems. The federal govern-ment, while increasing efforts to maintain macroeconomic and political stability, emphasized formation of economic institutions typical of a modern market and democratic society that were oriented to the peculiarities of Russia. The civil, taxation, budgetary, labour, land, water, and town planning codes, new pension legislation, and legislation on bankruptcy were accepted. Initiatives were undertaken to debureaucratize (that is, to deregulate), to perfect inter-budgetary relations (the federal budget, provinces, and municipal government) and currency legislation, and to reform natural monopolies, among other activities. However, for a number of reasons, progress was not made in several important areas. All stages of reforming market legislation were not accompanied by an efficient law-enforcement practice in reformed spheres.

During the years following the 2009 global financial crisis, Russia has been developing a new long-term development model for its economy, with a new structurally innovative background. There is a need for measures aiming at improvement of the investment climate, increasing labour efficiency, reducing the dependence of the budget from volatile revenues, supporting non-resource-based export, coping with limits posed by the infrastructure, enhancing the efficiency of governance, continuing anticorruption efforts, refining the pattern of partnership between state and business, and developing small and medium-sized enterprises (SMEs).

Influence on maritime activity

The comprehensive situational analysis of economic activities in the coastal zones of Russian seas in all provincial segments with cartographical support is suitable for assessing coastal zone economic assimilation in terms of the amount of the basic economy fields, levels of economic assimilation, the character and types of relationship with the sea, density of population, and ecological state. Analysis conducted at the beginning of the twenty-first century showed inefficient use of resource potential and a critical environmental state in practically all subjects of the federation, as well as the need for change in resource use management and interaction among all participants of the process – that is, stakeholders, local population, and authorities.

The following negative trends have become most evident in the maritime field.

- Some 95 per cent of Russian foreign trade cargo is transported by foreign shipowners, mainly the outflow of the most advanced merchant ships under foreign flags resulting from a poorly chosen privatization policy. As a result of the division of the fleet among the republics of the former USSR, Russia lost most of its transport ships on the Baltic, Caspian, and Black Seas. Currently, there are no Russian refrigerator ships or light carriers, and there is a significant decrease in the number of passenger ships and ice-class tankers in these seas, and technical conditions of the rest of the fleet, ports, and coastal infrastructure and repair facilities are, in the authors' opinion, extremely unsatisfactory.
- The volume of marine resources available under sovereign rights of the Russian state granted to foreign enterprises is too large, in the authors' view. As a result, about 50 per cent of the bioresources extracted in the Russian EEZ are exported as raw material and semi-finished products manufactured aboard foreign-flagged ships. At the same time, about a quarter of the fish products consumed in Russia are imported. The scale of poaching of bioresources in the Russian EEZ is often comparable to the volume of legal catch, and on products of high value the illegal catch exceeds the volume of the legal catch.
- The basic assets of all maritime economy branches are outdated, and their update is carried out mainly by the purchase of foreign equipment. Therefore, in the authors' view, the potential of the Russian ship-building, machine-building, and instrument-making capacity for the maritime activity is not being realized.
- The presence of private and foreign capital in strategically important sectors of maritime economy (such as in seaports and in the extraction of hydrocarbons and other resources) continues to grow.
- Research fleet opportunities and numerical strength are sharply reduced because of insufficient funding.
- As a result of a significant change in the costs of ship operations and icebreaker support, potential opportunities of the Northern Sea Route are poorly used to support sustainable livelihoods in the Russian Arctic zone and for transit cargo transportation.
- Russia's status as a large, leading sea power, which was based on the USSR's earlier engagement in marine research, participation in international maritime organizations, and by foreign policy and strategic support of maritime activities, is, in the view of the authors, gradually being lost.

Principal causes of these negative factors are as follows.

- There was a critical reduction in the level of state participation in management of the maritime economy complex: the numerous proprietors who arose as a result of privatization had insufficient financial resources and pursued purposes that were quite often incompatible with state interests; at the same time, their activity is poorly regulated through the economic policy of the country.
- Tax, customs, credit, tariff, and rent regulators are poorly equipped to stimulate effective use of marine resources and sea expanses in the interest of the state.
- The legislative base for maintenance of maritime activity is developing slowly.
- Internal funds of enterprises of the maritime economy are poorly used for expanding and modernising basic assets for prospecting, search for fisheries, or scientific research.

On the positive side, Russia has traditionally had a high level of qualified experts in the natural sciences, including marine sciences, an effective system of higher education, and accessible foreign experience in integrated approaches to marine resource use management.

Overall, the successor of the USSR, the Russian Federation, appears to be cut off from the powerful coastal infrastructure created in recent years. After the USSR collapsed, the Russian state put in place new geographic, economic, and political conditions that have not improved the level of marine activity. Moreover, as concerns Russia's access to sea, the country appears to have moved northeast towards Eurasia, leaving ports, bases, railways, roads, and pipelines behind. At the same time, the coastal population in the Arctic zone and the north of the Far East has decreased significantly. Together, these factors have resulted in reduced national interest and activity in the world's oceans. Russia has become geographically isolated from its foreign trade and global communication partners. Consequently, in the authors' opinion, this has increased the direct and indirect costs of economic and military strategic activity, and complicated foreign economic relations.

The difficulties that have arisen as a result of deep political, social, and economic transformations are surmountable. However, these demand not only time and appropriate financial resources, but also a well-developed general strategy of development of national marine activity. Such a strategy should include measures in the following categories:

- *vital* – that is, defence, trade, fishery, other economic activities directed to specific needs of the country;
- *status-related* – that is, measures aimed at maintaining the political position of Russia in the world, for example showing the flag, a designation of presence, or a demonstration of force; and
- *new long-term perspectives* – such as coastal and marine research and development, the formation of positions according to the future interests of Russia in the world's oceans, and strategic investments.

The maritime policy process: Its development and results

The prerogative of defining priority objections and the substance of the national maritime policy belongs to the president of the Russian Federation. The president, according to the Constitution, undertakes measures to secure the sovereignty of the Russian Federation in the world's oceans, to protect and secure the interests of a person, society, and the state in the field of maritime affairs. The president also provides guidance of the national maritime policy. The Security Council of the Russian Federation, as a special constitutional deliberative body headed by the president of the Russian Federation, reveals threats, determines vitally important demands of the society and the state, and works out major directions of the safety strategy of Russia in the world's oceans.

Issues of national maritime policy can be also considered at the sessions of the State Council of the Russian Federation, a deliberative body headed by the president of the Russian Federation, which was created according to presidential decree. The State Council aims to sustain and use the potential of the regional supreme officials. Issues of maritime activity are also supervised by the president's plenipotentiaries in federal districts of the Russian Federation, who can present their proposals in the field of maritime policy.

The Federal Assembly (Parliament) of the Russian Federation, consisting of two chambers, is responsible for legislative regulation of maritime activity. For these purposes to be achieved, the Commission on the National Ocean Policy functioned within the Council of Federation (upper chamber of Parliament) from 2004 to 2011. The main tasks of the Commission were: to monitor legislation in the field of the maritime affairs; to elaborate proposals for projects of federal acts for the Council of Federation, aimed at increasing the effectiveness of maritime economic activities; and to interact with federal bodies of the executive power involved in maritime affairs.

Draft laws go to the State *Duma* (the lower chamber of Parliament), where they are examined by the appropriate committees and commissions, or are elaborated on by the initiators, after which they have to be affirmed at plenary sessions. Draft laws accepted by the State Duma go on for approval to the Council of Federation. In cases of disagreement, conciliation commissions are created. When a draft law is approved in the upper chamber of Parliament, it goes to the president of the Russian Federation for signature.

Provisions stipulated by federal laws are formalized by the normative legal acts developed and accepted by the government of the Russian Federation, as well as by the appropriate federal executive authorities.

The 'World Ocean' Programme

The necessity of overcoming the negative consequences of maritime economic liberalization and uncontrolled privatization of the maritime economy basic production assets was evident in the mid-1990s. It was clear that Russia's participation in developing resources and expanses of the world's oceans was closely connected to improving management, including state regulation, together with purposeful scientific and technical development of marine activity in the country. For these purposes, the federal target programme (FTP) 'World Ocean' was approved by Act of the government of the Russian Federation in 1998. The concept of the programme was approved by decree of the president of the Russian Federation in 1997. All interested federal and provincial executive agencies, as well as many research organizations, took part in the development of this programme.

Adoption of FTP 'World Ocean' at the nationwide level was aimed at changing the existing narrowly focused sectoral and local-provincial approaches to conducting marine activity. These approaches were realized through several dozen state programmes with branch or regional orientation. Since 1998, FTP 'World Ocean' had become the basis of a nationwide system of regulation and management of Russia's marine activity aimed at increasing its integration and effectiveness.

The programme dealt with, at the federal level, the following issues:

- the study of the world's oceans;
- the use of marine resource potential;

- the development of transport communications;
- Russia's presence on marine expanses;
- the maintenance of sovereign rights and jurisdiction in Russia's coastal waters and international marine waters;
- control over the marine environment; and
- emergencies of a natural and man-made character, as well as other specific problems related to providing safe and sustainable livelihoods in both coastal regions and throughout Russia as a whole.

Significant problem areas that affect interests of the federal centre, subjects of the Russian Federation, local authorities, organizations, and enterprises participating in development of marine activity, were included in this programme. These issues determined the programme's inter-sectoral and inter-regional character. Comprehension of the value of the environment as a whole and its interrelations with marine problems were critical. Understanding the role of marine policy for protecting and improving coastal and marine ecosystem conditions became an important factor in the formation of the marine policy of Russia.

The basic purpose of FTP 'World Ocean' was the comprehensive resolution of issues of studying, developing, and effectively using resources and expanses of the world's oceans in the interests of economic development, safeguarding national security, and marine frontier protection. This programme included eleven sub-programmes, the state customers of which were thirteen federal executive agencies. The Ministry of Economic Development of the Russian Federation acted as the state customer coordinator. General coordination functions in the programme were carried out by the Interagency Commission on FTP 'World Ocean' (a federal government commission) and its Scientific-Expert Council from 1998 to 2004. Since 2005, the most critical issues of substance, implementation and continuation of FTP 'World Ocean' have been considered by the Marine Board of the government of the Russian Federation.

Taking into account the variety and complexity of the problems to be solved in the framework of FTP 'World Ocean', as well as its long-term character (it was designed to run for a fifteen-year term), the programme was divided into three stages, with appropriate purposes and tasks stipulated for each stage. Realization of the first stage of the programme stabilized the basic parameters of maritime activity in Russia and began the revitalization process. The purpose of the second stage was the development of sufficient reserves in economic, political, social, legal, nature protection, scientific, technological, and other spheres of marine activity in Russia that allowed Russia to satisfy its current needs and provided for its long-term interests and requirements. During the third stage, according to its national development strategy, Russia's position on the world scene and new structures of activity in the world's oceans would be strengthened. In the process of programme implementation, some corrective amendments had to be made, and the next tasks and solutions specified.

The programme was financed from two sources: federal budget and off-budget sources. Because resources are always limited, there arose the need to focus them on the areas of activity in the world's oceans that are most important from the perspective of priorities of social and economic development of the country and its long-term national interests. Therefore the number of areas in the programme was reduced as compared to the concept to optimize the use of means and resources.

In the aftermath of 1998 crisis, under the pressure from effects on the Russian economy, the implementation of the programme was further modified owing to limited budget. Funding of two sub-programmes, on mineral and biological resources, was from outside the framework of FTP 'World Ocean'; funding of another two sub-programmes, on development of technology and transport, was significantly reduced as compared to the original budget. The attempts to raise funds for certain sub-programmes from non-budget sources – which sometimes, as was the case of sub-programme 'Shelf', were meant to establish the major share of the budget of such sub-programme – ended up in failure. As a result, the implementation of six out of eleven sub-programmes was stopped in 2004–05.

Implementation of FTP 'World Ocean' focused on strengthening the economic and raw material potential of the country and the country's defence capabilities. Other priorities were sustaining and expanding scientific research in the world's oceans, expanding the sea transport cargo base, and increasing Russian seaport load levels, fisheries profitability, and, as a consequence, payments to the budget and employment levels through the creation of new job opportunities in coastal provinces.

Geological and geophysical research in various marine regions expanded fundamental knowledge of the sea and ocean bottom structure. This research also enlarged the available information on mineral resource distribution and development prospects.

Polymetallic nodules exploration solidified specific nodule resources in the Russian exploration region in the Pacific Ocean. In the Mid-Atlantic ridge crest, several prospective rich metal sulphide deposit sites were located. An estimate of ore and metal resources contained in them was made.

In support of definition and validation of the outer limits of the continental shelf beyond the 200 nautical mile EEZ of the Russian Federation within the legal framework of the 1982 Convention on the Law of the Sea (UNCLOS), existing geological, geophysical, and hydrographic information was processed. This information included data on recently conducted field studies to eliminate 'white spots' within the region of the Mendeleyev rise. This surveying provided a comprehensive picture of bathymetric and morphological aspects of the sea bottom, and in-depth structure of the crust, structure, and thickness of the sediment layer of the Arctic basin within the Russian sector of the Arctic Ocean. Geological and geophysical materials were prepared for assessment, according to international law, of a modified Russian claim for the outer limit of its continental shelf beyond the 200 nautical miles.

Study of ocean and marine ecosystem dynamics, ocean productivity, along with the ecosystem research on fish stock distribution and reproduction and trends in productive zone formation, carried out in the seas under Russia's jurisdiction, allowed the state to estimate the potential of ocean ecosystems and promoted improvement of estimation of the fish stocks.

The system of state control over fisheries and fishing bioresources was developed and put into commercial operation.

The results achieved within the framework of FTP 'World Ocean' research on climate change allowed regional assessments of risks associated with the observed and forecast climate change by 2030, 2050, and 2100, as well as the listing of Russian marine areas with extreme manifestations of climate change and critical change in frequency and magnitude of extreme climatic phenomena. Recommendations on mitigation of risks to economic activities in the Russian seas, regional scenarios of development of marine activities, and infrastructure under different scenarios of the climate change have been produced.

The concept of an expert system for forecasting of extreme oceanic phenomena (wind and waves, non-tropical cyclones, extreme sea-level rise, and tsunami) and assessing their impact on marine activities has been developed.

The outcomes include: an integrated system of region-specific monitoring of the Russian seas in the areas of development and transportation of hydrocarbons; recommendations on the monitoring system of composition, structure, and parameters of its functioning; the choice of satellites to monitor the status of the water surface; and documentation of monitoring costs.

The Strategic Action Plan on Protection of the Environment of the Arctic Zone of the Russian Federation (AZRF) was developed and adopted.

Proposals were produced regarding national legal tools on the protection of economic interests of the Russian Federation in ice-covered sea areas adjacent to the Russian EEZ overlaying its continental shelf in the Arctic ocean, in accordance with applicable international law and national laws of the Arctic states.

The project of the Russian multipurpose space system 'Arctic' was developed to meet the needs of hydrometeorology, communication, broadcasting, navigation, environmental monitoring, the safety of people, and nature use in the Arctic. The work on the development of top-priority high elliptical orbit hydrometeorological space subsystem 'Arctic-M' is ongoing within the framework of the Federal Space Programme of Russia.

Proposals were prepared on establishing the system providing for efficient management of traffic along the Northern sea route in areas of high concentration of vessels in the Arctic.

Proposals were elaborated on a multilayer prevention framework in the healthcare system to address the problem of early-age mortality among the communities of the Russian Arctic exposed to negative environmental and social impacts, to prevent socially significant non-infectious diseases as well as to improve demographic parameters of the region.

Pilot measures were taken to clean up polluted areas in the Arctic Zone of the Russian Federation, including sites in the Nenetsky Autonomous District, the Republic of Sakha (Yakutia), the Arkhangelsk Province, and on the Vrangel Island. Clean-up actions were effected in the Spitsbergen areas at the sites of the Russian presence.

The Russian science centre infrastructure was established, and the construction of objects of communication and its infrastructure have been completed on the Spitsbergen archipelago.

Proposals on the applicability of international law tools in respect of economic and research activities of Russian citizens and legal persons in areas governed by the Spitsbergen Treaty of 1920 were drafted. Proposals as to the Russian position on applicability of the Norwegian national law to economic activities in marine areas around Spitsbergen were formulated, with due regard to the official positions of other countries.

The new winter and transport complexes with a snow-ice runway at the 'Progress' station in the Antarctic were established, ensuring the continuous building of knowledge on the Antarctic environment. Thus core drilling was undertaken on the unique sub-glacial Vostok Lake within the integrated research programme (3,450 metres in February 2012). As a result of isotope analysis of ice samples, the existing knowledge on temperature pattern model for the central Antarctic during the last 400,000–800,000 years has been improved. The analysis of surface water of the sub-glacial Vostok Lake with molecular biology methods also allowed the identification of the DNA of bacteria previously unknown to the global research community.

The multifunction version of the Unified State System of Information on the World Ocean Conditions (USSIWOC) was put in operation (online at http://esimo.ru/portal/). The content of this system is provided by thirty-six organizations acting under the auspices of twelve Russian ministries and agencies. The system integrates more than 200 databases, reflecting more than 400 parameters of the world's ocean conditions (over 3,500 units of input data), including information on:

- the marine environment;
- the characteristics and position of cargo and fishing vessels, and port activities;
- the catch and processing of marine biological resources;
- the resources of the Russian Federation continental shelf and shelves of the world's oceans;
- shipyards operations;

- emergencies in the Russian seas and coastal areas; and
- economic and social conditions in the coastal subjects of the Russian Federation.

The technology developed for USSIWOC was utilized in the course of the establishment of the Ocean Data Portal of UNESCO Intergovernmental Oceanographic Commission (online at http://www.oceandataportal.org/).

Some key projects on development of Russian maritime activities were implemented.

Owing to the implementation of FTP 'World Ocean', it was possible to provide the minimal necessary level of marine field research and to protect scientific staff levels.

The programme also provided the framework for continuation of work on adoption of the tools of integrated coastal zone management (ICZM) and their adaptation to the conditions of Russia, with the following outcomes:

- relevant proposals for the Strategy for Development of Maritime Activities of the Russian Federation by 2030;
- guidelines on the development of programmes of integrated development of coastal land and water areas, and model performance indicators to monitor implementation in respect of this programmes (the guidelines were submitted to coastal subjects of the Russian Federation and the Ministry for Regional Development of the Russian Federation);
- documents on methodology for and recommendations on establishing of a national legal framework on the development of marine spatial planning, as an efficient instrument of functional zoning and strategic assessment of possible use of marine areas connected with land spatial planning in the Russian Federation and compatible with relevant European Union instruments;
- proposals concerning the development of management framework of the coastal zone of the Kaliningrad Province including:

  ○ improvement of the mechanism of delegation, from the federal-level executive bodies to subject-of-federation executive bodies, powers related to the protection of shoreline;

  ○ the use of subventions to implement the federal-level competence in that field; and

  ○ transborder cooperation with the southeast Baltic states on shoreline protection and coordination for the development of coastal areas; and

- recommendations regarding the science-based planning of economic activities within unique coastal landscapes (spits, beaches, dune complexes, lagoons), with the view to considering limitations associated with the specific vulnerability of natural objects.

Therefore FTP 'World Ocean' represented a tool with which to solve systemic problems, to ensure inter-sectoral cooperation in respect of maritime activities, and to test pioneer management approaches.

The initial goal, however, was far from integrating all maritime activities within the framework of one programme. For this reason, several FTPs were implemented simultaneously, including such independent FTPs as 'Modernization of the Russian Transport System' with its marine component, 'Development of Civil Marine Technology', 'Increasing the Efficiency of Use and Development of Resources Potential of Fisheries Complex', 'Establishing of the Black

Sea Fleet Base in the Russian Federation', and 'The State Border of the Russian Federation'. The programmes 'Development of the Kaliningrad Province', 'Economic and Social Development of the Far East and Areas East of the Baikal Lake', 'Social and Economic Development of the Kuril Islands (the Sakhalin Province)', and 'The Russian South' are related to the development of coastal areas. Moreover, certain efforts with maritime dimension have been, and still are, taken in different years within the framework of other FTPs. FTP 'World Ocean' did, however, succeed in carrying out a targeted programme approach to research, development, and use of spaces and resources of the world's oceans, which called for interdisciplinary approaches and joint efforts of different departments.

The Marine Doctrine

The Marine Doctrine of the Russian Federation for the period to 2020, authorized by the president of the Russian Federation in 2001, reveals the essence, content, and method of implementing a national marine policy. The Marine Doctrine is a major component of state policy of the Russian Federation. The doctrine sets out the concepts necessary for undertaking practical tasks in the world's oceans.

Balancing the interests of all states of the world community and taking into account globalization processes, Russia determined its national interests in the world's oceans and declared them in the Marine Doctrine. The purposes of the national marine policy of the Russian Federation are ensuring and protecting state sovereignty, sovereign rights, and freedoms of the high seas in the world's oceans. According to the doctrine, the national interests of the Russian Federation in the world's oceans are:

- the inviolability of Russian Federation sovereignty beyond its land territory to its internal maritime waters and the territorial sea, as well as to the air space over them, and to the seabed and subsoil;
- safeguarding the sovereign rights and jurisdiction of the Russian Federation in the EEZ and over the continental shelf – that is, the exploration, exploitation, and conservation of natural resources (both living and non-living resources located on the seabed, in its subsoil, and the waters superjacent to the seabed), as well as:

  - the management of these resources;
  - the generation of energy from water, currents, and wind;
  - the creation and use of artificial islands, installations, and structures;
  - marine scientific research; and
  - the protection and conservation of the marine environment;

- protecting high seas freedoms in the interest of the Russian Federation, including freedom of navigation, overflight, laying submarine cables and pipelines, fishing, and scientific research; and
- protecting human life at sea, preventing marine environmental pollution, maintaining control over vital sea communications functioning, and creating conditions promoting benefits from marine economic activities to the population of the Russian Federation, especially its coastal regions, and also to the state as a whole.

The Marine Doctrine of the Russian Federation provides the following criteria for evaluation of the national marine policy:

- the opportunity to implement the national marine policy's short-term and long-term tasks;
- the degree of realization of sovereign rights in the EEZ, over the continental shelf, and also high seas freedoms for merchant, fishing, research, and other Russian specialized fleets; and
- the ability of Russian maritime military component to protect marine territory and state interests in the world's oceans.

The Marine Doctrine of the Russian Federation stipulates that the subjects of national marine policy are the state and society. The state implements national marine policy through the bodies of state power of the Russian Federation and the subjects of the Russian Federation. Society participates in the formation and implementation of national marine policy through representative bodies of the Russian Federation and its subjects, institutions of local government, and public associations working under the jurisdiction of the Constitution and the legislation of the Russian Federation.

The Marine Doctrine structure reflects the character of national marine activity. According to existing theoretical premises, Russian national marine policy is carried out under two broad categories:

1 *functional* – that is, types of marine activity (transport, fishing, naval, etc.) depending on Russia's economic opportunities and its role in international relations; and
2 *regional* – that is, marine activities that take into account Russia's position in the world, as well as geographical and other regional features.

The following are the basic activities of the national marine policy:

- the identification of priorities of the national marine policy on short- and long-term prospects;
- the continuous upgrading of the content of national marine activity;
- management of the components of state marine potential, branches of the economy, and science related to maritime activity;
- the creation of a favourable legal regime, including economic, information, scientific, personnel, and other elements of the national marine policy; and
- evaluation of the efficiency of national marine policy and its subsequent updating.

The subjects of national marine policy are guided by the principles of national marine policy formulated in the Marine Doctrine. These principles are common for both functional and regional directions of national marine policy.

The Marine Board

The Marine Board of the government of the Russian Federation, headed by the deputy chairman of the government of the Russian Federation (the deputy prime minister), is a permanent coordinating body that was organized in 2001. The Marine Board brings together the actions of federal executive agencies, executive agencies of subjects of the Russian Federation, and the organizations engaged in maritime activity of the Russian Federation for the purpose of implementing Russia's maritime policy. The Marine Board members are heads of federal executive agencies, executive agencies of subjects of the Russian Federation, and maritime scientific, commercial, and other maritime-oriented organizations, including non-governmental organizations (NGOs). The membership of the Board is approved by the government of the Russian Federation. Representatives from other organizations can be invited to the Board's sessions.

Interagency commissions on functional aspects of maritime policy are created to draft legislative and other normative legal documents, as well as for preliminary consideration of draft decisions of the Marine Board. Additionally, working groups can be created. Experts from the interested federal executive agencies, executive agencies of subjects of the Russian Federation, state institutions, and scientific, commercial, and public organizations are also involved in the activities of these commissions and working groups.

Since the mid-2000s, the Marine Board has been managing to resolve one of the major problems: to establish connections with federal districts and subjects of the Russian Federation. Such connections are exercised through special coordinating advisory bodies, known as provincial maritime councils. Interactions with provincial maritime councils, which include representatives of provincial governments, local coastal communities, and other organizations, allow the taking into account of the demands and peculiarities of coastal provinces.

The record of the Marine Board experience suggests that this coordination body is the major practical focal point for maritime activity in the Russian Federation. It carries out preparation of scientific, political, and economic recommendations for adjusting and implementing the maritime policy of the state.

Integrated approaches to maritime management

Adaptation and introduction of modern management practices, including ICZM, into the coastal and maritime management in Russia started in the mid-1990s. At the federal level, it was achieved through federal marine research programmes and, since 2000, through the FTP 'World Ocean'. At the provincial level, it was mainly initiated by international projects and programmes (such as the Black Sea and the Caspian environmental programs).

The following had been accomplished to the year of 2010:

- assessments of Russia's coastal resource potential and the situation regarding its use;
- a study of international experience in the development and realization of national and local ICZM programmes;
- publication of a series of conceptual and methodological papers on ICZM;
- the drawing up of an ICZM curriculum and teaching of that curriculum in the national higher school system;
- the working out of an article-by-article summary of several versions of federal draft law on ICZM; and
- the taking of first steps towards introducing ICZM principles and approaches into the process of elaboration of local programmes for coastal development and use of coastal resources, by means of coordination of efforts of federal and international projects.

The first and only specialized international conference on ICZM in Russia (in 2000, in St Petersburg, with 130 participants from twenty countries) helped to identify problems hampering its development in countries with transition economies, endorsed recommendations for further actions with the assistance of leading foreign experts, and emphasized the special importance of administrative, economic, and scientific circles of these countries being familiar with global practice regarding the development and execution of ICZM approaches and programmes, and the role of the marine sciences in coastal zone management.

Summing up some general results of the ocean and coastal management in Russia, then, it is important to mention that:

- real introduction of ICZM approaches is a very hard task and will take many years;
- the essence of these approaches and instruments used by this methodology must be well understood;
- it is important that introduction of ICZM methods into management practices is executed both top-down (from the federal level) and bottom-up (from the local level), with the unity of approaches used; and
- the process of implementation must include all the ICZM instruments and approaches.

Despite the fact that the ICZM realization in Russia was implied by the FTP 'World Ocean' concept adopted by the decree of the president of the Russian Federation in 1997, these approaches still have not been adopted, remaining unknown in administrative circles at all levels and among potentially concerned entities (business, the local population, NGOs) whose participation in the decision-making process was of critical importance. In the authors' views, this underscores the fact that Russia rapidly entered the market economy and did not pay sufficient attention to assimilation of modern managerial practices on marine natural resource use. Despite the existence of 'points of growth' along the country's coasts (north, south, west, east), it remains difficult for ICZM 'to make its way' in current political and administrative processes.

Other aspects

With regards to domestic implementation of international agreements, the Russian Federation has established zones of offshore jurisdiction in accordance with UNCLOS. Russia has also ratified such basic maritime international conventions as, for example, the 1973 International Convention for the Prevention of Pollution from Ships, as amended by the Protocol of 1978 (MARPOL), the 1972 Convention on the Prevention of Marine Pollution by Dumping Wastes and Other Matters (London Convention), as modified by the Protocol of 1978, and the 1992 Convention on Biological Diversity (CBD). Provisions of MARPOL and its mandatory annexes, and main provisions of its optional annexes, as well as provisions of the London Convention, are reflected in national legislation. With regards to the Global Programme of Action for the Protection of the Marine Environment from Land-based Activities (UNEP, 1995), there is no national action plan to date. However, related activity on controlling land-based sources of pollution is carried out at the subnational level, for example within the framework of the Baltic Marine Environment Protection Commission (HELCOM), the Black Sea Environmental Programme, and the Caspian Environment Programme.

At the federal level, Russia has fourteen state natural reserves that include a marine component. Their total marine area is about 10 million hectares.

The current stage and long-term outlook

The Strategy for Development of Marine Activities

The Strategy for Development of Marine Activities of the Russian Federation by 2030, adopted by the government of the Russian Federation in 2010, determined major challenges and perspectives for the development of marine activities of the country, outlined strategic goals and targets, and established milestones in the development of marine activities. In the authors' opinion, they enabled Russia to move to a new stage in addressing many problems.

The Strategy is the first national document of its kind that recognizes such goals as economic development and improvement of living standards in the country as priorities of marine activities. The Strategy defines as one of the major challenges the need for an integrated approach to complement the existing sectoral approach to marine activities planning. One of the strategic goals is the realization of the integrated approach to the development of specific coastal land and marine areas by recognizing them as special objects of governance. This therefore predetermines the following strategic tasks:

- development of a legal framework for the establishment and implementation of programmes of integrated development of specific coastal land and marine areas (with the responsible executive being the Ministry of Economic Development of the Russian Federation); and
- establishment and implementation of programmes of integrated development of coastal land and marine areas as a specific element of integrated strategies and programmes of social and economic development of the coastal subjects of the Russian Federation and programmes of development of municipalities adjacent to sea (the responsible executives being subjects of the Russian Federation).

The Strategy stipulated further directions of the future development of the Russian Federation, which include, among others, such goals as:

- the introduction and elaboration of integrated (intersectoral) management at all levels;
- the extension of the marine component of integrated development programmes of land areas adjacent to sea and for coastal waters to the limits at which the Russian Federation exercises jurisdiction;
- coordination of these programmes with the management programmes for rivers catchment areas; and
- the use and development of tools of marine spatial planning.

It is noteworthy that the Strategy is silent on issues relating to:

- providing for the safety of the population of coastal areas and mitigation of the consequences of sea-related disasters;
- the development of sea-related tourism and recreational services;
- combating piracy; and
- the training of personnel for marine activities.

Unlike other important areas of marine activities, no specific quantitative milestones, including regional ones, were defined for the development of mariculture.

The Strategy also does not reflect such important principles of marine activities as:

- the involvement of all stakeholders in planning and implementation of marine activities;
- the application of adaptive management;
- the use of best available science and information;
- the precautionary approach and preventive measures; or
- biodiversity protection.

Given that certain milestones of the Strategy were not achieved during the first implementation stage (2010–12), the Marine Board recently resolved to produce proposals to amend the Strategy with regard to implementation tools and planned milestones for the second stage, so that they account for the existing sectoral strategies, and to establish uniform methodology of assessment of the results of the Strategy implementation to provide for realistic assessment of the implementation of the national marine policy.

Table 5.2 The main instruments containing conceptual, doctrinal, or strategic provisions defining different aspects of marine activities of the Russian Federation

Title	Year of adoption
Overarching instruments	
Marine Doctrine of the Russian Federation by 2020	2001
Concept of long-term social and economic development of the Russian Federation by 2020	2008
Strategy for Development of Marine Activities of the Russian Federation by 2030	2010
Strategies for Social and Economic Development of Coastal Federal Districts	2009–2011
Strategies for Social and Economic Development of Coastal Subjects of the Russian Federation	2007–2013
Departmental/sectoral instruments	
Environmental Doctrine of the Russian Federation	2002
Fundamentals of the State Policy for the Use of Mineral Resources and Subsoil	2003
Concept of the Development of Inland Water Transport of the Russian Federation	2003
On State Regulation in International Trade Relations	2003
Strategy for the Development of Shipbuilding Industry by 2020 and further on	2007
Transport Strategy of the Russian Federation by 2030	2008
Strategy for National Security of the Russian Federation by 2020	2009
Strategy for the Development of Fisheries of the Russian Federation by 2020	2009
Russian Energy Strategy by 2030	2009
Climate Doctrine of the Russian Federation	2009
Concept of the Development of Diving in the Russian Federation	2009
Main Focus Areas of the State Policy in Strengthening Energy Efficiency in Electroenergetics Sector Based on the Use of Renewable Energy Sources by 2020	2009
Military Doctrine of the Russian Federation	2010
Strategy for the Development of the Geological Sector of the Russian Federation by 2030	2010
Food Security Doctrine of the Russian Federation	2010
Strategy for Hydrometeorological and Accompanying Activities by 2030	2010
Concept of Development of Especially Protected Nature Areas of the Federal Level by 2020	2011
Strategy for Innovation Based Development of the Russian Federation by 2020	2011
Fundamentals of the State Policy of the Russian Federation for the Naval Activities by 2020	2012
Fundamentals of the State Policy for the Environmental Development of the Russian Federation by 2030	2012
Fundamentals of the Russian Federation Policy for the Development of Science and Technology by 2020 and further on	2012
Foreign Affairs Concept of the Russian Federation	2013
Specialized instruments	
Fundamentals of the State Policy of the Russian Federation in the Arctic by 2020 and further on	2008
Strategy for the Development of Activities of the Russian Federation in the Antarctic by 2020 and further on	2010
Strategy for the Development of the Arctic Zone of the Russian Federation and Ensuring National Security by 2020	2013

The process of revising the Marine Doctrine

Adoption of the Marine Doctrine of the Russian Federation in 2001 was an important contribution to raising awareness of marine activities and recognition of its importance for the development of the country. The adoption of the Doctrine marked the beginning of ongoing process of formation of strategic planning of marine activities, coordinated with national-level decisions, to this end.

However, for a number of reasons, not all of the goals contained in the Marine Doctrine have been achieved as expected. Meanwhile, the preceding period was marked with the adoption of new conceptual documents related to different aspects of marine activities of the Russian Federation (Table 5.2).

The integral analysis of the situation in light of more than a decade of experience of its implementation demonstrates that the Marine Doctrine requires modification, which means reviewing the whole document. The requests to this end were issued by the Security Council and the Marine Board.

So far, the structure of a revised Marine Doctrine of the Russian Federation by 2030 has been elaborated and a new draft has been prepared. Proposals regarding advancement of the national marine policy from functional and regional perspectives were developed, informed by the existing documents of strategic planning. The draft suggests renovation of both the structure and content.

The building of ships and vessels, and establishing the technological background for realization of all provisions of the Marine Doctrine, are distinguished as specific items within the draft. A new separate item covers marine pipelines, since the past decade has been marked with the rapid development of underwater systems of oil and gas transport, which is of strategic importance for the Russian Federation in connection both with national consumption of hydrocarbons extracted from the continental shelf and with international trade. It is suggested that the Marine Doctrine shall be supplemented with provisions on the development of sea-related tourism and recreational services, providing for the safety of the population of coastal areas and mitigation of the consequences of sea-related disasters, and combating piracy. Inland water transport is specifically addressed as being increasingly integrated into logistic chains around the globe, especially in coastal areas of the world's oceans both in small and long-range inshore coasting. More emphasis is placed on mariculture.

The region-specific national marine policy is supplemented by long-term goals related to the conservation of natural, cultural, and historical heritage. The Antarctic has been distinguished as the specific focus area.

The concept of administration of marine activities has been replaced by the concept of governance of marine activities. The key elements of such governance will be further developed in the draft Federal Law on the Governance of Marine Activities of the Russian Federation (see next section). Generalized criteria of efficiency of marine activities are presented.

Thus the Marine Doctrine is anticipated to provide guidance on general trends of development of marine activities, which are interrelated by nature, including clear definition of the economic development and improvement of living standards as the priority of marine activities (in line with the Strategy for National Security of the Russian Federation), involvement of business as a key stakeholder of the national marine policy along with the state and society, and transition from at large branch-specific (sectoral) approach to the management of marine activities to the holistic (integrated) approach based on modern management technology, in accordance with the Strategy for Development of Marine Activities.

Comprehensive annual assessment of the status of the marine component of the national security of the Russian Federation is scheduled. The Marine Board reports on results of such assessments to the president of the Russian Federation.

The draft Federal Law on Governance of Marine Activities

As we have noted earlier, the regulatory framework of national marine activities is sector-specific and use of resources is based on sectoral laws. The interests of sectoral operators are conflicting and the existing system of management does not provide reconciliation tools with which to ease these conflicts and, in the authors' view, lacks sufficient environmental emphasis.

It has to be said that whereas the Russian sea-related laws comprise quite well-developed instruments, they nonetheless stem from sectoral (departmental) efforts. The functions of national governance of marine activities in the Russian Federation are distributed among a range of the ministries, for which integration is generally of low priority. At the moment, neither of the existing federal bodies of executive power is in the position to provide for the balanced development of marine activities and the implementation of integrated and effective marine policy.

On the basis of these considerations, the Marine Board requested the development of a draft Law on Governance of Marine Activities of the Russian Federation to be enacted in 2015. Hopefully, this will allow for the establishment of an efficient governance tool adjusted to the current conditions of complex and interrelated marine activities.

In the authors' opinion, this is achievable provided that:

- the proposed law takes as its top priority the goal of economic development and improvement of living standards for marine activities, and identifies all potential advantages of integrated marine policy in the mid- to long-term perspective for fostering economic growth, competitiveness, investment potential, employment, comfortable conditions for living and its standards, and environmental protection, as well as its value for reducing the damage from natural and man-made catastrophes, and its contribution to the security of the country in terms of the natural competitive advantages of sea and coastal land areas;
- the important role that the contribution of the business community makes to the efforts by the state and society in the implementation of the national marine policy is ensured; and
- the transition from a branch-specific (sectoral) approach to planning and management of marine activities to the holistic (integrated) approach is ensured.

This draft law shall prioritize the development of the constitutional provision on 'joint competence', which defines the tools of participation of the subjects of the Russian Federation in the exercise of powers of the Russian Federation on determining the status of the territorial sea, its EEZ, and the continental shelf of the Russian Federation in the context of the definition of the competence of the federal centre and subjects of the Russian Federation, as applied to marine and coastal resources.

Other priorities are the formation of effective and understandable mechanisms concerning the identification and resolving of controversies emerging between the federal centre, the subjects of the Russian Federation, and the municipalities – that is, establishing the necessary working mechanisms for formation and practical implementation of the national marine policy.

The law proposition shall reflect principles of marine policy, such as the ecosystem approach, suggesting that the marine ecosystem should be viewed as a whole, with all processes therein being interrelated, involving participatory processes, as well as accountability, adaptive management, application of best available science and information, the precautionary approach, preventive measures, and the protection of biodiversity.

These principles have been developed by and introduced into national legislation by the leading sea powers; they are also part of international law. Some of them are reflected in Russian national legislation, such as in the Federal Law on the Environment Preservation, as well as international conventions ratified by the Russian Federation, such as the Convention on Biological Diversity. However, the majority of these principles have yet to be included in marine management directives, including the Strategy for Development of Marine Activities and the revised Marine Doctrine.

Integrated approaches to maritime management: Current stage

Pursuant to the task stipulated by the Strategy for Development of Marine Activities, the Ministry of Economic Development of the Russian Federation produced and disseminated among the Russian Federation coastal subjects in 2013 the guidelines on development of programmes of integrated development of coastal land and marine areas, as well as guiding documents and recommendations on marine spatial planning.

It is noteworthy that, following the adoption of the Strategy for Development of Marine Activities, relevant provisions on establishment and implementation by the coastal subjects of the Russian Federation of integrated programmes of coastal land and marine areas development were brought within the framework of implementation of strategies approved by the government of the Russian Federation, including the strategies of social and economic development by 2020 of the North-West, and the Southern and the Urals federal districts, as well as their implementation plans adopted by the government of the Russian Federation (2011–12). Analogous provisions were introduced into the Strategy for Development of the Arctic Zone of the Russian Federation and Ensuring National Security by 2020, decreed by the president of the Russian Federation, and the implementation plan of this Strategy adopted by the government of the Russian Federation in 2013.

In addition, the government of the Russian Federation introduced a draft Law on Marine Spatial Planning into its schedule for proposed legislation in 2014.

FTP 'World Ocean', 2016–31

The development of the new concept of FTP 'World Ocean' 2016–2031 by the Ministry of Economic Development of the Russian Federation, on request of the Marine Board, is under way. The process involves interested federal bodies of executive power. The priorities of the programme will be measured with the real opportunity of the state, and it is anticipated that such priorities will be associated with integrated research in the Russian seas and key areas of the world's oceans. They are supposed to ensure long-term interests and to provide scientific data supporting marine activities. This suggests expansion of marine surveys and renovation of the fleet of research vessels. Other priorities are supporting Russian presence in the Antarctic and improving marine activities management.

Conclusions

Russian experts, despite differing opinions on some specific estimates, agree that, within the next ten years, the scale, volume, and variety of maritime activity in the country will increase overall. This increase will be stimulated by the economic development of the country, the increasing score of Russia's connections with the external world, and the predictable expansion of Russia's participation in various forms and types of international cooperation. Development of marine activity in support of Russia's marine policy requires formation of priorities, modern administrative and managerial approaches to implementation, and the upgrading of coastal infrastructure, as well as manufacturing and infrastructure supporting maritime activity (that is, shipbuilding and ship repair, and all kinds of servicing) and information systems and technologies.

The system of long-term decision-making, along with the legislative and normative base, together ensure that a sound state marine policy has been developed in the Russian Federation and that it continues to be improved. Its conceptual bases have been developed according to the

1982 UN Convention on the Law of the Sea (UNCLOS), which Russia ratified in 1997, as well as to the Russian Federation's participation in other international maritime institutions, treaties, and agreements, and provide for appropriate national legislation development.

The adoption of the Marine Doctrine, followed by the adoption of the Strategy for Development of Marine Activities, marked the establishment of new national level corner-stones, goals, and targets of the national marine policy. The Marine Board became the main body responsible for formulation of short-term and current tasks. Therefore the decision-making framework required for the implementation of solid national marine policy is, on the whole, established and still being strengthened. It is FTP 'World Ocean' that, for a long time, used to be one of the major vehicles of implementation of such decisions and proved the effectiveness of the FTP approach to research, development, and use of spaces and resources of the world's oceans, involving interdisciplinary approaches and joint efforts of different departments.

A number of strategic documents recently approved by the president and the government of the Russian Federation, applicable on the national level or the level of federal districts, suggest that the coastal subjects of the Russian Federation should develop and implement programmes of integrated development of coastal land and water areas. Methodological guidelines to this end were produced on the federal level and submitted to these subjects.

Next steps in the strengthening of national marine policy will include drafting of a revised version of the Marine Doctrine of the Russian Federation (now by 2030), drafting propositions of laws on governance of marine activities of the Russian Federation and marine spatial planning in the Russian Federation, and drafting a new FTP 'World Ocean' for 2016–31.

There are reasons to hope that continuous efforts over many years to foster marine activities and to improve their management, as well as to tackle a number of cross-cutting challenges of relevance for the whole country and its economy, will result in the establishment of economic, legal, and institutional frameworks for optimum development of multiple competing sectors based on the resources of sea and its coasts, and accounting for federal, regional, and local interests, as defined in national marine documents, as well as documents of other levels and different sectors.

It appears that reaching these milestones in strengthening the management of national marine activities and its coastal regions – involving a transition from a specific (sectoral) approach to the management of marine activities for the holistic (integrated) manner, on the basis of up-to-date management technology – will be a valuable contribution for solving the tasks laid out in the strategic policy instruments related to the sea in the Russian Federation.

Note

1 A sailor, and renowned Russian scientist in the field of maritime affairs and policy, Dr Sinetsky passed away on 30 March 2014.

References

Aibulatov, N. A., Andreeva, E. N., Mikhaylichenko, Y. G., and Vylegjanin, A. N. (2005) 'Nature Resource Use in the Russian Sea Coastal Zone', *Proceedings of the Russian Academy of Science, Geography Series*, 4: 13–26 [in Russian].

Andreeva, E. E., Mikhaylichenko, Y. G., and Vylegjanin, A. N. (2003) 'Towards a Coastal Area Management Act of the Russian Federation', *Journal of Coastal Conservation*, 9(1): 19–24.

Kalinina, L. I., Lukyanova, S. A., and Solovyeva, G. D. (1992) 'Mapping of Russian Abrasion Coasts', *Bulletin of the Moscow University, Ser. 5, Geography*, 3: 46–50 [in Russian].

Mikhaylichenko, Y. G. (2004) 'Adaptation and Mastering of Integrated Coastal Management in Russia', *Proceedings of the Russian Academy of Science, Geography Series*, 6: 31–40 [in Russian].

Mikhaylichenko, Y. G. (2006) 'Development of an Integrated Coastal Zone Management System for the Black and Caspian Seas', *Mitigation and Adaptation Strategies for Global Change*, 11(3): 521–37.

Mikhaylichenko, Y. G. (2009) 'Marine and Coastal Resources', *State Management of Resources*, 1: 82–95 [in Russian].

Sinetsky, V. P. (2001) *System Analysis of Marine Problems of National Security Maintenance*, Moscow: SOPS [in Russian].

Sinetsky, V. P. (2007) *Search for Regularities*, Moscow: SOPS [in Russian].

United Nations Environment Programme (UNEP) (1995) *Global Programme of Action for the Protection of the Marine Environment from Land-Based Activities*, 5 December, Doc. No. UNEP (OCA)/LBA/ IG.2/7, online at http://daccess-dds-ny.un.org/doc/UNDOC/GEN/K96/000/14/PDF/K9600014. pdf?OpenElement [accessed 29 October 2014].

Voitolovsky, G. K. (2009) *Prospection on Systemic Sea Use: Introduction to Marinistics*, Moscow: Kraft+ [in Russian].

Conventions and treaties

1972 Convention on the Prevention of Marine Pollution by Dumping of Wastes and Other Matter ('London Dumping Convention') (1973) 11 ILM 129.

1973 International Convention for the Prevention of Pollution from Ships, as amended by the Protocol of 1978 (MARPOL)

1982 United Nations Convention on the Law of the Sea (UNCLOS), 10 December 1982, UN Doc. A/ CONF.62/122, (1982) 21 ILM 1261.

1992 Convention on Biological Diversity (CBD), UN Doc. DPI/130/7, (1992) 31 ILM 818.

1996 Protocol to the 1972 Convention on the Prevention of Marine Pollution by Dumping of Wastes and Other Matter (the 'London Protocol') (1996) 36 ILM 1.

6

STRATEGY FOR MARINE ENVIRONMENT CONSERVATION AND DEVELOPMENT IN THE UNITED KINGDOM

Hance D. Smith, Iwan Ball, Rhoda C. Ballinger, Tim Stojanovic, and Tara Thrupp

Introduction

In a strict historical sense, it is true to say that the United Kingdom (UK) has, until very recently, not possessed what may be termed an integrated oceans policy. Nonetheless, as one of the world's leading maritime nations, it has developed extensive, largely sectorally based, maritime policies, especially since the Industrial Revolution of the late eighteenth/early nineteenth centuries, which policies have closely reflected its emerging economic, political, and strategic interests. From the Battle of Trafalgar in 1805 to the outbreak of the First World War in 1914, the UK was the world's leading maritime nation; the profound influence of this period in shaping the country's maritime traditions, outlook, and related institutions remains to this day, despite the radical changes that occurred in the second half of the twentieth century in all aspects of marine affairs.

The main focus of the discussion here is to document in detail the development of UK marine and maritime policy over the past two decades, during which time it has become a major policy area characterized by increasing degrees of policy integration. However, an attempt has also been made, where appropriate, to place due emphasis upon the longer historical time scales, which are still very relevant to the present stage of development, especially concerning the individual groups of sea uses. While the global dimension of UK maritime policies will continue to develop, special interest in the present context attaches to current developments relating to the coasts and seas within the exclusive fishery zone, as defined in 1976. There are notable continuing initiatives in integrated management both of the coast – including the continuation of integrated coastal zone management (ICZM), and new marine plans covering parts of the territorial sea – and offshore, through the development of a succession of marine plans, offshore marine protected areas (MPAs), a revised offshore licensing system, and a UK Marine Policy Statement (HM Government, 2011).

Basic information and an overview of national ocean policy

Basic information

Constitutional and administrative evolution

The UK has evolved throughout the modern era of European history to become a somewhat complex political entity. There are four countries involved – England, Wales, Scotland and Northern Ireland – together with the Isle of Man and the Channel Islands. The parliaments of Scotland and England were united in 1707 – arguably, the true origin of the United Kingdom of Great Britain. The whole of the island of Ireland became part of the Union in 1801, and remained so until the declaration of independence of the twenty-six counties to become the Republic of Ireland in 1922, after the Easter Rising of 1916. The six counties of Northern Ireland remained part of the United Kingdom of Great Britain and Northern Ireland, to give it its full title. Northern Ireland was self-governing, with its own parliament (Stormont) from 1922 until the onset of the 'Troubles' in 1972, when direct rule from Westminster replaced it. Reconstituted in 1999, it was suspended in October 2002, but restored as a devolved government in April 2007.

Meanwhile, a high degree of administrative devolution was granted to Scotland in the 1880s and a lesser, but significant, administrative devolution to Wales in the 1960s. This was followed up in 1999 by granting of political devolution to Scotland (where the Scottish Parliament has limited tax-raising powers not so far used) and, to a lesser extent, Wales. The Silk Commission, established by the UK government in 2011, has since undertaken a review of the financial and constitutional powers of Wales, resulting in some increased powers being incorporated into a draft Wales Bill (December 2013). Scotland has maintained a separate legal system, which has resulted in separate Scottish legislation by the UK Parliament on certain matters for many years – notably, for sea and salmon fisheries.

Following the Government of Wales Act 2006, the Welsh government gained limited powers to create primary legislation across twenty devolved areas, including local government, environment, flood defence, and economic development. The Scottish Parliament also received further powers to legislate on devolved issues (Scotland Act 1998); an independence referendum in September 2014 maintained the Union, although there was a political decision taken as a consequence to extend devolved powers during 2015. Reserved matters, which remain the province of Westminster, include defence, trade, and industry, together with aspects of energy and transport regulation.

The UK joined the European Economic Community (EEC) in 1973 and became a Member State of the European Union in 1992. This has added a strong supranational dimension to aspects of marine policy – notably, concerning fisheries, waste management, pollution control, flood risk management, and nature conservation management. To illustrate current political uncertainties, at time of writing (April 2014) a referendum on continued EU membership has been promised for 2017 if a Conservative government is returned at the 2015 general election and there is growing political pressure within the UK to leave the European Union.

The Isle of Man and the Channel Islands (principally Jersey, Guernsey, Alderney, and Sark) remained something of a medieval constitutional anomaly before, during, and subsequent to the political and constitutional evolution outlined above. These remain outside the UK and European Union. Their parliaments are directly responsible to the Crown through the Privy Council, which approves all legislation. As islands with large offshore finance industries, this is of some considerable significance in the present maritime context.

Key characteristics of the United Kingdom

The basic geographic and demographic information on the UK is summarized in Table 6.1. With a gross domestic product (GDP) estimated at £1,531 billion in 2013 (ONS, 2014a), the UK remains one of the most developed countries in the world. The highly indented coastline gives a relatively high ratio of coastline length to land area. Over recent decades, there has been sustained immigration from many parts of the world, focused especially on southeast England and, to a lesser extent, the industrial towns of northern England, the English Midlands, and west central Scotland. The UK has one of the largest populations in the European Union, accounting for 13 per cent of the total (ONS, 2005), and is one of the most highly urbanized countries in the world. The population of the UK by end 2013 has been estimated at some 64.1 million (ONS, 2014b). Nonetheless, many peripheral rural areas have been associated with population decline since as early as the mid-nineteenth century.

Maritime boundaries

The UK's well-known location off the coast of northwest Europe renders it one of the most quintessentially maritime countries in the world. The maritime boundaries are summarized in Figure 6.1. In common with other states, these boundaries have evolved relatively quickly from the traditional 3 mile limit of the territorial sea, beginning in the 1960s. The delimitation of the UK continental shelf (UKCS) dates from the Continental Shelf Act 1964, pursuant to the 1958 Geneva Convention on the Law of the Sea.

Table 6.1 Geographic, demographic, and economic information about the United Kingdom

Geographic	
Length and breadth of the UK	Just under 1,000 km from north to south, and just under 500 km across at the widest point
Area of land	242,514 km²
Marine area	867,000 km²
Length of coastline	19,488 km
Biogeographic sea regions	Northern North Sea region
	Southern North Sea region
	Eastern English Channel region
	Western English Channel and Celtic Sea region
	Atlantic South West Approaches region
	Irish Sea region
	Minches and West Scotland region
	Scottish Continental Shelf region
	Faroe-Shetland Channel region
	Rockall Trough and Bank region
	Atlantic North West Approaches region
Demographic	
Population	62.0 million
% population change 1991–2003	5.7
Population density (people per km²)	244/km²
Population within 10 km of coast	16.9 million
Economic	
GDP (at market prices) (£ bn)	1,301.4
Gross value added (at basic prices) (£ bn)	1,259.6

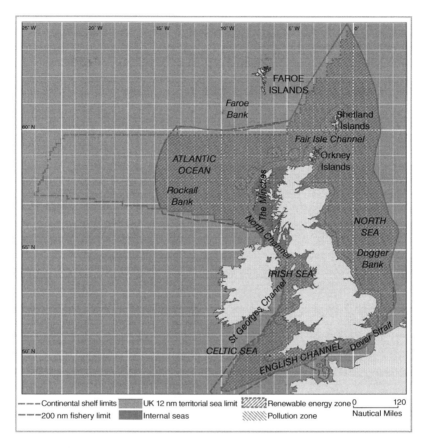

Figure 6.1 Maritime boundaries of the United Kingdom

The UK now has a 12 mile territorial sea defined by the Territorial Sea Act 1987, and introduced a 200-mile exclusive fishery zone in 1976, followed by provision for the designation of an exclusive economic zone (EEZ) under section 41 of the Marine and Coastal Access Act 2009. Owing to the basic coastal configuration of northwest Europe, it has been necessary to adopt a system of median lines in agreement with neighbouring states. The primary driving force for adoption of these has in effect been the development of offshore oil and gas resources, for which delineation of the continental shelf has been imperative: a few hydrocarbon reservoirs – most notably, the now defunct giant Frigg gas field – have at least the potential of straddling international boundaries. The first wave of boundary delimitation occurred between 1965 and 1971 in the North Sea, and included UK boundaries with Norway, Denmark, the Federal Republic of Germany, and the Netherlands. In the late 1970s, the boundary with France in the western English Channel was agreed, in 1988, that with Ireland in the Irish Sea, the St George's Channel, and Western Approaches, and in 1993, that with Belgium. Out to the northwest, the Continental Shelf Act 1964 recognized Rockall as an island generating its own limits. This was never agreed to by Iceland, Ireland, and Denmark, and the seaward boundaries were subsequently modified in negotiation with these countries. The agreement with Denmark for the boundary in the Faroe–Shetland Channel was concluded in 1999. Internally, both the

Isle of Man and Channel Islands (Jersey and Guernsey) have jurisdiction within their respective territorial seas. The Channel Islands sea area is an enclave within French waters.

In May 2006, the UK, Ireland, France, and Spain lodged a joint submission with the United Nations Commission on the Limits of the Continental Shelf (CLCS) claiming an area of continental shelf of approximately 800,000 km² (about the size of Ireland) in the area of the Celtic Sea and Bay of Biscay (CLCS, 2006). Three years later, in May 2009, the Commission concluded that the submission fulfilled the requirements and accepted the recommendations (CLCS, 2008a; 2009). The claim was the first of several to be made by the UK, some on behalf of the Overseas Territories.[1] In May 2008, the UK made another claim, this time for the extension of its territorial rights around Ascension Island in the South Atlantic as part of the British Overseas Territory of St Helena (CLCS, 2008b). The UK presented its claim to the CLCS in August 2008 and, after consideration, the Commission made recommendations. In April 2010, the Commission concluded that there was no morphological, geological, geophysical, or geo-chemical basis for the claim, and therefore the UK could not establish the outer limits of its shelf to include the area around Ascension Island (CLCS, 2010).

Marine sectors and resources

As regards the status and utilization of marine and maritime resources, the sea plays a leading role in the national economy. In 2002, UK marine-based industries were estimated to have had an annual turnover of around £70 billion, generating around 5 per cent of the then GDP and employing at least 1 million people directly (Pugh and Skinner, 2002) (see Table 6.2). However, such figures are fraught with uncertainty because of the precise accounting methodology involved, especially as a range of multiplier effects have to be considered, involving industries dependent on marine industries – notably, in shipping and ports, fisheries, and offshore hydrocarbons. Thus a 2005–06 *minimum* estimate was a £46 billion annual turnover, 4.2 per cent of GDP, and 890,000 jobs (Pugh, 2008).[2]

Table 6.2 Principal sectors of the marine economy: Turnover and value added

Sector	1999–2000 (£ m)	
	Turnover	Value added
Oil and gas	20,597	14,810
Leisure and recreation	19,290	11,770
Defence	6,660	2,531
Business services	4,535	1,080
Shipping	5,200	2,400
Shipbuilding	3,172	1,574
Equipment	2,326	1,358
Fisheries	2,447	825
Environment	1,050	435
Ports	1,690	1,183
Construction	500	190
Research	609	292
Telecommunications	500	190
Safety	316	129
Crossings	155	87
Aggregates	131	69
Education	49	25
Total	**69,227**	**38,948**

Source: Pugh and Skinner (2002)

The largest marine sectors by gross output are oil and gas, shipping (including ports, shipbuilding, and financial services), leisure and recreation, defence, equipment and materials, and fisheries. All major sea-use categories are important in a variety of ways. A classification of basic categories of use that can be used for conflict and management analyses is reproduced in Figure 6.2.

At a time when the region is embarking upon one of its long stages of development, an appreciation of economic and technological features of the key marine industries and associated marine activities is essential. In the field of mineral and energy resources, although the UKCS is now a mature hydrocarbon province that has been in production for nearly fifty years, it remains the most important of the all of the UK marine-related economic activities, meeting almost half of the UK's primary energy needs: in 2012, oil production was 490 million tonnes of oil equivalent, with an oil import dependency of 36 per cent; net gas production was 35 million tonnes of oil equivalent, with gas import dependency of 50 per cent. The industry currently employs some 450,000 people (Oil and Gas UK, 2014), as well as continuing to rank as a major global

			Sea and Land									Land only	
			Transport	Strategic	Minerals & energy	Living resources	Waste disposal	Leisure & recreation	Education & research	Conservation	Coastal engineering	Settlement	Manufacturing & services
TECHNICAL MANAGEMENT	Information Management	Environmental monitoring											
		Surveillance of uses											
		Information technology											
	Information Assessment	Environment											
		Technology											
		Economic											
		Social											
		Risk											
	Professional Practice	Natural/social sciences											
		Surveying											
		Engineering											
		Accountancy											
		Planning											
		Law											
GENERAL MANAGEMENT	**Technical management coordination**												
	Organization management												
	Policy												
	Strategic planning												

Figure 6.2 Basic sea-use matrix for conflict and management analysis

oil and gas province. The UK remains among the top twenty-five global producers of both oil (ranking twentieth) and gas (ranking twenty-third) despite the sharp decline in production since 2011 (Oil and Gas UK, 2013). Despite the province having entered a long-term declining production phase, there remains considerable potential for several decades ahead, which will mean both industrial restructuring and changes to government regulation (Wood, 2014).

The UK government's energy White Paper (DTI, 2007) set a long-term aim of a low-carbon economy, cutting CO_2 emissions by 60 per cent by 2050, along with a UK national target of producing 15 per cent of electricity from renewables by 2015, thereby stimulating substantial interest in harnessing marine renewable energy resources. The subsequent Energy Act 2004 introduced a number of powers in respect of renewable energy; most notably, it established a regulatory regime for offshore renewable energy installations beyond the territorial sea, in the UK's renewable energy zone (REZ), in accordance with rights set out in the 1982 United Nations Convention on the Law of the Sea (UNCLOS). The push for increasing decarbonization of energy generation has continued (HM Government, 2009), together with provisions for carbon capture and storage in line with the Climate Change Act 2008, most recently in the Energy Act 2013. Further, the UK also signed up to the EU Renewable Energy Directive (Directive 2009/28/EC), which sets out a strategic approach to meet national renewable energy requirements. The targets would mean an ambitious sevenfold increase (on 2008 levels) in renewable energy consumption, the largest commitment of any EU Member State.

Major investment has been directed towards developing offshore wind power. By the beginning of 2014, there were 3,653 megawatts (MW) of offshore capacity, the latest major investment to go on stream being the 630 MW London Array, the world's largest wind farm. In December 2013, wind power – both off- and onshore – provided a record 10 per cent of UK electricity demand. By late 2013, there were some 32,000 wind energy jobs overall. Meanwhile, significant experimental and development work continues in the field of both tidal and wave power technologies, in which the UK is also richly endowed with resources (Marine Energy Programme Board, 2014; Renewable UK, 2013).

Despite some £2 billion invested in the wind industry and recent government approval for the building of a new nuclear power station at Hinckley Point in Somerset, there currently remains profound political uncertainty surrounding the future pattern of investment in the energy industries. This is further complicated by the realization of enormous potential for the production of relatively inexpensive gas and oil in large quantities, especially onshore, through the introduction of fracking.

In 2012, over 22 per cent of the sand and gravel used in England and Wales was supplied by the marine aggregate industry, while most of the sand used in south Wales for construction comes from marine sources. In 2012, dredging licenses covered an area of 711 km^2, which equates to just over 0.08 per cent of the UKCS; around 13.6 per cent of the licensed area is being dredged. Some 16.79 million tonnes were dredged (BMAPA, 2014). Subject to continuing research and consideration of environmental impacts, the industry estimates that known existing resources will last for another half-century at present levels of extraction (BMAPA, undated). There are currently two distinct parts of the dredging fleet: an inshore group of dredgers, working for local markets; and much larger (in tonnage and number) dredgers, supplying the wider markets, especially in southeast England and adjacent continental Europe. It is likely that future production will move progressively further offshore into deeper water with a new generation of large dredgers.

While the UK remains one of the primary capital investors in the global shipping industry and London is the premier focus of the world's sea trade, labour supply has largely migrated to developing countries. To give some idea of the complexity of the industry, at the end of 2012 there were 675 UK direct-owned trading vessels (20.4 million gross tons); of these, 49 per cent were UK registered – only 34 per cent of the gross tonnage. However, the UK parent-owned fleet consisted of 845 ships, totalling 25.2 million gross tons, while there were 1,151 UK-managed trading ships, totalling 47.7 million gross tons (DfT, 2013). With the end of the recession that followed the 2008–09 global financial crisis, there has been a slight decrease in these figures. In the past fifteen years, the UK registered deadweight tonnage has increased six-fold, while the UK-owned tonnage has trebled, and UK parent-owned and UK-managed tonnage have both increased by almost 80 per cent. In addition, London remains one of the world's foremost maritime services centres, with a large cluster of marine insurance, ship finance, law, classification societies, shipbroking, education, and consultancy organizations (UK Chamber of Shipping, undated).

The UK ports industry is the largest in Europe and handled almost 500 million tonnes of freight in 2012. Over 95 per cent of imports and exports by volume, and 75 per cent by value, pass through the ports. The main categories of cargoes include raw materials (foodstuffs, oil and gas, mineral ores, metals), manufactured and semi-manufactured goods, and passengers (both short sea and cruise). These are associated with the major shipping categories of 'roll on, roll off' (ro-ro), liquid and dry bulk, container traffic, and cruise ships. In addition, there are specialized port functions associated with the offshore oil and wind, fisheries, and marine leisure industries. Altogether, there are approximately 120 commercial ports in the UK (UK Major Ports Association, undated). Ports directly employ 73,500 people, with a further 47,000 employed in ancillary industries (BPA, undated). There is a particular concentration of port and shipping activity in southeast England, because Europe has supplanted other global regions as the UK's principal trading partner. Notable recent developments have included the extension of the container terminal provision at Felixstowe and the opening of the London Gateway on the Thames. Port activity closely reflects the long-term development of the national economy, with forecast increases in container traffic of 182 per cent and 101 per cent in ro-ro traffic by 2030 (DfT, 2012).

Since the end of the Cold War in 1989, the UK has increasingly become a regional, rather than global, maritime power, operating as a member of the North Atlantic Treaty Organization (NATO). However, it retains substantial naval power and warship-building capacity, the latter mainly located at Barrow-in-Furness (submarines), and Portsmouth, Rosyth, and Glasgow (surface ships). The premier naval ports include Portsmouth, Devonport (Plymouth), and Faslane on the Gairloch (nuclear submarines). Substantial defence cuts during the recent recession have, however, led to decommissioning of part of the fleet, and corresponding reductions and consolidation in port activity.

The fishing industry has been subject to increasingly difficult times over the past several decades as the ability of the resources to sustain ever-increasing pressure exerted by both technological development and changing economic circumstances has been declining (House of Lords Committee on the European Union, 2008). Since the accession of the UK to the EEC in 1973, fisheries have been managed within a European framework. The first EU Common Fisheries Policy (CFP) came into force in 1983. Since then, there have been decadal reviews and updates of the CFP, the latest beginning with a Green Paper in 2009 (European Commission, 2009a) and being concluded in 2013. The serious pressures were acknowledged by the UK government after the 2002 CFP reform, with future policy measures proposed in 2004 (Cabinet Office, 2004) and a response published in 2005 (Defra, 2005a; Morley, 2005). Central throughout has been the need to reduce the catching power of the fleet through decommissioning (especially after 2002), the need to manage discards, and the need to introduce a regional dimension to management; these last two have been incorporated into the latest iteration of the CFP, which came into force on 1 January 2014. It stipulates that, between 2015 and 2020, catch limits should be sustainable and maintain the long-term viability of fish stocks (European Commission Fisheries, undated).

The UK also has a substantial salmon farming industry, which is located on the west coast of Scotland and in Shetland. Salmon production started from a low base in the early 1980s and reached over 162,000 tonnes in 2012 (Marine Scotland, undated). The industry is subject to global competition, especially from Norway, and has exhibited cyclical development. It is also vulnerable to disease and associated with environmental impacts, all of which present considerable management challenges. A second, much smaller sea farming industry produces mussels, and there are initiatives to produce other finfish – notably, halibut and cod.

Coastal and marine leisure and tourism remain very important economic activities around the UK, despite the demise of many traditional seaside resorts and the very competitive market for activity holidays. With increased mobility, a lengthening of the tourist season and an increase in active pursuits including water sports, rural areas continue to witness some of the greatest increases in tourism and leisure activity. Tourism agencies have been active in proposing development strategies for marine recreation, and recent years have seen a steady growth in both informal and organized marine recreational activities. Water-based recreation has inevitably been affected by the recession, but is now expanding again. In 2013, the best estimates for participation were 14.1 million adults, including 8.3 million enjoying leisure on the beach, and approximately half of the 3.4 million taking part in boating activity doing so on the coast. Although the complex structure of the industry makes it difficult to determine total economic value, industry revenue for the leisure, superyacht, and small commercial marine components in 2012–13 was £2,905 billion, including 37.2 per cent generated by exports (BMF, undated).

In the field of waste disposal, dumping of industrial waste and sewage sludge at sea, was phased out in the 1990s following intense pressure from other North Sea states. Offshore disposal of maintenance dredging spoil from port operations still occurs, with large port developments resulting in major dredging. Other improvements include the recent installation of secondary sewage treatment (and, in some cases, tertiary treatment) for all coastal towns to comply with the EU Urban Wastewater Directive (Directive 91/271/EEC), and a tightening of general pollution control standards as a result of regional, supranational, and national legislation. However, beach litter remains a major problem along many coasts, with significant inputs from beach users, fishing, and shipping activities (Defra, 2005b). There are also a large number of landfill sites within coastal areas that are potentially at risk from higher sea levels, rising groundwater levels, and increased rainfall (Environment Agency, 1999).

There is a large and diverse marine research and education community based on government laboratories (oceanography and fisheries) and the universities. There is also an important commercial consultancy sector actively engaged in research in marine science and technology, some of which is specifically linked to particular sectors, such as ports and the offshore oil industry. The UK Marine Science Strategy 2010–25 sets the general direction for future efforts in marine science and research, outlining high-level priority areas for scientific effort and addressing cross-cutting barriers to aid delivery of the science (Defra, 2010b).

Finally, coastal and marine conservation has been a rapidly growing activity. There are particularly high concentrations of designated areas in the west and north of the UK.[3] In Wales, over 70 per cent of the coastline is designated as being of national or international importance for nature conservation and landscape qualities. Marine species appearing in UK waters are protected under a number of international, European, and national legislative policies. Following reinterpretation of the extent of the EC Habitats Directive (Directive 92/43/EC) offshore and the conclusions of the Review of Marine Nature Conservation (RNMC) (Defra, 2004a), the case for greater protection of marine habitats and species continued to increase, because it was acknowledged that the current network of protected sites did not provide a comprehensive and ecologically coherent framework for the protection of the marine environment.

The RMNC established the Irish Sea Pilot project in 2002 as a means of testing an ecosystem-based management approach of ecologically distinct 'regional seas'. At the end of the pilot study period, sixty-four recommendations, including one that government should promote the 'regional seas' approach to marine and coastal management, were made (Vincent et al., 2004). The joint UK government response to the RMNC was published in December 2005. Presented in two parts, *Safeguarding Sea Life* (Defra, 2005b) sets out the government's and devolved administrations' shared policies for conserving marine biodiversity and what this means in relation to each of the RNMC's key recommendations. The report sets out new strategic goals for marine nature conservation, proposes the development of marine ecosystem objectives, and describes how the four administrations will use wider marine management regimes, including spatial planning, to help them to meet these objectives.

In 2007, the Offshore Marine Conservation Regulations were brought into force. These Regulations constitute the legal basis for the implementation of the EC Habitats Directive and the EC Birds Directive (Directive 92/43/EEC), and apply to the UK's offshore area beyond 12 nautical miles. The Regulations facilitate the formation and protection of special protection areas (SPAs) and special areas of conservation (SACs), known as 'European offshore marine sites', forming a network of sites known as Natura 2000.

The Marine and Coastal Access Act 2009, Marine (Scotland) Act 2010, and Marine (Northern Ireland) Act 2013 underpin legislative policy, and introduce new tools for marine and coastal conservation and management in the UK, providing a means for improved management and protection of marine species and habitats. Marine conservation is one of the principal issues addressed under the Acts, with the aim of expanding the current network of marine protected areas (MPAs) in UK waters, with national sites to complement the current European Natura 2000 network of sites that extend offshore. This will be facilitated through a series of provisions, including the amendment of marine boundaries to existing sites of special scientific interest (SSSIs), the improvement of national marine nature reserves, and the designation of marine conservation zones (MCZs) in England and MPAs in Scotland, offering more effective protection and covering wider areas. The MCZs will be designated in England by the Secretary of State and in Scotland, by the Scottish Government. In Wales, following the Welsh Government's consultation on options for highly protected MCZs and considerable opposition to proposed sites, the Welsh Government has withdrawn all proposed sites and has commissioned an assessment of the effectiveness of all current MPAs. However, marine planning, improved fisheries practices, and the regulation and enforcement of legislation under the Marine and Coastal Access Act 2009 will further aid the protection of marine species and habitats.

Patterns of sea use

In analysing conflicts among uses, distinctions may be based on overall intensities of use. It is useful to distinguish first 'urban' sea areas adjacent to large conurbations and generally within internal waters and the territorial sea. Within these areas, engineered shorelines, including extensive coast protection, sea defence, and harbour works are common, with a high level of commercial shipping and port activities, maritime leisure activities, and waste disposal. Beyond lie two regions of 'rural' seas, with a contrasting mix of generally less-intensive uses, including offshore hydrocarbon extraction, wind farms, aggregate dredging, fishing grounds, aquaculture, specialized marine leisure, shipping routes, military exercise areas, and coastal and offshore conservation zones. The more intensively used rural region consists of the southern North Sea, roughly south of 55°N, together with the eastern part of the English Channel; the less intensively used region comprises the remainder of the seas within UK maritime boundaries, as shown in Figure 6.1. Reference should also be made to the government's successive assessment of the state of the UK seas (Defra, 2005c; 2010a) for an overview of the main threats associated with human use facing the UK's biogeographic sea regions.

Given the size and highly developed nature of the UK economy, including its maritime sectors, and the extensive sea areas within UK maritime boundaries, it is scarcely surprising that oceans and coasts have been a policy area of sustained importance. The continued increase in maritime activities with associated conflicts of use has heightened concentration on coastal and marine policy, outlined at appropriate points in the following text.

Despite the obvious importance of the marine environment and its associated human activities, until recently ocean policy has been underplayed. This 'ocean policy' deficit was most apparent in the UK's first Sustainable Development Strategy (SDS) published in 1994 (HM Government, 1994). However, even later government documents relating to sustainable development, including the most recent SDS (HM Government, 2005) pay somewhat scant regard to the marine environment, although there is increasing consideration of a few related topic areas, including coastal risks from sea-level rise and the harnessing of offshore renewable energy. The emergence of marine policy as a clearly defined area began with the Marine Stewardship Report (Defra, 2002a) and continued, more recently, under the EC's Marine Strategy Framework Directive (MSFD) (Directive 2008/56/EC), which came into force in 2008. In 2009, Defra released a public consultation on draft regulations, followed by a response in June 2010. The Directive was transposed into UK legislation in July 2010, putting in place a clear legal framework to ensure the successful implementation of the MSFD in the UK. Meanwhile, the three country-specific Marine Acts remain one of the most crucial developments for UK ocean policy, representing a huge step towards delivering sustainable development of the marine and coastal environment by providing means for the improved management and protection of both inshore and offshore waters.

A brief overview of the nature and evolution of national ocean policy

Evolution and current status

The starting point for understanding national UK maritime policy remains the country's strategic and economic interests. For many centuries, the UK and its antecedent countries (mainly England) have rightly regarded themselves as maritime nations. In terms of maritime economic development, it is fair to argue that Holland, in the sixteenth and seventeenth centuries, was the first truly maritime industrial nation. However, from the end of the seventeenth century until the outbreak of the First World War in 1914, England – or rather, Britain – was the supreme maritime power, although this was not clearly consolidated until the Battle of Trafalgar in 1805 and the ensuing *Pax Britannica* during which the Royal Navy was the world's supreme military power at sea. As the world's first industrial nation from the 1780s until the end of the nineteenth century, this British maritime outlook was greatly strengthened. Even when Britain was overtaken as an industrial power by the United States and Germany in the early twentieth century, her maritime interests remained largely undiminished, notably in areas such as shipping and shipbuilding, fisheries, and marine science, as well as the military.

In the second half of the twentieth century and notably in the negotiations leading up to UNCLOS, the UK tended to maintain its historic position as a maritime power, for example with regard to the enclosure of the oceans and deep-sea mining. It is only in the last thirty years or so that greater political attention has become more focused on the seas surrounding the country itself. This helps to explain the involvement of the UK in the 'Cod Wars' with Iceland,[4] for example, as well as non-declaration of an EEZ until long after the UNCLOS provisions were enacted. Indeed, the UK's acknowledged strengths in marine research have provided considerable influence in negotiating maritime policy at both European and international levels (see 'Research and education' later in the chapter).

The UK Marine Policy Statement and marine plans

In the post-UNCLOS world now emerging, it is fair to say that the UK possesses a whole range of policy initiatives that are marine-related. The UK Marine Policy Statement (HM Government, 2011) is intended to provide a more integrated, cohesive framework for the sectoral system. This relatively short document has been adopted by all of the UK administrations: Westminster, the Scottish Government, the Welsh Government, and the Northern Ireland Executive. The policy statement makes a commitment to 'promote sustainable economic development'. Resulting marine plans and licensing decisions are supposed to be in conformity with the policy. Marine planning is currently being developed under the provisions of the several Marine Acts (Defra, 2010c). Provisions have been made for sequential development of ten regional plans in England and eleven regional plans in Scotland. At the time of writing (April 2014), draft plans are under way for the south coast of England and for Scotland. The Welsh Government has also recently issued a consultation, the *Statement of Public Participation on a single Welsh National Marine Plan* (Welsh Government, 2014), including an inshore (territorial sea) and offshore component, to be in place by the end of 2015. It is likely that there will also be a single marine plan for territorial waters around Northern Ireland.

Sectoral policies

Consistently signposted in the UK Marine Policy Statement is a set of sector-based policies with a strong legislative base. Over the long time scale since the Industrial Revolution (1780–1830), the development of sectoral policies can be followed in large measure from accompanying legislation and reflect, to a substantial degree, broad fifty-or sixty-year stages of maritime economic development. Major pieces of legislation tend to lag behind the pace of development, with periodic 'consolidating' Acts. The major sectors involved are shipping and ports, and fisheries, with much smaller numbers of key Acts in other sectors. Appendix 7.1 summarizes the most important pieces of legislation for key maritime sectors.

As a starting point for more detailed discussion of the policy development process in the following section, it is worth identifying the key areas of maritime policy development (Table 6.3). The majority of these are primarily economic in nature, and include: shipping and ports, for which the emphasis is on cooperation at the international level, both globally and in Europe; offshore mineral and energy development, which has a strong national focus; and sea fisheries and fish farming, over which the European Union is now the dominant influence. In the field of environmental policy, marine waste disposal and pollution control policy has been developed in conjunction, first, with regional international organizations operating under regional conventions that have a thirty-year track record and include the 1972 Convention for the Prevention of Marine Pollution by Dumping from Ships and Aircraft (the Oslo Convention), the 1974 Convention for the Prevention of Marine Pollution from Land-based Sources (the Paris Convention), and the 1992 Convention for the Protection of the Marine Environment of the North East Atlantic (the OSPAR Convention), and latterly with the European Union, which has focused on marine issues (other than fishing) only in the last few years.

Table 6.3 Major marine policy areas showing selected policy reviews in key sectors

Policy area	Recent policy reviews
Fisheries	*Review of Marine Fisheries and Environmental Enforcement* (Defra, 2004)
	Net Benefits: A Sustainable and Profitable Future for UK Fishing (Prime Minister's Strategy Unit, 2004)
	Turning the Tide: Addressing the Impact of Fisheries on the Marine Environment (RCEP, 2004)
	The Future for UK Fishing: Environment, Food and Rural Affairs Select Committee Sixth Report of Session 2004–05 (EFRA, 2005)
	Securing the Benefits: The Joint UK Response to the Prime Minister's Strategy Unit Net Benefits Report on the Future of the Fishing Industry in the UK (Defra, 2005)
	Charting a New Course (Defra, 2006)
	Fisheries 2027: A Long-term Vision for Sustainable Fisheries (Defra, 2007)
Ports and shipping	*A New Deal for Transport: Better for Everyone – The Government's White Paper on the Future of Transport* (DETR, 1998)
	British Shipping: Charting a New Course (DETR, 1998, modified 2003)
	Review of Trust Ports (DETR, 1998)
	The Future of the UK Shipping Industry: Environment, Transport and Rural Affairs Select Committee Twelfth Report of Session 1998–99 (ETRA, 1999)
	The Government's Response to the ETRA Committee's Twelfth Report of Session 1998–99 on The Future of the UK Shipping Industry (DETR, 1999)
	Identification of Marine Environmental High Risk Areas (MEHRAs) in the UK (DETR, 1999)
	Modernising Trust Ports: A Guide to Good Governance (DETR, 2000)
	Modern Ports: A UK Policy (DETR, 2000, modified 2003)
	Freight on Water: A New Perspective (Freight Study Group, 2002)
	The Government's Response to the Freight Study Group Report Freight on Water (Defra/DfT, 2002)
	Ports: Transport Committee's Ninth Report of Session 2002–03 (Transport Committee, 2003)
	The Government's Response to the Transport Committee's Ninth Report of Session 2002–03 on Ports (DfT, 2004)
	The Work of the Maritime and Coastguard Agency: Transport Committee Fourteenth Report of Session 2003–04 (Transport Committee, 2004)
	The Government's Response to the Transport Committee Fourteenth Report of Session 2003–04 (DfT, 2004)
	Ports Policy: Your Views Invited (DfT, 2006)
	Ports Policy Review Interim Report (DfT, 2007)

Marine energy	*Energy: The Changing Climate* (RCEP, 2000) *The Energy Review: A Performance and Innovation Unit Report* (Cabinet Office, 2002) *Future Offshore: Strategic Framework for the Offshore Wind Industry* (DTI, 2002) *Our Energy Future: Creating a Low Carbon Economy – Energy White Paper* (DTI, 2003) *Renewables Innovation Review* (DTI/Carbon Trust, 2004) *Creating a Low Carbon Economy: First Annual Report on the Implementation of the Energy White Paper* (DTI/Defra, 2004) *Nature Conservation Guidance on Offshore Windfarm Development* (Defra, 2005) *Renewable Energy 2005/6 Review of the Renewables Obligation: Preliminary Consultation* (DTI, 2005) *Creating a Low Carbon Economy: Second Annual Report on the Implementation of the Energy White Paper* (DTI/Defra, 2005) *Energy: Its Impacts on the Environment and Society* (DTI, 2005)
Marine nature conservation	*Review of Marine Nature Conservation: Interim Report* (Defra, 2001) *Natura 2000 in UK Offshore Waters* (JNCC, 2002) *State of Nature: Maritime – Getting onto an Even Keel* (English Nature, 2002) *Marine Nature Conservation and Sustainable Development: The Irish Sea Pilot – Report to Defra by the Joint Nature Conservation Committee* (JNCC, 2004) *JNCC Response to Consultees on Natura 2000 in UK Offshore Waters* (JNCC, 2004) *Review of Marine Nature Conservation Working Group Report to Government* (Defra, 2004) *Our Coasts and Seas: Making Space for People, Industry and Wildlife* (English Nature, 2005) *Safeguarding Sea Life: The Joint UK Response to the Marine Nature Conservation Review* (Defra, 2005)
Coastal protection and flood defence	*National Appraisal of Assets at Risk from Flooding and Coastal Erosion, Including the Potential Impact of Climate Change* (Defra, 2001) *National Assessment of Defence Needs and Costs for Flood and Coastal Erosion Management (NADNAC)* (Defra, 2004) *Making Space for Water: Developing a New Government Strategy for Flood and Coastal Erosion Risk Management in England – Consultation* (Defra, 2004) *High Level Targets for Flood and Coastal Erosion Risk Management* (Defra, 2005) *Making Space for Water: Taking Forward a New Government Strategy for Flood and Coastal Erosion Risk Management in England – First Government Response* (Defra, 2005)

(continued)

Table 6.3 (continued)

Policy area	Recent policy reviews
Water quality	*Quality Status Report of the Marine and Coastal Areas of the Irish Sea and Bristol Channel* (Defra, 2000)
	First Consultation Paper on the Implementation of the Water Framework Directive (2000/60/EC) (Defra, 2001)
	Second Consultation Paper on the Implementation of the Water Framework Directive (2000/60/EC) (Defra, 2002)
	Third Consultation Paper on the Implementation of the Water Framework Directive (2000/60/EC) (Defra, 2003)
	Directing the Flow: Priorities for Future Water Policy (England) (Defra, 2002)
	A New Diagnosis for the Long-Term Health of England's Waters (Defra, 2004)
	Integrated Pollution Prevention and Control: A Practical Guide (4th edn, Defra, 2005)
	Future Water: The Government's Water Strategy for England (Defra, 2008)
	Water Resources in England and Wales: Current State and Future Pressures (Environment Agency, 2009)
Marine heritage	*Protecting Our Marine Historic Environment: Making the System Work Better – Joint Consultation* (DCMS, 2004)
Coastal management	*Managing Coastal Activities: A Guide for Local Authorities* (Defra, 2004)
	ICZM in the UK: A Stocktake – Final Report (Defra, 2004)
Integrated/Cross-cutting	*Safeguarding Our Seas: A Strategy for the Conservation and Sustainable Development of Our Marine Environment – First Marine Stewardship Report* (Defra, 2002)
	Seas of Change: The Government's Consultation Paper to Help Deliver Our Vision for the Marine Environment (Defra, 2002)
	Regulatory Review of Development in Coastal and Marine Waters (DfT, 2002)
	The Government's Response to its Seas of Change Consultation to Help Deliver our Vision for the Marine Environment (Defra, 2004)
	The Marine Environment: Environment, Food and Rural Affairs Select Committee Sixth Report of Session 2003–04 (EFRA, 2004)
	Marine Environment: Government's Reply to the Committee's Report – Tenth Report of Session 2003–04 (EFRA, 2004)
	Delivering the Essentials of Life: Defra's Five Year Strategy (Defra, 2004)
	Charting Progress: An Integrated Assessment of the State of UK Seas (Defra, 2005)
	Cleaner Coasts/Healthier Seas: Working for a Better Marine Environment – Environment Agency Strategy for 2005–2011 (Environment Agency, 2005)
	Managing Our Marine Resources: Licensing under the Marine Bill (Defra, 2008)
	Statement of Public Participation for the UK Marine Policy Statement (Defra, 2009)
	Consultation on a Marine Planning System for England (Defra, 2010)

Coastal and marine conservation of environment and heritage has a long home-grown history, but environmental conservation has also been more recently and substantially driven by regulation emanating from the European Union, including the Habitats and Birds Directives. This is mainly the result of the cross-boundary issues associated with conservation, such as marine pollution and overfishing. The Marine Strategy Framework Directive (MSFD) will constitute the environmental element of future European maritime policy, driving future UK legislation towards achieving and maintaining good environmental status of its waters by 2020. Under the MSFD, the UK is required to develop a marine strategy for its waters in coordination with other countries from the same sea region.

More recently, there has been the development of what may be termed 'integrated coastal and marine management', ultimately aimed at bringing together all of the other policy areas. This is the newest field and is currently aimed primarily at integrated coastal management (ICM) and marine spatial planning (MSP). In both of these subfields, there has been considerable European influence, for example through the EU Recommendation on Integrated Coastal Zone Management (ICZM) of 2002, the European Spatial Development Perspective (ESDP) (Committee on Spatial Development, 1999), and the Framework European Marine Strategy (European Commission, 2002). The 2002 Recommendation was reviewed during the latter part of 2006 and early 2007, and in 2009 the UK government set out its approach to ICZM in its report *A Strategy for Promoting an Integrated Approach to the Management of Coastal Areas in England* (Defra, 2009a), along with similar reports in other home nations. A proposed framework Directive on Marine Planning and Integrated Coastal Management (European Commission, 2013) has the potential to draw together these two streams of policy. However, there are concerns that the ICM component of this is likely to be squeezed out as revisions to the draft Directive progress under increasing scrutiny, despite considerable consternation from various non-governmental organizations (NGOs), such as the World Wildlife Fund for Nature (WWF), and many of the southern European Member States, which already have to comply with ICM legislation in the form of the ICZM Protocol to the 1976 Convention for the Protection of the Mediterranean Sea Against Pollution (the Barcelona Convention). It is also probably true to say that coastal integration has been overshadowed by recent developments in marine planning.

In recent years, maritime policy has started to leave behind the sectorally based management approach with which it has historically been associated. In 2006, the European Commission published a Green Paper on integrated maritime policy, *Towards a Future Maritime Policy for the Union: A European Vision for the Oceans and Seas* (European Commission, 2006b), outlining maritime policy for the European Union. The policy considers the European Marine Strategy and the 2002 ICZM Recommendation under its broad framework, although arguably it can be seen as being weighted towards sustaining and increasing the economic benefits of the oceans rather than taking into account important environmental considerations. Whilst environmental pressures are addressed in the policy Green Paper, it has been argued that emphasis has been placed on the broader maritime economic sector and its benefits, and somewhat neglects the environmental consideration as a mere 'pillar' demanding less concern.

During the late 1990s and early 2000s, climate change and its linkages to the marine environment became more widely accepted, and thus formed a separate key area in marine policy development. The links were noted in the 2006–07 review of the ICZM Recommendation, with marine resources, marine and coastal tourism, and coastal zones being featured in the second European Climate Change Programme (ECCP II) in 2005. The European Commission published its 2007 Green Paper *Adapting to Climate Change in Europe: Options for EU Action* in response to the growing need to overcome the challenges faced by climate change (European Commission, 2007c). The Green Paper features sections on marine fisheries and marine ecosystems and biodiversity and draws attention to the 2006 Biodiversity Communication (European Commission, 2006a) and the annexed 'EU Action Plan to 2010 and Beyond and Indicators'. The Marine Strategy constituting EU maritime policy will integrate climate change and adaptation actions into marine policies and plans. Both the European Climate Change Programme and the Green Paper are of key importance to Europe's, and therefore the UK's, coastal zone.

Recent developments in UK policy stem from the broader framework of the EU's Integrated Maritime Policy. The Blue Paper *An Integrated Maritime Policy for the European Union* (European Commission, 2007b) outlines the need for 'joined-up' policymaking and a move away from isolated, sectorally based management. This is a continuation of the views expressed in the 2002 Recommendation on ICZM. Its main purpose is to develop a new approach and to establish a platform for the exchange of views and ideas, and it is expected to serve as a useful political instrument in maritime management (European Commission, 2009b). The policy sets out cross-sectoral tools to aid implementation of this new approach, including MSP, integrated surveillance, and improvements in marine knowledge.

The Blue Paper also sets out an ambitious action plan that aims to improve the maritime economy, whilst protecting and restoring the marine environment. In September 2012, the Commission adopted *Blue Growth: Opportunities for Marine and Maritime Sustainable Growth* (European Commission, 2012). The objective for the coming years is to promote sustainable development to come from the sea as the European economy gradually emerges from recession.

Historically, there has been no overall integrated policy for marine protection at the EU level and measures have traditionally been taken on a sector-by-sector basis. The European Marine Strategy, which represents the environmental dimension of the EU Sustainable Development Strategy as far as the marine environment is concerned, is intended to address the current patchwork of policies, legislation, programmes, and action plans. The Thematic Strategy on the Protection and Conservation of the Marine Environment promotes good environmental status of the European Union's marine waters by 2021, and protection of the resource base upon which marine-related economic and social activities depend.[5] The recent MSFD will establish European marine regions on the basis of geographical and environmental criteria. Each Member State, in cooperation with other relevant states within a marine region, will be required to develop strategies for its marine waters. These strategies will contain a detailed assessment of the state of the environment, a definition of 'good environmental status' at the regional level, and the establishment of clear environmental targets and monitoring programmes. The marine strategy is consistent with the Water Framework Directive (Directive 2000/60/EC), which requires that surface waters and ground waters achieve good ecological status by 2015.

Factors in the development of the UK's marine policy

While the broad thrust of UK policy and associated legislative development is closely related to the development of the maritime economy, a number of other contemporary influencing factors are worth noting. These include: international influences in the case of shipping, which are partly related to the activities of the International Maritime Organization (IMO); the influence of the European Union, most notably in the CFP, which involves derogations from UK sovereignty in fisheries matters; and coastal and marine environmental matters covered by a number of directives, notably the Habitats, Birds, and Water Framework Directives. In a wider environmental context, the regional conventions already noted earlier have been an important influence, together with the series of North Sea Ministerial Conferences instigated in Bremen in 1984.

Partly through these influences and partly because of continuing increases in use intensity, especially at the coast, there has been a notable development of coastal management initiatives based on public–private–voluntary sector partnership approaches to coastal environmental management. Of particular note are coastal forums at local and national levels in the constituent countries of the UK, and a range of previous programmes to produce estuary management plans, including the 'Estuaries Initiative' in England, established by English Nature in 1993, and 'Focus on Firths' in Scotland, established by Scottish Natural Heritage in 1995, as well as a collection of other non-statutory local coastal management initiatives led by a variety of organizations and the antecedent Heritage Coasts programme. Altogether, these have produced some eighty local initiatives or projects (Stojanovic and Barker, 2008). Whilst some have now been discontinued, there remain a considerable number of active coastal partnerships, towards which the Marine Management Organisation (MMO) in England has been turning to assist with stakeholder engagement in marine planning.

While, at the international and European levels, policy initiatives have primarily been a function of national government, there has sometimes been considerable pressure exerted by industry (notably shipping and fisheries) and the voluntary sector (notably the conservation interests). At subnational and local levels generally, a whole range of organizations play key roles, representing voluntary local wildlife trusts and national conservation societies (such as the Marine Conservation Society, the Royal Society for the Protection of Birds, the National Trust, and National Trust for Scotland), industrial interests (notably port authorities), the regional offices of government agencies, and local government. Overall, NGOs have played a significant role in campaigning for a more integrated approach to marine policy.

As regards conflicts in the development of national policies, the most difficult situations arise in the case of fisheries, in striking a balance between local and national interests, on the one hand, and European interests, on the other. These involve constant friction between industry and government at both national and European levels. A second type of conflict involves conflict of uses, notably between fisheries, on the one hand, and offshore hydrocarbon and aggregate extraction, on the other, competing for the same sea areas. The potential for conflict has grown considerably in recent years, as space for activities becomes overcrowded and resources decline, forcing sectors into new domains that, in most cases, already support several other marine activities. This is mediated in part in the case of the offshore oil industry by Oil and Gas UK (formerly known as the UK Offshore Operators Association, or UKOOA), a private sector trade association negotiating directly with the fishing industry. In 2006, UKOOA, working collaboratively with the Scottish Fishermen's Federation (SFF) and the National Federation of Fishermen's Organizations (NFFO), published a code of practice in an attempt to reduce conflicts between near-shore fishing activities and the offshore oil and gas industry. The code outlines oil and gas activities that may interact with, or disrupt, fishing practices and provides guidelines for avoiding such issues (Oil and Gas UK, 2006). The guidelines are designed to protect both sectors in an attempt to bring some harmonization to maritime relationships.

Increased development pressures upon the marine environment and potential for multiple-use conflicts, for example arising as a result of the current rapid expansion of offshore wind energy, fishing, dredging, and shipping activities, allied to international and European commitments to biodiversity conservation, have provided an impetus for government to explore MSP at both national and regional sea levels.

The policy development process

It will be apparent from the discussion so far that policy development is both diverse and of very long standing in most cases. It is beyond the scope of this chapter to trace the detailed origin and evolution of these areas. A more useful approach is to highlight the nature and timing of the current phase of marine policy development overall (see Table 6.3). Key factors in understanding policy development include: the transition from the post-war phase of economic development to the present phase, a gradual process largely concentrated in the past two to three decades; the growing importance of environmental policy, an important landmark being the Stockholm Conference on the Human Environment of 1972; and the transformed strategic position of the UK dating from the demise of the Soviet Union in 1989. As regards timing, the period from the early 1970s to the early 1990s encompasses most of the significant international political and economic changes influencing policy development. In addition to the Stockholm Conference, the UNCLOS negotiations between 1973 and 1982, the accession of the UK to the European Common Market (EEC) in 1973, the 1987 World Commission on Environment and Development (the Brundtland Commission), and the 1992 Convention on Biological Diversity (CBD) were all instrumental in contributing to the present international environment of marine policymaking. The period since the early 1990s has been one of more practical policy formulation and implementation.

The central theme now emerging in all policy areas is that of integration, an idea that transcends the individual policy areas per se and is spreading throughout the system. In this context, the influence of international and supranational policy relating to sustainable development since the Earth Summit of 1992 is important. In Europe, the Cardiff Process of 1998 put Article 6 of the EC Treaty into practice, requiring Council organizations to integrate environmental considerations into their respective activities. This does not mean that the sectoral policy approach is moribund; on the contrary, it remains vitally necessary. The approaches adopted will, however, in effect transform policymaking and set the scene for the next two or three decades – as far ahead as can reasonably be foreseen in economic and technological terms.

Initiation of policy

Significant marine sectoral policy development has taken place since the early 1990s. Major legislative and policy updating has been undertaken in the case of shipping, offshore hydrocarbons, and fisheries, for example (see Table 6.3). With respect to fisheries, apart from the governmental initiatives already discussed, the Royal Society of Edinburgh (RSE) and the Royal Commission on Environmental Pollution (RCEP) also produced reports (RSE, 2004; RCEP, 2004). In the new Common Fisheries Policy (CFP), it is worth noting that account has been taken with respect to its integration with the Integrated Maritime Policy (IMP) and MSFD.

In the case of both waste disposal and pollution, and coastal and marine conservation, a primary driver has been regulations deriving from the relevant EU directives, notably on Habitats (Directive 92/43/EC), Birds (Directive 79/409/EC, amended and codified by Directive 2009/147/EC), Urban Wastewater (Directive 91/271/EC, amended by Directive 98/15/EC), and Dangerous Substances in Water (Directive 76/464/EC), as well as by the more recent Water Framework Directive.

The approach to integrated policy has been somewhat different, although the influence of cross-cutting supranational policy, such as that relating to environmental impact assessment (EIA), under the EIA Directive (Directive 85/337/EEC, amended by Directive 97/11/EEC), strategic environmental assessment (SEA) (Directive 2001/42/EC),[6] and agreements and policy developments related to the Cardiff Process cannot be discounted. However, over the last decade, at more local levels along specific coastal stretches, there has been an emphasis on the coastal forums and estuary or firth programmes already noted (Stojanovic and Ballinger, 2009). Further, following the 2002 Bergen Declaration (agreed at the Fifth North Sea Ministerial Conference), commitments within the UK government's *Safeguarding our Seas* (Defra, 2002a) and *Seas of Change* (Defra, 2002b) documents, and recommendations within various UK sectoral policy reports (see Table 6.3), including the *Review of Marine Nature Conservation* (Defra, 2004a) and the subsequent government response, the UK Department for Environment, Food and Rural Affairs (Defra) and Marine Scotland have taken the lead in funding basic research projects on MSP, with pilot projects on marine planning in the Irish Sea (mentioned above), Dorset coast, Firth of Clyde, Shetland Islands, Sound of Mull, and Berwickshire coasts.

The country-specific Marine Acts also include provisions for management of fisheries within the territorial sea, MSP, licensing marine activities, improving marine nature conservation through measures such as the development of marine conservation zones (MCZs) and the establishment of the MMO, which is located centrally in relation to the UK as a whole in Newcastle upon Tyne.

Objectives

The UK Marine Policy Statement of 2011 sets out general issues for consideration in eleven topic areas: MPAs; defence and national security; energy production; ports and shipping; marine aggregates; marine dredging; telecommunications; fisheries; aquaculture; surface water and waste water; and tourism and recreation. The objectives of the policy areas noted in Table 6.3 are inevitably complex, although some features are worth noting. For policy and related legislation covering marine industries, sustainability of the respective industries is a major objective; this, of course, has not only economic, but also social and environmental, dimensions. It is also relevant in political and administrative terms.

For sustainable development of ocean areas and protection of biodiversity, there are environmental elements in the sectoral policy and legislation that are applicable. However, the main thrust now comes through administrative and legal approaches, including the Marine Acts. Again, this is realistic, although it may be a challenge to avoid conflicts of laws. In the UK, Acts of Parliament are amply cross-referenced as are the statutory instruments (SIs) – that is, the secondary regulations – that flow therefrom. Further, harmonization and consolidation is a key consideration in the European Union, where directives apply to all of the Member States. However, in most cases, harmonization remains a practicality, rather than a major objective.

Coordination of the government agencies involved in ocean affairs has been a long-standing aim of at least a few interests since the 1970s. It is also well established in a very few areas, such as search and rescue (SAR) at regional and national levels, in defence generally, and in fisheries protection. However, it has never been particularly well developed among central government departments. There is, for example, no equivalent of the *Missions Interministerielles* found in France. However, interdepartmental groups (IDGs) enable officials from across government to share best practice and to monitor the situation regarding particular areas of operations. For example, an IDG on Coastal Policy, chaired by Defra and including representatives from each central government department and the devolved administrations, was established to exchange information on government policy on coastal matters. However, this initiative, following the House of Commons Environment Select Committee's recommendations in 1992, was short-lived. Another example of coordination among government agencies was the creation of the Marine Consents and Environment Unit (MCEU), which held responsibility for the administration of a wide range of applications for statutory licences and consents to undertake works in tidal waters and at sea in UK waters and beyond (see 'Authority at the national level' later in the chapter), including marine development, offshore energy, coastal defence, dredging, and waste disposal. This function has since largely been taken over by the MMO in England, which is now the principal body in cross-government delivery, contributing to sustainable development of sea areas (MMO, 2014).

Major principles

It would be idle to pretend that there is application of a coherent set of prioritized principles in existing UK marine policy: the system is too diverse for that to be currently feasible. Nonetheless, there have been and remain some areas in which principles are historically and/or presently discernible, notably within sectoral legislation and general non-marine environmental legislation. Also notable are attempts to pull together a set of coherent principles for the marine environment through the Marine Stewardship Report (Defra, 2002a) and, following on from that, the 2009 high-level objectives report for the UK marine area, *Our Seas: A Shared Resource – High Level Marine Objectives* (Defra, 2009a), which both set out the government's vision for clean, healthy, safe, productive, and biologically diverse oceans and seas.

These principles are enshrined in the respective Marine Acts and focus on principles of good governance. The first principle of note is *stewardship*, which is the central theme of Defra's Marine Stewardship Report, a key harbinger of the emerging integrated policy, multiple-use, and participatory governance approaches. 'Stewardship' is defined in the report as 'entrusting people with a responsibility to care for the community they belong to', which means 'involving people in protecting the oceans and seas and using the resources they offer wisely' (Defra, 2002a: 7). The benefits of stewardship, as outlined in the report, include better decision-making, reduced reliance on regulation, generating a positive role for people and organizations, and greater inclusiveness.

In delivering its vision for the marine environment, the government recognizes that there is a need to reflect the economic and social needs of communities and individuals, as well as to protect marine habitats and species. *Sustainability* is therefore a core theme of the government's strategy and is primarily delivered through the regulatory framework, notably via the requirements for EIA and SEA. The latter in particular helps to deliver sustainable development by promoting integrated environmental and development decision-making, environmentally sustainable policies and plans, and consideration of best available techniques and alternatives (Defra, 2002a). More recently, the sustainability theme has featured as a high-level marine objective, set out in the context of five key principles for achieving a sustainable marine environment. This has been set out in the High Level Marine Objectives Report (Defra, 2009b).

The Marine Stewardship Report also endorsed the *ecosystem-based approach*, which it defined as 'the integrated management of human activities based on knowledge of ecosystem dynamics to achieve sustainable use of ecosystem goods and services and maintenance of ecosystem integrity' (Defra, 2002a: 6–7). This approach is also highlighted as a high-priority marine objective (Defra, 2009b). To this end, the government recognizes the need to better integrate marine protection objectives with sustainable social goals and economic growth, and to address conservation objectives alongside the full range of human activities and demands that are placed upon the marine environment (Defra, 2002a).

The UK government also has an obligation to undertake the ecosystem-based approach through several international legislative measures, including the CBD, the 1995 Jakarta Mandate on Marine and Coastal Biodiversity, and the 2001 Reykjavik Declaration on Responsible Fisheries in the Marine Ecosystem. An ecosystem-based approach to marine management also appears consistently in EU policy, being focal in the MSFD and in the 2009 Green Paper on reform of the CFP (European Commission, 2009a).

Finally, the ecosystem-based approach also features strongly in the designation of UK MCZs. The candidate MCZ sites were identified by means of four regional projects in England, covering the Irish Sea, the South West Approaches, the North Sea, and South East waters. In the autumn of 2011, 127 sites were proposed, but only thirty-one of these have been designated as a result of debates over the quality of the evidence base. Alongside the MCZs, special areas of conservation (SACs) and special protection areas (SPAs) will also continue to be identified in UK waters, under the EC's Habitats and Birds Directives, both of which embrace this approach.

Using *best available science*, or robust science, is another of the guiding principles put forward in the Marine Stewardship Report (Defra, 2002a) that underpins government policy for the marine environment. The document states that stewardship of the marine environment must be informed by the best available scientific evidence. It underlies government commitment to develop better integration of marine environmental monitoring and observation, and to continue to encourage appropriate collaboration with industry and NGOs.

Under the EC's Integrated Maritime Policy Blue Paper (European Commission, 2008), key actions highlighted include the Marine and Maritime Research Strategy, the first ever European strategy for promotion of marine research. The strategy will look to develop new capabilities and integrate current research efforts, with the aim of facilitating a holistic research effort in maritime affairs (European Commission, 2009c).

UK government departments and agencies, with the support of the Research Councils, deliver a high-quality marine science programme, which complements initiatives funded through other programmes such as the EC's Framework Research Programme and its successor, Horizon 2020. The initiatives of the North Sea Ministerial Conferences have also produced extensive marine science research aimed at establishing the status of the marine environment in the UK region generally, while the latest initiative with the same objective for UK seas through the UK Marine Monitoring and Assessment Strategy is the series of published 'State of the Seas' reports (Defra, 2005c; 2010a). The inputs of science are particularly significant in other key fields, including nature conservation, marine pollution, and coastal processes. The Marine Science Coordination Committee (MSCC), replacing the Inter-Agency Committee on Marine Science and Technology (IACMST), was devised to improve marine science coordination. The MSCC meets every six months and undertakes extensive consultation with the marine science community.

Scientific evidence may also help to identify areas in which a *precautionary approach* should be adopted, particularly in circumstances in which a range of scientific opinion exists. The precautionary approach is another principle that underpins the government's approach to marine stewardship. This means that decisions should err on the side of caution and that preventative measures should be taken if there are reasonable grounds for concern that damage may occur to the marine environment, 'even when there is no conclusive evidence of a causal relationship between the inputs and the effects' (Defra, 2002a: 7). This approach is also embedded as a major principle in national and EU policies, identified as a high-priority marine objective by the UK government (Defra, 2009b) and set out in the MSFD.

Integration is inherent to delivering the government's marine stewardship vision, as is *stakeholder involvement*. In pursuit of integration, the government is committed to: reviewing legislation affecting development in the coastal area, with a view to simplifying the regulatory system;[7] exploring how coordination among government departments can be improved, particularly with regard to the issuing of consents for development in the marine environment; exploring the role of MSP; building upon existing seabed mapping; and implementing the European Commission's 2002 ICZM Recommendation. The government made considerable progress at a relatively early stage on many of these issues, for example in commissioning the Marine Spatial Planning Pilot[8] and through the Irish Sea Pilot projects mentioned earlier (Vincent et al., 2004), which tested the recommendations of Defra's Review of Marine Nature Conservation on a regional scale and sought to integrate these with policies for other sectors, and in the Scottish Sustainable Marine Environment Initiative (SSMEI) (Scottish Government, undated).

Protection of biological diversity is another principle that is inherent in the stewardship approach envisaged by government, and is addressed in a specific chapter of *Safeguarding our Seas* (Defra, 2002a). The UK government is committed to continuing to play a leading role to help to conserve biodiversity in international forums at regional and global levels, such as the North Sea Conferences and at the OSPAR Commission. At the national level, it is committed, through the UK Biodiversity Action Plan, to taking forward the recommendations of the Marine Nature Conservation Review and Irish Sea Pilot, to implementing the ecosystem approach, and to applying the Habitats and Birds Directives out to the jurisdictional limit of UK seas.

There are other principles that are often present in national and EU policy documents. Sustainability of economic activity is undoubtedly central to much industry-based policy, albeit not usually explicitly stated. Similarly, timeliness is inherent in the evolution of legislation – perhaps especially so in the case of economic policy. Subsidiarity, which was introduced by the Maastricht Treaty, is, of course, an important guiding principle in the development of EU policy and directives. It states that decisions should be made as close as possible to the appropriate level of community interest, thus ensuring that no unnecessary management effort is undertaken. Accountability, adaptive management, and transparency are already built into policy and legislation to some extent, but measures are vulnerable to criticism at times on account of perceived absence of these principles.

Institutional arrangements

In the long history of marine policy development since the end of the eighteenth century, two mechanisms are particularly evident – namely, parliamentary committees and legally constituted committees of inquiry. The first, and more important, stems from processes of economic, technological, and social change that impact upon the political process. This results in consideration of specific topics, such as the state of the fisheries, by standing parliamentary committees with wide remits such as 'science and technology' and 'environment'. The principal contemporary committees are select committees of the House of Commons and committees of the House of Lords. These committees, which are relatively recent in nature, meticulously gather evidence from all of the stakeholders involved, including government itself where appropriate. In each case, their minutes of evidence, together with a report, are published. Most are important historical documents.

The Sixth Report on the Marine Environment of the House of Commons Select Committee on Environment, Food and Rural Affairs (EFRA) sought to examine the effectiveness and urgency with which government was pursing policies for the protection of the marine environment, and what institutional or other barriers existed that might hinder the implementation of policies in this area (EFRA, 2004). The EFRA Committee also conducted an inquiry into the future of the UK fishing industry (EFRA, 2005), already mentioned, which assessed many of the recommendations made in *Net Benefits*, the government's Strategy Unit report of March 2004 (Cabinet Office, 2004).

Government acts on each report in various ways. If major legislation is deemed necessary, either a Green Paper (consultative) or a White Paper is published, followed by a parliamentary Bill (a draft Act), which is then debated in both Houses of Parliament (the Commons and the Lords). Modifications are made in the course of debate, and the final Bill is then passed for royal assent and becomes law on a specified date. Behind this process, a whole range of consultations within government, and between government and private, voluntary, and civil society interests, may take place. A relevant division within the responsible government department will oversee the process. In matters of urgency – usually economic or political – this process can happen relatively quickly; in more evolutionary situations, such as coastal management, it may take between ten and twenty years. For example, the House of Commons Environment Committee's Report on Coastal Zone Protection and Planning (House of Commons Select Committee on the Environment, 1992) was published in 1992. However, the implications are still being worked out and are relevant to various pieces of legislation (including the Marine and Coastal Act 2009), although some recommendations were followed up within a five-year period, including a discussion paper from the Scottish Office in March 1996 (Scottish Executive Central Research Unit, 2001).

The second starting point for the policymaking process is accident or disaster. This may be followed by a public inquiry, which will normally make recommendations, and may also be the subject of a parliamentary committee inquiry in the overall field of the kind noted above. Most commonly, these situations occur in the wake of shipwrecks or offshore installation accidents. Notable examples include the *Piper Alpha* disaster of 1988, which became the subject of the Cullen Inquiry, as a result of which safety regulations under the Mineral Workings and Offshore Installations Act 1971 were extensively modified. A second example was the loss of the tanker *Braer* and the spilling of 84,000 tonnes of crude oil into the sea at the southern extremity of Mainland, Shetland, in early 1993, which led to the Donaldson Inquiry and the publication of *Safer Ships, Cleaner Seas* (DoT, 1994). In the nature of things, most of these inquiries and any resulting measures are predominantly sectoral, although a few in recent times have been primarily environmental in scope.

Table 6.4 summarizes the institutional basis of marine affairs and policy in the UK government in 2010. Departments with major responsibilities, such as the Department of Trade and Industry (DTI) – with marine-related functions now split between the new Department for Business, Innovation and Skills (BIS) and of Energy and Climate Change (DECC) – and Defra have several sections responsible for different aspects of administration and related policymaking. Most of the real administrative and management tasks are undertaken by executive and other agencies responsible to respective departments. These are summarized in detail in the *Civil Service Yearbook* (Cabinet Office, 2014). In practice, certain areas such as shipping, defence, and foreign policy are exclusively reserved to the UK government, whereas most other areas have varying degrees of devolved responsibility to the administrations in Scotland, Wales, and Northern Ireland. Arrangements for Jersey, Guernsey, and the Isle of Man have already been noted earlier in the chapter.

Table 6.4 UK government departments and their key delivery agencies and non-departmental public bodies (NDPBs) of significance to the marine environment*

Department for Environment, Food and Rural Affairs (Defra)

Executive agencies	*Executive NDPBs*
Centre for Environment, Fisheries and Aquaculture Science (CEFAS)	Commission for Rural Communities
	Consumer Council for Water
Food and Environment Research Agency (FERA)	Natural England
Animal Health	Environment Agency
	Joint Nature Conservation Committee (JNCC)
	Sea Fish Industry Authority (SFIA)
	Sustainable Development Commission
	Marine Management Organisation (MMO)

(continued)

Table 6.4 (continued)

Department for Environment, Food and Rural Affairs (Defra)

Advisory NDPBs	Other bodies
Advisory Committee on Hazardous Substances (ACHS)	Association of National Park Authorities
Advisory Committee on Pesticides (ACP)	British Waterways (public corporation)
Advisory Committee on Releases to the Environment [of genetically modified organisms] (ACRE)	Pesticides Forum
	Regional Flood Defence Committees (committees of the Environment Agency)
Air Quality Expert Group	Review of Fishing Vessel Licensing (review body)
Pesticide Residues Committee	Review of Marine Nature Conservation Working Group (review body)
Royal Commission on Environmental Pollution	Science Advisory Group (ad hoc advisory group)
Science Advisory Council	Sustainable Development Task Force (task force)
	UK Biodiversity Group

Department for Energy and Climate Change (DECC)

Executive NDPBs	Advisory NDPBs
Nuclear Decommissioning Authority	Committee on Radioactive Waste Management
Committee on Climate Change	Renewables Advisory Board

Department for Transport (DfT)

Executive agency	Advisory NDPBs (public corporations)
Maritime and Coastguard Agency (MCA)	Northern Lighthouse Board
	Trinity House Lighthouse Service (THLS)
	Trust Ports

Department for Business, Innovation and Skills (BIS)

Executive NDPBs	Advisory NDPBs
Nuclear Decommissioning Authority	Council for Science and Technology
Research Councils (NERC/BBSRC/EPSERC)	Renewables Advisory Board
	Sustainable Energy Policy Advisory Board

Ministry of Defence (MoD)

Executive agencies	Advisory NDPB
Defence Support Group (on-vote)	Defence Scientific Advisory Council
MoD Police and Guarding Agency (on-vote)	
Defence Science & Technology Laboratory (DSTL) (trading fund)	
Met Office (trading fund)	
UK Hydrographic Office (trading fund)	

Department for Culture, Media and Sport (DCMS)

Executive NDPBs	Advisory NDPBs
English Heritage	Advisory Committee on Historic Wreck Sites
UK Sport	Advisory Committee on National Historic Ships
Sport England	VisitEngland
National Maritime Museum	

Department for Communities and Local Government

Executive agencies	Advisory NDPB
Planning Inspectorate	Boundary Commissions for England and Wales
Ordnance Survey	

HM Revenue and Customs (border and frontier protection, environmental taxes)

* *Mainly relating to England*

A final important dimension of institutional arrangements concerns the division of land/sea responsibilities (Figure 6.3), which is of particular importance in the management of coastal areas. The oldest element of this is the Crown Estate, which owns around 55 per cent of the foreshore and the seabed to the limit of the territorial sea, together with mineral rights (excluding oil and gas) on the continental shelf beyond. The Crown Estate is legally a trustee body and de facto landowner, with extensive environmental management responsibilities reminiscent of those of landowners on land before the enactment of local government legislation in the late nineteenth century.

With the exception of the Northern Isles of Scotland, the land-use planning powers exercised by local government under national legislation do not extend into the marine environment, other than in relation to some estuaries and harbours that fall within local authority boundaries, for example within parts of the Severn Estuary. Certain local authorities exercise informal powers of this kind – most notably, the Highland Region (in Scotland), with regard to salmon farming. It should be noted that land-use planning can nonetheless be a considerable influence on adjacent sea areas. These administrative issues are addressed in greater detail in the next section.

The nature of the policy and legislation established

The nature of the resulting policy

UK marine policy is legislatively based in the first instance, with a fairly comprehensive framework of national law tied as appropriate into relevant international conventions and EU law. As already noted, the international conventions are particularly important for designation of the continental shelf and maritime boundary delimitation, shipping, ports, waste disposal, and pollution. EU law is an important influence on fisheries, waste disposal and pollution, conservation, and integrated measures. In this context, the Common Fisheries Policy (CFP) is an exclusive competence of the European Union, whereas the others are mixed competencies between the Union and national governments.

The practical implementation of marine policy is primarily administratively based, especially through government agencies responsible to cognate central government departments (see Table 6.4). The majority of these agencies are management agencies, for example those concerned with conservation (largely a devolved responsibility) and environment. There are also occasional sectoral development agencies, for example the Sea Fish Industry Authority (Seafish). Regional development agencies, now abolished, have historically also undertaken development functions.

Implementation of principles: Detailed assessment

As noted above, the explicit implementation of principles is difficult to gauge, although a detailed historical analysis would doubtless reveal practical application of these to varying degrees inherent in the policy- and law-making processes already referred to. The degree of success in attaining policy objectives has undoubtedly been variable, depending in no small measure on applicable sectors.

- *Organizational integration* Integration among sectors is inevitably limited in an administrative system organized along sectoral lines. Practical integration in emergency services and fisheries protection, for example, has already been noted. Perhaps more important historically is the current phase of thinking across these boundaries. Parliamentary committee

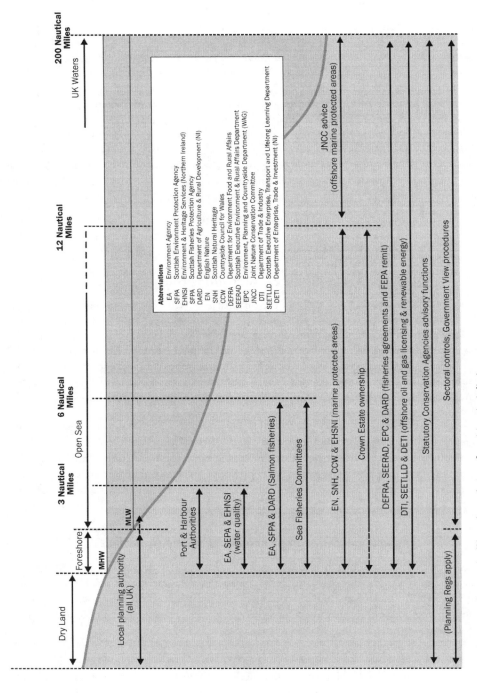

Figure 6.3 Onshore and offshore boundaries of maritime jurisdiction

inquiries undoubtedly encourage this, as do current policy statements such as the Marine Stewardship Report. Integration is also endorsed across many aspects of UK and EU policy, including the Marine Acts and the EU Integrated Maritime Policy (IMP). Integrative progress continues to be made, albeit at an early stage of development.

- *Environment–development integration* Previously, environment–development integration was encouraged (mainly on land) by means of regional development agency funding mechanisms. Specific marine instances are rare: an example is fisheries charting and gear technology development in Seafish. *Spatial integration* is now emerging in the contexts of the application of ICZM in the wake of both the national stocktakes and the EC evaluation (European Commission, 2007a), the provisions for MSP in the Marine Acts, and the emerging regional arrangements in the CFP. *Science–management integration* undoubtedly exists in a number of areas, including marine weather forecasting, defence, and waste management. Its spectacular failure has hitherto been in fisheries management, where the application of science has singularly failed to prevent overfishing. Overall, implementation lies partly at the level of thought, and less so at administrative and political levels.

- *Precautionary principle/approach* It would be wrong to say that the precautionary approach has been adopted in either fisheries or aquaculture management, for which the primary objective to date has been to maintain the industries. In the case of land-based and marine pollution, measures taken may be judged to fall into this category in a historical sense: these have been pursued using a mixture of domestic legislation, international conventions, and EU directives. A similar approach applies to marine conservation.

- *Ecosystem-based management* Transboundary agreements and arrangements for managing shared fish stocks are numerous, both within the CFP and largely through the CFP with adjacent states – most notably, the Scandinavian states outside the European Union. However, it would be idle to pretend that these amount to ecosystem-based management; more realistically, these are attempts to maintain industries and shifting fish stocks, such as the current issues surrounding the mackerel fishery. The necessary data and even scientific understanding of the complex marine ecosystems adjacent to Britain and Ireland remain incomplete in being able to support ecosystem-based management, even if the human activities side of the equation could be adequately managed. Having said that, the UK is very well endowed with national ocean science capabilities in all of the fields necessary for marine policy. This includes the earth sciences, oceanography, marine biology, meteorology, and fisheries science. There is extensive UK participation in regional and global marine science programmes through the International Council for the Exploration of the Sea (ICES) and the Intergovernmental Oceanographic Commission (IOC).

- *Public participation and community-based management* Public participation is already well established for marine projects that have a land component, such as coastal installations covered by the town and country planning legislation, which are subject to extensive consultation among stakeholders under the aegis of local authorities. Indeed, most of these and offshore large-scale projects are subject to environmental impact assessment (EIA) processes under European legislation, which requires extensive public consultation in line with the 1998 Convention on Access to Information, Public Participation in Decision-making and Access to Justice in Environmental Matters (the Aarhus Convention). In difficult cases, such arrangements are supplemented by planning or public inquiries. This was the case, for example, in a number of the offshore oil installation construction projects on the coast of Scotland in the 1970s and 1980s, and more recently with respect to offshore wind farm proposals around the UK coast. Particularly notable examples of locally focused legislation are the Zetland County Council Act and Orkney County Council Act, both of 1974, which gave the Shetland Islands and Orkney Islands Councils, respectively, substantial additional planning and management controls over offshore oil installations within 3 miles of the baseline of the territorial sea, in addition to land planning powers. Offshore, the Marine Acts will go some way toward extending public participation similar to that present in land legislation.

Authority at the national level

The organizational and policy framework for coastal and marine management at national (UK) level is extremely complex. A plethora of government departments, agencies, and other bodies are involved in an array of terrestrial and marine sectors, and are responsible for the management and regulation of many different activities. These are listed in Table 6.4 and a matrix of the responsibilities of key bodies is provided in Figure 6.4. The matrix is much simplified, and whilst an attempt has been made to be as accurate as possible, it is merely intended to illustrate the complexity of the situation and the number of government bodies/agencies involved in the management of the marine environment. This fragmented administrative structure has evolved in response to piecemeal, largely sectorally driven, legislation, which is often viewed as an impediment to effective coordination of policies and cooperation among bodies.

There are four principal sets of institutions involved in the role of national government in the management of the marine environment:

- UK central government departments;
- the devolved administrations (the Welsh Government, the Scottish Government, and the Northern Ireland Assembly) and their administrative arms;
- specialist bodies (executive agencies and non-departmental public bodies, or NDPBs) responsible for the practical implementation of management; and
- local government.

As previously noted, the UK government remains the cornerstone of a sectoral management system, with extensive legislation in all fields. Defra is the lead government department on most marine environmental matters with responsibility for, among other things: sustainable development and the environment; marine fisheries; designating European marine sites under the Habitats and Birds Directives, and listing protected species; regulating the disposal of waste at sea and the placement of structures in the marine environment – including offshore energy, formerly alongside the Department of Trade and Industry (DTI) and now with the Department for Energy and Climate Change (DECC); and responsibility for flood defence and coast protection in England and Wales. This remit is broadly confined to England, but Defra also has a strong UK coordinating role on some aspects of marine environmental affairs.

The DECC (and formerly the DTI) is responsible for regulating the exploration and production of oil and gas over the UK continental shelf and also for regulating renewable energy generation.[9] Other marine-related responsibilities lie with the Department for Transport (DfT) (ports and shipping), Department for Culture, Media and Sports (DCMS) (marine heritage), Ministry of Defence (MoD) (naval dockyards, naval vessels, and military firing ranges), and the Department for Communities and Local Government (DCLG), this last of which has overall responsibility for developing planning policy and legislation, and issuing planning policy guidance, although planning responsibilities will overlap with the Marine Management Organisation (MMO), which is responsible for planning up to mean high-water spring tide.

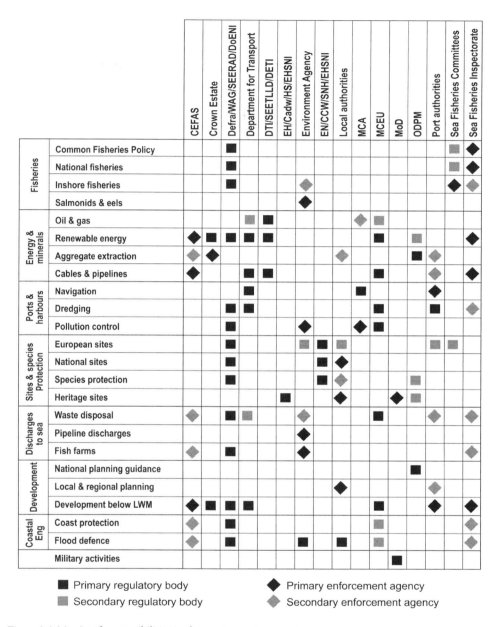

Figure 6.4 Matrix of responsibilities in the marine environment

Source: Based on Defra (2004a)

Efforts to coordinate and integrate management efforts have been ongoing for some time. For example, the Marine Consents and Environment Unit (MCEU) was established in April 2001 as a jointly managed cross-departmental unit of the Marine Environment Branch of Defra (formerly within the former Ministry of Agriculture, Fisheries and Food, or MAFF) and the Ports Division (Casework Branch) of the DfT, with the aim of providing a central facility for receipt and administration of applications to undertake works in tidal waters and at sea. This was replaced by the MMO under the Marine and Coastal Access Act 2009, which holds powers under the Act to improve integration of marine objects policies and plans (MMO, 2014). The MMO works in conjunction with the Welsh Government and Marine Scotland in several key areas. Some of the key challenges for the MMO include establishing a mandate and developing the institutional capacity to deliver marine planning and licensing (Fletcher et al., 2014). However, the resources within some of the devolved governments – notably, the Welsh Government – are perhaps even more limited now than under the previous arrangements.

The main role of government departments and their agencies is to implement government policy and to advise ministers. In doing so, these often work alongside local authorities, NDPBs, and other government-sponsored organizations. Sometimes referred to as 'quangos', NDPBs are national or regional public bodies that work independently of the ministers to whom they are accountable. The main NDPBs and their sponsoring departments are shown in Table 6.4. Executive NDPBs have executive, administrative, commercial, or regulatory functions. They work within a governmental framework and the degree of operational independence varies. Advisory NDPBs provide independent and expert advice to ministers and their departments on particular matters.

The national and subnational division of authority and interaction

The role of the devolved administrations

The administration of marine environmental control is complicated by the fact that the UK has devolved certain important powers over domestic matters. Furthermore, the UK system of devolution is asymmetric – that is, there are different levels of devolved responsibilities and there is no common pattern. Scotland, Wales, and Northern Ireland all have different forms of devolution, set out in the Scotland Act 1998 and Scotland Act 2012, the Government of Wales Act 1998 and the Government of Wales Act 2006, and the Northern Ireland Act 1998, respectively. The key departments, divisions, and agencies of the devolved administrations are shown in Table 6.5, although it should be noted that there has been a very recent change in Wales, in that Natural Resources Wales has replaced the Countryside Council for Wales and Environment Agency (Wales). A non-legislative framework of concordats – that is, agreements between government departments and the devolved administrations, under memoranda of understanding (Leeke et al., 2003) – commits the administrations to the principle of good communication with each other, especially where one administration's work may have some bearing upon the responsibilities of another.

The devolution settlements in Scotland and Northern Ireland have features in common, for example both have a legislature, which can pass primary, as well as secondary, legislation in areas that are not reserved for Westminster. The Scottish Parliament and Scottish Government in particular are responsible for a wide range of statutory functions in Scotland, including the environment, ports and harbours, fisheries, the natural and built heritage, economic development and tourism, and culture and sport. Similarly, the Northern Ireland Assembly, when it is sitting, has the power to make laws on all domestic matters that have been devolved by Westminster, including the environment, agriculture, fisheries and food, development, local government, some areas of trade and industry, and most areas of transport (Leeke et al., 2003). As noted above, the Welsh Government now has powers to make some primary legislation to meet distinctive Welsh needs on issues that have been devolved, including some that are significant for coastal and marine management (including most aspects of agriculture, fisheries, and

Table 6.5 Departments, divisions, and agencies of the devolved administrations with responsibility for the marine environment

Scottish Government	
Marine Scotland, formed from the merging of Fisheries Research Services (executive agency), Scottish Fisheries Protection Agency (executive agency), and Scottish Government Marine Directorate	Directorate for Transport Transport Strategy Division Aviation, Ports, Freight and Canals Division
Scottish Government Directorate for Learning Tourism, Culture and Sport Group Historic Scotland (executive agency)	Scottish Executive Development Department Planning Scottish National Heritage Scottish National Environment Protection Agency
Welsh Government	
Department for Environment, Sustainability and Housing Agriculture and Fisheries Policy Division Planning Division Environment Division Animal and Plant Health Division Planning Inspectorate (a joint agency with ODPM)	Department for Economy and Transport Economic Policy Division Transport Policy Division Local Government, Public Service and Culture Department Welsh Historic Monuments (Cadw) (executive agency) Natural Resources Wales
Northern Ireland Executive	
Department of the Environment (DoE) Planning and Natural Environment Division Climate and Waste Division Planning and Environmental Policy Group (PEPG) Planning Service (executive agency) Northern Ireland Environment Agency	Department of Agriculture and Rural Development (DARD) Animal Identification, Legalisation and Welfare Policy Branch (central policy group) Fisheries Division (central policy group) Agri-Food and Biosciences Institute Rivers Agency (executive agency)
Department for Regional Development (DRD) Ports and Public Transport Division Water Policy Division	Department of Enterprise, Trade and Investment for Northern Ireland (DETI) Energy Division (oil/gas/renewables minerals) Tourism, Agency Liaison and Equality Northern Ireland Tourist Board (executive agency)

food, the environment, including controlling water quality and food defence, and determining policies on town and country planning, economic development, including running the Welsh tourist board 'Visit Wales', culture, including ancient monuments and historic wrecks, and some aspects of transport.

Whilst, until now, administrative devolution in Wales, unlike Scotland, has been limited and comparatively land-oriented, this is gradually changing, with the newly constituted Marine Division in the Welsh Government steering the marine plan process for both inshore and off-shore Welsh waters. However, the UK government remains ultimately responsible for offshore activities, including ports and shipping, defence, and energy. For example, the DfT retains responsibility for shipping policy, pollution from ships, offshore safety, and most harbours in Wales (Boyes et al., 2003). Even in fisheries, the Welsh Government's remit covers only the

Wales territorial sea, but not the waters beyond. It should be noted that there has been little consistency in the way in which additional powers have been conferred, resulting in different provisions for the exercise of ministerial competencies within many Acts of Parliament. Each Act differs in the extent to which the Welsh Assembly is given powers, and some functions are exercisable wholly by the Welsh Government, partly by the Welsh Government and partly by Westminster, or by the Welsh Government together with ministers. Because there are no general principles in this respect, there is an extremely complex legislative picture (Ball and Smith, 2004).

Issues associated with the existing regulatory and administrative regime

The tendency to develop ad hoc solutions to problems as and when these arise has resulted in a sectoral approach to the management of the marine environment. Most of the principal uses of the marine environment are governed by Acts of Parliament (see Appendix 7.1), which often fail to recognize the interrelationship with legislation in other sectors. There is also a raft of secondary legislation that is of relevance to coastal and marine management, which exhibits little vertical integration (that is, at different spatial scales) or horizontal integration (that is, across sectors), leading to fragmented control of the marine environment (Ball and Smith, 2004). The complexity of the plethora of regulations has inevitably resulted in omissions and overlap in the regulatory framework, some of which may be attributed to limited reference to the coast or marine environment within more terrestrially focused legislation (Defra, 2004b). The piecemeal approach to amending legislation that has been described implies that much of the primary legislation governing some sectors – most notably, fisheries (see Appendix 7.1) – has been in place for many years and may no longer be fit for purpose. Furthermore, the complex structure of the regulatory regime makes it slow to react to new activities or pressures, such as offshore energy projects. The structure of the regulatory regime also has implications for enforcement agencies, as described under 'Enforcement' later in the chapter. In order to tackle this situation, there has recently been a campaign within government to reduce 'red tape', including the removal of defunct measures.

The domestic implementation of international agreements

As noted by Bell and McGillivray (2001), the relationship between UK and international law has two dimensions: the extent to which international law affects policymaking at the national level; and the contribution of the UK to policy and legal developments at the international level. With regard to the former, it is important to note that, in the UK, international agreements become part of national law only once given effect to by Parliament, usually through legislation.[10] Following devolution, the UK government is the contracting party that signs international agreements, and is consequently primarily responsible for their implementation and enforcement. The UK government undertakes the international negotiation, but implementation rests with devolved administrations if it is a devolved issue. The legislation to implement international obligations is typically implemented by an Act of Parliament or a ministerial order. However, the option exists both in theory and in practice for the government to delegate the task to the devolved administrations in particular cases, although international accountability would remain with the state (Cardiff University, 2001).

With regard to the second dimension, the UK has played a leading role in negotiating key maritime conventions, including UNCLOS and the earlier Geneva Conventions on the Law of the Sea relating to maritime limits. However, it is now often the case that the European

Commission negotiates on behalf of Member States, which sometimes makes it difficult to assess the particular stance taken by the UK (Bell and McGillivray, 2001). In the case of EU competence, the Commission speaks for the European Union. If not, the presidency speaks to the agreed position on behalf of the Member States, but there is more of an option for Member States to speak individually. Table 6.6 shows the status of domestic implementation of some key agreements related to the marine environment based on the specifications for papers submitted to the Ocean Policy Summit, held in Lisbon, 10–14 October 2005. In addition, it is worth mentioning that the UK is signatory to several other international maritime conventions, including the 1995 Agreement for the Implementation of the Provisions of the United Nations Convention on the Law of the Sea relating to the Conservation and Management of Straddling Fish Stocks and Highly Migratory Fish Stocks (the 1995 Fish Stocks Agreement, or UNFSA 1995).

The UK is also party to several important international conventions and agreements that are regional in scope, and which apply to the European/North Atlantic region, but are not part of the EU system. These include:

- the Convention on Future Multilateral Cooperation in Northeast Atlantic Fisheries, agreed to in 1959, and entered into force in 1963;
- the 1982 Paris Memorandum of Understanding on Port State Control (the Paris MoU);
- the Agreement for Co-operation in Dealing with Pollution of the North Sea by Oil (the Bonn Agreement) originally in 1969, and superseded by a new agreement in 1983, which entered into force in 1989;
- the 1992 Convention for the Protection of the Marine Environment of the North East Atlantic, which entered into force in 1998, and replaces the earlier Oslo (1972) and Paris (1974) Conventions; and
- the International Conferences for the Protection of the North Sea (commenced in Bremen in 1984).

Enforcement

The current organization of enforcement in the marine environment is complex. A review of the regulatory responsibilities and enforcement mechanisms relevant to marine nature conservation in the UK was undertaken as part of the Irish Sea Pilot project (Boyes et al., 2003), but has since undergone radical changes. The Marine and Coastal Act 2009 began a modernization of enforcement in UK marine policy. It sets out a core set of inspection and investigation powers across marine fisheries, licensing, and nature conservation (MMO, 2014). Under the Act, the MMO possesses powers of enforcement in UK offshore waters (excluding Scotland), and UK inshore waters in collaboration with Marine Scotland and the Welsh Government, in Scotland and Wales respectively.

In summary, enforcement provisions include a variety of measures that are sector-specific, and involve a multitude of government departments and agencies (see Figure 6.4). For example, with respect to sea fisheries, Defra's Sea Fisheries Inspectorate (SFI) had previous responsibility for enforcing the CFP and some national requirements throughout English and Welsh territorial waters and the exclusive fishing zone beyond the territorial sea (Defra, 2004a). This remit has now largely been reassigned to the MMO, which now coordinates fisheries enforcement, in addition to marine licensing and nature conservation enforcement, for all fishing activity within the UK fishing limit and UK vessels operating outside those waters (MMO, 2014).

For inshore waters in England and Wales, the twelve Sea Fisheries Committees (SFCs) originally established in the nineteenth century and funded by local authorities were replaced by

Table 6.6 Implementation of international agreements

International agreement	Implementation
UNCLOS	*Status*: The UK acceded to the Convention in August 1997. See Figure 6.1 for zones of offshore jurisdiction established in accord with the Convention. *Lead department*: FCO (Aviation, Maritime and Energy department) *Other interested departments and organizations*: CEFAS, Defra, DfT, DTI, MoD, NERC
CBD	*Status*: The UK ratified the Convention in June 1994. Also, in 1994, the government launched the UK Biodiversity Action Plan (UK BAP), a national strategy, which identified broad activities for conservation work over the next twenty years and established fundamental principles for future biodiversity conservation. To date, BAPs have been published for 1,150 species and 65 habitats. Local BAPs have also been identified and 163 so far have been developed. The UK has also identified several national focal points to coordinate the implementation of the Convention's thematic work programmes and cross-cutting issues. *Lead department*: Defra *Other interested departments and organizations*: CEFAS, Countryside Council for Wales, DoENI, EA, JNCC, Natural England, Scottish Natural Heritage, SEERAD, WAG
1973 International Convention for the Prevention of Pollution from Ships, as amended by the Protocol of 1978 (MARPOL)	*Status*: The UK has acceded to all annexes to MARPOL, including the 1997 Protocol (Annex VI), which provides for the establishment of international regulations for the prevention of air pollution from ships. *Lead department*: DfT (MCA) *Other interested departments and organizations*: Defra, DTI, EA, FCO, nature conservation agencies
1972 Convention on the Prevention of Marine Pollution by Dumping Wastes and other Matter (London Convention)	*Status*: The UK is a member of the London Convention and has ratified the 1996 Protocol, which entered into force on March 2006. *Lead department*: Defra *Other interested departments and organizations*: CEFAS, DfID, DfT, DTI, DoENI, EA, FCO, MoD, SEERAD, WAG
Global Programme of Action for the Protection of the Marine Environment from Land-based Sources (GPA)	*Status*: National and regional programmes of action are identified as the key modes for implementing the GPA. The UK has taken a leading role in supporting the implementation and development of the GPA and associated Washington Declaration, which reflects the importance that the GPA has for making a major contribution to the government's environmental and development objectives. Particular emphasis is placed on the role played by the regional seas conventions, including OSPAR, in delivering the Programme's objectives. To this end, the government has pledged to continue to work through the OSPAR Commission to prioritize additional hazardous substances for further action and to develop a Europe-wide framework for minimizing the risks from chemicals. OSPAR also sets a target of cessation of discharges, emissions, and losses of hazardous substances by 2020. *Lead department*: Defra *Other interested departments and organizations*: DfID, DoENI, EA, FCO, SEERAD, WAG

Note: CEFAS – Centre for Environment, Fisheries and Aquaculture Science (Defra); Defra – Department for Environment, Food and Rural Affairs; DfID – Department for International Development; DfT – Department of Transport; DoENI – Department of Environment Northern Ireland; DTI – Department of Trade and Industry; EA – Environment Agency; FCO – Foreign and Commonwealth Office; JNCC – Joint Nature Conservation Committee; MCA – Maritime and Coastguard Agency; MoD – Ministry of Defence; NERC – Natural Environment Research Council; SEERAD – Scottish Executive Environment and Rural Affairs Department; WAG – Welsh Assembly Government ten Inshore Fisheries Conservation Authorities (IFCAs) in April 2011, pursuant to the Marine and Coastal Access Act 2009. The two SFCs that Wales shared with England were combined into a single IFCA for Wales. The SFCs regulated and enforced their own, and some national, requirements for fisheries and environmental purposes in inshore waters (Defra, 2004a), and their functions were continued by the IFCAs. The Environment Agency acts in a similar capacity in instances in which no SFC/IFCA was established – notably, in a number of estuaries – and licenses fishing for salmonids and other migratory species, such as eels. Defra contracts the Royal Navy to carry out fisheries enforcement at sea by means of the Royal Navy's Fisheries Protection Squadron, while the Ministry of Defence Police has seagoing capabilities and powers, and its staff are sworn in as British fisheries officers (Boyes et al., 2003). The country conservation agencies, Natural England and Natural Resources Wales (formerly its CCW component), offer scientific advice, but do not have enforcement powers at sea.

In Scotland, the lead management agency in fisheries enforcement is Marine Scotland, following the Marine (Scotland) Act 2010. Marine Scotland brings together the former functions of the Fisheries Research Services, Scottish Fisheries Protection Agency, and the Scottish Government Marine Directorate. Similarly, in Northern Ireland, the Department of Agriculture and Rural Development (DARD) has responsibility for regulating fisheries within territorial waters and for enforcing fisheries legislation in inshore coastal waters, while the Northern Ireland Sea Fisheries Inspectorate carries out enforcement onshore and at sea (Boyes et al., 2003).

The Review of Marine Fisheries and Environmental Enforcement (Defra, 2004a) acknowledged that there was very little enforcement of environmental requirements, and made several recommendations for improving the organizational, legal, and financial framework for fisheries enforcement in England and Wales. Principal among these was that, in the medium to long term, there should be a single agency with responsibility for wider marine fisheries and environmental management. It recognized that this would require primary legislation and, with the subsequent adoption of the Marine and Coastal Access Act 2009 and the establishment of the MMO, fisheries and environmental management in UK waters is being drawn together into a single agency.

In the case of shipping standards, apart from the global role of the classification societies and International Maritime Organization (IMO) conventions, a major means of implementation is through the ports signatory to the Paris MoU, which deals with port state control of substandard ships. Ships using UK ports are subject to rigorous inspection and can be detained if in breach of the regulations. Breach of fishery regulations and rules relating to waste disposal and pollution are punishable by the courts in the form of fines and confiscation of gear. Port authorities may also be fined for negligence in pollution incidents, as happened, for example, in the case of Milford Haven in the wake of the *Sea Empress* oil spill of 1996.

There are, however, several factors complicating the effectiveness and enforcement of legislation in the marine environment, particularly beyond controlled waters. Some of the problems are organizational, in that bodies may lack appropriate powers or there is a reliance on those whose major concerns are with other sector-specific matters. Because the functions of enforcement bodies are often defined by legislation, there may be constraints upon their ability to cooperate with others in the marine environment if their statutory and duties are expressed in narrow sectoral terms. For example, prior to the Transport and Works Act 1992, most harbour authorities had no statutory powers to promote nature conservation and were solely concerned with the interests of commercial navigation (Ball and Smith, 2004). Other difficulties arise from distinct jurisdictional boundaries between land and sea, and as a result of the different spatial boundaries that apply to different regulations and consenting regimes (see Table 6.7). Consequently, different sectors are controlled to a varying extent and distance from the shore

Table 6.7 Onshore and offshore boundaries of key legislation

Act or powers	Distance from shore to which the law applies
Town and Country Planning Act 1993	Mean low–water mark
Local harbour powers	Boundaries of harbour
Land Drainage Act 1991	3 nautical miles ('controlled waters')
Water Resources Act 1991	
Sea Fisheries Act 1981	6 nautical miles
Transport and Works Act 1992	12 nautical miles (limits of territorial waters)
Electricity Act 1989 (since extended)	
Petroleum Act 1998	200 nautical miles (UK waters)
Food and Environmental Protection Act 1985	
Marine and Coastal Access Act 2009	200 nautical miles (UK waters)
Marine (Scotland) Act 2010	

Source: EFRA (2004); Defra (2010)

and jurisdictions, which leads to several confusing tiers of enforcement responsibilities, as illustrated in Figure 6.3. There are also perceived to be basic problems in enforcing legislation in a dynamic environment. For example, the jurisdiction of the SFCs (and their successor IFCAs) extended only to 6 nautical miles from the coast, whereas important fish stocks are clearly not confined to this zone (Ball and Smith, 2004). There are also inconsistencies between the devolved administrations in the implementation of legislation, which is sometimes further complicated by differences in jurisdictional boundaries.

Research and education

The UK possesses a large infrastructure of marine laboratories in fisheries and oceanography, supplemented by the marine research capabilities of the Armed Forces and the Meteorological Office (Met Office). Several universities also have substantial marine research capabilities (see Table 6.8). These participate in the full range of national and international research programmes. Since the early 1900s, there has been a commendable trend in government to publish position or policy papers on both sectoral and integrated aspects of coastal land marine policy, the most recent being *Charting Progress 2: The State of the UK Seas* (Defra, 2010a; 2010b).

The Natural Environment Research Council (NERC) delivers independent research, survey, training, and knowledge transfer in the environmental sciences. The Council's Strategy for Marine Science and Technology (SMST) has been one of seven sectoral strategies covering the environmental sciences. Current marine thematic programmes involve successful partnerships between NERC centres/surveys, universities, industry, and other non-academic bodies (see Table 6.9). Much of the recent spend, however, has been on core strategic programmes at the Centre for Coastal and Marine Sciences, encompassing Plymouth Marine Laboratory, Proudman Oceanographic Laboratory, and Dunstaffnage Marine Laboratory, at the National Oceanography Centre, Southampton, and at the Sea Mammal Research Unit (NERC, undated).

Building an ocean ethic has fallen primarily to the voluntary sector – notably, the WWF, the Royal Society for the Protection of Birds (RSPB), the Marine Conservation Society, the Whale and Dolphin Society, the National Trust, and wildlife trusts – through publications and management of coastal properties for which they are responsible in a number of cases. The school system, for example through the National Curriculum, cannot be said to be marine-oriented to any great extent.

Table 6.8 Leading maritime-related academic research and education institutions in the UK

University	Department
Bournemouth University	School of Applied Sciences
Cardiff University	School of Earth and Ocean Sciences
Heriot-Watt University	School of Life Sciences/Institute of Petroleum Engineering
Imperial College	Department of Earth Science and Engineering/Department of Civil and Environmental Engineering
Napier University	School of Life Sciences
Queens University Belfast	School of Biology and Biochemistry
Southampton Solent University	School of Engineering, Construction and Maritime
University College London	Departments of Civil and Environmental/Geomatic/Mechanical Engineering
University of Aberdeen	Department of Zoology/Department of Geography and Environment
University of East Anglia	School of Environmental Sciences
University of Essex	Department of Biological Sciences
University of Glamorgan	Faculty of Health, Sport and Science
University of Glasgow	Department of Naval Architecture and Marine Engineering
University of Greenwich	Greenwich Maritime Institute
University of Highlands and Islands	Scottish Association for Marine Science (UHI)
University of Hull	Department of Biological Sciences/Scarborough Centre for Environmental and Marine Sciences/Institute of Estuarine and Coastal Studies
University of Liverpool	Department of Earth and Ocean Sciences/School of Environmental Sciences
University of Newcastle	School of Marine Science and Technology, Marine Sciences and Coastal Management
University of Plymouth	Department of Biological Sciences/Institute of Marine Studies
University of Portsmouth	School of Earth and Environmental Sciences
University of Southampton	School of Ocean and Earth Science/School of Engineering, Ship Science
University of St Andrews	School of Geography and Geosciences Sustainability Institute and Scottish Oceans Institute
University of Stirling	Institute of Aquaculture
University of Strathclyde	Department of Naval Architecture and Marine Engineering
University of Ulster	School of Biological and Environmental Studies, Coastal Studies Research Group
University of Wales, Bangor	School of Ocean Sciences

Source: Adapted from Pugh and Skinner (2002)

Financing

By far the most important source of funding is the Consolidated Fund of HM Treasury. The UK does not subscribe to the kind of hypothecated taxation that might be used to fund marine policy projects directly. There is one exception, the Aggregate Sustainability Levy Fund, which is financing a substantial marine research programme through Defra. By means of taxation, hundreds of millions of pounds sterling are devoted every year to central administration, agency operation, enforcement, scientific research, and higher education.

Table 6.9 NERC-funded maritime-related research centres

NERC research centres
British Antarctic Survey
British Geological Survey
Centre for Ecology and Hydrology
National Oceanography Centre

NERC collaborative centres
National Centre for Atmospheric Science
National Centre for Earth Observation
Plymouth Marine Laboratory
Scottish Association for Marine Science
Sea Mammal Research Unit
Centre for Population Biology
Centre for Earth Observation Instrumentation
National Institute for Environmental Science
Tyndall Centre for Climate Change Research
UK Energy Research Centre

Data centres
British Oceanographic Data Centre (BODC)
British Ocean Sediment Core Research Facility
Dundee Satellite Receiving Station (AVHRR, CZCS, SeaWIFS)
Global Population Dynamics Database (includes marine species)
MarLIN (MarineLife Information Network for Britain and Ireland)
National Geoscience Data Centre
National Oceanography Centre, Southampton-SeaDOG (Deep Ocean Geophysics: British Mid-Ocean Ridge Programme)
NERC Earth Observation Data Centre
NERC Earth Observation Data Acquisition and Analysis Service
OceanNET (national portal to data and information about the marine environment set up by BODC on behalf of the Inter-Agency Committee on Marine Science and Technology)
Polar Data Centre

Source: NERC (undated)

Implementation, evaluation, and long-term outlook

A review of problems, issues, or obstacles addressed by the policy/programme

From an environmental point of view, the principal immediate problems in marine policy centre round degradation of the ecosystem resulting from overfishing and conflicts of use in intensively used sea areas generally, but not always inshore. Major related issues include the need to secure a fully representative system of coastal and marine protected areas (MPAs) for conservation, allied to an effective system of development control offshore. The present sectorally based administration and management does this to some extent; proposed developments in integrated coastal zone management (ICZM), Europe's Integrated Maritime Policy (IMP), and the Marine Acts will establish a long-term framework for future marine management, and lay the foundations for effectively tackling related problems and issues over the next ten years or so.

At the national level, it is likely that the UK will retain most of the key marine policy powers mentioned in this chapter. However, there is likely to be progressive adjustment in the long run, gradually involving devolved administrations to a greater extent, especially in areas in which they already possess competencies. This has already been demonstrated somewhat through changes under the Marine Acts, in that Scottish bodies now hold complete powers for marine licensing, fisheries, nature conservation, and marine planning in its waters, and similarly, but to a lesser extent, the Welsh Government has acquired specific powers in Welsh waters.

From a governance perspective, the recent period in UK marine policy can be characterized as a transition from a relatively hierarchical to a more networked and place-based system. This is illustrated by growing involvement of public, private, and voluntary sector interests in marine policy development, as well as in the development of regional marine plans. It may be that such plans will increasingly reflect the spatial turn in terrestrial planning and provide spatial allocation of resources. There is a risk that these trends in governance may lead to 'governing by consultation'. However, the role of government remains important, especially with respect to planning, licensing, and enforcement. The role of the private sector and civil society is arguably also important to deal with the complexity of human activities in UK coasts and the seas beyond. But this approach brings new challenges of equitable representation, consultation fatigue, and conflicting agendas. The issues associated with the aborted Welsh Government Marine Conservation Zone Programme aptly illustrate this.

Monitoring, evaluation, and adjustment

The monitoring and evaluation systems involved in UK marine policy are multifarious and beyond the scope of this chapter to consider in any detail. However, it remains true to say that an integrated marine policy is just beginning to emerge, with corresponding indicators similarly emerging, for example in the 'Charting Progress' series alluded to above, as well as provisions made in association with the Marine Strategy Framework Directive (MSFD). The MSFD, IMP, and the Marine Acts all provide considerable impetus in this direction.

The short-term outlook is bright. The arrival of an integrated approach in the fields of coastal management, marine spatial planning, and fisheries management from both European and UK legislation discussed in this chapter will, in a sense, complete the process of sectoral evolution that began in the late eighteenth century.

In the longer term – that is, two to three decades in economic and technological terms – the system of sectoral and integrated policy measures and related legislation should be robust. However, it is likely to be held together more effectively by the emergence of an integrated approach at the level of basic thinking, which will apply to the sectors as well as the integrated parts of the whole enterprise.

Recommendations for improving national ocean policy

In both short- and long-term future development of marine policy and related legislative and other measures, two particular guiding ideas are of special significance. The first of these concerns the regional approach. In promoting national-level action in relation to the marine environment, it is crucially important to be mindful of the international (both European and global in a legal sense) and supranational EU dimensions, and to aim for an equitable division of effort in ensuring that decision-making takes place in the most suitable forums at the most suitable level. This regional approach also has to be borne in mind when considering subnational-level action: much of the most effective measures in the marine environment can be enacted at a relatively local level, as on land. The activities of port authorities and fishing communities, as well as local authorities, are cases in point. It may take some time for this to evolve further.

The second guiding idea concerns the future relationships among the public, private, and voluntary sector institutions, together with civil society more generally, both with regard to the economic sectors of the marine economy and the new integrated approaches. Hitherto, in most quarters, there has historically been an implicit 'statist' approach to marine policy, as in many other areas of public policy. It is widely expected that the state will take practically all of the responsibility in such matters; indeed, the very term 'public policy' infers this. However, in a process that has gradually intensified over the past three decades, the voluntary sector and civil society have been providing much of the thinking and not a little political pressure in the field of UK marine policy, while the private sector – and the public, through the tax system – has to meet the cost. There is thus emerging a gradual awareness of the need for a partnership approach involving all of these interests as a more explicit, rather than implicit, state of affairs. As with the permeation of integrated ideas through management of the economic sectors, this could lend a whole new dimension in promoting a culture of integrated marine policy based on continuous dialogue – a continuation of the UK maritime heritage in a twenty-first-century world.

Acknowledgement

Figures were produced with the assistance of Mr Alun Rogers, Cartographic Unit, Cardiff University School of Earth and Ocean Sciences.

Notes

1 The Overseas Territories are: Anguilla; Bermuda; British Antarctic Territory; British Indian Ocean Territory; British Virgin Islands; Cayman Islands; Falkland Islands; Gibraltar; Montserrat; St Helena, Ascension, and Tristan da Cunha (formerly St Helena and Dependencies); Turks and Caicos Islands; Pitcairn Island; South Georgia and South Sandwich Islands; and Sovereign Base Areas in Cyprus.
2 See also Saunders et al. (2010); Austen et al. (2011).
3 See, e.g., Scottish Executive (1997); Scottish Government (undated).
4 The first 'Cod War' took place in 1958 when Iceland extended its coastal fishing limit from 4 miles to 12 miles. The second Cod War started in 1972 when Iceland extended its coastal non-fishing limit to 50 miles. It ended with an agreement between the two countries that limited British fishing to restricted areas within the 50 mile limit. The third Cod War (November 1975–June 1976) almost brought the two countries to the brink of real war and was resolved through the intercession of the North Atlantic Treaty Organization (NATO) and an agreement that limited British trawlers to a maximum number of twenty-four within the 200 mile limit.
5 The European Marine Strategy was one of seven thematic strategies resulting from the Community's Sixth Environmental Action Programme (EAP). It was composed of three documents: a communication, representing the European Marine Strategy; a legislative proposal for a Marine Framework Directive; and a Commission (regulatory) impact assessment.
6 The SEA Directive (Directive 2001/42/EC) applies to plans and programmes within its scope (fisheries, energy, industry, transport, tourism, waste disposal, including disposal of dredgings, and water management), which set the framework for future development consent of projects listed in Annexes I and II of the EIA Directive (Directive 85/337/EEC) prepared after 21 July 2004. Ahead of the Directive, the Department of Trade and Industry (DTI) had, since 1999, been undertaking a rolling programme of SEA on the UKCS prior to the release of blocks for oil and gas licensing. The SEA programme for oil and gas has now been merged with that for renewables.
7 See, e.g., Defra (2013).
8 As part of the Government's Marine Stewardship initiative, Defra commissioned a research consortium to investigate options for developing, implementing, and managing marine spatial planning in the UK. Following completion of a literature review to identify international experiences in marine spatial planning and their applicability to the UK (ABPMer et al., 2005), a pilot project was established in the Irish Sea to simulate the development of a regional and local plan.

9   In Wales, developments under 50MW are administered by the Welsh government; the Scottish Executive and Department of Enterprise, Trade and Investment Northern Ireland (DETI) are responsible for administering wind farm applications in their respective territorial waters.

10 This is the case even when the treaty is ratified, because ratification is a matter for the central government, not Parliament.

References

ABP Marine Environmental Research Ltd (ABPmer), O'Rourke, T., Risk & Policy Analysis, Geoltek, Anderson, H., and Coastal Management for Sustainability (2005) *Marine Spatial Planning Literature Review*, online at http://www.abpmer.net/mspp/docs/finals/MSPliteraturereview_Final.pdf [accessed 30 April 2014].

Austen, M. C., Malcom, S. J., Frost, C., Hattam, S., Mangi, G., Stentiford, S., Benjamins, M., Burrows, M., Butenschon, C., Duck, D., Johns, G., Meriino, N., Ieszkowska, A., Miles, I., Mitchell, E., Pimm, E., and Smyth, T. (2011) 'Marine', in UK National Ecosystem Assessment, *UK National Ecosystem Assessment: Technical Report*, Cambridge: UNEP-WCMC.

Ball, I., and Smith, H. (2004) *Legislative Reform for the Welsh Marine Environment: A Report to WWF Cymru*, Cardiff: Marine and Coastal Environment Group, Cardiff University.

Bell, S. and McGillivray, D. (2001) *Environmental Law: The Law and Policy Relating to the Protection of the Environment*, 5th edn, Oxford: Blackstone Press.

Boyes, W. L., Warren, H., and Elliott, M. (2003) *Regulatory Responsibilities and Enforcement Mechanisms Relevant to Marine Nature Conservation in the United Kingdom: Report to the Joint Nature Conservation Committee*, Hull: Institute of Estuarine and Coastal Studies, University of Hull.

British Marine Aggregates Producers Association (BMAPA) (undated) 'Area Dredged: Marine Aggregate Summary Statistics 1998–2013', online at http://www.bmapa.org/issues/area_dredged.php [accessed 2 November 2014].

British Marine Aggregates Producers Association (BMAPA) (2014) *Strength from the Depths: Seventh Sustainable Development Report for the British Marine Aggregate Industry*, London: BMAPA.

British Marine Federation (BMF) (undated) 'Who We Represent', online at http://www.britishmarine.co.uk/About-BMF/Who-we-Represent [accessed 31 March 2014].

British Ports Association (BPA) (undated) 'Market Overview', online at http://www.britishports.org.uk/uk-ports-industry/market-overview [accessed 31 March 2014].

Cabinet Office (2004) *Net Benefits: A Sustainable and Profitable Future for UK Fishing*, London: Prime Minister's Strategy Unit, Cabinet Office.

Cabinet Office (2014) *Civil Service Yearbook 2013/2014*, 50th edn, online at http://www.civilserviceyearbook.co.uk [accessed 8 May 2014].

Cardiff University (2001) *Analysis of Options for Improving the Planning and Management of the Wales Territorial Sea: Report to the Countryside Council for Wales (CCW)*, Cardiff: Cardiff University.

Commission on the Limits of the Continental Shelf (CLCS) (2006) *Joint Submission by France, Ireland, Spain and the United Kingdom of Great Britain and Northern Ireland*, CLCS/52, online at http://www.un.org/depts/los/clcs_new/submissions_files/submission_frgbires.htm [accessed 1 July 2010].

Commission on the Limits of the Continental Shelf (CLCS) (2008a) *Statement by the Chairman of the Commission on the Limits of the Continental Shelf on the Progress of Work in the Commission*, Twenty-second Session, 11 August–12 September, New York, CLCS/60, online at http://www.un.org/depts/los/clcs_new/commission_documents.htm#Statements by the Chairman of the Commission [accessed 1 July 2010].

Commission on the Limits of the Continental Shelf (CLCS) (2008b) *Submission by the United Kingdom of Great Britain and Northern Ireland*, CLCS/60, online at http://www.un.org/depts/los/clcs_new/submissions_files/submission_frgbires.htm [accessed 1 July 2010].

Commission on the Limits of the Continental Shelf (CLCS) (2009) *Statement by the Chairman of the Commission on the Limits of the Continental Shelf on the Progress of Work in the Commission*, Twenty-third Session, 2 March–9 April, New York, CLCS/62, online at http://www.un.org/depts/los/clcs_new/commission_documents.htm#Statements by the Chairman of the Commission [accessed 1 July 2010].

Commission on the Limits of the Continental Shelf (CLCS) (2010), *Statement by the Chairman of the Commission on the Limits of the Continental Shelf on the Progress of Work in the Commission*, Twenty-fifth Session, 15 March–23 April, New York, CLCS/66, online at http://www.un.org/depts/los/clcs_new/commission_documents.htm#Statements by the Chairman of the Commission [accessed 1 July 2010].

Committee on Spatial Development (1999) *European Spatial Development Perspective: Towards Balanced and Sustainable Development of the Territory of the European Union*, online at http://ec.europa.eu/regional_policy/sources/docoffic/official/reports/pdf/sum_en.pdf [accessed 2 November 2014].

Department for Environment, Food and Rural Affairs (Defra) (2002a) *Safeguarding Our Seas: A Strategy for the Conservation and Sustainable Development of Our Marine Environment – First Marine Stewardship Report*, London: Defra.

Department for Environment, Food and Rural Affairs (Defra) (2002b) *Seas of Change: The Government's Consultation Paper to Help Deliver Our Vision for the Marine Environment*, London: Defra.

Department for Environment, Food and Rural Affairs (Defra) (2004a) *Review of Marine Nature Conservation: Working Group Report to Government*, London: Defra.

Department for Environment, Food and Rural Affairs (Defra) (2004b) *Review of Marine Fisheries and Environmental Enforcement*, London: Defra.

Department for Environment, Food and Rural Affairs (Defra) (2005a) *Securing the Benefits: The Joint UK Response to the Prime Minister's Strategy Unit Net Benefits Report on the Future of the Fishing Industry in the UK*, London: Defra.

Department for Environment, Food and Rural Affairs (Defra) (2005b) *Safeguarding Sea Life: The Joint UK Response to the Review of Marine Nature Conservation*, London: Defra.

Department for Environment, Food and Rural Affairs (Defra) (2005c) *Charting Progress: An Integrated Assessment of the State of UK Seas*, London: Defra.

Department for Environment, Food and Rural Affairs (Defra) (2009a) *A Strategy for Promoting an Integrated Approach to the Management of Coastal Areas in England*, London: Defra.

Department for Environment, Food and Rural Affairs (Defra) (2009b) *Our Seas: A Shared Resource – High-Level Marine Objectives*, London: Defra.

Department for Environment, Food and Rural Affairs (Defra) (2010a) *Charting Progress 2: An Assessment of the State of UK Seas*, London: Defra.

Department for Environment, Food and Rural Affairs (Defra) (2010b) *UK Marine Science Strategy: Shaping, Supporting, Co-ordinating, and Enabling the Delivery of World Class Marine Science for the UK, 2010–2025*, London: Defra, on behalf of the Marine Science Co-ordination Committee.

Department for Environment, Food and Rural Affairs (Defra) (2010c) *Marine Planning System Newsletter*, Issue 2, March, online at http://archive.defra.gov.uk/environment/marine/documents/legislation/201003msp-news.pdf [accessed 2 November 2014].

Department for Environment, Food and Rural Affairs (Defra) (2013) *A Coastal Concordat for England*, online at http://www.gov.uk/government/publications/a-coastal-concordat-for-england [accessed 31 March 2014].

Department of Trade and Industry (2007) *Meeting the Energy Challenge: A White Paper on Energy*, London: HMSO.

Department of Transport (DoT) (1994) *Safer Ships, Cleaner Seas: Report of Lord Donaldson's Inquiry into the Prevention of Pollution from Merchant Shipping*, London: HMSO.

Department for Transport (DfT) (2012) *National Policy Statement for Ports*, online at http://www.gov.uk/government/uploads/system/uploads/attachment_data/file/3931/national-policy-statement-ports.pdf [accessed 31 March 2014].

Department for Transport (DfT) (2013) *Maritime and Shipping Statistics*, online at http://www.gov.uk/government/collections/maritime-and-shipping-statistics [accessed 31 March 2014].

EFRA *See* Select Committee on Environment, Food and Rural Affairs

Environment Agency (1999) *The State of the Environment of England and Wales: Coasts*, London: HMSO.

European Commission (2002) *Communication from the Commission to the Council and the European Parliament: Towards a Strategy to Protect and Conserve the Marine Environment*, COM (2002) 539 final, 2 November, online at http://www.eea.europa.eu/policy-documents/com-2002-539-final [accessed 2 November 2014].

European Commission (2006a) *Communication from the Commission: Halting the Loss of Biodiversity by 2010 – And Beyond: Sustaining Ecosystem Services for Human Well-being*, 22 May, COM(2006) 216 final, online at http://eur-lex.europa.eu/legal-content/EN/TXT/?uri=CELEX:52006DC0216 [accessed 2 November 2014].

European Commission (2006b) *Green Paper: Towards a Future Maritime Policy for the Union – A European Vision for the Oceans and Seas*, 7 June, SEC (2006) 689, online at http://eur-lex.europa.eu/legal-content/EN/TXT/?qid=1414934785754&uri=CELEX:52006DC0275(02) [accessed 2 November 2014].

European Commission (2007a) *Communication from the Commission: Report to the European Parliament and Council – An Evaluation of Integrated Coastal Zone Management (ICZM) in Europe*, 7 June, COM (2007)

308 final, online at http://eur-lex.europa.eu/legal-content/EN/TXT/?uri=CELEX:52007DC0308 [accessed 2 November 2014].

European Commission (2007b) *Communication from the Commission to the European Parliament, the Council, the European Economic and Social Committee and the Committee of the Regions: An Integrated Maritime Policy for the European Union*, 10 October, COM (2007) 575 final, online at http://eur-lex.europa.eu/ LexUriServ/LexUriServ.do?uri=COM:2007:0575:FIN:EN:PDF [accessed 2 November 2014].

European Commission (2007c) *Green Paper on Adapting to Climate Change in Europe: Options for EU Action*, 29 June, COM(2007) 354, online at http://europa.eu/legislation_summaries/environment/tackling_ climate_change/l28193_en.htm [accessed 2 November 2014].

European Commission (2008) *Communication from the Commission to the Council, the European Parliament, the European Economic and Social Committee and the Committee of the Regions: A European Strategy for Marine and Maritime Research – A Coherent European Research Area Framework in Support of a Sustainable Use of Oceans and Seas*, 3 September, COM (2008) 534 final, online at http://ec.europa.eu/research/participants/ portal/doc/call/fp7/fp7-ocean-2011/30283-com(2008)534-a_european_strategy_for_marine_and_ maritime_research_en.pdf [accessed 2 November 2014].

European Commission (2009a) *Green Paper: Reform of the Common Fisheries Policy*, 22 April, COM(2009)163 final, online at http://eur-lex.europa.eu/LexUriServ/LexUriServ.do?uri=COM: 2009:0163:FIN:EN:PDF [accessed 2 November 2014].

European Commission (2009b) *Report from the Commission to the Council, the European Parliament, the European Economic and Social Committee and the Committee of the Regions: Progress Report on the EU's Integrated Maritime Policy*, 15 October, COM (2009) 540 final, online at http://eur-lex.europa.eu/ LexUriServ/LexUriServ.do?uri=COM:2009:0540:FIN:EN:PDF [accessed 2 November 2014].

European Commission (2009c) *The European Strategy for Marine and Maritime Research*, online at http:// www.eu.hermes.net/policy/SPP2009/03_Saab_EU_Maritime_Policy.pdf [accessed 1 October 2009].

European Commission (2012) *Communication from the Commission to the European Parliament, the Council, the European Economic and Social Committee and the Committee of the Regions: Blue Growth – Opportunities for Marine and Maritime Sustainable Growth*, 13 September, COM (2012) 494 final, online at http:// ec.europa.eu/maritimeaffairs/policy/blue_growth/documents/com_2012_494_en.pdf [accessed 2 November 2014].

European Commission (2013) *Proposal for a Directive of the European Parliament and of the Council Establishing a Framework for Maritime Spatial Planning and Integrated Coastal Management*, 12 March, COM(2013) 133 final, online at http://eur-lex.europa.eu/LexUriServ/LexUriServ.do?uri=COM:2013:0133:FIN:EN:PDF [accessed 2 November 2014].

European Commission Fisheries (undated) 'Managing Fish Stocks', online at http://ec.europa.eu/fisheries/ cfp/fishing_rules/index_en.htm [accessed 31 March 2014].

Fletcher, S. R., Jefferson, R., Glegg, G., Rodwell, L., and Dodds, W. (2014) 'England's Evolving Marine and Coastal Governance Framework', *Marine Policy*, 45: 261–8.

HM Government (1994) *Sustainable Development: The UK Strategy*, London: HMSO.

HM Government (2005) *Securing the Future: UK Government Sustainable Development Strategy*, London: HMSO.

HM Government (2009) *The UK Renewable Energy Strategy*, London: HMSO.

HM Government (2011) *UK Marine Policy Statement*, London: HMSO.

House of Commons Select Committee on the Environment (1992) *Coastal Zone Protection and Planning: Second Report of the House of Commons Select Committee on the Environment*, London: HMSO.

House of Lords Committee on the European Union (2008) *The Progress of the Common Fisheries Policy, Vol. I: Report*, Twenty-first Report of Session 2007–08, London: HMSO.

Leeke, M., Sear, C., and Gay, O. (2003) *An Introduction to Devolution in the UK*, Parliament and Constitution Centre Research Paper 03/84, London: House of Commons Library.

Marine Energy Programme Board (2014) *Maximising the Value of Marine Energy to the United Kingdom*, online at http://www.renewableuk.com/en/publications/reports.cfm/Maximising-the-Value-of-Marine-Energy-to-the-UK [accessed 31 March 2014].

Marine Management Organisation (MMO) (undated) 'About Us', online at http://www.marinemanage-ment.org.uk/about/ [accessed 31 March 2014].

Marine Scotland (undated) 'Aquaculture', online at http://www.scotland.gov.uk/Topics/marine/Fish-Shellfish [accessed 31 March 2014].

Morley, E. (2005) 'Speech by Elliot Morley MP, Minister of State, Climate Change and Environment', Presented at the Coastal Futures 2005 Conference, 19 January, SOAS, University of London.

Natural Environment Research Council (NERC) (undated) 'Marine Science', online at http://www.nerc. ac.uk/funding/marinesci/index.shtml [accessed 20 August 2005; page no longer available].

Office for National Statistics (ONS) (2005) *UK 2005: The Official Yearbook of the United Kingdom of Great Britain and Northern Ireland*, London: HMSO.

Office for National Statistics (ONS) (2014a) 'National Population Projections Based on Data Published on 6/11/13', online at http://www.ons.gov.uk [accessed 31 March 2014].

Office for National Statistics (ONS) (2014b) 'GDP 2013', online at http://www.ons.gov.uk [accessed 31 March 14].

Oil and Gas UK (2006) 'Guidelines Launched to Improve Relations between the Oil and Gas Industry and Nearshore Fishermen', Press release, 27 July, online at http://www.oilandgasuk.co.uk/news/news. cfm/newsid/248 [accessed 2 November 2014].

Oil and Gas UK (2013) *Economic Report 2013*, online at http://www.oilandgasuk.co.uk/cmsfiles/modules/ publications/pdfs/EC038.pdf [accessed 31 March 2014].

Oil and Gas UK (2014) *Economic Report 2014*, online at http://www.oilandgasuk.co.uk/cmsfiles/modules/ publications/pdfs/EC038pdf [accessed 9 November 2014].

Pugh, D. (2008) *Socio-economic Indicators of Marine-related Activities in the UK Economy: Project OSR07–04 Final Report*, London: HMSO/The Crown Estate.

Pugh, D., and Skinner, L. (2002) *A New Analysis of Marine-related Activities in the UK Economy with Supporting Science and Technology*, IACMST Information Document No. 10, London: IACMST.

Renewable UK (2013) *Wind Energy in the UK: State of the Industry Report 2013*, online at http://www. renewableuk.com/en/publications/reports.cfm/state-of-the-industry-report-2012-13 [accessed 31 March 2014].

Royal Commission on Environmental Pollution (RCEP) (2004) *Turning the Tide: Addressing the Impact of Fisheries on the Marine Environment – 25th Report of the Royal Commission on Environmental Pollution*, London: HMSO.

Royal Society of Edinburgh (RSE) (2004) *Inquiry into the Future of the Scottish Fishing Industry*, Edinburgh: Royal Society of Edinburgh.

Saunders, J., Tinch, R., and Hull, S. (2010) *Valuing the Marine Estate and UK Seas: An Ecosystem Services Framework*, London: The Crown Estate.

Scottish Executive (1997) *NPPG-13 Coastal Planning*, Discussion paper, 1 August, online at http://www. scotland.gov.uk/Publications/1997/08/nppg13-coastal [accessed 27 September 2010].

Scottish Executive Central Research Unit (2001) *Indicators to Monitor the Progress of Integrated Coastal Zone Management: A Review of Worldwide Practice*, Edinburgh: Central Research Unit.

Scottish Government (undated) 'Local Planning Pilots (SSMEI)', online at http://www.scotland.gov.uk/ Topics/Environment/Wildlife-Habitats/protectedareas/SSMEI [accessed 1 October 2010].

Select Committee on Environment, Food and Rural Affairs (EFRA) (2004) *The Marine Environment: Environment, Food and Rural Affairs Select Committee Sixth Report of Session 2003–04*, HC76, London: HMSO.

Select Committee on Environment, Food and Rural Affairs (EFRA) (2005) *The Future for UK Fishing: Environment, Food and Rural Affairs Select Committee Sixth Report of Session 2004–05*, HC 122, London: HMSO.

Stojanovic, T., and Ballinger, R. C. (2009) 'Integrated Coastal Management: A Comparative Analysis of Four U.K. Initiatives', *Applied Geography*, 29(1): 49–62.

Stojanovic, T., and Barker, N. B. (2008) 'Improving Governance through Local Coastal Partnerships in the UK', *The Geographical Journal*, 174(4): 344–60.

UK Chamber of Shipping (undated) 'The Status and Economic Impact of UK Shipping', online at http:// www.ukchamberofshipping.com/information/ [accessed 31 March 2014].

UK Major Ports Association (undated) 'Welcome', online at http://www.ukmajorports.org.uk [accessed 31 March 2014].

Vincent, M. A., Atkins, S. M., Lumb, C.M., Golding, N., Lieberknecht, L. M., and Webster, M. (2004) *Marine Nature Conservation and Sustainable Development: The Irish Sea Pilot – Report to Defra by the Joint Nature Conservation Committee*, Peterborough: Joint Nature Conservation Committee.

Welsh Government (2014) *Consultation Document: The Statement of Public Participation for the Welsh National Marine Plan*, 3 February, Doc. No. WG20730, online at http://wales.gov.uk/docs/desh/ consultation/140203marine-planning-public-participation-statement-en.pdf [accessed 2 November 2014].

Wood, Sir I. (2014) *UKCS Maximising Recovery Review: Final Report*, London: HM Government (the Wood Review).

Appendix 7.1 Selected key UK maritime legislation

Offshore Petroleum Activities (Conservation of Habitats) Regulations 2001 (SI 2001/1754)
Offshore Chemicals Regulations 2002 (SI 2002/1355)
Offshore Installations (Emergency Pollution Control) Regulations 2002 (SI 2002/1861)
Offshore Petroleum Activities (Oil Pollution Prevention and Control) Regulations 2005 (SI 2005/2055)
Merchant Shipping (Implementation of Ship-source Pollution Directive) Regulations 2009 (SI 2009/1210)

Marine renewable energy

Coast Protection Act (CPA) 1949, s. 34
Food and Environment Protection Act 1985, s. 5
Electricity Act 1989, ss 36 or 37
Town and Country Planning Act 1990, ss 57 or 90
Water Resources Act 1991, s. 109
Transport and Works Act 1992
Sustainable Energy Act 2003
Energy Act 2004
Climate Change and Sustainable Energy Act 2006
Climate Change Act 2008
Energy Act 2008
Planning and Energy Act 2008
Energy Act 2010
Conservation (Natural Habitats &c.) Regulations 1994 (SI 1994/2716), as amended by SI 1997/3055, SI 2000/192, and SI 2001/1754
Harbour Works (Environmental Impact Assessment) Regulations 1999 (SI 1999/3445)
Electricity Works (Environmental Impact Assessment) (Scotland) Regulations 2000 (SSI 2000/320)
Electricity Works (Environmental Impact Assessment) (England and Wales) Regulations 2000 (SI 2000/1927)
Transport and Works (Applications and Objections Procedure) (England and Wales) Rules 2000 (SI 2000/2190)
Electricity Act 1989 (Requirement of Consent for Offshore Wind and Water Driven Generating Stations) (England and Wales) Order 2001 (SI 2001/3642)
Electricity Act 1989 (Requirement of Consent for Offshore Wind and Water Driven Generating Stations) (Scotland) Order 2002 (SSI 2002/407)

Military activities

Dockyard Ports Regulations Act 1865
Military Lands Act 1892, Pt II, as amended by the Military Lands Act 1900, s. 2
Land Powers (Defence) Act 1958, s. 7
Wildlife and Countryside Act 1981
Protection of Military Remains Act 1986
Countryside and Rights of Way Act 2000
Conservation (Natural Habitats &c.) Regulations 1994 (SI 1994/2716), as amended by SI 1997/3055, SI 2000/192, and SI 2001/1754

Ports, shipping, and navigation

Harbours, Docks and Piers Clauses Act 1847
Coast Protection Act 1949
Harbours Act 1964
Hovercraft Act 1968
Carriage of Goods by Sea Act 1971
Health and Safety at Work Act 1974

Transport Act 1981

Public Health (Control of Diseases) Act 1984

Dangerous Vessels Act 1985

Food and Environment Protection Act 1985

Pilotage Act 1987

Aviation and Maritime Security Act 1990

Ports Act 1991

Carriage of Goods by Sea Act 1992

Transport and Works Act 1992

Merchant Shipping Act 1995

Merchant Shipping and Maritime Security Act 1997

Marine Safety Act 2003

Marine and Coastal Access Act 2009

Marine (Scotland) Act 2010

Merchant Shipping (Prevention of Oil Pollution) Order 1983 (SI 1983/1106), as amended by SI 1985/2002, SI 1991/2885, and SI 1993/1580

International Oil Pollution Compensation Fund (Parties to Convention) Order 1986 (SI 1986/2223)

Dangerous Substances in Harbour Areas Regulations 1987 (SI 1987/37)

Merchant Shipping (Prevention and Control of Pollution) Order 1987 (SI 1987/470), as amended by SI 1987/664, SI 1990/2595, SI 1992/2668, SI 1997/2569, and SI 1998/254

Merchant Shipping (Prevention of Pollution by Garbage) Order 1988 (SI 1988/2252), as amended by SI 1993/1581, SI 1997/2569, and SI 1998/254

Merchant Shipping (Registration of Ships) Regulations 1993 (SI 1993/3138), as amended by SI 1994/541, SI 1998/1915, SI 1998/2976, and SI 1999/3206

Merchant Shipping Act 1995 (Appointed Day No. 1) Order (SI 1995/1210)

Merchant Shipping Act 1995 (Appointed Day No. 2) Order (SI 1995/3107)

Merchant Shipping (Prevention of Pollution) (Law of the Sea Convention) Order 1996 (SI 1996/282)

Merchant Shipping (Prevention of Oil Pollution) Regulations 1996 (SI 1996/2154), as amended by SI 1997/1910 and SI 2000/483

Merchant Shipping and Maritime Security Act 1997 (Commencement No. 1) Order (SI 1997/1082)

Merchant Shipping and Maritime Security Act 1997 (Commencement No. 2) Order (SI 1997/1539)

Merchant Shipping (Prevention of Oil Pollution: Substances other than Oil) (Intervention) Order 1997 (SI 1997/1869)

Merchant Shipping (Oil Pollution Preparedness, Response and Co-operation Convention) Regulations 1998 (SI 1998/1056), as amended by SI 2001/1639

Merchant Shipping (Prevention of Pollution by Garbage) Regulations 1998 (SI 1998/1377)

Merchant Shipping (Control of Pollution) (SOLAS) Order 1998 (SI 1998/1500)

Merchant Shipping (Port State Control) Regulations 1995 (SI 1995/3128), as amended by SI 1998/1433, SI 1998/2198, SI 2001/2349, and SI 2003/1636

Merchant Shipping (Safety of Navigation) Regulations 2002 (SI 2002/1473)

Merchant Shipping and Fishing Vessels (Port Waste Reception Facilities) Regulations 2003 (SI 2003/1809)

Merchant Shipping (Vessel Traffic Monitoring and Reporting Requirements) Regulations 2004 (SI 2004/2110)

Merchant Shipping (Implementation of Ship-source Pollution Directive) Regulations 2009 (SI 2009/1210)

Marine leisure and recreation

Public Health Acts Amendment Act 1907, ss 82–84

Public Health Act 1936, s. 231, as amended by Public Health Act 1961, s.76

Countryside Act 1968

Local Government Act 1972, s. 236

Local Government (Miscellaneous Provisions) Act 1976, s. 17
Local Government, Planning and Land Act 1980, s. 185
Countryside and Rights of Way Act 2000
Marine and Coastal Access Act 2009
Marine (Scotland) Act 2010
Bathing Waters (Classification) Regulations 1991 (SI 1991/1157)
Merchant Shipping (Small Commercial Vessels and Pilot Boats) Regulations 1998 (SI 1998/1609)
Merchant Shipping (Vessels in Commercial Use for Sport or Pleasure) Regulations 1998 (SI 1998/ 2771)
Merchant Shipping (Vessels in Commercial Use for Sport or Pleasure) (Amendment) Regulations 2000 (SI 2000/482)

Sea fisheries

Sea Fisheries Regulation Act 1966
Sea Fish (Conservation Act) 1967
Fisheries (Shellfish) Act 1968
Salmon and Freshwater Fisheries Act 1975
Fisheries Limits Act 1976
Fisheries Act 1981
Salmon Act 1986
Sea Fisheries (Wildlife Conservation) Act 1992
Environment Act 1995 (inserting s. 5A into the Sea Fisheries Regulation Act 1966)
Sea Fisheries (Shellfish) (Amendment) Act 1997
Sea Fisheries (Shellfish) Amendment (Scotland) Act 2000
Territorial Waters Order in Council 1964
Fisheries Limits Act 1976 (Commencement) Order 1976 (SI 1976/2215)
Fisheries Act 1968 (Commencement No. 1) Order 1969 (SI 1969/1551)
Various sea fisheries regulations, and several orders and bylaws

Marine nature conservation

National Parks and Access to the Countryside Act 1949
Conservation of Seals Act 1970
Endangered Species (Import and Export) Act 1976
Wildlife and Countryside Act 1981
Wildlife and Countryside (Amendment) Act 1985
Food and Environment Protection Act 1985, Pt II
Environmental Protection Act 1990
Sea Fisheries (Wildlife Conservation) Act 1992
Environment Act 1995
Countryside and Rights of Way Act 2000
Marine and Coastal Access Act 2009
Marine (Scotland) Act 2010
Conservation (Natural Habitats &c.) Regulations 1994 (SI 1994/2716), as amended by SI 1997/3055, SI 2000/192, and SI 2001/1754
Offshore Marine Conservation (Natural Habitats &c.) Regulations 2007 (SI 2007/1842)
Offshore Marine Conservation (Natural Habitats &c.) (Amendment) Regulations 2009 (SI 2009/7)
Offshore Marine Conservation (Natural habitats &c.) (Amendment) Regulations 2010 (SI 2010/491)

Marine archaeology and cultural heritage

Protection of Wrecks Act 1973
Ancient Monuments and Archaeological Areas Act 1979
Protection of Military Remains Act 1986

Planning (Listed Buildings and Conservation Areas) Act 1990 (does not apply below LWM)

Town and Country Planning Act 1990 (does not apply below LWM)

Transport and Works Act 1992

Merchant Shipping Act 1995, ss 224 and 236

Merchant Shipping and Maritime Security Act 1997, s. 24

National Heritage Act 2002

Various planning guidance and non-statutory codes of practice

Waste disposal and pollution control

Food and Environment Protection Act 1985

Environmental Protection Act 1990, ss 146–147 and 162, and Sch. 6

Water Resources Act 1991, Pt III

Environment Act 1995

Pollution Prevention and Control Act 1999

Environmental Protection (Duty of Care) Regulations 1991 (SI 1991/2839)

Waste Management Licensing Regulations 1994 (SI 1994/1056)

Pollution Prevention and Control (Scotland) Regulations 2000 (SSI 2000/323)

Pollution Prevention and Control (England and Wales) Regulations 2000 (SI 2000/1973), as amended by SI 2001/503, SI 2002/275, SI 2002/1559, and SI 2002/1702

Water Environment (Water Framework Directive) (England and Wales) Regulations 2003 (SI 2003/3242)

Pollution Prevention and Control (Northern Ireland) Regulations 2003 (SR 2003/46)

Hazardous Waste (England and Wales) Regulations 2005 (SI 2005/894)

Waste and Air Pollution (Miscellaneous Amendments) Regulations 2007 (SI 2007/3476)

Coastal defence and shoreline management

Coast Protection Act 1949

Food and Environment Protection Act 1985, Pt II

Town and Country Planning Act 1990

Land Drainage Act 1991, as amended by the Land Drainage Act 1994

Water Resources Act 1991

Environment Act 1995

Water Act 2003

Town and Country Planning (General Permitted Development) Order 1995 (SI 1995/418)

Various planning guidance and non-statutory measures

7

BRAZIL'S NATIONAL OCEAN POLICY

Milton L. Asmus, Etiene Villela Marroni, and Guilherme G. Vieira

Introduction

Basic information and overview of national ocean policy

Brazil is a significant presence in the South American Atlantic. With a coastline of approximately 8,500 km and 25 per cent of the national population – which generates 70 per cent of the Brazilian gross national product (GNP) – living in coastal counties, Brazil has a significant impact on the adjacent territorial waters and beyond. From a global perspective, socio-economic activities in the coastal zone of Brazil have a great potential to influence not only the national economy and society, but also the environmental state of the South Atlantic.

The geographical importance of the coast was sufficient reason for the Brazilian government to establish a policy concerning the coastal zone, territorial sea, and the living, as well as non-living, resources of its continental shelf in 1980. With the recent growth of the Brazilian economy (IBGE, 2014), there has also been an increase in the possibility for new investment in several sectors, one of which involves a whole segment of ocean-related businesses and professionals. Although Brazil has been naturally provided with a coastline with great possibilities for navigation, fishery resources, and other natural resources, it was through prospecting for oil that the Brazilian legal continental shelf became a target of interest for domestic and foreign investors (Marroni, 2013). In this chapter, the Brazilian ocean policy, known as the National Policy for Sea Resources (*Política Nacional para os Recursos do Mar*, PNRM), is analysed in terms of its historical development, and the political and institutional context that establishes the basis for development of implementation plans and programmes. This includes the recent establishment of the Brazilian marine space (also called 'Blue Amazon') as a national priority and actions related to marine spatial planning (MSP).

The chapter describes the past and present processes of coastal planning in Brazil, its mechanism towards the establishment of an ocean policy, and the way in which it is intended to be integrated into the national socio-political system. The PNRM was a significant advancement in organizing the multiple uses of the coastal and marine zones of Brazil. Established in the 1980s, but has been reviewed biannually and implemented through new proposed action plans. The Brazilian National Plan of Coastal Management (PNGC) is also analysed, which highlights the role of coastal municipalities in the planning and management processes. The role of local communities and municipalities in developing public policies and management processes is still being developed in Brazil, and may be enhanced with a better evaluation model and increasing local participation (Marroni and Asmus, 2013).

The coastal zone of Brazil

Brazil has a coastal zone of approximately 514,000 km², including the terrestrial and aquatic sectors (see Figure 7.1). The Brazilian coast is a diversified space as a result of the variety of environmental characteristics and socio-economic activities in the area. Characterized as a zone of multiple uses, the distinct economic activities in the coastal zone have led to serious conflicts in terms of the use of the land, which have been made more serious because of the intensity of human use (Scherer et al., 2011). The diverse natural environment and uses of the coastal zone constitute a complexity that demands effective planning and regulation.

The terrestrial limit of the coastal zone involves selected counties according to the criteria established by the maritime policy of the country, with 397 counties distributed along 7,367 km of coast (a total of 8,698 km if bays and indentations are considered), and a total area of approximately 388,000 km². The Brazilian territorial sea extends out to 12 nautical miles. In terms of latitude, the Brazilian coast extends from 4° 30'N to 33° 44'S in inter-tropical and subtropical zones with very different environments of high ecological relevance. The Brazilian coast consists of a variety of ecosystems and habitats, including coastal reefs, mangroves, coastal lagoons, barrier islands, marsh areas and swamps, tidal zones, and beaches and dunes (MMA, 2008a).

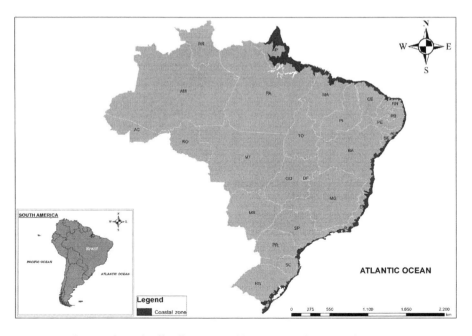

Figure 7.1 Coastal zone of Brazil, officially composed by 397 coastal municipalities

Note: The two-character names are the abbreviations for the Brazilian states.

Source: Adapted from Marroni and Asmus (2013)

The importance of environmental assets has been highlighted in Brazil – especially the ecological functions of coastal ecosystems. Previously, these ecosystems had direct value only as natural resources, for example as fish or fuel wood. More recently, the importance of the functions of the coastal ecosystems has been recognized, not only as a result of the often serious impacts and damage resulting from improper human use, but also because of the use of bio-economic models of these ecosystems (MMA, 2008a).

Approximately 74 million people live in the coastal zone. The mean population density in this area is 105 inhabitants per square kilometre – five times the national mean (20 inhabitants/km²). Population concentrations vary along the Brazilian coast; there are regions with low demographic density, such as Amapá state, with 6.1 inhabitants/km² in the coastal zone, while the states of Pernambuco (803.0 inhabitants/km²) and Rio de Janeiro (656.5 inhabitants/km²) have the highest coastal population concentrations. In general, the coastal population is smaller in the north (4.0 million) and south (3.3 million), and larger in the northeast (17.4 million) and southeast (16.0 million). Given the current population growth trend, problems associated with high coastal concentrations are expected to increase (Marroni and Asmus, 2005). The construction of houses in areas of high environmental sensitivity (such as dunes, mangroves, and estuaries) and the lack of basic sanitation, along with agricultural and urban activities, degrade the natural environment through organic pollution, sediment deposit, and destruction of natural habitats. Economic activities, historically, have been concentrated along the coast. These activities include oil extraction and refining, ports, agriculture, aquaculture, mineral extraction, fishing, cattle-raising, reforestation, salt production, and summer resorts and tourism. For historical and structural reasons, economic growth has been based on industrialization in the coastal region, inducing population and urban growth. This strong association between urbanization and industrialization characterizes the processes of territorial, population, and economic dynamics in Brazil, and, as a consequence, in its coastal zone (Fernandes, 2013; MMA, 2008b).

Since the 1960s, an export-oriented policy has determined the settlement pattern of important industrial areas in the coastal zone. Harbour installations and the formation of important industrial complexes, such as Cubatão, Santos and São Vicente (Estate of São Paulo), and Rio Grande (Estate of Rio Grande do Sul), characterize the area. Brazilian industrial exports are mainly in sectors of high pollution potential, such as pulp, paper, and chemicals. These industries are usually located in metropolitan regions of the coastal zone. New growth in these sectors as a result of increased demand for oil (70 per cent of which comes from offshore extraction), as well as the position of Brazil in the world economy, which reinforces the logistical role of the coastal zone, would have significant repercussions on the coastal zone (World Bank, 2014).

Currently, the focus of interest on the use of the Brazilian continental shelf has turned to the extraction of oil and gas. Brazil has extracted oil from its continental shelf since the mid-1990s, with an increase in its production since 2010. Meanwhile, the country has made progress in developing new technologies for prospecting and extracting oil and gas in deep waters on the continental shelf and beyond its limits, with wells that exceed 1,800 metres into the seafloor. The new system of exploration and extraction of offshore oil and gas has established a complex support structure and new activities, and is affecting and competing with other economic uses, such as inshore and deep-sea fishing (Marroni, 2011). The establishment of drilling platforms for oil and gas, and the intense maritime route used by boats to support production systems, demanded immediate action from the Brazilian environmental protection agency (*Instituto Brasileiro do Meio Ambiente e dos Recursos Naturais Renováveis*, IBAMA), determining exclusive fishery zones in an attempt to prevent damage to operating structures and accidents among the fleets. This initiative has generated conflicts between activities and has led the fishery system, as a whole, to become the biggest reason for the establishment of mitigating and compensatory measures related to oil and gas exploration and exploitation (Marroni and Asmus, 2013).

Historical antecedents and the importance of local governance in forming public policies

The preoccupation of the Brazilian government with the use of maritime resources and coastal spaces started in the 1970s, with new environmental concerns in state planning. In 1973, the Special Secretary of Environment of the Presidency was created, establishing a meaningful step in the institutional history of this process. A year later, the Inter-Ministry Commission for Sea Resources (*Comissão Interministerial para os Recursos do Mar*, CIRM) was formed. Its main goal was the coordination of themes that would lead to a national policy for the coastal region of Brazil, issued through Decree No. 74.577 of 12 September 1974. However, only with the latest Brazilian Constitution, sanctioned in 1988, have the legal questions related to the environment started to take shape and to have some effect (Constituição da República Federativa do Brasil, 1988).

Before offering a brief historical presentation of policies, plans, and programmes related to oceanic areas in Brazil, it is important to explain and analyse its federal Constitution, which outlines the essential functions and rules that define and frame the competencies among the federal government, states, and counties. In Chapter VI, article 225, of the Brazilian Constitution, the importance of the environment is acknowledged:

> [E]verybody has the right to an ecologically balanced environment, a good of common use by the people and essential for a healthy quality of life, imposing for the Public Government and for the population the duty of defending and preserving it for both present and future generations.

Article 225(4) establishes that the coastal zone is a national 'patrimony'. For this reason, it is applied as a law aiming at 'preservation' of the coastal zone. However, although the coastal zone is thus defined as a national patrimony, it cannot be defined as either federal or state property. The latter is where the state performs its domain. The national patrimony constitutes a good of public interest. It is characterized as an eminent domain, as practised by the legal regulation of its use. The coastal zone is a patrimony that interests all Brazilians and, for this reason, its protection, as well as the protection of its natural resources, ought to respect its regional peculiarities. Rules should be modified in accordance with respective federal, state, and county contexts, and if applied to coastal management, it is inferred that the state (nation) will be responsible for the fulfilment of general rules. The states should respect the rules promulgated at the federal level by establishing complementary rules or standards. It is the responsibility of the municipalities to protect local interests, legislating when some peculiarity demands or justifies more specific rules. It is clear that federal or state rules do not include sufficient standards to meet the demands or necessities of different localities.

The implementation of policies pertinent to the oceanic area of Brazil (see Table 7.1) is oriented toward the rational use of resources to guarantee the quality of life of the coastal population and effectively to protect existing ecosystems. In this sense, the improvement of human resources that will carry out work themes related to coastal and marine environments is a major factor in the implementation of coastal and maritime policies. Specific policies for certain coastal spaces are necessary, with the aim of improving life conditions of the population according to regional peculiarities. Otherwise, the resolution of serious problems will be relegated to future Brazilian generations.

Improving the life conditions of the community is the real function of 'local governance'. However, in the past, people were generally unused to being involved with environmental causes

Table 7.1 The evolution of federal ocean policies in Brazil

Year	Event
1973	Establishment of the office of Special Secretary of Environment of the Presidency. This represents the beginning of an environmental component in the planning process.
1974	Inter-Ministry Commission for Sea Resources (CIRM) established. It represented the concern of the Brazilian government with the use of coastal and marine resources.
1980	National Policy for Sea Resources (PNRM) instituted as result of work done by the CIRM. This policy is implemented through multi-annual plans and programmes elaborated within the CIRM.
1981	Brazilian National Policy on the Environment (PNMA) established. It aims at the preservation, improvement, and recovery of environmental quality.
1982	CIRM opens a Sub-Commission on Coastal Management, which organizes a seminar in Rio de Janeiro.
1983	International Seminar on Coastal Management, organized by CIRM.
1984	II Brazilian Symposium on Sea Resources, with representatives from many Brazilian universities.
1985	II ENCOGERCO (Brazilian Summit on Coastal Management) defines the institutional model (attempting to be decentralized and participatory).
1987	CIRM establishes the National Programme of Coastal Management (GERCO), specifying the zoning methodology and the institutional model for its application.
1988	Federal Constitution of 1988 highlights legal and institutional process of Brazilian coastal management, as it declares the coastal zone a national patrimony (art. 225(4)).
	National Plan of Coastal Management (*Plano Nacional de Gerenciamento Costeiro*, PNGC) established by Law No. 7.661 of 1988, with the political and judicial support of the CIRM and the National Environmental Council (CONAMA).
1990	Resolution CIRM 001/90 approves first version of the PNGC, defining the methodological basis, the institutional mode, and the tools to support GERCO.
	Law No. 8.028 creates the Environment Secretary of the Presidency.
1992	Special Secretary of Environment (SEMAM) transformed into the Ministry of Environment.
1992	V ENCOGERCO takes place.
	Group of Coordination of Coastal Management, created within CIRM, begins to update the first version of the PNGC.
1994	VI ENCOGERCO evaluates the impact of sectorial federal policies.
1996	VII ENCOGERCO evaluates the need for inter-institutional and inter-sectorial coordination.
1997	Resolution CIRM 005 establishes PNGC II, which creates the Group of Integration of the Coastal Management and the Subgroup for Integration of the State Programmes.
1998	Federal Action Plan for the Coastal Zone of Brazil aims at reorienting federal actions in the coastal zone towards promotion of inter-institutional cooperation to articulate public policies in the coastal zone.
2001	CIRM establishes new guidelines for action in its plans and programmes, striving to foster technological development and international action by Brazil in matters related to the sea.
2003	Coastal Agency created, a public interest non-governmental organization (NGO), with the aim of coordinating actions to integrate society and government in coastal management initiatives.
2004	ENCOGERCO evaluates fifteen years of implementation of the PNGC.
2005	New revised version of the Federal Action Plan for the Coastal Zone of Brazil (PAF) issued.
2005	PNRM updated to meet scientific and technological needs in relation to the study of the seas and oceans.
2013	Conference entitled 'Oceans and Society' evaluates and celebrates twenty-five years of the PNGC in Brazil.
2013	New National Policy for the Conservation and Sustainable Use of the Brazilian Marine Biome proposed in Congress.
2013	CIRM establishes working group to deal with the uses of the ocean and marine spatial planning (MSP).

through any participatory social process, leaving the responsibility to the exclusive domain of the government. More recently (especially in the twenty-first century), the public has started to get involved, organizing in groups to deal with local problems. This development led to calls for state action through plans and strategies set up by professionals in the system or by the public itself. It is this 'local governance', democratically organized through community participation, which encourages the bureaucracy or government institutions to act, aiming for the improvement of the quality of life of coastal citizens (CIRM, 2013; Marinha do Brasil, 2011).

It is imperative that the study of public policy address not only the rational use of the coastal ocean environment, but also the management of a whole range of economic and social policies, looking at the proper maintenance and use of ocean spaces under Brazilian jurisdiction (Marroni, 2013). The implementation of activities related to marine resources has been undertaken in a decentralized manner by several actors in different ministries, states, municipalities, research institutions, the private sector, and the scientific community, according to their respective competencies, as identified in Table 7.1. The CIRM, consistent with the guidelines established in the National Policy for Sea Resources (PNRM), works as the 'organizer' between the public and private sectors, integrating professionals and institutions by means of meetings, projects, and policies.

In the last decade, Brazil has seen initiatives to facilitate dialogues between the government and the public on how to strengthen the policies that protect the country's marine and coastal environment. For example, the Brazilian Agency for Coastal Management (the Coastal Agency) was established in 2002 as a 'non-profit association formed by people and organizations interested in contributing to the sustainable development of the coastal and marine zone of Brazil, in ways that ensure its integrity and environmental quality, and maintain its natural and cultural heritage' (Poleti et al., 2012: 1546) and is a public interest non-governmental organization (NGO) formally recognized by government (*Organização da Sociedade Civil de Interesse Público*).

The Coastal Agency conducted four National Meetings on Coastal Management (*Encontro Nacional de Gerenciamento Costeiro*, ENCOGERCO) in 2002, 2004, 2006, and 2009. The ENCOGERCO has been recognized as Brazil's foremost event for the discussion of technical and political developments related to coastal management and sustainable development of marine and coastal areas. The events saw wide participation from government and NGOs, academia, and the private sector.

The difficulty in achieving a national event in Brazil results from geographical limitations, which prompted ENCOGERCO to establish the online platform *Fórum do Mar* (Sea Forum) (Fórum do Mar, 2012) in 2010. The Fórum was established as a space for discussion to establish a broad and participatory platform for contact and dialogue among government, civil society, and social actors in light of their shared responsibilities. The Coastal Agency serves as the executive secretariat of the Fórum do Mar.

Since its creation, Fórum do Mar has encouraged and coordinated classroom- and Internet-based discussions on policy issues regarding the development of marine and coastal systems, especially on the aspects of integrated marine and coastal management, sustainable development, ocean energy sources, and the 'blue economy'. In addition, Fórum do Mar participates in national committees on environment and development, such as the Brazilian Panel for Biodiversity (*Painel Brasileiro de Biodiversidade*, PainelBio) (Merico, 2013). In the future, it is expected that Fórum do Mar will have a broader impact on issues and be able to act in conjunction with international institutions to stimulate Brazilian policies to align with global recommendations and agreements toward the sustainable use of the ocean.

The policy development process

The Brazilian National Policy on the Environment (*Política Nacional do Meio Ambiente*, PNMA), instituted by Law No. 6.938 of 31 August 1981, mainly aimed to implement policy by means of 'a harmonization of the social-economic development with the preservation of the environment quality and its ecological balance' (article 4°), 'considering the environment as a public patrimony' and 'aiming at the population's use' of the environment (article 2°). This law organized the national system of environmental control in the country, delegating to the Environment National Council the responsibility of composing a legislative body to carry out the actions foreseen in the PNMA.

The PNRM, instituted in 1980 by means of Presidential Decree No. 5.377 of 23 February 2005, has the CIRM as its executive organ. The coordinating institution for the CIRM is the Brazilian Navy. The CIRM aims to be a connecting element among the various federal sectorial policies concerning the marine and coastal environment. It also sets the essential arrangements for the promotion of integration of the territorial sea and continental shelf to the Brazilian space and rational exploration of the oceans. This policy is implemented through multi-annual plans and programmes developed by the CIRM. Such plans include the Sectorial Plan for Sea Resources (*Plano Setorial Para os Recursos do Mar*, PSRM), the Plan for the Survey of the Brazilian Continental Shelf (*Plano de Levantamento da Plataforma Continental Brasileira*), and the National Plan of Coastal Management (*Plano Nacional de Gerenciamento Costeiro*, PNGC) (CIRM, 2005).

It is worth mentioning the interface of basic regulatory instrument Decree No. 99.540 of 21 September 1990, even though it goes beyond the coastal zone. This decree establishes the Coordinating Commission of the Ecologic-Economic Zoning of the National Territory, which has the responsibilities for planning, coordinating, and evaluating procedures that involve land and marine activities.

The Inter-Ministry Commission for Sea Resources (CIRM)

As noted above, the implementation of the PNRM through multi-annual plans and programmes is developed within the CIRM. The CIRM is a 'multidisciplinary' unit supervised by the Brazilian Navy. It has a secretariat, known as SECIRM, which gathers and executes all programme activities. However, the unit has its own autonomy concerning the hiring or creation of groups and subgroups of technical and administrative workers. Implementation of activities connected to sea resources is not centralized. The Commission connects staff from within the different ministries, states, counties, research institutions, scientific community, and private entities according to their respective competencies consistent with the decisions of the PNRM. Therefore the CIRM works as a coordinator between the public and the private sectors, organizing meetings, projects, and policies in which scientific professionals are invited to participate. The competencies and attributes of the CIRM directly assist the president in the execution of the PNRM. The CIRM proposes the general rules of the policy, monitors the results and, when necessary, suggests possible changes to the president, presents opinions and suggestions, and establishes connections with other ministries, state governments, and the private sector to attain the necessary support for the execution of the plans and programmes of PNRM for which they share common objectives. Box 7.1 outlines the main functions of the CIRM (Marroni and Asmus, 2013).

Box 7.1   Functions of the CIRM

- To formulate the rules proposed for the execution of the PNRM
- To plan activities related to sea resources
- To coordinate the development of multi-annual and annual plans and programmes
- To develop activities related to the sea and the Antarctic
- To follow up on the results and to propose changes in the execution of the Brazilian Program for the Antarctic (*Programa Antártico Brasileiro*, PROANTAR)
- To produce reports and suggestions related to issues concerning sea resources
- To follow the results and propose adjustments for the PNRM

Source: Modified from CIRM (2003)

The focal point for the CIRM is the PNRM, which was instituted on 12 May 1980. The PNRM consolidates the work of the CIRM in undertaking measures for the integration of the territorial sea and the continental shelf to Brazilian space, and for the rational exploration of marine resources.

Decree No. 6.107 of 2 May 2007 determines the following composition for the CIRM: Office of the Presidency of the Republic; Ministry of Defence; Ministry of Foreign Affairs; Ministry of Transportation; Ministry of Agriculture, Livestock and Food Supply; Ministry of Education; Ministry of Health; Ministry of Development, Industry and Foreign Trade; Ministry of Mines and Energy; Ministry of Science, Technology and Innovation; Ministry of the Environment; Ministry of National Integration; Fisheries and Aquaculture; Ministry of Tourism; Ministry of Sports; Ministry of Planning, Budget and Management; Secretariat of Ports; Navy Command; SECIRM; subcommittees; executive committees; and working groups (CIRM, 2003; 2012).

The current composition of the CIRM can be considered to be a large institutional framework for the purposes of developing a new perspective on Brazilian marine resources. The upgrade was effected by means of 'modernization' of the PNRM, which provided for greater reciprocity between government agencies, research, and funding for the development of research on living and non-living resources in the sea (Marroni, 2013).

The National Policy for Sea Resources (PNRM): History and characterization

The PNRM was a significant advance in organizing the multiple uses of the coastal and marine zones of Brazil. Established in the 1980s, the PNRM has been reviewed biannually. The frequency of reviews is justified by dynamic changes in the Brazilian population and in the marine environment. Even though it was created during the Brazilian military regime, the PNRM has been changed and adapted to the current regime, and provides a strong instrument for governance in the ocean and coastal areas. Presidential Decree No. 5,382 reformulated the policy through PSRM VIII, which sets out the basic rules that guide all initiatives in the fields of teaching, research, exploration, and exploitation of sea resources. One of the goals of the PNRM is to introduce measures aimed at promoting the:

> . . . integration of the territorial sea and the continental shelf to Brazilian land and to the rational exploitation of the oceans, involving living resources, minerals and the

energy of the water column, seabed and seabed sub-floor that are of interest for the sustainable development of the country.

(VIII PSRM, 2012)

The policy also outlines the level of participation of public institutions, community organizations, and private entities in undertaking these measures.

The basic principles of the PNRM (see Box 7.2) harmonize the policy, wherever possible, with the PSRMs. In order to implement these principles, it is necessary to ensure significant cooperation and use of technical and scientific staff throughout the country, whose specializations in fields related to the ocean and its resources require enhancement through appropriate qualifications. It is also important to update Brazilian legislation concerning ocean resources to ensure their sustainable development and, most importantly, to stimulate and extend technical-scientific and policy exchange.

Box 7.2 Principles of the PNRM

- Harmonization with national policy
- Non-centralized and participatory implementation
- Coordination of financial resources
- Definition of project priorities
- Inclusion of the private sector
- Protection of biodiversity and genetic patrimony
- Supervision of governmental action
- Collaboration with international programmes

Source: Adapted from the PNRM

It should be noted that PNRM updates by means of the PSRMs are also influenced by international treaties and agreements. It was in multilateral forums – particularly in the United Nations, in which Brazil has increased its international performance – which led to increased participation in ocean issues and provided more visibility to the country in the international arena. By incorporating the PNRM into the Brazilian institutional system, the upgraded plans should be formulated according to the specific policy needs of the country, taking into account national and international circumstances (Marroni, 2013).

The strategies for implementation of the PNRM are established under the 'conditions' for its execution. First, the policy will be consolidated through multi-annual and annual plans and programmes. To implement such plans, the CIRM, in consultation with the Secretary of Planning of the Presidency, should aim to coordinate programmes and to use available resources more efficiently (CIRM, 2011b). The PNRM's multi-annual plans also need to include objectives, strategies, proposed programmes, financial estimates, and sectorial programmes in specific projects (that is, basic work documents). Any planning to establish strategies for action will have to consider human resource development, marine research, science, and technology, and exploration and sustainable use of sea resources. It is also of fundamental importance to review the strategies for action with the aim of maximizing the execution and management of the current plans and programmes resulting from the broader policy.

The nature of the policy and established legislation

Sectorial Plans for Sea Resources (PSRMs)

The PSRM constitutes an 'upgrade' of the PNRM and is valid for a couple of years. Planning of all activities related to sea resources (see Table 7.2) follow the guidelines of the PSRM.

According to the Sectorial Plan (VIII PSRM, 2012), the importance of living marine resources derives not only from their exploitation for food production, but also from their contribution to biodiversity and genetic patrimony as relates to their relevance for biotechnology. The focus on marine biodiversity was an innovation in the updates proposed since the VI PSRM. In addition to constituting a new framework for the PNRM, biotechnology provides new opportunities for research and human resource development, and can generate more precise information on the potential economic value of Brazil's coastal-marine resources. In this sense, the policy aims to establish strategies of action consisting of human capacity-building, research, exploitation, and the sustainable use of marine resources. The constant revision of strategies for action is of fundamental importance, always aiming toward maximizing the processes of implementation and management of plans arising from broader political programmes (Marroni, 2014).

The PSRM also updates legislation concerning the exploration of non-living marine resources both on and below the seabed, such as petroleum and natural gas. As finite resources, they must be exploited in a rational and sustainable manner. There is also a strong recommendation to conduct scientific research to find alternative renewable energy sources to replace non-renewable resources. In addition to updating the PNRM, the PSRM is subject to, and in agreement with, basic instruments of international law to which Brazil is a signatory. These legal instruments define the global legal framework within which the country should carry out sustainable use of ocean resources.

The PSRM is also subordinate to internal legislation, such as the 1988 Constitution. The Constitution defines the territorial sea, the resources of the exclusive economic zone (EEZ) and the continental shelf as 'goods' of the nation, and the coastal zone as national patrimony. It is also important to note that the PSRM is based on the precautionary principle in relation to exploitation and use of ocean resources. The PSRM also adopts the principles of integrated coastal and marine management, and incorporates the concepts of sustainability (ecological, economic, and societal), and non-centralized and participatory planning (see Table 7.3).

Table 7.2 Main plans and programmes of the national policy

National Policy for Sea Resources (PNRM)				
PSRM	PROANTAR	LEPLAC	PROMAR	GERCO
Sectorial Plan for Sea Resources (*Plano Setorial Para os Recursos do Mar*)	Antarctic Programme (*Programa Antártico Brasileiro*)	Plan for the Survey of the Brazilian Continental Shelf (*Plano de Levantamento da Plataforma Continental Brasileira*)	Programme for Promotion of a Maritime Mentality (*Programa de Promoção da Mentalidade Marítima*)	National Programme of Coastal Management (*Programa Nacional de Gerenciamento Costeiro*)

Source: Adapted from the PNRM

Table 7.3 Sectorial Plans for Sea Resources

Plan	Period	Features
I PSRM	1982–85	Structured research activities and prospecting relating to sea resources, iterating the meaningful interests of the Brazilian society in the use of these resources within the national productive system.
II PSRM	1986–89	Established objectives for overcoming socio-economic difficulties within the country; outlined the scientific and technical capacity of the organizations and human resources involved in the projects.
III PSRM	1990–93	Validation of II PSRM; study of the effects of confirmation, by Brazil, of the 1982 United Nations Convention on the Law of the Sea (UNCLOS) establishing as a main target the investigation and rational exploration of resources in the exclusive economic zone (EEZ).
IV PSRM	1994–98	Validation of III PSRM; implementation of the Programme for the Study of Sustainable Potential of Live Resources Capture of the Exclusive Economic Zone (*Programa de Avaliação do Potencial Sustentável de Recursos Vivos na Zona Econômica Exclusiva*, REVIZEE), which gained a new institutional momentum.
V PSRM	1999–2003	PNRM update, with precepts aimed at sustainable development, preparation of human resources, and incentives to research.
VI PSRM	2004–08	Update of IV PSRM, adapting it to the conjuncture foreseen for its period.
VII PSRM	2008–11	Recognition of the oceans' role in global climate change; highlighting of the need for cooperation among government, academia, civil society, and private sector actors to manage the sustainable use of sea resources.
VIII PSRM	2012–15	Introduction of a novel model for the integrated and participative management of sea resources, involving several ministries, scientific society, and private sector actors; stimulates integration of actions; highlights the importance of data availability for society; defines the conservation of sea resources a priority; stimulates the development of human resources and the international cooperation – and all with an especial focus on the natural resources of coastal zones.

Source: Adapted from Marroni and Asmus (2013)

It is important to understand the theoretical and leadership basis of the PNRM, and how PSRM mechanisms work. The multi-annual plans assess the adequacy and 'modernization' of the PNRM. These plans provide the executive branch of government with operational guidelines for a period of four years. A political update through multi-annual plans and programmes is developed by the Executive Secretary of the CIRM.

Some issues included in the sectorial plans stem from obligations resulting from treaties and agreements signed by Brazil on the protection and rational use of coastal and marine resources. For example, the 1982 United Nations Convention on the Law of the Sea (UNCLOS), which the Brazilian National Congress ratified in December 1988, deals with, among other topics, mining of the deep seabed and straddling fish stocks and highly migratory fish stocks, which, in association with Chapter 17 of Agenda 21 (see Figure 7.2), have formed the basis for various programmes in the CIRM (Agenda 21 Brasileira, 2002; MMA, 2006; 2008a).

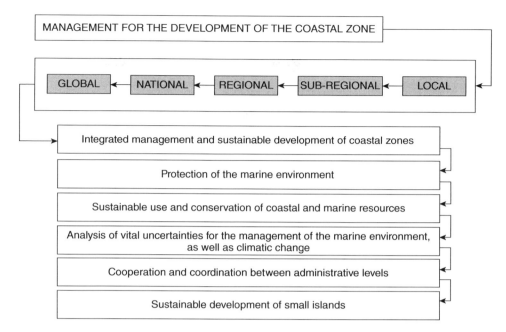

Figure 7.2 Programme areas for the management of the Brazilian coastal zone, according to Agenda 21

Source: Adapted from Agenda 21 Brasileira (2002)

The frequent updates of the PNRM renew the specific plans and programmes for each area that guides the policy as a whole. The non-centralized execution of the PNRM, with the participation of states, cities, and civil NGOs, encourages proposals for the resolution of conflicts in specific areas of Brazilian coastal and marine zones. Executing public policies and plans has become a significant part of the analytical process because the execution and management of these plans involves specialized technical staff to manage financial resources for each area. To be successful, they must analyse and evaluate results regarding the benefits for the population and rationalize the areas covered under the PNRM. Table 7.4 lists some of the plans and programmes executed by the PNRM.

Implementation programmes

The implementation of Brazilian oceanic and coastal policy under the PNRM is accomplished by means of national programmes coordinated, in most cases, by the CIRM and the Ministry of Environment. The programmes of implementation are regulated and adapted through the development of consecutive PSRMs. Several of these programmes are described briefly below.

REVIMAR (*Avaliação, Monitoramento e Conservação da Biodiversidade Marinha*, or Evaluation, Monitoring and Conservation of Marine Biodiversity), established on 14 September 2005, is coordinated by the CIRM Secretariat (SECIRM) and supported by the Ministry of Environment. REVIMAR's role is to evaluate, monitor, and promote the conservation of marine biodiversity by means of an ecosystem-based approach, aiming at the establishment of a scientific basis and integrated actions that can inform policy, strategies, and conservation actions for the sustainable use of living resources.

Table 7.4 Main ongoing programmes from the National Policy for Sea Resources

Abbreviation/Acronym	Programme
REVIMAR	Evaluation, Monitoring and Conservation of Marine Biodiversity
AQUIPESCA	Action for Aquaculture and Fisheries
REMPLAC	Programme for Evaluation of Mineral Potential of the Brazilian Legal Continental Shelf
GOOS/Brazil	Brazilian System of Ocean Observing and Climate
PROARQUIPÉLAGO and PROTRINDADE	Programme for Scientific Research in the Oceanic Islands
BIOMAR	Action for Marine Biotechnology
PPG-MAR	Capacity Building of Human Resources in Marine Sciences
PROAREA	Programme for the Prospecting and Exploration of Mineral Resources in the International Area from South and Equatorial Atlantic
PROMAR	Programme for Promotion of a Maritime Mentality
PROANTAR	Brazilian Antarctic Programme

Source: Adapted from the PNRM

In recent years, Brazil, through the Ministry of Environment (*Ministério do Meio Ambiente*, MMA) and Brazilian Institute of Environment and Renewable Resources (*Instituto Brasileiro do Meio Ambiente e dos Recursos Naturais Renováveis*, IBAMA), has been promoting initiatives for the integrated and participatory management of its marine zone, seeking, among other goals, to increase knowledge on population dynamics and assessment of fish stocks. Among the initiatives, REVIZEE (*Programa de Avaliação do Potencial Sustentável de Recursos Vivos na Zona Econômica Exclusiva*, or Programme for the Study of Sustainable Potential of Live Resources Capture of the Exclusive Economic Zone), which began in 1995 and officially ended in September 2006, can be highlighted and considered to be the largest integrated effort undertaken in the country to survey the potential sustainable catch of living resources in the EEZ (CIRM, 2012).

AQUIPESCA (*Ação Aquicultura e Pesca*, or Action for Aquaculture and Fisheries) is coordinated by the Brazilian Navy and the Ministry of Fisheries and Aquaculture. It seeks to articulate, in an inter-ministry cooperative environment, the implementation of priority actions in the Plan for Sustainable Development of Fisheries and Aquaculture, in order to qualify fishermen, to adjust fishing effort, and to encourage mariculture. Its main objective of developing and promoting national fisheries and aquaculture production in a sustainable manner is based on the strategic need to promote sovereign exploitation of fishery resources in the country's EEZ (AQUIPESCA, 2011).

The Programme for Assessment of Mineral Potential of the Legal Brazilian Continental Shelf (*Programa de Avaliação da Potencialidade Mineral da Plataforma Continental Jurídica Brasileira*, REMPLAC) is coordinated by the CIRM and the Ministry of Mines and Energy. The programme was established by CIRM Resolution No. 004 on 3 December 1987. This programme aims to make the basic systematic geological and geophysical survey of the continental shelf detail, in appropriate scale, sites of geo-economic interest, and perform the analysis and evaluation of mineral deposits. From mid-2005, the programme received a financial boost with high priority assigned to it by the Department of Geology, Mining and Mineral Processing of the Ministry of Mines and Energy (MME) and the Directorate of Geology and Mineral Resources of the Geological Survey of Brazil (CPRM) (CIRM, 2011a).

GOOS/Brazil (Brazilian System of Ocean Observing and Climate) is coordinated by the Brazilian Navy, with executive coordination by the SECIRM. The CIRM approved the GOOS/Brazil Pilot Project on 30 April 1997. The Brazilian System of Ocean Observing and Climate is part of the Global Ocean Observing System (GOOS), which was established by the Intergovernmental Oceanographic Commission (IOC) of the United Nations Educational, Scientific and Cultural Organization (UNESCO), in cooperation with the World Meteorological Organization (WMO) and the United Nations Environment Programme (UNEP), in consideration of the provisions of UNCLOS and Agenda 21 (GOOS Brazil, 2011).

The Brazilian Oceanic Islands Programme comprises PROARQUIPELAGO (*Programa Arquipélago de São Pedro e São Paulo*, or Archipelago Programme of São Pedro and São Paulo) and PROTRINDADE (*Programa de Pesquisas Científicas na Ilha da Trindade*, or Scientific Research Programme at Trindade [and Martin Vaz Island]). PROARQUIPELAGO is coordinated by the Brazilian Navy and established by CIRM Resolution No. 001/96 of 11 June 1996, with the main objective of ensuring the continued habitability of the remote region of the archipelago, which provides the country with an additional EEZ of 450,000 km^2 (CIRM, 2012).

PROTRINDADE is coordinated by the CIRM and was established by CIRM Resolution No. 3 of 15 May 2007. The programme aims to:

- construct and maintain, with the consent of the Brazilian Navy, accommodation facilities for researchers carrying out scientific research at the existing facilities on the island of Trindade;
- provide the means and support needed for the transportation of resident researchers; and
- promote and manage the development of scientific research in the Trindade and Martin Vaz islands and adjacent sea area (CIRM, 2011b).

BIOMAR (*Biotecnologia Marinha*, or Programme for the Survey and Evaluation of Biotechnological Potential of Marine Biodiversity) is coordinated by SECIRM, with the support of the Ministry of Science, Technology and Innovation. The programme was established by Brazilian Navy Decree on 14 September 2005. BIOMAR aims to promote the sustainable use of the biotechnological potential of marine organisms in coastal and transitional ocean areas under national jurisdiction. It also focuses on the development of knowledge, technology transfer, and innovation in the human, environmental, agricultural, and industrial sectors (BIOMAR, 2011).

The PPG-Mar (*Programa de Pós-graduação em Ciências do Mar*, or Research and Graduate Programme in Marine Sciences) is coordinated by SECIRM and has the strong participation of several universities. The programme was established by Decree No. 232 of 14 September 2005. The PPG-Mar aims to strengthen the building of human capacity to promote knowledge about the components, processes, and resources of marine environments and transition zones (Krug, 2012).

PROAREA (*Programa de Prospecção e Exploração de Recursos Minerais da Área Internacional do Atlântico Sul e Equatorial*, or Programme for the Prospecting and Exploration of the Mineral Resources from International Areas of South and Equatorial Atlantic) is coordinated by the CIRM/SECIRM, together with the Ministry of Foreign Affairs and the Geological Service of Brazil. The programme was established by CIRM Resolution No. 003 of 16 September 2009, and aims to identify and assess the mineral potential of areas located beyond the limits of national jurisdiction with economic, political, and strategic importance to Brazil. PROAREA is expected to become an important way in which to increase Brazilian presence in the south and equatorial Atlantic, and the Brazilian position in the international arena. PROAREA activities are divided into the following projects:

- the creation of a database for integration and systematization of information regarding the mineral resources in maritime space;
- the evaluation of the mineral potential of cobaltic crusts of the Rio Grande Rise; and
- the geological survey and assessment of the biotechnological potential of the hydrothermal deposits of mid-ocean ridge in the South and Equatorial Atlantic.

The regions covered by PROAREA are characterized by the presence of iron, copper, zinc, gold, silver, lithium, and silicon (CIRM, 2011c; 2011d). These projects are linked to the formulation of the Brazilian proposal for exploration and exploitation of deep seabed minerals for submission to the International Seabed Authority (ISBA). The next step in this programme will be the exploration of polymetallic sulphates along the Mid-Atlantic Ridge and near the Archipelago of São Pedro and São Paulo, which was approved in 2010.

PROMAR is the Programme for Promotion of a Maritime Mentality, coordinated by the Brazilian Navy and SECIRM. This programme aims to stimulate, through continuing actions, the development of a maritime mentality among the Brazilian population. PROMAR deals with issues involving greater knowledge of the ocean and its resources, the importance of the ocean to Brazil, the responsibility of rational and sustainable exploitation, and awareness of the need for preservation (Krug, 2012).

PROANTAR (*Programa Antártico Brasileiro*, or Brazilian Antarctic Programme) was created in January 1982. In early December 1982, Brazil made a hydrographic, oceanographic, and meteorological expedition in areas within the northwest sector of Antarctica and selected the location where it would install the future Brazilian Station. The success of Operation Antarctica I resulted in international recognition of the Brazilian presence in Antarctica allowing, on 12 September 1983, the acceptance of Brazil as a consultative party in the Antarctic Treaty. On 6 February 1984, the Brazilian Antarctic Station 'Comandante Ferraz' (EACF) was established on King George Island, 130 km from the Antarctic Peninsula (CIRM, 2012).

The National Plan of Coastal Management (PNGC)

Law No. 7.661 of 16 May 1988 established the National Coastal Management Plan (PNGC) as part of the National Policy for Sea Resources (PNRM) and the National Policy on the Environment (*Política Nacional do Meio Ambiente*, PNMA). The PNGC establishes the principles that lead to coastal management concepts and definitions, objectives and rules, as well as instruments, competences, and resources. However, although it defines some aspects of management, the PNGC cannot be characterized as the only juridical norm that guides government and citizens' actions in the coastal zone.

On 18 April 1990, Decree No. 99.213 created the Coordination Group of Coastal Management (COGERCO). On 27 June 1995, however, this decree was cancelled by Decree No. 1.540, which set out COGERCO's current constitution. The updated and self-applicable plan was submitted for consideration to the CIRM, which was responsible for approving it in conjunction with the National Council for the Environment (CONAMA), a unit of the Environment Ministry. Following this determination, it was approved by CIRM Resolution No. 1 of 21 November 1990. The PNGC was superseded by the revised National Plan of Coastal Management (PNGC II), following the acceptance of CIRM Resolution No. 5 of 3 December 1997. PNGC II was promulgated after it was submitted to CONAMA and approved by CIRM Resolution No. 5. As a result, coastal management, as a whole, now has a more focused methodology in plans and action strategies that involve the nation, states, and counties (PNGC II, 1997).

The new plan contains principles that will direct coastal management, as well as norms and rules for its implementation in states and counties. It should be noted that PNGC II, like the PNGC, does not establish the zoning of uses and activities, nor does it determine specific measures to be adopted or analysed in the coastal zone. Such specifications are the responsibility of the other policies to be established by the nation, states, and counties. PNGC II may be regarded as a 'way of doing things' that contemplates actions and instruments capable of minimizing the existing use conflicts on the Brazilian coast.

One of the most important aspects of the plan is the selection of a management process that is integrated, non-centralized, and participatory. This procedure theoretically focuses on community participation in the decision-making process, setting non-centralized management as a target to be reached. Integration is characterized as a link between all of the steps of the action plan, and induces a new period of democratic dynamism between the public, government, and the other segments of society. Through this process, the division of responsibilities and tasks is made possible, and favours the involvement of the community in the process of regional development.

PNGC II establishes that development should have the support of three levels of government: federal, state, and county. States will have autonomy in the execution of regional proposals, as long as these are consistent with the principles of the national plan. The county, as a consequence, will execute the state proposal, although it is the responsibility of the state to indicate emergency areas to address. The Municipal Plan of Coastal Management (*Plano Municipal de Gerenciamento Costeiro*, PMGC), once legally established, should explain the work of the PNGC and the State Plan of Coastal Management (*Plano Estadual de Gerenciamento Costeiro*, PEGC), with the aim of implementing the Municipal Plan, including the responsibilities and the institutional procedures for its execution. The PMGC should have a close connection with the use and territorial occupation plans and other pertinent measures to county planning. PNGC II makes it clear that it is important to plan actions through definition of priorities and elaboration of annual operative plans at the federal, state, and city levels.

However, what seems to be decentralization from the central government is, in fact, also the division of social and sectorial responsibilities. The decentralization means a change of competences in many respects, especially social, but overall a transfer of autonomy and real decision-making power to the states, and more specifically to the municipalities. As a consequence of these transformations, local governments have an expanded presence in the state apparatus and in the definition of public policies. The presence of the municipality is reinforced as the best option for putting in place a management process. This is justified because it is the cities that can better gather the efforts in favour of local sustainability. If each city government were able to organize and develop its own management plan according to the national plan, the exchange of information and solutions at the state level would be made easier and optimized. Further, community involvement would become more prominent, paving the way for more regional analysis and proposals (Marroni and Asmus, 2005).

Presently, public government does not have the management capacity to provide immediate and effective answers that minimize problems at the local level. This limited capacity is a product of the horizontal authority among different parts of the government (secretaries, departments, and other sectors). Also, national support organizations have a double function in that they finance and provide technical assistance to most cities. For this reason, the decentralization of initiatives and decisions is necessary.

Because cities face scarcity of resources, in both economic and management capacity, as well as minimal levels of community intervention, actions should be planned that aim to use the available resources efficiently. PNGC II can be viewed as a continuous and sustainable process in which various social, political, and governmental sectors interact in relation to a common project with a potential for overcoming conflict. A successful action plan has the potential to reinforce the credibility of the governors, to facilitate the viability of integrated management, and to contribute to the strengthening of a democratic culture.

The Federal Action Plan for the Coastal Zone of Brazil (PAF)

A year after the establishment of PNGC II, CIRM Resolution No. 05/98 created the Federal Action Plan for the Coastal Zone (*Plano de Ação Federal Para a Zona Costeira do Brasil*, PAF). The PAF is an action instrument for setting a benchmark for programmatic activities of the state in the coastal territories. It aimed to articulate the activities and actions of the nation regarding the PNGC. To make this initiative possible at the federal level, PNGC II created a permanent forum of inter-institutional discussion in the country: the Group of Integration of the Coastal Management (GI–GERCO). This unit is coordinated by the CIRM, with the mission of promoting the articulation of current federal actions in the coastal zone from the approved PAF (CIRM, 2008).

The need to ensure that PNGC actions were compatible with sectorial public policies concerning the coastal zone led to the elaboration of a document oriented toward the various activities of the federal government concerning the Brazilian coast: the PAF. The aim of this initiative is to ensure that integration meets sustainability parameters. The concept of PAF, which establishes partnerships and articulates work on the national scale among the regional units in order to integrate actions and optimize results, was proposed by GI–GERCO to resolve these issues (MMA, 2008b). Box 7.3 sets out the objectives of PAF regarding the evolution of these relationships of cooperative action and institutional partnership projects.

Box 7.3   Objectives of PAF

- To orient the federal government actions in the coastal zone by means of participative planning and integrated implementation
- To identify opportunities for the optimization of established capacity
- To promote institutional cooperation
- To promote the development of strategic actions for the articulation of public policies that occurs in the coastal zone

Source: Adapted from Asmus et al. (2004)

The structure of PAF programmes was outlined in specified lines of action. From criteria adopted to prioritize actions of the plan, some priority, converging, and complementary actions were identified and grouped into four programme areas, with respective lines of action.

In 2005, CIRM Resolution No. 7 established the revision of PAF. The revised PAF included new lines of action, which were classified into three main blocks: territorial planning of the coastal zone; conservation and protection of natural and cultural heritage; and monitoring and control.

For implementing the requirements of each action line, projects were defined, with specific focus, goals, areas of expertise, activities, and institutional arrangements. The strategic basis for the implementation of PAF involved strengthening inter-sectorial coordination of the components in GI-GERCO, attention to regional demands, and advances in the regulatory and standardization processes in the coastal zone (CIRM, 2005).

Evaluation, outcomes, and long-term outlook

The analysis of Brazilian ocean policy should be considered within the country's historical and political context. The establishment of the National Policy for Sea Resources (PNRM) in 1981 occurred at the end of a military government regime characterized by strong nationalist tendencies. This characteristic, associated with the preoccupation of Brazil with participating in UNCLOS, directed its oceans policy towards geopolitical and national security interests. At that time, there was a strong preoccupation in the country with evaluating and guaranteeing the right for exploration and possession of its marine resources, and establishing internationally accepted limits of its territorial sea. There was not, however, a national awareness or concern in exploiting its marine resources in a sustainable way, and even less in community involvement through democratic participation in the processes of managing those resources. The first Brazilian policy for oceans can be considered to be a national landmark that defines its marine territory in the context of UNCLOS.

The historical and evolutionary aspect of the PNRM is well characterized by the way in which it has been reviewed and implemented through successive Sectorial Plans for Sea Resources (PSRMs) (presently in their eighth edition). Such evolution demonstrates the transition of Brazil's political landscape and its national priorities since the early 1980s. New principles have gradually been incorporated into the most recent plans. Examples of these principles are the integrated focus on sustainability in establishing targets of exploration of marine resources and an increasing national preoccupation with preserving marine and coastal biodiversity. Likewise, the principles of the National Plan for Coastal Management (PNGC), such as coastal management, based on the characteristics of ecosystems and involvement of local communities, have been included in its most recent version.

An example of the evolution of the basic principles of Brazilian ocean policy is the involvement of institutions that elaborate, implement, and evaluate it. In this respect, the role of the Inter-Ministry Commission of for Sea Resources (CIRM) in the coordination and implementation of policy is important. The CIRM, in addition to representing the various ministries that form it, is presided over by the Brazilian Navy through its Secretariat (SECIRM). SECIRM is a unit with a traditional role related to the integration of Brazilian territory rather than aspects of sustainable development of the ocean. This decision by the Brazilian government demonstrates that it was more focused on geopolitical principles than conservation during the establishment of its national ocean policy. The CIRM, however, played an important role in the following years as a coordinating unit for marine policy and the various plans related to it. On the other hand, implementing the many sectorial plans of the policy involves an increasing variety of institutions with activities predominantly comprising environmental research and management. These institutions include the federal and state agencies for environmental control, universities and research institutes, and other representative groups in civil society. In this second phase, principles connected to community participation and sustainable development were gradually incorporated.

The evolution of Brazil's national ocean policy is summarized in Table 7.5.

Recently, there have been new concerns and trends arising in Brazilian ocean policy. They represent a growing national interest for the ocean areas within and beyond national jurisdiction.

Table 7.5 Evolution of Brazil's national ocean policy

CIRM 1974	CIRM 2001
National Policy for Sea Resources (PNRM)	
First version (1980)	Revised version (2005)
Main purpose: To outline plans and programmes for Brazilian maritime space in accordance with international agreements	*Main purposes*: To offer a new international perspective on prospecting, exploring, and exploiting natural resources from the sea
	To meet the need to review the principles and objectives of PNRM 1980
Sectorial Plans for Sea Resources (PSRMs)	
I PSRM (1982–85)	V PSRM (1999–2003)
II PSRM (1986–89)	VI PSRM (2004–08)
III PSRM (1990–93)	VII PSRM (2008–11)
IV PSRM (1994–98)	VIII PSRM (2011–15)
Plans, programmes, and related actions	
1982	PROANTAR	Antarctic Programme (Brazilian Navy/CIRM)
1987	REMPLAC	Programme for Assessment of Mineral Potential of the Legal Brazilian Continental Shelf (CIRM/MME)
	GERCO	National Programme of Coastal Management (CIRM/Ministry of Environment)
1988	PNGC	National Plan of Coastal Management
1989	LEPLAC	Plan for the Survey of the Brazilian Continental Shelf (CIRM/Ministry of Environment)
1990 (Final version)	PNGC	National Plan of Coastal Management (CIRM/Ministry of Environment)
1996	PROARQUIPELAGO	Archipelago Programme of São Pedro and São Paulo (Brazilian Navy/CIRM)
1997	PNGC II	Revised National Plan of Coastal Management (CIRM/MMA)
	Programa Train-Sea-Coast	
	GOOS/Brazil (First Version)	Brazilian System of Ocean Observing and Climate (OMI/COI/UNESCO/SECIRM/Institute of Marine Studies Almirante Paulo Moreira, or IEAPM – with the representation of the following ministries: Education; Mines and Energy; Environment; Science, Technology and Innovation; Livestock and Supply)

	REMPLAC	Update to the Programme for Evaluation of Mineral Potential of the Brazilian Legal Continental Shelf (CIRM/MME)
	PROMAR	Programme for Promotion of a Maritime Mentality (Brazilian Navy/SECIRM)
1999	REVIZEE	Programme for the Study of Sustainable Potential of Live Resources Capture of the Exclusive Economic Zone (Brazilian Navy/SECIRM/Brazilian Institute of Environment and Renewable Natural Resources/National Council for Scientific and Technological Development – with the following ministries: Foreign Relations; Education; Science, Technology and Innovation; Livestock and Supply)
2004	GERCO	Update to the National Programme of Coastal Management (CIRM/Ministry of Environment)
2005	BIOMAR	Programme for the Survey and Evaluation of Biotechnological Potential of Marine Biodiversity (CIRM/Ministry of Science, Technology and Innovation)
	PPG-MAR	Research and Graduate Programme in Marine Sciences (SECIRM/Ministry of Education/Federal University of Rio Grande)
	REVIMAR	Evaluation, Monitoring and Conservation of Marine Biodiversity (CIRM/Brazilian Institute of Environment and Renewable Natural Resources)
	AQUIPESCA	Action for Aquaculture and Fisheries (SECIRM/Ministry of Fisheries and Aquiculture)
2007	PROTRINDADE	Scientific Research Programme at Trindade [and Martin Vaz Island] (Brazilian Navy/CIRM)
2009	PROAREA	Programme for the Prospecting and Exploration of the Mineral Resources from International Areas of South and Equatorial Atlantic (CIRM/Ministry of Foreign Relations)
Still running and based on PNRM	PROANTAR GERCO PROMAR REMPLAC LEPLAC GOOS/BRAZIL PROARQUIPELAGO Train-Sea-Coast	(DOALOS/UNESCO/Federal University of Rio Grande/Federal University of Rio de Janeiro/Federal University of Pernambuco)

Source: Marroni (2013)

Such interest can be interpreted in light of aspects involving economic, strategic, and foreign policy, as well as those related to the country's image in the international context. From an economic standpoint, it is undeniable that the increasing exploration and exploitation of oil and gas in deep waters of the Brazilian EEZ has produced a substantial interest in advancing action in terms of the sustainable use of these resources. This has produced some reflections on policy through, for example, the incorporation into PSRMs of some clear indications on the need for the country to move forward with scientific and technological innovations related to exploratory and economic activities in deep waters.

The gradual accumulation of knowledge on the Brazilian continental shelf and continental margin in recent decades has allowed Brazil to reassess their limits; in 2004, it presented a petition to the UN Commission for the Limits of the Continental Shelf (CLCS) regarding an extension of its continental shelf area. Such a request would expand the Brazilian EEZ by 950,000 km², reaching the total of 4,489,919 km² of ocean area, and is referred to in Brazil as the 'Blue Amazon'(Figure 7.3).

Brazilian interest in the open ocean component of policy can also be identified in the operationalization of implementation programmes. PROAREA is an example of the growing involvement of Brazil in areas of economic and strategic interests in marine areas beyond its jurisdiction (CIRM, 2011c). Similarly, the CIRM has recently established a new working group focused on the study of shared use of the Brazilian ocean, with two thematic subgroups: legal issues, and marine spatial planning. The expectation is that this new initiative will establish novel guidelines and recommendations for future ocean policy in Brazil (CIRM, 2013).

Recommendations for improving the national oceans policy

There are recommendations that can be made for improving the oceans policy of Brazil. Aspects to be improved include organizational character, implementation processes, community involvement, and government priorities.

It is not easy to have a clear idea of the hierarchy of the national policy for the ocean in Brazil. It is formally established and structured with a legal basis, but the instruments that define and implement ocean policy sometimes do not fall into a hierarchical structure. It is common practice that a public policy is implemented through specific or multi-thematic programmes. In the case of the National Policy for Sea Resources (PNRM), there are incongruences about its hierarchical organization, which raises questions when it is analysed or when one tries to use it to contextualize the actions regarding desired uses of the Brazilian ocean coast – and this indicates the need for better normative instructions from the federal government in order to define its hierarchical structure clearly.

An important component in any analysis of the PNRM is to highlight the need for a larger and more logical documentation of its legal, normative, and operational aspects. It would be very useful if all of the official documents about PNRM were to be centralized and available in a governmental unit such as the CIRM.

Concerning its implementation, there is a general feeling in the oceans community (especially scientific) that the Brazilian ocean policy has not been totally implemented. It is, in fact, difficult to achieve a more objective and perceptive evaluation of progress achieved as a result of the fact that targets and success indicators are rarely taken into consideration in the implementation of its plans and programmes. The use of social, economic, and ecological indicators as elements of evaluation of the implementation of marine policies – and, more specifically, for coastal management – has been found useful for the establishment of sustainable development in coastal zones (Murawski et al., 2008). In Brazil, unfortunately, such indicators have not been

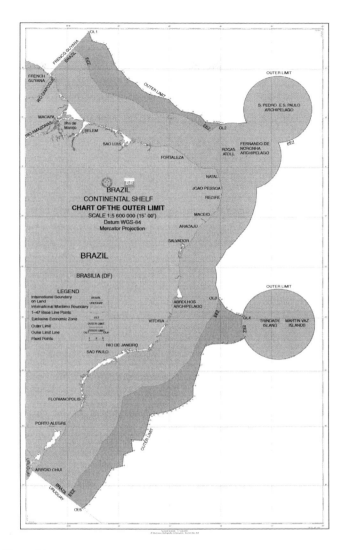

Figure 7.3 Brazil's maritime space

Source: Government of Brazil (2013)

well established, although there are official plans to implement them through the Environmental Quality Report (one of the instruments of the PNGC). It is important that the Brazilian government, along with the universities and research institutes, establish suitable indicators for adequate evaluation of its policies, plans, and programmes.

When considering the conceptual picture and the general process of implementation of Brazilian ocean policy, it is clear that its development has been characterized by a very official (or predominantly government) 'top–down' style (Asmus et al., 2004), although to a lesser extent now than historically. It would be beneficial if there were increasing participation of universities and organizations from civil society in the whole process of development and application of the policy in the future. This process can be participatory, especially during the identification and analysis of major issues and problems, providing opportunities for all stakeholders to shape the policy and participate in the implementation of strategic actions after the policies are put in place. In addition, increased community participation and participation of qualified research institutions would tend to decentralize and regionalize Brazilian policy, which would in turn increase efficiency in its implementation. Similarly, in a decentralized system, it becomes much easier to monitor the policy outcomes. The same levels of progress in biodiversity conservation, for example, can have very distinct meanings in such an extensive and diverse coast as that of Brazil, which indicates that a decentralized system of monitoring would be suitable.

Besides the need for a more participatory approach, it would be beneficial if the new versions of the PNRM were consistently to include the new internationally agreed principles in the management and governance of ocean, islands, and coastal zones. Examples of principles that have being incorporated into Brazilian policy include, among others, the integrated approach, intergenerational and intra-generational equity, participatory governance, and stewardship. Certainly a better application and more inclusion of these principles could produce a new age in Brazilian environmental policy and lead to the social, ecological, and economical sustainability of its marine resources.

There are high expectations among the Brazilian scientific community and society in general that the policies for the ocean and coastal zone will rank higher in the priorities set out in the government's multi-annual plans. Brazil should acknowledge its territorial sea and coastal and ocean zones as priorities by taking concrete and participatory actions to enhance social well-being.

References

Agenda 21 Brasileira (2002) 'Bases para discussão', Brasília, online at http://www.agenda21.org.br/ [accessed 12 November 2013].

Aquicultura e Pesca (AQUIPESCA) (2011) 'CIRM: Comissão Interministerial para os Recursos do Mar', online at http://www.mar.mil.br/secirm/proarqui.htm [accessed 30 September 2012].

Asmus, M. L., Kitzmann, D., and Laydner, C. (2004) *Gestão Costeira no Brasil: Estado Atual e Perspectivas*, Montevideo: Encuentro Regional Cooperación en el Espacio Costero.

Biotecnologia Marinha (BIOMAR) (2011) 'CIRM: Comissão Interministerial para os Recursos do Mar', online at http://www.mar.mil.br/secirm/biomar.htm [accessed 30 September 2012].

Comissão Interministerial para os Recursos do Mar (CIRM) (1980) *Política Nacional Para os Recursos do Mar: Diretrizes Gerais (PNRM)*, Brasília: CIRM.

Comissão Interministerial para os Recursos do Mar (CIRM) (2003) *O Modelo Brasileiro para o Desenvolvimento das Atividades Voltadas para os Recursos do Mar*, online at http://www.secirm.mar.mil.br [accessed 17 September 2013].

Comissão Interministerial para os Recursos do Mar (CIRM) (2005) *Plano de Ação Federal da Zona Costeira do Brasil*, online at http://www.secirm.mar.mil.br [accessed 17 September 2013].

Comissão Interministerial para os Recursos do Mar (CIRM) (2008) *Plano de Ação Federal Para a Zona Costeira do Brasil (PAF)*, Brasília: CIRM/GIGERCO and MMA/SIP/DEMAI/GERCO.

Comissão Interministerial para os Recursos do Mar (CIRM) (2011a) 'Programa de Avaliação da Potencialidade Mineral da Plataforma Continental Jurídica Brasileira (REMPLAC)', online at http://www.mar.mil.br/secirm/remplac.html [accessed in 30 September 2012].

Comissão Interministerial para os Recursos do Mar (CIRM) (2011b) 'Organização da CIRM', online at http://www.mar.mil.br/secirm/cirm-org.htm#criacao [accessed 11 October 2012].

Comissão Interministerial para os Recursos do Mar (CIRM) (2011c) *Programa de Prospecção e Exploração de Recursos Minerais da Área Internacional do Atlântico Sul e Equatorial (PROAREA)*, online at http://www.cprm.gov.br/geomar/digeom/proarea.pdf [accessed 30 September 2012].

Comissão Interministerial para os Recursos do Mar (CIRM) (2011d) 'PROAREA Realiza Coleta de Rochas na Elevação de Rio Grande', *INFOCIRM*, 23(2): 2.

Comissão Interministerial para os Recursos do Mar (CIRM) (2012) *Atuação da CIRM no Mar e na Antártica*, online at http://www.mar.mil.br/secirm/ [accessed 15 March 2014].

Comissão Interministerial para os Recursos do Mar (CIRM) (2013) 'Resolução No. 001/Uso Compartilhado do Ambiente Marinho', online at http://www.mar.mil.br/secirm/p-ataseresolucoes. html [accessed 12 March 2014].

Constituição da República Federativa do Brasil (1988) 'Senado Federal, Brasília', online at http://www. senado.gov.br/legislacao/const/con1988/CON1988_13.07.2010/index.shtm [accessed 10 October 2004].

Fernandes, L. F. C. (ed.) (2013) *O Brasil e o Mar no Século XXI: Relatório aos Tomadores de Decisão do País*, Brasília: CEMBRA.

Fórum do Mar (2012) 'O Que É?', online at http://www.forumdomar.org.br/page.php?p=7 [accessed 28 July 2014].

GOOS/Brazil   *See* Sistema Brasileiro de Observação dos Oceanos e Clima

Government of Brazil (2013) 'LEPLAC: Plano de Levantamento da Plataforma Continental Brasileira', online at http://www.mar.mil.br/dhn/dhn/quadros/ass_leplac.html [accessed 12 November 2013].

Instituto Brasileiro de Geografia e Estatística (IBGE) (2014) *Brasil em Síntese*, online at http://brasilemsin-tese.ibge.gov.br [accessed 17 March 2014].

Krug, L. C. (ed.) (2012) *Formação de Recursos Humanos em Ciências do Mar: Estado da Arte e Plano Nacional de Trabalho – VIII Plano Setorial para os Recursos do Mar*, Pelotas: Editora Textos.

Marinha do Brasil (2011) 'Vertentes da Amazônia Azul', online at http://www.mar.mil.br/menu_v/ama-zonia_azul/html/vertentes.html [accessed 21 November 2011].

Marroni, E. V. (2011) *O Pré-Sal e a Soberania Marítima do Brasil: Uma Nova Geopolítica para o Atlântico Sul*, Anais do IV Seminário Nacional de Ciência Política – Teoria e Metodologia em Debate, Programa de Pós-Graduação em Ciência Política da Universidade Federal do Rio Grande do Sul (UFRGS), online at http://www.ufrgs.br/sncp/4SNCP/GT_PolIntern/EtieneVillela.pdf [accessed 15 October 2012].

Marroni, E. V. (2013) *Política Internacional dos Oceanos: Caso Brasileiro Sobre o Processo Diplomático para a Plataforma Continental Estendida – Tese de Doutorado*, Programa de Pós-Graduação em Ciência Política, Universidade Federal do Rio Grande do Sul (UFRGS).

Marroni, E. V. (2014) 'The Importance of Public Policy for Blue Amazon Marine Spatial Planning, Brazil', *Development Studies Research*, 1(1): 161–7.

Marroni, E. V., and Asmus, M. L. (2005) *Gerenciamento Costeiro: Uma Proposta para o Fortalecimento Comunitário na Gestão Ambiental*, Pelotas: União Sul-Americana de Estudos da Biodiversidade.

Marroni, E. V., and Asmus, M. L. (2013) 'Historical Antecedents and Local Governance in the Process of Public Policies Building for Coastal Zone of Brazil', *Ocean & Coastal Management*, 76: 30–7.

Merico, L. (2013) 'Desafios da UICN para a Conservação no Brasil', in IUCN-Sur (ed.) *Conservación Ahora*, Quito: IUCN-Sur.

Ministério do Meio Ambiente (MMA) (2006) *Macrodiagnóstico da Zona Costeira e Marinha*, Brasília: MMA.

Ministério do Meio Ambiente (MMA) (2008a) *Brazilian Coastal and Marine Biodiversity*, Brasília: MMA.

Ministério do Meio Ambiente (MMA) (2008b) *Macrodiagnóstico da Zona Costeira e Marinha do Brasil: Instituto Brasileiro do Meio Ambiente e dos Recursos Naturais Renováveis (IBAMA)*, Brasília: MMA.

Murawski, S., Cyr, N., Davidson, M., Hart, Z., Balgos, M., Wowk, K., and Cicin-Sain, B. (2008) 'Policy Brief: Ecosystem-based Management and Integrated Coastal and Ocean Management and Indicators for Progress', Policy Brief on EBM/ICM and Indicators for Progress Prepared for the Fourth Global Conference on Oceans, Coasts, and Islands, Pre-conference version, 30 March.

Plano Nacional de Gerenciamento Costeiro (PNGC II) (1997) *Plano Nacional de Gerenciamento Costeiro II (PNGC II)*, online at http://www.mma.gov.br/estruturas/sqa_sigercom/_arquivos/pngc2.pdf [accessed 15 October 2012].

Plano Setorial Para os Recursos do Mar (VIII PSRM) (2012) 'Sumário Executivo', online at http://www. mar.mil.br/secirm/document/doc_psrm/sum_viiipsrm.pdf [accessed 10 January 2013].

Poleti, A. E., Filet, M., Sanches, M., and Scherer, M. E. G. (2012) 'Da Agência Costeira ao Fórum do Mar: Um Instrumento de Integração da Gestão Costeira e Oceânica', in I Congreso Iberoamericano de Gestión Integrada de Áreas Litorales, Cadiz, *Libro de Comunicaciones y de Posters*, Cadiz: Universidade de Cadiz.

Scherer, M., Asmus, M. L., Filet, M., Sanches, M., and Poleti, A. E. (2011) 'El Manejo Costero en Brasil: Análisis de la Situación y Propuestas para una Posible Mejora', in J. Farinós (ed.) *La Gestión Integrada de Zonas Costeras: Algo Más Que una Ordenación del Litoral Revisada?*, Valência: IIDL/Publicacions de la Universitat de València.

Sistema Brasileiro de Observação dos Oceanos e Clima (GOOS/Brazil) (2011) 'CIRM: Comissão Interministerial para os Recursos do Mar', online at http://www.mar.mil.br/secirm/goos.htm [accessed 30 September 2012].

World Bank (2014) 'Brazil Overview', online at http://www.worldbank.org/en/country/brazil/overview [accessed 12 March 2014].

8

JAMAICA'S OCEANS AND COASTAL POLICY

Laleta Davis Mattis and Peter Edwards

Introduction

Over the past thirty years, there has been much discussion on a global scale relating to integrated coastal zone management (ICZM). The call for a comprehensive regime to manage coastal and oceans resources has not gone unheeded. In fact, coastal states and regions have risen to the challenge, and recognized that dependence on coastal resources must result in their protection and proper management. The recognition that coastal states and regions are part of whole ecosystems, and that there is a need for holistic management that is cognizant of the interconnectedness of people and resources, has been catalytic in prompting nations to move towards ICZM. For small island developing states (SIDs), the application of ICZM is all the more critical owing to their peculiar vulnerabilities and the vicissitudes associated with being so classified. For purposes of this chapter, SIDs can be characterized as low-lying coastal countries that share similar sustainable development challenges. These challenges include a small population, a lack of resources, remoteness, susceptibility to natural disasters, an excessive dependence on international trade and vulnerability to global developments, and a lack of economies of scale (DESA, undated). Jamaica is a small island developing state.

In May 1994, the island states of the Caribbean adopted the Barbados Programme of Action (UNGA, 1994). Following from Agenda 21,[1] the Global Conference on the Sustainable Development of Small Island Developing States, held in Bridgetown, Barbados, on 25 April–6 May 1994, was the first conference that sought to translate the provisions of Agenda 21 into a programme of action. Small island developing states share with all nations a critical interest in the protection of coastal zones and oceans against the effects of land-based sources of pollution (UNGA, 1994: Pt I, para. 3(3)). The Barbados Programme of Action provides that:

> Sustainable development in Small Island Developing States depends largely on coastal and marine resources, because their small land area means that those States are effectively coastal entities. Population and economic development, both subsistence and cash, are concentrated in the coastal zone. The establishment of the 200-mile exclusive economic zone has vastly extended the fisheries and other marine resources available to them. Their heavy dependence on coastal and marine resources emphasizes the need for appropriate and effective management.
>
> *(UNGA, 1994: Pt IV, para. 25)*

This chapter explores Jamaica's recent attempts at ICZM. Despite the changes in nomenclature regarding oceans management, throughout this chapter the phrase 'integrated coastal zone management' is used interchangeably with 'integrated coastal and oceans management'.

The setting

Jamaica is the third largest island of the group of islands in the Caribbean known as the West Indies. The West Indian islands are actually the summits of a submarine range of mountains, which, in prehistoric times, may have formed one large land mass connecting Central America to Venezuela in South America (Morris, undated). Jamaica is located within the Greater Antilles at latitude 18°21'23"N and longitude 78°20'43"18W. Jamaica is flanked by Cuba approximately 145 km to the north, Haiti approximately 161 km to the east, Central America to the west, and South America to the south. Jamaica's terra firma is approximately 10,831 km² in area. Jamaica is the largest of the English-speaking West Indian islands. It has a coastline approximately 895 km in length (Earthtrends, 2003). It lies on the direct sea routes from the United States and Europe to the Panama Canal. Jamaica belongs to the Central American region of the Western Hemisphere. It comprises one main island and some sixty small rocks, islets, and cays, generally situated about 60 miles south of the main island. In accordance with the 1982 United Nations Convention on the Law of the Sea (UNCLOS), Jamaica declared itself an archipelagic state in 1996 by virtue of the Maritime Areas Act, 1996. Straight baselines connect the main island to a rock, Southwest Rock, and to a series of small cays. The Pedro Banks, an area of shallow seas extending generally east to west for over 160 km, lie southwest of Jamaica. A cluster of cays are associated with the Banks. To the southeast lie the Morant Cays, 51 km from Morant Point, the easternmost point of Jamaica (US Library of Congress, undated). The Cays and Southwest Rock are, in turn, connected by straight lines to Blower Rock.

The highest area in Jamaica is that of the Blue Mountains. These eastern mountains are formed by a central ridge of metamorphic rock running northwest to southeast, from which many long spurs jut to the north and south. For a distance of over 3 km, the crest of the ridge exceeds 1,800 metres. The highest point is Blue Mountain Peak, at 2,256 metres. The Blue Mountains rise to these elevations from the coastal plain in the space of about 16 km, producing one of the steepest general gradients in the world (US Library of Congress, undated). The climate of Jamaica is mainly tropical, with the most important climatic influences being the northeast trade winds and the island's orographic features (mainly the central ridge of mountains and hills) (CSGM, 2012: 2–3).

The coastline of Jamaica is one of many contrasts. The northeast shore is severely eroded by the ocean. There are many small inlets along the rugged coastline, but no coastal plain of any extent. A narrow strip of plains along the north coast offers calm seas and white-sand beaches. Behind the beaches is a flat raised plain of uplifted coral reef. The southern coast of the island, for the most part, has a wider coastal plain that stretches inland. Geogenic sediments typically comprise the beaches on the south coast, the majority being black sand and pebble beaches. In many instances, these beaches are backed by cliffs of limestone where the plateaus end.

As of July 2014, Jamaica had a population of approximately 2.9 million, most of which lives along the coast (CIA, undated). Approximately two-thirds of the island's population is concentrated in major towns and cities situated along the coast. Historical factors, including a reliance on exports, necessitated a reliance on the sea. Jamaica's location made it strategic as a trans-shipment port; thus shipping, as an industry, has been a historical imperative. Vocations such as fishing also led to a dependence on the use of coastal resources. More recently, tourism emerged as a significant foreign-exchange earner, and this is an industry that relies almost solely on the coastal resources of sea and land. Land-use development along the coast required more regulation, because tourism development was focused along the coasts.

Jamaica: Coastal development issues by coastal region

In Jamaica, like most coastal states, there is a heavy dependence on the use of coastal and oceans resources. Correspondingly, there is a great diversity in uses and use patterns, and an attendant diversity in the number of users. Jamaica's coastal and oceans resources include wetlands and mangroves, swamp forests, estuarine areas, cays, coral reefs, coastal lagoons, and seagrass beds. These ecosystems provide the economic base for activities such as tourism, fisheries, shipping, industries, and housing. The sad reality is that these coastal and ocean resources face the prospect of depletion and degradation, thereby undermining the very economic base on which Jamaicans depend.

The east coast

The coastal area of Kingston and St Andrew stretches from the Fresh River to the west to Bull Bay (Nine Miles) to the east, and includes the Kingston waterfront and the Palisadoes Peninsula on which the Norman Manley International Airport and Port Royal are located. Existing development in the area includes a variety of commercial and industrial activities with competing yet complementary uses, for example an electricity-generating power plant, an oil refinery, a port and container terminal, fishing beaches, and housing (National Council on Ocean and Coastal Zone Management, 2001: 75). Kingston Harbour is the most heavily used port, offering several facilities for docking and berthing, in addition to the container terminal for trans-shipment operations. Kingston Harbour and Hunts Bay act as receiving water bodies for both industrial and domestic wastes.

Fishing is a vibrant economic activity along this coast. There are public recreational beaches as well. However, owing to the effects of pollution from industry and sewage, recreational bathing is limited in the Kingston Harbour area. Other formal recreational activities include boating and yachting (National Council on Ocean and Coastal Zone Management, 2001: 75).

In the parish of St Thomas, agriculture is the principal land activity, and crops cultivated include coconuts, sugar cane, and bananas. Smaller areas are used for intensive mixed farming and cattle grazing. There are five public bathing beaches and fourteen fishing beaches, in addition to manufacturing and food processing facilities, along the coast.

The parish of Portland is noted for high-end tourism and is probably one of the tourism parishes in which coastal development has progressed on a more conservative path. Recently, however, a marina was constructed to attract small craft and to boost the tourism industry in this part of the island. The east coast of Jamaica is home to one of Jamaica's three Ramsar sites,[2] the Palisadoes–Port Royal, which was designated on 22 April 2005 (Ramsar, 2014).

The north coast

The north coast is noted for its vibrant tourism industry, and offers a wide range of facilities and activities. Hotels, for the most part, are located directly along the shoreline. St Ann is the prime tourist resort area on the island. Ports and harbours accommodate both cargo and cruise ship piers. The Master Plan for Tourism Development identifies Ocho Rios and Montego Bay as resort centres (with more than 2,000 rooms each), along with Negril (Richards, 2008: 12).

As Jamaica focuses on tourism as a major foreign-exchange earner, the obligation to balance development and the sustainable management of natural resources becomes more apparent – especially in the context of a climate in which hotel development has the coastline as its preferred location. The policies of the government of Jamaica encourage further growth in this sector, with targets of a 5.5 per cent per annum increase in stopover visitors, a 8.4 per cent per annum increase in visitor expenditure, and a 4 per cent per annum increase in room stock (Richards, 2008). One of the most significant recent coastal developments on Jamaica's north coast is the construction of the Falmouth cruise ship pier, which saw the reclamation and development of approximately 14 acres of coastal resources and land. The development saw massive remove of coral reefs and other marine flora and fauna. Development is interspersed with marine parks, for example the Montego Bay Marine Park and the Ocho Rios Marine Park.

The south and west coasts

The south coast is characterized by a combination of coastal and non-coastal wetlands. Large portions of the coastal area are also covered by deciduous forest. The Black River, which drains the Morass and is the largest river in Jamaica, is used for waste dilution, especially dunder from sugar factories, and supports an important artisanal fishery, which produces an estimated J$3 million annually (National Council on Ocean and Coastal Zone Management, 2001: 103).

The Black River Lower Morass and the Portland Bight Wetlands and Cays are Ramsar sites, added on 7 October 1997 and 2 February 2006, respectively (Ramsar, 2014).

The west coast of Jamaica has also seen its fair share of coastal development, including both hotel and road infrastructure. Significant growth areas include Trelawney, Hanover, and Westmoreland, particularly with regards to the increase in the number of hotel rooms. On the west coast alone, there are more than 10,000 acres of morass land, the largest part of which is called the Great Morass. The Negril Marine Park is also located along the west coast.

Redefining Jamaica as a coastal state

Under the Exclusive Economic Zone Act, 1991, Jamaica actively sought to address the management of resources within the exclusive economic zone (EEZ). Additionally, pursuant to the Maritime Areas Act, 1996, Jamaica was redefined as an archipelago. This redefinition expanded Jamaica's marine space to include a significant amount of marine area. As may be appreciated, the close proximity of Jamaica to neighbouring states has mandated negotiations for delineation of national EEZs.

To date, Jamaica has concluded delimitation agreements with Cuba and Colombia. The agreement between Jamaica and Colombia creates a joint management regime for the management of resources within that area. A fisheries management agreement is still anticipated. The Jamaica and Cuba delimitation agreement was signed in 1994 and entered into force in 1995; it delineates the EEZ and continental shelf between the two states. The method of delimitation was equidistance, which the parties held to offer an equitable settlement (CARICOM, undated). Delimitation talks with the Cayman Islands, Honduras, and Nicaragua are ongoing. Delimitation negotiations with Haiti are also anticipated. An area of disagreement between Jamaica and Nicaragua is the ownership of South West Rock, located at latitude 16°47'N and longitude 78°12'W. At the initiative of the Council, a lighthouse was erected on this cay as part of Jamaica's claim, which includes traditional rights.

Towards a legislative framework for integrated coastal zone management in Jamaica

In 1976, following the 1972 United Nations Conference on the Environment in Stockholm, Jamaica attempted to address the need for an umbrella environmental management agency. The Natural Resources Conservation Department (NRCD) was formed by merging several existing entities – namely:

- the Watershed Protection Commission established under the Watersheds Protection Act, 1963;
- the Wildlife Protection Committee established under the Wild Life Protection Act, 1991;
- the Beach Control Authority under the Beach Control Act, 1956;
- the Natural Resources Planning Unit;
- the Marine Advisory Committee; and
- the Kingston Harbor Quality Monitoring Committee.

The NRCD was established to protect general environmental quality and, as such, was organized and given additional technical capability in the areas of aquatic resources (water quality, wetlands, and oceanography) and wildlife management. The focus on oceanography and water quality led to some emphasis on the management of the coast and coastal resources. Nonetheless, prior to Jamaica's ratification of UNCLOS in 1983, there was no policy with an integrated approach to the management of coastal and oceans resources. It was only the coming into force of UNCLOS that prompted Jamaica to move beyond traditional concepts of near-shore coastal resources management and to consider the possibility of exercising jurisdictional rights over a 200 nautical mile EEZ. It started to look more critically at its marine resources and potential resources in the wider Caribbean Sea. Supported by Agenda 21 and the Barbados Programme of Action, Jamaica embarked on the path of integrated coastal zone management (ICZM).

The Natural Resources Conservation Authority Act, 1991, recognized the need for a comprehensive integrated approach to environmental protection and sustainable natural resources management. The Act creates a statutory regulatory body, the Natural Resources Conservation Authority (NRCA), and provides for the management, conservation, and protection of the natural resources of Jamaica. The 1991 Act recognizes the concept of an ecosystem approach to the management, conservation, and protection of natural resources. Further, the Act provides for the coordination of the efforts of all agencies and departments of government that carry out any functions similar to those of the Authority. To date, the Authority is the principal agency with responsibility for holistic environmental management, including that of coasts and oceans.

The Authority was given primary management responsibility for coastal areas through the Coastal Zone Management Unit, and was responsible for developing and implementing an ICZM programme. Policy statements on resource use and management, and on the development of criteria, standards, and guideline for ICZM, were the main areas of emphasis. Guidelines were prepared for activities such as: the planning and execution of coastal and estuarine dredging works and disposal of dredged material; the planning, construction, and maintenance of facilities for shoreline protection and enhancement; marinas and small craft harbours; the deployment of benthic structures; and the construction, maintenance, and monitoring of underwater pipelines and cables. The emphasis on policy and legislation therefore, even though referring to resources within the sea, was aimed at the protection of specific resources and paid little or no attention to any integrated approach to the use and management of those resources. The 1995 Jamaica

National Environmental Action Plan refers to ICZM as a management tool that is important to Jamaica. The document suggests that the Jamaican government should develop, within the subsequent two years, a comprehensive coastal zone management plan incorporating all of the necessary measures to facilitate more effective management of the natural resource base (NRCA/PIOJ, 1995: 23–7).

In 1997, the NRCA, focusing on the development of policy and legislation, and the government of Jamaica, with the assistance of the Swedish International Development Agency (SIDA), conscientiously began a series of national consultations for the development of a policy on ICZM. The Coastal Zone Planning/Management Project emphasized an integrated approach to coastal zone management, and produced a coastal planning atlas and a manual for integrated coastal planning and management, prepared by the NRCA in collaboration with the Uppsala University in Sweden (NRCA, 1999: 60).

In 1996, the Jamaica Cabinet endorsed and established an interagency committee to address the competing and complementary issues and players in Jamaica's coastal and oceans resources. The Department of Marine and Aviation was established within the Ministry of Foreign Affairs and Foreign Trade, with the primary objectives of: coordinating the development of an integrated marine policy covering Jamaica's coastal and maritime zones, including the territorial sea and the EEZ; developing policy in relation to the law of the sea and the management of the oceans; coordinating the development of an air policy; and developing an overall national strategy aimed at increasing international transportation services.

The primary participant in the Interagency Advisory Committee included government agencies having jurisdiction over ports, border enforcement, shipping, fisheries, and environmental management. The Committee evolved into the Council for Oceans and Coastal Zone Management in 1998, with the mandate to lead the policy formulation process. With the assistance of the Commonwealth Secretariat, the Marine and Aviation Unit, with the strong support of the NRCA, led the ocean and coastal management policy formulation process.

The effective use and management of coastal and oceans resources is critical to any effort at planning Jamaica's development for this generation and the next. Jamaica recognized the need for the effective management of coastal resources through legislation quite early. Coastal management legislation, in the form of the Beach Control Act, 1956, addressed the need for marine protected areas (MPAs), as well as the preservation of coral reefs and mangroves. Under the Wild Life Protection Act, 1991, corals, mangroves, dolphins, and other forms of marine life are protected. Marine parks regulations also seek to protect the marine environment and to provide a regulatory regime within which the resources within a marine park may be utilized.[3] Development orders address marine areas as one category, which require special planning considerations. The Town and Country Planning (Trelawny Parish) Provisional Order 1980, for example, addresses coastal erosion, pollution, water-borne effluent, quarrying, and sand mining. The Order declares that severe coastal erosion and deleterious alteration to the marine environment by natural and artificial means requires control of development that might affect the coast and coastal waters, and provides for the repair of damage that is evident or imminent. It also refers to the need to effect coastal modification works for shoreline protection. The Order provides that the coast and coastal waters are to be protected against pollution by controlling adjoining development. Development proposals are to be examined, with regard to prevention or control of pollution. Sectoral in origin, the application of these early pieces of legislation provided limited scope for the effective management of Jamaica's coastal and oceans resources.

As noted above, the Interagency Advisory Committee was created in 1996 under the Exclusive Economic Zone Act, 19991, to examine the issue of resource-use licences within the EEZ, as well as broader issues dealing with managing the coastal zone. This Committee evolved into the National Council on Ocean and Coastal Zone Management, with a mandate that included definition of national policy for the effective use and management of Jamaica's coastal and oceans resources. Housed in the Ministry of Foreign Affairs, the terms of reference of the Council include:

- Formulating a comprehensive coastal and ocean development management policy and developing an action plan for the management of Jamaica's EEZ, archipelagic waters, territorial waters and coastal and marine resources
- Becoming the advisory body for coastal and ocean use and development and monitoring the implementation of the policy
- Coordinating training and human resource development initiatives to enable Jamaica to make use of its living and non-living resources and to fulfill its obligations under international conventions relating to the marine environment
- Making recommendations for cooperation with regional and global organizations dealing with marine environmental issues
- Making recommendations for further capacity strengthening in the area of marine management including the development plan for:
  i.   an interim institutional focal point
  ii. an appropriate institution responsible for marine affairs.
  (National Council on Ocean and Coastal Zone Management, 2001: 6–7)

The Council, chaired by the minister with portfolio responsibility for foreign affairs, began the arduous task of formulating a policy to address the management of coastal and ocean resources. It was imperative for the Council to recognize the various regulatory coastal interests, because agencies have mandates under several pieces of legislation (see Table 8.1). Principal among them are the NRCA, the National Environment and Planning Agency (NEPA), and the Town and Country Planning Authority. The Port Authority of Jamaica and the Maritime Authority of Jamaica exercise port state control. The National Solid Waste Management Authority is critical in reducing and eliminating debris from entering the marine environment. These multijurisdictional interests in coastal and ocean resources mandated the creation of a high-level, high-profile body, the chair of which, in the person of the minister, was answerable to the Jamaican Cabinet. Further, this entity needed to comprise critical personnel who, based on their operational capacities, would be able to pool resources and make cooperative decisions for the benefit of Jamaica. Only an organization of this nature could command the attention and respect due to Jamaica's ocean resources. Hence the Council comprises the principal heads of all agencies of government whose activities impact on the coastal zone, members of academia, an environmental non-government organization (NGO), and others with expertise in coastal and oceans issues (see Table 8.2).

The policy formulation process

Development of a regime for the implementation of ICZM begins with an analytical process to set objectives for the development and management of the coastal zone. Integrated coastal zone management should seek to attain a compromise between the broad spectrum of interest groups in the process of setting objectives, planning, and implementation, and aim at balancing those interests in the overall use of the country's coastal zones. The process includes policymaking, enforcement, boundary definitions, zoning, monitoring, and the development of guidelines, standards, appropriate control measures, incentives, and disincentives (Clarke, 1992: 167).

Table 8.1 Government agencies with responsibilities for coastal and ocean matters in Jamaica

Agency	Coastal-related mandate, function and/or activity
Airports Authority of Jamaica	Management of Norman Manley and Sangster International Airports
Attorney General's Department	Preparation of legal instruments
Chief Parliamentary Council	Review of documentation
Civil Aviation Authority	Regulation and control of airports; development of guidelines in the vicinity of airports
Council on Ocean and Coastal Zone Management	Definition of national policy; promotion of coordination of administrative and operational functions; ensuring compliance with international treaties and protocols
Environmental Control Division	Water quality monitoring and testing
Fisheries Division	Fisheries management
Forestry Department	Watershed management
Jamaica Constabulary Force	Enforcement of law and order
Jamaica Fire Brigade	Control of fires (including forest fires)
Jamaica National Heritage Trust	Cultural heritage, including intangible cultural heritage, flora, fauna, buildings, landmarks, and artefacts of historical or archaeological importance
Jamaica Tourist Board	Recreational areas/cruise ship terminals
Marine and Aviation Affairs Department, Ministry of Foreign Affairs	Coordination of the development of marine and aviation policy
Maritime Authority of Jamaica	Marine transportation regulatory agency
Mines and Geology Division	Regulation of mining and quarrying
Ministry of Land and the Environment	Policy related to environmental matters and land use
Ministry of National Security and Justice	Policy on matters of national justice
Ministry of Tourism and Sports	Tourism policy and planning
National Water Commission	Treatment of domestic water supply and wastewater and sewage
Office of Disaster Preparedness and Emergency Management	Natural hazards contingency planning, and disaster response and recovery
Office of the Harbour Master	Regulation and safety of shipping in harbours and ports
Petroleum Corporation of Jamaica	Petroleum shipments
Planning Institute of Jamaica	Initiation and coordination of planning for economic, financial, social, cultural, and physical development
Port Authority of Jamaica	Management of ports and port operations
Scientific Research Council	Pure and applied research
Shipping Association of Jamaica	Provision of skilled labour to shipping industry operations and management of relevant parts of the industry
Tourism Product Development Company Ltd	Standards and regulation of tourism product
Urban Development Corporation	Development and planning
Water Resources Authority	Groundwater quality and extraction

Jamaica's ICZM policy formulation process involved and encompassed dialogue that emphasized the interface between government efforts and the role of coastal communities. The process began with Cabinet endorsement of the policy process and identification of the primary issues involved. The process was led by the National Council on Ocean and Coastal Zone Management. Environmental NGOs and fishing communities also played critical roles in articulating the issues. National consultations sought to further advance the need for a new

Table 8.2 Composition of the National Council for Ocean and Coastal Zone Management

---

Ministry of Foreign Affairs and Foreign Trade (Chair)
Attorney General's Department
Caribbean Coastal Area Management Foundation (environmental NGO)
Caribbean Maritime Institute
Centre for Marine Sciences, University of the West Indies, Mona
Division of Mines and Geology, Ministry of Mining and Energy
Fisheries Division, Ministry of Agriculture
Jamaica Defence Force–Coast Guard
Jamaica National Heritage Trust
Maritime Authority of Jamaica
Ministry of Land and Environment
Ministry of Tourism and Sport
Ministry of Transport and Works
Ministry of Water and Housing
National Commission on Science and Technology
National Council on Science and Technology
National Environment and Planning Agency (representing NRCA)
National Land Agency
Planning Institute of Jamaica
Port Authority of Jamaica
Shipping Association of Jamaica
Survey Department, National Land Agency

---

regime and to set the stage for the official policy formulation process. The Council established subcommittees to deal with three key areas: legal, financial, and natural resources.

- The Legal Subcommittee was tasked with identifying the critical legal issues that needed to be addressed and to provide an ad hoc forum for matters from the Council to be referred to for advice and comment.
- The Natural Resources Subcommittee was to look at the critical sustainable development issues and to make recommendations for their sustainable use within an ICZM context. This group also operated in an ad hoc manner, accepting instructions from the Council.
- The Financial Subcommittee was to look at the financial implications for the government of Jamaica and the agencies acting under it, and specifically to advise on funding, logistical arrangements, and human resources allocation.

The policy formulation process was preceded by, and interspersed with, coastal conferences that sought to identify pertinent coastal issues. These conferences also sought to sensitize Jamaicans and coastal groups to the need for sustainable use of coastal resources. Their primary purpose was to garner support for the formulation of a comprehensive strategy for the management and conservation of oceans resources.

The National Policy on Ocean and Coastal Zone Management

In 2002, the Jamaica Cabinet endorsed the Green Paper on Coastal and Oceans Management for Jamaica (National Council on Ocean and Coastal Zone Management, 2001). The National

Policy on Oceans and Coastal Zone Management entered into force and the process of implementation began. Broadly speaking, Jamaica's Ocean and Coastal Zone Management Policy aims to enhance national institutional capacities for ICZM, to integrate planning and management of the fisheries, agriculture, and forestry sectors into coastal area management, and to prevent and control environmental degradation in coastal areas. More specifically the policy seeks, among other things, to:

1   Implement effective domestic integrated coastal management strategies.
2   Incorporate integrated coastal management goals in local level strategies, as an element of improved commitment to sustainable development at the highest levels.
3   Encourage community-based participatory approaches in coastal and management planning, and in conservation of critical habitats, and develop an integrated decision-making process including all sectors to promote compatibility and a balance of uses.
4   Facilitate and catalyze [*sic*] private sector action to contribute to effective management of coastal resources.

[. . .]

7   Improve assessment, management and exploitation practices to conserve and sustain marine living and non-living resources.

[. . .]

10   Develop practical measures to expand the base of appropriate technology that supports sustainable management of Jamaica's marine resources.
11   Promote the development and application of innovative economic instruments for generating necessary revenue and financing in combination with effective monitoring and enforcement of resources and environmental regulations.
12   Strengthen the human capability towards a better management of the marine environment[.]
13   Strengthen institutions responsible for marine affairs by defining responsibilities, mandates and coordination.

[. . .]

16   Consistently include in components of the Action Plan provisions to involve stakeholders, in particular local authorities and communities and relevant social and economic sectors, both public and private.

[. . .]

19   Develop and harmonize existing laws and policies[.]
(Ministry of Foreign Affairs and Foreign Trade et al., 2000: 5–6)

Each objective needs to be evaluated and the parameters for evaluation must be clearly articulated. Setting evaluation criteria enhances transparency and accountability, and fosters a spirit of cooperation between government and communities. The policy includes a process for evaluation of the implementation process and sets out the benchmarks from which success will be measured. These benchmarks are:

1　Effective integrated coastal management practices incorporated in local environmental management programs.

2　Coastal environmental and economic goals integrated in local planning strategies.

3　Successful collaborative ventures addressing problems pertaining to inter agency management.

4　Measurable improvement in the quality of the coastal waters, as a result of comprehensive approach to prevention, reduction and control of pollution.

5　Local economies able to meet the obligations or spirit of the national policy and action plan to prevent, reduce and control marine pollution.

6　Enhanced sustainable use and conservation of marine living resources throughout the region.

7　Enhanced protection of critical habitats, biological diversity and normal ecosystem function, on which sustainability of commercially important marine resources depends.

8　Reduced adverse effects of maritime transport and tourism industries on marine resources.

9　Research, exchange of information, technology and expertise.

10　Broadened application of technology to support sustainable development of the marine environment.

11　Strengthened human and institutional capability.

12　New approaches developed to improve decision making at the highest levels.

13　Broaden participation at all levels in integrated coastal and marine management.

14　Public and private sector participation and partnership.

15　Communities, business and governments involved collaboratively in action to assure the sustainability of the marine environment.

16　Effective legal and policy framework[.]

17　Effective enforcement of laws and regulations[.]

(Ministry of Foreign Affairs and Foreign Trade et al., 2000: 7)

Critical sustainable development issues

The policy identifies sectoral issues that contribute to the web of competing interests within the coastal zone. These include, among others, fisheries, mariculture, forestry, watershed management, agriculture, recreation, tourism, mining, quarrying, coastal industrial development, ports, airports, shipping, climate change, population concentration, and access to coastal resources. Aspects of some of these sectors are explored in more detail next.

Fisheries

Jamaica's waters are considered to be overfished. Overfishing by all methods, as well as the harvesting of juveniles, has resulted in an overall decline in the fishable stock. Management mechanisms are therefore necessary for the sustenance and, in some cases, restoration of these resources. Fish-harvesting practices in Jamaica range from the use of fish traps, gill nets, drag nets, and trawlers, to spear fishing and the use of dynamite and chemicals (National Council on Ocean and Coastal Zone Management, 2001: 11). The principal challenges associated with the effective management of fisheries resources include inadequate personnel and equipment at the Fisheries Division, insufficient institutional arrangements among enforcement agencies and the prevalence of unsustainable fishing methods, the (in)effectiveness of fisheries management organizations, and lack of a legal framework within which to manage fisheries resources, as well as insufficient research, including stock assessments (National Council on Ocean and Coastal Zone Management, 2001: 11). The Fisheries Division in the Ministry of Agriculture and Fisheries has expanded the number of fish sanctuaries and has included, as part of the sustainable management structure, fisheries conservation management areas with no-fishing zones to facilitate the reproduction of critical fish populations. The Fisheries Division is also addressing illegal, unreported, and unregulated fishing, a critical threat to Jamaica's queen conch (*Strombus gigas*), which is protected under the 1973 Convention on International Trade in Endangered Species of Wild Flora and Fauna (CITES).

Recreation and tourism

Tourism is the largest contributor to Jamaica's foreign-exchange earnings. There is a focus on increasing the number of rooms to increase foreign-exchange earnings. However, this will place a greater strain on coastal resources and highlights the need for the implementation of sustainable tourism practices. The Tourism Master Plan identifies these needs and recommends a diversion from 'sun, sea, and sand' to ecotourism, heritage, and cultural tourism. Cruise shipping has also grown in importance, with a corresponding need for port reception facilities. In February 2011, a new cruise shipping pier was opened in Falmouth, Trelawny, in part to receive the largest cruise ship in the world. The project involved extensive dredging and relocation of live corals.

Conflicts over public access remain a critical component of beach management and the tourism sector. A beach policy was drafted to address, among other things, issues relating to access to beaches and oceans resources by Jamaicans. Currently, the foreshore and the floor of the sea, the water column, and all resources therein are vested in the Crown (Ministry of Land and the Environment, 2000). The current revised policy, dubbed the Beach Access and Management Policy, is expected to be submitted to the Jamaica Cabinet during 2014.

Agriculture, Forestry, and Watershed Management

A long history of monoculture agriculture on the low-lying plains – namely, sugar and banana plantations – as well as small-scale farming on the steep hillsides has presented significant challenges to environmental management in Jamaica. Issues associated with the agricultural sector include poor land management practices and the inappropriate use of chemicals and fertilizers, which impact negatively on the marine environment. Poor watershed management practices have plagued upland areas, which have had corresponding negative effects on the marine environment. The clearing of land for various types of development and for fuel and charcoal production has contributed to the degradation of critical watersheds. Non-point-source pollution from agrochemicals and unplanned residence in the hinterland also contribute to the degradation of the marine environment. A Green Paper on watershed management in Jamaica remains under review at time of writing (NEPA, 2004). The policy was reviewed in 2012, but has yet to be formalized. Jamaica is also now addressing its obligations under the 1994 United Nations Convention for Combating Desertification (UNCCD), with a focus on watershed management.

Climate Change

Jamaica has ratified both the 1992 United Nations Framework Convention on Climate Change (UNFCCC) and the 1997 Kyoto Protocol, giving prominence to Jamaica's commitment to coastal zone management. Whilst Jamaica is not a significant contributor to global greenhouse gas (GHG) emissions, it is formulating a policy on climate change. The Climate Change Policy Framework and Action Plan Green Paper has been the subject of island-wide consultation and lists as its objectives:

I   To mainstream climate change considerations into sectoral and financial planning and build the capacity of sectors to develop and implement their own climate change adaptation and mitigation plans.

II To support the institutions responsible for research and data collection at the national level on climate change impacts to Jamaica to improve decision-making and prioritization of sectoral action planning.

III To improve communication of climate change impacts so that decision makers and the general public will be better informed.

(Ministry of Water, Land, Environment and Climate Change, 2013: 9)

Coastal Industrial Development

Jamaica's strategic position within the Americas means that it is a major trans-shipment centre. It is no surprise that industrial development, with a focus on international trade, has expanded along the coast. Some of the challenges inherent in such expansion include the need to rezone certain areas, to assess the cumulative effects of such activities, and to expand port facilities.

Marine Protected Areas

With an expanded jurisdictional area, marine protected areas (MPAs) have become more critical to Jamaica. The use of marine parks has always been a challenge, and the question of 'no take' or multiple-use areas become increasingly important. Jamaica has three MPAs: Negril Marine Park, Ocho Rios Marine Park, and Montego Bay Marine Park. The Portland Bight Protected Area includes the body of water between Portland Ridge and Hellshire Hills.

Cays

Jamaica's coastline is dotted with both nearshore and offshore cays, which are used for recreational purposes such as boating and watersport activities, as well as for commercial purposes such as fishing. Since the beginning of 2008, the National Council on Ocean and Coastal Zone Management has been focusing on the state of Jamaica's cays. It established the Cays Management Committee (CMC) with the primary objective of examining sustainable management of the cays and their ecosystems, and making recommendations for the formulation of a policy and legislative framework for their management. The policy is currently being finalized. Key objectives include:

- to protect the ecological integrity of Jamaica's islands and cays, and their associated ecosystems;
- to manage Jamaica's islands and cays so as to facilitate access on a qualified basis, and to promote the sustainable use of the islands and cays and their associated ecosystems in such a way as to contribute to the social, economic, and cultural well-being of people and communities;
- to protect marine and cultural heritage resources, recognizing their value to the cultural and economic development of Jamaica;

- to incorporate climate change mitigation and adaptation measures, as a part of the management of Jamaica's islands and cays, and to ensure that coastal hazard risks are accounted for in decision-making processes; and
- to facilitate sustainable financing to support the sustainable management of cays (Jamaica Information Service, 2008).

Pollution

The primary sources of marine pollution are industrial effluent, sewage effluent, oil spills, particles from quarrying of aggregates, improperly treated sewage by direct discharge or leakage through soakaway pits, pesticides and fertilizers, and solid waste dumped into gullies and rivers that eventually empty into the ocean. The National Solid Waste Management Authority, the principal body with the mandate for the sustainable management of solid waste, is preparing legislation in the form of regulations to address in a more holistic manner, among other things, domestic waste, the grant of licences for persons wishing to operate solid waste disposal facilities, and the management of waste disposal sites and the generators of solid waste. The enforcement of these regulations will contribute greatly to Jamaica's obligations under the 1983 Convention for the Protection and Development of the Marine Environment of the Wider Caribbean Region (Cartagena Convention). A further endorsement to Jamaica's commitment to the sustainable management of solid waste is reflected in a recent pronouncement by the minister responsible for water, land environment and climate change that Jamaica is far advanced in its plans to accede to the 1999 Protocol Concerning Pollution from Land-Based Sources and Activities (LBS Protocol) (Thompson, 2013).

The boundary issue and guiding principles

Any discussion on ICZM intrinsically involves a description of boundary issues. Where does the coastal zone begin and where does it end? There is also an innate proposition that such boundary issues may not be fixed in law, nor can they necessarily be drawn along physical lines; rather, such boundary issues will depend on the peculiar circumstances of each coastal state. For the purposes of Jamaica's oceans policy, there is the recognition that the coastal and oceanic environment is a complex web of interacting natural and man-made systems, and that socio-cultural, economic, environmental, legal, and scientific issues pertaining to these systems defy easy spatial or geographic delineation. While an ocean policy might define boundaries to extend as far seaward as necessary, there is a primary focus on the land–sea interface. Integrated coastal area management represents a flexible, cross-sectoral, interagency, and multijurisdictional effort to respond to the many and varied issues affecting the biological and physical resource base within this wider coastal and oceanic environment (National Council on Ocean and Coastal Zone Management, 2001: 11). An ocean policy should also recognize that it would be more appropriate to allow the particular issue under consideration to determine how the boundaries of the programme or project area are to be drawn, bearing in mind the need for practicality in extending coastal zone management concepts (National Council on Ocean and Coastal Zone Management, 2001: 11).

An ocean policy should also identify principles by which decisions are to be guided. Chief among them are the precautionary principle, the 'polluter pays' principle, intergenerational equity, and the global and regional division of environmental impacts. Each of these principles in turn is described further in the context of Jamaica's ocean policy.

Jamaica's ocean policy does not explain the application of the principle, neither is there any specific legislation that directly addresses it. The practical application of the precautionary principle will depend on the issue at hand and what is at stake. In coastal planning, for example, the use of mitigation measures to prevent likely negative environmental effects may arguably discharge the burden of the principle. For example, when the broader sustainable development mandate results in decisions that recognize elements of uncertainty, the need for housing may require that a decision be made in the interest of providing housing. The need to mitigate environmental damage may therefore be the preferred option. The most appropriate use of the precautionary principle is perhaps found in the request for environmental impact assessments (EIAs) by the Natural Resources Conservation Authority (NRCA) or National Environment and Planning Agency (NEPA) for coastal development projects.

Jamaica has adopted the 'polluter pays' principle in legislation regarding sewage and industrial discharges, and air emissions from fixed air emission sources. Currently, there are legislative provisions that address discharges to the coast. The Natural Resources (Permits and Licences) Regulations, 1996, provide for the licensing of any discharge to and into the ground. The application of the 'polluter pays' principle is, however, not a specific component of the Regulations.

The principle of intergenerational equity challenges decision-makers to plan both for this generation and for the next. The implication is that traditional five- and ten-year plans may be an insufficient planning time frame. The principle connotes justice and fairness for generations yet unborn, whose options for the effective utilization of natural resources must be accounted for in current decision-making processes. The principle is not enshrined in law in Jamaica, but it is endorsed by the various processes geared at ensuring the sustainable use of oceans resources. Practical application of the principle is reflected in the requirement for EIAs for coastal development and the imposition of conditions to ensure the maintenance of environmental integrity.

No island is merely an island, but a part of a unique ecosystem shared by neighbouring states. The global recognition that the exercise of one state's sovereignty should not prejudice the rights of other states is an age-old principle borrowed from tort law. As part of the wider Caribbean basin, Jamaica is obliged to ensure that activities within its jurisdiction do not negatively affect the marine environment of the wider Caribbean. Jamaica's Ocean and Coastal Zone Management Policy seeks to identify ways in which these activities – in particular, prospects for ICZM – can be evaluated within the context of global sustainability.

Policy implementation

In seeking to ensure the proper allocation of resources, Jamaica's Ocean and Coastal Zone Management Policy recommends actions for each sustainable development issue identified and assigns institutional responsibility to implement these actions. Within the context of the Council, agencies will provide high levels of accountability for their respective statutory obligations under the overarching objectives of the policy.

Since ICZM must also be supported by academia, the Ocean and Coastal Zone Management Policy calls for the establishment of a centre of excellence as a primary contributor to the implementation process. The centre is to be housed at the University of the West Indies, and will provide an avenue for research and development. It is, however, recognized that the University of the West Indies already provides programmes, both at the undergraduate and graduate levels, in marine affairs. The Caribbean Maritime Institute specifically conducts tertiary training in maritime affairs. Discussions to date relating to the implementation of the policy have not addressed the establishment of such a centre, primarily owing to financial constraints. The Centre for Marine Sciences at the University of the West Indies, however, plays an advisory role in relation to the Council. The University of the West Indies also boasts several programmes that seek to foster advanced learning in coastal zone management.

Jamaica's response to international and regional agreements

The global call for the management of coasts and oceans is echoed in international agreements. Jamaica has recognized its role in protecting the global environment and has given its commitment to many of these agreements. Jamaica is party to numerous international conventions, and has implemented legislation to address some of the provisions of these conventions and agreements.

Although Jamaica is a party to the 1973 International Convention for the Prevention of Pollution of Ships, as amended by the Protocol of 1978 (MARPOL), to date there is no specific legislation to address ship-source pollution. However, the Maritime Authority of Jamaica is currently preparing a Marine Pollution Act that will govern ship-source pollution standards in Jamaican waters. Although there is no specific legislation relating to the 1972 Convention on the Prevention of Marine Pollution by Dumping of Wastes and Other Matter, the Beach Control Act, 1956, and the Exclusive Economic Zone Act, 1991, contain provisions for its implementation. Jamaica has, however, passed specific legislation to satisfy its obligations under the 1989 Basel Convention on the Control of Transboundary Movements of Hazardous Wastes and their Disposal. The Natural Resources (Transboundary Movement of Hazardous Wastes) Regulations, 2002, offer comprehensive coverage of export and transit vessels carrying hazardous waste. The Regulations specifically ban the import of hazardous wastes into Jamaica.

In the absence of specific Jamaican legislation, there is a Caribbean regional cooperative effort to respond to oil pollution, as called for under the 1990 International Convention on Oil Pollution Preparedness, Response and Co-operation. This is also a requirement under the 1983 Protocol Concerning Co-operation in Combating Oil Spills in the Wider Caribbean Region (1983 Cartagena Protocol). A contingency plan enumerates the roles of three principal agencies – the Office of Disaster Preparedness and Emergency Management, the Jamaica Defense Force Coast Guard, and the National Environment and Planning Agency – in responding to oil spills. The person causing the spill is expected to initiate clean-up measures. With the limited resources available to the relevant agencies, the response is likely to be confined to nearshore coastal areas.

The Endangered Species (Conservation and Regulation of Trade) Act, 2000, governs the implementation of CITES, and codifies all of its elements and principles. Jamaica has drafted a strategy and action plan concerning biological diversity (NRCA/NEPA, 2003), as called for under the 1992 Convention on Biological Diversity (CBD), the implementation of which is ongoing. Existing legislation – specifically, the Natural Resources Conservation Authority Act, 1991 – is deemed to contain sufficient provisions to ensure the fulfilment of Jamaica's obligations under the CBD. In order to meet its obligations under the 2000 Cartagena Protocol on Biosafety to the Convention on Biological Diversity (1983 Cartagena Protocol), Jamaica embarked on a series of national consultations and is currently reviewing a draft policy framework, with a view to putting in place a definitive policy and a legislative framework to address biosafety (NEPA, 2007). The main elements of this framework are a regulatory/legislative system, a decision-making system that includes risk assessment and management, and mechanisms for public participation and information. To date, the legislative framework has yet to be put in place.

As noted earlier in the chapter, Jamaica has designated four wetlands of international importance under the 1971 Rasmar Convention: the Black River Lower Morass; Portland Bight Wetlands and Cays; Palisadoes-Port Royal; and Mason River, an inland plaustrine wetland (Ramsar, 2014). The Black River Morass is the largest freshwater wetland ecosystem in Jamaica and the Caribbean. The site includes mangrove swamps, permanent rivers and streams, freshwater swamp forest, and peatlands. It is a biologically diverse and extremely complex natural wetland ecosystem, which supports diverse plant and animals communities that include rare, endangered, and endemic species. Portland Bight (or bay) includes some 8,000 hectares of coastal mangroves, among the largest contiguous mangrove stands remaining in Jamaica, as well as a salt marsh, several rivers, offshore cays, coral reefs, seagrass beds, and open water. The site constitutes a critical feeding and breeding location, as well as a general habitat for internationally threatened species. Palisadoes–Port Royal is located on the southeastern coast just offshore from the capital Kingston. The site contains cays, shoals, mangrove lagoons, mangrove islands, coral reefs, seagrass beds, and shallow water, thus hosting a variety of underrepresented wetland types. Its historic and cultural values are very high, because not only does the site include forts on the dunes, but it also includes part of the city of Port Royal, said to have been the largest city in the Americas, which sank in an earthquake in 1692 and is now a unique archaeological treasure. A management plan is in place, and the University of the West Indies operates research facilities at this site.

Jamaica also passed two critical pieces of legislation to give effect to specific provisions of UNCLOS: the Exclusive Economic Zone Act, 1991, and the Maritime Areas Act, 1996. In addition, the very notion of an oceans policy comes under the rubric of the general consensus to comply with the obligations of the Convention. There are outstanding matters, however. For example, Jamaica has not ratified the 1995 Agreement for the Implementation of the Provisions of the United Nations Convention on the Law of the Sea of 10 December 1982 relating to the Conservation and Management of Straddling Fish Stocks and Highly Migratory Fish Stocks (the 1995 Fish Stocks Agreement, or UNFSA 1995). Jamaica therefore has not capitalized on the benefits to be derived under this agreement.

The Caribbean Planning for Adaptation to Global Climate Change project was developed in response to the UNFCC and the Kyoto Protocol. The overall objective of the project is to support Caribbean states in preparing to cope with the effects of global climate change, especially sea–level rise in coastal and marine areas. As discussed earlier, the current Green Paper on Climate Change addresses not only Jamaica's obligations under the Convention and Protocol as they relate to Jamaica's common, but differentiated, responsibilities, but also to the responsibility of Jamaicans to adapt to global climate change.

The Preamble to the 1983 Cartagena Protocol, under the auspices of the 1983 Cartagena Convention, calls for effective action to be taken at the national level to organize and coordinate preventative, mitigation, and clean-up activities. Article 2 of the Protocol applies to oil spill incidents that have resulted in, or which pose a significant threat of, pollution to the marine and coastal environment of the wider Caribbean region, or which adversely affect the related interests of one or more of the contracting parties. The Caribbean Oil Pollution Preparedness Response and Co-operation Plan and the National Guidelines for Disaster Relief and Response set out Jamaica's programmes for combating oil spills. These programmes reflect the country's general obligations under the Protocol.

Although Jamaica has not ratified the 1990 Cartagena Protocol Concerning Specially Protected Areas and Wildlife (SPAW Protocol), the current legislative regime governing protected areas and associated ecosystems addresses the more salient resource issues. The Natural Resources Conservation Authority Act, 1991, with its mechanism for marine and terrestrial protected areas, the Fishing Industry Act, 1975, with provisions for the declaration of fish sanctuaries, and the Forestry Act, 1996, with its provisions for the establishment of protected areas are some examples of current legislative provisions relevant to the SPAW Protocol.

Jamaica has, however, developed a programme of action that will satisfy its obligations under the 1999 Protocol Concerning Pollution from Land–Based Sources of Marine Pollution. The objectives of the programme are:

- To protect human health
- To reduce the degradation of the coastal and marine environment
- To promote the conservation and sustainable use of the coastal and marine environment
- To develop a framework for the acquisition, analysis and use of data for decision-making in the protection of the marine environment
- To develop an institutional and planning framework to improve the planning process in order to ensure protection of the coastal and marine environment for sustained economic development.

(NEPA, 2003: 6)

Areas of focus include sewage, oil hydrocarbons, sediments, nutrients, pesticides, solid waste, and marine debris.

Issues and recommendations

Achieving ICZM is never an easy task. The multijurisdictional issues alone are enough to act as a deterrent to any concerted effort. However, ICZM has moved beyond a 'concept' to become an active method of addressing competing interests with one goal: effectively managing coastal and oceans resources. Jamaica has been fortunate in that there has been consistent political will to meet Jamaica's obligations to protect and manage marine resources. As a result, much has been accomplished in Jamaica. Probably the single most important initiative is the creation of the National Council on Ocean and Coastal Zone Management. The Council forms the nucleus of the process of implementation of ICZM. The success of ICZM in Jamaica therefore depends on the effective operation of the Council.

Given the challenges experienced by the Council to date, one recommendation is for the Council, through the Ministry of Foreign Affairs or through any other competent government agency, to seek funding from donor agencies to implement specific projects. Specific agencies or community groups would undertake these projects, with the Council having oversight over the implementation of the project or the allocation of resources. To date, the major project over which the Council exercised initiative was the provision of a decompression chamber operated by the University of the West Indies.

During the discussions on Jamaica's Ocean and Coastal Zone Management Policy, one position was whether the Council should enjoy legal standing as a regulated body. Answering this question was deferred, because it was felt that it would be more appropriate to take this decision when the operations of the Council become more apparent. To date, this issue has not been resolved.

The individual mandates of the agencies comprising the Council and the attendant budgetary arrangements have also hindered opportunities to focus on priority issues and to pool resources. In drafting the policy, it was envisaged that the Council member agencies' common interests in the management of coastal and ocean resources would allow them to identify priority issues and to allocate action items in their respective budgets. The realities of budgetary constraints have rendered this a difficult goal to achieve. However, with the Council members playing a more active 'intrusive' role in the oversight of coastal and oceans projects, this goal is not as elusive as it may appear. The effective application of this objective will require a concerted effort on the part of the agencies to identify common goals in the interest of integrated initiatives. At present, there is no focus, but discussions are taking place on the need for the coordination of activities within the marine sector. Since the National Council on Ocean and Coastal Zone Management has no legal standing in law, this has perhaps stymied its ability to influence decision-making at critical crossroads in the implementation of oceans and coastal policy.

One recommendation is for the Council to be given legal recognition, not necessarily as a statutory body enjoying legal personality, but as a body with specific functions and duties for the purpose of integrating decision-making relating to coastal and ocean resources. The point is, however, that the Council, within the hierarchy of the political directorate, is strategically well placed, with a sphere of influence that commands the attention of the major players in the coastal and oceans sector. By reporting to the Cabinet through a minister who chairs the meetings, the Council enjoys access to political decision-makers, and is therefore in a strategic position to ensure the kind of coordination required for the implementation of the policy.

The role of the Ministry of Foreign Affairs and Foreign Trade in providing the necessary expertise to handle environmental matters properly also needs refinement. This issue was partially resolved when the NRCA was identified as the lead agency with primary responsibility for implementing the policy. Since the Ministry has responsibility for ensuring that Jamaica meets its obligations under UNCLOS, the Council is properly placed. With the NRCA as lead agency, the policy will be able to utilize the expertise of an agency that already has coastal zone management as part of its mandate. Secretariat support for the Council, however, needs to be addressed. The review document refers to the formation of a Marine and Aviation Unit within the Ministry to provide administrative services to the Council (National Council on Ocean and Coastal Zone Management, 2001: 11). However, in late 2003 the Marine and Aviation Unit merged with the International Economic Affairs Department to become the Economics Affairs Department. In the short term, as contemplated by the policy, the Council should be provided with the secretariat support necessary to achieve its objectives. The International Economic Affairs Department does operate in the capacity of secretariat, but in a purely administrative mode.

There is no doubt that the creation of the Council enhances potential collaboration and effective decision-making regarding coastal and ocean affairs. Discussion with a Council member reveals a passion for the management of Jamaica's coastal and ocean resources, and the clear need for a Council despite the challenges that it faces.

The Ocean and Coastal Zone Management Policy recognizes that legislation will be required to implement ICZM effectively in Jamaica. At the time that the policy was drafted, however, the belief was that the current legislative policy framework, while somewhat fragmented, was sufficient to move ICZM forward in Jamaica. The vision therefore was to develop specific legislation to address ICZM. The legislation envisioned would also give the Council a functional role as an entity to be consulted in matters of national importance as they relate to coastal and ocean resources. The modalities of such consultations and the classification of 'matters of national importance' would need to be further articulated.

One of the fears expressed by the Council and by others during the consultation phase was the delay in the implementation of the policy. Although that fear has been realized to some extent, it need not paralyse the work of the Council. Implementation, as contemplated under the policy, has been slow and there has been little systematic attempt to implement the five-year plan contemplated in the policy. Critical elements of the policy must be identified with a view to implementation.

Commendably, the Council has consistently met since the adoption of the oceans and coastal zone management policy. The implementation of the Ocean and Coastal Zone Management Policy has its own inherent challenges, resulting primarily from the non-binding nature of the decisions of the Council and perhaps from the uncertainty surrounding the reporting role of the Council to the Cabinet or even to Parliament. What are certainly noteworthy and commendable are the consistent dedication of the Council to ocean and coastal zone matters and the very serious approach of the Ministry of Foreign Affairs and Foreign Trade in chairing the Council and maintaining uninterrupted scheduled meetings. Whilst it is recognized that the various agencies and departments of government represented on the Council do have their own portfolios and accompanying budgets and strategic plans, there appears to be limited input from the Council with respect to these strategic governance issues pertaining to the sustainable management of oceans and coastal resources.

The Council has also addressed a variety of ancillary issues. Matters relating to the delimitation of Jamaica's exclusive economic zone (EEZ) have received the ongoing attention of the Council. The Council has also addressed the need for decompression facilities to accommodate divers. This acquisition was not spurred solely by the provision of facilities for Jamaican divers, but for others found in Jamaican waters whether legal or otherwise. Establishment of safety guidelines for commercial divers has also been on the Council's agenda. After extensive research, a set of drafting instructions for diver safety regulations have been formulated that will form the basis of national consultations among interested and relevant groups. The Council has also sought to put a coastal conference back on the national agenda. This goal is yet to be realized.

There is a need to engage political decision-makers on the importance of management of ocean resources and to provide sufficient financial resources to implement the policy. In the absence of any legal requirement to do so, Council members should agree among themselves to identify all coastal developments within their purview. This mechanism would provide a knowledge base on what is happening around Jamaica's coasts and invite discussion that would, hopefully, influence decision-makers.

New initiatives

Jamaica's Ocean and Coastal Zone Management Policy is currently under review. The aim of the review is to evaluate the state of implementation and to make recommendations for more effective means of decision-making. A critical factor in the success of the implementation of the policy is reflected in the adequacy of the current legal framework for the implementation of ICZM. The five-year action plan contemplates the comprehensive review and reconciliation of national legislation, policy, and programmes that affect the coastal zone, with a view to making them compatible with national needs and concerns relating to ICZM (National Council on Ocean and Coastal Zone Management, undated). It also addresses protection and compensation of local knowledge on coastal management. The action plan also calls for a programme that will support activities leading to the preparation of instructional material, the formal and informal training of citizens, and the training of trainers. Although the coastal conference represents a step in this direction, the action plan calls for a more systematic public education programme. The current review of the policy addresses these issues, and advances recommendations for the incorporation of Council decisions and input in the strategic programmes within the respective agencies represented on the Council. The review also focuses on legislative provisions to give effect to the role of the Council. The Cays Policy also seeks to facilitate and promote strict adherence to the Ocean and Coastal Zone Management Policy.

Commensurate with this review, the Cays Policy represents a decision to take a critical look at Jamaica's islands and cays, especially within the context of the country's archipelagic status. The Council therefore established a Cays Management Committee. Jamaica has sixty-five rocks, islands, and cays comprising the Jamaican archipelago. No comprehensive legislative or governance framework addresses the sustainable management of cays. The Council therefore gave the Cays Management Committee the primary role of making recommendations for the development of a comprehensive policy and administrative arrangement for the management of Jamaica's islands and cays and associated ecosystems. In so doing, the CMC was required:

- to review the state of the islands and cays and associated ecosystems, identifying the issues affecting them and the existing arrangements and efforts being made to address those issues, and to provide a detailed report thereon;
- to develop policies related to the issues affecting the islands and cays and associated ecosystems, and to formulate strategies for implementation; and
- to make recommendations for the drafting and implementation of appropriate legislation, where necessary.

The objectives of Jamaica's cays and associated ecosystems policy is to protect their ecological integrity, and to enhance management measures to promote their sustainable use and thus contribute to the social, economic, and cultural well-being of people and communities. Further objectives include: protecting marine and cultural heritage resources, by recognizing their value to the cultural and economic development of Jamaica and incorporating climate change mitigation and adaptation measures in their management regime; and ensuring that coastal hazard risks are accounted for in decision-making processes. The Cays Policy also seeks to facilitate the retention of the islands and cays as Crown property held on public trust, to encourage the prohibition of the sale of islands and cays to private persons, and to incorporate elements of conservation easements over select cays and to facilitate sustainable financing to support the management of cays. The Cays Policy is seen as a 'stand-alone' policy as well as a subset of the broader Oceans and Coastal Zone Management Policy.

Conclusion

Jamaica has made significant strides in implementing ICZM. This management model is ideal given the size, geographic, and oceanic features of the country. Integrated coastal zone management calls for cooperation at the highest level. The Council provides that cooperative mechanism. However, much more needs to be done. The five-year implementation plan that accompanied the policy was never systematically pursued despite the high level of accountability displayed by the Council to date. Perhaps current management mechanisms within the government machinery require review. For example, within the Council itself, each entity represented should be required to include the strategies of the policy as part of its own strategic and operational activities. The nature of the accountability of the Council should also reflect ongoing and defined reporting to the Cabinet through the responsible minister.

Integrated coastal zone management is not an option, neither is it an experiment; it is an imperative. Within this model is encapsulated the critical environmental and sustainable development obligations, inclusive of climate change and adaptation, controlling land-based sources of marine pollution, developing sustaining fisheries management models – and the list goes on. Of credible note is the fact that the Ministry of Water Land Environment and Climate Change and the Ministry for Foreign Affairs and Foreign Trade have recognized not only the value of the policy, but also the imperatives in implementation. The very recent mandate to review current oceans policy and to look specifically at the sustainable conservation and management of Jamaica's islands and cays, forming the archipelago, reflects a commitment to sound environmental management within the context of sustainable development. The commitment to ratify the LBS Protocol, coupled with the recent thrust towards enhancing Jamaica's protected areas system plan inclusive of financial sustainable measures, is a leap in the right direction. Jamaica has done much, talked much, and consulted much. The engine has been engaged . . . Let Jamaica navigate sustainably.

Notes

1 Agenda 21 is a comprehensive plan of action to be taken globally, nationally, and locally by organizations of the United Nations system, governments, and major groups in every area in which humans impact the environment. It represents a global consensus and political commitment at the highest level on development and environment cooperation. It was adopted in 1993. Chapter 17 of Agenda 21 provides that the marine environment – including the oceans and all seas and adjacent coastal areas – forms an integrated whole that is an essential component of the global life-support system and a positive asset that presents opportunities for sustainable development (UNGA, 1992).

2 Designated wetlands of international importance under the 1971 Convention on Wetlands of International Importance (the Ramsar Convention).

3 The Natural Resources (Marine Parks) Regulations, 1992, were made pursuant to the Natural Resources Conservation Authority Act, 1991. To date, there are three established marine parks: Montego Bay Marine Park, Negril Marine Park, and Ocho Rios Marine Park. Other marine areas are protected under a different nomenclature, for example the Portland Bight Protected Area and the Palisados Port Royal Protected Area. The latter two are inclusive of adjoining land areas.

References

Caribbean Community (CARICOM) (undated) 'Delimitation of Maritime Boundaries within CARICOM', Background Paper for *Development of Relevant Rules for Delimitation of Maritime Boundaries, Including Practical Illustrations of the Operations of such Rules*, online at http://www.caricom-fisheries. com/website_content/publications/documents/Delimitation_of_Maritime_Boundaries_within_ CARICOM.pdf [accessed 6 March 2014] (no longer available).

Central Intelligence Agency (CIA) (undated) 'Demographic Statistics', *The World Factbook*, online at https://www.cia.gov/library/publications/the-world-factbook/geos/jm.html [accessed 6 May 2014].

Clarke, J. R. (1992) *Integrated Management of Coastal Zones*, FAO Fisheries Technical Paper No. 327, Rome: FAO.

Climate Studies Group Mona (CSGM) (2012) *State of the Jamaican Climate 2012: Information for Resilience Building (Full Report)*, produced for the Planning Institute of Jamaica, Kingston, online at http://idbdocs.iadb.org/wsdocs/getDocument.aspx?Docnum=38578258 [accessed 11 July 2014].

DESA *See* United Nations Department of Economic and Social Affairs, Division for Sustainable Development

Earthtrends (2003) 'Coastal and Marine Ecosystems: Jamaica', online at http://earthtrends.wri.org/pdf_ library/country_profiles/coa_cou_388.pdf [accessed 14 August 2006] (no longer available).

Jamaica Information Service (JIS) (undated) 'Committee to Develop Policy Document for Cays', online at http://jis.gov.jm/committee-to-develop-policy-document-for-cays/ [accessed 17 November 2014].

Ministry of Foreign Affairs and Foreign Trade/Natural Resources Conservation Authority (NRCA)/ Coastal Water Quality Improvement Project (CWIP) (2000) *Toward Developing a National Policy on Ocean and Coastal Zone Management*, online at http://www.nepa.gov.jm/symposia_03/Policies/ OceanandCoastalZoneManagementPolicy.pdf [accessed 2 November 2014].

Ministry of Land and the Environment (2000) *Towards a Beach Policy for Jamaica: A Policy for the Use of the Foreshore and the Floor of the Sea*, Kingston: Government of Jamaica.

Ministry of Water, Land, Environment and Climate Change (2013) *Green Paper: Climate Change Policy Framework and Action Plan*, online at http://www.japarliament.gov.jm/attachments/440_Climate%20 Change.pdf [accessed 30 April 2014].

Morris, M. (undated) 'Geography and Politics of Jamaica', online at http://www.discoverjamaica.com/ geography.htm [accessed 5 December 2006].

National Council on Ocean and Coastal Zone Management (undated) *Towards Ocean and Coastal Zone Management in Jamaica: Proposed Action Plan (First Five Years)*, online at http://www.nepa.gov.jm/ symposia_03/Policies/OceanandCoastalZoneManagementActionPlan.pdf [accessed 11 July 2014].

National Council on Ocean and Coastal Zone Management (2001) *Towards Developing an Ocean and Coastal Zone Management Policy in Jamaica*, Green Paper No. 9/01, Kingston: Ministry of Foreign Affairs and Trade.

National Environment and Planning Agency (NEPA) (2003) *Draft National Programme of Action (NPA): Jamaica's National Programme of Action for the Protection of the Coastal and Marine Environment from Land Based Sources and Land Based Activities*, online at http://www.nepa.gov.jm/symposia_03/Policies/JAMAICANPA.pdf [accessed 7 July 2014].

National Environment and Planning Agency (NEPA) (2004) *Watershed Policy for Jamaica*, Kingston: NEPA.

National Environment and Planning Agency (NEPA) (2007) *Draft National Biosafety Policy*, Revised Draft 002, online at http://www.nepa.gov.jm/documents/Draft-Biosafety-Policy.pdf [accessed 7 July 2014].

Natural Resources Conservation Authority (NRCA) (1999) *Jamaica National Environmental Action Plan: JaNEAP 1999–2002*, online at http://www.mona.uwi.edu/cardin/virtual_library/docs/1200/1200.pdf [accessed 4 July 2014].

Natural Resources Conservation Authority/National Environment and Planning Agency (NRCA/NEPA) (2003) *National Strategy and Action Plan for Biodiversity in Jamaica*, online at http://www.cbd.int/doc/world/jm/jm-nbsap-01-p5-en.pdf [accessed 30 April 2014].

Natural Resources Conservation Authority/Planning Institute of Jamaica (NRCA/PIOJ) (1995) *Jamaica National Environmental Action Plan 1995–1998*, online at http://www.mona.uwi.edu/cardin/virtual_library/docs/1199/1199.pdf [accessed 7 July 2014].

Ramsar (2014) *National Report on the Implementation of the Ramsar Convention on Wetlands*, online at http://www.ramsar.org/sites/default/files/documents/2014/national-reports/COP12/cop12_nr_jamaica.pdf [accessed 17 November 2014].

Richards, A. (2008) 'Development Trends in Jamaica's Coastal Areas and the Implications for Climate Change', online at http://www.pioj.gov.jm/portals/0/sustainable_development/jamaica_climate_change_paper.pdf [accessed 6 March 2014].

Thompson, K. (2013) 'Jamaica to Ratify LBS Protocol: Pickersgill', *Jamaica Observer*, 3 October, online at http://www.jamaicaobserver.com/news/Jamaica-to-ratify-LBS-protocol——Pickersgil_15182892 [accessed 7 July 2014].

United Nations Department of Economic and Social Affairs, Division for Sustainable Development (DESA) (undated) 'Who Are the SIDs?', online at http://www.un.org/esa/sustdev/sids/sids.htm [accessed 2 August 2005] (no longer available).

United Nations General Assembly (UNGA) (1992) *Report of the United Nations Conference on Environment and Development*, UN Doc. A/CONF.151/26/REV.1, vols I–III, online at http://www.un.org/Docs/journal/asp/ws.asp?m=A/CONF.151/26/Rev.1%20 [accessed 3 November 2014].

United Nations General Assembly (UNGA) (1994) *Report of the Global Conference on the Sustainable Development of Small Island Developing States, Bridgetown, Barbados, 25 April–6 May 1994*, UN Doc. A/CONF.167/9, online at http://www.sidsnet.org/docshare/other/BPOA.pdf [accessed 2 August 2005].

United States Library of Congress (undated) 'Caribbean Islands: Jamaica – Geography', online at http://countrystudies.us/caribbean-islands/20.htm [accessed 12 August 2006].

9

DEVELOPMENT OF NATIONAL OCEAN POLICY IN JAPAN

Hiroshi Terashima and Moritaka Hayashi

Basic information and an overview of national ocean policy

The importance of the oceans to Japan is self-evident from its geographical location. Equally, the oceans have played a significant role in Japanese culture and in Japan's economy. The long history of the nation's use and misuse of oceans and coastal areas has led to a proliferation of related laws and institutions. Most of these were put in place individually in particular sectors without coordination with others. Given such a sectoral approach, Japan has been slow to develop an overall ocean policy. A new era, however, has started for ocean policy development, with the adoption by the *Diet* (Japanese Parliament) of the Basic Act on Ocean Policy in 2007[1] and the subsequent announcement by the government of an ocean policy paper entitled *Basic Plan on Ocean Policy*.[2]

Maritime zones claimed

The total land area of Japan is 377,835 km², and the area covered by its territorial sea and internal waters is about 430,000 km². Its exclusive economic zone (EEZ) covers some 4,050,000 km². The length of its coastlines is approximately 35,000 km. The population of Japan is approximately 127 million, as of 1 May 2014. Coastal cities and communities occupy around 30 per cent of the land area, and about half of the population lives in coastal cities and communities.

The EEZ claimed by Japan overlaps partially with the EEZs claimed by its neighbouring countries. Japan has not concluded delimitation agreements with its neighbours, except a partial provisional fishery line with the Republic of Korea. Major obstacles to concluding agreements include the territorial problems relating to several islands with China, the Republic of Korea, and Russia. It should be noted, however, that Japan did conclude an agreement with the Republic of Korea on the delimitation of the continental shelf, excluding the area surrounding the disputed island of Takeshima.[3]

Significance of the oceans to the nation

Surrounded completely by oceans, for many centuries Japan has depended heavily on the oceans for food, transportation, industry, recreation, trade, and the exchange of people and cultures.

The seas around Japan consist of three large marine ecosystems – that is, the *Kuroshio* and *Oyashio* currents, and the Sea of Japan. They are rich in fauna and flora, with some 3,100 species of fish and 5,500 species of algae having been identified. About 40 per cent of the animal protein intake by the Japanese is provided by fish and fishery products. In 2003, Japan was the largest importer of fish and fishery products in the world, accounting for 18 per cent in value and 11 per cent in quantity of world imports (Fisheries Agency, 2006: 31). Imports into Japan, however, have declined slightly since then in quantity, and now Japan is the second largest importer after China (Fisheries Agency, 2009). Japan also depends on the oceans for its international trade. In 2013, sea transportation accounted for 99.7 per cent of the total of 963 million tons, and 76.6 per cent in value of exports and imports (Japanese Shipowners' Association, 2013). Japan has recognized the importance of the oceans by designating the third Monday of July a national holiday called 'Ocean Day'.

Ocean and coastal use and conflicts

Extensive development of Japan's coastal areas started in the 1960s, when its economy began to expand rapidly. Many industrial facilities – particularly large-scale petrochemical processing complexes and accompanying infrastructure – were built along the coasts and on reclaimed land areas. Inevitably, this, together with the increasing concentration of the population in coastal cities, has caused serious environmental damage in the coastal zones, as well as tragic human suffering. A most serious example is the outbreak of *Minamata* disease, first discovered in 1956 in Minamata City, Kumamoto Prefecture, caused by methyl mercury in the waste water of a chemical factory. The highly toxic chemical contaminated adjacent coastal waters and caused severe neurological syndrome, including death, to thousands of people who had eaten seafood from the area.

Another notable development is the construction of seawalls and other artificial improvements of the coastline against typhoons, tsunami, and high tides. As a result, natural beaches have disappeared, causing the loss of mudflats, seagrass beds, and other ecosystems vital in preserving marine and coastal biodiversity. Public access to beaches, natural beauty, and other amenities has also been reduced. Approximately 34 per cent of Japan's shoreline has been transformed for ports, harbours, and other development purposes. Only about 53 per cent remains as natural sandy and rocky shores.

These and other developments have led to a number of conflicts and disputes among such key stakeholders as industrial and land developers, construction companies, fishermen and fishery cooperatives, shipping companies, conservationists, and recreation seekers, as well as government authorities. Other players include *Keidanren* (the Japan Business Federation)[4] and active local governments. Several disputes have ended up in courts.

Major legislation

Typically, government response to these problems has been piecemeal. Special laws or provisional measures were adopted, specifically aimed at environmental improvement of certain areas such as bays and semi-enclosed seas. Periodic amendments have been made to existing laws of more general application.

Until the adoption of the Basic Act on Ocean Policy, to be discussed below, the main existing laws relating to the oceans and coastal zone could be grouped under several headings, as follows:[5]

- *Legislation on planning* – National land sustainability and development; national land-use planning; natural parks; development of the national capital region; development of remote islands; development of Hokkaido; development of the Ogasawara Islands; city planning; factory locations
- *Law of the sea* – territorial sea; the EEZ; the continental shelf
- *Land preservation and management of public goods* – Government property; coastal areas; rivers; improvement of fishing harbours and grounds; improvement of infrastructures with over-riding priority
- *Fisheries* – Basic fisheries policy; conduct of fishing; preservation and management of marine living resources; promotion of development of marine fishery resources; fishing boats; exercise of sovereign rights with regard to fisheries and other activities
- *Mineral resource development* – Mining; mining safety; development of petroleum resources and inflammable natural gas; stockpiles of petroleum; petroleum pipeline industry; extraction of gravel
- *Shipping* –Marine transport; domestic shipping industry; maritime traffic safety; personnel for ships; aids to navigation; port regulations; prevention of collision at sea; maritime accident inquiry; life-saving in coastal accidents; pilotage; hydrographic activities; security of ships; port facilities
- *Space utilization* – Reclamation of publicly owned water surface; city planning; urban parks; factory locations
- *Energy* – Stockpiles of petroleum; electricity industry; promotion of energy resources development; restriction of nuclear source material and nuclear reactors; prevention of radioactive damage caused by radioactive isotopes
- *Waste disposal* – Waste disposal and cleaning; regional offshore environmental improvement
- *Environmental protection* – Basic policy on the environment; environmental impact assessment (EIA); natural parks; prevention of water pollution; prevention of marine pollution and maritime disasters; special measures for the conservation of environment of the Seto Inland Sea.

With the exception of the legislation relating to the law of the sea, these laws essentially address sectoral issues that generally have been dealt with by an individual government ministry or agency. Only after the turn of the century did some laws start to address problems requiring inter-sectoral collaboration involving several ministries. Two examples are the Law concerning the Special Measures for Regeneration of the Ariake and Yatsushiro Seas,[6] and the Law concerning the Promotion of Nature Restoration.[7] Implementation of the latter requires cooperation among the Ministry of Land, Infrastructure, Transport and Tourism (MLIT), the Fisheries Agency, and the Ministry of the Environment.

The Law on the EEZ and the Continental Shelf,[8] adopted at the time of Japan's ratification of the 1982 United Nations Convention on the Law of the Sea (UNCLOS) in 1996, consists of merely four articles that essentially define the extent of the EEZ and the continental shelf, and confirm the application of Japanese laws to various activities therein. Although the law is supplemented by other laws relating mainly to fisheries activities in the EEZ, these laws are not sufficient and are unclear with regard to activities such as scientific research, environmental protection, or the activities of unidentified vessels (Hayashi, 1997).

Relevant ministries and agencies

The principal ministries and agencies relevant to ocean and coastal areas are:

- the Ministry of Defense;
- the Ministry of Public Management, Home Affairs, Post and Telecommunications;

- the Ministry of Foreign Affairs;
- the Ministry of Education, Culture, Sports, Science and Technology (MEXT);
- the Ministry of Agriculture, Forestry and Fisheries, including the Fisheries Agency;
- the Ministry of Economy, Trade and Industry, including the Agency for Natural Resources and Energy;
- the Ministry of Land, Infrastructure, Transport and Tourism, including the Japan Coast Guard, the Meteorological Agency, and the Geographical Survey Institute; and
- the Ministry of the Environment.

Policy development process

In follow-up to the 1992 United Nations Conference on Environment and Development in Rio de Janeiro, Japan started the slow process of designing national ocean policy with a more multipurpose and comprehensive approach. In 1993, the government prepared a National Action Plan for Agenda 21 for submission to the United Nations (Government of Japan, 1993). The Action Plan recalled that the Fourth Comprehensive National Development Plan, adopted in 1987, had stated that local communities should lead the development of comprehensive plans for coastal zone use in order to ensure the preservation and safety of the coastal environment, to actively seek the possibility of multiple use, and to promote attractive local development through integrated use of wider coastal areas. The National Development Plan further stated that the government would actively support local communities in their preparation and adoption of such plans. The Action Plan confirmed these points and recorded the government's intention to pursue the National Development Plan, with a major emphasis on that commitment.

In the Fifth Comprehensive National Development Plan, adopted in 1998 and entitled *The Grand Design for the National Land in the 21st Century* (MLIT, 1998), the government laid down a long-term direction for coastal zone development. This Fifth National Development Plan went a step further than the last and called upon local communities to lead the preparation of comprehensive 'coastal zone management plans' by regarding the coastal zone as a natural system. The commitments of the government were also strengthened. The government was required to prepare a set of guidelines for the preparation of these management plans and to assist, if necessary, in the preparation of local community plans in those areas in which two or more communities are involved. The Grand Design also stated that, in order to create a better environment, it was necessary to deal with coastal areas from wider perspectives, with long-term objectives, and to rebuild and improve the environment through step-by-step plans. At the same time, the Grand Design stressed the need to pursue its management goals in an integrated manner, by coordinating the individual activities and projects of a variety of players, and by encouraging communities to build partnerships.

Pursuant to the Grand Design, the government released its Guidelines for the Development of Integrated Coastal Zones Management (ICZM) Plans in 2000 (MLIT, 2000).The purpose of the Guidelines was to provide direction to local communities and various non–governmental organizations (NGOs) in preparing for integrated management of coastal zones on the basis of the Grand Design. The Guidelines laid down the basic principles and approaches of ICZM, and identified the main elements to be incorporated in the management plans, as well as other major points to be taken into account in formulating and implementing such plans. The management plans were to be prepared by 'ICZM Commissions', with the local government (prefecture or city) playing the main role, together with representatives from (among others) other administrative bodies, private industry, local civil society, and non–profit organizations. Unfortunately, however, these Guidelines were not much utilized because they were not accompanied by institutional and financial measures.

All of these initiatives were, however, limited to the management of coastal zones, and did not cover the wider ocean and coastal areas. The first indication of the government's interest in a broader oceans and coastal policy is found in a request that the Minister for Education, Culture, Sports, Science and Technology made to the Council for Science and Technology, an expert advisory group under the Ministry, for advice on the basic concepts and policy to advance ocean development with a long-term perspective. In August 2002, the Council's Subdivision on Ocean Development submitted its report entitled *Japan's Ocean Policy in the Early 21st Century* (MEXT, 2002).

The report first stressed the utmost importance of changing policy perspective from the benefits to be reaped from the ocean to the realization of sustainable use of the ocean. It proposed the formulation of a new ocean policy that promotes the three basic pillars of 'preservation', 'utilization', and 'understanding' of the ocean in a well-balanced manner. Second, it pointed out the need to implement the ocean policy in a strategic manner with an international perspective. And third, the report stressed the need to develop and implement the ocean policy in a comprehensive manner, including social and human science elements, in partnership among relevant ministries and agencies. The report then set out, under each of the three pillars, the essential principles on the basis of which to formulate the ocean policy and specific measures for its implementation. Further, the report discussed the need to improve the foundations on which the ocean policy is to be pursued, underlining the importance of capacity-building, the free flow of information, and international cooperation.

In 2005, the *Diet* took a step to facilitate coastal and ocean management by amending the Comprehensive National Land Development Act in the form of the National Land Sustainability Plan Act,[9] and adding to its scope the utilization and preservation of marine areas, including the EEZ and the continental shelf. In June 2006, the MLIT, which is in charge of several administrative functions relating to ocean and coastal areas, issued *The MLIT Policy Outline on Ocean and Coastal Areas* (MLIT, 2006). This document set out eight basic policy directions, including the promotion of integrated management of the ocean and coastal areas, followed by ninety-five specific measures that the government intends to implement.

Side-by-side with these governmental initiatives, several private bodies have also made recommendations to improve ocean and coastal zone management and policy. For example, *Keidanren* circulated a proposal entitled *A Grand Design for Oceans in the 21st Century* in June 2000 (Keidanren, 2000). The proposal suggested that the three dimensions of the ocean – that is, surface, water column, and seabed – should be utilized in a well-balanced manner by means of research, wiser use, and protection. *Keidanren* urged industry, the government, and academic circles to collaborate, and called on relevant government agencies to work together under an integrated framework with a view to pursuing ocean development in an integrated way. It is apparent that the *Keidanren* proposal had a strong impact on the 2002 report of the Council for Science and Technology (MEXT, 2002).

Another private group, the Japanese Association for Coastal Zone Studies, issued *2000 Appeal: Proposals for the Sustainable Use and Preservation of the Coastal Zone Environment* (JACZS, 2000). It proposed, among other things, the enactment of a 'Law on the Integrated Coastal Zone Management', and empowering and requiring local governments to develop and implement their own ICZM plans, with central government sharing the required financial resources.

In March 2002, the Nippon Foundation released a fundamental set of general proposals on ocean policy entitled *Oceans and Japan: A Proposal on the Ocean Policy of Japan for the 21st Century* (Nippon Foundation, 2002). The report proposed the development of the ocean policy on the basis of a 'Basic Ocean Law' and set out basic policy elements, as well as restructured government organizations, which were to be centrally supervised by a 'Minister for the Ocean'.

Other proposals included: review and improvement of relevant legislation; the rational management of fishery resources, and the coordination between fisheries and the other uses of the sea; the adoption of specific measures for the overall management of the EEZ and the continental shelf under Japanese jurisdiction; the strengthening of youth education; and the adoption of an interdisciplinary approach in ocean research and education.

In 2005, a Committee on Ocean Science of the Science Council of Japan issued a report entitled *The Need for Integrated Promotion of Science Relating to Oceans: A Proposal for Comprehensive Ocean Policy-Making* (Science Council of Japan, 2005). The report pointed to the proliferation of ocean-related administration in several ministries and agencies, with no overarching basic law, and the lack of comprehensive policy and national strategy on ocean issues. The Committee called for the establishment of a special deliberative body, over and above the existing ministries and agencies concerned, to promote integrated studies relating to oceans.

The Nippon Foundation proposal was further developed by the Ocean Policy Research Foundation (OPRF) in its proposal, announced in late 2005, entitled *Oceans and Japan: A Proposal for the Ocean Policy of the 21st Century* (OPRF, 2005).The proposal stressed, among other things, the need for the drafting of 'National Policy Guidelines' and a 'Basic Ocean Law', as well as the restructuring of administrative organization by creating a post of 'Cabinet Minister in charge of Oceans'. These proposals were followed by specific recommendations on such key areas as the management of the EEZ and the continental shelf, maritime security, the protection and preservation of the marine environment, the development of resources, the promotion of an ICZM system, the strengthening of natural disaster prevention and reduction, the collection of information and data, and the promotion of research and education. The OPRF promoted the proposal widely and submitted it to policymakers, including then Chief Cabinet Secretary Shinzo Abe and high officials of the ruling Liberal Democratic Party (LDP) in November 2005. The OPRF, along with the Nippon Foundation, then made a formal recommendation to the LDP to consider the proposal. The LDP agreed to work for submission of a Basic Ocean Bill at the following session of the *Diet*.

These developments have led to new initiatives on basic ocean policy and legislation on the part of some law-makers. Thus, in April 2006, a multi-partisan 'Basic Ocean Law Study Group' was established, consisting of *Diet* members from the LDP, the New *Komeito* Party (which was a coalition partner of the LDP), and the Democratic Party, together with scientists and experts from various ocean-related fields, as well as observers from relevant ministries and agencies. The OPRF, represented by Hiroshi Terashima, one of the authors of this chapter, served as secretariat for the Study Group, facilitating its establishment and operations.

The Basic Ocean Law Study Group produced, in December 2006, Guidelines for Ocean Policy and an outline of a draft Basic Ocean Law. The Guidelines pointed out the urgent need for new systems with which to address ocean issues and especially the need for a 'Basic Ocean Act' in support of comprehensive ocean management. The Group called for the Act to state clearly the basic philosophy and principles underlying ocean policy, and the obligations and responsibilities of national and local public entities, industry, and the general public. It also called for the government to prepare a 'Basic Ocean Plan', stating the specific policy measures necessary to implement the comprehensive management and sustainable development of the ocean, and the establishment of comprehensive ocean policy headquarters in the Cabinet, as well as the appointment of a minister for the ocean to ensure the effective and continuous implementation of ocean policy.

The LDP, the New *Komeito* Party, and the opposition Democratic Party have since followed up on the outcome of the Study Group, preparing a draft Basic Act on Ocean Policy based on its ideas, which Act the *Diet* passed on 20 April 2007 and which entered into force on 20 July 2007.

The Basic Act and the Plan on Ocean Policy

The Basic Act on Ocean Policy lays down as its fundamental philosophy the harmonization of ocean development and use with the conservation of the marine environment, securing the safety and security on the oceans, the improvement of scientific knowledge of the oceans, the sound development of ocean industries, and the comprehensive governance of the oceans, as well as international partnership with regard to oceans. Regarding new institutions, the Act provides for the establishment of Headquarters for Ocean Policy, with the Prime Minister as its head, and the Chief Cabinet Secretary and the Minister in Charge of Ocean Policy as deputies. The rest of the Cabinet ministers serve as members. The Minister in Charge of Ocean Policy is to be appointed by the Prime Minister to assist in ocean policy matters. Among the principal functions of the Headquarters are to draft and promote the execution of the Basic Plan on Ocean Policy, and to pursue the overall coordination of activities of the various administrative bodies concerned with the basis of the Basic Plan.

The Prime Minister appointed initially the Minister for Land, Infrastructure, Transport and Tourism as the Minister in Charge of Ocean Policy. More recently, the Prime Minister has appointed a minister specially to be in charge of ocean policy, together with some other cross-sectoral issues. The Ocean Policy Headquarters has also established an Advisory Council consisting of ten experts from various ocean-related fields, with the function of advising the head of the Headquarters.

The Cabinet decision was made on the first Basic Plan on Ocean Policy in March 2008, after its draft had gone through the process of public comment in February. It was intended to cover a period of around five years, and consisted of a general introduction and three parts.

- Part 1 set out the basic directions of Japan's ocean policy.
- Part 2 dealt with the policy measures that the government should pursue in a comprehensive and systematic manner. The measures enumerated in Part 2 consisted of the following twelve areas:

  - the development and use of marine resources;
  - the protection and preservation of the marine environment;
  - the development of the EEZ and the continental shelf;
  - securing efficient and stable maritime transport;
  - safety and security at sea;
  - the promotion of scientific research and survey;
  - the promotion of research and development in ocean science and technology;
  - the promotion of ocean industries;
  - the promotion of integrated management of coastal zones;
  - the conservation of remote islands;
  - the promotion of international cooperation and coordination; and
  - the enhancement of citizens' understanding of the oceans and human resource development.

- Part 3 addressed other matters that were necessary for pursuing ocean policy in a comprehensive and systematic manner.

During the initial half-decade of implementation, the Basic Act and Plan on Ocean Policy have already produced a number of important achievements in the multi-sectoral management of ocean spaces and conservation and development of marine resources. First, the government worked out more detailed plans for the implementation of the Basic Plan in several areas. Thus, it adopted in 2009 a plan for the development of marine energy and mineral resources. The government also adopted sets of basic guidelines for the conservation and management of remote islands,[10] for the exploration for mineral resources and conduct of marine scientific research in the EEZ and the continental shelf,[11] on the establishment of marine protected areas (MPAs),[12] and for the promotion of marine renewable energy development.[13]

Second, several new laws have been adopted under the Basic Act and Plan. For example, an Act for the suppression and punishment of acts of piracy was adopted,[14] which gave the Maritime Self-Defense Force and Coast Guard the authority to conduct anti-piracy operations on the high seas, particularly in the area off Somalia, and to undertake the first prosecution of four Somali pirates in Japan.[15] The Mining Act was revised in order to make it applicable to the mining activities on the continental shelf. New laws were also adopted for the protection of coastal areas that are important for establishing baselines for maritime zones, and for amending the Maritime Transportation Act and the Act on Seafarers, as well as on the passage of foreign-flagged ships through the territorial sea and internal waters (Terashima, 2010; 2013).

Third, the government, through the Headquarters for Ocean Policy, has implemented several other measures that require actions by several ministries and agencies, as well as close coordination among them. One notable example is the preparation and submission of the scientific data concerning the limit of Japan's outer continental shelf to the Commission on the Limits of the Continental Shelf in 2008.

In April 2013, the Cabinet adopted a new Basic Plan on Ocean Policy, which is intended to cover another period of roughly five years. The new Basic Plan follows the basic structure of the first one, but describes particularly the policy measures in a more detailed and concrete manner.

- In Part 1, which sets out the basic directions of policy measures, the Basic Plan emphasizes the following six measures:

  o the promotion and creation of ocean industries;
  o securing maritime security and safety;
  o the promotion of scientific research, and the integration of maritime information and facilitation of public access thereto;
  o the development of human resources and strengthening of technology;
  o the integrated management of marine areas and coastal zones; and
  o other measures, including those for disaster preparedness and environmental protection, taking into account the Great East Japan Earthquake, and those in response to the changing Arctic resulting from climate change.

- In Part 2, which contains the policy measures that the government should implement, some examples of detailed measures include the following:

  o the development of the technology necessary for commercial mining of methane hydrates and polymetallic sulphides by 2018, with a view to starting projects for commercial developments led by the non-governmental sector within five years after 2023;
  o the development and testing of floating wind farm systems for energy production;
  o the promotion of strategic growth of internationally competitive ocean industry for resource development, including the development of floating logistic hubs for transporting workers and materials for at-sea facilities of production, storage, and offloading of liquefied natural gas (LNG);

- ○ the further promotion of MPAs, with the aim of placing 10 per cent of the coastal and maritime areas under appropriate conservation and management by 2020;
- ○ the promotion of research and development in ocean science and technology, particularly for responding to the policy needs of prediction of, and adaptation to, global warming and climate change, the exploitation of ocean energy and mineral resources, the conservation of marine ecosystem and sustainable development of living resources, the development of renewable ocean energy, and responses to natural disasters;
- ○ the development of a comprehensive legal system for the management of the EEZ and the continental shelf;
- ○ the promotion of integrated management of coastal zones;
- ○ continued monitoring of water, bottom soil, and marine life for radioactive contamination; and
- ○ the further development of human resources that support the country as an ocean state.

Evaluation and outlook

In summary, Japanese ministries traditionally addressed the management of coastal zones and enclosed sea areas in a sectoral and uncoordinated manner. This has made it difficult to prevent resource depletion, ecosystem deterioration, and environmental degradation in those areas. It is only in recent years that initiatives aimed at vertical and horizontal integration have started, through governmental guidelines and special laws. National policy on coastal zone management, however, has remained largely in the form of administrative guidance or guidelines addressed to local governments. There was no national legal framework under which such guidelines were to be implemented. These approaches started to change, however, with the implementation of the 2007 Basic Act on Ocean Policy.

As far as the formulation of national oceans policy is concerned, Japan has also been slow in starting a formal process. The slow action on the part of the Japanese government has prompted private organizations to put forward their own recommendations. One such recommendation, a carefully drafted proposal by a non-governmental think tank, finally moved politicians and assisted them in launching the collaborative process of law-making among *Diet* members, experts, and private stakeholders as well as relevant ministries and agencies. Such a process of initiating and drafting a law was truly unprecedented in Japan's legislative history.

These efforts have finally borne fruit, with the adoption by the *Diet* of the Basic Act on Ocean Policy and the release of the Basic Plan on Ocean Policy. Japan has thus entered a new phase for ocean policy, equipped with a new government mechanism to formulate and implement it in a comprehensive and systematic manner. Some progress has already been made in the last several years in implementing the Basic Act and the Basic Plan on Ocean Policy through active initiatives of the Headquarters for Ocean Policy.

The new Basic Plan has overcome the weakness of the original Plan, which tended to describe measures in terms that were too general and abstract, without specific objectives, target years, and road maps. Part 2 of the Plan sets out in a specific manner the measures necessary for achieving the objectives under each of the twelve basic policy measures defined by the Basic Act on Ocean Policy. In Part 3, the new Plan further stipulates those measures that are required for pursuing policy measures in a comprehensive and systematic manner, such as strengthening of the functions of the Advisory Council and the Secretariat of Ocean Policy Headquarters in implementing policy plans in a more comprehensive and effective way, as well as clarifying the roles of local governments, communities, and civil society, and promoting active information-sharing with the public. In conclusion, with the adoption of this new Basic Plan of Ocean Policy in 2013, Japan has consolidated its most recent stage in ocean policymaking that had started in 2008.

Notes

1 Act No. 33 of 27 April 2007, online at http://law.e-gov.go.jp/cgi-bin/idxselect.cgi [accessed 12 June 2008].
2 Available online at http://www.kantei.go.jp/jp/singi/kaiyou/kihonkeikaku/080318kihonkeikaku.pdf [accessed 12 June 2008].
3 Agreement Concerning the Establishment of a Boundary in the Northern Part of the Continental Shelf Adjacent to the Two Countries, signed on 30 January 1974, entered into force on 22 June 1978. See Oda (1974).
4 *Keidanren* is a major business organization, with membership comprising of some 1,300 representative companies of Japan, 120 nationwide industrial associations, and 47 regional economic organizations.
5 The list is based on Kisugi (2004: 33), supplemented by other materials.
6 Law No. 120 of 29 November 2002, online at http://law.e-gov.go.jp/htmldata/H14/H14HO120.html [accessed 1 April 2007].
7 Law No. 148 of 11 December 2002, online at http://law.e-gov.go.jp/htmldata/H14/H14HO148.html [accessed 1 April 2007]. The Law requires the government, local governments, and communities, as well as private organizations and the general public, to cooperate in restoring and preserving natural environments such as rivers, wetlands, and mudflats, with a view to regaining and preserving the ecosystems and other natural features that have been damaged or destroyed.
8 Law No. 74 of 14 June 1996, online at http://law.e-gov.go.jp/cgi-bin/strsearch.cgi [accessed 1 April 2007].
9 Law No. 89 of 29 July 2005, online at http://law.e-gov.go.jp/htmldata/S25/S25HO205.html [accessed 1 April 2007].
10 See online at http://www.kantei.go.jp/jp/singi/kaiyou/ritouhoushin.pdf [accessed 18 March 2014].
11 See online at http://www.kantei.go.jp/jp/singi/kaiyou/CS/honbusiryou.pdf [accessed 18 March 2014].
12 See online at http://www.kantei.go.jp/jp/singi/kaiyou/annual/H24/H24-2.pdf [accessed 18 March 2014], p. 29.
13 See online at http://www.kantei.go.jp/jp/singi/kaiyou/energy/torikumihousin.pdf [accessed 18 March 2014].
14 See Hayashi (2010; 2013: 257).
15 See Furuta and Tsuruta (2013).

References

Fisheries Agency (2006) 'FY 2005 Trends in Fisheries: Fisheries Measures for FY 2006 – Outline', online at http://www.jfa.maff.go.jp/hakusyo/kaigai/fy2005.pdf [accessed 1 April 2007] [in Japanese].
Fisheries Agency (2009) 'FY 2009 Trends in Fisheries: Fisheries Measures for FY 2010', online at http://www.jfa.maff.go.jp/j/kikaku/wpaper/h21/pdf/h121pdf [accessed 10 August 2010] [in Japanese].
Furuta, K., and Tsuruta, J. (2013) 'The *Guanabara Case*: The First Prosecution of Somali Pirates under the Japanese Piracy Act', *International Journal of Marine and Coastal Law*, 28: 719.
Government of Japan (1993) *National Action Plan for Agenda 21*, online at http://env.go.jp/en/earth/iec/agenda [accessed 1 April 2007].
Hayashi, M. (1997) 'Japan: New Law of the Sea Legislation', *International Journal of Marine and Coastal Law*, 12: 570–80.
Hayashi, M. (2010) 'Japan: Anti-Piracy Law', *International Journal of Marine and Coastal Law*, 25: 143.
Hayashi, M. (2013) 'Japan's Anti-Piracy Law and UNCLOS,' in H. N. Scheiber and J.-H. Paik (eds) *Regions, Institutions, and Law of the Sea: Studies in Ocean Governance*, Leiden: Martinus Nijhoff.
Japanese Association for Coastal Zone Studies (JACZS) (2000) *2000 Appeal: Proposals for the Sustainable Use and Preservation of the Coastal Zone Environment*, online at http://www.jaczs.com/jacz2000.pdf [accessed 1 April 2007] [in Japanese].

Japanese Shipowners' Association (2013) *The Current State of Japanese Shipping*, online at http://www.jsanet.or.jp/data/pdf/data2_2013a.pdf [accessed 10 January 2014] [in Japanese].

Keidanren (2000) *A Grand Design for Oceans in the 21st Century*, online at http://www.keidanren.or.jp/japanese/policy/2000/028.html [accessed 9 July 2006] [in Japanese].

Kisugi, S. (2004) 'The Current Status of Japan's Coastal Areas', in Ocean Policy Research Foundation (OPRF) (ed.) *Kaiyo Hakusho* [*White Paper on Oceans*], Tokyo: OPRF [in Japanese].

Ministry of Education, Culture, Sports, Science and Technology (MEXT) (2002) *Japan's Ocean Policy in the Early 21st Century*, online at http://www.mext.go.jp/b_menu/shingi/gijutu/gijyutu0/toushin/020801b.htm [accessed 1 September 2005] [in Japanese].

Ministry of Land, Infrastructure and Transport (MLIT) (1998) *The 5th Comprehensive National Development Plan: Grand Design for the 21st Century – Promotion of Regional Independence and Creation of Beautiful National Land*, online at http://www.mlit.go.jp/kokudokeikaku/zs5-e/index.html [accessed 1 July 2006].

Ministry of Land, Infrastructure and Transport (MLIT) (2000) *Guidelines for the Development of Integrated Coastal Zones Management Plans*, online at http://www.mlit.go.jp/kokudokeikaku/enganiki/shishin.html [accessed 1 July 2006] [in Japanese].

Ministry of Land, Infrastructure and Transport (MLIT) (2006) *The MLIT Policy Outline on Ocean and Coastal Areas*, online at http://www.mlit.go.jp/kisha/kisha06//01/010621.html [accessed 1 April 2007] [in Japanese].

Nippon Foundation (2002) *Oceans and Japan: A Proposal on the Ocean Policy of Japan for the 21st Century*, online at http://nippon.zaidan.info/seikabutsu/2001/00888/mokuji.htm [accessed 1 September 2005] [in Japanese].

Ocean Policy Research Foundation (OPRF) (2005) *Oceans and Japan: A Proposal for the Ocean Policy of the 21st Century*, online at http://www.sof.or.jp/topics/2005/pdf/20051120_01.pdf [accessed 1 July 2006] [in Japanese].

Oda, S. (1974) 'Conclusion of the Japan–Korea Continental Shelf Agreements,' *Juristo*, 559: 98 [in Japanese].

Science Council of Japan, Liaison Committee on Ocean Science Study (2005) *The Need for Integrated Promotion of Science Relating to Oceans: A Proposal for Comprehensive Ocean Policy-Making*, online at http://www.scj.go.jp/ja/info/kohyo/pdf/kohyo-19-t1031-3.pdf [accessed 1 April 2007] [in Japanese].

Terashima, H. (2010) 'Japan's Ocean Policymaking', Paper presented at the International Conference on National Oceans Policymaking, Taipei, 4–5 August.

Terashima, H. (2013) 'Japan's Establishment as a New Ocean State and the Act on Ocean Policy', *Journal of Island Studies*, 3(1): 76–88 [in Japanese].

10

DEVELOPMENT OF A NATIONAL OCEAN POLICY IN MEXICO[1]

*Porfirio Alvarez-Torres, Antonio Díaz-de-León-Corral,
Gustavo Pérez-Chirinos, Juan Carlos Aguilar, Roberto Rosado,
Fausto Efrén Burgoa, Sofía Cortina, Mariela Ibáñez, Gaëlle Brachet,
Norma Patricia Muñoz Sevilla, Evelia Rivera-Arriaga, and Isaac
Andres Azuz Adeath*

Introduction

This contribution presents recent developments in ocean and coastal policy in Mexico. The Mexican case study needs to be conceived of in the context of the great diversity of coastal and marine ecosystems, as well as the various productive coastal capabilities, numerous social structures, dissimilar levels of development, different types of use and user, unrelated economic interests, and wide variety of views and/or sectoral policies. Preparing a specific ocean and coastal policy is an urgent need that has to be fulfilled taking into account the multi-sectoral nature of these zones, their economic importance, and the increasing number of people living along the coast. This contribution briefly describes the steps taken to develop this policy and identifies the challenges, limitations, strengths, and opportunities for its implementation. In conclusion, the authors propose several recommendations for the successful execution of such a policy.

Basic information and an overview of national ocean policy

Mexico's marine area is larger than its terrestrial area and comprises approximately 11,600 km of coastline, a continental shelf of 393,253 km^2, 1,567,000 hectares of estuaries, and an insular surface of 5,083 km^2. The territorial sea embraces nearly 291,585 km^2 and the exclusive economic zone (EEZ) extends to 2,997,679 km^2 (Arriaga Cabrera et al., 1998; Burke et al., 2001; Contreras and Castañeda, 2004; De la Lanza, 2004; EarthTrends, undated). There are seventeen coastal states, divided into four regions as follows: Gulf of Mexico and Caribbean Sea on the Atlantic Ocean; the California Current Region; the Gulf of California; and the America's Central Pacific Coastal Zone (Sherman and Alexander, 1989).

The total population within a 100 km coastal strip is estimated at 14,572,188, which is 14.9 per cent of the total national population. However, the coastal zone has the highest population growth rate, at 2.8 per cent annually, and the trend for migration to the coast is increasing. It is estimated that the coastal zone has 35,626 coastal settlements, but that 0.1 per cent of them host more than 50,000 inhabitants (Díaz–de-León–Corral and Alvarez, 2004).

Four ocean areas bound Mexico. The convergence of warm and cold ocean currents and rich nutrient blooms results in high biodiversity and endemism (Salazar Vallejo and González,

1993). The ocean areas within Mexican jurisdiction reach 4,000 metres depth in the Sigsbee Trench in the Gulf of Mexico and more than 2,000 metres in the Guaymas Basin in the Pacific Ocean. Numerous volcanoes and hydrothermal vents embrace large colonies of benthic abyssal organisms with high rates of productivity in the Gulf of California.

Within the Mexican coastal region, there are highly productive ecosystems, such as coastal lagoons, mangroves, wetlands, salt marshes, coral reefs, sand dunes, and bays, among others, supporting these important offshore trophic chains (INE, 1995). Coastal ecosystems hold numerous endemic species. Of the twenty-nine animal phyla recognized for the marine environment worldwide, fourteen are endemic to Mexico. Mexico is among the top ten mega–diverse countries, with around 10 per cent of the world's biological diversity (Rodríguez et al., 2003). Mexico occupies second place for reptile, mammal, and ecosystem diversity; third place for endemism; fourth place for amphibians, vascular plants, and total biodiversity; and twelfth for bird diversity. Of the total of the world's endemic species, Mexico has 62 per cent of amphibians, 56 per cent of reptiles, and 32 per cent of mammals (Mittermeier et al., 1997: 503).

An overview of existing oceans and coasts policy mechanisms

Several pillars support the Mexican approach to ocean policy, as follows.

Identification of seventy priority coastal and oceanic regions

The National Commission for Knowledge and Use of Biodiversity (*Comisión Nacional para el Conocimiento y Uso de la Biodiversidad*, CONABIO) has developed the database and bank of these priority regions, which were identified owing to their biodiversity, environmental quality (ecological integrity, endemic species, and richness), economic importance (commercial species, fishing zones, tourism, and strategic resources), and risk and vulnerability criteria (pollution, landscape changes, introduction of exotic species, and impacts from land-based activities). Of these seventy regions, fifty-eight have high diversity, forty-one are under threat, thirty-eight are under pressure from some economic activity, and eight are of geographic importance, but there is no biological information about these last.

Identification of major coastal and oceanic economic activities

Mexico's coastal zone is cross-cut by several economic activities, including: fisheries, aquaculture, and mariculture; tourism; industry; energy production; and ports and transportation. Table 10.1 outlines the economic activities within the coastal regions.

Cross-cutting activities and objectives

There are numerous overlapping issues when economic activities and impacts take place in priority coastal and marine regions. The activities listed in Table 10.1 are of importance to the regional and national economy. However, these activities compete for available natural resources and space, inducing a permanent tension and conflicts among uses and users, and between environmental conservation and economic development objectives.

Coastal and marine priority issues

The results of these intensive productive activities in the coastal zone include fishery overexploitation and decreased fisheries catch, high levels of water pollution, habitat deterioration and loss, diminished landscape quality, unregulated urban growth, and loss of public space and

Table 10.1 Coastal regions and productive activities

Coastal region	Main productive activities
Gulf of Mexico	Energy production and related activities
	Ports and transportation
	Fisheries
Caribbean Sea	Tourism
California Current Region	Fisheries
	Tourism
Gulf of California	Tourism
	Fisheries
Central Pacific	Ports and transportation
	Fisheries
	Tourism

recreational areas. Each of these factors can be used as an indicator for identifying unsustainable natural resource use. The overlapping of economic activities and conservation areas, and the resultant consequences, highlights the need for a coordinated approach to management of the ocean and coastal zones in order to guarantee sustainable development of both realms.

Ocean and coastal policy integration and coordination

On 8 June 2006, Mexico presented and published an integrated policy for the environmental sector in regard to oceans and coasts entitled the *National Environmental Policy for the Sustainable Development of Oceans and Coasts: Strategies for its Conservation and Sustainable Use* (*Política Ambiental Nacional para el Desarrollo Sustentable de Océanos y Costas: Estrategias para su Conservación y Uso Sustentable*, PANDSOC) (SEMARNAT et al., 2006a). The main purposes for PANDSOC were to align and develop policy integration within the environment sector, and to ensure that each federal agency does not continue to work in isolation given its duties and faculties under the 1976 Organic Law of Federal Public Administration (*Ley Orgánica de la Administración Pública Federal*), sectoral plans, and the National Constitution. Previously, there was poor coordination or a lack of coordination with other federal agencies except for specific problems or cases that required such coordination. Each economic and social sector promoted its own projects and activities. For example, the Ministry of Tourism (*Secretaría de Turismo*, SECTUR) fostered the development and implementation of tourism projects; the Ministry of Agriculture, Livestock, Rural Development, Fisheries and Food (*Secretaría de Agricultura, Ganadería, Desarrollo Rural, Pesca y Alimentación*, SAGARPA) supported fishing and agricultural activities in coastal areas; and the Ministry of Environment and Natural Resources (*Secretaría de Medio Ambiente y Recursos Naturales*, SEMARNAT) implemented models for preservation and conservation of natural resources, focusing on sustainable development actions.

In addition to this poor, or lack of, coordination between federal agencies, there was also poor communication among actors at the federal, state, and local levels. This resulted from different visions and competences in the coastal region and the country as a whole. This lack of coordination increased when the different sectors could not reach a consensus on productive activities and conservation strategies, thereby risking the sustainability of efforts made in each region. Thus there was a need to establish a coordinated and integrated strategy to synchronize local and federal objectives concerning productive activities, while considering sustainability criteria in order to ensure the adequate use of available natural resources and to preserve ocean and coastal ecosystems.

Table 10.2 summarizes the institutions and organizations that should be involved in the development of a state policy for oceans and coasts in Mexico. Each of these stakeholders has its own vision for the sustainable management of ocean and coastal zones, as well as direct competence for natural resources.

Policy development process

The project to establish PANDSOC was undertaken at three levels, as follows.

1 As a social and academic demand from the National Council for Sustainable Development, an advisory council to SEMARNAT
2 At a coordination level under a project called 'Agenda del Mar' (Oceans Agenda)

This initiative, led by the President's Office, aimed to create a discussion forum for all of the perspectives and activities promoted among the public federal administration and other sectors

Table 10.2 Stakeholders in an integrated ocean and coastal strategy for Mexico

	Institution
Federal	Ministry of Agriculture, Livestock, Rural Development, Fisheries and Food (*Secretaría de Agricultura, Ganadería, Desarrollo Rural, Pesca y Alimentación*, SAGARPA)
	National Fisheries Commission (*Comision Nacional de Acuacultura y Pesca*, CONAPESCA)
	National Fisheries Institute (*Instituto Nacional de Pesca*, INP)
	Ministry of Tourism (*Secretaría de Turismo*, SECTUR)
	National Trust Fund for Tourism Development (*Fondo Nacional de Fomento al Turismo*, FONATUR)
	Ministry of the Navy (*Secretaría de Marina*, SEMAR)
	Ministry of Communications and Transport (*Secretaría de Comunicaciones y Transportes*, SCT)
	Ministry of Economy (*Secretaría de Economía*, SE)
	Ministry of Social Development (*Secretaría de Desarrollo Social*, SEDESOL)
	Ministry of Energy (*Secretaría de Energía*, SENER)
	Mexican State Oil Corporation (*Petróleos Mexicanos*, PEMEX)
	Federal Electricity Commission (*Comisión Federal de Electricidad*, CFE)
	Ministry of Environment and Natural Resources (*Secretaría de Medio Ambiente y Recursos Naturales*, SEMARNAT)
	National Commission for Natural Protected Areas (*Comisión Nacional de Áreas Naturales Protegidas*, CONANP)
	National Water Commission (*Comisión Nacional del Agua*, CONAGUA)
	National Forestry Commission (*Comisión Nacional Forestal*, CONAFOR)
	National Ecology Institute (*Instituto Nacional de Ecología*, INE)
	National Congress
Local	State governments
	Municipal governments
	Local Congress
Social	Non-government organizations (NGOs)
	Coastal communities
Academia	Universities and research institutions

with a stake in ocean and coastal activities. Representatives from the federal agencies listed in Table 10.2 participated in this work. Agenda del Mar started in 2000, with a team that presented development projects and proposals for the regulation of some coastal and ocean activities.

3   At an operational level, led by SEMARNAT

SEMARNAT led the project to prepare PANDSOC, which was published in 2006. It aimed to establish an inter- and intra-institutional coordination mechanism for efficiently address-ing current and emerging management and planning issues in the ocean and coastal zones. SEMARNAT exercises its legal mandate in coastal areas through the Maritime and Coastal Federal Zone Administration (*Zona Federal Marítima Terrestre y Areas Costeros*, ZOFEMATAC), and by protecting, conserving, assessing, and promoting sustainable exploitation of natural resources, ecosystems, and biodiversity.

Agenda del Mar transitioned to a decentralized model, granting state governments partial, but limited, stewardship over the coastal zone. PANDSOC enhanced sound coordination at the operational level, which was a key factor in the success of the implementation process. It referred to the objectives, strategies, and results achieved by SEMARNAT.

The background process and previous efforts to construct PANDSOC

The first attempt to develop an integrated oceans and coastal policy was made in 2001. Nineteen branches of SEMARNAT participated in an exercise to identify the main activities and dif-ficulties faced by the Ministry regarding the coastal zone. However, this strategy ended in a diagnosis, but without any treatment.

In 2004, the need for coordination was evident, as several coastal issues arose that had to be addressed by more than one department within the Ministry. At this point, the 2001 diagnosis was reconsidered, and several actions were undertaken under the leadership of the Ministry's Undersecretary for Planning and Environmental Policy, through its Directorate General for Environmental Policy, Sectoral and Regional Integration (*Dirección General de Política Ambiental Integración Regional y Sectorial*, DGPAIRS). The main tasks focused on were broadening the scope of the Ministry's institutional group, updating the diagnosis document, establishing a position paper on oceans and coastal issues, and generating a discussion forum for coordinating actions concerning the coastal zone from every department within the Ministry.

PANDSOC contains the main strategies and guidelines that define the environmental policy regarding the oceans and coasts of Mexico. This public policy entered into force as a joint work-ing programme within SEMARNAT. PANDSOC served as the basis for discussing all federal policies related to integrated management of oceans and coasts through the Agenda del Mar project in the President's Office.

The main objectives of PANDSOC and its strategies are:

- to develop a strategy for the integrated management of oceans and coasts;
- to strengthen coordinated actions between and within coastal and marine-related institutions;
- to promote social and economic welfare through environmental and biodiversity conserva-tion and the sustainable use of coastal and marine natural resources;
- to strengthen the institutional (legal, normative, and administrative) framework for man-agement of oceans and coasts; and
- to develop an information system specific to oceans and coastal issues.

To achieve these objectives, PANDSOC embraced and promoted the following principles.

- *Ecosystem-based management* – An integrated oceans and coastal management strategy needs to focus on the analysis of the ecosystem as a unit. This is the main principle of the environmental policy promoted by SEMARNAT.
- *Multiple-use management* – Multiple economic activities should be developed, encouraged, and managed in the ocean and coastal zones.
- *Sustainability* – All activities in the oceans and coastal zones must be undertaken in a sustainable manner to maintain such activities and to ensure that future generations can use and enjoy marine natural resources.
- *Precautionary approach* – Even if some zones require intensive restoration, it is important to impose the precautionary principle in order to promote the conservation of the well-preserved zones and to deter further environmental damage. In cases of poor or lack of scientific evidence, a precautionary approach must be used.
- *Integrated approach* – Successful implementation of the policy will require the participation and coordination of all relevant actors.
- *Protection of biodiversity* – This principle complements the ecosystem-based perspective.
- *Adaptive management* – This principle will be implemented to ensure learning and adjustment, and consideration of both the spatial and temporal perspectives.
- *Participatory governance* – Social participation is a basic element for successful implementation of any ocean and coastal management strategy.

PANDSOC was implemented within a framework of transparency, access to information, and accountability, and promoted by all federal agencies.

One of the most important elements of PANDSOC is the consensus reached on the geographic definition of the coastal zone. The coastal zone region is defined as 'the zone influenced by coastal affairs and activities'.

PANDSOC is divided into five main subjects that assemble the most relevant topics for the Ministry. For the purpose of facilitating the organization and integration of the information, every section was outlined in the following order: a brief diagnosis of the subject; a description of the Ministry's competences; evaluation of the main problems in addressing the subject (the most relevant section in terms of policymaking); and evaluation of implemented policies, identification of areas of opportunity, and establishment of guidelines to address the identified problems.

The five main subjects are as follows.

1 *Legal framework and economic instruments* – This subject was divided into two main topics:

   a   the legal framework examination; and
   b   an exhaustive review of the economic instruments implemented to achieve environmental policies.

2 *Ecosystems and biodiversity* – This subject includes the information about species (both endangered and protected), ecosystems diversity, and relevant conservation areas, as well as natural protected areas and priority marine regions.

3 *Economic activities and their environmental impact* – This section was divided into the most important activities undertaken in the coastal and marine zones in terms of their production, as well as environmental impact.

4   *Enforcement mechanisms and stakeholder participation* – This subject contains the main elements to strengthen the enforcement system and dealt with all aspects in which civil participation was essential.

5   *Research agenda, education, and public outreach* – This section was created with the aim of setting an agenda for further research in accordance with identified needs in the other subjects. A second objective was to gather and integrate available information within the Ministry and other agencies about ocean and coastal issues.

Finally, to meet its objectives successfully, the project was divided into four stages.

- *Stage 1* (short-term, 2004–05)

  o   Build coordination mechanisms between the environmental sector and other areas related to issues concerning oceans and coasts.
  o   Develop a position paper.
  o   Prepare PANDSOC, based on the sectoral diagnosis analysis.
  o   Promote public consultation on the contents of the proposed PANDSOC.

- *Stage 2* (medium–term, 2005–06)

  o   Obtain consensus on the proposed PANDSOC among all federal agencies and authorities linked to ocean and coastal issues.
  o   Strengthen institutional capacity-building capabilities to deal with coastal and ocean issues.
  o   Draft PANDSOC.

- *Stage 3* (medium–term, 2006)

  o   Build consensus for PANDSOC among the public and promote participation in the decision-making process.
  o   Publish and implement PANDSOC.

- *Stage 4* (long–term)

  o   Evaluate, adapt, and improve PANDSOC.
  o   Create a Mexican Intergovernmental Commission for Oceans and Coastal Affairs (*Comisión Intersecretarial para el Manejo Sustentable de Mares y Costas*, CIMARES)
  o   Develop the cross–cutting National Policy for Oceans and Coasts of Mexico (*Política Nacional de Mares y Costas de México*, PNMC) (CIMARES, 2012).

Implementing PANDSOC

Once the second stage of the strategy was initiated, PANDSOC was incorporated into the Agenda del Mar project. This resulted in a scenario in which all federal executive agencies jointly participated in oceans and coastal management. Various federal agencies and their branches were linked to all aspects of coastal zone management. Thus it was necessary to identify areas of conflict, cooperation opportunities, and scenarios and prospects to enhance national development. Coordination of these federal agencies allowed the PANDSOC objectives to be defined according to their own interests, while at the same time maintaining PANDSOC's main goal of integrated management of ocean and coastal areas.

Successful coordination was based on the following strategies:

- integration of formal working groups to ensure that every relevant actor was considered and had competence to participate in the decision-making process;
- establishment of a specific agenda and objectives for each working group;
- definition of a National Policy for Oceans and Coasts, the most important objective of which being to institutionalize and guarantee that such a policy includes all sectoral activities and will be implemented and enforced at the national level; and
- implementation of PANDSOC being reinforced by both formal and informal collaboration between the involved actors.

In considering the issue of agency coordination, it is important to recognize that each actor or sector has different interests in the coastal zone, most of them without a range of order or priority.

Table 10.3 Authority of federal agencies related to ocean and coastal zones

Agency	Competences
Ministry of the Interior (*Secretaría de Gobernación*, SEGOB)	Administration of islands in the federal jurisdictional territory
	Coordination and response on hazards and disasters
Ministry of the Navy (*Secretaría de Marina*, SEMAR)	Execution of sovereignty and enforcement of law in the coastal, islands, and economic zones
	Coordination and elaboration of cartography and oceanic charts
	Coordination of research expeditions in federal zones
	Coordination and response on marine hazards and disasters
Ministry of Social Development (*Secretaría de Desarrollo Social*, SEDESOL)	Coordination of urban planning and development
	Human habitat coordination
Ministry of Environment and Natural Resources (*Secretaría de Medio Ambiente y Recursos Naturales*, SEMARNAT)	Coordination of conservation, restoration, and preservation of natural resources policies
	Administration of the federal coastal and maritime zone
	Creation of natural protected areas and coordination of productive activities in these areas
Ministry of Agriculture, Livestock, Rural Development, Fisheries and Food (*Secretaría de Agricultura, Ganadería, Desarrollo Rural, Pesca y Alimentación*, SAGARPA)	Coordination of public policies addressing fisheries, livestock, agriculture, and rural development
Ministry of Energy (*Secretaría de Energía*, SENER)	Coordination of energy extraction and distribution policies
Ministry for Agrarian Reform (*Secretaría de la Reforma Agraria*, SRA)	Coordination of the policies addressing agriculture entities
Ministry of Tourism (*Secretaría de Turismo*, SECTUR)	Coordination of tourism policy, in particular for coastal zone activities and cruise ships
Ministry of Communications and Transport (*Secretaría de Comunicaciones y Transportes*, SCT)	Coordination of maritime transportation and communication (ports, navigation permits, coastal infrastructure, etc.)
Ministry of Economy (*Secretaría de Economía*, SE)	Coordination of export/import policies

Table 10.3 summarizes the responsibilities or competence of federal agencies in the oceans and coastal zone, indicating converging, and sometimes overlapping, responsibilities and competences of federal agencies therein.

Main objectives

The Mexican federal government indicated its determination to move towards the alignment of economic growth and sectoral development, accompanied by a continuous improvement in the environmental performance, by streamlining actions among sectors that use ocean and coastal natural resources. The objective of its oceans policy is to promote sustainable development, while ensuring the quality of life of coastal communities and a pollution-free environment, as well as protecting and conserving the coastal and marine heritage. As noted earlier, the main objectives of PANDSOC are to promote governance and to improve sustainable use of coastal and ocean resources, to strengthen critical habitats, to promote adequate ocean and coastal use planning, to promote sustainable development of coastal areas, and to strengthen control of land-based sources of pollution. This oceans policy acted as a catalyst for sea and coastal use planning activities in an area of the world that is environmentally and economically complex. Implementing this oceans policy certainly presented tremendous challenges.

Principles and pillars

To facilitate effective cooperation for conservation and enhancement of Mexico's oceans and coastal ecosystems, the federal government was determined to participate actively, together with civil society, to achieve sustainable development and to increase government accountability regarding the enforcement of environmental laws. In this context, as noted, PANDSOC incorporates key policy principles, including sustainability, ecosystem-based management, adaptive management, the precautionary approach, multiple-use management, an integrated approach, transparency and accountability, and participatory governance. All of these principles are aimed at enhancing the participatory process and conflict resolution for all government levels, civil society, private entrepreneurs, and non-governmental organizations (NGOs).

Considering both ecosystem-based management and integrated management approaches as the key components to achieving environmental sustainability, PANDSOC was based on four pillars, as follows.

1   To enhance the quality of life of coastal communities
2   To promote harmony between economic and social development and sustainable conservation of the oceans and coasts
3   To achieve social equity and poverty alleviation
4   To promote sustainable values and ethical appropriation

Contents and policy guidelines

In keeping with its objectives, PANDSOC focuses on four major issue areas:

- the sustainable development of the coastal zone;
- the conservation of coastal and marine resources and ecosystems;
- mandates and rights for Mexicans to the coastal and marine zones; and
- a coastal and marine zones integrated management approach.

PANDSOC comprised the following sections:

1 Legal and institutional frameworks
2 Environmental policy tools and instruments
3 Ecosystems and biodiversity
4 Sectoral activities and human settlements
5 Governance and public participation
6 Information systems (capacity-building, research, development, education, and public awareness).

PANDSOC policy guidelines can be summarized as follows.

1 Build institutional capacity for the coastal and marine zones.
2 Develop policies in consistency to the national policy for sustainable development and the environmental principles within each agency's jurisdiction.
3 Promote inter-institutional coordination through cross-cutting issues and approaches.
4 Create regional councils for coastal and marine development.
5 Provide benchmarks for sustainable development in the coastal and marine zones.
6 Generate a clearing house on coastal and ocean data and information.
7 Improve environmental law framework and to develop innovative policy tools.
8 Maximize the social and economic benefit.
9 Assess security, risks and vulnerability to climate change.
10 Evaluate of the policy results and outcomes.

Current examples of application in Mexico

Application of ecosystem-based management stresses the importance of environmental issues as a cross-cutting sector. Currently, Mexico has directed its efforts to adopt the ecosystem-based approach (SEMARNAT et al., 2007) through ocean-use planning initiatives in the Gulf of California (SEMARNAT et al., 2006b), the Gulf of Mexico and the Caribbean Sea, which was completed in 2012 (SEMARNAT et al., 2006c), and the Northern Mexican Pacific Ocean, which is to be completed in 2014 (SEMARNAT, undatedb). Further, it completed the Transboundary Diagnostic Analysis and Strategic Action Programme for the Gulf of Mexico Large Marine Ecosystem (GoM LME) project in 2013 (GoM LME, undated), developed the Regional Programme of Action for the Yucatan Peninsula for the Protection of the Marine Environment from Land-based Sources of Pollution (SEMARNAT et al., 2006d), and will soon undertake efforts on the Central-Southern Pacific coast of Mexico (SEMARNAT, undateda) and in the Caribbean Large Marine Ecosystem Project (CLME Project, undated).

Difficulties and recommendations for PANDSOC implementation

Given the scope of this policy, it certainly faced some problems during the design and implementation phases. Some of the conflicts that arose and the strategies with which they were addressed were as follows:

1   Differences between sectoral objectives and the difficulty integrating the provided criteria

As noted earlier, most agencies have different interests in the ocean and coastal zones, and are seeking to fulfil their own sectoral agenda. To overcome the coordination problem, policy initiatives focused on the objectives for which consensus could be built between the different agencies to achieve long-term benefits.

2   Established power relationships that might lead to conflicts when encouraging other sectors to adopt specific policy roles or when granting legal attributes or authority

This was one of the most difficult issues to overcome, given the current legal framework. The establishment of a unique coordination agency within the highest hierarchy to generate consensus among different actors is one of the most important recommendations that will address this problem.

3   Resistance from some social stakeholders who are current beneficiaries of the lack of coordination (or the status quo).
4   Different visions on priorities and strategies from local governments, with the divergence ranging from general and particular visions
5   Lack of administrative capacity of most municipal authorities.

The following points are recommended to overcome the difficulties described above:

- *Higher coordination office* – The President's Office or the Ministry of the Navy should act as the coordination entity, because these might provide assertive and effective harmonization, and have the capacity to manage and generate consensus among the diverse interests of the different government agencies.
- *Institutionalization of the ocean and coastal policy* – It is important to build infrastructure and a culture that supports the methodology, practices, and procedures for oceans and coasts, even after those who originally defined them are gone from the institutions or no longer in the arena. Institutionalization of oceans and coastal policy will overcome the challenge of requiring government bureaucracies to reach a consensus on every problem area.
- *Decisions must be translated into public policies* – Considering the momentum of Mexico's newly initiated decentralization policy towards the coastal zone, there was, and is, an urgent need for public policymakers to incorporate the legal framework, along with the concerns of coastal communities, users, and stakeholders, into the public policymaking process.
- *Operationalization* – This is the process of translating conceptual policies into specific, observable actions, with results and outcomes that can be measured. It is therefore recommended that a task force group be established to oversee and survey the policy implementation process.
- *Participation of the scientific community* – Decision-making for coastal and ocean zones should be based on accurate, up-to-date, and sound scientific information. In addition, the scientific community is willing to support the recent initiative to establish a national coordinated network for monitoring and observation of the oceans such as (or similar to) the Global Ocean Observing System (GOOS).
- *Involvement of state and municipal governments* – To be successful, this project must involve state and municipal governments in the management of oceans and coasts. Local governments have the best opportunity to implement, survey, enforce, and evaluate the ocean and coastal policy.

- *Stakeholder involvement* – Another key actor is the public, particularly from coastal municipalities, since it is the public that must to promote stewardship and ownership of the oceans and coasts. In addition, SEMARNAT considers public involvement to be mandatory in the policy- and decision-making processes.
- *Implementation of environmental policy instruments* – Policies protecting and managing coasts and oceans can be grouped into three categories: governance, regulatory (limits to access or use), and economic (incentives or disincentives).

    o Policies that support localized management mostly revolve around decentralization of authority to local governments and communities, use of protected areas and integrated coastal management regimes, various types of regulation governing use of an area or the resource, and appropriate economic incentives, such as user fees, trust funds, or compensation payments.

    o Policies that support global (national and international) protection and management of coasts and oceans include international or national marine parks, transnational or national integrated coastal management programmes, legal frameworks that recognize local management regimes, long-term lease agreements and management rights, valuation tools to raise awareness, property rights, adjustment of various national laws, bans on import/export of vulnerable species, pollution taxes, conservation tax write-offs, market entry fees, debt-for-nature swaps, and carbon emission taxes, among others. These instruments seek to ensure the sustainability of coasts and oceans.

- *Clearing house on coastal and ocean zones* – Equal access to sound and accurate information about ocean and coastal issues is essential for policy- and decision-making, and for translating such policies into tangible benefits for the environment and for society. Information diffusion and availability is a social and ethical requirement for sustainable development, for promoting collaborative actions, encouraging expertise exchange and active participation and involvement, establishing and disseminating guidelines, workflow models, and best practices, and carrying out joint projects.
- *Policy evaluation* – Measuring the impact of a policy or programme usually involves measuring the counterfactual – that is, the number of positive outcomes that would have been observed amongst the eligible population if the programme were not in place. Sometimes, evaluations need to focus only on either the process *or* the impact of the policy or programme. More commonly, evaluation of a policy will involve *both* elements, and thus both types of evaluation.
- *Public observatories* – These would facilitate the follow-up and benchmarking of policy progress at the local and regional levels, with strong public and informed participation.

Recent developments in ocean and coastal policy in Mexico (2007–14)

During the past seven years, Mexico has made significant progress in marine and coastal policy. Under the leadership of the Directorate General for Environmental Policy, Sectoral and Regional Integration (*Dirección General de Política Ambiental Integración Regional y Sectorial*, DGPAIRS), land and sea use (*ordenamiento ecológico*) or LSUP has been central to the sustainable management that promotes green development in a cross-cutting fashion and adaptation policies to climate change.

Sea-level rise is one of the most significant climate-change-derived challenges facing the coastal zone of Mexico. The LSUP provides the key tool for local governments for addressing risk and vulnerability issues. The ordinance is a binding instrument, which grants legal certainty for promoting green investments, enhancing sustainable activities, and prohibiting damaging practices. Therefore, coastal municipalities with a published LSUP model are able find the blueprint with which to identify and address climate change hotspots. As a result, the ordinance has transformed the way in which risk and vulnerability have been treated at local and national levels in the policy agendas, and how resources are mobilized to prevent and alleviate coastal disasters.

In marine areas within the territorial sea and the Mexican exclusive economic zone (EEZ), LSUP ordinances are soft-law agreements for actors and users of the ocean space. Establishment of LSUPs in four marine areas have been initiated, involving the three government levels, academia, NGOs, organized social groups, stakeholders, and the private sector, in order to provide a general environmental, social, and economic assessment of the threats that each marine zone faces, such as conflict among users, too many users, marine pollution (including from ballast water), illegal fishing, and issues related to oil and gas exploration and extraction. To date, two LSUPs have been decreed (for the Gulf of California, and the Gulf of Mexico and Caribbean), a third is under public consultation (for the Northern Pacific), and the Southern Pacific Marine LSUP is under development (SEMARNAT et al., 2006b; 2006c).

Another major step was made with the Temporary Employment Programme (*Programa de Empleo Temporal*, PET) (SEDESOL, undated). This innovative action was created to alleviate the economic suffering caused by natural disasters, particularly in the fisheries sector. The PET has evolved into a powerful incentive to promote numerous actions that support the coastal environment, such as mangrove and wetlands hydrological systems restoration, marine mammal and turtle protection, and beach, river, *cenotes*,[2] and wetlands clean-up actions. The PET has also proven its potential when directed towards increasing coastal resilience to climate change.

The strengthening of capacity of local institutions through specific funding directed to this purpose has helped to create environmental and LSUP departments, given direction to numerous states and municipalities, and has equipped and trained state ministries. This funding has also contributed to the creation of state environmental protection agencies, enhanced legal frameworks, and the development of air quality networks and vehicle verification processes.

A major achievement has been the development of new environmental public policies that allow for a better understanding of the ocean dynamics and complexity through intra-governmental coordination, which also resulted in improved assessment and management approaches to environmental emergencies (see Box 10.1). Additional achievements include:

- the creation of the Intergovernmental Commission for Oceans and Coastal Affairs (*Comisión Intersecretarial para el Manejo Sustentable de Mares y Costas*, CIMARES) (SEMARNAT, undated*f*);
- the design and approval of the National Policy for Oceans and Coasts of Mexico (*Política Nacional de Mares y Costas de México*, PNMC) (CIMARES, 2012);
- funding allocation through the PET using environmental criteria, which was used to finance thirty-one out of thirty-two federal offices of SEMARNAT, as well as four other environmental entities:

    ○ the National Water Commission (*Comisión Nacional del Agua*, CONAGUA) (SEMARNAT, undated*d*);
    ○ the National Forestry Commission (*Comisión Nacional Forestal*, CONAFOR) (SEMARNAT, undated*e*);

- o   the Commission of Natural Protected Areas (*Comisión Nacional de Áreas Naturales Protegidas*, CONANP) (SEMARNAT, undated*e*); and
- o   the Federal Attorney for Environmental Protection (*Procuraduria Federal de Proteccion al Ambiente*, PROFEPA) (SEDESOL, undated; SEMARNAT, undated*g*);

- the development of a bilateral marine agenda by means of a memorandum of understanding involving SEMARNAT and the United States Environmental Protection Agency (EPA);[3]
- implementation of the trilateral Mexico/United States Global Environment Facility (GEF)/United Nations Industrial Development Organization (UNIDO) Gulf of Mexico Large Marine Ecosystem (GoL LME) programme, which was granted the 'Gulf of Mexico Guardian' Award by the US EPA in June 2013, as well as the Ecological Award by SEMARNAT (GoM LME, undated);
- the Decree of the General Territory Land Use Planning Programme, which provides for the first time, a baseline for territorial planning and as result of which there was a three-fold increase of LSUPs for coastal territories, at 43 per cent, and of 34 per cent for marine LSUPs (SEMARNAT, undated*h*).

Box 10.1   Policy documents (2006–14)

2006 – PANDSOC was published and entered into force.

2007 – In February 2007, President Felipe Calderón launched the National Strategy for Ecological Ocean and Coastal Use Planning and Management (*Estrategia Nacional para el Ordenamiento Ecológico del Territorio en Mares y Costas*) in Mazatlan, Mexico (SEMARNAT et al., 2007), a major strategy covering all Mexican ocean areas and the most pressured coastal zones in the country, by investment, infra-structure, or urban development.

2008 – During the presentation of the National Strategy, President Calderón directed the establishment of CIMARES, which was installed in office in December 2008 according to the Presidential Decree of June 2008 that created it. Since its inception, the Commission – composed of ten ministries and four decentral-ized agencies, including the Mexican State Oil Corporation (*Petróleos Mexicanos*, PEMEX) and the Federal Electricity Commission (*Comisión Federal de Electricidad*, CFE) – meets two times a year to approve important agreements and policy instru-ments (SEMARNAT, undated*f*).

2011 – The National Policy for Oceans and Coasts of Mexico: Integrated Management of the Most Dynamic Regions of the Country (CIMARES, 2012) and the National Strategy for Management of Mangrove Ecosystems (*Estrategia Nacional para la Atención del Ecosistema de Manglar*) were established (SEMARNAT, 2012), while the National Strategy for Marine and Coastal Biodiversity in Mexico (*Estrategia Nacional de Atención a la Biodiversidad Marina y Costera de México*) is cur-rently undergoing a public consultation process (SEMARNAT, 2011).

2013 – SEMARNAT finished the policy paper *Coastal Dunes Ecosystem Management: Ecological Criteria and Strategies* (SEMARNAT, 2013), and the National Oceans Strategy was presented to the GEF (Díaz-de-León-Corral, 2013).

In view of the challenges currently faced by Mexico in its coastal and ocean ecosystems, it is important to acknowledge that the decision-making process for the coastal and marine zones in Mexico requires a strong partnership between the state and the private sector in order to carry out best practices and strategies, including those that address externalities affecting the environment and natural resources.

Government institutions and governance practices have been strengthened by means of integrated coastal zone management (ICZM) approaches and increasing public participation. Land and sea use meetings at coastal local levels are building strategic paths through conflict resolution, enhancing transparency, accountability, and effective channels of information and communication. All of these add up to bottom-up decision-making in coastal areas.

Mexico has made a great effort in the last ten years or so to increase its institutional capacity, and the public involvement and awareness in ocean and coastal matters. The idea is to take advantage of ocean and coastal economic opportunities, while improving the livelihoods of inhabitants and conserving and improving the natural ocean and coastal patrimony. These efforts are expected to increase and to be consolidated in the years to come.

Notes

1 The contents of this document are based on the authors' opinions and do not necessarily represent the position of the Ministry of Environment and Natural Resources of Mexico.
2 *Cenotes* are sinkholes formed by the dissolution of carbonate bedrock, exposing groundwater or seawater: see, e.g., Schmitter-Soto et al. (2002).
3 Memorandum of Understanding between the Environmental Protection Agency of the United States of America and the Ministry of Environment and Natural Resources of the United Mexican States Concerning Environmental Cooperation in Coastal and Aquatic Ecosystems, signed 24 September 2012, Washington DC, online at http://www2.epa.gov/international-cooperation/memorandum-understanding-mou-between-epa-and-semarnat-cooperation-coastal [accessed 11 July 2014].

References

Arriaga Cabrera, L., Vázquez Domínguez, E., González Cano, J., Jiménez Rosenberg, R., Muñoz López, E., and Aguilar Sierra, V. (eds) (1998) *Regiones Prioritarias Marinas de México* [*Priority Marine Regions of Mexico*], México: Comisión Nacional para el Conocimiento y Uso de la Biodiversidad.
Burke, L., Kura, Y., Kassem, K., Revenga, C., Spalding, M., and McAllister, D. (eds) (2001) 'Coastal Zone: Extent and Change', in *Pilot Analysis of Global Ecosystems: Coastal Ecosystems*, online at http://pdf.wri.org/page_coast_005_extent.pdf [accessed 27 November 2006].
Caribbean Large Marine Ecosystem (CLME Project) (undated) 'About CLME', online at http://clmeproject.org/briefclme.html [accessed 10 July 2014].
Comisión Intersecretarial para el Manejo Sustentable de Mares y Costas (CIMARES) (2012) *National Policy for Oceans and Coasts of Mexico: Integrated Management of the Most Dynamic Regions*, online at http://web2.semarnat.gob.mx/temas/ordenamientoecologico/cimares/Documents/pnmc%20_ingles_final.pdf [accessed 10 July 2014].
Contreras, E. F., and Castañeda, L. O. (2004) 'La Biodiversidad de las Lagunas Costeras [The Biodiversity of Coastal Lagoons]', *Ciencias*, 76(Oct–Dec): 46–56.
De la Lanza, E. G. (2004) 'Gran Escenario de la Zona Costera y Oceánica de México [The Great Coastal Zone and Oceanic Landscape of Mexico]', *Ciencias*, 76(Oct–Dec): 4–13.
Díaz-de-León-Corral, A. (2013) 'Estrategia Mexicana para Mares y Costas [Mexican Strategy for the Sea and Coasts]', Presentation to Diaologo GEF-Mexico, 13 November, México, online at http://www.semarnat.gob.mx/archivosanteriores/dialogonacional/Documents/PLENARIAS/Sesion%20I.%20Planes,%20Programas%20y%20Estrategias/Estrategias%20Nacionales/Mares%20y%20Costas.%20Antonio%20Diaz%20de%20Leon.pdf [accessed 10 July 2013].
Díaz-de-León-Corral, A., and Álvarez, P. (2004) 'Océanos, Costas y Gestión de Recursos Marinos [Oceans, Coasts, and the Management of Marine Resources]', *Economía Informa* 328: 36–45.

EarthTrends (undated) 'Coastal and Marine Ecosystems: Mexico', online at http://earthtrends.wri.org [accessed 27 November 2006] (no longer available).

Gulf of Mexico Large Marine Ecosystem Project (GoM LME) (undated) 'Video of GoM LME Achievements', online at http://gomlme.iwlearn.org/en/news-1/video-of-gom-lme-achievements-1 [accessed 10 July 2014].

Instituto Nacional de Ecología (INE) (1995) *Áreas Naturales Protegidas: Economía e Instituciones* [*Protected Natural Areas: Economics and Instititions*], México: INE.

Mittermeier, R. A., Robles, G. P., and Mittermeier, G. C. (1997) *Megadiversidad: Los Países Biológicamente Más Ricos del Mundo* [*Megadiversity: Earth's Biologically Richest Nations*], México: Cemex, S.A. de C.V.

Rodríguez, P., Soberón, J., and Arita, H. T. (2003) 'El Componente Beta de la Diversidad de Mamíferos de México [The Beta Component of Mammalian Diversity in Mexico]', *Acta Zoológica Mexicana*, 89: 241–59.

Salazar Vallejo, S. I., and González, N. E. (1993) 'Panorama y Fundamentos para un Programa Nacional [Overview of and Basis for a National Programme]', in S. I. Salazar Vallejo and N. E. González (eds) *Biodiversidad Marina y Costera de México*[*Marine and Coastal Biodiversity in Mexico*], México: CONABIO/ CIQRO.

Schmitter-Soto, J. J., Comín, F. A., Escobar-Briones, E., Herrera-Silveira, J., Alcocer, J., Suárez-Morales, E., Elías-Gutiérrez, M., Díaz-Arce, V., Marn, L. E., and Steinich, B. (2002) 'Hydrogeochemical and Biological Characteristics of Cenotes in the Yucatan Peninsula (SE Mexico)', *Hydrobiologia*, 467: 215–28.

Secretaría de Desarrollo Social (SEDESOL) (undated) 'Programa de Empleo Temporal (PET) [Temporary Employment Programme]', online at http://www.sedesol.gob.mx/es/SEDESOL/Empleo_Temporal_ PET [accessed 25 June 2014].

Secretaría de Medio Ambiente y Recursos Naturales (SEMARNAT) (undated*a*) 'Bitácora Ambiental del Programa de Ordenamiento Ecológico Marino y Regional del Pacífico Centro Sur [Environmental Log Book of the Central and South Pacific Marine Ecological and Regional Land and Marine Use Planning Programme]', online at http://www.semarnat.gob.mx/temas/ordenamiento-ecologico/bita-cora-ambiental/bitacora-del-programa-de-ordenamiento-ecologico-0 [accessed 10 July 2014].

Secretaría de Medio Ambiente y Recursos Naturales (SEMARNAT) (undated*b*) 'Bitácora de Ordenamiento Ecológico Marino y Regional del Pacífico Norte [Ocean Use Planning Process of the Northern Mexican Pacific Ocean]', online at http://www.semarnat.gob.mx/temas/ordenamiento-ecologico/ bitacora-ambiental/bitacora-de-ordenamiento-ecologico-marino-y-regional [accessed 10 July 2014].

Secretaría de Medio Ambiente y Recursos Naturales (SEMARNAT) (undated*c*) 'Comision Nacional de Areas Naturales Protegidas (CONANP) [National Commission for Natural Protected Areas]', online at http://www.conanp.gob.mx/index.php [accessed 10 July 2014].

Secretaría de Medio Ambiente y Recursos Naturales (SEMARNAT) (undated*d*) 'Comision Nacional del Agua (CONAGUA) [National Water Commission]', online at http://www.conagua.gob.mx/inicio. aspx [accessed 10 July 2014].

Secretaría de Medio Ambiente y Recursos Naturales (SEMARNAT) (undate*e*) 'Comision Nacional Forestal (CONAFOR) [National Forestry Commission]', online at http://www.conafor.gob.mx/ web/ [accessed 10 July 2014].

Secretaría de Medio Ambiente y Recursos Naturales (SEMARNAT) (undated*f*) 'Histórico CIMARES [History of CIMARES]', online at http://web2.semarnat.gob.mx/temas/ordenamientoecologico/ cimares/Paginas/Cimares.aspx [accessed 10 July 2014].

Secretaría de Medio Ambiente y Recursos Naturales (SEMARNAT) (undated*g*) 'Procuraduria Federal de Proteccion al Ambiente (PROFEPA) [Federal Attorney for Environmental Protection]', online at http://www.profepa.gob.mx/ [accessed 10 July 2014].

Secretaría de Medio Ambiente y Recursos Naturales (SEMARNAT) (undated*h*) 'Programa de Ordenamiento Ecológico General del Territorio (POEGT) [General Territory Land Use Planning Programme]', online at http://www.semarnat.gob.mx/temas/ordenamiento-ecologico/programa-de-ordenamiento-ecologico-general-del-territorio-poegt [accessed 10 July 2014].

Secretaría de Medio Ambiente y Recursos Naturales (SEMARNAT) (2011) *Estrategia Nacional de Atención a la Biodiversidad Marina y Costera de México* [*National Strategy for Marine and Coastal Biodiversity in Mexico*], online at http://web2.semarnat.gob.mx/temas/ordenamientoecologico/cimares/Documents/ enabmc_consultapublica.pdf [accessed 10 July 2014].

Secretaría de Medio Ambiente y Recursos Naturales (SEMARNAT) (2012) *Estrategia Nacional para la Atención del Ecosistema de Manglar* [*National Strategy for Management of Mangrove Ecosystems*], online at http://web2.semarnat.gob.mx/temas/ordenamientoecologico/cimares/Documents/ENAEM_ver.edicion.pdf [accessed 10 July 2014].

Secretaría de Medio Ambiente y Recursos Naturales (SEMARNAT) (2013) *Coastal Dunes Ecosystem Management: Ecological Criteria and Strategies*, México: SEMARNAT.

Secretaría de Medio Ambiente y Recursos Naturales (SEMARNAT)/Subsecretaría de Planeación y Política Ambiental/Dirección General de Política Ambiental, Integración Regional y Sectorial (DGPAIRS) (2006a) *Política Ambiental Nacional para el Desarrollo Sustentable de Océanos y Costas: Estrategias para su Conservación y Uso Sustentable* [*National Environmental Policy for the Sustainable Development of Oceans and Coasts: Strategies for its Conservation and Sustainable Use*], online at http://www.invemar.org.co/redcostera1/invemar/docs/RinconLiterario/2013/JC-213.pdf [accessed 27 November 2006].

Secretaría de Medio Ambiente y Recursos Naturales (SEMARNAT)/Subsecretaría de Planeación y Política Ambiental/Dirección General de Política Ambiental, Integración Regional y Sectorial (DGPAIRS) (2006b) *Programa de Ordenamiento Ecológico Marino del Golfo de California* [*Gulf of California Marine Ecological Use Planning Programme*], online at http://www.semarnat.gob.mx/archivosanteriores/temas/ordenamientoecologico/Documents/documentos%20decretados/decretos_2010/decreto_poemgc.pdf [accessed 27 July 2007].

Secretaría de Medio Ambiente y Recursos Naturales (SEMARNAT)/Subsecretaría de Planeación y Política Ambiental/Dirección General de Política Ambiental, Integración Regional y Sectorial (DGPAIRS) (2006c) *Ordenamiento Ecológico Marino y Regional del Golfo de México y Mar Caribe* [*Gulf of Mexico and Caribbean Sea Marine Ecological and Regional Land and Marine Use Planning*], online at http://seduma.tamaulipas.gob.mx/wp-content/uploads/2013/04/pronostico_consulta_oemr_gmmc.pdf [accessed 27 July 2007].

Secretaría de Medio Ambiente y Recursos Naturales (SEMARNAT)/Subsecretaría de Planeación y Política Ambiental/Dirección General de Política Ambiental, Integración Regional y Sectorial (DGPAIRS)/Dirección de Integración Regional (2006d) *Programa de Acción Regional para el Control de las Fuentes Terrestres de Contaminación Marina en la Península de Yucatán (PAR-Yucatán)* [Regional Programme of Action for the Yucatan Peninsula for the Protection of the Marine Environment from Land-based Sources of Pollution], online at http://web2.semarnat.gob.mx/temas/ordenamientoecologico/Documents/cimares/reuniones_titulares/par_yucatan.pdf [accessed 27 July 2007].

Secretaría de Medio Ambiente y Recursos Naturales (SEMARNAT)/Subsecretaría de Planeación y Política Ambiental/Dirección General de Política Ambiental, Integración Regional y Sectorial (DGPAIRS) (2007) *Estrategia Nacional para el Ordenamiento Ecológico del Territorio en Mares y Costas* [*National Strategy for Ecological Ocean and Coastal Use Planning and Management*], online at http://web2.semarnat.gob.mx/temas/ordenamientoecologico/cimaresold/Documents/integracion/estrategia_nacional_oe_mares_costas.pdf [accessed 10 July 2014].

Sherman, K., and Alexander, L. M. (eds) (1989) *Biomass, Yields and Geography of Large Marine Ecosystems*, Boulder, CO: Westview Press.

11

DEVELOPMENT OF A NATIONAL OCEAN POLICY IN THE UNITED STATES

Biliana Cicin-Sain, Gerhard F. Kuska, Caitlin Snyder, and Kateryna Wowk[1]

Introduction: Basic information and an overview of national policy

The United States is a major maritime nation. Its ocean and coastal areas are priceless assets that support the nation's economy, security, health and well-being, and long-term resilience. The United States claims an exclusive economic zone (EEZ) of 11.5 million km^2 – the world's largest (FAO, 2005). The US EEZ is 25 per cent larger than the US land mass area of 9.2 million km^2, and its coastline extend for 19,924 km (CIA, 2014).

The oceans and coasts of the United States directly support marine transportation, fisheries and aquaculture, energy production, recreation, biotechnology, and other uses. US coastal shoreline counties (including the Great Lakes) accounted for 41 per cent of the nation's gross domestic product (GDP) in 2010. This economic activity contributed about 44 million jobs and US$2.4 trillion in wages (NOAA, 2012). While these figures certainly matter, it is important to consider that these are only the market values of ocean and coastal resources. The non-market values, such as that of public access to the beach or an unobstructed ocean view, are estimated at over $100 billion a year (NOEP, undated). Further, it is becoming increasingly clear that the services that these ecosystems provide, including coastal storm protection, carbon sequestration, and the regulation of climate, natural hazards, disease, wastes, and water quality, are of significant value that has not yet been captured (NOAA, undated*a*). Ocean and coastal ecosystems and sectors are clearly a vital part of the US economy.

The United States signed, but has not acceded to, the 1982 United Nations Convention on the Law of the Sea (UNCLOS). However, the United States claims the following maritime zones consistent with UNCLOS:

- a territorial sea 12 nautical miles from baseline;[2]
- a contiguous zone 24 nautical miles from baseline;[3]
- an EEZ 200 nautical miles from baseline;[4] and
- a continental shelf that was specified before UNCLOS.[5]

US coastal states generally have authority over between zero and 3 statute (geographical) miles offshore, with the exception of Texas and Florida, the authority of which extends over

3 marine leagues (about 10 statute miles) into the Gulf of Mexico (Cicin-Sain and Knecht, 2000).

In the 1970s, the United States was a world leader in the creation of sectoral marine laws addressing issues such as fisheries, offshore oil and gas, water quality, and marine mammal protection, and enacted the world's first coastal management legislation, the Coastal Zone Management Act of 1972. General increases in public awareness of environmental concern (such as the 1969 Santa Barbara oil spill), the 1969 Stratton Report on US ocean policy (Stratton Commission, 1969), and a proactive Congress combined to give rise to significant ocean–related legislation and regulation from 1969 through the 1970s, resulting in enactment of an extensive body of generally sectoral laws. These include the Marine Protection, Research, and Sanctuaries Act of 1972, the Clean Water Act of 1972, the Marine Mammal Protection Act of 1972, the Magnuson–Stevens Fishery Conservation and Management Act of 1976, and the Outer Continental Shelf Lands Act Amendments of 1978, among others.

The Council on Environmental Quality (CEQ), the National Oceanic and Atmospheric Administration (NOAA), the Environmental Protection Agency (EPA), and the State Department's Bureau of Oceans and International Environmental and Scientific Affairs (often known simply as 'Oceans, Environment, and Science', or OES) were also created during this time period. Key Cabinet-level departments and agencies that deal with the ocean and coasts were expanded to include the Departments of State, Commerce, Interior, Justice, Transportation, Homeland Security, Defense, and the EPA (USCOP, 2004; Cicin-Sain and Knecht, 2000). However, no Cabinet-level 'Department of Oceans and Coasts' or 'Department of Natural Resources and Environment' was created, nor was any federal department or agency with oversight of all ocean and coastal issues. Over 140 statutes, regulations, and policies govern the use of US oceans, coasts, and the Great Lakes, and ocean and coastal issues are split among at least eleven of the fifteen cabinet departments, plus four independent agencies and several commissions (Kuska, 2005).

While path-breaking in many ways, the ocean laws and programmes created in the 1970s were also flawed, in the sense that they were largely based on single-sector approaches to ocean governance. Few, if any, effective mechanisms were made available to reconcile conflicts among multiple ocean and coastal uses and agencies, to encourage area-wide planning and management, to provide a vision for governance and for future uses, and to allow for cross-sectoral national ocean policy (Cicin-Sain and Knecht, 2000).

This is not to say that important successes have not been made in US ocean policy in the past thirty years. In fact, much has been achieved. Coastal management programmes have been established in thirty-four of the thirty-five coastal states and territories. Great strides have been made in marine mammal and endangered species protection. In fisheries, the 200 nautical mile EEZ was 'Americanized', giving fishing priority to American fishers (although, despite the ban on foreign fishing fleets, many US fish stocks continue to be threatened by foreign fleets). Significant strides have also been made in the control of point sources of pollution and, under the EPA's National Estuary Program, management planning has been completed for twenty-eight estuaries in eighteen states and Puerto Rico. A network of thirteen national marine sanctuaries and twenty-eight national estuarine research reserves is also in place under NOAA, along with ten National Seashores under the Department of Interior's (DOI) National Park Service, as well as several national monuments in marine areas. Further, the Papahānaumokuākea Marine National Monument, located in the Northwestern Hawaiian Islands, was created by presidential proclamation in 2006 and covers 362,000 km², making it one of the largest marine protected areas (MPAs) in the world (Papahānaumokuākea Marine National Monument, undated). In 2009, the National System of MPAs also was officially established under NOAA and DOI, to advance the conservation and sustainable use of the nation's vital natural and cultural marine resources. It now includes 437 MPAs, covering an area of 191,030 square miles, including federal MPAs, federal–state partnerships, and MPAs in states and territories. Until recently, the offshore oil and gas programme also generated significant energy supplies with a good safety record – until that record was significantly tarnished in 2010 as a result of the *Deepwater Horizon* oil spill in the Gulf of Mexico.

While notable achievements have been made, marine ecosystems and US ocean policy have nonetheless suffered significant problems. Serious declines have occurred in many fisheries. Also, while significant successes have been achieved in the management of point sources of marine pollution, it has been much more difficult to manage non-point sources of marine pollution (such as agricultural and urban runoff), which is a factor in 90 per cent of all incidents in which water quality is determined to be below standard (EPA, 1998). While extensive anecdotal evidence exists about successes achieved in coastal management, the absence of measurable coastal indicators in this area has made it difficult to sustain the political support necessary for continued development and growth of the programme. Although significant planning has taken place in American estuaries, few plans have been implemented and enforced. Serious conflicts have occurred between marine mammals and fisheries, and have not been adequately resolved, and conflicts between marine mammals, defence activities (sonar) and marine transportation (speed) continue to present a challenge in many areas, although there has been recent advancement with this issue, particularly surrounding the endangered northern right whale in the Northeast (NOAA, 2013).

Newer uses of the EEZ, such as marine aquaculture, marine biotechnology, and offshore wind power, are only beginning to be addressed and effective policy frameworks have not yet been implemented. Thus, in some cases, important new opportunities for economic development have been delayed. Newer challenges, such as sea-level rise, also are only beginning to be addressed in a systematic fashion.

Internationally, the United States has yet to accede to important treaties such as UNCLOS (the 'constitution' for the world's oceans) and the 1992 Convention on Biological Diversity (CBD), which provides guidance on protection of biodiversity, and access to and exploitation of the world's genetic resources.

In general, Congress and the Administration have played an oversight role on US ocean policy mainly issue by issue, and law by law, paying little attention to how well the various parts, issues, and laws fit together. The result has been conflicts among users, agencies, and levels of government, and significant declines in resources as narrowly focused policies continue unabated and economic opportunities in the ocean are forgone. However, the sectoral approach to ocean management is changing, with significant strides occurring in recent years (Cicin-Sain, 2012).

Movement toward integrated ocean policy

Over the past twenty years, non-governmental organizations (NGOs) and academic groups have articulated the need to go beyond the sector-by-sector approach to ocean policy in the United States to address multiple-use conflicts, to preserve ecosystems, and to take advantage of new economic opportunities in the ocean, underscoring that the US approach to oceans governance is 'less than the sum of its parts' (Cicin-Sain and Knecht, 2000: 7). In particular, an academic network, the Ocean Governance Study Group, organized in the late 1980s and led by the Universities of Delaware, California, and Hawaii, brought together analysts and practitioners to review the functioning of the US ocean governance system and to discuss alternatives for policy reform (Cicin-Sain, 1992). Ocean academics and activists began calling for a comprehensive look at the US system of ocean management, and for the convening of a national ocean commission to examine all aspects of US ocean policy and to develop recommendations for ocean governance reform. However, although a number of Bills were introduced in Congress to create such a commission, they were not enacted.

The call for comprehensive reform was given serious attention in 2000, when two ocean commissions were created to examine all aspects of US ocean policy. The first was the privately funded Pew Oceans Commission (POC), which appointed a number of distinguished leaders to lead an inquiry panel into US ocean policy, examining, in particular, environmental issues related to ocean degradation, resource decline and their underlying causes. The Pew Oceans Commission issued its report in June 2003, after holding many public meetings and commissioning several policy studies (POC, 2003).

Around the same time, and with a heightened awareness of the need to forward US ocean policy, Congress enacted the Oceans Act of 2000 on 25 July, which was signed into law by the President on 7 August 2000. The Oceans Act created the US Commission on Ocean Policy (USCOP) and tasked it with conducting research and giving recommendations to Congress and the President for a coordinated and comprehensive national ocean policy that would promote: protection of life and property; stewardship of ocean and coastal resources; protection of the marine environment and prevention of marine pollution; enhancement of maritime commerce; expansion of human knowledge of the marine environment; investments in technologies to promote energy and food security; close cooperation among government agencies; and US leadership in ocean and coastal activities. The Commission released its final report, *An Ocean Blueprint for the 21st Century*, on 20 September 2004 (USCOP, 2004).

As will be discussed later in the chapter, President George W. Bush took actions to implement the recommendations of the USCOP during 2004–08. The Obama Administration also took interest in the ocean, and in 2009 began working on a national policy to enhance stewardship of the ocean, coasts, and the Great Lakes, and to promote the long-term conservation and use of these resources, as well as a recommended framework for effective coastal and marine spatial planning (CEQ, 2013). This work culminated in 2010 in an integrated policy for the stewardship of oceans, coasts, and the Great Lakes – the first comprehensive national ocean policy for the United States.

The policy development process

Within the past thirty years, various attempts have been made at collaboration among, or to promote collaboration among, the activities of various executive branch departments and agencies involved with ocean and coastal issues by means of formalized federal interagency groups. In addition to the Commission on Marine Science, Engineering and Resources (the 'Stratton Commission', so-called after its chair Julius Stratton), the Marine Resources and Engineering Development Act of 1966 created a high-level, interagency council within the Executive Office of the President to coordinate federal ocean and coastal activities. The National Council on Marine Resources and Engineering Development (Marine Council) coordinated ocean and coastal activities from June 1966 until June 1971. The establishing legislation stipulated that the Council would be terminated after the Commission submitted its final report. In fact, the Nixon Administration maintained the Council until after NOAA was established and operating. After the sunset of the Council in the early 1970s, no new high-level interagency coordinating body for ocean policy existed for the next thirty years (Kuska, 2005).

In 1997, the Marine Board of the National Research Council published a report calling for a national marine council. In 2003, the Pew Oceans Commission published its report, with recommendations for a new ocean policy, including a permanent interagency council to help coordinate 'at least [six] departments of the federal government and dozens of federal agencies in the day-to-day management of ocean and coastal resources' (POC, 2003: 14).

The US Commission on Ocean Policy was tasked to coordinate with the states, a scientific advisory panel, and the public to develop a national oceans report. The report was to give 'equal consideration to environmental, technical feasibility, economic, and scientific factors' (USCOP, 2004: 56). The Commission's report addressed a wide range of issues, including: an assessment of facilities (people, vessels, computers, satellites); a review of federal activities; a review of federal laws; a review of the supply and demand for ocean and coastal resources; a review of the relationship among federal, state, and local governments, and the private sector; a review of the opportunities for the investment in new products and technologies; recommendations for modifications to federal laws and/or the structure of federal agencies; and a review of the effectiveness of existing federal interagency policy coordination.

The Commission was composed of sixteen members nominated by Congress and appointed by the President, including leaders with backgrounds in shipping and ports, public aquaria, natural and social sciences, the offshore oil industry, fisheries, and state government. Interestingly, there were no appointees from environmental NGOs. The Commission worked with a Science Advisory Panel, required by the Oceans Act to assist in preparing its report and to ensure 'that the scientific information considered is based on the best available data' (Oceans Act of 2000, Appendix A, para. A3). The Commission held sixteen public meetings around the country and conducted eighteen regional site visits, receiving input from 447 witnesses on national ocean policy issues and how they should be resolved, and carried out a number of supporting analyses. As called for in the Oceans Act, the final report contains comments received from the state governors on the Commission's preliminary report (comments were received from thirty-seven governors, five tribal leaders, and a multitude of other organizations and individuals) (USCOP, 2004).[6]

The first page of USCOP's final report, *An Ocean Blueprint for the 21st Century,* showed a map of the US EEZ (as noted, the largest in the world), and the Commission resoundingly pointed out that 'the United States is an Ocean Nation' (USCOP, 2004: 1). In its letters to the President and to the Congress delivering the report, the Commission emphasized that:

> A comprehensive and coordinated national ocean policy requires moving away from the current fragmented, single-issue way of doing business and toward ecosystem-based management. This new approach considers the relationships among all ecosystem components, and will lead to better decisions that protect the environment while promoting the economy and balancing multiple uses of our oceans and coasts.
>
> *(USCOP, 2004)*

The Commission, as well, evoked a new vision of the oceans as 'one in which our oceans and coasts are clean, safe, sustainably managed, and preserved for the benefit and enjoyment of future generations' (USCOP, 2004). At the international level, the Commission called for the immediate accession to UNCLOS.

This was a landmark assessment of US ocean policy: the first comprehensive effort, thirty-five years after the Stratton Commission issued the first blueprint for US management of its oceans in 1969 (Stratton Commission, 1969). Completion of the report represented a massive effort: it is 522 pages long, with seven appendices, plus six lengthy appendices published separately. A broad array of issues in US ocean policy is addressed, albeit to different degrees of depth. There are 212 specific recommendations in the report, directed to the Congress, the

Executive Branch leadership, federal agencies, and regional bodies. The report also includes recommendations related to international affairs.

Policies recommended by the US Commission on Ocean Policy

The major recommendations of USCOP relate to guiding principles for ocean governance and for improving the system of ocean governance, at both the national and regional levels. There are many recommendations, as well, for each sector of US ocean policy. Given the large number of recommendations found in the report, only major recommendations for change are highlighted here.

Guiding ocean principles

The Commission first established an overarching set of principles to guide national ocean policy. These are articulated in a clear and succinct manner, and provide an excellent core for comprehensive governance. This is an important advance in US ocean policy, in so far as past guiding principles were tied to specific ocean sectors or programmes. The newly articulated principles represent overall guidance on how the United States should manage all uses of ocean and coastal ecosystems in an integrated manner.

The principles include: sustainability; stewardship; ocean–land–atmosphere connections; ecosystem-based management; multiple-use management; preservation of marine biodiversity; best available science and information; adaptive management; understandable laws and clear decisions; participatory governance; timeliness; accountability; and international responsibility. Definitions of the principles are noted in Table 11.1.

National ocean governance reform

In a major innovation, the Commission recommended new institutions for national ocean policy coordination to encourage eleven Cabinet-level departments and four independent agencies with important roles in ocean and coastal policy to undertake coordinated and joint action. These recommendations included the following.

- An National Ocean Council (NOC) within the Executive Office of the President, chaired by an Assistant to the President, and composed of Cabinet secretaries of departments and administrators of independent agencies with relevant ocean- and coastal-related responsibilities, should be established. The NOC should provide high-level attention to ocean, coastal, and Great Lakes issues, develop and guide the implementation of appropriate national policies, and coordinate the many federal departments and agencies with responsibilities over oceans.
- The Assistant to the President should chair the NOC, advise the Office of Management and Budget (OMB) and the agencies on appropriate levels of funding for important coastal- and ocean-related activities, and prepare a biennial report on oceans, as mandated by the Oceans Act of 2000.
- Two committees should be set up under the NOC: a Committee on Ocean Resource Management (to be chaired by the chair of the existing Council on Environmental Quality), and a Committee on Ocean Science, Education, Technology, and Operations (to be chaired by the chair of the existing Office of Science and Technology Policy).

Table 11.1 Definitions of guiding principles for national ocean policy in the United States

Principle	Definition
Sustainability	Ocean policy should be designed to meet the needs of the present generation without compromising the ability of future generations to meet their needs.
Stewardship	The principle of stewardship applies both to the government and to every citizen. The US government holds ocean and coastal resources in the public trust – a special responsibility that necessitates balancing different uses of those resources for the continued benefit of all Americans. Just as important, every member of the public should recognize the value of the oceans and coasts, supporting appropriate policies and acting responsibly, while minimizing negative environmental impacts.
Ocean–land–atmosphere connections	Ocean policies should be based on the recognition that the oceans, land, and atmosphere are inextricably intertwined, and that actions that affect one earth system component are likely to affect another.
Ecosystem-based management	US ocean and coastal resources should be managed to reflect the relationships among all ecosystem components, including humans and non-human species and the environments in which they live. Applying this principle will require defining relevant geographic management areas based on ecosystem, rather than political, boundaries.
Multiple-use management	The many potentially beneficial uses of ocean and coastal resources should be acknowledged and managed in a way that balances competing uses, while preserving and protecting the overall integrity of the ocean and coastal environments.
Preservation of marine biodiversity	Downward trends in marine biodiversity should be reversed where they exist, with a desired end of maintaining or recovering natural levels of biological diversity and ecosystem services.
Best available science and information	Ocean policy decisions should be based on the best available understanding of the natural, social, and economic processes that affect ocean and coastal environments. Decision-makers should be able to obtain and understand quality science and information in a way that facilitates successful management of ocean and coastal resources.
Adaptive management	Ocean management programmes should be designed to meet clear goals and provide new information to continually improve the scientific basis for future management. Periodic re-evaluation of the goals and effectiveness of management measures, and incorporation of new information in implementing future management, are essential.
Understandable laws and clear decisions	Laws governing uses of ocean and coastal resources should be clear, coordinated, and accessible to the nation's citizens to facilitate compliance. Policy decisions and the reasoning behind them should also be clear and available to all interested parties.
Participatory governance	Governance of ocean uses should ensure widespread participation by all citizens on issues that affect them.
Timeliness	Ocean governance systems should operate with as much efficiency and predictability as possible.
Accountability	Decision-makers and members of the public should be accountable for the actions they take that affect ocean and coastal resources.
International responsibility	The United States should act cooperatively with other nations in developing and implementing international ocean policy, reflecting the deep connections between US interests and the global ocean.

Source: USCOP (2004)

- A President's Council of Advisors on Ocean Policy should be created, consisting of representatives from state, territorial, tribal, and local governments, and academic, public interest, and private sector organizations, to ensure a formal structure for non-federal participation in the NOC on ocean and coastal policy matters.
- A small Office of Ocean Policy should be created, to provide staff support to all of these bodies.
- To get the process of national ocean governance reform started immediately, pending Congressional action, the Commission recommended that the President put this structure in place by means of executive order.

Other major critical actions called for by the Commission

Among its 212 recommendations, the Ocean Commission points particularly to the following set of 'Critical Actions' to provide the foundation for a comprehensive national oceans policy:

Improved Governance
- Improve the federal agency structure by strengthening NOAA and consolidating federal agency programs according to a phased approach.
- Develop a flexible, voluntary process for creating regional ocean councils, facilitated and supported by the National Ocean Council.
- Create a coordinated management regime for activities in federal offshore waters.

Sound Science for Wise Decisions
- Double the nation's investment in ocean research, launch a new era of ocean exploration, and create the advanced technologies and modern infrastructure needed to support them.
- Implement the national Integrated Ocean Observing System (IOOS) and a national monitoring network.

Education – A Foundation for the Future
- Improve ocean-related education through coordinated and effective formal and informal efforts.

Specific Management Challenges
- Strengthen coastal and watershed management and the links between them.
- Set measurable goals for reducing water pollution, particularly for nonpoint sources, and strengthen incentives, technical assistance, enforcement, and other management tools to achieve those goals.
- Reform fisheries management by separating assessment and allocation, improving the Regional Fishery Management Council system, and exploring the use of dedicated access privileges.
- Accede to the United Nations Convention on the Law of the Sea to remain fully engaged at the international level.

Implementation
- Establish an Ocean Policy Trust Fund based on unallocated revenues from offshore oil and gas development and new offshore activities, that is dedicated to supporting improved ocean and coastal management at federal and state levels.

(USCOP, 2004: 25)

Overall, the Commission's report was a landmark achievement: it is comprehensive in its coverage; it presents a moderate and balanced perspective (promoting both ecosystem protection and appropriate use for the benefit of the nation); and it provides a blueprint to guide national ocean policy for years to come. The participatory process involved in the preparation of the report re-energized the nation on oceans, mobilized many groups and individuals around ocean issues, and convinced even sceptics about the need for ocean policy reform. The stage was set for achieving significant improvements in America's management of its oceans and coasts.

With such a lengthy and wide-ranging report, there are, of course, some sections that are stronger than others. The Commission did a particularly good job in laying out the general framework for national ocean governance reform and in arguing for (and actually costing out) very substantial increases in funding for oceans and coasts. It was likewise very strong in making the case for a significant investment in formal and informal marine education programmes ('promoting lifelong ocean education') and was very detailed and effective in the prescriptions on an integrated ocean observing system (Cicin-Sain, 2012).

The Commission should also be commended for its call to revitalize US international leadership on oceans and especially for its very strong stance in urging immediate US accession to UNCLOS (the first formal action taken by the Commission being a unanimous resolution calling for accession to UNCLOS).

On the other hand, the Commission's report did not provide sufficient guidance on some important governance issues, relying much on the proposed National Ocean Council to address these issues. In fact, forty-four of the Commission's recommendations, and some of the most important ones, were aimed at the National Ocean Council.

One area in which insufficient guidance had been given was the very important question of regional ocean governance. This is a complex topic, given the presence of multiple federal agencies and multiple states in all regions, and great diversity in the presence or absence of past inter-sectoral cooperative efforts and cross-cutting institutions in different regions of the country. The Commission called for development of a 'flexible and voluntary process for the creation of regional ocean councils', and prescribed that 'states, working with relevant stakeholders, should use this process to establish regional ocean councils, with support from the National Ocean Council' (USCOP, 2004: 87). While rightly calling for building on existing regional institutions and recognizing the diversity found in different regions of the country, the Commission failed to point to a catalysing entity that could get the regional dialogue and coordination processes started, leading to an ambiguous leadership situation at the regional level (Cicin-Sain, 2012).

Another topic on which one would have expected the Commission to provide detailed guidance concerns the absence of a policy framework for new uses of federal waters (3–200 nautical miles offshore), such as offshore aquaculture, offshore wind power, and marine biotechnology. While the Commission recognized the need for establishing a coordinated offshore management regime, instead of developing the framework of such a system it called for the National Ocean Council to ensure that each current and emerging activity in federal waters would be administered by a lead federal agency in coordination with other applicable authorities (Cicin-Sain, 2012).

There is also insufficient detail on federal agency reorganization – a topic that many had hoped would have been at the centre of the Commission's recommendations. The report called for a phased approach, starting first with establishing an organic act for NOAA to codify its existence and mission, and then for a period of review by the Assistant to the President, with subsequent recommendations of opportunities for agency consolidation.

The Commission also avoided the politically hot topic of the offshore oil moratorium, which, at the time, prohibited development of oil and gas resources in much of the nation's outer continental shelf by means of the imposition of spending moratoria on the Department of the Interior's Minerals Management Service (MMS), which was responsible for the nation's natural gas, oil, and other mineral resources on the outer continental shelf, in the congressional appropriations process. While one can understand why the Commission would not want to deal with this very conflictual issue, avoidance of recommendations on the topic perpetuated management via the congressional appropriations process – which is not a good governance approach – and neglected to address questions of energy supply in a time of steeply rising energy costs. The Commission also steered clear of major maritime issues, such as the 1920 Jones Act (on merchant marines) and its impact on the US maritime sector (Cicin-Sain, 2012).

Initial implementation of the US Commission on Ocean Policy recommendations

The Bush Administration's response to the US Commission on Ocean Policy

As mandated by the Oceans Act of 2000, the Administration of President George W. Bush responded to the Commission's report and recommendations on 17 December 2004, issuing Executive Order 13366 to establish a Cabinet-level Committee on Ocean Policy (COP),[7] and releasing the US Ocean Action Plan (USOAP) (White House, 2004). The plan proposed many specific actions, as well as a mechanism to further evaluate and address the Commission's recommendations. It also outlined principles for implementation, including:

- a focus on achieving meaningful results;
- ensuring the continued conservation of coastal and marine resources, while at the same time ensuring that the American public enjoys and benefits from those same resources;
- employing the best science and data to inform decision-making;
- continuing to work towards an ecosystem-based approach;
- encouraging innovation and employing economic incentives over mandates; and
- working with states, tribal and local governments, communities, and interested individuals to advance mutual objectives, and ensuring that programmes are effectively coordinating on ocean and coastal activities.

The six sections of the USOAP focused on improving and supporting US marine and Great Lakes environments in several areas.

Leadership and coordination

To improve leadership and coordination, the Administration proposed two key items: first, it proposed to codify the existence of NOAA within the Department of Commerce through an organic act; second, the USOAP announced the intent to create a governance structure beneath the Cabinet-level Committee on Ocean Policy (COP), including a supporting committee and subcommittee structure (Figure 11.1). The Chairman of the CEQ was designated chair of the COP under Executive Order 13366.

Knowledge and understanding

The USOAP proposed development of an ocean research strategy to oversee coordination of key research priorities, including earth and ocean observation, new research platforms (satellites, ships, etc.), water quality monitoring, and mapping, as well as to increase research on connections between oceans and human health, and to promote ocean education.

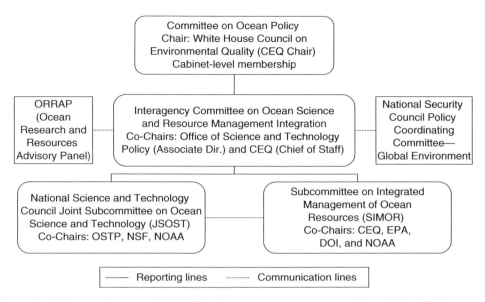

Figure 11.1 The core elements of the US federal ocean governance structure from 2005 to 2009

Source: White House (2004)

Use and conservation

The USOAP addressed improvement of fisheries (emphasizing support for individual fishing quotas, wider representation on regional fishery management councils, better data collection, guidelines on the use of science, and international coordination), called for protection of coral reefs and protection of marine mammals, sharks, and sea turtles (via international cooperation, by-catch reduction, and better enforcement), proposed legislation and guidelines for offshore aquaculture, argued for improved coordination of marine managed areas, such as MPAs, supported offshore energy development, and called for the protection of shipwrecks with national heritage implications.

Coastal and watershed management

To improve coastal watersheds, the Administration proposed specific programmes that would focus on assisting local and state management, addressing effects from farming, increasing and restoring wetlands, preventing the spread of invasive species, and reducing pollution from runoff and airborne deposition.

Transportation

For marine transportation, the goals were to facilitate the coordination, development, and efficiency of shipping through interagency cooperation, capacity-building, short-sea shipping, reduced taxes, and improved navigation, and to reduce vessel pollution (through a programme to reduce effects from federal ships and ports, and by EPA rules to decrease vessel air emissions).

International science and policy

The Administration's international priorities included acceding to UNCLOS, ratifying marine engine pollution reduction requirements under the 1973 International Convention for the Prevention of Pollution from Ships, as amended by the Protocol of 1978 (MARPOL), and strengthening the 1972 Convention on the Prevention of Marine Pollution by Dumping of Wastes and Other Matter (London Convention). Through various partnerships, the Administration also sought to improve earth and ocean observing, management of marine-based ecosystems, protection of coral reefs, and land-based pollution reduction programmes on global scales.

It should be noted that development of the USOAP was intended to be budget-neutral: no new funding was anticipated for the successful implementation of the actions outlined. This was in contrast to the recommendations of the US Commission on Ocean Policy, which recommended an increase in the first year for implementing its recommendations of US\$1.5 billion, with ongoing annual costs of nearly \$3.9 billion (USCOP, 2004).

Beginning in January 2005, the Chair of the Committee on Ocean Policy – and the President's Advisor on Oceans – began to implement the USOAP recommendations. The Interagency Committee on Ocean Science and Resource Management Integration (ICOSRMI) began meeting on a bimonthly basis in January 2005. This senior-level group of federal leaders reported directly to the Committee on Ocean Policy and was composed of the second-tier federal agency leadership (that is, at the level of Undersecretary/Assistant Secretary). Its functions were specified to include:

> . . . 1) coordinate and integrate activities of ocean-related Federal agencies and provide incentives for meeting national goals; 2) identify statutory and regulatory redundancies or omissions and develop strategies to resolve conflicts, fill gaps, and address new emerging ocean issues for national and regional benefits; 3) guide the effective use of science in ocean policy and ensure the availability of data and information for decision making at national and regional levels; 4) develop and support partnerships among government agencies and nongovernmental organizations, the private sector, academia, and the public; 5) coordinate education and outreach efforts by Federal ocean and coastal agencies; 6) periodically assess the state of the Nation's oceans and coasts to measure the achievement of national ocean goals and 8) [*sic*] make recommendations to the Committee on Ocean Policy on developing and carrying out national ocean policy, including domestic implementation of international ocean agreements.
>
> *(White House, 2004: 7–8)*

Additionally, the subcommittees and associated interagency mechanisms met regularly (at least monthly) to coordinate the oversight of USOAP implementation (MMS, 2006). Although new funding was not originally anticipated under the USOAP, some new funding was proposed in the President's budget request and by international entities (such as the Coral Triangle Initiative) (DOS, 2007).

The Administration issued updates of its progress in implementing the USOAP on several occasions, including the President's statement that, with the exception of US accession to UNCLOS, which required Senate Advice and Consent, all of the USOAP recommendations had been implemented as of September 2008 (White House, 2008). The Bush Administration listed its major accomplishments in ocean conservation and management as:

- Accomplished conservation in marine environments over the past eight years on par with what we have achieved on land over the past 100.

- Released an Ocean Action Plan in 2004 and created the first ever Cabinet Committee on Ocean Policy. All 88 actions recommended in the Ocean Action Plan have been met or are on track, making our oceans, coasts and Great Lakes cleaner, healthier and more productive.

- Designated nearly 140,000 square miles of coral reef ecosystems and surrounding waters in the northwestern Hawaiian Islands, which contain more than 7,000 species, many of which are found nowhere else on earth, such as the Papahānaumokuākea Marine National Monument, giving the area the nation's highest form of marine environmental protection.

- Designated three areas of the Pacific Ocean, covering more than 195,500 square miles, as marine national monuments: the Marianas, Pacific Remote Islands, and Rose Atoll Marine National Monuments.

- Announced the expansion of the Monterey Bay National Marine Sanctuary by 775 square miles to include the Davidson Sea Mount.

- Protected our oceans under US control by taking action to end overfishing and conserve habitats.

- Increased funding for National Oceanic Atmospheric Administration by US$770 million.

(White House, undated)

Implementation progress 2005–07, as measured by the Joint Ocean Commission Initiative

Following the release of the US Commission on Ocean Policy and the Pew Oceans Commission reports, the two commissions formed a collaborative effort – the Joint Ocean Commission Initiative (JOCI) – to advance the goals put forth therein. The JOCI works with stakeholders at the local, state, regional, and national levels to achieve ocean policy reform. To assess the advancement of ocean policy reform, the JOCI issued a yearly 'US Ocean Policy Report Card' for the period 2005–07, which graded the progress of ocean policy in a number of areas and offered suggestions for improvement. As is shown in Table 11.2, the report cards for 2005–07 showed great room for improvement in the majority of the subject areas.

National ocean governance reform

As shown in Table 11.2, at a national level, the JOCI assessed ocean governance reform in the 2005–07 period as being slow. The JOCI acknowledged the establishment of the ocean governance structure and the Ocean Action Plan, along with a series of accomplishments, including the establishment of the Papahānaumokuākea Marine National Monument in June 2006, efforts at the United Nations to stop destructive fishing practices on the high seas in November 2006, reauthorization of the Magnuson-Stevens Fishery Conservation and Management Act in January 2007, and release of an Ocean Research Priorities Plan and Implementation in January 2007 with an initial funding request. Further, a number of accomplishments were not noted in the JOCI's assessment, including efforts to protect important recreational fish species launched in October 2007, a Marine Debris Initiative launched in November 2007, and initial support and new funding committed to the Coral Triangle Initiative in Southeast Asia in December

Table 11.2 Comparison of US Ocean Policy Report Cards, 2005–07, issued by the Joint Ocean Commission Initiative

Subject	2005	2006	2007
National ocean governance reform	**D+**	**C−**	**D**
	Development of the US Ocean Action Plan and establishment of the Committee on Ocean Policy are significant actions, but to date the tangible results have been limited given the scope of the challenges facing our nation. Despite pending legislation and efforts of the Committee, legislative and administrative reforms addressing organizational deficiencies in NOAA and mandatory interagency coordination and integration of ocean-related programs have been inadequate. Moreover, the steps taken to date do not embody the governance reform principles put forth by the Joint Ocean Commission Initiative.	*Notable Progress* • Declaration of Northwestern Hawaiian Islands Marine National Monument • Expanded federal interagency planning and coordination • Increased opportunities for stakeholder input on federal plans • Legislative deliberation on mission, role, and organization of the National Oceanic and Atmospheric Administration (NOAA) *Improvements Needed* • Enact legislation that would: adopt a statement of national ocean policy; codify a permanent federal coordinating committee for oceans in the White House; reform NOAA; establish a comprehensive offshore management regime; and create a regional ocean governance framework • Expand protection for ecologically or culturally important marine areas	*Notable Progress* • House deliberation on comprehensive ocean governance reform • Progress by the House and Senate Commerce Committee on ocean legislation • National stakeholder process to strengthen the Coastal Zone Management Act *Improvements Needed* • Reform national ocean governance by enacting legislation that creates a national ocean policy, codifies NOAA, and strengthens federal coordination • Pass pending ocean legislation, including ocean observing, ocean exploration, coastal land conservation, and ballast water management • Reauthorize and strengthen the Coastal Zone Management Act, National Marine Sanctuaries Act, and Oceans and Human Health Act • Create a national framework to help initiate and coordinate regional efforts
Regional and state ocean governance reform	**B−**	**A−**	**A−**
	Promising ocean governance efforts are underway in a number of regions and states. The Joint Initiative encourages more regional collaboration and calls on additional states to demonstrate a commitment to ocean governance reform. The federal government should do more to facilitate and support ocean governance reform efforts in regions and states and should strive for better coordination among federal agencies at the regional level.	*Notable Progress* • New (2006) regional and state initiatives, including the Gulf of Mexico, West Coast, New York, and Washington • Progress on existing (pre-2006) regional and state initiatives, including the Great Lakes, Northeast, California, Florida, and Massachusetts *Improvements Needed* • Create a national framework to support regional collaborations and approaches • Implement additional regional and state ocean governance efforts and increase progress on existing initiatives	*Notable Progress* • Progress establishing and implementing state ocean legislation in MA, NJ, and NY and noteworthy progress in AK, CA, FL, HI, LA, OR, and WA • Significant progress in Gulf of Mexico and West Coast regions *Improvements Needed* • Strengthen existing initiatives, including expanding state commitment and federal support • Implement regional initiatives in Southeast and Mid-Atlantic

International leadership	**F** While some positive steps have been taken regarding international leadership on ocean issues, our continued failure to become a party to the United Nations Convention on the Law of the Sea hampers our ability to enhance and protect our national security interests and to demonstrate international leadership. Despite overwhelming support from a diverse array of interests, the Senate has yet to schedule the convention for a floor vote, and more vigorous support from the Administration is needed.	**D−** *Notable Progress* • Presidential statement calling for an end to destructive fishing practices on the high seas • US leadership on fisheries and whale conservation efforts in the United Nations • Enactment of high seas monitoring and compliance provisions in the Magnuson-Stevens Fishery Conservation and Management Reauthorization Act of 2006 (MSA) *Improvements Needed* • Accede to the United Nations Convention on the Law of the Sea	**C+** *Notable Progress* • Presidential support for the Law of the Sea Convention • Senate Foreign Relations Committee approval of the Convention • Active support for the Convention by a bipartisan coalition of industry, military, and environmental leaders • Administration support for international ocean policy issues *Improvements Needed* • Senate approval of the Law of the Sea Convention
Research, science, and education	**D** Doubling the ocean research budget and significantly increasing the support for ocean science and education are fundamental to improving our understanding and management of the oceans and coasts. The lack of an integrated ocean observing system capable of providing decision makers with important information compromises our nation's capacity to manage the oceans. The absence of an ocean and coastal stewardship ethic and a sluggish effort to coordinate the public education and outreach activities needed to enhance such an ethic hamper support for reform and funding.	**D+** *Notable Progress* • Administration's Ocean Research Priorities Plan and Implementation Strategy • Enacted legislation addressing tsunami preparedness and marine debris prevention and reduction • Enhanced role for science–based management in reauthorized MSA • New interagency working group leading development of national strategy on ocean education • Consideration of legislation on ocean exploration, ocean and coastal observing, coral reef conservation, ocean and coastal mapping, ballast water management, and coastal land protection	**C−** *Notable Progress* • Administration focus on implementing the Ocean Research Priorities Plan and Implementation Strategy • Continued efforts to develop ocean and coastal observing systems • Expanded federal support and coordination on ocean education • Congressional deliberation on ocean science legislation

(continued)

Table 11.2 (continued)

Subject	2005	2006	2007
		Improvements Needed • Address chronic under-funding of ocean science and education • Increase recognition of the ocean's role in climate change • Reestablish a congressional science and technology advisory entity	*Improvements Needed* • Fund implementation of the Ocean Research Priorities Plan and Implementation Strategy • Pass pending ocean science legislation on ocean observing, ocean acidification research, ocean exploration, and coastal and ocean mapping • Reestablish a congressional science and technology advisory office
Fisheries management reform	C+ Broad bipartisan support has been garnered for a Senate bill to reauthorize the Magnuson-Stevens Fishery Conservation and Management Act, and the Joint Initiative applauds the effort to reach out to Commissioners and other stakeholders during the development of the bill. The Joint Initiative appreciates the Administration's thoughtful consideration of fisheries management reform in its bill and subsequent input to the Senate bill. The House should build on and strengthen the Senate bill to reflect the full suite of fisheries management principles articulated by the Joint Initiative and work with the Senate to make reauthorization of the Magnuson-Stevens Act a reality this year.	B+ *Notable Progress* • Congressional and Administration support leading to passage of the Magnuson-Stevens Fishery Conservation and Management Reauthorization Act of 2006 *Improvements Needed* • Provide rulemaking and funding to implement newly enacted provisions	C+ *Notable Progress* • Initial steps implementing the Magnuson-Stevens Reauthorization Act • Fewer stocks overfished or experiencing overfishing • Progress toward establishing limited access privilege programs • US leadership on international fisheries and habitat conservation *Improvements Needed* • Expedite implementation and funding for Magnuson-Stevens Reauthorization Act reforms • Increase emphasis on incorporating science into decision-making • Improve recreational fisheries monitoring and management • Increase commitment to international fisheries conservation

New funding for ocean policy and programs	F	F	D+
	Funding for essential ocean programs, outlined above, remains woefully insufficient and is far outpaced by current and future challenges. Failure to provide even the modest funding increases recommended by the Commissions, compounded by funding rescissions in important ocean programs, jeopardizes the economic and ecological benefits our nation receives from its oceans and coasts. New investment must be made so that we can address ocean and coastal issues effectively.	*Notable Progress* • Significant Senate funding support for ocean programs in NOAA, National Science Foundation, and National Aeronautics and Space Administration despite overall lack of new federal investment in oceans • Increased state funding for ocean programs in a number of states, such as California *Improvements Needed* • Develop an integrated budget for federal ocean and coastal programs • Resolve chronic House under-funding of ocean programs and address severe funding reductions and uncertainty associated with Fiscal Year 2007 Continuing Resolution • Include oceans in ongoing initiatives to address climate change and the President's American Competitiveness Initiative • Establish an Ocean Trust Fund to support state and federal programs	*Notable Progress* • House joined the Senate in increasing funding support for NOAA • Presidential funding support for Ocean Research Priorities Plan and Implementation Strategy • Ocean research recognized as part of national competitiveness initiative *Improvements Needed* • Increase funding for ocean research, management, and infrastructure, including ocean and coastal observing systems • Establish an integrated budget for federal ocean programs • Establish a dedicated ocean trust fund for state and federal programs

(continued)

Table 11.2 (continued)

Subject	2005	2006	2007
Links between oceans and climate change	n/a	n/a	C *Notable Progress* • Expanded state efforts to mitigate and adapt to impacts of climate change on coastal communities and resources • Increased recognition by Congress of the role of oceans in climate change and the impacts of this change on oceans and coasts *Improvements Needed* • Enact legislation that incorporates ocean science, management, and education into a strategy to mitigate and adapt to climate change • Expand ocean research, observing, modeling, and information delivery systems • Increase federal support of state and regional efforts to address ocean-related impacts of climate change

Note: The grading scheme is arranged from A (best grade) to F (failing grade).

Source: Extracted from JOCI (2005; 2006b; 2007)

2007. Despite these accomplishments, the JOCI grades for ocean governance reform remained low owing to the lack of legislation that:

> . . . develops a national ocean policy, codifies and reforms NOAA, establishes a permanent interagency coordinating structure in the White House, provides a structure for federal agencies to support and participate in regional partnerships, and institutes coordinated and comprehensive management of offshore waters.
>
> *(JOCI, 2007: 3–4)*

Indeed, while several noteworthy accomplishments were achieved, the agencies were challenged from the beginning to advance actions that were under way or to develop new actions only with existing funds. The fact that no new funding was contemplated under the USOAP was in direct conflict with the recommendations of the US Commission on Ocean Policy and played heavily into the JOCI assessment.

In March 2006, the JOCI received a request from ten Senators to prepare a report listing the top ten actions that Congress ought to take in order to implement the recommendations of the US Commission on Ocean Policy and the Pew Oceans Commission, and the top funding priorities for implementation of the recommendations. In the resulting June 2006 report, *From Sea to Shining Sea: Priorities for Ocean Policy Reform*, the JOCI noted the following ten priority actions:

1 Adopt a statement of national ocean policy. . . .
2 Establish the NOAA in law and work with the Administration to identify and act upon opportunities to improve federal agency coordination on ocean and coastal issues. . . .
3 Foster ecosystem–based regional governance. . . .
4 Reauthorize an improved Magnuson-Stevens Fishery Conservation and Management Act. . . .
5 Enact legislation to support innovation and competition in ocean-related research and education consistent with key initiatives in the Bush Administration's Ocean Research Priorities Plan and Implementation Strategy. . . .
6 Enact legislation to authorize and fund IOOS. . . .
7 Accede to the United Nations Convention on the Law of the Sea. . . .
8 Establish an Ocean Trust Fund in the U.S. Treasury as a dedicated source of funds for improved management and understanding of ocean and coastal resources by federal and state governments. . . .
9 Increase base funding for core ocean and coastal programs and direct development of an integrated ocean budget. . . .
10 Enact ocean and coastal legislation that has already progressed significantly in the 109th Congress. . . .

> *(JOCI, 2006a: 8–10)*

In January 2007, a comprehensive oceans Bill, the Ocean Conservation, Education, and National Strategy for the 21st Century Bill (OCEANS 21), was introduced into the House of Representatives. With seventy co-sponsors, OCEANS 21 was a bipartisan Bill that sought to address a number of the key recommendations of the US Commission on Ocean Policy. OCEANS 21 provisions sought to:

- establish a national oceans policy;
- strengthen and formally authorize NOAA and create an Under Secretary of Commerce for Oceans and Atmosphere to serve as its administrator;
- create a National Oceans Advisor, a Committee on Ocean Policy, and the Council of Advisors on Oceans Policy;
- establish an Ocean and Great Lakes Conservation Trust Fund; and
- call for US$1.3 billion annually for the development and implementation of regional eco-system management plans that address regional priority problems.

However, no agreement could be reached between the House and Senate and the Bush Administration in several areas, including the contents of a national ocean policy, the position of a national oceans advisor, the establishment of a trust fund, and the authorization of significant new funding resources. The Administration was not supportive of the funding levels proposed by the Commission (approximately $4 billion in new funding to supplement the existing funding of approximately $9.5 billion in federal funding for non-military ocean and coastal funding). Ultimately, OCEANS 21 did not see congressional action.

Regional and state ocean governance reform

Regional and state ocean governance reform received the highest grades from the JOCI, because states and regional organizations worked to advance ocean and coastal policies through a number of initiatives. Regional initiatives include the Gulf of Mexico Alliance, the West Coast Governors' Agreement on Ocean Health, the Great Lakes Regional Collaboration, the Northeast Regional Ocean Council, the Mid-Atlantic Regional Council for the Ocean, and the Governors' South Atlantic Alliance. The policy of the Bush Administration was to develop regional partnerships in areas in which states were willing to come together around specific ocean and coastal priorities, as defined by those states. In some cases, for example Gulf of Mexico and the Northeast, collaboration between state and federal entities played an important role in the development of the regional initiatives. In other instances, for example the West Coast Governors' Agreement, states acted alone to form the partnership and then extended an invitation to federal partners to participate in the regional initiative. Other partnerships, such as the Mid-Atlantic Regional Council for the Ocean, formed over a longer period of time; a few regions (Pacific, Caribbean, and Alaska) did not take an affirmative step during this time to form a similar type of regional partnership.

Several regions across the country also established ocean-observing partnerships to provide the necessary coordination of observations, modelling activities, and data management as part of the development of an integrated ocean-observing system. For example, the Mid-Atlantic region formed an ocean-observing regional association in 2004 and a coordinated regional observatory in 2007 (MARACOOS, undated).

In collaboration with states, federal agencies also focused their programmes and efforts on state-identified priorities to coordinate and leverage resources. As a key partner in these efforts, the NOAA funded Sea Grant programmes undertook multi-year regional research planning assessments with seven regional projects: Alaska/Aleutian Islands; West Coast; Gulf of Mexico; Great Lakes; South Atlantic; Mid-Atlantic; and the Gulf of Maine. These collaborative projects among the regions sought to identify priority regional research and information needs, and then to develop a research action plan to meet those needs.

States took individual action as well, with several passing legislation or improving governance efforts between 2006 and 2008. Notably, in 2008, the Commonwealth of Massachusetts became the first US state to pass a comprehensive Oceans Act that governs all development activity in state waters and places the authority for ocean activities within a single office. New Jersey and New York also passed oceans legislation, while California, Florida, and Washington actively developed ocean governance initiatives, and Alaska, Hawai'i, Louisiana, and Oregon established councils, authorities or other mechanisms (JOCI, 2007). It should be noted that some states had been active in ocean conservation and management for many years preceding the Ocean Commission report and its subsequent implementation. Further, sub-regional efforts had been under way in different parts of the country, including Chesapeake Bay, Gulf of Maine, and Long Island Sound (JOCI, 2007).

International leadership

The United States is among the last major coastal nations that have not acceded to UNCLOS. The USOAP included a goal of acceding to the Convention, and in 2007 President Bush, along with industry, military, and environmental leaders, called for the Senate to approve it. However, despite holding several public hearings on the Convention, the Senate failed to act. The Senate Committee on Foreign Relations held hearings on UNCLOS on 27 September and 4 October 2007, and passed it with an overwhelming seventeen-to-four majority on 31 October 2007 (the Committee again held hearings on 23 May 2012 and 14 June 2012). However, UNCLOS continues to await a full Senate vote. The failure of the United States to accede to the Convention is a major reason behind the poor JOCI grade in this category.

Research, science and education

The grades for research, science, and education slowly improved over the three report cards, yet, as will be detailed later in the chapter, much remains to be done to educate the public about the importance of the oceans and their role in human, economic, and environmental health. In 2006, the Bush Administration released the Ocean Research Priorities Plan and Implementation Strategy, which set forth research priorities for the ocean and coasts, and stressed the importance of educational activities. The JOCI also stressed the need to advance the IOOS, because it can lead to increased understanding of ocean and coastal processes, especially in terms of climate change. Although efforts to advance the IOOS during this period were hampered by poor funding levels, as is explained later significant advances were made in this area.

Fisheries management reform

Fisheries management reform received a large grade boost in 2006 with the passage and enactment into law of the Magnuson–Stevens Fishery Conservation and Management Reauthorization Act of 2006, which attempts to address continuing problems with US fisheries – namely, continued overfishing and a lack of annual catch limits for many stocks, questions of the balance of membership of regional fishery management councils, data gaps for many fish stocks, and a lack of ecosystem-based management provisions. The goals of the 2006 Act are to end overfishing, to promote market-based management approaches, to improve the strength and role of science in the development of fishery management plans (FMPs), and to enhance international cooperation. Annual catch limits on overfished stocks were required by 2010, and on all other stocks, regardless of status, by 2011. Initial implementation of the Act's provisions began in 2007, with NOAA Fisheries initiating formal rule-making on a number of issues, including catch limits.

Funding for ocean policy and programmes, 2005–08

Underfunding of federal ocean programmes is a chronic problem, as reflected by the poor grades given to this subject by the JOCI. In 2005 and 2006, even though NOAA's budget request saw increases since 2001, the Administration's overall requests for funding increases were limited for ocean-related programmes. The USOAP released in December 2004 was intended to be a budget-neutral initiative, meaning that no new resources were necessary to implement it (White House, 2004). Following the initial implementation, new initiatives emanating from the USOAP called for new funding, including (as noted earlier) for the release of the Ocean Research Priorities Plan and Implementation Strategy in January 2007. The Administration requested US$143 million in new funding above the previous request to advance ocean science and research ($80 million), to protect and restore coastal and marine areas ($38 million), and to end overfishing and ensure sustainable use of ocean resources ($25 million). In 2007, the House of Representatives supported a funding increase for NOAA; however, the increases were not reflected in the fiscal year 2008 Omnibus Appropriations Bill. The Administration's budget remained flat, at $3.9 billion from 2005 to 2008.

Links between oceans and climate change

The links between oceans and climate change are a subject that the JOCI added in 2007, because there had been a growing emphasis on the impacts of climate change phenomena on the ocean, as well as the ability of the ocean to mitigate climate change. The JOCI noted that a number of states were taking steps to address climate change, and Congress considered several Bills dealing with the impacts of climate change and increasing understanding of climate change impacts, such as sea-level rise and coral reef degradation. Despite these developments, additional steps were recommended, including greater federal support for state and regional initiatives addressing climate change, expanded funding for ocean research, monitoring, and modelling, and full funding for the implementation of the Bush Administration's Ocean Research Priorities Plan and Implementation Strategy. Overall, the JOCI reports reflected the fact that, while progress had been made in many areas, much work remained to be done.

The creation of President Obama's National Ocean Policy, 2009–13

President Obama established the Interagency Ocean Policy Task Force (Task Force) on 12 July 2009 to develop recommendations for a national ocean policy and for effective coastal and marine spatial planning. To complete its recommendations, the Task Force, led by the chair of the Council of Environmental Quality and composed of twenty-four senior policy-level officials from the executive departments, agencies, and offices, reviewed federal, state, and foreign policies and models, past and pending legislation, and the recommendations contained in the reports of the US Commission on Ocean Policy and the Pew Oceans Commission.

The efforts of the Task Force culminated in a set of final recommendations released on 19 July 2010 (CEQ, 2010). Simultaneously, President Obama signed an executive order that adopted many of the recommendations.[8] The executive order established the first comprehensive, integrated National Policy for the Stewardship of the Ocean, Our Coasts, and the Great Lakes (see Box 11.1), and set a mechanism and methodology for comprehensive, sustainable planning and management for oceans and coasts, even though, as detailed later on, challenges with comprehensive planning persist. The executive order also established a Cabinet-level National Ocean Council (NOC), representing an interagency assembly led by CEQ and the OSTP to oversee implementation of the policy.

Box 11.1   Summary of the National Ocean Policy of the United States (2010)

Purpose

- Establishes a national policy to ensure the protection, maintenance, and restoration of the health of ocean, coastal, and Great Lakes ecosystems and resources
- Preserves maritime heritage
- Supports sustainable uses and access
- Provides for adaptive management to enhance understanding of, and capacity to respond to, climate change and ocean acidification
- Coordinates with national security and foreign policy initiatives
- Calls for regional plans that will enable a more integrated, comprehensive, ecosystem-based, flexible, and proactive approach to planning and managing sustainable multiple uses across sectors, and improve the conservation of the ocean, coasts, and the Great Lakes

Principles

- Policies, programmes, and activities of the United States should be managed and conducted in a manner that seeks to prevent or minimize adverse environmental impacts to the ocean, coasts, and Great Lakes ecosystems and resources, including cumulative impacts, and to ensure and improve their integrity. They should be managed and conducted in a manner that does not undermine efforts to protect, maintain, and restore healthy and biologically diverse ecosystems and the full range of services that they provide.
- Decisions affecting the ocean, coasts, and the Great Lakes should be informed by and consistent with the best available science. Decision-making will also be guided by a precautionary approach, as reflected in the Rio Declaration of 1992, Principle 15 of which states: ' . . . Where there are threats of serious or irreversible damage, lack of full scientific certainty shall not be used as a reason for postponing cost-effective measures to prevent environmental degradation.'
- Actions taken to protect the ocean, coasts, and the Great Lakes should endeavour to promote the principles that environmental damage should be avoided wherever practicable and that environmental costs should be internalized, taking into account the approach that those who cause environmental damage should generally bear the cost of that damage.
- Human activities that may affect ocean, coastal, and Great Lakes ecosystems should be managed using ecosystem-based management and adaptive management, through an integrated framework that accounts for the interdependence of the land, air, water, and ice, and the interconnectedness between human populations and these environments. Management should include monitoring and have the flexibility to adapt to evolving knowledge and understanding, changes in the global environment, and emerging uses.
- Current and future uses of ocean, coastal, and Great Lakes ecosystems and resources should be managed and effectively balanced in a way that:

    o   maintains and enhances the environmental sustainability of multiple uses, including those that contribute to the economy, commerce, recreation, security, and human health;

○ harmonizes competing and complementary uses effectively;

○ integrates efforts to protect, maintain, and restore the health, productivity, and resiliency of ocean, coastal, and Great Lakes ecosystems and the services that they provide; and

○ recognizes environmental changes and impacts, including those associated with an increasingly ice-diminished Arctic, sea-level rise, and ocean acidification.

• There should be support of disciplinary and interdisciplinary science, research, monitoring, mapping, modelling, forecasting, exploration, and assessment to continually improve understanding of ocean, coastal, and Great Lakes ecosystems. These efforts should include improving understanding of physical, biological, ecological, and chemical processes and changes.

Nine national priority objectives of the National Ocean Policy

Implementation of the National Ocean Policy is to focus around nine priority objectives that the United States will pursue 'to address some of the most pressing challenges facing the ocean, our coasts, and the Great Lakes' (CEQ/OSTP, 2010: 2).

1 **Ecosystem–Based Management:** Adopt ecosystem-based management as a foundational principle for comprehensive management of the ocean, our coasts, and the Great Lakes.

2 **Coastal and Marine Spatial Planning:** Implement comprehensive, integrated, ecosystem-based coastal and marine spatial planning (CMSP) and management in the United States.

3 **Inform Decisions and Improve Understanding:** Increase knowledge to continually inform and improve management and policy decisions and the capacity to respond to change and challenges. Better educate the public through formal and informal programs about the ocean, our coasts, and the Great Lakes.

4 **Coordinate and Support:** Better coordinate and support Federal, state, tribal, local, and regional management of the ocean, our coasts and the Great Lakes. Improve coordination and integration across the Federal Government, and as appropriate, engage with the international community.

Areas of Special Emphasis

1 **Resiliency and Adaptation to Climate Change and Ocean Acidification:** Strengthen resiliency of coastal communities and marine and Great Lakes environments and their abilities to adapt to climate change impacts and ocean acidification.

2 **Regional Ecosystem Protection and Restoration:** Establish and implement an integrated ecosystem protection and restoration strategy that is science-based and aligns conservation and restoration goals at the Federal, state, tribal, local, and regional levels.

3 **Water Quality and Sustainable Practices on Land:** Enhance water quality in the ocean, along our coasts, and in the Great Lakes by promoting and implementing sustainable practices on land.

4   **Changing Conditions in the Arctic:** Address environmental stewardship needs in the Arctic Ocean and adjacent coastal areas in the face of climate-induced and other environmental changes.

5   **Ocean, Coastal, and Great Lakes Observations, Mapping, and Infrastructure:** Strengthen and integrate Federal and non-Federal ocean observing systems, mapping, sensors, and data collection platforms into a national system and integrate that system into international observation efforts.

(CEQ/OSTP, 2010: 2)

Governance changes

The National Ocean Policy is intended to help agencies coordinate and cooperate in ocean and coastal management, with an emphasis on the need to balance ocean health and community prosperity, to level the playing field for all stakeholders, to make decisions based on the best available science, and to respect the unique character of each US region. The policy is premised on the building blocks of the NOC and the nine national priority objectives. The NOC represents a strengthened governance structure to provide for sustained, high-level, and coordinated attention to ocean and coastal issues. This Council restructures the Committee on Ocean Policy (Figure 11.2) and provides for a stronger mandate and direction.

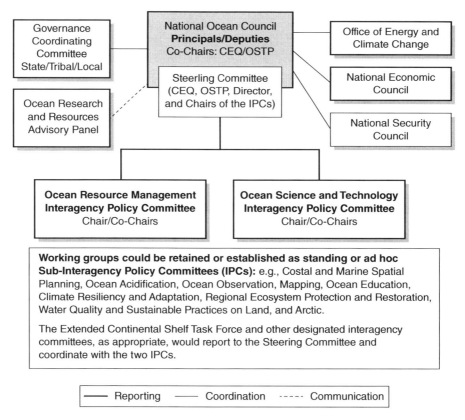

Figure 11.2 The structure of the National Ocean Council

Source: OPTF (2010)

As noted above, the co-chairs of the NOC include the chair of CEQ and the director of OSTP. The NOC further comprises:

- the Secretaries of State, Defense, the Interior, Agriculture, Health and Human Services, Commerce, Labor, Transportation, Energy, and Homeland Security;
- the Attorney General, the Administrator of the Environmental Protection Agency, the Director of the Office of Management and Budget, the Under Secretary of Commerce for Oceans and Atmosphere (NOAA Administrator), the Administrator of the National Aeronautics and Space Administration, the Director of National Intelligence, the Director of National Science Foundation, and Chair of the Joint Chiefs of Staff;
- the National Security Advisor and the Assistants to the President for Homeland Security and Counterterrorism, Domestic Policy, Energy and Climate Change, and Economic Policy;
- an employee of the federal government designated by the Vice-President; and
- such other officers or employees of the federal government as the co-chairs of the Council may, from time to time, designate.

The NOC co-chairs are granted the authority and responsibility to:

- advise the President on the implementation of the national policy;
- facilitate the development of strategic action plans and identify progress toward meeting defined goals and objectives;
- facilitate implementation of coastal and marine spatial planning;
- coordinate reporting and accountability;
- coordinate the development of an annual budget guidance memorandum on ocean priorities consistent with the goals and objectives of the National Ocean Policy;
- bring any presidential ocean actions or priorities to the NOC, as appropriate, for action; and
- coordinate with the Secretary of State and the heads of other relevant agencies on matters related to the policy issues that arise within international organizations, such as the Intergovernmental Oceanographic Commission (IOC), International Whaling Commission (IWC), Arctic Council, International Maritime Organization (IMO), and regional fishery management organizations.

The NOC is tasked, at the principal level, with the overall responsibility for the implementation of the National Ocean Policy (including coastal and marine spatial planning), the periodic updating and setting of national priority objectives, and the review and provision of annual direction on National Ocean Policy implementation objectives, based on Administration priorities and recommendations. The NOC is also to act as a forum for dispute resolution and decision-making issues that cannot be resolved at lower levels.

The Governance Coordinating Committee, comprising eighteen members from states, federally recognized tribes, and local governments, serves as the formal entity for state, tribal, and local government representatives to deliberate and coordinate with the NOC. The Ocean Research and Resources Advisory Panel (ORRAP), consisting of individuals from the national academies, state governments, academia, and ocean industries, continues to provide for independent advice and guidance on areas as requested by the NOC.

National Ocean Policy implementation and the 2012 Joint Ocean Commission Initiative Report

Overall implementation of the National Ocean Policy is to rely on existing legal authorities, to build upon and inform current plans, projects, and decision-making, and to require cooperation with state, local, and tribal governments. A key component of the policy centres on the need to ensure that decision-making is based on science, including a clear understanding of communities and ecosystems, and that objective measures are used for ocean, coastal, and Great Lakes health.

At the outset, the NOC recognized that a greater level of specificity was needed to instruct federal agencies on *how* to reach the nine priority objectives of the National Ocean Policy, including a timeline and milestones to measure success. For the initial years following establishment of the policy by executive order, the NOC set out to translate the policy into on-the-ground, specific actions that federal agencies could undertake to advance it and to improve the health and well-being of marine ecosystems and coastal communities. The NOC sought and received thoughtful input from national, regional, and local stakeholders from all ocean sectors, tribal, state, and local governments, the private sector, academia, and the public. As drafts of a National Ocean Policy Implementation Plan were being developed and circulated for comment, the JOCI provided an additional assessment through a 2012 US Ocean Policy Report Card (Table 11.3).

Unsurprisingly, the 2012 Report Card showed mixed results. While notable progress again was made, progress was lacking in certain key needs, sources of dedicated funding could not be secured, and enhanced stakeholder and interagency coordination required greater attention.

A significant step forward was taken in the spring of 2013 with the release of the National Ocean Policy Implementation Plan (NOC, 2013a), which was the subject of extensive public comments. The Plan stressed the importance of incremental change, pilot projects, support for local and regional capacity and self-determination, and the fundamental need for more and better information. Most likely reorganized to provide more effective communications aspects and clearer direct linkages to national priorities, the Implementation Plan specifies how federal actions on the nine priority objectives will benefit five major areas: the ocean economy; safety and security; coastal and ocean resilience; local choices; and science and information. The Implementation Plan gives a higher-level snapshot of the major actions and outcomes under each area, while a Technical Appendix provides greater details on specific planned actions – grounded in the nine priority objections – that federal agencies will take in the near-to-medium term to advance the National Ocean Policy, and to meet the challenges facing US oceans, coasts, ocean sectors, and coastal communities.

Additional efforts that have advanced in recent years, which support a major component of National Ocean Policy and the Implementation Plan, involve advancing regional marine planning around the country.

Coastal and marine spatial planning

In response to the executive order, the Task Force recommendations provide for a framework for effective CMSP that is fair and objective, regionally and community focused, and guided by ecosystem-based management (CEQ, 2010: Pt 4). This comprehensive, integrated, regional and ecosystem-based approach to management will attempt to bring together federal, state, and tribal partners to balance conservation, economic activity, user conflict, and sustainable use of the ocean, coastal, and Great Lakes resources. Decision-making is to be underpinned by sound science and the best available information to reduce conflicts, to improve planning and

Table 11.3 US Ocean Policy Report Card, 2012, issued by the Joint Ocean Commission Initiative

Subject	Grade	Comments
National support and leadership	**C** Good groundwork laid but need for better communication, expanded stakeholder engagement, and tangible results	*Notable Progress* NOC and GCC of local and state-level advisors established NOP Draft Implementation Plan released National workshop on regional ocean planning processes *Improvements Needed* Stronger outreach efforts to engage private sector and stakeholders and citizens in the NOP implementation process Expanded engagement of GCC to increase input from states, tribes, and local governments and of ORRAP Interagency review of ocean-related policies to reduce duplication and inefficiencies and resolve interagency conflicts Constructive engagement from Congress to improve and support National Ocean Policy implementation
Regional, state, and local leadership and implementation	**A–** Regional ocean partnerships continue to make progress but need more support from states and federal agencies	*Notable Progress* Multi-state regional ocean partnerships demonstrate leadership with stakeholder engagement, science and research, federal agency support, and leveraging resources to advance priorities Significant progress on advancing priorities across the country *Improvements Needed* Maximum flexibility for states and regions working to implement actions consistent with the NOP High-level support from states and federal agencies for advancing regional priorities Meaningful engagement of stakeholders and the public on implementation of priorities at the national and regional scales
Research, science, and education	**C** Some progress but funding and program cuts, as well as delayed implementation of critical tools, weakened ocean science, research, and education	*Notable Progress* NOC ocean data portal created, complemented by regional portals for access to regional, state, and local data Strong regional efforts to coordinate regional ocean and coastal research, observing, mapping, and restoration priorities *Improvements Needed* Fully develop and support the IOOS Release updated Ocean Research Priorities Plan and move toward achieving the goals outlined in this plan Reverse decreases or elimination of ocean education funding

	Grade	Description	Progress / Needs
Funding	**D–**	Ocean programs continue to be chronically underfunded, highlighting the need for a dedicated ocean investment fund	*Notable Progress* NOAA ROP Grants provided modest funding *Improvements Needed* Enact dedicated funding to provide support for improved ocean and coastal management, science, and education Rebalance NOAA's funding portfolio so satellite programs and ocean and coastal programs each have adequate resources Increase funding for regional ocean partnerships Prioritize funding of activities identified in the Ocean Research Priorities Plan and the National Research Council's Report on Infrastructure for Ocean Research and Societal Needs in 2030
Law of the Sea Convention	**F**	Strong support from Administration and private sector leaders but no successful senate vote yet	*Notable Progress* Strong support for UNCLOS from the Secretaries of Defense and State and other senior leaders within the Administration Active support for the Convention by a bipartisan coalition of national security, industry, and nongovernmental leaders Senate Committee on Foreign Relations hearing on UNCLOS *Improvements Needed* Senate advice and consent to accede to UNCLOS

Source: Extracted from JOCI (2012)

regulatory efficiencies, to decrease costs and delays, to engage affected communities and stakeholders, to enhance public engagement, and to preserve critical ecosystem functions.

The national goals of CMSP are to:

1   Support sustainable, safe, secure, efficient, and productive uses of the ocean, our coasts, and the Great Lakes, including those that contribute to the economy, commerce, recreation, conservation, homeland and national security, human health, safety, and welfare;

2   Protect, maintain, and restore the Nation's ocean, coastal, and Great Lakes resources and ensure resilient ecosystems and their ability to provide sustained delivery of ecosystem services;

3   Provide for and maintain public access to the ocean, coasts, and Great Lakes;

4   Promote compatibility among uses and reduce user conflicts and environmental impacts;

5   Streamline and improve the rigor, coherence, and consistency of decision-making and regulatory processes;

6   Increase certainty and predictability in planning for and implementing new investments for ocean, coastal, and Great Lakes uses; and

7   Enhance interagency, intergovernmental, and international communication and collaboration.

(CEQ, 2010: Pt 3)

The CMSP process is to include the US territorial sea, the EEZ, and the continental shelf, and to extend landward to the mean high-water line. The Great Lakes also are to be included, extending from the ordinary high-water mark and including the lakebed, subsoil, and water column to the limit of the US–Canada international boundary, and including Lake St. Clair and the connecting channels between lakes. Inland bays and estuaries on the coast and in the Great Lakes would be included, but privately owned lands would be excluded from the geographic scope. For the purposes of planning, the United States is divided into nine regional planning areas (see Figure 11.3), based on large marine ecosystems and allowing for the incorporation of existing state or regional ocean governance bodies, which include the Northeast, Mid-Atlantic, South Atlantic, Great Lakes, Caribbean, Gulf of Mexico, West Coast, Pacific Islands, and Alaska/Arctic. If desired each region may voluntarily establish a corresponding regional planning body, with federal, state, and tribal representatives, to develop regional goals, objectives, and ultimately regional CMSP.

Since the release of the Task Force recommendations, CMSP has faced significant opposition from some portions of Congress and industries, which view the process as an overreach of federal authority and are concerned that a lack of transparency will result in a process that, overall, favours some uses – including conservation – over others (NOC, 2011). While many industry representatives are mobilizing to ensure that they have a seat at the planning table, there are some that object to CMSP as just an additional layer of bureaucracy that will limit state and industry rights to access. Staff at the NOC has reached out to both Congress and industry to try to assuage these concerns, and to assure states and industry that their views will be represented through a transparent process. Further, recognizing that some stakeholders took issue with the 'spatial' component of CMSP (that is, by equating 'spatial' with 'ocean zoning'), the NOC took a deliberative approach to rebrand CMSP simply as 'regional marine planning' (RMP) (NOC, 2013b). However, the rebranding did little to counter opposition. It is likely that, until plans are

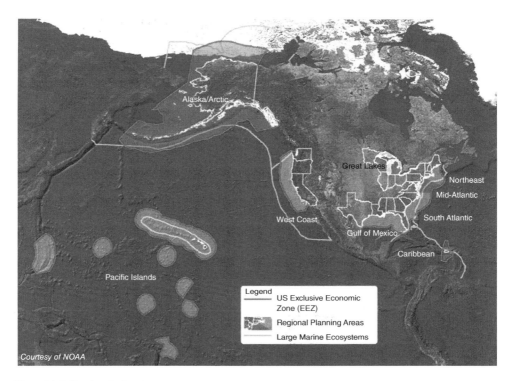

Figure 11.3 Regional planning areas

Source: CMSP (undated)

crafted and implemented through a trusted process, these concerns will persist at varying degrees throughout the regions.

Rather than prescribing a 'one size fits all' approach, the NOC has noted that it will work with those regions that are ready to move forward with regional planning bodies (RPBs) by supporting those bodies, their priorities, and the development of their plans, and will work with other regions through existing mechanisms. This is important, because it is clear that some regions do recognize the value of this process and want to move forward, as is evidenced by the fact that four have voluntarily established RPBs (the Northeast, Mid-Atlantic, Caribbean, and the Pacific Islands) to begin marine planning.

The NOC and federal agencies also are seeking to support RMP in other ways. For example, in eight of the nine regions, regional ocean partnerships (ROPs) have been voluntarily established through governors' agreements, in collaboration with federal agencies, tribes, local governments, and stakeholders (NOAA, undated*b*). The ROPs are as follows:

- the Caribbean Regional Ocean Partnership;
- the Council of Great Lakes Governors;
- the Governors' South Atlantic Alliance;
- the Gulf of Mexico Alliance;
- the Hawai'i Ocean Resources Management Plan;
- the Mid–Atlantic Regional Council on the Ocean;
- the Northeast Regional Ocean Council;

- the Pacific Regional Ocean Partnership; and
- the West Coast Governors Alliance on Ocean Health.

Although their methods and approaches may differ, the ROPs are all working to address similar challenges and share a belief that multi-sector, multi-state management decisions will result in an improved ocean environment and coastal economy. Federal funding has been supplied to support these efforts through the Regional Ocean Partnership Funding Program; in early 2012, NOAA awarded US$6.18 million to regional partners, with further awards in September 2012 of $3.14 million, and in 2013, of $3.7 million (NOAA, undated*b*).[9] Unfortunately, the fiscal year 2014 Omnibus Appropriations Bill zeroed funding for Regional Ocean Partnerships (HR, 2014). Nevertheless, new sources of funding are being sought, including from the private sector, and the RPBs being established around the country continue to provide a new, complementary platform for federal, state, and tribal authorities to interact, share, and collaborate. In at least one instance – the Mid-Atlantic Council for the Ocean – the existing governors' agreement is working closely with the new Mid-Atlantic RPB to build on the existing infrastructure and experiences, and to find synergies for each of the entities and their members and stakeholders. Tying in the regional ocean and coastal data product providers in each region (for example the IOOS regional associations) will serve the critical decision-making by the RPB and governors' agreement. It is encouraging to see that the federal government is committed to working with the states in the regions to develop a multi-sectoral approach to governance, regardless of what governance structure a region may adopt.

As with the previous Bush Administration, notable advances have been made under the Obama Administration, not least of which has been the creation of a comprehensive National Ocean Policy. The significance of this key effort should be recognized. However, the true work to improve the conditions of US ocean communities, economies, and ecosystems is only beginning, and significant challenges remain. Funding for ocean policy initiatives year after year has yet to be secured. Although small in number, there remains significant opposition to the ocean policy from both some congressional representatives and industry groups. Further, the policy makes clear that implementation is to rely on existing legal authorities, and to build upon and inform current plans, projects, and decision-making.

Strong championship is needed in support of further implementation and funding support for the National Ocean Policy. All of this, of course, is occurring within the context of threatened coastal and ocean resilience, increasingly at-risk coastal communities, and habitats that are disappearing, along with the societal benefits that they provide. For example, one of the 2013 milestones in the National Ocean Policy Implementation Plan was to document the status and trends of coastal wetlands using the most recent data from 2004 to 2009. The resulting report found that wetlands loss around the United States has accelerated over the last five years, from 60,000 to 80,000 acres per year (Dahl, 2013). This demonstrates that we do not have the luxury of time on our side. Agencies and partners are making progress under the National Ocean Policy, but without strong leadership, support, and funding to meet the challenges along US oceans and coasts, sustainability and a healthy ocean economy will not be realized.

Concluding observations

In the United States, the seeds of opportunity for far-reaching ocean policy reform have been sown. Two commissions have produced blueprints for the nation, echoing similar themes and solutions. There is much resolve among the ocean policy community to persevere to make important change happen. Implementation of a more coordinated and coherent approach to ocean governance in the United States, following the recommendations of the US Commission on Ocean Policy, began in the period 2005–09 with President Bush's Ocean Action Plan, during which time some progress was made. Additional advances have been made under President Obama's Administration in the period 2009–13 and are ongoing. Several Bills have been introduced in the Congress to implement various aspects of the Commission's recommendations, although few have yet been enacted. Perhaps most importantly, at the regional, state, and local levels, efforts continue to achieve practical solutions to real-world issues. However, the reforms made so far are relatively modest and do not yet address many of the major recommendations of the US Ocean Commission. Nevertheless, it is encouraging that implementation of some recommendations has come to fruition in the past ten years.

We should note that the Joint Ocean Commission has played a key role in providing an external look at the policy, evaluating its progress or lack thereof, and in prodding both the Congress and the Administration to accelerate their efforts to implement the National Ocean Policy. Additionally, professional and industry associations, along with NGOs and academia, have actively participated in policy development and implementation efforts. This underlines the importance of 'external parties' as important vehicles for keeping political attention focused on a national ocean policy and for mobilizing the broader ocean community.

The National Ocean Policy has been marked by slow implementation, especially in terms of the creation of RPBs and development of marine spatial plans. There are major problems and obstacles that hinder the full implementation of an integrated national ocean policy in the United States, which need to be addressed to enhance the likelihood of future success, as follows:

1   *Absence of a legislative basis* – In both the Bush and Obama Administrations, the basis for national ocean policy has been executive action, not legislative action. This means that the next administration can wipe out the changes made through the implementation of a national ocean policy in recent years 'at the stroke of a pen'. In fact, President Obama revoked President Bush's executive order on 19 July 2010 when he issued his own Executive Order for a National Ocean Policy.[10]

In the long term, a legislative basis to the National Ocean Policy must be obtained. In this way, the policy will have the force of law and will likely have much more enduring implementation success. A legislative solution has not yet been realized, in part because of political divisions in Congress and increases in partisanship in both Houses of Congress. Additionally, compromise has eluded Congress as a result of concerns and influence from the more liberal and conservative wings of the parties.

2   *Absence of industry support* – In contrast to other countries in which ocean industry groups have supported the development of an integrated national ocean policy, in the United States industry groups have generally either been lukewarm toward the National Ocean Policy or have expressed outright opposition to it, especially regarding the marine spatial planning aspects. The National Ocean Policy has been viewed, by some, as too environmentally inclined and not sufficiently focused on promoting economic development opportunities. Further, the term 'stewardship' has become associated, for some groups, with negative implications for industry and the economy. The underlying meaning of stewardship and related terms refer to responsible planning for the future – something

that is critical for any nation, industry, business, or other entity with a long-term outlook. Society will need to restore, acknowledge, and embrace the concept of balancing long- and short-term decision-making, to gain the broader support necessary to advance ocean policy development and implementation.

3   *Insufficient funding and staffing support for the policy* – Implementation of the National Ocean Policy has also been plagued by inadequate funding and staffing support. Very little new funding has been made available to support the implementation of the policy, and the National Ocean Council operates with only a handful of people. Strategies to implement new approaches have been left largely to government agencies, where budgets are shrink- ing in response to economic challenges and growing national debt. A combination of public and private investments will be necessary to achieve the goals of a well-understood and well-managed ocean and coastal domain.

4   *Competition for issue prioritization* – Ocean policy has not been an executive or legisla- tive priority in the past several administrations. There are many high-priority issues that compete on a daily basis for attention and resources, including areas such as the economy, defence, health care, and education. One indication of the relative importance of this policy issue in a presidential administration is the level of attention accorded by the senior staff. Despite the importance of oceans and coasts to the economic and overall well-being of the nation, ocean policy has been largely relegated to the environmental and scientific arms of the White House in the past two administrations rather than made the concern of the more powerful economic, domestic policy, or national security entities in the White House structure. The relative low level of importance placed on ocean policy in the past several administrations is further expressed through the lack of funding prioritiza- tion for ocean-policy-related issues and programmes. Even in times of crisis, such as the *Deepwater Horizon* oil spill, attention is focused for a period of time and then returned to other policy areas deemed more critical for the nation. Simultaneously, ocean science is equally undervalued and underfunded because of its low prioritization relative to other research areas and policy issues. And yet the value of knowledge (both basic and applied) is critical for effective and efficient decision-making. It is unlikely that ocean policy and ocean science will be a high priority in coming administrations, and this will be a chal- lenge to overcome for the executive and the legislative branches, as well as society in general, if any consistent and meaningful ocean policy and management advances are to occur.

5   *Geography of decision-making versus implementation* – Much discussion occurs in the fed- eral and state capitals, while implementation occurs in the field, far from offices and committee rooms. While the focus of reform is often placed on the coordination of central authorities, the breadth and nuance of the issues across the country require a regional approach to development and implementation of ocean and coastal policies and programmes. The required emphasis on regional approaches, however, is chal- lenging because of the variety of participants, the federal–state jurisdiction/sovereignty questions, the varying political forces, the competing policy priorities, the disconnect between regional decision-makers and regional ocean and coastal information provid- ers, and the overall lack of funding and support available to address these challenges in an appropriate and meaningful way. A significant shift in focus and resource mobiliza- tion toward the regions will be necessary.

The United States must, in the long run, enact an oceans law and policy patterned on the major recommendations of the US Commission on Ocean Policy, and it must provide sufficient funding and staffing support for the implementation of the policy. To achieve this, major and persistent mobilization of civil society groups will be essential.

Notes

1 The authors write in their personal capacities. Any views expressed herein do not necessarily reflect the views of the Mid-Atlantic Regional Association Coastal Ocean Observing System (MARACOOS), the National Oceanic and Atmospheric Administration (NOAA), or the Gerald J. Mangone Center for Marine Policy, University of Delaware.
2 Executive Proclamation 5928 of 27 December 1988 on the Territorial Sea of the United States.
3 Proclamation 7219 of 2 September 1999 on the Contiguous Zone of the United States.
4 Proclamation 5030 of 10 March 1983 on the Exclusive Economic Zone of the United States of America.
5 Presidential Proclamation 2667 of 28 September 1945 on the Policy of the United States with Respect to the Natural Resources of the Subsoil of the Sea Bed and the Continental Shelf.
6 Several of the authors of this chapter worked directly with the US Commission on Ocean Policy. Biliana Cicin-Sain served on the Scientific Advisory Panel of the US Commission on Ocean Policy and later, on the Scientific Advisory Committee of the Joint Ocean Commission Initiative; Gerhard Kuska worked as a staff member of the Commission, especially focusing on the ocean governance aspect of the report, and later, with NOAA, then as Director of Ocean and Coastal Policy in the Executive Office of the President, implementing the Ocean Action Plan, which was President George W. Bush's response to the US Commission on Ocean Policy's recommendations. Kateryna Wowk is working at NOAA, in part implementing the recommendations of the US Ocean Commission by means of President Obama's National Ocean Policy.
7 Executive Order 13366 of 17 December 2004 on a Committee on Ocean Policy.
8 Executive Order 13547 of 19 July 2010 on the Stewardship of the Ocean, Our Coasts and the Great Lakes.
9 However, it should be noted that, at the time of writing, despite the Administration's request for 2014 of US$5 million for regional ocean partnership grants, zero funding was appropriated for the programme.
10 See n. 8.

References

Central Intelligence Agency (CIA) (2014) 'United States', in *CIA Factbook 2014*, online at https://www.cia.gov/library/publications/the-world-factbook/geos/us.html [accessed 1 February 2014].

Cicin-Sain, B. (1992) *Ocean Governance: A New Vision – The Work Program of the Ocean Governance Study Group*, Newark, DE/San Diego, CA/Honolulu, HI: University of Delaware Sea Grant College Program/University of California Sea Grant College Program/University of Hawaii at Manoa Sea Grant College Program.

Cicin-Sain, B. (2012) 'The Evolution of US Marine Policy: Current Status and Future Prospects', Presentation at University of Delaware, 1 October.

Cicin-Sain, B., and Knecht, R. (2000) *The Future of US Ocean Policy*, Washington, DC: Island Press.

CMSP   *See* National Oceanic and Atmospheric Administration, Coastal and Marine Spatial Planning

Council on Environmental Quality (CEQ) (2010) *Final Recommendations of the Interagency Ocean Policy Task Force, July 19, 2010*, online at http://www.whitehouse.gov/files/documents/OPTF_FinalRecs.pdf [accessed 4 July 2014].

Council on Environmental Quality (CEQ) (2013) 'National Ocean Policy Implementation Plan', online at http://www.whitehouse.gov/administration/eop/ceq/initiatives/oceans [accessed 1 February 2014].

Council on Environmental Quality and the Office of Science and Technology Policy (CEQ/OSTP) (2010) *Federal Ocean and Coastal Activities Report to the US Congress for Calendar Years 2008 and 2009*, Report prepared by the, 16 November, online at http://www.whitehouse.gov/sites/default/files/microsites/ostp/20101116-focar.pdf [accessed 4 July 2014].

Dahl, T. E. (2013) *Status and Trends of Wetlands in the Conterminous United States 2004 to 2009*, online at http://www.fws.gov/wetlands/Documents/Status-and-Trends-of-Wetlands-in-the-Conterminous-United-States-2004-to-2009.pdf [accessed 4 July 2014].

Department of State (DOS) (2007) *Senior Officials' Meeting of the Coral Triangle Initiative on Coral Reefs, Fisheries and Food Security*, 7 December, online at http://2001–2009.state.gov/g/oes/rls/rm/2007/96747.htm [accessed 31 October 2014].

Environmental Protection Agency (EPA) (1998) *Clean Water Act Section 303(d) Lists: Overview of TMDL Program*, Washington, DC: EPA.

Food and Agricultural Organization of the United Nations (FAO) (2005) *Fishery Country Profile: The United States of America*, online at ftp://ftp.fao.org/Fi/DOCUMENT/fcp/en/FI_CP_US.pdf [accessed 1 November 2013].

House of Representatives (HR) (2014) 'Explanatory Statement Submitted by Mr. Rogers of Kentucky, Chairman of the House Committee on Appropriations Regarding the House Amendment to the Senate Amendment on H.R. 3547', online at http://docs.house.gov/billsthisweek/20140113/113-HR3547-JSOM-FM-B.pdf [accessed 1 February 2014].

Joint Ocean Commission Initiative (JOCI) (2005) *US Ocean Policy Report Card*, online at http://www.jointoceancommission.org/resource-center/2-Report-Cards/2006-02-01_2005_US_Ocean_Policy_Report_Card.pdf [accessed 1 November 2013].

Joint Ocean Commission Initiative (JOCI) (2006a) *From Sea to Shining Sea: Priorities for Ocean Policy Reform – Report to the United States Senate*, online at http://www.jointoceancommission.org/resource-center/1-Reports/2006-06-13_Sea_to_Shining_Sea_Report_to_Senate.pdf [accessed 4 November 2014].

Joint Ocean Commission Initiative (JOCI) (2006b) *US Ocean Policy Report Card*, online at http://www.jointoceancommission.org/resource-center/2-Report-Cards/2007-01-01_2006_Ocean_Policy_Report_Card.pdf [accessed 1 November 2013].

Joint Ocean Commission Initiative (JOCI) (2007) *US Ocean Policy Report Card*, online at http://www.jointoceancommission.org/resource-center/2-Report-Cards/2008-02-27_2007_Ocean_Policy_Report_Card.pdf [accessed 1 November 2013].

Joint Ocean Commission Initiative (JOCI) (2012) *US Ocean Policy Report Card*, online at http://www.jointoceancommission.org/resource-center/2-Report-Cards/2012-06-06_2012_JOCI_report_card.pdf [accessed 1 February 2014].

Kuska, G. F. (2005) 'Collaboration toward a More Integrated National Ocean Policy: Assessment of Several US Federal Interagency Coordination Groups', PhD thesis, University of Delaware, Newark, DE.

Mid-Atlantic Regional Association Coastal Ocean Observing System (MARACOOS) (undated) 'About MARACOOS', online at http://maracoos.org/about_maracoos [accessed 1 November 2013].

Minerals Management Service (MMS) (2006) 'Special Interagency OAP Issue', *MMS Ocean Science*, 3(5), online at http://www.boem.gov/uploadedFiles/BOEM/Newsroom/Publications_Library/Ocean_Science/mms_ocean_06_sep_oct.pdf [accessed 1 November 2013].

National Ocean Council (NOC) (2011) 'Coastal and Marine Spatial Planning: Public Comments Received 1/24/2011–4/29/2011', online at http://www.whitehouse.gov/sites/default/files/microsites/ceq/cmsp_comments_and_attachments_1.24.11-4.29.11.pdf [accessed 1 November 2013].

National Ocean Council (NOC) (2013a) *National Ocean Plan Implementation Plan*, online at http://www.whitehouse.gov/sites/default/files/national_ocean_policy_implementation_plan.pdf [accessed 4 July 2014].

National Ocean Council (NOC) (2013b) *Marine Planning Handbook*, online at http://www.whitehouse.gov//sites/default/files/final_marine_planning_handbook.pdf [accessed 1 February 2014].

National Ocean Economics Program (NOEP) (undated) 'Environmental and Recreational (Non-Market) Values: Overview', online at http://oceaneconomics.org/nonmarket/ [accessed 1 February 2014].

National Oceanic and Atmospheric Administration (NOAA) (undateda) 'NOAA's State of the Coast, http://stateofthecoast.noaa.gov/economy.html [accessed 1 February 2014].

National Oceanic and Atmospheric Administration (NOAA) (undatedb) 'Regional Activities', online at http://cmsp.noaa.gov/activities/index.html [accessed 1 February 2014].

National Oceanic and Atmospheric Administration (NOAA) (2012) 'Spatial Trends in Coastal Socioeconomics', online at http://coast.noaa.gov/digitalcoast/data/stics [accessed 1 February 2014].

National Oceanic and Atmospheric Administration (NOAA) (2013) 'Ship Strike Reduction Rule Proves Effective Protecting North Atlantic Right Whales', Press release, 6 December, online at http://www.nmfs.noaa.gov/mediacenter/2013/12/04_12_shipstrikereduction_final_rule.html [accessed 1 February 2014].

National Oceanic and Atmospheric Administration, Coastal and Marine Spatial Planning (CMSP) (undated) 'National Framework', online at http://www.msp.noaa.gov/framework/index.html [accessed 1 November 2013].

Ocean Policy Task Force (OPTF) (2010) *Final Recommendations of the Interagency Ocean Policy Task Force*, 19 July, online at http://www.whitehouse.gov/files/documents/OPTF_FinalRecs.pdf [accessed 1 November 2013].

Papahānaumokuākea Marine National Monument (undated) 'About', online at http://www.papahanau-mokuakea.gov/about/ [accessed 1 November 2013].

Pew Oceans Commission (POC) (2003) *America's Living Oceans: Charting a Course for Sea Change*, Washington, DC: POC.

US Commission on Marine Science, Engineering and Resources (Stratton Commission) (1969) *Our Nation and the Sea: A Plan for National Action – Report of the Commission on Marine Science, Engineering and Resources*, Washington, DC: US Government Printing Office.

US Commission on Ocean Policy (USCOP) (2004) *An Ocean Blueprint for the 21st Century: Final Report*, online at http://govinfo.library.unt.edu/oceancommission/documents/full_color_rpt/welcome.html [accessed 1 November 2013].

White House (undated) 'Protecting Our Nation's Environment', online at http://georgewbush-white-house.archives.gov/infocus/environment/ [accessed 1 September 2013].

White House (2004) *US Ocean Action Plan*, 17 December, online at http://www.nauticalcharts.noaa.gov/ocs/hsrp/admin/mar2005/OceanActionPlan.pdf [accessed 1 September 2013].

White House (2008) 'President Bush Discusses Ocean Action Plan', Press release, 26 September, online at http://georgewbush-whitehouse.archives.gov/news/releases/2008/09/20080926-4.html [accessed 1 September 2013].

12

TOWARD AN INTEGRATED POLICY FOR THE OCEAN IN PORTUGAL[1]

Mário Ruivo, Tiago Pitta e Cunha, Márcia Marques, and Raquel Ribeiro

Introduction

The ocean is an extremely important asset for Portugal owing to its geography and biophysics. If managed in a sustainable and integrated way, it might become – even more than it currently is – a primary source of socio-economic, scientific, and cultural development. The maritime areas under Portuguese sovereignty/national jurisdiction offer promising potential for the exploitation of mineral and energy resources, biodiversity, development of a biotechnology industry, and tourism and associated recreational activities. Traditional ocean uses, such as fisheries, aquaculture, fish processing, port infrastructure, shipbuilding, ship repair, shipping, and telecommunications, can be further developed and restructured. An integrated policy for the ocean also requires a significant investment in the maritime components of national defence, environmental protection, marine areas conservation, and a reinforcement of marine science, technology capabilities, and specialized human resources.

Basic information

Land area and maritime areas under national sovereignty/jurisdiction

With an exclusive economic zone (EEZ) approximately eighteen times larger than its land area, the maritime area under Portugal's jurisdiction is one of the largest in the European Union, covering more than 1,700,000 km². A proposal to extend the Portuguese continental shelf was submitted to the United Nations' Commission on the Limits of the Continental Shelf (CLCS) on 11 May 2009, and if approved, Portugal's continental shelf will be approximately forty times larger than its land area (see Figure 12.1). Including the mainland, Azores, and Madeira, Portugal has a coastline of 2,587 km (Table 12.1) and a total land area of 92,212.02 km² (Table 12.2).

The national EEZ and continental shelf cover a large area of the northeast Atlantic region and have very specific oceanographic characteristics. Processes and phenomena in this area include the North Atlantic Oscillation (NAO), coastal upwelling, and the strong influence of Mediterranean water on the Atlantic Ocean. These features are particularly important for scientific research in ocean–atmosphere interactions, and assessment and management of natural

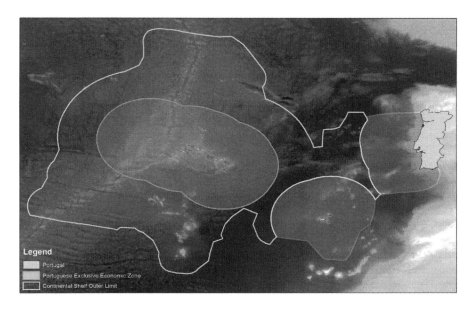

Figure 12.1 Portuguese EEZ and proposed extended continental shelf

Source: Task Force for the Extension of the Continental Shelf (*Estrutura de Missão para a Extensão da Plataforma Continental*, EMEPC)

Table 12.1 Length of the Portuguese coastline in kilometres and nautical miles

Territory	Length of the coastline	
	Km	*Nautical miles*
Mainland	1,242	500
Azores	943	473
Madeira	402	187
Total	**2,587**	**1,160**

Note: 1 nautical mile = 1,852 metres

Source: CAOP (2012)

Table 12.2 Total Portuguese land area

Territory	Land area (km²)
Mainland	89,088.93
Azores	2,321.96
Madeira	801.12
Total	**92,212.02**

Source: CAOP (2012)

hazards and impacts of anthropogenic origin. Furthermore, the seabed and subsoil subjacent to the national EEZ is extensive and diverse. These areas include the continental margin, vast abyssal plains, canyons, seamounts and islands, hydrothermal vents, and a large portion of the Mid-Atlantic Ridge.

Some of the most interesting ecosystems in the ocean are located in these areas. They constitute a valuable national heritage and offer potential economic resources that must be assessed, managed, and preserved for future generations. Moreover, the tectonic, seismic, and hydrothermal characteristics of the Mid-Atlantic Ridge and the Azores archipelagic region offer exceptional opportunities for a better understanding of biosphere–geosphere–hydrosphere processes, climate change, and natural hazards (including earthquakes and tsunamis). An intensified and coordinated research and monitoring effort at the national and international levels is necessary to increase understanding of the large marine ecosystems (LMEs) in these regions, as well as of associated natural processes and anthropogenic impacts. These actions will further contribute to the acquisition of data and information for management and sustainable development.

Population

At the beginning of the twenty-first century, Portugal had a population of 10,356,117 (Table 12.3). It is estimated that 49.6 per cent of the Portuguese population lives within 10 km of the sea. This trend accentuates the risk of conflicts between the different ocean users (for example tourism, fisheries, aquaculture, maritime transportation and recreational boating, discharge of pollutants, industries), and has resulted in claims for effective protection and conservation of biodiversity and coastal ecosystems.

The maritime economy contributed approximately 2.5 per cent to the national gross value added in 2010, only considering direct effects, and comprised 2.3 per cent of national employment (DGPM, 2013). It is expected that, by 2020, the maritime economy will contribute around 3 per cent to the national economy.

A considerable portion of commodities and energy arrive by sea through the country's ports. Tourism is one of the biggest industries in Portugal. The fishing community – one of the largest in Europe – still draws its livelihood from the ocean. However, indicators point to a considerable decline in traditional fisheries' activities, maritime transportation, and shipbuilding in recent decades. It is commonly accepted that the status and utilization of marine resources in Portugal need to be properly assessed, as well as the activities associated with the oceans that spin off into various subsectors of economic activity (Pitta e Cunha, 2011; COTEC Portugal, 2012).

In this context, institutional initiatives and arrangements have arisen the last few years, although some of them were fragmented and somehow experimental. The next section covers the evolutionary course of Portugal's efforts toward an integrated national ocean strategy and governance.

Table 12.3 Demographic data

Territory	Population		Population density (residents/km²)		Effective growth rate (%)
	2001	*2011*	*2001*	*2011*	
Portugal	10,356,117	10,562,178	112.4	114.5	2
Mainland	9,869,343	10,047,621	110.8	–	–
Azores	241,763	246,772	104.1	106.3	2.1
Madeira	245,011	267,785	312.2	334.3	9.3

Source: INE (2002; 2013)

The policy development process

Portugal has come a long way in developing its current ocean policy. This development process was influenced by global and regional milestones, and also resulted from Portugal's membership of the European Union, taking into account its national priorities oriented towards sustainable development.

The rationale behind a national policy for the ocean is anchored in the need to manage maritime areas under national sovereignty/jurisdiction by means of a comprehensive, long-term stable policy based on an inter-sectoral, interdisciplinary, and effective transversal integration of ocean affairs. The goal is to overcome fragmented sectoral approaches by providing an integrated framework based on the inspired insight, expressed in the 1982 United Nations Convention on the Law of the Sea (UNCLOS), that 'the problems of ocean space are closely interrelated and need to be considered as a whole', so as to reap the full benefits of this natural resource in a sustainable way.

This section describes the Portuguese path toward integrated governance of ocean affairs since 1998 until the adoption of the National Ocean Strategy (2013–2020) presently in force, highlighting the most relevant features and events in the period.

Evolutionary pathway

After testing a governance model based upon broad mandates assigned to two Ministries of the Sea (in 1983 and 1991) in an attempt to institutionally integrate all oceans' sectors under a sole governmental structure, the Portuguese government initiated a process for the formulation of a national strategy for the ocean in 1998. The intention, among other relevant objectives, was to define effective mechanisms and institutional arrangements, to reinforce capabilities for sustainable development and effective management of maritime areas under national sovereignty/jurisdiction, to reduce sectoral conflicts in the uses of the sea, and to invest in new promising activities to increase competitiveness, innovation, and economic growth. These initiatives were to be solidly based on reliable scientific knowledge, data, and information.

This process was fuelled by several major projects and actions in which Portugal was directly involved (see Figure 12.2), including Lisbon's hosting of the 1998 World Exposition (EXPO'98), which triggered a positive debate on ocean affairs and humankind. Themed 'The Oceans: A Heritage for the Future', EXPO'98 was designed to promote public awareness and better understanding of the ocean, and had a strong national impact on schools, civil society, the public and private sectors, non-governmental organizations (NGOs), and the media. Other key actions included Portugal's submission to the United Nations General Assembly, through the Intergovernmental Oceanographic Commission (IOC) of the United Nations Educational, Scientific and Cultural Organization (UNESCO), of the proposal to declare 1998 'International Year of the Ocean' and the presentation by the Independent World Commission on the Oceans, chaired by Mário Soares, former President of Portugal, of the report *The Ocean, Our Future* (IWCO, 1998a), in conjunction with the *1998 Lisbon Declaration: Ocean Governance in the 21st Century – Democracy, Equity and Peace in the Ocean* (IWCO, 1998b), which advanced ideas aimed at improving global ocean governance from a holistic point of view. These actions had a significant effect on national and international efforts to promote ocean affairs.

In this context, several initiatives were approved – namely, those published in a special edition of the national Official Gazette, *Diário da República,* symbolically printed in blue to integrate with commemorations of the 1998 International Year of the Ocean and the theme of EXPO'98. In particular, a resolution of the Council of Ministers formulated guidelines for

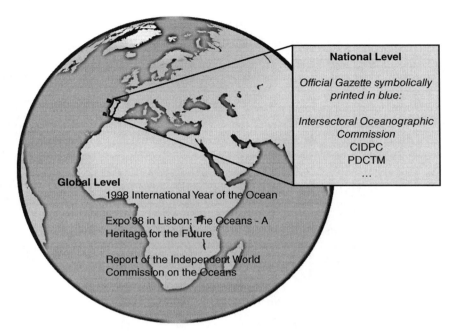

Figure 12.2 1998 as the trigger for ocean-related issues at the national and global levels

the development of a national maritime strategy based on UNCLOS and other relevant instruments.[2] This led to the creation of the Intersectoral Oceanographic Commission (*Comissão Oceanográfica Intersectorial*, COI) within the framework of the National Ministry for Science and Technology (*Ministério da Ciência, Tecnologia e Inovação*, MCTI),[3] the launching of the Programme to Promote Ocean Sciences and Technologies (*Programa Dinamizador das Ciências e Tecnologias do Mar*, PDCTM) within the framework of the Foundation for Science and Technology (*Fundação para a Ciência e a Tecnologia*, FCT),[4] and the establishment of the Interministerial Commission for the Delimitation of the Continental Shelf (*Comissão Interministerial para a Delimitação da Plataforma Continental*, CIDPC) beyond 200 nautical miles in accordance with the provisions of UNCLOS.[5]

Additional institutional steps took place in the first decade of the twenty-first century. After a long period of isolation, which came to an end in 1974, Portugal started a process of turning outwards, of being actively involved in international fora and the global ocean agenda. Portugal expressed willingness to host two international organizations dealing with ocean affairs: the European Centre for Information on Marine Science and Technology (EurOcean); and the European Maritime Safety Agency (EMSA).

EurOcean, with an office currently located in Lisbon under a protocol with the FCT, was launched in 2002 as a joint French–Portuguese initiative, the main functions being facilitating access and compiling appropriate data and information on marine sciences and technologies, and other products of relevance for participating members and other users, to enhance communication between European research organizations. Presently, this NGO has fourteen full members from different European countries and three cooperating members (EurOcean, undated). This centre has been bridging the competent structures existing at national and European levels by means of the networking of activities.

EMSA was created in 2002 in the aftermath of the *Erika* disaster and has had its headquarters in Lisbon since 2006. The Agency provides technical assistance to the European Commission in the development and follow-up of EU legislation on maritime safety, including monitoring visits in Member States. Moreover, EMSA organizes training activities and provides technical advice, assists Member States when affected by pollution from ships, and facilitates cooperation on establishing a European vessel traffic monitoring and information system. The Agency has been gradually assigned more maritime transport-related tasks, enhancing technical assistance and providing expertise and services to Member States in the maritime domain (EMSA, undated).

In response to ocean governance requirements, institutional adjustments, new mechanisms, and new approaches were required to reinforce national capacities in ocean affairs.

The Maritime Authority System was established in 2002,[6] with a mandate to ensure maritime safety, surveillance, and police patrolling of maritime waters under national jurisdiction in accordance with UNCLOS.

In 2003, the prime minister created a high-level working group to further advance the holistic approach to ocean and maritime affairs, the Strategic Oceans Commission (*Comissão Estratégica dos Oceanos*, CEO),[7] which was composed of representatives of nine ministries with responsibilities in ocean affairs, and of the regional governments of Azores and of Madeira, as well as experts from diverse sectors of civil society. The CEO was assigned the task of submitting a report providing the rationale for the definition of a national strategy for the ocean, guided by the principle of sustainable development. The aim of such a strategy was to strengthen Portugal's role in ocean affairs, including maximizing benefits of the rational use of marine resources, and to promote more effective management and exploitation of maritime areas under national sovereignty/jurisdiction (CEO, 2004).

Following up on this process, the Portuguese government created the Task Force for Maritime Affairs (*Estrutura de Missão para os Assuntos do Mar*, EMAM) in 2005, under the authority of the then Ministry of National Defence and Sea Affairs.[8] The mission of EMAM was to submit a report offering guidelines for the implementation of an integrated national policy for the ocean, and to ensure the interdepartmental coordination of maritime affairs in order to facilitate and promote a concerted action of all public and private entities dealing with marine and maritime activities. This process resulted in the first such strategy, the National Ocean Strategy (2006–2016), which was approved by the Council of Ministers on National Ocean Day, 16 November 2006 (EMAM, 2006). The Task Force was supported by a consultative council, comprising representatives of the different ministries with responsibilities in maritime affairs, as well as representatives from the private sector.

In compliance with its mandate, EMAM was involved in the preparation of the Joint Contribution (Portugal, Spain, and France) to the European Commission's Green Paper on Maritime Policy (European Commission, 2006), which contribution was submitted to the European Commission on 4 April 2005. The Task Force undertook the gathering of national contributions and positions in collaboration with representatives of relevant ministries.

In the course of the efforts undertaken to extend the Portuguese continental shelf, the Portuguese Task Force for the Extension of the Continental Shelf (*Estrutura de Missão para a Extensão da Plataforma Continental*, EMEPC) was also created under the authority of the then Secretary of State for National Defence and Sea Affairs/Ministry of National Defence.[9] The mission of EMEPC was to prepare a proposal for the extension of the national continental shelf beyond 200 nautical miles, in accordance with the provisions of UNCLOS, and to submit it to the CLCS by 13 May 2009 (EMEPC, undated). Owing to the complexity of the technical and scientific tasks, EMEPC's mandate had to be extended from the original deadline of April 2006 to 2016.[10] The Task Force continues its mission of collecting and updating data to reinforce the goal of delimiting the Portuguese continental shelf. It is presently under the authority of the Ministry of Agriculture and Sea (*Ministério da Agricultura e do Mar*, MAM). A consultative commission, comprising representatives of various ministries, as well as specialists from relevant research institutions, has been created to follow up EMEPC's activities and to contribute to the success of its mission.

In 2006, a law defining the extension of the maritime areas under national sovereignty/jurisdiction was enacted, within the remit of UNCLOS, regulating the powers of the state within national maritime areas and in the high seas.[11] It is worth noting that, within this overall process, Portugal succeeded in negotiating the creation and management of the first marine protected area (MPA) in the high seas. Negotiated under the 1992 Convention for the Protection of the Marine Environment of the North-East Atlantic (OSPAR) in 2006, the MPA was designated for the protection of the hydrothermal vent field 'Rainbow', located in close proximity to the Portuguese EEZ in the area of Azores (OSPAR Commission, 2013).

Based on the need for a more coordinated action in ocean affairs, a major step was taken in 2007 when the Council of Ministers approved the creation of an Interministerial Commission for Ocean Affairs (*Comissão Interministerial para os Assuntos do Mar*, CIAM).[12] The Commission functioned under the leadership of the Minister of National Defence, with the support of EMAM, whose mandate was then extended until 31 December 2009. The Commission was composed of several ministers in relevant ocean affairs portfolios, as well as representatives from the regional governments of the Azores and Madeira. The Commission had the opportunity to integrate representatives of the private sector and NGOs as required. The main objectives of CIAM were to coordinate, follow up, and evaluate the implementation of the National Ocean Strategy (2006–2016), to promote the enforcement of international commitments undertaken by Portugal, and to encourage investment from the private sector in maritime-related activities, so as to promote the development of a sound, forward-looking, maritime-based economy. Following a long debate on the functions and responsibilities of CIAM, it was finally decided that the prime minister should chair the Commission,[13] with a view to facilitating effective interministerial and cross-sectoral coordination in ocean affairs.

Recognizing that the implementation of a national ocean strategy as a national goal requires the involvement of relevant sectors of civil society, the first meeting of CIAM in 2007 launched a new ad hoc entity devoted to ocean affairs to act as an informal bridge between civil society and the government: the Standing Forum for Maritime Affairs (*Fórum Permanente Para os Assuntos do Mar*, FPAM). Despite logistical limitations, the FPAM contributed to the promotion of an inter-sectoral and multidisciplinary approach to ocean governance by providing a platform for the expression of diverse perspectives and points of view (Portal MarOceano, undated).

Responding to the commitment to complying with international principles, particularly at the European level, it was under the Portuguese presidency of the European Union that the EU Integrated Maritime Policy was adopted in Lisbon in October 2007 by the European Commission, and endorsed by the Heads of State and Government in the European Council in December 2007 (European Commission, 2009).

In the context of recent governmental and institutional changes, the private sector and maritime clusters organized themselves, and appeared in the national arena in the 2009 Oceano XXI exposition (Oceano XXI, undated) and in the 2010 Business Forum for Ocean Economy (*Fórum Empresarial para a Economia do Mar*, FEEM) (FEEM, undated). These initiatives contributed to a more effective integration of Portuguese ocean strategies (Figure 12.3).

Within the EU context, Brussels adopted its strategy for the Atlantic region: the European Union Maritime Strategy for the Atlantic Ocean Area (European Commission, 2011). It was

Figure 12.3 European and national context in the initial tenure of the National Ocean Strategy (2006–2016)

Source: Compiled by the authors from European Commission (undated) and Museus Portugal (undated)

launched during the Lisbon Atlantic Conference in 2011 and aimed at generating a public discussion encompassing all parties concerned (Atlantic Area Transnational Programme, 2011).

After 2012, new adjustments were made in the national institutional framework for ocean affairs in Portugal. During the tenure of the nineteenth constitutional government, a Secretary of State for the Sea was created under the jurisdiction of the Ministry of Agriculture, Sea, Environment and Spatial Planning (*Ministério da Agricultura, do Mar, do Ambiente e do Ordenamento do Território*, MAMAOT), which has since been restructured into two ministries: the Ministry of Environment, Spatial Planning and Energy (*Ministério do Ambiente, Ordenamento do Território e Energia*, MAOTE); and the Ministry of Agriculture and Sea (*Ministério da Agricultura e Mar*, MAM). A new Directorate General for Maritime Policy (*Direção-Geral de Política do Mar*, DGPM) was also created within MAM, and given the mission to develop, evaluate, and update the National Ocean Strategy (including developments in European policies). The DGPM also proposed a national ocean policy, which included a plan and a management strategy for the use and activities of maritime spaces, as well as a mandate to monitor and participate in the development of the EU Integrated Maritime Policy, and to promote national and international cooperation related to the ocean.[14] The DGPM provides support to the Interministerial Commission for Ocean Affairs (CIAM),[15] and currently has the following objectives: to implement and update the National Ocean Strategy; and to set annual goals for the implementation of the action plan in conjunction with the national state budget and the multi-annual medium- and long-term budgets.[16] Moreover, CIAM, under the chairship of the prime minister, is now formally open to the involvement of personalities of recognized merit in ocean affairs, reinforcing civil society's representation.

A second Lisbon Atlantic Conference took place in 2013, two years after the official announcement of the EU Maritime Strategy for the Atlantic Area. This event provided not only a 'peer review', but also took stock of the progress made so far – in particular, the approval of the Action Plan for a Maritime Strategy in the Atlantic Area. It facilitated identification of future opportunities within the framework of the European Union for national ocean strategies and dedicated action plans (Lisbon Atlantic Conference, 2013). In this regard, Portugal presented its proposal of a partnership agreement (*Acordo de Parceria 2014–2020*) to the European Union on 31 January 2014, with a strong focus on ocean affairs (Governo de Portugal, 2014).

In the national arena, two major legal instruments were adopted in 2014:

- the Framework Law for National Maritime Spatial Planning and Management (*Bases da Política de Ordenamento e de Gestão do Espaço Marítimo Nacional*), which aims to provide an adequate structure for the use of the national maritime areas in order to ensure the objectives of sustainable development;[17] and
- the current National Ocean Strategy (2013–2020), which was justified by the need to update the 2006 Strategy, taking into account the European Union's strategies, policies, and financial cycles.

Portuguese ocean policy

National Ocean Strategy (2006–2016)

The National Ocean Strategy (2006–2016*)* was approved in 2006[18] to promote the integration and coordination at the governmental level of cross-cutting and multidisciplinary policies dealing with ocean affairs. The main pillars of the strategy were its articulation of plans, strategies, and programmes pertaining to the coastal zone and adjacent waters, the participation of stakeholders, and the build-up of solid and credible scientific information. Furthermore, as noted earlier, CIAM was established in 2007 to facilitate short-term implementation.

This strategy was supported by eight strategic actions, which identified the cross-cutting measures needed to create favourable conditions for the development of an integrated policy based on the sustainable use of the sea. The implementation of the strategic actions, in articulation with other national strategies, would ensure the enforcement of the three strategic pillars on which the actions were based: knowledge; spatial planning policy; and active promotion and defence of national interests.

National Ocean Strategy (2013–2020)

Several developments in recent years prompted the adoption of the National Ocean Strategy (2013–2020) (see Figure 12.4):

- a new EU multi-annual financial framework (MFF) for 2014–2020;
- the 2007 approval of the EU Integrated Maritime Policy, a year after the launch of the first Portuguese National Ocean Strategy;
- the Europe 2020 strategy, launched in 2010 as the European Union's growth strategy for the coming decade;
- the 'Blue Growth' Agenda (2012) – the European Union's long-term strategy to support sustainable growth in the marine and maritime sectors as a whole, and the maritime contribution to achieving the goals of the Europe 2020 strategy for smart, sustainable, and inclusive growth; and
- the new Common Fisheries Policy (CFP) made effective by the European Council and Parliament on 1 January 2014.

The National Ocean Strategy (2013–2020) was adopted in 2014.[19] The Strategy 'was subjected to extensive public debate . . . having received over one hundred [written] contributions from civil society, in the Academy, but also from public and private entities, which have helped to improve and enrich the document' (DGPM, 2013: 7). The main difference from the previous

Figure 12.4 Revision of the National Ocean Strategy (2006–2016), leading to the adoption of National Ocean Strategy (2013–2020)

Source: Compiled by the authors from European Commission (undated)

Strategy (2006) is the adoption of an action plan (*Plano Mar-Portugal*), which sets out concrete objectives, a timeline for meeting them and the resources needed to carry them out, and provisions for the assessment of progress made.

The National Ocean Strategy (2013–2020) reaffirms 'the national maritime identity in a modern, proactive and entrepreneurial framework', inserting 'Portugal, on a worldwide level, [as] a leading maritime nation and an undisputed partner of the IMP [Integrated Maritime Policy] and of the EU maritime strategy, in particular for the Atlantic area' (DGPM, 2013: 62). On this basis, it is necessary to create conditions for 'attracting investment, both national and international, in all sea economy sectors, promoting growth, employment, social cohesion and territorial integrity, and, until 2020, promoting an increase of the sea economy contribution for the GDP' (DGPM, 2013: 62). Furthermore, it aims at strengthening 'national scientific and technological capacity, stimulating development of new areas of action that promote the knowledge of the ocean and effectively, efficiently and sustainably enhance its resources, uses and activities as well as the ecosystem's services' (DGPM, 2013: 62).

Implementation, evaluation, and long-term outlook

The National Ocean Strategy (2013–2020) is accompanied by the action plan *Plano Mar-Portugal*, 'which includes the programs to be run and developed, in order to achieve specific objectives and produce the desired effects, being subject to proper monitoring, evaluation, review and update mechanisms' (DGPM, 2013: 7). The action plan is intended to be integrative (see Figure 12.5), and is 'aimed at the economic, social and environmental valorization of the national maritime space through the implementation of sectorial and cross-sectorial projects' (DGPM, 2013: 7).

The transformation of Portuguese institutions to a more rational and effective ocean governance model reflects, to a certain degree, the 'experimental nature' of the process, as evaluated by Mário Ruivo in a recent presentation (Ruivo, 2014). The strategic and structural instruments have been progressively shaped by more open interaction of public administration with stakeholders and interest groups, and reinforced by a better understanding of public opinion in ocean affairs. The coming years will indicate whether the recent developments and the current

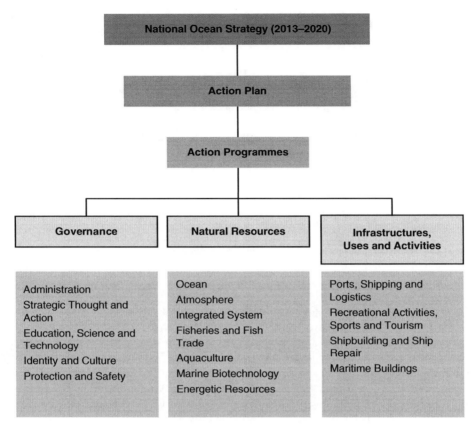

Figure 12.5 Main areas of the action programmes from *Plano Mar-Portugal*

Source: Based on DGPM (2013)

National Ocean Strategy provide a sustainable course of action for the next decades, allowing Portugal to integrate marine and maritime affairs as a major contribution to reinforcing its economic, social, and environmental dimensions. Finally, Portugal should continue to contribute to marine and maritime affairs with a view toward promoting more responsible ocean governance, to enhance ocean sustainability worldwide.

Notes

1 The opinions expressed in this document are those of the authors.
2 Resolution of the Council of Ministers No. 83/98 of 10 July 1998, *Diário da República* No. 157, Series I–B.
3 Resolution of the Council of Minister No. 88/98 of 10 July 1998, *Diário da República* No. 157, Series I–B.
4 Resolution of the Council of Ministers No. 89/98 of 10 July 1998, *Diário da República* No. 157, Series I–B.
5 Resolution of the Council of Ministers No. 90/98 of 10 July 1998, *Diário da República* No. 157, Series I–B.
6 Law-Decree No. 43/2002 of 2 March 2002, *Diário da República* No.52, Series I–A.

7 Resolution of the Council of Ministers No. 81/2003 of 17 June 2003, *Diário da República* No. 138, Series I-B.

8 Resolution of the Council of Ministers No. 128/2005 of 10 August 2005, *Diário da República* No. 153, Series I-B, and Resolution of the Council of Ministers No. 110/2006 of 7 September 2006, *Diário da República* No. 173, Series I.

9 Resolution of the Council of Ministers No.9/2005 of 17 January 2005, *Diário da República* No. 11, Series I-B.

10 Resolution of the Council of Ministers No. 26/2006 of 14 March 2006, *Diário da República* No. 52, Series I-B; Resolution of the Council of Ministers No. 55/2007 of 4 April 2007, *Diário da República* No. 67, Series I; Resolution of the Council of Ministers No. 32/2009 of 16 April 2009, *Diário da República* No. 74, Series I; Resolution of the Council of Ministers No. 3/2011 of 12 January 2011, *Diário da República* No. 8, Series I.

11 Law No. 34/2006 of 28 July 2006, *Diário da República* No. 145, Series I.

12 Resolution of the Council of Ministers No. 40/2007 of 12 March 2007, *Diário da República* No. 50, Series I.

13 Resolution of the Council of Ministers No. 119/2009 of 30 December 2009, *Diário da República* No. 251, Series I.

14 Regulatory Decree No. 17/2012 of 31 January 2012, *Diário da República* No. 22, Series I.

15 The Task Force for Maritime Affairs (*Estrutura de Missão para os Assuntos do Mar*, EMAM), the former technical structure of support to CIAM, has been disbanded and its mission has been split and transferred to DGPM and EMEPC.

16 Resolution of the Council of Ministers No. 62/2012 of 13 July 2012, *Diário da República* No. 135, Series I.

17 Law No. 17/2014 of 10 April 2014, *Diário da República* No. 17, Series I.

18 Resolution of the Council of Ministers No. 163/2006 of 12 December 2006, *Diário da República* No. 237, Series I.

19 Resolution of the Council of Ministers No. 12/2014 of 12 February 2014, *Diário da República* No. 30, Series I.

References

Atlantic Area Transnational Programme (2011) 'Lisbon Atlantic Conference and Stakeholder Day', online at http://www.coop-atlantico.com/presentation/maritime-strategy-for-the-atlantic-ocean-area/lisbon-atlantic-conference-and-stakeholder-day/lisbon-atlantic-conference-and-stakeholder-day [accessed 15 May 2014].

Carta Administrativa Oficial de Portugal (CAOP) (2012) *CAOP 2012.1* [*Official Administrative Map of Portugal, 2012.1*], online at http://nsnig.igeo.pt/produtos/cadastro/caop/versao20121.htm [accessed 12 November 2014].

Comissão Estratégica dos Oceanos (CEO) (2004) *Relatório da Comissão Estratégica dos Oceanos* [*Report of the Strategic Oceans Commission*], online at http://www.eurocean.org/np4/file/128/RelatorioCEO_Parte_I.pdf and http://www.ordemengenheiros.pt/fotos/editor2/eng.naval/2relatorioceo_parte_ii.pdf [accessed 4 November 2014].

COTEC Portugal (2012) *Blue Growth for Portugal: Uma Visão Empresarial da Economia do Mar* [*A Corporate View of the Blue Economy*], online at http://www.cotecportugal.pt/index.php?option=com_content&task=view&id=2165&Itemid=420 [accessed 7 July 2014].

Direção-Geral de Política do Mar (DGPM) (2013) *National Ocean Strategy (2013–2020)*, online at http://www.dgpm.mam.gov.pt/Documents/ENM_Final_EN_V2.pdf [accessed 4 November 2014].

Estrutura de Missão para a Extensão de Plataforma Continental (EMEPC) (undated) 'A Submissão Portuguesa [The Portuguese Submission]', available online at http://www.emepc.pt/index.php?option=com_content&task=view&id=578&Itemid=285 [accessed 4 November 2014].

Estrutura de Missão para os Assuntos do Mar (EMAM) (2006) *National Ocean Strategy (2006–2016)*, online at http://www.emam.mdn.gov.pt/National_Ocean_Strategy_Portugal_en.pdf [accessed 4 November 2014].

EurOcean (undated) 'About EurOcean', online at http://www.eurocean.org/np4/41 [accessed 4 November 2014].

European Commission (undated) 'National Helpdesks', online at http://ec.europa.eu/ecip/help/national_helpdesks/index_en.htm [accessed 4 November 2014].

European Commission (2006) *Green Paper: Towards a Future Maritime Policy for the Union – A European Vision for the Oceans and Seas,* COM (2006) 275 final, online at http://eur-lex.europa.eu/search.html?t ype=expert&qid=1415136520672 [accessed 4 November 2014].

European Commission (2009) *Communication from the Commission to the European Parliament, the Council, the European Economic and Social Committee and the Committee of the Regions: An Integrated Maritime Policy for the European Union,* COM (2007) 540 final, online at http://eur-lex.europa.eu/legal-content/EN/ TXT/?uri=CELEX:52007DC0575 [accessed 4 November 2014].

European Commission (2011) *Communication from the Commission to the European Parliament, the Council, the European Economic and Social Committee and the Committee of the Regions: Developing a Maritime Strategy for the Atlantic Ocean Area,* COM (2011) 782 final, online at http://eur-lex.europa.eu/legal-content/EN/ TXT/?qid=1401902717268&uri=CELEX:52011DC0782 [accessed 4 November 2014].

European Maritime Safety Agency (EMSA) (undated) 'About Us', online at http://emsa.europa.eu/about. html [accessed 4 November 2014].

Fórum Empresarial para a Economia do Mar (FEEM) (undated) 'About Us', online at http://feemar. weebly.com/quem-somos.html [accessed 4 November 2014].

Governo de Portugal (2014) *Portugal 2020: Acordo de Parceria 2014–2020* [*Partnership Agreement 2014–2020*], online at http://www.portugal.gov.pt/media/1489775/20140730%20Acordo%20Parceria%20UE.pdf [accessed 4 November 2014].

Independent World Commission on the Oceans (IWCO) (1998a) *The Oceans, Our Future,* New York: Cambridge University Press.

Independent World Commission on the Oceans (IWCO) (1998b) *The 1998 Lisbon Declaration: Ocean Governance in the 21st Century – Democracy, Equity and Peace in the Ocean,* New York: Cambridge University Press.

Instituto Nacional de Estatística (INE) (2002) *Censos 2001 Resultados Definitivos* [*Final Results of Census 2001*], online at http://censos.ine.pt/xportal/xmain?xpid=CENSOS&xpgid=ine_censos_publicacao_ det&contexto=pu&PUBLICACOESpub_boui=377750&PUBLICACOESmodo=2&selTab=tab1&p censos=61969554 [accessed 4 November 2014].

Instituto Nacional de Estatística (INE) (2013) *Anuário Estatístico 2012* [*Statistical Yearbook of Portugal*], online at http://www.ine.pt/xportal/xmain?xpid=INE&xpgid=ine_publicacoes&PUBLICACOESpub_boui =209570943&PUBLICACOESmodo=2 [accessed 4 November 2014].

Lisbon Atlantic Conference (2013) 'Presentation', Conferência do Atlântico, 4–5 December, online at http://www.fem.pt/lac2013/Ingles/princ_ing.htm [accessed 16 May 2014].

Museus Portugal (undated) 'Início [Start]', online at http://www.museusportugal.org/default.aspx [accessed 4 November 2014].

Oceano XXI (undated) 'Home', online at http://www.oceano21.org/ [accessed 4 November 2014].

OSPAR Commission (2013) *2012 Status Report on the OSPAR Network of Marine Protected Areas,* online at http://www.ospar.org/documents/dbase/publications/p00618/p00618_2012_mpa_status%20report. pdf [accessed 4 November 2014].

Pitta e Cunha, T. (2011) *Portugal e o Mar* [*Portugal and the Sea*], Lisbon: Fundação Francisco Manuel dos Santos.

Portal MarOceano (undated) 'Fórum Permanente para os Assuntos do Mar [Standing Forum for Maritime Affairs]', online at http://www.maroceano.pt/forum-permanente-para-os-assuntos-do-mar [accessed 4 November 2014].

Ruivo, M. (2014) 'Mudanca de Paradigma da Relacao dos Sistemas Humanos com o Oceano: Implicacoes Estrategicas e Estruturais [A Paradigm Change in the Relationship between Human Systems and the Ocean: Strategic and Structural Implications]', Presentation to Academia de Marinha, 21 January, online at http://www.maroceano.pt/governacao-instituicoes-e-estrategias/2451-mudanca-de-para- digma-da-relacao-dos-sistemas-humanos-com-o-oceano [accessed 4 June 2014].

13

TOWARD A NATIONAL OCEAN POLICY IN NEW ZEALAND

*Prue Taylor**

Introduction: Basic information and an overview of national ocean policy

In July 2000, the New Zealand government initiated a process for developing a National Oceans Policy (NOP). The stated intention was to develop a policy for the integrated and consistent management of the oceans within New Zealand's jurisdiction. The policy should take a whole-of-government approach, cover all aspects of oceans management (including the effects from land), and extend out to the edge of the exclusive economic zone (EEZ) and the continental shelf beyond.

In 2003, the government decided to suspend the NOP process, pending resolution of some politically sensitive, overlapping issues, including ownership of the public foreshore and seabed, and public access to the coast.[1] The government announced a restart of the process in 2005, but little progress was made. The NOP Secretariat did not produce a 'draft' NOP document for public comment and the momentum that had built up during the initial consultation phase was lost. Little happened to further the development of the NOP between 2006 and 2008. In 2008, a new national government was elected to office. The priority of this government is to improve the regulatory regime for the environmental impact of human activities within the EEZ. It has not expressly indicated an interest in the NOP process (Ministry for the Environment, undated*a*). As a consequence, this chapter can provide only basic information and an overview of the NOP development process conducted between 2000 and 2005.[2]

At first glance, it would appear that New Zealand has much more to learn from this oceans policy comparative study than other participant nations. This observation is, however, misleading. New Zealand has a relatively advanced policy and legislative regime in place for the achievement of integrated coastal management (ICM). This regime was created by the Resource Management Act 1991 and has been developing over the last fourteen years. The main purpose of this regime is to provide for the 'sustainable management' of the nation's 'coastal marine area' (CMA) – that is, from the inland boundary out to the 12 nautical mile limit.[3] However, this regime is by no means perfect and, as will be seen, inherent and emerging problems are part of the motivation for the development of an NOP.

Basic information

New Zealand is an island nation. It lies in the southwest expanse of the Pacific Ocean, 2,580 km east of Melbourne, Australia, 5,430 km north of the South Pole, and 9,370 km west of

Santiago, Chile. It comprises the North and the South Islands, and a number of smaller islands, with a total land area of 267,707 km².[4] The coastline is immense and diverse; it stretches for some 15,134 km (making it the tenth longest in the world) (Statistics New Zealand, 2008: 2) and includes subtropical beaches, mangroves, estuaries, sounds, and fiords. New Zealand's island geography is such that no human settlement is more than 130 km from the sea (New Zealand Ocean Policy Secretariat, 2001: 15); hence the linkages between the people, the land, and the oceans remain strong. The oceans are of social, economic, spiritual, and economic significance to New Zealanders (New Zealand Ocean Policy Secretariat, 2001: 15).

As of August 2010, New Zealand's culturally diverse human population was approximately 4.37 million (Statistics New Zealand, 2010b). At the 2006 census, 14.6 per cent of the population identified as belonging to the Māori ethic group and 265,974 identified as Pacific Islanders (Statistics New Zealand, undated*a*). Māori are the indigenous people of New Zealand and, as such, have special rights and interests in, together with a unique cultural and spiritual connection to, the marine environment. The Treaty of Waitangi was signed in 1840 by the Crown and selected Māori chiefs. It forms the basis of the current relationship between the government and Māori, and is the primary framework for the negotiation of rights.

Approximately three-quarters of New Zealanders live in 'main urban areas' (that is, areas with a population of 30,000 or more), with the largest of these population centres located on the coast (Statistics New Zealand, 2008: 92).[5] As in most modern nations, there is a range of diverse and competing interests and users exerting pressure on the marine environment. As a significant amount of New Zealand's total biodiversity exists here, it is critical that policy and law exists to manage this pressure.[6] New Zealand has an extensive and integrated regime in place for managing the terrestrial environment, and to a lesser extent, the CMA, but not for managing the marine environment. Developing such a framework will not be an easy task. New Zealand has the fourth largest EEZ in the world (405 million hectares). This is more than fifteen times the area of the country's land mass (see Table 13.1).

Maritime boundaries

New Zealand's interpretation of its maritime boundaries and the jurisdiction associated with them is set out in Table 13.1. Since the creation of this table, New Zealand and Australia have settled outstanding issues concerning the delimitation of their EEZ and continental shelf boundaries. In this respect, the outer limits are now finalized. Where the EEZs of Australia and New Zealand overlap, the median line forms the boundary. The continental shelf boundary makes adjustments to take into account various islands under the separate jurisdiction of New Zealand and Australia.[7] New Zealand also has rights and responsibilities in the southern ocean sector of the Ross Dependency in Antarctica.[8]

As noted earlier, New Zealand's EEZ is massive; its continental shelf is even more expansive. The portion of the shelf that extends beyond the EEZ adds approximately a further 1.5 million km² (Table 13.1). Very little is known about these huge areas. New Zealand is currently conducting a comprehensive ocean seafloor survey and discovering a whole new world, including volcanoes the size of those on land. Figure 13.1 sets out New Zealand's maritime boundaries.

New Zealand claims the following:

- a continental shelf that extends 200 nautical miles or to the extent of the continental margin;
- an EEZ that extends 200 nautical miles from the coast;
- a contiguous zone that extends 24 nautical miles from the coast; and
- a territorial sea that extends 12 nautical miles from the coast.

Table 13.1 New Zealand's maritime boundaries

Zone/boundary	Location	Area	Rights/Obligations under the 1982 United Nations Convention on the Law of the Sea (UNCLOS) and domestic law
Land	Landward and above the line of mean high-water springs (MHWS)		New Zealand has full 'sovereignty' over its land (which is beyond the scope of UNCLOS).
Foreshore	Seaward of MHWS to mean low-water springs (MLWS), generally the line of low water of the coast (i.e. those areas that are normally wet)		New Zealand has full 'sovereignty' over its foreshore (which, again, is beyond the scope of UNCLOS). This is a territory defined under the Resource Management Act 1991 (RMA).
Coastal marine area	The foreshore, seabed, and coastal water, and the air space above the water extending from MHWS (with a few exceptions) to the limits of the territorial sea (12 nautical miles)		This is a territory defined under the RMA. New Zealand has full 'sovereignty' within the CMA that extends between MHWS and MLWS (because this is beyond the scope of UNCLOS). Seaward of MLWS to the 12 nautical mile limit, New Zealand's sovereignty is subject to rights and duties established by UNCLOS and to other rules of international law. Other states have rights such as 'innocent passage' of their vessels.
Baselines	Normally, the line of MLWS, but with exceptions for rivers, bays, islands, fiords, harbour works, etc.		
Internal waters	Waters on the landward side of the baseline of the territorial sea		Part of New Zealand's 'sovereign territory', which means that New Zealand has full 'sovereignty' over its internal waters.
Territorial sea	Seaward of the baseline out to 12 nautical miles		New Zealand has full 'sovereignty' over its territorial sea, subject to the rights and duties established in UNCLOS and to other rules of international law. Other states have rights such as 'innocent passage' of their vessels.
Contiguous zone	Between the outer limits of the territorial sea (12 nautical miles) to 24 nautical miles		In addition to 'sovereign rights' conferred over this area as part of the EEZ, New Zealand may exercise such control as is necessary to prevent and punish infringements in its territory or territorial sea of its customs, immigration, tax, and sanitary laws.

(continued)

Table 13.1 (continued)

Zone / boundary	Location	Area	Rights/Obligations under the 1982 United Nations Convention on the Law of the Sea (UNCLOS) and domestic law
Exclusive economic zone (EEZ)	Seaward of the outer limits of the territorial sea, including the contiguous zone, to an outer limit of 200 nautical miles from the baselines (i.e. the breadth of the EEZ is normally 188 nautical miles)	New Zealand's EEZ is the fourth largest in the world, with an area of 405 million hectares. This amounts to more than fifteen times the area of its land mass.	New Zealand has 'sovereign rights' – a more limited jurisdiction than sovereignty – for the purposes of exploring and exploiting, conserving, and managing natural resources of the waters, seabed, and subsoil. It also has 'jurisdiction' with regard to the establishment of artificial islands, installations, and structures, marine scientific research, and the protection and preservation of the marine environment. New Zealand must also have due regard for the rights of other states. Other states have certain freedoms, including navigation, overflight, laying cables in the EEZ, etc.
Continental shelf	The seabed and subsoil of submarine areas beyond the territorial sea (12 nautical miles) to the outer edge of the continental margin or to 200 nautical miles from the baselines (whichever is greatest)	Although the outer limits of New Zealand's continental shelf limits are not yet finalized, the area that extends beyond New Zealand's EEZ to the limits of its continental shelf is likely to include up to 1.5 million km^2 or more. This area alone equates to about six times the area of its land mass.	New Zealand has 'sovereign rights' (as for the EEZ) for the purpose of exploring and exploiting the natural resources of the seabed and subsoil (including immobile organisms that live on or under the seabed or subsoil). In areas in which the continental shelf extends beyond 200 nautical miles from the baselines, the water itself above the continental shelf is not within New Zealand's jurisdiction and is part of the high seas.
High seas	Water column beyond the outer limits of coastal states' EEZ		Open to all states subject to due regard for the interests of other states. All states have 'freedom of the high seas', which includes freedom of navigation, overflight, the laying of cables and pipelines, construction of artificial installations, fishing, and scientific research.
The Area	Seabed and subsoil beyond the limits of national jurisdiction (i.e. seaward of the outer limit of the continental shelf)		Vested in humankind as a whole and administered by the International Seabed Authority (ISA). No state can claim or exercise sovereignty or sovereign rights over the Area.

Source: Ministry for the Environment (undated*b*)

Figure 13.1 New Zealand's maritime boundaries

Source: New Zealand Oceans Policy Secretariat (undated) (copy on file with the author)

The utilization of marine and maritime resources

In 2003, the government commissioned a report on the economic opportunities associated with the oceans to inform NOP development (Centre for Advanced Engineering, 2003: 1). Much of the information given in this section is taken from that report; hence the economic interpretation of 'utilization'. This also accounts for the report's interpretation of the NOP initiative as:

> . . . explicitly searching for the best ways to pursue economic opportunities within New Zealand's oceanic territory, generally in accordance with accepted principles of sustainable development . . . to inform the crafting of a policy approach that will enable innovation to expedite value creation from whatever opportunities present themselves, subject to straightforward sustainability tests.
>
> *(Centre for Advanced Engineering, 2003: 1)*

This report also opines that:

> Government will be expected to ensure that a well-balanced enabling policy, and statutory framework with clear and efficiently administered property rights, are provided. . . . Among the critical components identified as being necessary for a successful oceans management are provision for:
>
> - integrated approaches;
> - knowledge-based industries;
> - property rights that enable access to opportunities;
> - regulatory frameworks that attract investment;
> - acknowledgement of the process of technology advancement and that what is known today will be different in future; and
> - conservation rather than preservation.
>
> *(Centre for Advanced Engineering, 2003: iii–v)*

In stark contrast, a report on the management of New Zealand's marine environment, written by the New Zealand Parliamentary Commissioner for the Environment (PCE), identified a wide range of other values associated with the marine environment, which included climate control, education, and intrinsic values (PCE, 1999: 11). This report warned of the danger of focusing solely on economic value and noted that:

> If New Zealand is to promote the sustainable development of its marine assets, it must recognise the diverse range of commercial interests and factors in the marine arena, and clearly establish where and how they relate to each other. Improving our understanding of the interrelationships between commercial, environmental and other values in marine resources, and developing robust frameworks for the comparative assessment and integration of ecological and economic data, are urgent priorities . . . [9]
>
> *(PCE, 1999: 11)*

Because most of the available data focuses on quantifying the economic utilization of the marine environment, the next section is largely confined to this narrow form of valuation.

In the late twentieth century, the oceans became of the focus of a number of new industries. These include:

- fisheries (commercial, recreational, and customary);
- aquaculture;
- the urbanization of the coast;
- increased seaborne trade of goods;
- the installation of high-capacity communication cables;
- the development of gas fields and the transfer of electricity by undersea cables;
- ecotourism (such as marine life tours);
- luxury craft design and manufacture; and
- recreational/leisure uses.

Commercial fisheries

The commercial fisheries industry has grown significantly in the last ten years. The marine ecosystems of the EEZ are diverse, containing more than 1,000 known fish species, 61 per cent of which occur in coastal rock pools. Despite having the fourth largest EEZ in the world, New Zealand contributes only 1 per cent of global fish production. Currently, two-thirds of the EEZ is considered to be 'commercially barren' (Statistics New Zealand, 2010a: 5).

Some 15,000 people are employed in the fisheries industry (including aquaculture). Commercial marine fishing contributes to around 80 per cent of the total value from New Zealand fisheries production. In 2009, fish exports contributed over NZ$1.42 billion in export earnings (Ministry for Primary Industries, 2010). New Zealand's commercial catch was 409,449 tonnes for the 2009–10 fishing year (Ministry for Primary Industries, 2010).

Commercial fisheries are controlled via a quota management system (QMS), which creates management areas and identifies quota species.[10] The QMS is not subsidized by government. There are currently ninety-seven species (or species groups) subject to the QMS, with the EEZ divided into ten quota management areas (Ministry for Primary Industries, undated). The Ministry of Fisheries sets commercial catch limits annually, based on government and industry research and submissions from the industry and interest groups.[11] Commercial fishers own or lease an individual transferable quota, which entitles them to fish a portion of the catch limit of a fish stock. It is the right to harvest that is owned and only New Zealand residents can own such rights. The task of harvesting can be contracted to overseas companies.

Māori are key stakeholders in the commercial fishing industry. As a result of a settlement between the Crown and Māori, initiated in the late 1980s, Māori own around 40 per cent of the commercial fishing quota.[12]

Recreational fishing

Recreational fishing is the catching of fish for non-commercial purposes. Fish caught cannot be sold. Recreational fishing is an important component of New Zealand fisheries. New Zealanders are avid fishers, with more than 500,000 going fishing at least once a year (DOC, 2002). The government has set allowable recreational catch levels, which, together with the total allowable commercial catch (TACC), forms the total allowable catch (TAC). Customary fishing (see next) also forms part of the TAC.

The Ministry of Fisheries administers the recreational fishing regulations. New Zealand is divided into three recreational management areas: north, central, and south. Each area has different rules that apply to different fish stocks. These rules govern catch limits, fish/shellfish size, etc. Fishery officers monitor recreational catch and can impose significant penalties for violation of regulations.

Customary fishing[13]

The Treaty of Waitangi (Fisheries Claims) Settlement Act 1992 provided a legislated settlement of fishing claims with Māori. This settlement addressed the right of Māori to a commercial stake in New Zealand's fishing industry (see above) and the right of Māori to non-commercial customary fishing.

Customary fishing regulations were developed by the Crown and Māori, with the aim of providing for the traditional rights of customary fishing and preserving the sustainability of the fisheries. These regulations cover only non-commercial customary fishing; permission to take fish under the customary fishing rights does not include the right to sell or trade fish for any form of payment. These regulations apply to marine fish and not to freshwater fish.[14]

Customary fishing regulations also provide for the establishment of *mataitai* reserves within which Māori can manage all non-commercial fishing. These areas are administered by Tangata

Kaitiaki/Tiaki,[15] individuals, or groups who can authorize customary fishing. The Tangata Kaitiaki/Tiaki grant fishing rights, size and quantity taken, fishing method, and area in which the fishing is to occur. They provide catch statistics to the Ministry of Fisheries.[16]

Aquaculture/marine farming

Aquaculture has emerged as a fast-growing industry over the last decade. Several coastal areas lend themselves to this form of commercial activity. On the basis of 2005 figures, aquaculture or marine farming was worth around NZ$250 million per annum, with approximately 1,200 marine farms nationally (Ministry for the Environment, 2005a).

In 2004, the government completed a comprehensive reform of the sector (Ministry for the Environment, 2005a). This was necessary because the high demand for space revealed problems with the pre-existing resource planning and fisheries legislation. Local authorities were dealing with consents and permits on an ad hoc, 'first in, first served' basis. There was little strategic overview of how marine farming fitted into overall coastal management. The cumulative environmental effects of farms were leading to poor environmental outcomes. All parties experienced delays and high consent/permit-processing costs. A moratorium (now lifted) was imposed during the period of reform.[17]

The Aquaculture Reform Act 2004 (legislation implementing the reform) came into effect on 1 January 2005. The stated purpose of the reform was 'to enable the sustainable growth of aquaculture and ensure the cumulative environmental effects are properly managed while not undermining the fisheries regime or Treaty of Waitangi settlements'.[18]

The main aspects of this reform, governed by the Resource Management Amendment Act 2004 and the Aquaculture Reform Act 2004, are that:

- it creates a single process for aquaculture planning and consents, governed by the 'sustainable management' purpose of the Resource Management Act 1991 (RMA);
- regional authorities have clearer direction and responsibilities for managing all environmental effects of aquaculture, including effects on fisheries[19] and other marine resources;
- marine farms can occur only in zoned areas known as 'aquaculture management areas' (AMAs);
- the creation of AMAs can be initiated by regional authorities, the industry, or individuals; and
- certainty is provided by settling claims for Māori commercial aquaculture (see next).

The government is committed to providing Māori with 20 per cent of the marine farming space within every region of New Zealand under the Treaty of Waitangi (Fisheries Claims) Settlement Act 1992. Consequently, 20 per cent of all existing space created since September 1992, and 20 per cent of all space created in the future, will be allocated to Māori for commercial use. If space is not available, the financial equivalent must be made available by government, under the Māori Commercial Claims Settlement Act 2004 and Te Ture Whenua Māori Amendment Act 2004.[20]

Public interest and biodiversity protection

New Zealand's coastal and marine ecosystems are highly diverse. Some 8,000 marine species have been identified, with up to seven new species being identified every two weeks. Scientists estimate that up to 80 per cent of the New Zealand's indigenous biodiversity is found in the sea (DOC, 1998: 47).

The total area included within marine reserves comprises 7 per cent of New Zealand territorial sea. However, only 0.3 per cent of New Zealand's total marine environment is protected by marine reserves.[21] The Marine Reserves Bill, if enacted, will govern the bulk of this area, and provides for free access and marine protection for areas within the territorial sea and beyond, to include areas within the EEZ.[22] Marine *reserves* are 'no-take' areas, protecting them from direct human impacts such as fishing, energy exploration, and pollution. Government has set a target of having 10 per cent of New Zealand's marine environment protected by 2010.[23] The establishment of marine reserves is often contested by fishers (recreational, commercial, and customary), but supported by those with interests in tourism, non-extractive recreation, education, and research. Although marine reserves are protected from direct human impacts such fishing and pollution, they remain highly vulnerable to land-use practices that result in sedimentation (some 390 million tonnes of sediment per annum), the dumping of human waste (including storm water and heavy metals), and run-off from nutrients such as inorganic fertilizers (DOC, 2002).

New Zealand also has a number of marine parks and other protected areas, which are not administered by the Department of Conservation (DOC) and generally receive a lesser protection.[24] Marine parks, for example, are gaining support as a mechanism for achieving broad environmental objectives, while still allowing recreational fishing. After a lengthy period of consultation, the government finally introduced the Marine Protected Areas: Classification, Protection Standard and Implementation Guidelines, in 2008 (Ministry of Fisheries and DOC, 2008). This was accompanied by the Marine Protected Area Policy and Implementation Plan, which was designed to give national guidance for the selection and management of a range of marine habitats and ecosystems (DOC and Ministry of Fisheries, 2005).

Marine mammals abound in New Zealand's marine ecosystems. Almost half of the world's cetaceans have been reported in New Zealand's waters. The main threats to marine mammals are habitat degradation and disturbance, global climate change, by-catch, and accumulation of pollutants. The Marine Mammals Protection Act 1978 provides for the protection and management of marine mammals, including the establishment of marine mammal sanctuaries. There are currently six in existence, primarily for the protection of Hector's dolphins, sea lions, and the Southern Right whale. The increase in deep-sea fishing over the last ten years has led to a marked increase in the by-catch of many marine mammals.

New Zealand is a founding member of the International Whaling Commission (IWC) and supported the establishment of the Southern Ocean Whale Sanctuary, which includes all of its EEZ south of 40°S (DOC, undated*a*).

Maritime trade and transport

New Zealand's economy is heavily orientated toward international trade. The export of goods and services accounts for just fewer than 33 per cent of total outputs (Statistics New Zealand, 2008: 331). Shipping is crucial to this trade. Eighty-five per cent of New Zealand exports by value and over 99 per cent of exports by volume are carried by sea (Statistics New Zealand, 2008: 425). There are thirteen commercial ports in New Zealand and they are largely owned by local government. Tourism is also served by shipping: in 2007, some 61,035 visitor arrivals and departures by cruise ship were recorded.[25]

There is also a frequent inter-island rail/ferry service that crosses the strait between the two main islands. Various domestic ships carry bulk cargos and livestock around coastal waters. New Zealand also has to import oil, by ship, to meet domestic oil demand. Around 51 per cent of New Zealand's total energy consumed is from oil, the great majority of which is imported. This needs to be shipped through coastal waters (Statistics New Zealand, 2008: 311).

New Zealand is a signatory to the International Maritime Organization (IMO) and International Labour Organization (ILO) treaties. The Maritime Transport Act 1994 regulates safety, liability, and marine environment protection. It is administered by Maritime New Zealand.

Biosafety is a major concern, because it is threatened by organisms attaching to hulls and contained in ballast water and debris from defouling. Over the last 100 years, an average of 1.4 new marine species per year are known to have established themselves (DOC, 2002). The Biosecurity Act 1993 is the main piece of legislation used to implement biosecurity measures. In 1998, the government adopted an Import Health Standard for Ships' Ballast Water (New Zealand Biodiversity, undated*a*).

Energy and mining sectors

As noted above, New Zealand is heavily dependent on imported oil to meet its energy needs. Gas comprises only 3.3 per cent of its total energy supply. Seventy-seven per cent is produced by one offshore field, which is in decline and may cease to be a source by the end of the decade.

All large-scale mining is currently onshore, but this may change dramatically in the near future.[26] However, there is a small industry dealing in the extraction and removal of sand, gravel, and stones. The government issues permits under the Crown Minerals Act 1991 in respect of strategic minerals such as petroleum. The RMA management regime for the CMA deals with other minerals.

Tourism and recreation

Tourism (both international and domestic travellers) is a fundamental component of New Zealand's economy. International travel is a top foreign-exchange earner, at around NZ$5.1 billion per annum. In 2007, there were 2.45 million arrivals (Statistics New Zealand, 2008: 271). Access to iconic coastal landscapes, pristine waters, and bush–clad fiords, and easy access to marine wildlife (such as dolphins, seals, whales, and penguins) are crucial to this industry.

More generally, the sea is a vital aspect of tourism and recreation in New Zealand. Most New Zealanders live in towns and cities with easy access to harbours and beaches. New Zealanders often holiday by the sea in their holiday homes, and enjoy boating, swimming, scuba diving, kayaking, water skiing, and fishing. One in every four households in Auckland City (New Zealand's largest city, bounded by two harbours) has a boat of some kind. Auckland has twice hosted the America's Cup yacht race and is known as the 'City of Sails'.

The marine recreation and tourist industry is diverse, and includes:

- boat design and construction;
- sail making and marine engineering;
- the construction/operation of marinas;
- recreational fishing supplies and services (such as guided trips);
- sports equipment and accessories; and
- wildlife centres and tourism ventures, such as whale watching and swimming with dolphins.

Recreational uses have been changing over time from largely passive and low-impact activities to those that are more active and high-impact. For example, the increase in the number of jet skis and related accidents has resulted in regulation. Local authorities are increasingly being urged to provide facilities for the offloading and disposal of effluent from commercial and recreational boats to deal with pollution concerns.

Upstream users (including coastal development)

The impacts of upstream users on the marine environment are dramatic in New Zealand. The adverse effects are perhaps most evident in respect of inshore coastal environments. In New Zealand, the major land uses and processes affecting the marine environment are:

- discharges from industrial facilities released into rivers and waterways flowing to the sea;
- sewage discharges (most are treated in some manner) and storm water (a mix of rainfall and pollutants);
- run-off from dairy farms into rivers and waterways, releasing nutrients (nitrogen and phosphorus) (PCE, 2004), and other agricultural and horticultural discharges;
- sedimentation (from road construction, subdivisions, forestry); and
- hydro dams changing the rates and timing of freshwater flows.

Development of New Zealand's coastlines for residential purposes is proceeding quickly. There is currently a lot of concern about the effect of this development on coastal landscape and amenity values.

New frontiers

Government and industry are working to identify natural compounds from marine organisms that may have significant commercial importance (that is, bio-prospecting). Sponges, mussels, and sea horses have received particular attention for their potential medicinal properties. Intellectual, cultural, and genetic property rights issues with respect to indigenous plants and animals are not easily resolved in New Zealand, owing to the Treaty of Waitangi.

In 2001, the government closed nineteen seamounts (around 100,000 km²) to bottom trawling. In 2004, the government announced further policy initiatives to extend bans until effective management frameworks could be put in place. The government is generally supportive of a UN moratorium on bottom trawling on the high seas, and in 2006 adopted a bottom trawling strategy.[27]

Oceans and coasts: An important policy area

Why are oceans and coasts an important policy area in New Zealand? In many ways, the facts and figures given throughout the last sections speak for themselves. In 1991, New Zealand put in place a policy and legal framework for the integrated coastal management of the CMA (from the inland boundary out to a 12 nautical mile limit).[28] However, it has long been recognized that this was only a starting point. A comprehensive and unifying framework was also needed for a much larger area, out to the limits of the EEZ and continental shelf. Furthermore, reforms were needed in respect of the CMA. Currently, some eighteen statutes apply to the marine environment, of which fourteen apply to the CMA (Harris, 2004: 238). New Zealand is also a signatory to thirteen international conventions applicable to the marine environment. The legislative coverage is complicated, piecemeal, and contradictory (DOC, 2002). There are in excess of fourteen agencies and a growing plethora of government strategies. A much more cohesive government approach to marine management is urgently required. To illustrate this, Figure 13.2 sets out how the key legislation fits together. Tables 14.2, 14.3, and 14.4 also show legislation and administrative responsibilities.

Before proceeding, it is useful, at this point, to give a very brief overview of how the RMA applies to the CMA, because it is one of New Zealand's core environmental planning statutes.

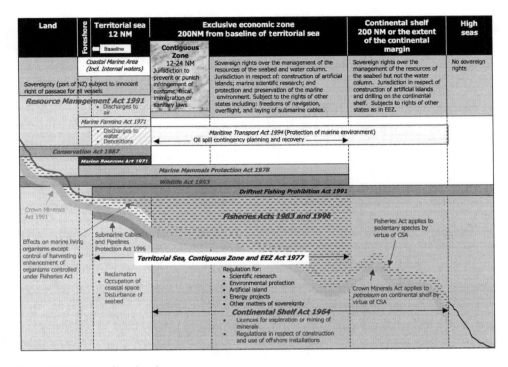

Figure 13.2 Scope of key legislation

Source: Enfocus Ltd et al. (2002: 17)

Table 13.2 Statutory framework covering the coastal marine environment

Legislation	Agencies with functions	Coverage and purpose
Biosecurity Act 1993	Ministry of Fisheries (primary) Department of Conservation	Prevent the entry of, and control and manage, unwanted marine organisms, among other things
Conservation Act 1987	Minister of Conservation Department of Conservation	To protect indigenous fauna and flora, including habitat, and to protect historic heritage, among other things
Continental Shelf Act 1964	Ministry of Transport Maritime New Zealand	To define navigational rules and the placement of structures on the seabed or waters within the continental shelf
Control of Dogs Act 1996	Territorial authorities	Control of dogs on beaches

(continued)

Table 13.2 (continued)

Legislation	Agencies with functions	Coverage and purpose
Crown Minerals Act 1991	Crown Minerals Group (MEcD) Minister of Energy Minister of Conservation Regional councils (see RMA)	Mining permits (EEZ and CMA); granting access where no exclusive occupation under RMA; resource consents process
Fisheries Acts 1983–96	Ministry of Fisheries	Managing the gathering and use of organisms in EEZ and CMA, including marine farming and control of disease and pests
Foreshore and Seabed Endowment Revesting Act 1991	Minister of Conservation	To revest land within the CMA in Crown control, and to regulate its vesting and disposal
Foreshore and Seabed Act 2004	Minister of Conservation Department of Conservation Regional councils	Clarifies Crown ownership of CMA; provides for some Māori interests in CMA
Marine Mammals Protection Act 1978	Department of Conservation	To protect marine mammals by means of a variety of devices
Marine Reserves Act 1971	Minister of Conservation Department of Conservation	To protect seabed and waters for scientific and conservation purposes
Maritime Transport Act 1994	Ministry of Transport Maritime New Zealand	To manage oil spills; to control navigation; to regulate pollution in EEZ
Resource Management Act 1991	Minister of Conservation (restricted coastal activities and coastal plans) Regional Councils (managing coastal planning and resource consent process) Ministry for the Environment (regulations and overall control of Act)	To control development and use of resources within the CMA by means of resource consent process (may be extended to the EEZ and continental shelf); to set marine pollution regulations
Territorial Sea, Contiguous Zone and Exclusive Economic Zone Act 1977	Minister of Fisheries Minister of Energy (Crown Minerals)	Fisheries management; control of mining; control of resources generally
Treaty of Waitangi (Fisheries Claims) Settlement Act 1992	Ministry of Fisheries Treaty of Waitangi Fisheries Commission	To regulate the quota system and compensate *iwi* for treaty claims forgone through changes in statute
Wildlife Act 1953	Department of Conservation Ministry of Fisheries	To protect marine life

Source: Adapted and updated from Harris (2004: 238)

Table 13.3 Environmental statutes

Statute	Coverage (inside CMA)	Coverage (outside CMA)
Biosecurity Act 1993	Yes (Ministry of Agriculture and Forestry, Ministry of Fisheries, Department of Conservation)	Only to 24 nautical miles (44.44 km)
Continental Shelf Act 1964	Yes (Ministry of Transport)	Yes
Crown Minerals Act 1991	Yes (Ministry of Economic Development)	Yes (only parts – see below)
Dog Control Act 1996	Yes (district councils)	No
Foreshore and Seabed Endowment Revesting Act 1991	Yes (Minister of Conservation)	No
Hazardous Substances and New Organisms Act 1996	Yes (Environmental Risk Management Authority)	No
Marine Mammals Act 1978	Yes (Department of Conservation)	Yes
Marine Reserves Act 1971	Yes (Department of Conservation)	Yes (if current amendments are enacted)
Maritime Transport Act 1994	Yes (oil pollution) (Ministry of Transport)	Yes
Territorial Sea, Contiguous Zone and Exclusive Economic Zone Act 1977	Yes (Ministry of Transport)	Yes
Wildlife Act 1953	Yes (Department of Conservation)	Yes

Source: Harris (2004: 259)

Table 13.4 Protection statutes applying to coastal and marine areas

Statute	Foreshore coverage	Inside territorial sea/CMA	Outside 12 nautical miles
Conservation Act 1987	Yes	Yes	Yes (coverage under Coastal Marine Strategies created under the Conservation Act 1987 and marine mammal conservation plans)
Fisheries Act 1996 (Mahinga Mataitai and Taiapure Reserves)	Yes	Yes	Not currently
Marine Mammal Protection Act 1978	Yes	Yes	Yes
Marine Park legislation (Hauraki Gulf and Sugarloaf)	Yes	Yes	Not relevant
Marine Reserves Act 1971	Yes	Yes	Not currently, but Marine Reserves Bill may include areas outside the territorial sea/CMA
National Parks Act 1980	Yes	No	No
Reserves Act 1977	Yes	No	No
Wildlife Act 1953	Yes	Yes	Yes, for seabirds and protected marine species

Note: Detailed tables containing information on and an analysis of legislation, government strategies, and international instruments were produced as part of the Stocktake Report (Enfocus Ltd et al., 2002).

Source: Harris (2004: 260)

Under the RMA, regional councils set and enforce controls by means of regional coastal plans. These plans are approved by the Minister of Conservation and must 'give effect' to national-level planning documents created by government, such as the New Zealand Coastal Policy Statement (NZCPS),[29] which are intended to give national guidance.[30] The regional coastal plans and the NZCPS are created for the purpose of achieving 'sustainable management' of the CMA. In essence, this regime enables use and development, subject to environmental criteria being met, to the degree that environmental effects will be judged 'minor'.[31] An activity (such as use, occupation, disturbance, or discharge) with more than 'minor' environmental effects is prohibited unless it is expressly allowed by a consent/permit issued by the regional council (or the Minister of Conservation),[32] or by a rule in a plan.[33] Because of the linkages between land use and impacts on the CMA (and vice versa), regional coastal plans may form part of regional plans dealing with land use (creating a broader concept of the 'coastal environment').[34] Regional coastal plans should also 'give effect' to regional policy statements and be consistent with the plans of adjacent regional councils.[35] Other land-use planning documents created under the RMA, such as district plans, must not be inconsistent with coastal plans.[36]

Ocean policy: Part of a national sustainable development strategy?

New Zealand does not have a national sustainable development strategy and has no clear agenda to develop one. In 2003, a document entitled *Sustainable Development for New Zealand: Programme of Action* was released (Department of the Prime Minister, 2003). This document has a limited scope and does not address ocean policy.

A brief overview of the nature and evolution of a national ocean policy in New Zealand

Status

As stated in the introduction, New Zealand does not yet have a national ocean policy (NOP), but was partway through the process of creating one when the process was suspended. Despite an official 'restart' in 2005, no significant developments have occurred since this date because development of an NOP has failed to qualify as a priority area for central government. According to available information, the NOP was to have been developed in three stages and was intended to answer the question: 'How should we govern our oceans?'

Stage One was completed in 2001. An extensive public consultation process eventually led to the development and adoption of a vision statement (see 'The policy development process: Objectives'). Stage Two was intended to develop options and preferences based on a variety of reports and working papers. This process was nearing its completion when, at the end of 2003, a political decision was taken to suspend the process, pending resolution of two related issues: ownership of the public foreshore and seabed; and public access to the coasts, across private land. The first of these two issues was, to some extent, resolved with the passing of the Foreshore and Seabed Act 2004. This Act vests ownership of the public foreshore and seabed in the Crown and denies Māori customary title to these areas.[37] Paradoxically, with a new national government coming into power after the 2008 elections, this Act is being repealed and will be replaced with new legislation. It is not yet known if this new Act will substantially clarify ownership, public access, and the process for claiming customary title, although this is the intention (Finlayson, 2010).[38] The second issue of public access to the coast, across private land, has not yet been fully resolved. Broadly speaking, the government proposed passing law granting rights of public access to water bodies (including the coastline) across private land. There is no indication of when this issue might be resolved.[39]

As of August 2010, the government has not indicated whether it will resume the NOP process or even initiate an entirely new process. Its stated priority is better regulation of the environmental impact of activities within the EEZ, including deep-sea mining (Ministry for the Environment, undated*a*).

Factors that gave rise to a national ocean policy initiative

Increasing human pressure on the marine environment (both extractive and non-extractive) has created an urgent need to resolve the perennial problems of integrating development interests with the objective of comprehensive environmental protection. Two legislative Acts are indicative of this. The first is the Aquaculture Reform Act 2004, which was created to better manage the environmental impacts of the rapidly burgeoning marine farming industry, without stifling its growth. The government has also drafted (but not yet enacted) the Marine Reserves Bill, which aims to increase the total area of New Zealand's marine environment protected by the creation of 'no take' areas. These responses, together with numerous strategy documents, have their merits, but they also contribute to the ad hoc nature of the pre-existing regulatory and policy regime for the marine environment.

In short, the main factor giving rise to the NOP initiative in New Zealand has been the realization that the current framework has reached the end of its useful life and that something new is now needed.[40] In the late 1980s and early 1990s, much effort was spent on reforming the policy and regulatory framework for the terrestrial environment. The outcome was the comprehensive Resource Management Act 1991 (RMA), which created an integrated management regime for land, air, and water (including the CMA). This regime is based on the guiding principle of 'sustainable management'. What is now needed is a similar reform exercise for the marine environment. Arguably, this is a much more difficult task, because relatively little is known about many of its complex ecosystems.

The paragraphs that follow summarize the main issues arising out of the current oceans management system, as identified by the New Zealand Oceans Policy Secretariat (Enfocus Ltd et al., 2002: 1–6).[41] The common theme is the lack of integration between legislation, policy, decision-making, and activities in the marine environment. In this respect, the issues identified are very similar to those identified in 1999 by the Parliamentary Commissioner for the Environment's report *Setting Course for a Sustainable Future: The Management of New Zealand's Marine Environment* (PCE, 1999). One of the main recommendations of that report was addressed to the prime minister. It recommended the establishment of a Coastal and Oceans Task Force for the purpose of developing a strategy comprising goals and principles, and actions and policies, for the future sustainable management of New Zealand's marine environment (PCE, 1999: 99).

No overriding goal and differing philosophies

There is no overriding and consistent goal for management of the marine environment beyond 12 nautical miles (for example some expression of 'sustainability'). In addition, there are inconsistent statutory approaches to reconciliation of competing interests, environmental protection, and public participation in decision-making. For example, there is a lack of integration between the Fisheries Act 1996 (which controls commercial fishing), the RMA (which controls CMA and, in part, marine farming), and the Marine Reserves Act 1971 (regarding preservation areas). The Fisheries Act also has a less transparent and participatory decision-making process than most other legislation for the marine environment.

Statutory gaps

Statutory gaps are a particular problem *beyond* the limits of the territorial sea. There are, for example, limited statutory requirements for environmental impact assessment (EIA) or public participation in decision-making. This is the case with the Continental Shelf Act 1964: it does not provide for EIA, public participation, or monitoring. There is no formal process for dealing with conflicting uses, such as protection of seamounts versus mining or bottom trawling. Thus far, decisions have been made in an ad hoc manner, and at the discretion of ministers and officials. Advances in technology and scientific knowledge will increase pressure to resolve or manage these conflicts. There are very significant gaps in the ability of the government to provide for environmental protection in the EEZ (Ministry for the Environment, 2005b). As noted above, this is currently a priority area for the New Zealand government.

Gaps also exist within the territorial sea, although regional councils could be considered lead agencies, with the RMA providing most of the regulatory control. However, there are questions about whether this act allows for good decision-making between competing uses, whether it deals adequately with land use impacts and cumulative effects, and whether there is adequate protection for marine cultural heritage of national significance.

Implementation difficulties

Statutory and non-statutory tools for implementation are not being fully utilized. The RMA provides for strategic and integrated decision-making from the national level down to the regional and local (territorial) levels. However, there are questions about:

- the lack of national level involvement, including the need for more collaboration between central government and local government agencies with overlapping areas of jurisdiction, along with a need for more central government leadership or guidance through the application and monitoring of clear and rigorous standards under the New Zealand Coastal Policy Statement, or by other means;
- capacity at the regional and local levels, including enforcement capacity, and lack of funding, necessary skills, and information;
- problems with the land–sea interface resulting from increasing land-based sources of marine pollution and uncertainty about responsibilities around the mean high-water springs (MHWS) mark boundary (inland boundary for CMA); and
- tools with which exist to address Treaty of Waitangi issues and to incorporate Māori values into some aspects of oceans management, along with a lack of experience of how and when to use such tools.

Beyond concerns with the RMA, the following are indicative of broader implementation problems.

- The Fisheries Act focuses on management of fish stocks and not the linkages between commercial fishing and environment protection. A Strategy for the Managing the Environmental Effects of Fishing (Ministry of Fisheries, 2005) signals an effort to address this failure. The strategy describes how to set limits to manage the environmental effects of fishing, but does not itself set those limits.
- The is a lack of clarity over what statutory mechanisms to use in situations in which a range of mechanisms exist. For example, there is no overall strategy on how to use existing law to protect marine sites. A Marine Protected Areas Policy and Implementation Plan has been developed to assist (DOC and Ministry of Fisheries, 2005).

- In the area of biosecurity, there is a lack of capacity to deal with prevention and incursions. A limited national biosecurity strategy was released in 2008, in an attempt to better link the various applicable laws and agencies (New Zealand Biodiversity, undated*a*).
- The ability to manage areas of jurisdiction, beyond the 12 nautical mile limit, in an efficient and transparent manner is questionable.

Information needs

There is no shared understanding and approach to what data is collected and for what purposes, and only inadequate accessibility and coordination of information already collected. Furthermore, existing information is poor, dispersed between many agencies, and incomplete. This detrimentally affects the quality of current decision-making. Owing to the complexity of ecosystems, lack of information is likely to be a problem in the short and long terms. Tools for managing a lack of information, such as the precautionary principle, risk management, and adaptive management, are inconsistently interpreted and applied. Education, compliance, and enforcement are also negatively affected by lack of information.

The specific strengths and weaknesses of current regime

Some of the key strengths of the current regime include the following.

- *Single uncontested jurisdiction* – New Zealand has clear sovereignty over areas claimed via international law of the sea treaties and conventions. An outstanding issue, the extent of the continental shelf, has recently been resolved with Australia. Internally, jurisdiction is not complicated by federalism.
- *Integrated coastal management* – The current regime for management of resource use (excluding fisheries) by a single statute (RMA) and single agencies (regional councils) across the land–water interface out to the 12 nautical mile limit is a major advantage. This model is thought to have the potential to address most coastal management issues once there is more capacity, greater interagency cooperation, and national guidance or leadership.
- *Treaty of Waitangi and indigenous rights* – The Treaty provides a formal framework for addressing and considering Māori rights and interests, and relationships with the government. It has proven helpful in the context of fisheries and aquaculture, but there is continuing discontent and uncertainty over resolution of the foreshore and seabed issue.[42] Treaty principles are increasingly being incorporated into domestic legislation.
- *Allocation of fisheries* – The quota management system (QMS) is considered by some to be relatively efficient. However, it is compromised by lack of reliable information about some fisheries stock, failure to provide for environmental protection, limited coverage of stock, and problems dealing with competing uses (including non-commercial fishers and other extractive and non-extractive users such as shipping and tourism).

Some of the main weaknesses (in addition to those already listed) include the following.

- *Proliferation of goals and strategies* – In the absence of a unifying goal, officials are left applying goals from a selection of international agreements such as UNCLOS and Agenda 21. There is a proliferation of non-statutory strategies such as the New Zealand Biodiversity Strategy (New Zealand Biodiversity, undated*b*).
- *Limits of the Treaty of Waitangi* – There is considerable uncertainty about the scope of Māori title and rights to many areas of jurisdiction *beyond* the public foreshore and seabed. Future decisions will have a considerable impact on an NOP.

- *Spatial management units* – Most management regimes are based on geo-spatially concentric and geological boundaries under UNCLOS. These boundaries ignore ecosystem processes, making management based on the effect of activities on ecosystems very difficult. This if further exacerbated by the sectoral approach to fisheries management.

As noted, many of the issues identified by the NOP Secretariat had already been scoped by the Parliamentary Commissioner for the Environment in 1999 (PCE, 1999). However, two features of the PCE report lead to a fundamentally different type of analysis. First, the PCE is an independent parliamentary officer, whereas the Secretariat comprises government officials. The PCE therefore has the opportunity to be more objective and critical. Second, the PCE's report took a comprehensive 'systems approach', or 'systems thinking approach', which went far beyond merely acknowledging the complexity of marine ecosystems. In the words of that report:

> This systems focus is fundamental to the long-term sustainable management and development of New Zealand's marine environment. The systems approach needs to extend beyond the context of the ecological system. It has to extend to the way we think about and ultimately organize our management of marine assets. We need to assess the merits of a 'systems thinking' approach to our legislation, to our recognition of the rights and values of *tangata whenua* [Māori], to our research, to our development of property rights and responsibilities, and to our policy formulation.
>
> *(PCE, 1999: 3)*

Adopting this 'system's approach', the PCE identified the following key shortcomings (amongst others).

- A lack of communication and a grave lack of trust among major stakeholders are severely inhibiting sustainable management of the marine environment.
- Māori, are *kaitiaki* (guardians), holders of customary rights, and major commercial quota holders. As such, they are strategic partners in sustainable management of the marine environment. There is inadequate understanding and recognition of their full range of rights under the Treaty of Waitangi.
- Adequate government investment in knowledge about marine ecosystems and species is critical. This investment has been waning in recent years.
- The fisheries management rights regime is immature and cannot, by itself, ensure sustainable management of ecosystems.
- Sustainability is threatened by a lack of an overarching framework or strategy that guides positive collaboration between stakeholders, and integrates diverse interests and values into environmental management solutions (PCE, 1999: 4–5).

Stakeholders involved

To date, the key stakeholders have been:

- central government agencies (departments and ministries);
- local government (territorial and regional authorities);
- Māori (as holders of various customary rights, commercial quota, and as *kaitiaki*);
- non-government organizations (NGOs);
- Crown research institutes and university researchers;

- sector groups and business associations; and
- the general public (via first-round public consultation).

Obstacles to development of NOP

There will be an array of barriers to the development of an NOP, not all of which are currently evident. Those most apparent, include:

- lack of national guidance, in the form of a national sustainable development strategy;
- differing interpretations of 'sustainability', including fundamental disagreement over weak versus strong sustainability (PCE, 2002; SANZ, 2009);
- the resolution of public access to the coast;
- Māori rights and interests within and beyond the statutorily defined foreshore and seabed;
- reconciling environmental protection objectives with commercial interests of extractive and non-extractive users;
- parallel ten-year reviews of the New Zealand Coastal Policy Statement (NZCPS), with the development process for an NOP crossing over with the review of the NZCPS, which has been significantly delayed;[43]
- the continued ad hoc development of new legislation and strategy documents;
- the absence of strong central government leadership and commitment, together with inadequate ministry budgets to fund the process;[44]
- the complexities of taking an integrated ecosystems-based approach and moving beyond the historical spatial approach;
- a lack of support for a 'high-level policy' resulting from uncertainty about what it will actually achieve at a local level; and
- major deficiencies in current scientific knowledge, including lack of government investment and a focus on species of commercial value rather than on ecological information needed as a foundation to understanding the marine environment.

The policy development process

Initiation of the policy

New Zealand's marine environment is of enormous importance, but debates about how to manage it have raged for decades. Several factors led to New Zealand adopting a fresh approach, including the United Nations designating 1998 the 'International Year of the Oceans', at which time the New Zealand government began to consider the role of an oceans policy. In December 1999, the Parliamentary Commissioner for the Environment (PCE) presented his report to Parliament and considerably strengthened the case for an oceans policy. As previously noted, the PCE is an officer of Parliament and therefore independent of government. His or her statutory role includes reviewing the effectiveness of environmental management systems and making recommendations for change. The PCE believed that the marine environment was a key priority area.

In his report, the PCE described a 'chronic discontent' with the way in which things were being managed. He acknowledged that considerable attempts had been made to address these concerns and that New Zealand's fisheries management regime had therefore been under almost constant revision in the twenty-one years leading up to the 1999 report – including, for example, introduction of a world-leading quota management system (QMS) (PCE, 1999: 2).

However, tensions remained, and it was time to look at management of marine resources from *first principles*. The 1999 report recommended, among other things, that a task force be set up to examine critical issues affecting the marine environment and to develop a long-term strategic framework for their management.

The PCE summarized 'a notable consistency' in the debate amongst New Zealanders about priorities for action, including:

- the sustainability of fishing;
- the operations and outcomes of New Zealand's fisheries QMS;
- Treaty of Waitangi rights and entitlements;
- the decline and degradation of species and ecosystems;
- by catch of non-target fish species, marine mammals, and seabirds;
- biosecurity and the risks to New Zealand of alien marine organisms;
- provision for recreational fishing;
- the lack of comprehensive scientific data for most of New Zealand's marine resources and ecosystems;
- the small proportion of New Zealand's coastal seas under protective status as marine reserves;
- the lack of any deep-sea protection; and
- the impacts of land-based management on marine environments (PCE, 1999: 2).

In July 2000, the New Zealand Cabinet[45] agreed to the development of an NOP as the basis for an overhaul of the system. This was to be a 'cross-government', or 'whole of government', exercise involving officials and ministers from all key government departments, working together (see 'Objectives'). The NOP was to cover all aspects of oceans management within the jurisdiction of New Zealand, including effects from land, and extending out to the edge of the EEZ and the continental shelf beyond. While the policy would take account of relevant international obligations, it was not intended to explicitly address issues associated with New Zealand's role in the Southern Ocean or New Zealand's role in the wider South Pacific region (New Zealand Ocean Policy Secretariat, 2001: 10).[46]

The authority to manage development of the policy was delegated by Cabinet to an ad hoc group of six Cabinet ministers. At the time of its inception, the portfolios of these ministers collectively covered government responsibilities for any likely environmental and economic outcomes in relation to the oceans, as well as the Treaty of Waitangi implications, presumably reflecting the need for the policy to incorporate each of these aspects.[47] An Oceans Policy Secretariat was also set up to coordinate and administer policy development (see 'Institutional arrangements').

The ad hoc ministerial group then established a Ministerial Advisory Committee on Oceans Policy (MACOP) to assist the ministers in their role of overseeing the development of the oceans policy and, in particular, to undertake a public consultation process. Eight members of MACOP were appointed by Cabinet, including a former diplomat, a former governor-general, two academics, and four company directors from various industries such as tourism and oil and gas exploration. Members of MACOP were appointed 'not as representatives of sector interests of specific values systems', but because they had the 'skills and experience necessary to engage in constructive conversations with New Zealanders about what they want to happen with the marine environment' (New Zealand Ocean Policy Secretariat, 2001: Appx B).[48]

The terms of reference for MACOP were to gain a clear understanding of the vision, values, goals, and objectives of New Zealanders, as a basis for a management framework in the oceans context. The members were appointed in March 2001 and held office during the pleasure of the ad hoc ministerial group. It concluded its function following submission of its final report to the ministers, which they were required to produce by the end of September 2001 (New Zealand Ocean Policy Secretariat, 2001).

Objectives

Prior to its suspension, the NOP was being developed in three stages. The first stage defined the 'vision', including the goals, values, and principles to guide the process (assisted by the MACOP work referred to above). The next stage was to define the strategic and operation policy framework for this vision. Some further consultation on this framework would then take place, and finally, the NOP would be implemented. Figure 13.3 illustrates this process.

Stage One was completed in 2001m when MACOP reported to Cabinet on the outcome of a wide public consultation process. This enabled Cabinet to enunciate a vision for the NOP. The consultation process included seventy-one public meetings held all over New Zealand, including twenty-four *hui* (meetings on a Māori *marae* or in a Māori meeting house) and three *fono* (meetings specifically for Pacific Islanders).[49] An oceans policy website was also established, supplemented by media releases and consultation resource packs. As a result of all of this work, some 1,160 written submissions were received, which MACOP then analysed and distilled into a report to help Cabinet to develop a vision for the NOP.

The Advisory Committee noted that Māori had a well-established set of values relevant to an NOP. These included *kaitiakitanga* (an obligation to protect the physical and spiritual well-being of valued things within their control) and *whanaungatanga* (the view that relationships are everything, and that humans are an equal and not superior part of the natural world). Māori consistently sought recognition as a partner (via the status accorded them in the Treaty of Waitangi), rather than as a mere interest group. The values that should drive an oceans policy were identified as follows:

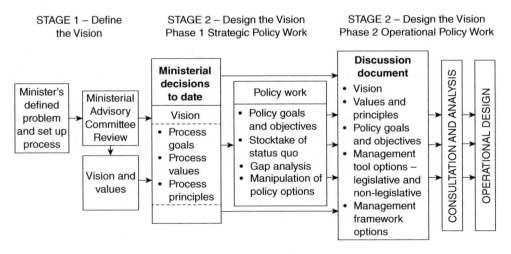

Figure 13.3 The oceans policy process

Source: New Zealand Oceans Policy Secretariat (2005)

- to set clear goals;
- to integrate separate management processes;
- to provide open and transparent decision-making that allows for informed participation;
- to provide fair and equitable means to balance competing aspirations;
- to reflect the range of values held in relation to the marine environment;
- to strike a balance between the need for adaptability and consistency;
- to provide for the optimal realization of economic benefits without compromising the quality of the environment;
- to ensure that management decisions are informed by adequate knowledge and due caution is exercised; and
- to promote a collective sense of responsibility (New Zealand Ocean Policy Secretariat, 2001: 8).

On the basis of the MACOP report, Cabinet identified a vision statement for New Zealand's oceans:

Healthy Oceans: New Zealanders understand marine life and marine processes and, accordingly take responsibility for wisely managing the health of the ocean and its contribution to the present and future social, cultural, environmental and economic well-being of New Zealand.[50]

Ministers distilled from the MACOP work the following principles to inform some of the further NOP development work that followed (Enfocus Ltd et al., 2002: 45):

- to integrate decision-making and administration across different activities, consistent with common objectives (such as vision and goals);
- to take into account:
  - the biophysical characteristics of the marine environment; and
  - the range of values held by New Zealanders in relation to that environment;
- to determine when, how, and by whom decisions are made and implemented;
- to build in compliance and enforcement models and to encourage voluntary compliance;
- to address the Crown's responsibilities under the Treaty of Waitangi in relation to the marine environment;
- to manage information about the marine environment and to assist in identifying information needs; and
- to monitor the implementation of the systems and make changes as required.

Stage Two of the policy process focused on developing various options and preferences for the mechanisms and management approaches necessary to achieve the vision. The first phase of Stage Two involved the commissioning of a number of reports and analyses. This included the release of eleven draft working papers for comment, to NOP stakeholders, by the Secretariat during March 2003.[51] These papers looked at particular issues and how they were currently being managed, and concluded with ways in which an NOP might improve the situation. Meetings were held by the Secretariat to receive feedback, which was then summarized in a report used to help to formulate the draft NOP. General feedback on the report included a view that it was difficult to understand how an NOP (that is, a high-level policy) might actually change things to make the current system less confusing.

As far as can be determined, matters have not progressed beyond this point. Phase Two of Stage Two involved the writing of a 'discussion document'. It is understood that this may have been drafted in some form, after feedback was received on the working papers, but it has never been presented to Cabinet (or to the public). It is further understood that, after such a long delay, this document would have to be redrafted and updated following further work by government agencies.[52]

Stage Three would have implemented the policies, processes, and mechanisms identified in Stage Two, and any necessary legislation would have been passed and institutional changes effected. Stage Three was also to include monitoring mechanisms to assess progress towards the vision and identify any further work required.

Major principles

Many of the major principles applicable to an NOP are evident in pre-existing law or strategies. On this basis, it is highly likely that they will be reflected in either an NOP or any implementing strategy or law. The following list identifies the most commonly utilized principles and links them to some relevant legislation.

- *Adaptive management* – Legislation regulating marine farming
- *Indigenous rights* – RMA; legislation regulating marine farming and fisheries[53]
- *Protection of marine diversity* – Marine Reserves Bill
- *Intergenerational and intragenerational equity* – RMA; Local Government Act 2002
- *Stewardship* – *RMA (including recent amendments)*; Māori fisheries legislation; Foreshore and Seabed Act 2004
- *Subsidiarity* – RMA; Local Government Act 2002
- *Sustainability* – RMA; Fisheries Act 1996; Local Government Act 2002
- *Timeliness, transparency, the precautionary principle, the 'polluter pays', principle, participatory governance, multiple-use and ecosystem-based management, best available science, and accountability* – RMA; Local Government Act 2002

Institutional arrangements

The Hon. Pete Hodgson (Minister of Fisheries, 2002–04) led the initial NOP process and the Ministry for the Environment hosted it. Policy development has been led by the Secretariat, comprising officials from key oceans-related departments, including the Ministry for the Environment, the Department of Conservation, the Ministry of Economic Development, the Ministry of Fisheries, and Te Puni Kokiri (the Ministry of Māori Affairs). Other central and local government agencies are considered partners in the policy development process. Figure 13.4 provides an overview of the project structure and the ministers and government agencies involved.

The Ministry for the Environment, as lead agency for the NOP project,[54] commissioned a 'stocktake', in two parts, to identify and record information on how the oceans are currently managed in New Zealand. The stocktake comprised two elements: to document the status quo and the problems associated with it; and to investigate international approaches to ocean policy development. The information in both parts was then to be analysed against the vision and the policy goals and objectives. This exercise was intended to help to indicate whether the status quo was consistent with the vision and the policy goals and objectives, or how it differs and why (Enfocus Ltd et al., 2002: 45).

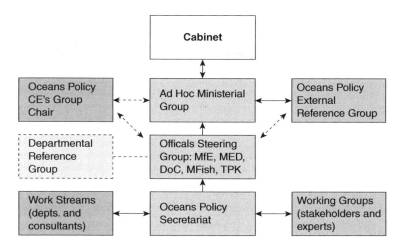

Figure 13.4 National Ocean Policy project structure

Source: Ministry for the Environment (undated*c*)

The first part of the stocktake is complete and it provides a detailed record of legislation (national and international) and strategies that apply in New Zealand. It also provides an overview of some of the strengths and weaknesses of the current approach to managing the oceans, and considers themes across legislation and tensions within the current regime. A summary was given earlier.

Summary of the position to 2010

As noted previously, the NOP stalled in July 2003 and was reactivated only in November 2005, but no significant developments have occurred since that date. One might conclude from this that the political process has taken over. The Secretariat has been disbanded for the time being, with members returning to their own ministries and departments. This was of concern to the Parliamentary Commissioner for the Environment (PCE) and while, in oral communications with the author, the then PCE indicated that his office might review the situation in 2006, owing to the lack of progress, this did not subsequently occur. The current PCE has not been very active in the area of marine environmental protection. There has also been some criticism of the policy process itself, with a suggestion that the long hiatus might provide an opportunity to reassess the policy process. Clearly, a lot of the momentum has been lost and it will be a very difficult (if not impossible) task to restart it.

Developments to 2013

Following the disastrous oil spill in the Gulf of Mexico, the Environmental Defence Society, an NGO, renewed the call for oceans reform in New Zealand (Mulcahy et al., 2011). Noting that the legal and institutional frameworks were developed in the 1970s, and calling attention to the new boom in oil and gas exploration, the Society warned that New Zealand was unprepared to manage the risks (Mulcahy et al., 2011). On 5 October 2011, *RV Rena* hit a reef off the coast of New Zealand, releasing its load of heavy oil and freight into the sea. *RV Rena* was a cargo ship, but its demise led to one of the most serious oil spills in New Zealand's history. Despite the obvious wake-up call, the current government has not restarted the NOP process nor has it indicated any intention to do so.[55] Instead, a sequence of new and amending legislation has been enacted, with the primary objective of maximizing the contribution of mining to the economy (Ministry of Economic Development, 2012). This practice does nothing to address to government's own acknowledgement of 'sporadic' policy initiatives on marine issues and the urgent need to get the right frameworks in place (Ministry for the Environment, 2011). Overall, oceans governance in New Zealand remains 'sector-based and fragmented among a range of policies, programmes and agencies with marine responsibilities' (Institute for Governance and Policy Studies, Victoria University of Wellington, undated). This approach to policy development was previously identified as an obstacle to the development of an NOP.

This section outlines recent legislation and other developments.[56]

The Resource Management Act 1991 and related legislation and policy

After a very long delay, the Minister for Conservation finally announced an intention to adopt the 2010 reviewed New Zealand Coastal Policy Statement (NZCPS). It came into effect on 3 December 2010 and updates the preceding 2004 management strategy for the coastal marine area (CMA). In 2011, the government made progress on its repeal of the contentious Foreshore and Seabed Act 2004. Section 11 of the Marine and Coastal Area (Takutai Moana) Act 2011 states that 'neither the Crown nor any other person owns, or is capable of owning, the common marine and coastal area'. Particular provision is made under section 43 for the acknowledgement of special customary interests, including 'customary marine title'.

From 2010 onwards, the government addressed the issue of management of the environmental effects of extractive activities, such as petroleum and mineral exploration and mining, within the EEZ and extended continental shelf. The Exclusive Economic Zone and Continental Shelf (Environmental Effects) Act 2012 became law in September 2012. A new agency, the Environmental Protection Authority,[57] is responsible for implementing the regulations made under this Act, including an environmental impact assessment (EIA), procedure together with the limited public participation processes.[58] Section 10 of the 2012 Act includes an overarching sustainable management purpose. A parallel sequence of amendments was made to the Crown Minerals Act 1991, the intent of which was to make the permitting process more efficient and effective for applicants (mining companies) (New Zealand Petroleum and Minerals, 2013).[59]

At the same time as the government's legislative agenda for economic growth has rapidly progressed, legislation for marine ecosystem and mammal protection has largely languished.[60] The Marine Reserves Act 1971 remains unreformed. However, on a more positive note, the Subantarctic Islands Marine Reserves Act 2014 was recently enacted, creating three new marine reserves off the coast of New Zealand. Furthermore, in March 2013, the Minister for Conservation announced an intention to create five new marine reserves, boosting the total area protected from 33,574 to 51,102 hectares.[61]

Finally, the Local Government Act 2002 was amended in 2012, removing 'sustainable development' from a statement of local governments' statutory purpose. This erodes their legal competence to deliver of the broad agenda of sustainability.[62]

Marine spatial planning

The Auckland Council (New Zealand's largest metropolitan local government authority) recently announced an intention to create a marine spatial plan for the Hauraki Gulf (Hauraki

Gulf Forum, 2011), following best international practice. A fourteen-member stakeholder working group is tasked with drafting the spatial plan, which will take two years to create. Its objective is to provide guidance for the policies and processes of the various authorities and agencies with a role to play in protection of the Gulf's values (Sea Change, 2014). This will be the first attempt at marine spatial planning in New Zealand.

The Blue Economy

On the face of it, New Zealand has embraced the idea of the 'Blue Economy' following Rio+20. A government website places emphasis on 'sustainable fisheries, protection of the marine environment and the elimination of harmful fisheries subsidies' (Ministry of Foreign Affairs and Trade, undated). However, thus far, no innovative initiatives have emerged that are specifically linked to the Blue Economy.[63] In particular, the interactions between climate change and changes to the ocean environment have received very little attention.[64]

Notes

* This report has been written by Prue Taylor, Deputy Director, New Zealand Centre for Environmental Law and Senior Lecturer, School of Architecture and Planning, University of Auckland, New Zealand. Research assistance was provided by Lisa Sutherland. Peer review was provided by Mr Tony Seymour, Coastal Planning Team Leader, Northland Regional Council. Thank you to Ms Lisa Sheppard, Manager Oceans, Ministry for the Environment, for her comments on the final draft. The information in this report (including all references and web links) was accurate as of September 2010. See the final section 'Developments to 2013' for an update.

1 The foreshore and seabed issue was, at least initially, resolved with the passage of the Foreshore and Seabed Act 2004. By virtue of s. 13(1) of that Act, absolute property in the public foreshore and seabed vests in the Crown. The 'public foreshore and seabed' extends from mean high-water springs out to the limits of the Territorial Sea. Very little of the foreshore and seabed was in private ownership prior to the passage of this Act. However, as explained in the section 'Brief Overview of the Nature and Evolution of a National Ocean Policy in New Zealand', this Act is now in the process of being repealed. See also the section in this chapter, 'Developments to 2013'. Public access to coastal areas, across private land, remains an issue in New Zealand.

2 Oceans Policy Secretariat documents, related to the NOP, were available online at http://www.mfe. govt.nz/issues/oceans/previous-work/index.html [accessed 20 August 2010] (no longer available).

3 Resource Management Act 1991, s. 2. New Zealand claims 12 nautical miles as its Territorial Sea.

4 This figure excludes offshore islands under 20 km².

5 As at the 2006 Census, 86 per cent lived in urban areas, with 72 per cent of the resident population in 'main urban areas' (Statistics New Zealand, 2008: 92).

6 A ten-year international survey of marine life recently revealed a very large endemic population of marine life in New Zealand waters (University of Auckland Library, 2010).

7 Treaty between the Government of Australia and the Government of New Zealand of 25 July 2004 establishing Certain Exclusive Economic Zone and Continental Shelf Boundaries (2004) 2441 UNTS 235.

8 The Ross Dependency EEZ is a further 2.3 million km² (Helm, 1998:252).

9 The need to bring in wider values was confirmed during the initial broad public consultation stage of the policy development process. See 'Summary of Position to 2010'.

10 Created by the Fisheries Act 1996; see also the Fisheries Amendment Act 1999.

11 Each year a total allowable catch (TAC) is set for each species or group on the basis of the maximum sustainable yield (MSY), i.e. the estimated capacity of a population in a fisheries habitat necessary to sustain its numbers.

12 Treaty of Waitangi (Fisheries Claims) Settlement Act 1992. The settlement gave Māori around a third of the commercial fisheries quota and an entitlement to 20 per cent of the quota for any new species brought into the quota management regime. In return, all other claims to commercial fishing rights were extinguished (PCE, 1999: 13).

13 A wide range of values, in addition to fishing and utilization of marine resources, give particular meaning to the marine and coastal environment for Māori (PCE, 1999: 14–17).

14 Regulation 27 of the Fisheries (Amateur Fishing) Regulations 1986; Kaimoana Customary Fishing Regulations 1998; the Fisheries (South Island Customary Fishing) Regulations 1998.

15 Kaitiaki are considered to be stewards or guardians of customary fishing areas.

16 Tangata Kaitiaki/Tiaki currently covers practically the entire South Island and much of the North Island.

17 The Resource Management (Aquaculture Moratorium) Amendment Act 2002 imposed a moratorium on 25 March 2002, which expired on 31 December 2004.

18 Explanatory note to the Aquaculture Reform Bill 2004, p. 24.

19 Effects of aquaculture on fishing activity are taken into account through this process by means of a test under the Fisheries Act 1996 (whether there are any undue adverse effects on commercial, customary, or recreational fishing) and the Fisheries Amendment Act 2004.

20 It is likely that the current government will review this regime, responding to pressure from the aquaculture industry.

21 This figure does not include other marine protected areas (MPAs) such as marine parks. Some 99 per cent of the area protected by marine reserves comprises two large remote reserves around the Kermedec and Auckland Islands (DOC, undated*b*).

22 The Marine Reserves Act 1971 is limited in its scope to the Territorial Sea. The Marine Reserves Bill has not yet been enacted and is still before Parliament: see 'Developments to 2013'.

23 This figure is fairly conservative given international recommendations that a minimum of 20–50 per cent of a country's EEZ be zoned no-take areas (PCE, 1999: 22).

24 For example, the Hauraki Gulf Marine Park Act 2000 takes an integrated land-to-sea approach and attempts to provide for the management of the Hauraki Gulf across twenty-one Acts of Parliament, including the Resource Management Act of 1991 (RMA).

25 There were 32,018 arrivals and 29,017 departures. The year 2007 is the last in which Statistics New Zealand made data on arrivals and departures by sea available (Statistics New Zealand, undated*b*: Table 1.05). In 2003, the Parliamentary Commissioner for the Environment issued a report on the environment effects (including, but beyond, biosecurity risks) of cruise ships visiting New Zealand waters (PCE, 2003).

26 The Minister of Energy considered an application from a Chinese company to mine 3,617 km² of the seabed (*Waikato Times*, 9 March 2005, p. 1).

27 Department of Conservation (2008) 'Bottom Trawling Strategy', online at http://www.doc.govt.nz/conservation/marine-and-coastal/commercial-fishing/bottom-trawling-strategy/ [accessed 20 August 2010] (no longer available).

28 RMA, s. 2, reads as follows: ' "Coastal marine area" means the foreshore, seabed, and coastal water, and the air space above the water: (a) Of which the seaward boundary is the outer limits of the territorial sea: (b) Of which the landward boundary is the line of mean high water springs, except that where that line crosses a river, the landward boundary at that point shall be whichever is the lesser of – (i) One kilometre upstream from the mouth of the river; or (ii) The point upstream that is calculated by multiplying the width of the river mouth by 5.'

29 The 1994 New Zealand Coastal Policy Statement (NZCPS) (DOC, 1994) was reviewed and replaced by the 2010 NZCPS: see 'Developments to 2013'.

30 RMA, s. 67(2), as amended by the Resource Management Amendment Act 2003, s. 28.

31 RMA, s. 5(2).

32 RMA, s. 28(1)(c). This permitting power was recently repealed under the Resource Management (Simplifying and Streamlining) Amendment Act 2009, s. 25(1).

33 RMA, s. 12.

34 RMA, s. 64.

35 RMA, s. 66(2)(d), as amended by the Resource Management Amendment Act 2005.

36 RMA, s. 75(2)(c), as amended by the Resource Management Amendment Act 2003, s. 32.

37 Foreshore and Seabed Act 2004, s. 13(1). Note, however, that this Act provides for the recognition of 'territorial customary rights' and 'customary activities'. These can be operationalized via creation of reserves and customary rights orders under the Resource Management (Foreshore and Seabed) Amendment Act 2004. The United Nations Committee on Elimination of Racial Discrimination (CERD) considered the legislation under its early warning procedures. Questions were raised by the Committee regarding, among others, the distinction drawn between Māori customary rights and

private freehold rights, and the different treatment accorded them, and the appropriateness of redress, compared with the possibility of compensation, which is denied by the Act. The Committee reported that: 'Bearing in mind the complexity of the issues involved, the legislation appears to the committee, on balance, to contain discriminatory aspects against Māori, in particular, in its extinguishment of the possibility of establishing Māori *customary title* over the foreshore and seabed' (CERD, 2005: para. (f), emphasis added).

38 The Act was subsequently repealed: see 'Developments to 2013'.

39 In 2008, the Walking Access Act came into force, creating the New Zealand Walking Access Commission. It is tasked with promoting walking access across private land, to the coastline, rivers, lakes, and forests. However, it cannot require public access over private land.

40 In non-government circles, this problem was being extensively discussed prior to and since 1998 (Helm, 1998).

41 As part of Stage One of the policy development process, the Oceans Policy Secretariat commissioned a detailed study of what laws and policies were already in place and how effective they were. Tables 2–4 from this stocktake summarize major legislation, strategies, and international instruments that applied in New Zealand.

42 See n. 1 and n. 37.

43 The NOP Secretariat has suggested that an oceans plan could (subject to further assessment) supersede the NZCPS. This proposal was not supported by the reviewer of the NZCPS (Rosier, 2004: 22).

44 The Ministry for the Environment comments that adequate leadership did exist up until the process was put on hold in 2003.

45 Cabinet comprises selected ministers from key government departments and ministries. It makes all major decisions on national policy issues.

46 There is no indication, for example, that it will link to regional oceans programmes in the Pacific.

47 Ministries specifically represented on the committee as its members at its inception were energy, fisheries, research, science, and technology, foreign affairs and trade, conservation, local government, Māori affairs, commerce, and environment.

48 The information that the ministers sought to gain through this report was 'the views, values, principles and any shared vision raised in the consultation process, issues that need to be addressed and associated key risks, and recommended goals, objectives and principles for the management of New Zealand's oceans' (New Zealand Ocean Policy Secretariat, 2001: Appx B).

49 An initial *hui* was held to determine the best way in which to involve Māori. The meetings were held over a two-month period. Approximately 2,000 people attended these meetings. Summaries were placed on the website.

50 Ministry for the Environment (undated) 'Stage One: Defining the Vision', online at http://www.mfe. govt.nz/issues/oceans/previous-work/index.html [accessed date unknown] (no longer available).

51 The papers, meeting, and summary feedback reports are available at Ministry for the Environment (2003).

52 Author interview with Lisa Sheppard, representative of the Ministry for Environment, 27 April 2005. Sheppard noted that bringing people together from diverse government agencies and refocusing them after a break of years would be challenging.

53 But see earlier comments on the Foreshore and Seabed Act 2004, n. 1 and n. 37.

54 The total budget for the work by Ministry for the Environment between 2000 and 2003 was approximately NZ$1.6 million.

55 An Oceans Governance Project between government and academia has been initiated (Institute for Governance and Policy Studies, Victoria University of Wellington, undated).

56 The names of new government agencies involved in aspects of oceans management include the Ministry of Primary Industries (fisheries), the Ministry of Business, Innovation and Employment, and the Ministry of Energy and Resources (the latter two both concerned with mining in the marine environment).

57 Established by means of the Environmental Protection Authority Act 2011.

58 The government is currently considering draft Exclusive Economic Zone and Continental Shelf (Environmental Effects – Non-Notified Activities) Regulations 2013. The non-notification provisions that restrict public participation are contentious. See, e.g., Environmental Defence Society (2014) and the Exclusive Economic Zone and Continental Shelf (Environmental Effects-Permitted Activities) Regulations 2013.

59 These amendments came into effect in May 2013, under the Crown Minerals Amendment Act 2013. Amendments were also made to the Continental Shelf Act 1964.

60 Penalties for harm to marine mammals have been increased by the Conservation Natural Heritage Protection Act 2013.
61 Tourism New Zealand (2013) 'New Zealand to Create Five New Marine Reserves', online at http://www.newzealand.com/travel/media/press-releases/2013/3/nature_five-new-marine-reserves.cfm [accessed 20 February 2014] (no longer available).
62 Local Government Act 2002 Amendment Act 2012.
63 For a very different interpretation and approach, see Bargh (2014).
64 A new report has been released by the Office of the Prime Minister's Science Advisory Committee, entitled *New Zealand's Changing Climate and Oceans: The Impact of Human Activity and Implications for the Future* (PMCSA, 2013).

References

Bargh, M. (2014) 'A Blue Economy for Aotearoa New Zealand?', *Environmental Development and Sustainability*, 16: 459–70.

Centre for Advanced Engineering (2003) *Economic Opportunities In New Zealand's Oceans: Informing the Development of Oceans Policy*, Prepared for the New Zealand Oceans Policy Secretariat by the Centre for Advanced Engineering, online at http://www.mfe.govt.nz/publications/marine/economic-opportunities-new-zealands-oceans-informing-development-ocean-policy [accessed 18 August 2010].

CERD   *See* United Nations Committee on Elimination of Racial Discrimination

Department of Conservation (DOC) (undated*a*) 'Marine Mammals: DOC's Role', online at http://www.doc.govt.nz/conservation/native-animals/marine-mammals/docs-role/ [accessed 18 August 2010].

Department of Conservation (DOC) (undated*b*) 'Marine Reserves A–Z', online at http://www.doc.govt.nz/conservation/marine-and-coastal/marine-protected-areas/marine-reserves-a-z/ [accessed 20 August 2010].

Department of Conservation (DOC) (1994) *New Zealand Coastal Policy Statement 1994*, online at http://doc.org.nz/publications/conservation/marine-and-coastal/new-zealand-coastal-policy-statement/archive/new-zealand-coastal-policy-statement-1994/ [accessed18 August 2010].

Department of Conservation (DOC) (1998) 'Theme Three: Coastal and Marine Biodiversity', in *New Zealand's Biodiversity Strategy: Our Chance to Turn the Tide*, Consultation Document, online at http://www.biodiversity.govt.nz/picture/doing/nzbs/part-three/theme-three.html [accessed 18 August 2010].

Department of Conservation (DOC) (2002) 'Sea: Our Future – Protecting Special Places in the Sea', Information Kit (copy on file with the author).

Department of Conservation (DOC) and Ministry of Fisheries (2005) *Marine Protected Areas: Policy and Implementation Plan*, online at http://www.fish.govt.nz/NR/rdonlyres/85FB2343-1355-45EA-A329-88C269A2A84C/0/MPAPolicyandImplementationPlan.pdf [accessed 20 August 2010].

Department of the Prime Minister (2003) *Sustainable Development for New Zealand: Programme of Action*, online at http://www.msd.govt.nz/documents/about-msd-and-our-work/publications-resources/archive/2003-sustainable-development.pdf [accessed 18 August 2010].

Enfocus Ltd, Hill Young Cooper Ltd, and URS NZ Ltd (2002) *Oceans Policy Stocktake, Part 1: Legislation and Policy Review, Schematic Representation of Jurisdictional Boundaries of Key Statues*, Report prepared for New Zealand Oceans Policy Secretariat, online at http://www.mfe.govt.nz/publications/oceans/stocktake-report-dec02/index.html [accessed 18 August 2010] (no longer available).

Environmental Defence Society (2014) 'EDS Calls for Offshore Oil and Gas Regulations to be Dropped', Press release, 3 February, online at http://business.scoop.co.nz/2014/02/03/eds-calls-for-offshore-oil-and-gas-regulations-to-be-dropped/ [accessed 20 February 2014].

Finlayson, C. (2010) 'Repeal of Foreshore and Seabed Act Announced', 14 June, online at http://www.beehive.govt.nz/release/repeal+foreshore+and+seabed+act+announced [accessed 20 August 2010].

Harris, R. (ed.) (2004) *Handbook of Environmental Law*, 2nd edn, Wellington: Royal Forest and Bird Society.

Hauraki Gulf Forum (2011) *State of Our Gulf: Tikapa Moana – Hauraki Gulf State of the Environment Report*, online at http://www.aucklandcouncil.govt.nz/EN/environmentwaste/naturalenvironment/Pages/stateofthehaurakigulf.aspx [accessed 20 February 2014].

Helm, P. (1998) 'New Zealand's Ocean Future Opportunities and Responsibilities', in C. Wallace, B. Weeber, and S. Buchanan (eds) *Proceedings of the 1998 Sea Views Conference*, Environment and Conservation Organization of New Zealand, 11–14 February, Wellington.

Institute for Governance and Policy Studies, Victoria University of Wellington (undated) 'Oceans Governance: The New Zealand Dimension', online at http://igps.victoria.ac.nz/events/completed-activities/Emerging%20Issues%20Programme/Ocean%20Governance.html [accessed 20 February 2014].

Ministry for the Environment (undated*a*) 'Exclusive Economic Zone and Continental Shelf (Environmental Effects) Act 2012', online at http://www.mfe.govt.nz/issues/oceans/current-work/index.html [accessed 20 August 2010].

Ministry for the Environment (undated*b*) 'What is Ocean Policy? Jurisdictional Boundaries under International Law', online at http://www.mfe.govt.nz/issues/oceans/what/jurisdictional.html [accessed 9 February 2007] (no longer available).

Ministry for the Environment (undated*c*) 'Who's Involved?', online at http://www.oceans.govt.nz/whos-involved/stage-two-projects.html [accessed date unknown] (no longer available).

Ministry for the Environment (2003) 'Oceans Policy Stage Two Working Papers', online at http://mfe.govt.nz/publications/oceans/stage-two-papers-feb03/index.html [accessed 18 August 2010].

Ministry for the Environment (2005a) 'Aquaculture Reform 2004: An Overview', online at http://mfe.govt.nz/publications/rma/aquaculture-info-overview-jan05/index.html [accessed 18 August 2010].

Ministry for the Environment (2005b) *Offshore Options: Managing Environmental Effects in New Zealand's Exclusive Economic Zone*, online at http://www.mfe.govt.nz/publications/marine/offshore-options-managing-environmental-effects-new-zealands-exclusive-economic [accessed 5 November 2014].

Ministry for the Environment (2011) *The Natural Resources Sector Briefing to Incoming Ministers*, http://mfe.govt.nz/publications/about/briefing-incoming-minister-2011/nrs-bim-2011.pdf [accessed 20 February 2014].

Ministry for Primary Industries (undated) 'New Zealand's Quota Management System (QMS)', online at http://fs.fish.govt.nz/Page.aspx?pk=81 [accessed 20 August 2010].

Ministry for Primary Industries (2010) 'New Zealand Fisheries at a Glance', online at http://www.fish.govt.nz/en-nz/Fisheries+at+a+glance/default.htm [accessed 30 August 2010].

Ministry of Economic Development (2012) *Review of the Crown Minerals Act 1991 Regime: Discussion Paper*, online at http://www.med.govt.nz/sectors-industries/natural-resources/pdf-docs-library/oil-and-gas/crown-minerals-act-review/Review%20of%20the%20Crown%20Minerals%20Act%201991%20regime%20-%20Discussion%20paper.pdf [accessed 20 February 2014].

Ministry of Fisheries (2005) *Strategy for Managing the Environmental Effects of Fishing*, online at http://www.fish.govt.nz/NR/rdonlyres/B341907D-D284-4E3D-BC19-9628D3301902/1505/SMEEFPapa2.pdf [accessed 20 August 2010].

Ministry of Fisheries and Department of Conservation (DOC) (2008) *Marine Protected Areas: Classification, Protection Standard and Implementation Guidelines*, online at http://www.doc.govt.nz/publications/conservation/marine-and-coastal/marine-protected-areas/marine-protected-areas-classification-protection-standard-and-implementation-guidelines/ [accessed 20 August 2010].

Ministry of Foreign Affairs and Trade (undated) 'Rio + 20', online at http://www.mfat.govt.nz/Foreign-Relations/1-Global-Issues/Environment/1-Governance/Rio-20.php [accessed 20 February 2014].

Mulcahy, K., Peart, R., and Bull, A. (2011) *Safeguarding Our Oceans: Strengthening Marine Protection in New Zealand*, online at http://www.eds.org.nz/eresources/e-books.cfm#faq119262 [accessed 20 February 2014].

New Zealand Biodiversity (undated*a*) 'Border Requirements', online at http://www.biodiversity.govt.nz/seas/biosecurity/border/index.html [accessed 20 August 2010].

New Zealand Biodiversity (undated*b*) 'The New Zealand Biodiversity Strategy', online at http://www.biodiversity.govt.nz/picture/doing/nzbs/index.html [accessed 20 August 2010].

New Zealand Oceans Policy Secretariat (undated) 'Ocean Boundaries', online at http://www.mfe.govt.nz/issues/oceans/what/map.html [accessed 15 December 2006] (no longer available).

New Zealand Ocean Policy Secretariat (2001) *Healthy Sea, Healthy Society: Towards an Oceans Policy for New Zealand*, Report on consultation undertaken by the Ministerial Advisory Committee on Oceans Policy, online at http://www.mfe.govt.nz/publications/oceans/healthy-seas-healthy-society/healthy-sea-healthy-society-sep01.pdf [accessed 18 August 2010] (no longer available).

New Zealand Oceans Policy Secretariat (2005) 'The Oceans Policy Process', online at http://www.oceans.govt.nz/policy/policy-process.html [accessed 3 January 2005] (no longer available)

New Zealand Petroleum and Minerals (2013) 'New Crown Minerals Act Regime', online at http://www.nzpam.govt.nz/cms/about-nzpam/news/archive/new-crown-minerals-act-regime [accessed 20 February 2014].

Office of the Prime Minister's Science Advisory Committee (PMCSA) (2013) *New Zealand's Changing Climate and Oceans: The Impact of Human Activity and Implications for the Future*, online at http://www.pmcsa.org.nz/wp-content/uploads/New-Zealands-Changing-Climate-and-Oceans-report.pdf [accessed 20 February 2014].

Parliamentary Commissioner for the Environment (PCE) (1999) *Setting the Course for a Sustainable Future: The Management of New Zealand's Marine Environment*, online at http://www.pce.parliament.nz/assets/Uploads/Reports/pdf/Sustainable_Future_report.pdf [accessed 5 November 2014].

Parliamentary Commissioner for the Environment (PCE) (2002) *Creating Our Future: Sustainable Development for New Zealand*, online at http://www.pce.parliament.nz/publications/all-publications/creating-our-future-sustainable-development-for-new-zealand-3 [accessed 5 November 2014].

Parliamentary Commissioner for the Environment (PCE) (2003) *Just Cruising? Environmental Effects of Cruise Ships*, online at http://www.pce.parliament.nz/publications/all-publications/just-cruising-environmental-effects-of-cruise-ships-5 [accessed 5 November 2014].

Parliamentary Commissioner for the Environment (PCE) (2004) *Growing for Good: Intensive Farming, Sustainability and New Zealand's Environment*, online at http://www.pce.parliament.nz/publications/all-publications/growing-for-good-intensive-farming-sustainability-and-new-zealand-s-environment-3 [accessed 5 November 2014].

PMCSA *See* Office of the Prime Minister's Science Advisory Committee

Rosier, J. (2004) *An Independent Review of the New Zealand Coastal Policy Statement*, Report to the Minister of Conservation, online at http://www.doc.govt.nz/publications/conservation/marine-and-coastal/independent-review-of-the-new-zealand-coastal-policy-statement/ [accessed 18 August 2010] (no longer available).

Sea Change (2014) 'Sea Change: Tai Timu Pari Stakeholder Process', Press release, 6 January, online at http://www.seachange.org.nz/News-Events/Media-release-library/Sea-ChangeTai-Timu-Tai-Pari-stakeholder-process-moves-ahead/ [accessed 20 February 2014].

Statistics New Zealand (undated*a*) '2006 Census Data and Reports', online at http://www.stats.govt.nz/Census/2006CensusHomePage.aspx [accessed 20 August 2010].

Statistics New Zealand (undated*b*) 'Tourism and Migration 2007', online at http://www.stats.govt.nz/browse_for_stats/industry_sectors/tourism/tourism-migration-2007.aspx [accessed 30 August 2010].

Statistics New Zealand (2006) '2006 Census QuickStats National Highlights: Cultural Diversity', online at http://www.stats.govt.nz/census/2006-census-data/national-highlights/2006-census-quickstats-national-highlights.html [accessed 15 December 2006] (no longer available).

Statistics New Zealand (2008) *Official New Zealand Yearbook 2008*, 106th edn, Wellington: Statistics New Zealand.

Statistics New Zealand (2010a) *Fish Monetary Stock Account 1996–2009*, Wellington: Statistics New Zealand.

Statistics New Zealand (2010b) 'New Zealand's Population Climbs to 4.37 Million', Press release, 13 August, online at http://www.stats.govt.nz/browse_for_stats/population/estimates_and_projections/NationalPopulationEstimates_MRJun10qtr.aspx [accessed 20 August 2010].

Sustainable Aotearoa New Zealand Inc. (SANZ) (2009) *Strong Sustainability for New Zealand: Principles and Scenarios*, Auckland: Nakedize Ltd.

United Nations Committee on Elimination of Racial Discrimination (CERD) (2005) *Decision 1(66): New Zealand Foreshore and Seabed Act 2004*, 66th Session, 21 February–11 March, UN Doc. CERD/C/66/NZL/Dec.1, online at http://www2.ohchr.org/english/bodies/cerd/docs/CERD.C.66.NZL.Dec.1.pdf [accessed 18 August 2010].

University of Auckland Library (2010) 'Census of Marine Life', 6 August, online at http://blogs.library.auckland.ac.nz/science/archive/2010/08/06/1661.aspx [accessed 20 August 2010].

14

NORWAY'S MARINE POLICY

Toward comprehensive oceans management

*Alf Håkon Hoel and Terje Lobach**

Introduction

Geographical and economic features

With an ocean area of more than 2 million km² and 22,000 km of coastline, Norway is a maritime nation. The marine area of the country is situated between 55 and 84°N latitude, with the Svalbard archipelago to the north in the Barents Sea and Jan Mayen Island to the northwest in the Norwegian Sea. Offshore petroleum, fisheries, aquaculture, shipping, and various marine industries are fundamental to the country's economy, culture, and politics.

The ocean areas under Norwegian jurisdiction are in the North Sea (bordering on the territories of Sweden, Denmark, and the United Kingdom), in the Norwegian Sea (bordering the waters of the United Kingdom, Greenland, Iceland, the Faroe Islands, and areas beyond national jurisdiction), and in the Barents Sea (bordering the waters of Russia and areas beyond national jurisdiction) (Figure 14.1). The only country with territories in both polar areas, Norway has also established territorial waters off Bouvet Island in the Southern Atlantic.

The offshore petroleum industry is the most important industry in the country in terms of economics, accounting for 23 per cent the country's gross domestic product (GDP) and half of its exports earnings in 2012 (Norwegian Petroleum Directorate, 2013: 21). Fisheries and aquaculture are small in terms of GDP, but are significant in terms of export value – 7 per cent in 2012 (Ministry of Industry and Fisheries, 2013: 7) – and are important to coastal communities. With a total seafood production of 3.6 million tons in 2012, Norway ranked twelfth among the world's seafood-producing nations in 2011 (Ministry of Industry and Fisheries, 2013: 4). With a small population of 5.1 million, Norway has a limited domestic market and exports most of its marine products. It is the world's seventh largest exporter of oil and third largest exporter of natural gas (Norwegian Petroleum Directorate, 2013: 20), and the second largest exporter of fish products (Ministry of Industry and Fisheries, 2013: 5).

The role of the oceans in domestic politics

Vital to the economy of the country, the management of oceans and coastal areas is an important issue in domestic politics, as well as in foreign policy. Marine affairs are embedded in a broader security framework, the basis of which is the country's membership in the North Atlantic Treaty Organization (NATO), of which Norway is a founding member. Norway is not a

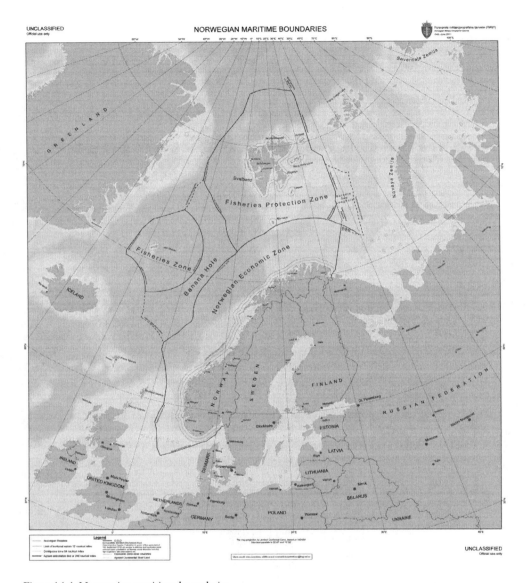

Figure 14.1 Norwegian maritime boundaries

Source: Norwegian Military Geographic Service

member of the European Union, but is closely associated with the internal market by means of the European Economic Area (EEA). Its position vis-à-vis the European Union – with regard to marine natural resources in particular – is an important issue in domestic politics (Ingebritsen, 1998).

Its ocean areas more than six times its land area, Norway is a major beneficiary of the post-war developments in international ocean law (Hoel, 2005). Its domestic oceans regime has evolved over time in response to these legal developments, as well as in response to scientific, technological, economic, and political changes. With a small and widely scattered population, a long coast, and vast oceans, conflicts among different types of use of the oceans have, until recently, not been as acute in Norway as they have been in more densely populated regions. Historically, the development of ocean policies has therefore largely comprised sector-based endeavours. Conflicts that have arisen, for example between fisheries and petroleum development, have been possible to reconcile.

Within the various policy sectors, regulatory frameworks have been institutionalized by means of the establishment of ministries and subsidiary bodies, scientific institutions, and legislation. The uses of the oceans are, in general, strictly regulated. The last decade brought with it a drive towards greater coordination between various marine policy sectors and integrated oceans management (Hoel and Olsen, 2012). The evolution of scientific knowledge, first-hand experience with problems related to pollution and overfishing, the requirements of international agreements, and increasing conflicts between different uses of the oceans are important driving forces. A critical juncture was a report to the *Storting* (the Norwegian Parliament) in 2002 that outlined principles for a more integrated and ecosystem-oriented marine policy (Ministry of the Environment, 2002), which set the stage for the development of plans for integrated management of the oceans.

In what follows, we first describe the substantive issue areas of fisheries, petroleum, and environment; then, we address the implementation of international agreements and account for the nature of polices and legislation. We conclude with observations on the current development of comprehensive oceans management strategies and recent legislation in Norway.

The policy sectors: Petroleum, fisheries, and environment

The most significant economic sectors in the marine realm in Norway are petroleum and fisheries. Shipping is also an important maritime activity, but since this mostly occurs in foreign and international waters and, to a large extent, under foreign flags, it will not be discussed here.[1] Environmental affairs constitute a policy sector in its own right, with its own issues and concerns.

The petroleum sector

Following the discovery of oil in the North Sea, Norway claimed jurisdiction over the continental shelf in 1963. Regular oil production started less than a decade later. Increasingly, production of natural gas has become important and now accounts for more than 40 per cent of the country's total petroleum production. Since 1980, when the first licences for activity north of 62°N were issued, the Norwegian Sea has become an important petroleum province. To the north of the Arctic Circle, in the Barents Sea, regular production started in 2007 when the Snøhvit ('Snow White') gas field came on stream. The North Sea continues to be the most important petroleum region in Norway. As of 2013, seventy-six fields were in operation on the Norwegian shelf.

The total estimates of known and unknown petroleum resources on the Norwegian continental shelf are 13.6 billion standard cubic meters potential energy (Sm^3 pe). Nearly half of this has been produced, most of it in the North Sea, and about two-thirds of the remaining estimated resources have been discovered (Norwegian Petroleum Directorate, 2013: 26).

The petroleum industry accounted for 30 per cent of the state's income in 2012 (Norwegian Petroleum Directorate, 2013: 21). The income from petroleum activity is partly allocated to the annual state budget and partly invested in the state's pension fund. The fund, which invests internationally, was established in 1990 to ensure macro-economic stability and to provide a financial reserve to meet the future demands on the pension system stemming from an ageing population. In 2012, the fund was some NOK4,000 billion (Norwegian Petroleum Directorate, 2013: 22), making it one of the world's largest sovereign wealth funds.

Total employment in the petroleum industry and related activities was some 200,000 in 2009 (Norwegian Petroleum Directorate, 2013: 21). Alongside the petroleum industry itself, a major services industry has developed, sustaining a large number of businesses providing goods and services to the petroleum sector in Norway, as well as abroad.

Fisheries

About twenty-five species are significant in commercial fisheries in Norway. There are practically no subsistence fisheries. Total marine fisheries were 2.3 million tons in 2012 (Ministry of Industry and Fisheries, 2013: 13). In addition, foreign fishing vessels land substantial quantities in Norwegian ports. Arctic cod (358,000 tons), Atlanto-Scandian herring (611,000 tons), capelin (268,000 tons), mackerel (176,000 tons), haddock (161,000 tons), and saithe (176,000 tons) are among the most important fisheries. These six species constitute some 75 per cent of the total catch.

As to marine mammals, harp and hooded seals and minke whales are harvested commercially. In 2012, 465 minke whales were taken. Sealing takes place in the Arctic on the ice. About 5,600 animals were taken in 2012 (Ministry of Industry and Fisheries, 2013: 21). While seals are caught mainly for their pelts; minke whales are hunted for their meat. The exploitation of marine mammals is of minor economic significance nationally, but important to the local communities engaged in the hunt (Kalland, 1995).

Many coastal communities depend on the fisheries for income, and are culturally and historically attached to them. Aquaculture has become increasingly important, in 2012 accounting for some 60 per cent of the export value from the fisheries sector (Ministry of Industry and Fisheries, 2013: 22). The importance of aquaculture is growing steadily in relation to marine fisheries, and Norway is the world's biggest producer of farmed salmon (1.2 million tons in 2012). Considerable efforts are now invested in research and development of the farming of marine species such as cod, halibut, and shellfish. The fish farming industry depends on pelagic fish species for fish feed. Fish meal and oil are important components of the diet of many species of farmed fish.

Of the 6,500 fishing vessels that are registered (2009), 4,000 are 11 metres in length or less (Ministry of Fisheries and Coastal Affairs, 2010a: 9). The large majority of vessels are small vessels fishing with nets, lines, and jigs. Trawlers, purse seiners, and longliners dominate the offshore fisheries. Norway is not a distant-water fishing nation, although a few vessels fish in the Antarctic for krill, a fishery managed with cooperation under the 1982 Convention for the Conservation of Antarctic Marine Living Resources (CCAMLR). Norwegian vessels also fish in international waters in the North Atlantic, which are managed by the regional fisheries management organizations (RFMOs) there, the North East Atlantic Fisheries Commission (NEAFC), the North Atlantic Fisheries Organization (NAFO), and the International Commission for the Conservation of Atlantic Tuna (ICCAT). Most of the Norwegian harvest is taken in the waters under Norwegian jurisdiction, providing income for some 12,000 fishers and about 10,000 workers at fish plants. About 5,800 people are employed in fish farming (Statistics Norway, 2014), bringing the total direct employment in the fishing industry to some 28,000. As in the petroleum sector, fisheries generate employment in related activities such as shipbuilding, repairs and gear production, and public administration, as well as sales and exports.

Exports of fish and fish products constitute 7 per cent of the value of exports from Norway (Ministry of Industry and Fisheries, 2013: 7). More than 90 per cent of the fish landed in Norway is exported. The most important market is the European Union, which receives about 60 per cent of the exports. By country, Russia is the single most important market, followed by France, Poland, and Denmark. The total export value of the fisheries production was about NOK61 billion in 2013 (US$10 billion) (Norwegian Seafood Council, 2014). The single most important species in terms of export value is salmon.

The ocean environment

Ranging from the temperate waters of the North Sea to the polar waters in the Arctic Ocean north of the Svalbard – a distance of more than 3,000 km – Norway's marine environment is highly diverse. A significant environmental force is the North Atlantic Current, which brings warm waters from the southwest Atlantic towards northern Europe. As a result of this ocean current, coastal Norway is 5–8°C warmer than other regions at the same latitude. The ocean currents combine with strong upwellings and nutrient-rich waters to create a highly productive ocean environment that sustains large populations of fish and marine mammals (Sakshaug et al., 2009).

The marine environment is generally healthy, but is exposed to climate-related stresses, as well as pollution (IMR, 2010). The major sources of pollution are found outside Norwegian waters, because Norway sits downstream of major sources in Europe. Emissions of nuclear materials from British plants are among the most threatening pollution sources,[2] but emissions of persistent organic pollutants (POPs) are also a cause for concern (AMAP, 2009). In the ocean areas under Norwegian jurisdiction, discharges of produced water in the petroleum drilling and dumping of toxic materials are the primary sources of pollution of the marine environment.

A recent concern is shipments of petroleum from ports in the Kola and Archangelsk regions in northwest Russia, which pass through the Barents Sea and along the north Norwegian coast. The frequency of petroleum shipments in the region is not yet high compared to that of the major petroleum port facilities in southern Norway. However, it is likely to increase in the years to come as a consequence of further development of petroleum operations in northwest Russia (Bambulyak and Frantzen, 2009).

Implementation of international agreements

Institutionalization of international norms

As a small state, Norway has a fundamental interest in a robust international legal order, which provides for stability in international relations. The growth of rule-based, as opposed to power-based, interactions among countries in oceans affairs is a definite characteristic of the international oceans regime that has developed over the last decades (Ebbin et al., 2005). This development serves the interest of smaller countries, in that it contributes to enhanced international stability and predictability of oceans governance.

Commencing with the 1958 United Nations Conference on the Law of the Sea (UNCLOS I), a broad framework regulating almost all uses of the oceans has since evolved. The most important agreement emanating from the 1958 Conference was the 1958 Convention on the Continental Shelf. This framework was significant when Norway, in 1965, concluded agreements with Denmark and the United Kingdom on the delimitation of the continental shelf in the North Sea, which settled the issue of ownership to the petroleum resources in that sea. The Third United Nations Law of the Sea Conference, UNCLOS III (1973–82), introduced the concept of the exclusive economic zone (EEZ), which set the stage for a major reconfiguration of rights to natural resources in the oceans and the development of a coastal state-based system of resource management regimes (Churchill and Lowe, 1999). Many fish stocks were not circumscribed by the EEZ of any one country, but were shared among two or more countries and, in some instances, also the high seas (Munro et al., 2004). These situations led to calls for the establishment of international arrangements for cooperation on transboundary stocks of living marine resources.

The institutionalization of the EEZ and delimitation issues

The current domestic ocean regime is constituted by several Acts, notably the 1976 Economic Zone Act,[3] the 1996 Petroleum Act,[4] and the 2008 Oceans Resources Act.[5] The 1976 Act extends jurisdiction over living marine resources to 200 nautical miles and reserves the utilization of those resources for vessels under Norwegian flag. In addition, other Acts elaborate the regulatory framework for petroleum resources, fisheries, and the environment. The Acts provide for enabling legislation to facilitate the adaptation of regulatory regimes to changing circumstances.

The EEZ off the Norwegian mainland was established on 1 January in 1977 (see Figure 14.1). In 1978, this was followed by the establishment of a 200 nautical mile fisheries protection zone around Svalbard. A 200 nautical mile fisheries zone around Jan Mayen was introduced in 1980. Norway has not established EEZs off its territories in the Antarctic.[6] Territorial waters were established off Bouvet Island in 2005.

The extension of jurisdiction over the shelf and the waters raised several jurisdictional issues with neighbouring countries. Since the establishment of EEZs in the late 1970s, Norway has, however, settled virtually all boundary conflicts with its neighbouring countries that arose as a consequence of extended state coastal jurisdiction. In the Barents Sea, talks were initiated in 1974 between Norway and the then Soviet Union with a view to resolving a boundary there. Talks continued for more than thirty-five years before an agreement on a maritime boundary 1,680 km long was reached between Norway and the Russian Federation, dividing a disputed area of 175,200 km² into two equally sized parts. The final agreement was signed in September 2010 and entered into force on 7 July 2011 (Government of Norway, Office of the Prime Minister, 2010).

A second jurisdictional issue concerns the fisheries protection zone around Svalbard. The 1920 Svalbard Treaty accords 'full and absolute' sovereignty over the archipelago to Norway. It also provides that nationals of other parties to the treaty retain equal access to economic activity on Svalbard, and states that its provisions pertain to the land territory and territorial waters. In terms of its 1976 Economic Zone Act, Norway considers that it is entitled to establish an EEZ around Svalbard. The sensitive nature of the region at the height of the Cold War and opposition from other countries to a Norwegian EEZ there resulted in a more cautious design: a non-discriminatory fisheries protection zone. Although this action was met with opposition,[7] the zone has been in effect and largely complied with since it was established in 1977.

Boundary disputes with Iceland and Denmark (that is, Greenland) have been settled. The Iceland–Jan Mayen boundary was settled in 1981 by arbitration, and the Greenland–Jan Mayen delimitation line was decided on in 1993 by the International Court of Justice. The boundary between Svalbard and Greenland was settled in 2006. The circumpolar continental shelf facing the Arctic Ocean was settled for Norway by a recommendation from the UN Commission on the Limits of the Continental Shelf (CLSC) in March 2009 (CLCS, 2009), the first in the central Arctic Ocean. The outer limits are in the process of being implemented.

International cooperation in fisheries

An important aspect of Norwegian fisheries is that the most important fish stocks are transboundary and shared with other countries. International cooperation in their management is therefore necessary. Since the introduction of extended jurisdiction in the 1970s, a comprehensive network of bi- and multilateral agreements has been negotiated with neighbouring countries to provide for the management of shared and straddling fish stocks and marine mammals. The most important of these agreements are with Russia and the European Union. There are bilateral arrangements also with Iceland, the Faroese Islands, Greenland, and others. Norway is also party to several regional fisheries management organizations, of which NEAFC, NAFO, and CCAMLR are the most important. The activity under these agreements and their performance are subject to annual review by the *Storting*, on the basis of reports from the government.[8]

The Joint Norwegian–Russian Fisheries Commission, established in 1975, meets annually to agree on total allowable catches (TACs) of shared stocks and on the allocation of quotas for the fisheries of shared stocks in the Barents Sea – namely, cod, haddock, and capelin. Since 2009, Greenland halibut is also considered a shared stock. The total quotas are divided between the two countries according to a fixed ratio. The arrangement includes mutual access for the two countries' vessels to each other's area of jurisdiction and exchange of quotas in other fisheries.[9] A part of the total quota is traded to third countries that fished in the area before the establishment of the EEZ in return for access for Norwegian vessels to their waters. The cooperation also entails joint efforts in fisheries research (Haug et al., 2008) and in enforcement of fisheries regulations (Hønneland, 2012). The 2010 boundary agreement contains provisions on fisheries and states that the fisheries cooperation between the two countries, as laid down in the fisheries cooperation agreements from 1975 and 1976, is to continue, maintaining the relative shares of the shared fish stocks.

Cooperation with the European Union on the management of shared fish stocks in the North Sea is based on a 1980 agreement. The Union has a Common Fisheries Policy (CFP) whereby the Member States have renounced their competence over fisheries management (Holden, 1994), the fundamentals of which were established around 1970. Bilateral cooperation on the management of the fisheries in the North Sea therefore is between Norway and the European Union, rather than the relevant Member States with coasts on the North Sea. The arrangement involves seven shared stocks and quota exchanges for several non-shared stocks. There are differences between Norway and the Union as to the approaches to resource management – among other things, on the discard of fish. A revision of the CFP that will enter into force in 2014 modifies the EU approach in this regard. The enforcement of fisheries regulations is also addressed by the cooperation agreement, but Norway has also entered into a series of bilateral agreements on control with all relevant EU Member States, because this is still within their competence.

Other fisheries agreements in the Northeast Atlantic area include agreements between Norway and Greenland, Iceland, and the Faroe Islands, respectively. International cooperation on the regulation of salmon fisheries is vested in the North Atlantic Salmon Conservation Organization (NASCO).[10]

The management of marine mammals is vested in several multilateral bodies: the International Whaling Commission (IWC); the North Atlantic Marine Mammals Commission (NAMMCO); and JointFish. The IWC does not set quotas, because controversies over whether whales should be considered natural resources and therefore subject to harvest have paralysed the organization (Andresen, 2004). Since 1993, Norway has therefore set unilateral quotas for its commercial take of minke whales on the basis of the work of the IWC Scientific Committee. NAMMCO adopts management measures for cetaceans and seals in the northern northeast Atlantic area.

Norway ratified the 1995 Agreement for the Implementation of the Provisions of the United Nations Convention on the Law of the Sea of 10 December 1982 Relating to the Conservation and Management of Straddling Fish Stocks and Highly Migratory Fish Stocks (the 1995 UN Fish Stocks Agreement, or UNFSA) in 1996. The provisions of the agreement have been influential in shaping the country's policies in fisheries management, in particular as regards the application of a precautionary approach and the enhancement of regional fisheries cooperation for fisheries on the high seas. The International Council for the Exploration of the Sea (ICES) (see 'Fisheries') has provided precautionary advice on catch levels since the late 1990s. In 2002, JointFish decided that, from 2004 onwards, multiannual quotas based on a precautionary approach should apply. Cooperation between Norway and the European Union are based on ICES advice and involve work on long-term management plans for certain stocks. Both the Norway–Russia and the Norway–EU cooperation has led to development of long-term management plans with associated harvest control rules for the most important fisheries.

As to the RFMOs, NEAFC has seen its role enhanced as a result of increased fisheries on the high seas and the impetus for regional fisheries cooperation provided by UNFSA (Hoel, 2010). In response to UNFSA, the members of NEAFC have strengthened its role in enforcement of the fisheries on the high seas in the region – that is, the NEAFC Regulatory Areas. Disagreement over allocation of quotas has, however, prevented NEAFC from taking an active stance on several high seas fisheries in the region.[11] A 1996 agreement between five coastal states in the northeast Atlantic to manage Atlanto-Scandic herring has been in effect until recently, when disagreement over allocation prevented agreement on a TAC. Similar problems have been hampering the management of other species, such as blue whiting and mackerel.

Norway participates in bringing the work of RFMOs elsewhere in the world in line with the principles laid down in UNFSA. This includes the Antarctic (the CCAMLR), the southeast Atlantic (the South East Atlantic Fisheries Organization, or SEAFO), and the northeast Atlantic (NAFO).

International cooperation on the marine environment

Beyond stating a very general duty to protect and conserve the marine environment, the 1982 United Nations Convention on the Law of the Sea (UNCLOS) does not contain specific obligations for states when it comes to the marine environment. The major global treaties regulating pollution of the seas are:

- the 1972 Convention on the Prevention of Marine Pollution by Dumping of Wastes and Other Matter, as amended by the 1996 Protocol (the London Convention), which regulates the discharge of waste from vessels into the ocean;
- the 1973 International Convention for the Prevention of Pollution from Ships, as amended by the Protocol of 1978 (MARPOL); and
- several other instruments under the International Maritime Organization (IMO), which stipulate the standards with which vessels engaged in international shipping have to comply.

Beyond these legal instruments, international cooperation on the protection of the marine environment is based on regional institutions and soft law.

In the northeast Atlantic, regional measures to protect the marine environment are developed under the auspices of the 1992 Convention for the Protection of the Marine Environment of the North-East Atlantic (OSPAR Convention). The work of the OSPAR Commission[12] is organized under five strategies, based on the Convention's five annexes, which deal with land-based pollution, dumping, ocean-based pollution, environmental assessment, and conservation of ecosystems and biodiversity. The annexes and measures adopted under OSPAR are the basis for domestic implementation. For Norway, for example, it is particularly relevant that OSPAR has evolved to include measures for removal of derelict offshore installations (1998) and treatment of wastewater from petroleum production (2001). Of particular importance to marine conservation – a major concern for Norway, given its vast oceans and resources – is the work of the Biodiversity Committee, which includes ecological quality objectives (EcoQOs), assessments of species and habitats in need of protection, and marine protected areas (MPAs) in its work.

As to air pollution and its consequences for the marine environment, global treaties of importance include the 1997 Kyoto Protocol to the 1992 United Nations Framework Convention on Climate Change, and the 1979 United Nations Economic Commission for Europe Convention on Long-range Transport of Air Pollution (LRTAP) and its protocols. The Kyoto Protocol, which has been extended beyond its original expiry period of 2012 to 2018, specifies permitted emission levels of greenhouse gases and time frames for achieving reductions for a group of industrialized countries. The petroleum industry is a major contributor to Norway's emissions of greenhouse gases and several measures have been adopted to reduce those emissions.

When it comes to the conservation of species and ecosystems, the 1992 Convention on Biological Diversity (CBD) is very general in its approach and relies on countries to develop plans for its implementation. Protected areas are a key measure in this regard. A specific initiative concerning the marine environment was adopted in 1995,[13] and a marine programme of work has been in effect for several years. Cooperation under the CBD has adopted ambitious goals for the establishment of marine protected areas, aiming for protection of 10 per cent of the oceans by 2020. This was supported by the 2012 Rio+20 conference, which added that conservation could be achieved 'through effectively and equitably managed, ecologically representative and well-connected systems of protected areas and other effective area-based conservation measures' (UNGA, 2012: para. 177).

The provisions of UNCLOS and UNFSA arguably appear more suited to actual biodiversity conservation than the CBD (Hoel, 2003). However, the work under the CBD has helped to bring attention to specific issues such as threats to coral reefs and the use of MPAs. Future developments may see the CBD becoming a more important arena for marine conservation. The same applies to cooperation under the 1973 Convention on International Trade in Endangered Species of Wild Fauna and Flora (CITES), which is seeking a role in the management of living marine resources.

International cooperation based on soft law arrangements

Environmental politics is characterized by the existence of a large number of 'soft law' arrangements that supplement legally binding agreements (Birnie and Boyle, 2002). Such processes and arrangements often provide vehicles for bringing issues quickly on the political agenda, and they have gained in importance over the years. Important examples include Agenda 21 – in particular, Chapter 17 on oceans – and the 2002 Johannesburg Plan of Implementation from the World Summit on Sustainable Development (WSSD), which provides specific guidance to governments in developing their ocean policy. Such soft law arrangements also exist at the regional level. In the northeast Atlantic region, the most important arrangements are the North Sea Conference (OSPAR Commission, undated) and the Arctic Council (Arctic Council, 2011). These do not engage in actual management or regulation, but serve as arenas for policy-shaping in their respective areas.

In relation to fisheries, the Food and Agriculture Organization of the United Nations (FAO) 1995 Code of Conduct of Responsible Fisheries, and the international action plans adopted to further the implementation of the code at national and regional levels, are particularly important soft law instruments. International plans of action have been developed for addressing overcapacity in fisheries, for by-catch of seabirds and sharks, and for targeting illegal, unreported, and unregulated (IUU) fishing (FAO Fisheries and Aquaculture Department, undated).

Of increasing importance in the Arctic parts of Norway's seas are the Arctic Council and its working groups. The Arctic Council was established in 1996 to provide a high-level forum for cooperation on economic, cultural, and environmental issues in the high north. The substance of the cooperation takes place in its six working groups. The Arctic Monitoring and Assessment Programme (AMAP) and the Protection of the Arctic Marine Environment (PAME) working groups have been particularly important in the marine context, producing comprehensive assessments of the status of climate change, shipping, petroleum development, and pollution, among other issues.

Policies and legislation

The political context of policies

All major policy sectors in the marine realm in Norway have strong international dimensions. As pointed out above, the most important marine pollution problems are brought from abroad by ocean currents. Major fish stocks and other living marine resources are shared with other countries. Owing to the small domestic market, fisheries products, as well petroleum production, have to be exported. The international aspects of marine policy in Norway can therefore hardly be overstated.

In recent years, the international debate concerning traditional uses of the oceans and their resources has increasingly included concerns with non-use aspects (Norse, 1993). For example, in recent decades, a high-profile issue in Norwegian ocean politics has been the harvest of marine mammals (Andresen, 2004). While there is a broad consensus at the domestic level, Norway's marine mammal policy has attracted considerable attention, as well as opposition, from other countries, as in the case of the European Union and its ban on the importation of seal products (Wegge, 2013). At the same time, Norway is generally regarded as being environmentally conscious and has been a leader in promoting tougher environmental standards in a number of international forums (Skjærseth, 2004). To the foreign eye, these traits may seem incompatible, but to Norwegians the two perspectives are mutually dependent upon each other, because conservation of resources presupposes their use and vice versa. Domestic politics relating to issues of oceans is therefore use-oriented, but at the same time is conservation-minded. This is demonstrated not least in fisheries policy, in which conservation and use has been combined in a largely successful manner (Gullestad et al., 2013).

Norway is a parliamentary democracy: the government remains in power as long as it commands the confidence of the majority in the *Storting*. Its political system includes a strong tradition for participation of organized interests in the formulation and execution of public policies (Olsen, 1983), a comparatively high degree of centralization of decision-making power, and a relatively consensual political process (Heidar, 2001), whereby the differences between political parties may be difficult to discern – at least as viewed from abroad. The implication for oceans policy is that policymaking in the marine sectors tends to be inclusive, involving interests that are likely to be affected by decisions.[14] The ministries are the hubs of decision-making, with little power devolved to regions.[15] In broad terms, questions of the use of the oceans and reconciliation of conflicting uses have, until recently, been relatively uncontroversial in Norway, at least when compared to the situation in more densely populated countries. The advent of petroleum–related activities in the North has, however, brought a higher level of controversy, with petroleum pitted against fisheries and the environment.

Marine issues are frequently at the centre stage of domestic political debate. Perhaps the most significant political issue in the country in recent decades, Norway's relationship with the European Union, has had fisheries as a chief topic. The country has twice negotiated agreements with the Union about membership (1972 and 1994), but the negotiation outcomes in the area of fisheries contributed to a 'no' in referendums. Also, marine issues increasingly command the attention of non-state actors, such as environmental non–governmental organizations (NGOs) and indigenous peoples' groups – a development that brings increased levels of controversy. In the last few years, it has been the issue of petroleum development that has been most contro-versial. The general elections in 2009 and 2013 both saw the issue of petroleum development in the North, and in the Lofoten region in particular, as one of the most important issues in the campaigns. In both elections, the outcome resulted in coalitions, under which the negotiation of parliamentary platforms resulted in compromises, holding a partial lid on petroleum develop-ments in the North.

Petroleum

The main objective of petroleum policy in Norway is to maximize the returns from the industry for the benefit of Norwegian society as a whole. National control over the development of the petroleum industry, the development of a domestic industry, and participation by the state in the activity are the key elements of the domestic policy. Taking a broad perspective, this policy has been successful, as evidenced by the existence of in a significant international oil company (Statoil), in which the Norwegian state is a majority shareholder, a major supply industry, and the establishment of the state pension fund.

The development of a comprehensive institutional framework has been critical to the achievement of national control over the industry. The Ministry for Petroleum and Energy and the Petroleum Directorate are the locus of this control.[16] A standing committee of the *Storting* also plays a significant role. The 1996 Petroleum Act provides for a licensing system that is the core of the regulatory regime, empowering the government to regulate all aspects of the indus-try. In addition, control of the domestic industry is exercised through the state-owned petro-leum company, Statoil, which was established in 1972.[17] Concern over the power of Statoil in the domestic petroleum system brought about establishment of the state's direct financial interest (SDFI) arrangement in 1985. Through SDFI, the state invests, takes risks, and earns revenue, as do commercial companies. In 2001, Statoil was partially privatized, with current state ownership at about 65 per cent.

Petroleum activities are subject to a strict regulatory regime: exploratory drilling has to be approved by the Petroleum Directorate and the Norwegian Environment Agency, while new fields and pipelines require the consent of the *Storting*. Also, operators are required to undertake environmental and socio-economic assessments (ESIAs) and to subject them to public hearings before new activities are authorized. Plans for the termination of operations have to be approved in advance of the start of regular operations.

The 1975 Petroleum Tax Act provides that the petroleum industry is taxed in the same man-ner as other businesses in Norway – that is, at a rate of 28 per cent on net income. However, an additional special tax of 50 per cent is levied, justified by the super-profitability stemming from the resource rent.

To maintain the level of revenue generated by the industry and the level of activity in associated industries, new fields have to be found and brought into operation continuously. As a consequence of this and high petroleum prices, producers are increasingly interested in exploiting reserves in northern waters. Results from exploratory drilling in the North have so far been mixed. With more than 100 wells drilled since the opening of petroleum activities in the North in 1980, only one gas field – the abovementioned Snøhvit – has come on stream. A small oil field, Goliat, is in the process of being developed, and several other discoveries are also likely to be developed in the future. Following the agreement with the Russian Federation on a boundary in the Barents Sea in 2011, new areas have become available to exploration in the Barents Sea.

Fisheries

The objectives for fisheries policy relate to sustainability, economic efficiency, and regional policy. These are issues that have to be reconciled through the policymaking process (Ministry of Fisheries and Coastal Affairs, 1997). Over the years, increasing weight has been accorded to resource conservation. This is an issue that is largely decided in international negotiations.

The point of departure for decisions on living marine resource management is the scientific advice on catch levels provided by the International Council for the Exploration of the Sea (ICES). Based on the work carried out in research institutions in the member countries, ICES reviews and analyses information about the status of fish stocks and the marine environment, and provides scientific advice on management measures to member states and RFMOs. The primary marine research institution in Norway is the Institute of Marine Research (IMR).

Quota levels for fish stocks in Norwegian waters are set in cooperation with other countries as far as shared fish stocks are concerned. The most critical conservation decisions – the amount of fish that can be removed from a given stock – is the basis for the domestic decision-making process. An open regulatory meeting, which includes industry representation, as well as other stakeholders, is tasked with advising the Fisheries Directorate and the Ministry of Industry, Trade and Fisheries on proposed regulatory arrangements for the various fisheries. The industry is thereby involved directly in the decision-making process, which serves to enhance the legitimacy of government regulations (Mikalsen and Jentoft, 2003).[18]

Broadly speaking, the status of the major fish stocks in the North is sound (IMR, 2010), although fishing pressure is too high on some stocks – notably, coastal cod and redfish. Cooperation between Norway and Russia may be characterized as well functioning (Hønneland, 2012). Cooperation with the European Union on management of joint stocks, such as herring and saithe, in the North Sea has had positive results, while other joint stocks in the North Sea, among them North Sea cod, are in a perilous condition.[19]

From the point of view of fisheries managers, it is a constant dilemma that the quota advice from the scientists varies from one year to the next. This results from natural variations in the fish stocks, errors in catch data, and/or revisions of the criteria and reference points that constitute the basis for the management advice provided by the scientists. In some years, this could have serious negative consequences for the fishing industry. Therefore, when fixing the quotas for the different stocks, fisheries managers also take into account the objectives of economic sustainability, industry stability, and other social and economic factors, in addition to biological sustainability of the stock.

The 2008 Marine Resources Act[20] and the 1999 Act on Participation in Fisheries[21] provide the basis for a comprehensive regulatory framework for the management of living marine resources at the domestic level. The approach taken is a combination of limits to access to fisheries, restrictions on catches in the form of quotas, and technical regulations determining the type of fishing gear to be used and fishing seasons and areas. Important aspects of the regime include discard bans and flexible closures of areas with juvenile fish. Vessels with heavy gear (trawls) generally have to fish beyond the 12 nautical mile limit. All vessels are required to hold a licence and, in most instances, also need a quota to be admitted to a particular fishery. There are virtually no open-access fisheries in Norway. Access to fisheries for larger vessels has been substantially restricted for more than forty years.

The enforcement of fisheries regulations occurs both at sea and when the fish is landed. The Coast Guard (a service in the Navy) is responsible for inspecting fishing vessels at sea and checking their catch against the vessel's logbook. The Fisheries Directorate carries out inspections at sea, but focuses most on territorial waters. Foreign vessels fishing in Norwegian waters – about 700 licences are issued – are also inspected. Because many foreign vessels land their catches abroad, control of such vessels fishing in Norwegian waters must be based on measures other than landing control. In order to monitor their activities, foreign vessels have to carry vessel monitoring systems, report in advance of and during fishing, and present themselves at designated checkpoints for possible control by the Coast Guard.

Ocean-going vessels under the Norwegian flag are required to use electronic reporting devices that enable the authorities to monitor their activity.[22] The landing of fish is controlled by the Fisheries Directorate, as well as the organizations buying the fish. The latter reports the landed quantity to the Directorate of Fisheries, which is responsible for compiling fisheries statistics and checking of vessels' landings against their quotas. Today, this system works almost in real time. The various authorities engaged in enforcement cooperate and enforcement resources are deployed strategically, following a risk-based assessment of where they are most needed (Directorate of Fisheries, 2010).

Increased IUU fishing over the last decade represented a challenge to enforcement in Norway, as elsewhere.[23] The covert nature of such activities, involving flags of convenience and trans-shipment on the high seas, makes enforcement difficult. A blacklist of vessels that have engaged in IUU activities in northeast Atlantic waters had already been established by the Norwegian authorities in 1994 and was the first in the world. The blacklist has been extended to cover IUU fishing in areas under the auspices of RFMOs to which Norway is a member and is continuously updated. Vessels on the list are perpetually prohibited from operating in Norwegian waters or entering its ports and will not be entitled to fly the Norwegian flag, irrespective of changes in ownership.[24] The blacklist is believed to have had considerable effect in deterring illicit activities on fishing grounds in northern Europe. Furthermore, a comprehensive port control system has been established by the NEAFC, which also covers fisheries conducted within the national waters of the parties. As a consequence of the various measures adopted against IUU fishing at domestic, as well as international, levels in the North, IUU activity has declined dramatically in recent years (Ministry of Fisheries and Coastal Affairs, 2010b).

The environment

In the Norwegian polity, environmental concerns are generally considered a cross-sector, or rather *super*-sector, issue. All ministries are required to check their policies against environmental standards, and the country's constitution explicitly states the right to a good environment (Ulfstein, 1999).[25] From this perspective, the environment is not a sector, but a general overarching concern. Looked after by a ministry with several designated agencies,[26] the environment nevertheless takes on sector policy qualities when different concerns and constituencies must be reconciled in a policymaking process.

In the marine environment, the primary international environmental agreements were referred to earlier in the chapter – in particular, OSPAR and the Kyoto Protocol. Petroleum-related activities on the Norwegian shelf are subject to a strict environmental regime. Organic compounds, oil, and chemicals used in petroleum production are the most significant discharges to the sea in Norwegian waters. Domestic regulation of such discharges is largely mandated by OSPAR and, in the North, included the objective of zero harmful discharges to the sea until a relaxation of policy in 2011. New fields are required to re-inject produced water rather than to discharge it into the ocean. Drilling operations in the North are also required to have zero discharges.[27]

Both the petroleum industry and the fisheries are significant contributors to air emissions. For carbon dioxide (CO_2), about a quarter of national emissions stem from the petroleum industry's production of energy at production platforms (Ministry of Petroleum and Energy, 2004: 57). In 1991, Norway introduced a CO_2 tax, which was aimed at reducing emissions and was applied to fossil fuels producing emissions of CO_2.[28] The 1981 Pollution Control Act imposes a number of regulations on all types of emission and discharge.

As pointed out already, the relationship between fisheries and petroleum activity is a major environmental issue. During more than four decades of petroleum activity on the Norwegian shelf, some seventy-five fields have been developed and 11,000 km of pipelines have been laid. In this period, one major accident incurring significant pollution (the 1977 *Bravo* incident) has occurred. The general experience from the North Sea is that fisheries and the petroleum industry can coexist, although the fishers have a mixed experience in this regard. The increase in petroleum-related activities in the North has, however, brought new awareness of the real and potential impacts of petroleum activities on fisheries. Several measures have been introduced to limit potential damage from eventual oil spills, including a programme to reduce the risk of spills from tankers (which has a higher probability that spills from platforms) and the adoption of an integrated management plan for the Barents Sea.

Development of a comprehensive policy

Initiation

The concept of integrated oceans management has been debated in Norwegian (and international) academia for more than three decades (Underdal, 1980; Østreng, 1982). Its introduction into public policy occurred in a report to the *Storting* in 2002 that signalled the need to develop a more comprehensive oceans policy to guide and reconcile the various uses of the oceans (Ministry of the Environment, 2002). Following that, a process to develop an integrated management plan for the Norwegian part of the Barents Sea and the sea areas off the Lofoten Islands by 2006 was initiated. In the same period, the government set in motion work to develop a more modern legislative framework for the oceans. A new Oceans Resources Act would provide for a more holistic approach to the management of living marine resources.

Major principles and their implementation

The implementation of the precautionary approach, the ecosystem-based approach, and the integrated oceans management approach are significant aspects of the current oceans policy in Norway. While the implementation of a precautionary approach to fisheries management has been in the works since the late 1990s, efforts to implement an ecosystem-based approach and integrated oceans management are more recent.

The precautionary approach

The precautionary approach is espoused in UNFSA 1995, which mandates its application in fisheries (Balton, 1996).[29] Applying a precautionary approach in fisheries management is meant to reduce the risk of overexploitation of fish stocks. This entails the setting of reference points that signal objectives for management and threshold levels for spawning stock size and mortality of fish. Fisheries management is then about ensuring that the fish mortality rates and the size of the spawning stock biomass are maintained at or above desired levels.

ICES has provided advice on catch levels based on the precautionary approach since the late 1990s. Further refinement of management principles in the ICES context now includes the introduction of the maximum sustainable yield (MSY) principle. JointFish decided in 2002 that, from 2004 onwards, multiannual quotas, based on a precautionary approach, should apply. A new management strategy adopted in 2003 includes a decision rule stipulating that total allowable catch (TAC) levels for any three-year period shall be in line with the precautionary reference values provided by ICES. Another important element of the strategy is limits to the annual adjustments that can be made to quota levels. In the North Sea, Norway and the European Union cooperate on long-term management plans for certain stocks, with management objectives and the time frame within which they are to be achieved. The development and implementation of long-term management plans represent an improvement upon earlier practices, because they provide decision-makers with long-term scenarios of the effects of their decisions.

The precautionary approach, like any management principle, is a work in progress. Experience over the last decade has brought lessons that scientists and managers try to include in development of the approach. One critical issue is how threshold levels for spawning stock sizes and rates of decline (fish mortality) should be established. Fundamentally, these are issues of risk, and therefore questions of values and societal choice as much as they are matters of science. The task will often be to strike a balance between the need to minimize risk of stock depletion, on the one hand, and the risk of depriving coastal communities of fishing opportunities, on the other.

The ecosystem-based approach

The 2002 Johannesburg Implementation Plan encourages countries to apply an ecosystem approach to the management of living marine resources by 2010 (DESA, 2002: para. 29(d)). It is important to distinguish between different types of situation for the application of the principle. The ecosystem-based approach, as applied broadly to oceans use and management, is different from the approach applied to the management of living marine resources. A useful distinction can be maintained between 'type I' (multiple-use) and 'type II' (sector-based) ecosystem-based management (Hoel, 2005). The relevant knowledge, regulatory arrangements, actors, and institutional frameworks vary between these settings. For fisheries, the principles for the application of an ecosystem-based approach are laid out in the Reykjavik Declaration on Responsible Fisheries in the Marine Ecosystem.[30]

In a general sense, the notion of an ecosystem-based approach implies that different types of use of the ocean environment have to be reconciled to realize the objective of sustainable use and conservation of the marine environment. While this understanding of the ecosystem-based approach to oceans management may appear uncontroversial, it is nonetheless political: decisions reconciling different uses and concerns will often favour certain interests at the expense of others. The approach taken in Norway by means of the integrated management plans is rather technical, framing the issue of oceans management as one of rational choice: the ordering of priorities, the selection of criteria and the collection of data to assess against them, and then deriving the optimal choice. While appealing from the point of view of public administration, casting the issue as a technical one does not remove the political tensions inherent in ecosystem-based management.

No universally agreed definition of the ecosystem-based principle or approach exists (Hilborn et al., 2004; Curtin and Prezello, 2010). The approach, as applied to fisheries, promoted by Article 5 of UNFSA and by the 2001 Reykjavik Declaration, focuses on the relationship between fisheries and the wider ecosystems of which fish resources are a part. In the application of an ecosystem-based approach, the impacts of fishing on parts of the ecosystem other than target stocks are to be considered in the regulations for the fishery (Pikitch et al., 2004). This includes the biotic, as well as the abiotic, environment. Conversely, impacts of environmental factors on fish, for example predation from other species, also have to be considered.

The implementation of an ecosystem-based approach to fisheries management therefore requires that the knowledge base for management also addresses the effects of fisheries on other parts of the ecosystem, as well as the impacts of, for example, climate change and predation of marine mammals on fish stocks (Brander, 2010). For most of the stocks that are subject to Norwegian fisheries management, the scientific knowledge on the single-stock level is sufficient for single-stock measures that take into account the precautionary principle. However, optimum utilization of different stocks requires knowledge about the interdependence of species, the effects of fisheries on their ecosystems, and relevant oceanic, environmental, and socio-economic conditions. Compiling this list is a comprehensive and expensive task, and for the foreseeable future a fully fledged ecosystems-based approach has yet to be developed on the basis of conventional data and methods. To facilitate implementation of this new approach, the Institute of Marine Research (IMR) operates on the basis of a number of ecosystem programmes (including the Barents Sea, Norwegian Sea, and North Sea), drawing on the work of several specialized research groups.

As to regulatory measures, flexible fishing areas that can be opened and closed according to conservation needs are likely to remain a critical feature of a regulatory system based on an ecosystem-based perspective. The long-standing practice of including cod's consumption of capelin when fixing quotas for the capelin fishery is also likely to be an important feature of an ecosystem-based regulatory regime (Hoel, 2005). This practice will have to be considered for other species relationships too, for example by maintaining the harvest of marine mammals that predate on fish. The impact of fishing on abiotic features of ecosystems is already addressed. Norway has prohibited bottom trawling in vulnerable areas in the NEAFC region of the high seas in the northeast Atlantic following the decision by the NEAFC in 2004 to close five sea-mounts on the high seas from 1 January 2005. Norway has also established protected areas in vulnerable ecosystems in its own waters. In these areas, fishing is restricted or prohibited to protect cold-water coral reefs among other ecological features.

Norway is also considering the establishment of coastal MPAs, with the aim of protecting unique nature types along its 22,000 km of coastline, and in 2013 three such areas were established. This is also in response to the Johannesburg Plan of Implementation, which advocates the establishment of MPAs by 2012, as well as ambitions under the CBD to increase the ocean area that is subject to area-based conservation measures (DESA, 2002: para. 31(c)). A recent review of existing regulatory measures by the Ministry of Industry, Trade and Fisheries indicates that more than half of the oceans under Norwegian jurisdiction are subject to area-based regulatory measures.[31]

Institutional arrangements for implementation of new policies

The new and comprehensive Oceans Resources Act, which was adopted by the *Storting* in 2008 and entered into force in 2009, and the concept of integrated management plans, first to be applied to the Barents Sea region in 2006 (revised in 2010), then the Norwegian Sea in 2009, and the North Sea in 2013, represent first steps in the direction of integrated oceans management in Norway. The concept of integrated management plans is important, because it stems from concerns similar to those driving the 'type I' ecosystem-based approach and aims at reconciling conflicting uses of the oceans. As fisheries, aquaculture, tourism, transportation, military activity, and other uses of the ocean increase, they tend to become more competitive, placing multiple stresses on ocean ecosystems.

An integrated management plan

A significant institutional development following from the increased level of activity in several industries – petroleum, in particular – is the integrated management plan for the Barents Sea.[32] This initiative grew out of the report to the *Storting* in 2002 on the environmental status of the country's oceans and policy development in that regard (Ministry of the Environment, 2002).[33] The rationale for the report and the policy laid out in it is that it is necessary to coordinate the various human uses of the oceans to ensure to that the total human impact on the marine environment is within the limits of the carrying capacity of marine ecosystems (Hoel and Olsen, 2012).

The management plan establishes a holistic framework for decision-making that takes into account the interests of fisheries, petroleum, and transportation, as well as the marine environment. The plan, which was adopted by the *Storting* in March 2006, identifies environmental impacts from various economic activities, as well as impacts of external factors such as climate change. The resulting *total load* on the environment is the basis for the specifications in the plan of where and when petroleum-related activities can take place in the Norwegian part of the southern Barents Sea and the coast off northern Norway. The integrated management plan therefore can be understood as a zonal arrangement, separating potentially incompatible uses of the oceans from each other in space and time.

The trade-offs between the concerns represented by different policy sectors can have far-reaching consequences. For now, petroleum-related activities are not permitted in areas that are considered vulnerable. The restrictions are temporary, and if petroleum activity in the future is to be permitted in the sea area off the Lofoten archipelago in winter, it may have implications for the fisheries in the area. On the other hand, too much emphasis on environmental protection may deprive society of revenue from petroleum activity. Public debate on the issue has intensified considerably over the last few years. In general, many politicians in the North are supportive of petroleum development because it is seen as important to the further development of the local communities and the maintenance of employment and settlements. The resistance to the activity is most of all voiced by environmental NGOs and ad hoc campaigns. As pointed out, however, the issue has also become important in national-level politics.

Work on the development of the plan entailed close cooperation between agencies from different sectors under the oversight of a high-level inter-ministerial group. In spite of the political nature of trade-offs between sectors, it is not envisaged, however, that the implementation of the integrated management plans will require changes to the current sector-based organization of the government authorities responsible for policy formulation and implementation. An important aspect of integrated oceans management in Norway is therefore that the management plans introduced another layer of decision-making rather than replaced existing structures. The management plan structures address the total impact of various stressors on marine ecosystems and attempt to reconcile conflicting concerns. The actual regulation of activities occurs on the basis of existing legislation.

Three working groups are established to implement the plans. An Advisory Group on Monitoring, a Forum on Environmental Risk Management, and a Management Forum involve representatives from several research institutes and government agencies. The groups report to the Inter-Ministerial Steering Group.

The management plan for the Barents Sea pertains only to the Norwegian part of the area. There is cooperation at the science and agency levels to develop the foundations for cooperation between Norway and the Russian Federation on this issue in the future. This has resulted in a comprehensive report on the state of the ecosystems (Stiansen and Filin, 2007).

The Marine Resources Act[34]

The Marine Resources Act was developed by a committee appointed by the government in 2003. The committee comprised ten members representing the fishing industry, the government prosecution, public administration (fisheries, environment, and foreign affairs), academia, and the Saami Parliament. It brings new elements into the legislation, as well as a modernization and clarification of existing legislation. Most importantly, the Act, as adopted by the *Storting* in 2008, consolidates all relevant provisions for the management of living marine resources into a single statute,[35] thereby facilitating its implementation. It also incorporates new principles for oceans governance, such as the ecosystem-based approach.

The Act represents a significant modernization of the current legislation. Its overall objective is to ensure a socio-economically profitable management of wild marine resources by sustainable use and long-term conservation of the resources (paragraph 1.1). It expands upon current legislation by making the new law applicable to all utilization of marine resources that live in the wild, including genetic material (paragraph 1.2). It explicitly states that the natural resources are the property of the state as long as they are in the wild. While property rights to living marine resources can be acquired upon capture, genetic material belongs to the state (paragraph 1.5) – a rule that will have implications for bio-prospecting. The Act also lists a series of principles and concerns that are to be taken into consideration in the management of living marine resources and genetic material. Among these are the precautionary approach, an ecosystem-based approach, optimum utilization and allocation of resources, effective control of harvesting, implementation of international law, transparency in decision-making, and regard for the Saami culture.

The Act also establishes the legal basis for the establishment of MPAs (paragraph 2.1) and requires that all catches have to be landed (paragraph 2.6). The latter provision is intended to deter discards of fish by ensuring that all by-catch is landed. The Act requires fishers to maintain and remove gear that is left in the sea (paragraph 2.10), to prevent 'ghost fishing' and accumulation of marine debris.

The Act also provides that, in implementing an ecosystem-based approach to the management of living marine resources, precise resource management objectives can be established (paragraph 3.1). A provision on marine bio-prospecting states that such activities will require a permit from the Fisheries Directorate and that a share of an eventual economic surplus shall be shared with the state. The Act sets out measures to be employed in the regulation of fisheries (Chapters IV and VII) and their enforcement (Chapters VIII, IX, XII, and XIII). It also sets the stage for the state to capture a share of resource rent through the establishment of 'management fees' whereby the industry will have to cover part of the cost of resource management (paragraph 10.1) and a provision for the collection of resource rent (paragraph 10.2).

In addition to the Marine Resources Act, a new Nature Diversity Act was adopted in 2009 that will have some implications for the management of marine ecosystems. The general objective of the Act is to conserve nature. It applies to all nature, including the territorial waters of the oceans under Norwegian jurisdiction. The principle of sustainable use is central to the law, and it is to be implemented through the establishment of conservation objectives for nature types and species, and through the enactment of certain environmental principles, including the precautionary principle, the 'polluter pays' principle, and the principle of responsible technologies and practices.

Conclusion

A major beneficiary of extended coastal state jurisdiction, the Norwegian ocean governance regime has adapted to changing circumstances and challenges. The central defining characteristic of this regime is its international orientation: most living marine resources are shared with other countries and have to be managed through international cooperation; significant marine pollution is brought into Norwegian waters with ocean currents; and, as a small country in terms of population, almost all of Norwegian fishery and petroleum production has to be exported.

While the introduction of the EEZ brought vast ocean areas under Norwegian jurisdiction, the drawing of new boundaries raised controversial issues in relations with other countries, few of which remain after the conclusion of a boundary with Russia in the Barents Sea in 2010. The differing views on principles for the drawing of boundaries and application of international treaties appear not to have hampered cooperation on resource management or compliance with Norwegian policies.

The uses of the ocean are generally regulated on a sectoral basis, with different sets of legislation applied according to type of use. Owing to the small population and the long coastline, conflicts deriving from competing uses of the oceans and the coast have not been as prevalent in Norway as they have been in more densely populated countries. This may be changing, however, in particular with the northwards expansion of the petroleum industry, the growth of aquaculture, increasing environmental awareness, and the development of other uses of the oceans.

From a fisheries perspective, it is important to sustain marine ecosystems, the living resources of which provide food and serve as a basis for income for present and future generations. Healthy, well-functioning, and productive ecosystems will provide optimal levels of production for harvesting. The idea that fisheries constitute a major environmental problem ignores the fact that fisheries are critically dependent upon the ocean environment and healthy fish stocks. In fact, other human activities in the oceans can pose threats to viable fisheries and food security if they are not properly managed. This situation calls for 'type I' ecosystem-based approaches to protect fisheries, as the case is with the management plan for the Barents Sea. Fisheries will, however, often impact on marine ecosystems in the process of harvesting. A management target would therefore be to obtain the maximum benefit from harvesting without reducing the future value of the resources and the marine environment.

In Norway, the domestic ocean governance regime has been subject to a major overhaul over the last decade. New legislation concerning the management of living marine resources has been introduced. The integrated management plan for the Barents Sea (Ministry of the Environment, 2006) is the first of several such plans to cover all oceans under Norwegian jurisdiction, and was followed by the adoption of the Norwegian Sea plan (Ministry of the Environment, 2009), a revision of the Barents Sea plan (Ministry of the Environment, 2011), and a plan for the North Sea (Ministry of the Environment, 2013). The driving force behind the integrated management plans was the development of petroleum activities in the North, near major fishing grounds. The need to implement international agreements in domestic policy, developments in marine science, a general increase in use of the oceans, and the need to modernize legislation all contributed to the initiation of a comprehensive policy and new legislation to effect integrated ocean management in Norway.

Notes

★ The views expressed in this chapter are those of the authors and not necessarily those of the Institute of Marine Research or the Directorate of Fisheries.

1 Some 5 per cent of the world's merchant fleet sails under the Norwegian flag. Of Norway's total exports, 11 per cent derive from the shipping industry. See the Norwegian Shipowners' Association (*Norges Rederiforbund*, http://www.rederi.no/) for more information.

2 In April 2004, the British government decided that the Sellafield plant should introduce new cleansing technologies, reducing emission of technetium-99 by more than 90 per cent.

3 *Lov om Norges økoonomiske sone*, No. 91 of 1976.

4 Originally, *Lov om undersjøiske naturforekomster* [Law on Offshore Resources], No. 12 of 1963, replaced with subsequent legislation, of which *Lov om petroleumsvirksomhet* [Law on Petroleum Activities], No. 72 of 1996 relating to petroleum activities (as amended) is the most recent.

5 *Lov om forvaltningen av viltlevende marine ressurser*, No. 37 of 2008, in force 1 January 2009.

6 Norway's Antarctic territories consist of Bouvet Island, Peter I Island, and Queen Maud Land. Bouvet Island is outside the Antarctic Treaty area. Norwegian jurisdiction is not disputed, and Norway declared a 12 nautical mile Territorial Sea limit there in 2004. Peter I Island and Queen Maud Land are situated in the Antarctic Treaty area south to the sixtieth parallel, and are subject to the Antarctic Treaty regime.

7 The Soviet Union and Iceland have protested the establishment of a fisheries protection zone around Svalbard. Finland and Canada have, in the past, supported the Norwegian stance. Several countries have reserved their rights under the Svalbard Treaty.

8 The latest is the Ministry of Trade, Industry and Fisheries (2012–13).

9 Based on a 1976 agreement on reciprocal access to fisheries in the two parties' EEZs: *Om samtykke til inngåelse av en avtale mellom Regjeringen i Kongeriket Norge og Regjeringen i Unionen av Sovjetiske Sosialistiske Republikker om gjensidige fiskeriforbindelser* [*On the Consent to Enter into an Agreement between the Government of the Kingdom of Norway and the Government of the Union of Soviet Republics on Reciprocal Fisheries*] (1976–77) St. prp. nr 74, online at https://www.stortinget.no/no/Saker-og-publikasjoner/ Stortingsforhandlinger/Saksside/?pid=1970-1981&mtid=109&vt=b&did=DIVL118314 [accessed 5 November 2014].

10 There has not been a commercial marine salmon fishery in the North Atlantic for two decades.

11 Mackerel is a case in point.

12 See http://www.ospar.org

13 The so-called Jakarta Mandate on Marine and Coastal Biological Diversity adopted in 1995. See Convention on Biological Diversity (undated).

14 This has been a legal requirement for public policymaking for more than forty years.

15 Coastal zone planning is an exception. The municipal authorities have substantial powers to plan and execute local policies regarding use of the coastal zone.

16 The Directorate is tasked with supervising the activity in the industry, and ensuring that it operates according to regulations and permits.

17 In addition, a private company, Saga, was established, along with a petroleum division of Norsk Hydro, which later acquired Saga.

18 In addition to the fishing industry, Saami interests and a representative of regional authorities are represented in the meetings. The proceedings are open to other interested parties, e.g. environmental organizations.

19 For a number of years, ICES has advised that the cod fisheries in the North Sea be closed.

20 *Lov om forvaltningen av viltlevende marine ressurser*, No. 37 of 2008

21 *Lov om retten til å delta i fiske og fangst (Deltakerloven)* [Law relating to the right to participate in fishery and hunting of marine mammals (the Participation Act)], No. 15 of 1999.

22 This applies to vessels over 15 metres in length.

23 In the mid-2000s, IUU fishing in the Barents Seas is estimated to have been 100,000 tons or more annually.

24 See Directorate of Fisheries (undated) for the current version of the list.

25 1814 *Nogres Riges Grundlov* [Constitution], § 100b.

26 These agencies include pollution control, nature management, and polar affairs.

27 Except for those resulting from the drilling of the top-hole section for surface casing.

28 In 2005, the carbon dioxide tax was NOK0.78 per Sm3 of gas or litre of diesel.

29 UNFSA, Art. 6 and Annex 2.

30 The Reykjavik Declaration is available online at ftp://ftp.fao.org/fi/DOCUMENT/reykjavik/ y2198t00_dec.pdf [accessed 20 February 2014].

31   2013 unpublished Ministry of Trade, Industry and Fisheries note, on file with author.
32   The title is not entirely appropriate, because a large part of the area is to the west of the Barents Sea proper, and the Russian part of the Barents Sea is not included at all.
33   The incumbent government had stated its intention to do this in its 2001 programme.
34   This section emphasizes new developments and does not discuss the draft law in a comprehensive manner.
35   Legislation pertaining to the right to participate in fisheries is contained in a separate Act: *Deltakerloven*, No. 15 of 1999.

References

Andresen, S. (2004) 'Whaling: Peace at Home, War Abroad', in J. B. Skjaerseth (ed.) *International Environmental Regimes and Norway's Environmental Policy*, Aldershot: Ashgate.

Arctic Council (2011) 'About the Arctic Council', online at http://www.arctic-council.org/index.php/en/about-us/arctic-council/about-arctic-council [accessed 5 November 2014].

Arctic Monitoring and Assessment Programme (AMAP) (2009) *Arctic Pollution 2009*, online at http://www.amap.no/documents/doc/arctic-pollution-2009/88 [accessed 25 March 2014].

Balton, D. (1996) 'Strengthening the Law of the Sea: The New Agreement on Straddling Fish Stocks and Highly Migratory Fish Stocks', *Ocean Development and International Law*, 27(1–2): 125–51.

Bambulyak, A., and Frantzen, B. (2009) *Oil Transport from the Russian Part of the Barents Region, Status per January 2009*, online at http://www.bioforsk.no/ikbViewer/page/tjenester/publikasjoner/publikasjon?p_document_id=85803 [accessed 5 November 2014].

Birnie, P., and Boyle, A. (2002) *International Law and the Environment*, Oxford: Oxford University Press.

Brander, K. (2010) 'Impacts of Climate Change on Fisheries', *Journal of Marine Systems*, 79: 389–402.

Churchill, R. R., and Lowe, A. W. (1999) *The Law of the Sea*, Manchester: Manchester University Press.

CLCS   *See* United Nations Commission on the Limits of the Continental Shelf

Convention on Biological Diversity (undated) 'Marine and Coastal Biodiversity', online at http://www.biodiv.org/programmes/areas/marine/default.asp [accessed 27 January 2007]

Curtin, R., and Prellezo, R. (2010) 'Understanding Marine Ecosystem Based Management: A Literature Review', *Marine Policy*, 34: 821–30.

DESA   *See* United Nations Department of Economic and Social Affairs

Directorate of Fisheries (undated) 'Norwegian Black List', online at http://www.fiskeridir.no/fiskeridir/english/blacklisted_vessels [accessed 27 January 2007].

Directorate of Fisheries (2010) 'Offentliggjør Prioriterte Områder for Ressurskontrollen', Press release, 11 May, online at http://www.fiskeridir.no/fiske-og-fangst/aktuelt/2010/0110/offentliggjoer-prioriterte-omraader-for-ressurskontrollen-i-2010 [accessed 20 February 2014].

Ebbin, S. A., Hoel, A. H., and Sydnes, A. K. (eds) (2005) *A Sea Change: The Exclusive Economic Zone and Governance Institutions for Living Marine Resources*, Dordrecht: Springer.

Food and Agriculture Organization of the United Nations (FAO) Fisheries and Aquaculture Department (undated) 'International Plans of Action', online at http://www.fao.org/fishery/code/ipoa/en [accessed 20 February 2014].

Government of Norway, Office of the Prime Minister (2010) 'Treaty on Maritime Delimitation and Cooperation in the Barents Sea and the Arctic Ocean Signed Today', Press release, 15 September, online at http://www.regjeringen.no/en/dep/smk/press-center/Press-releases/2010/treaty.html?id=614254 [accessed 20 February 2014].

Gullestad, P., Aglen, A., Bjordal, A., Blom, G., Johansen, S., Krog, J., Misund, O. A., and Røttingen, I. (2013) 'Changing Attitudes 1970–2012: Evolution of the Norwegian Management Framework to Prevent Overfishing and to Secure Long-Term Sustainability', *ICES Journal of Marine Science*, online at http://icesjms.oxfordjournals.org/content/early/2013/07/27/icesjms.fst094 [accessed 5 November 2014].

Haug, T., Misund, O. A., Gjøsæter, H., and Røttingen, I. (2008) 'Long-Term Bilateral Russian–Norwegian Scientific Cooperation as a Basis for Sustainable Management of Living Marine Resources in the Barents Sea', *Proceedings of the Twelfth Norwegian-Russian Symposium, Tromsø, 21–22 August*, IMR/PINRO Joint Report Series No. 5, Bergen: Institute of Marine Research.

Heidar, K. (2001) *Norway: Elites on Trial*, Boulder, CO: Westview Press.

Hilborn, R., Punt, A. E., and Orensanz, J. (2004) 'Beyond Band-Aids in Fisheries Management: Fixing World Fisheries', *Bulletin of Marine Science*, 74(3): 493–507.

Hoel, A. H. (2003) 'Marine Biodiversity and Institutional Interplay', *Coastal Management*, 30: 25–36.

Hoel, A. H. (2005) 'The Performance of Exclusive Economic Zones: The Case of Norway', in S. A. Ebbin, A. H. Hoel, and A. K. Sydnes (eds) *A Sea Change: The Exclusive Economic Zone and Governance Institutions for Living Marine Resources*, Dordrecht: Springer.

Hoel, A. H. (2010) 'Performance Reviews of RFMOs: A Passing Fad?', in D. Russell and D. VanderZwaag (eds) *Strengthening Transboundary Fisheries Management Arrangements*, Dordrecht: Brill.

Hoel, A. H., and Olsen, E. (2012) 'Integrated Oceans Management as a Strategy to Meet Rapid Climate Change: The Norwegian Case', *Ambio*, 41: 81–95.

Holden, M. (1994) *The Common Fisheries Policy*, Oxford: Fishing News Books.

Hønneland, G. (2012) *Making Fishery Agreements Work: Post-Agreement Bargaining in the Barents Sea*, Cheltenham: Edward Elgar.

Ingebritsen, C. (1998) *The Nordic States and European Unity*, Ithaca, NY: Cornell University Press.

Institute of Marine Research (IMR) (2010) *Havforskningsrapporten 2010* [*Ocean Research Report 2010*], online at http://www.imr.no/filarkiv/2010/07/havforskningsrapport2010_web.pdf/nb-no [accessed 20 February 2014].

Kalland, A. (1995) 'Marine Mammals and the Culture of Norwegian Coastal Communities', in A. S. Blix, L. Walløe, and Ø. Ulltang (eds) *Whales, Seals, Fish and Man*, Amsterdam: Elsevier.

Mikalsen, K., and Jentoft, S. (2003) 'Limits to Participation? On the History, Structure, and Reform of Norwegian Fisheries Management', *Marine Policy*, 27(5): 397–407.

Ministry of the Environment (2002) *Rent og rikt hav* [*Protecting the Riches of the Sea*], Stortingsmelding No. 12 (2001–02), online at http://www.regjeringen.no/nb/dep/kld/dok/regpubl/stmeld/20012002/stmeld-nr-12-2001-2002-.html?id=195387 [accessed 5 November 2014].

Ministry of the Environment (2006) *Helhetlig forvaltning av det marine miljø i Barentshavet og havområdene utenfor Lofoten (forvaltningsplan)* [*Integrated Management of the Marine Environment of the Barents Sea and the Sea Areas off the Lofoten Islands (Management Plan)*], Stortingsmelding No. 8, online at http://www.regjeringen.no/nb/dep/kld/dok/regpubl/stmeld/20052006/stmeld-nr-8-2005-2006-.html?id=199809 [accessed 27 March 2014].

Ministry of the Environment (2009) *Helhetlig forvaltning av det marine miljø i Norskehavet (forvaltningsplan)* [*Integrated Management of the Marine Environment of the Norwegian Sea (Management Plan)*], Stortingsmelding No. 37, online at http://www.regjeringen.no/nb/dep/kld/dok/regpubl/stmeld/2008-2009/stmeld-nr-37-2008-2009-.html?id=560159 [accessed 27 March 2014].

Ministry of the Environment (2011) *Oppdatering av forvaltningsplanen for det marine miljø i Barentshavet og havområdene utenfor Lofoten* [*Updated Management Plan for the Barents Sea and the Sea Areas off the Lofoten Islands*], Stortingsmelding No. 10, online at http://www.regjeringen.no/nb/dep/kld/dok/regpubl/stmeld/2010-2011/meld-st-10-2010-2011.html?id=635591 [accessed 27 March 2014].

Ministry of the Environment (2013) *Helhetlig forvaltning av det marine miljø i Nordsjøen og Skagerrak (forvaltningsplan)* [*Integrated Management of the Marine Environment of the North Sea and Skagerrak (Management Plan)*], Stortingsmelding No. 37, online at http://www.regjeringen.no/nb/dep/kld/dok/regpubl/stmeld/2012-2013/meld-st-37-20122013.html?id=724746 [accessed 27 March 2014].

Ministry of Fisheries and Coastal Affairs (1997) *Perspektiver på utvikling av norsk fiskerinæring* [*Perspectives on the Development of Norwegian Fisheries Sector*], Stortingsmelding No. 51, online at https://www.stortinget.no/no/Saker-og-publikasjoner/Stortingsforhandlinger/Lesevisning/?p=1997-98&paid=3&wid=c&psid=DIVL1520 [accessed 5 November 2014].

Ministry of Fisheries and Coastal Affairs (2010a) *Facts about Fisheries and Aquaculture 2010*, Oslo: Ministry of Fisheries and Coastal Affairs.

Ministry of Fisheries and Coastal Affairs (2010b) 'Svært Gode Resultater i Kampen mot Ulovlig Fiske [Very Good Results in Combating IUU Fishing]', Press release, 26 April, online at http://www.regjeringen.no/nb/dep/fkd/pressesenter/pressemeldinger/2010/Svart-gode-resultater-i-kampen-mot-ulovlig-fiske.html?id=601898 [accessed 20 February 2014].

Ministry of Foreign Affairs (2003) *Om lov om Norges territorialfarvann og tilstøtende sone* [*On the Act of Norway's territorial waters and adjacent zone*], Ot. Prp. No. 35 (2002–03), Oslo.

Ministry of Industry and Fisheries (2013) *Fakta om Fiskeri og Havbruk 2013* [*Facts on Fisheries and Aquaculture 2013*], online at http://www.regjeringen.no/upload/FKD/Brosjyrer%20og%20veiledninger/2013/L-0553N_WEB.pdf [accessed 20 February 2014].

Ministry of Petroleum and Energy (2004) *Petroleumsvirksomheten* [*Petroleum Activity*], Stortingsmelding No. 38, online at http://www.regjeringen.no/nb/dep/oed/dok/regpubl/stmeld/20032004/Stmeld-nr-38-2003-2004-.html?id=404848 [accessed 5 November 2014].

Ministry of Trade, Industry and Fisheries (2012–13) *Melding til Stortinget: Fiskeriavtalane Noreg har Inngått med Andre Land for 2013 og Fisket Etter Avtalene i 2011 og 2012* [*The Fisheries Agreements into which Norway Has Entered with Other Countries 2013 and the Fisheries According to the Agreements in 2011 and 2012*], Meld. St. No. 40, online at http://www.regjeringen.no/nb/dep/nfd/dok/regpubl/stmeld/2012-2013/meld-st-40-20122013.html?id=729136 [accessed 20 February 2014].

Munro, G., van Houtte, A., and Willmann, R. (2004) *The Conservation and Management of Shared Fish Stocks*, FAO Fisheries Technical Paper No. 465, Rome: FAO.

Norse, E. (ed.) (1993) *Global Marine Biological Diversity: A Strategy for Building Conservation into Decision Making*, Washington DC: Island Press.

Norwegian Petroleum Directorate (2013) *Fakta 2013 Norsk Petroleumsverksemd* [*Norwegian Petroleum Activity Facts 2013*], online at http://npd.no/Global/Norsk/3-Publikasjoner/Faktahefter/Fakta2013/FAKTA_2013.pdf [accessed 20 February 2014].

Norwegian Seafood Council (2014) 'Strong Growth in Seafood Exports', Press release, 6 January, online at http://en.seafood.no/News-and-media/News-archive/Press-releases/Strong-growth-in-seafood-exports2 [accessed 20 February 2014].

Olsen, J. P. (1983) *Organized Democracy: Political Institutes in a Welfare State: The Case of Norway*, Oslo: Universitetsforlaget.

OSPAR Commission (undated) 'North Sea Conferences', online at http://www.ospar.org/content/content.asp?menu=00590624000000_000000_000000 [accessed 5 November 2014].

Østreng, W. (1982) 'Norway's Law of the Sea Policy in the 1970s', *Ocean Development and International Law*, 11(4): 69–93.

Pikitch, E. K., Santora, C., Babcock, E. A., Bakun, A., Bonfil, R., Conover, D. O., Dayton, P., Doukakis, P., Fluharty, D., Heneman, B., Houde, E. D., Link, J., Livingston, P. A., Mangel, M., McAllister, M. K., Pope, J., and Sainsbury, K. J. (2004) 'Ecosystem-Based Fishery Management', *Science*, 305(5682): 346–7.

Sakshaug, E., Johnsen, G., and Kovacs, K. (eds) (2009) *Ecosystem Barents Sea*, Trondheim: Tapir Academic Press.

Skjærseth, J. B. (2004) 'Introduction', in J. B. Skjærseth (ed.) *International Regimes and Norway's Environmental Policy*, Aldershot: Ashgate.

Statistics Norway (2014) 'Akvakultur, 2013, Foreløpige Tall', online at http://www.ssb.no/jord-skog-jakt-og-fiskeri/statistikker/fiskeoppdrett/aar-forelopige [accessed 20 February 2014].

Stiansen, J. E., and Filin, A. A. (eds) (2007) *Joint PINRO/IMR Report on the State of the Barents Sea Ecosystem in 2006 with Expected Situation and Considerations for Management*, IMR/PINRO Joint Report Series No. 2, Bergen: Institute of Marine Research.

Ulfstein, G. (1999) 'Internasjonal miljøretts stilling i norsk rett [The Position of International Environmental Law under Norwegian Statute]', *Lov og Rett*, 7: 402–18.

Underdal, A. (1980) 'Integrated Marine Policy: What? Why? How?', *Marine Policy*, 4(3): 159–69.

United Nations Commission on the Limits of the Continental Shelf (CLCS) (2009) *Statement by the Chairman of the Commission on the Limits of the Continental Shelf on the Progress of Work in the Commission*, Twenty-third session, 2 March–9 April, online at http://daccess-dds-ny.un.org/doc/UNDOC/GEN/N09/307/58/PDF/N0930758.pdf?OpenElement [accessed 5 November 2014].

United Nations Department of Economic and Social Affairs (DESA) (2002) *Johannesburg Plan of Implementation*, online at http://www.un.org/esa/sustdev/documents/WSSD_POI_PD/English/WSSD_PlanImpl.pdf [accessed 20 February 2014].

United Nations General Assembly (UNGA) (2012) *Resolution 66/288 Adopted by the General Assembly on 27 July 2012: The Future We Want*, UN Doc. A/RES/66/288, online at http://www.uncsd2012.org/thefuturewewant.html [accessed 20 February 2014].

Wegge, N. (2013) 'Politics between Science, Law and Sentiments: Explaining the European Union's Ban on Trade in Seal Products', *Environmental Politics*, 22(2): 255–73.

15

THE PHILIPPINE NATIONAL MARINE POLICY

*Jay L. Batongbacal**

Introduction

The Philippines is a Southeast Asian archipelago comprising more than 7,500 islands bounded by the South China Sea, Sulawesi Sea, and the Pacific Ocean. The total land and inland waters area is estimated at 300,000 km², with a coastline extending 36,289 km. The Philippine exclusive economic zone (EEZ) covers an area of 2.2 million km², much of which overlaps with the EEZs of neighbouring Malaysia and Indonesia to the south, and China to the north. The coastal and marine areas host a wide range of uses, from providing basic subsistence to small fishers to accommodating the wide-ranging infrastructure of most of the country's major cities. Fisheries, tourism, and marine transportation are among the major economic sectors, undertaken in many forms, from small and medium-sized enterprises (SMEs) to large corporate operations. The Philippines also lies within a region considered to be the world's centre of marine biodiversity. Fragile and rich coral, seagrass, and mangrove ecosystems nurture rich and abundant tropical species of aquatic flora and fauna. All of these ecosystems are now suffering the stresses of direct human exploitation and indirect impacts from other coastal and marine resource uses.

Over half of the Filipino people live in coastal cities and municipalities, which have a very high population density of about 285 persons per km². Around 80 per cent lived in poverty in 2000, with fisher households having much lower incomes and a higher poverty incidence than other households in general (DENR et al., 2004). In 2010, the population was estimated at 94 million (NSO, 2010).

The origins of the Philippine National Marine Policy

The Philippines was one of the first countries in the Asia-Pacific to adopt a national marine policy document in 1994 to express the guiding principles for management and development of the country's ocean space, in which it recognized the importance of the coastal and marine environment to its economic development (CCMOA, 1994). The genesis of the National Marine Policy (NMP) may be traced to two activities undertaken in 1994 under then-President Fidel V. Ramos. The impending entry into force of the 1982 United Nations Convention on the Law of the Sea (UNCLOS) prompted the government to consider its impact on the country's national boundaries and maritime interests. President Ramos emphasized that the Philippines needed to refocus on the coastal zone and oceans of the archipelago, as well as to prepare to carry out its obligations under UNCLOS (PRIMEX et al., 1994: Annex F). There was agreement within government to get its 'marine house' in order (PRIMEX et al., 1994: Annex D). In July 1994, President Ramos issued Executive Order No. 186,[1] which reconstituted the former Cabinet Committee on the Law of the Sea that had originally reviewed UNCLOS prior to its ratification, and renamed it the Cabinet Committee on Maritime and Ocean Affairs (CCMOA), under the chairship of the Department of Foreign Affairs (DFA). Ramos also ordered the DFA to conduct interagency consultations on this issue.

These initiatives coincided with the National Marine Policy Research Project conducted by the Center for International and Strategic Studies of the Foreign Service Institute (FSI), the research and training arm of the DFA. The FSI commissioned several working papers and consolidated the results of the interagency consultations, which eventually identified the ocean's potential contributions to national economic development as the primary motivation for establishing an oceans policy (Pangilinan, 1994). After a number of internal discussions and technical committee meetings, the CCMOA presented a confidential NMP paper to President Ramos on 8 November 1994. The Cabinet approved the NMP document, but revised and drastically shortened it for public dissemination. The public version was a small pamphlet, about eight pages long, which described all too briefly the broad strokes of the country's marine policy (CCMOA, 1994).

The Philippine National Marine Policy

The NMP describes itself as a 'developmental and management program' to mobilize the extensive marine resources of the Philippines to achieve economic growth and development, and calls for a model of development that is 'faithful' to the archipelagic nature of the Philippines (CCMOA, 1994). Its contents may be broken down into three major categories of policy statements: official premises; key principles; and policy goals and objectives. These policy statements all revolve around four priority concerns:

- the extent of the Philippines' national territory;
- the protection of the marine environment/ecology;
- the management of the marine economy and technology; and
- maritime security.

Official premises

Certain key premises lay under the four priority concerns that established the programmes of action of the NMP. The first group of premises related to the extent of the national territory. They acknowledged that the international recognition of the territorial limits claimed by the Philippines under the 1898 Treaty of Paris 'remain[ed] an issue', but contended that the extended maritime jurisdictions of the country were already 'well established under municipal and international law', since they were based on customary norms of international law (CCMOA, 1994: 7–8).[2]

The protection of the marine environment/ecology[3] was premised on the proper management of coastal resources and its protection from marine and land-based sources of pollution. This complemented the related priority area on the management of the marine economy and technology, which recognized that coastal resources played an important role in economic development. It assumed that the country could produce competitive products and services in the marine sector by means of management of resources, upgrading of equipment and infrastructure, and development of information technology. This would also enhance the development of coastal communities (CCMOA, 1994: 8–10).

Finally, the NMP notably defined maritime security as 'a state wherein the country's marine assets, maritime practices, territorial integrity and coastal peace and order are protected, conserved, and enhanced' (CCMOA, 1994: 11). The security component reflected a historically rooted concern about the unique interests and vulnerabilities inherent in the Philippine archipelagic domain, and was tied to concerns over the national territory.

Two key observations may be noted from these premises. First, with respect to national territory, the government was definitely reluctant to acknowledge the international community's rejection of the Philippines' original negotiating position during the United Nations Conferences on the Law of the Sea that the Treaty of Paris limits[4] defined the country's national maritime territorial borders. This was owing to the domestic political implications of redrawing the country's national maritime boundaries in accordance with UNCLOS's rules. It was feared that it would be portrayed as an abandonment of territorial space and thus could seriously undermine the position of any administration with less than prodigious political will.

Unfortunately, no administration mustered the political will necessary to deliberately reconfigure the country's convoluted maritime zones into compliance with the Convention. It took a political crisis to spur this process in early 2008, when allegations of corruption surfaced concerning the Joint Marine Seismic Undertaking, a trilateral project between the Philippines, Vietnam, and China to undertake offshore seismic exploration in the waters of the Kalayaan Island Group. It was alleged that then-President Gloria Macapagal-Arroyo had sold out the country's interests in the Kalayaan Island Group in exchange for kickbacks on soft-loans and development assistance from China (Bower, 2010). A politically hostile House of Representatives passed a Bill establishing archipelagic baselines that included the entire Kalayaan Island Group and Bajo de Masinloc (Scarborough Shoal), triggering a contentious public debate among the legislators and the executive branch, which preferred to keep those islands separate.[5] A year later, in March 2009, Republic Act No. 9522 was passed,[6] which modified the technical description of the baselines to conform with the technical rules in Part IV of UNCLOS. The Kalayaan Island Group and Bajo de Masinloc remained outside the archipelagic baselines, although as a political compromise it was expressly placed under a so-called 'regime of islands' described in Article 121 of UNCLOS.

The second insight gleaned is the overriding concern for socio-economic development and the sustainable use of the country's ocean resources for such development. This utilitarian character of the NMP was a sign of the times: in 1994, the country was still recovering from political instability and economic stagnation under the Aquino Administration (1986–92), which replaced the dictatorial Marcos Administration. That same year, the Ramos Administration had just achieved progress in establishing a broad legislative coalition in Congress and succeeded in bringing three fronts of insurgency (Communists, military rebels, and Muslim secessionists) before the bargaining table, enabling the country to achieve sufficient political equilibrium in order for the socio-economic outlook to improve. The idea of ocean management readily assimilated the Ramos Administration's optimism that the nation could finally embark upon socio-economic progress.

Key principles

The NMP contained four key principles as the basis for present and future policies, as follows (CCMOA, 1994: 7–12):

1 Coastal marine areas are considered as the locus of community, ecology, and resources.
2 UNCLOS is to be implemented within the framework of the NMP.

3   Coordination and consultations between the CCMOA and the concerned and affected sectors is necessary.

4   Continuous updating and enhancement of the NMP is needed to make it more responsive to the national interest.

Whether or not these principles directly influenced policy- and decision-making since 1994 is probably the main question that should be asked in determining the effectiveness of the NMP. However, since the Philippines is an archipelago with long-standing maritime sectors, it is quite difficult to determine whether policies and decisions are made directly on the basis of consideration of the NMP, or emerged anyway in light of surrounding circumstances.

Priority goals and objectives

The NMP identified particular goals and objectives for the last three of these priority areas of concern. It is significant to note that no specific goals and objectives had been mentioned for the concern over the national territory.

   Beginning with the component on the protection of the marine ecology/environment, several goals were identified, inlcuding:

- the exploration, development, and management of offshore and oceanic resources based on the principle of sustainable development;
- the development and management of coastal resources within an integrated coastal zone management (ICZM) framework;
- the development and enhancement of national marine consciousness through a comprehensive information programme;
- the encouragement of the development of a marine research programme;
- the adoption of the 'polluter pays' principle to protect the marine environment; and
- the maintenance of high-quality maritime professional schools and similar institutions for training experts in maritime-related issues (CCMOA, 1994: 9).

In the management of the marine economy and technology, the following objectives were noted:

- the promotion of a viable marine fisheries management programme;
- the provision of a continuous and adequate supply of energy;
- the development of technological capabilities in the maritime sector;
- the promotion of investment programmes in marine areas;
- the harnessing of information technology to serve NMP goals;
- the enhancement of regional economic and technical cooperation in marine and ocean affairs; and
- the strengthening of trade policies supportive of maritime issues (CCMOA, 1994: 10).

Finally, maritime security was to be enhanced by means of:

- the promotion and enhancement of maritime security as a key component of national security;
- the provision of a stable and peaceful socio-political and administrative environment that fosters sustained profitability and growth for maritime industries;

- the protection and defence of the integrity of Philippine marine resources;
- ensuring preparedness for and effective response to natural calamities and man–made disasters; and
- the provision of leadership and guidance in the collection, processing, and distribution of strategic information supportive of the NMP (CCMOA, 1994: 11).

Widely varied degrees of compliance were indicated by various activities and achievements in the management of coastal and ocean resources in the Philippines. For example, international recognition was accorded to regulatory requirements under the country's environmental impact assessment (EIA) system and the Local Government Code[7] as being supportive of the first objective, the protection of the marine ecology/environment, in the course of pursuing offshore petroleum development. The first commercially significant petroleum production platform, the Malampaya Deepwater Gas to Power Project located some 50 km off the environmentally sensitive waters of the island province of Palawan, was awarded the World Business Summit Award for Sustainable Development Practices in 2002 by the International Chamber of Commerce (ICC) and the United Nations Environment Programme (UNEP) (ICC, 2004). Coastal resource management was heavily promoted by both the Department of Environment and Natural Resources (DENR) through the Coastal Resource Management Project in the late 1990s (CRMP, undated), and the Department of Agriculture-Bureau of Fisheries and Aquatic Resources (DA-BFAR) through the Fisheries Resource Management Project and the Fisheries for Improved Sustainable Harvest Project. These emphasized local-level and integrated management of fisheries and other coastal resources, and showed that the second objective was being pursued as well. At present, both of these agencies work together on the Philippines' implementation of the Coral Triangle Initiative, an internationally funded regional initiative with six countries aiming to help to manage and protect coral reefs in Southeast Asia (Republic of the Philippines, 2009).

However, objectives such as development and enhancement of the marine consciousness, harnessing information technology, or development of technological capabilities of the marine sector could not be tracked as easily. In other areas, such as promotion of investments in the marine sector, promotion of a viable marine fisheries programme, and protection and defence of Philippine marine resources, the activities were inherently embedded in the archipelagic nature of many of the country's local government units and industries. It is virtually impossible to tell whether such activities were the result of active awareness and concern for the NMP's directives, or simply 'business as usual'.

The Archipelagic Development Framework

Between 1994 and 2001, a convergence of events led to the formulation of a proposed alternative to the NMP. First, in the initial years after the approval of the NMP, the operation of the CCMOA through its Technical Committee encouraged the development of a network of mid-level government personnel and academic institutions that regularly interacted through policy discussions and consultations, and which appreciated the need to revise the NMP. Even after the untimely abolition of the Technical Committee in 2001, this network formed an informal and multidisciplinary core of specialists, who participated in subsequent initiatives to review and revise policy.

Second, the DENR hosted two major coastal management projects that included significant policy-review components. The Coastal Resource Management Project, funded by the US Agency for International Development (USAID), developed a draft National Coastal Resource Management Policy (CRMP, 2004: 38). The Regional Programme on Partnerships in Environmental Management for the Seas of East Asia, funded by the Global Environment

Facility (GEF), the United Nations Development Programme (UNDP), and the International Maritime Organization (IMO), produced the Sustainable Development Strategy for the Seas of East Asia (SDS-SEA) (PEMSEA, 2003). These projects provided a basis for regular discussion of proposals and refinement of programmes of action. These discussions, combined with the pool of mid-level technical personnel and academic counterparts who previously worked with the Technical Committee, provided a continuity and institutional memory with which to reform efforts, even though individuals shifted between projects and offices.

Third, the University of the Philippines established an Archipelagic and Ocean Studies Program in the late 1990s. The programme established a network of academics interested in various aspects of the marine component of the archipelago and linked them to concerned government offices. It sought to assist government in developing policies and programmes to enhance integrated management through an 'archipelagic' perspective (DENR et al., 2004: 6; Batongbacal, 2001b). This allowed more concrete and active linkages between the government and academia, and allowed academic discussion to flesh out the archipelagic perspective and to support government agencies more actively in policy discussions and capacity building.

The convergence of these policymaking discussions and the relatively constant and common core of expertise among the three initiatives provided an opportunity to create a more sophisticated policy framework. In 2002, UNDP granted assistance to the Philippines to develop a national coastal policy framework that would guide the planning and implementation of government programmes for the protection of the environment and sustainable development of natural resources. This activity was undertaken in anticipation of the adoption and implementation of PEMSEA's regional-level policy framework for coordinated and consistent management of coastal and marine resources of the East Asian Seas implemented through national-level policies. This provided the incentive and opportunity for the formulation of the Sustainable Philippine Archipelagic Development Framework (ArcDev) document that was circulated for national consultations in 2004 (DENR et al., 2004).

Rationale

ArcDev emphasized that, as a unity of land and water, the sustainable development of the Philippines must recognize the implications of its predominantly maritime environment and geographically fragmented terrain. It described the current development paradigm pursued by the country as being overwhelmingly land-based, indifferent to the contributions of the oceans, and unable to sustain coastal and marine resource uses. It asserted that the Philippines failed to express its archipelagic identity and consciousness in its laws, institutions, policies, and development programmes. Thus ArcDev was proposed to provide a more balanced, integrative, and comprehensive approach that 'sees the land and ocean as a seamless web of interactions between human beings and nature' (DENR et al., 2004: 7). It sought to establish clear linkages between land-based and ocean-based development, to democratize the opportunities for beneficial use of the oceans, to allow planners to maximize the potential of coastal communities as the focus of development efforts, and to enable those communities to act as responsible stewards of the environment (DENR et al., 2004: 7–8).

Imperatives for sustainable archipelagic development

Five key imperatives form the foundation of ArcDev. The first is the need to establish growth with equity, recognizing that poverty is a key driver of the unsustainable use of coastal and marine resources. Poverty alleviation is therefore a primary objective of the proposed integrated policy framework.

The second imperative is that of sustainable resource management: in order to sustain and renew the resources of the ocean and provide wealth and security to future generations, effective and responsible management of the marine environment is considered essential.

The third imperative is national integration and political unity. The geographic fragmentation of the country exacerbated the linguistic, cultural, religious, and political divisions among the Filipino people. ArcDev addressed this by emphasizing the need to develop maritime transportation systems and air services as a means of linking people, improving the delivery of services and facilitating socio-economic exchanges throughout the archipelago.

The fourth imperative is the need to secure a peaceful and stable external environment. The country's maritime domain is seen as an arena for interstate disputes over resources, sovereignty, and jurisdiction, resulting from threats of piracy, transboundary pollution, maritime transnational crime, and other threats to maritime security. Located at a crossroads to Southeast Asia and the South China Sea, the country's waters are also an important focus of regional and international security concerns.

The fifth imperative of good global citizenship and international cooperation underscores the need to harmonize national policies and international obligations that benefit Filipinos and the international community. In particular, the Philippines needed to ensure effective and responsible implementation of the international marine and environmental agreements such as UNCLOS, the 1992 Convention on Biological Diversity (CBD), and the 1995 Global Programme of Action for the Protection of the Marine Environment from Land-based Sources of Pollution.

Vision and mission

The Archipelagic Development Framework specifically expresses its vision thus:

> A secure and healthy life for all citizens of the archipelago, where peace, harmony and economic prosperity reign, where people equitably benefit from the nation's bounty, where everyone shares responsibility, in the spirit of stewardship and genuine cooperation, for nurturing and sustainably managing the use of our valuable coastal and marine ecosystems and natural resources, and where we, as one nation, cherish and take pride in our people, country, traditional values, and cultural heritage.
>
> *(DENR et al., 2004: 33)*

To achieve this vision, ArcDev declares its overall objective as '[t]o harness and strengthen partnerships among various stakeholders in our marine and coastal ecosystems, and to pursue our national interests consistent with sustainable development goals and responsibility as peaceful global citizens' (DENR et al., 2004: 33).

Principles and programmes

ArcDev enumerates seven key principles to guide policy actions and programmes:

- Promotion of archipelagic integration that recognizes the interaction of land, sea, air, and people in an archipelagic setting
- Meaningful and responsive participation, by all concerned stakeholders, in all phases of the planning, formulation, implementation, monitoring and evaluation, of an effective integrated ocean-management policy

- Application of resource-valuation methods and programmatic environmental impact assessments in decision-making
- Conservation, protection and management of both marine resources and the marine environment for the benefit and use of present and future generations of Filipinos
- Adoption of the precautionary principle and the principle of stewardship in the pursuit of economic activities in the coastal and marine environment
- Implementation of international instruments relevant to the management of the coastal and marine environments in accordance with national laws and policies
- Communication with stakeholders to enhance understanding of the oceans, ocean processes, marine resources and ecosystems, and to strengthen multi-sectoral participation and obtain scientific support for sustainable development.

(DENR et al., 2004: 34)

These core principles were embodied in three strategic programmes, which each identified five to six major objectives and can be summarized as follows:

1   *To establish a Philippine archipelagic ecosystem approach*

   i   Conserve, manage, and protect the integrity of the archipelago's healthy ecosystems in perpetuity
   ii   Develop and enhance degraded areas while preventing further degradation and over-exploitation of the archipelago's ecosystem and resources
   iii Sustain the value of ecosystems and equitably utilize their resources
   iv   Rationalize the management of foreshore areas and small islands, to maintain the integrity of their ecosystems
   v   Institutionalize adaptive ecosystem co-management through appropriate interdisciplinary approaches that utilize science-based and traditional ecological knowledge and wisdom in the pursuit of research, development, and extension
   vi   Enable communities and local governments to integrated complementary activities within an archipelagic framework

2   *To promote sustainable development and shared stewardship of the country's archipelagic heritage*

   i   Encourage and establish suitable industries in coastal and marine areas
   ii   Institutionalize programmes that integrate social development and economic incentives
   iii Improve the quality of life in coastal areas
   iv   Develop and sustain partnerships in environmental investments with a focus on coastal and marine environments
   v   Improve planning processes that will lead to the economic and environmental management of coastal and marine areas

3   *To institutionalize the Sustainable Philippine Archipelagic Development Policy*

   i   Establish a highly developed archipelagic development planning and management framework
   ii   Define and reach consensus on long-term priorities and medium-term strategies and programmes, focusing on the sustainable development of coastal and marine areas and the promotion of the territorial integrity of the archipelago
   iii Harmonize conflicts, address gaps, and rationalize overlaps in law and policy to streamline the legal and policy frameworks for archipelagic concerns

iv   Strengthen law enforcement and regulatory measures to ensure sustained protection of the environment and reduce threats to public order and to the safety of the country's citizens and coastal communities

v   Build and strengthen awareness of maritime issues and organize constituencies for the promotion of sustainable archipelagic development across different sectors and levels of governance

(DENR et al., 2004: 35–45)

Recommended implementation strategies and performance targets accompany each objective. Possible institutional arrangements were also proposed for consideration.

The integrated coastal management policy

Although ArcDev was submitted to consultation in 2004, successive changes in the administration of the DENR appear to have delayed final decision on the proposal. However, ArcDev undoubtedly influenced at least the DENR itself, providing continuity in the development of policy, and allowed further refinement and experimentation. In June 2006, then-President Gloria Macapagal-Arroyo signed Executive Order No. 533,[8] which adopted integrated coastal management (ICM) as a national strategy for the sustainable development of the country's coastal and marine environment. Notably, this strategy is relied upon 'to achieve food security, sustainable livelihood, poverty alleviation, and reduction of vulnerability to natural hazards, while preserving ecological integrity'.[9] This document directly addressed key commitments made under the SDS-SEA, establishing ICM and identifying certain elements and best practices as basic policy guidelines for both national- and local-level management bodies. It also maintained the thread of priorities established in the NMP and ArcDev.

The Order defines ICM as 'a dynamic process of planning and management involving stakeholders, and requiring the analysis of the environmental and socio-economic implications of development, the ecosystem processes, and the interrelationships among land-based and marine-related activities across jurisdictions'.[10] In the implementation of ICM, agencies were required to take into account:

* the establishment of interagency, multisectoral coordinating mechanisms;
* the formulation of long-term coastal strategies and action plans accompanied by programmes of action addressing priority issues and concerns;
* the conduct of public information programmes to increase public awareness, understanding, and responsibility for planning and implementing ICM programmes;
* the mainstreaming ICM programmes into existing planning and socio–economic development programmes;
* the establishment of capacity-building programmes and integrated environmental monitoring; and
* the allocating adequate financial and human resources, and promoting investments and sustainable financing mechanisms for environmental protection and conservation.

Section 4 of the Order also identifies several best practices to achieve ICM, including:

* coastal and marine use zonation as a management tool;
* sustainable fisheries and the conservation of living resources;

- the protection and rehabilitation of coral reefs, mangroves, seagrass, estuaries, and other habitats, particularly through marine protected areas (MPAs), nature reserves, and sanctuaries;
- the development of upland, watershed, catchment areas, and basin-wide management approaches;
- integrated waste management for all major sources of waste;
- integrated management of port safety, health, security, and environmental protection; and
- involvement of the private and business sectors as ICM partners.

The DENR is the lead agency for developing a national ICM programme in consultation with other agencies, sectors, and stakeholders. This includes principles, strategies, development priorities, action plans, national targets, and a national ICM coordinating mechanism. The programme is intended to provide guidance to local government units (LGUs) – that is, the provinces, cities, and municipalities – and local stakeholders in the development and implementation of local ICM programmes.[11] The LGUs were to play a prominent role in the national ICM programme because they were clearly identified as 'frontline agencies' for planning, formulation and implementation.[12] The national ICM programme is embedded within DENR's standing sectoral development plans and programmes (DENR and UNDP, 2007).

Four supporting mechanisms, encompassing ICM education, capacity-building, resource accounting and valuation, and a coastal and marine environment information system, were also mandated for development by the DENR.[13] While no new finances were provided in the Executive Order, all government departments were permitted to allocate part of their budgets to ICM activities. The LGUs may also raise their own revenues and secure independent sources of funding.[14]

The issuance of Executive Order No. 533 marked an important step in the right direction, since it is directed toward one of the key areas of concern of the NMP and ArcDev: the sustainable development and management of coastal areas. It should be noted that the ICM policy allowed the government to focus and coordinate the management in this key component of the archipelago, and particularly to dedicate resources and programmes to the front lines. Under existing Philippine laws, DENR's jurisdiction is primarily land-based; aside from mining activities, it has very limited control over what happens beyond the low-water line. By emphasizing its coordinative and supportive role for LGUs, the ICM policy provides the opportunity for local-level strengthening and enhancement of environmental management on the land from which a significant portion of the myriad coastal and marine management issues and problems arise. The LGUs were the ideal focus for the enhancement of management capabilities. Unlike the DENR, LGUs exercise fisheries management and specific environmental management jurisdictions over municipal waters, which extend 15 km seaward from the shoreline under the Local Government Code of 1991 and the Fisheries Code of 1998.[15] Local jurisdiction compensated for the limits of the DENR's jurisdiction, and allowed coastal and nearshore marine issues to be addressed from a more geographically comprehensive and multi-sectoral approach.

The DENR continues to undertake local capacity-building through projects such as: the Philippine Environmental Governance 2 (EcoGov) Project to strengthen collaboration between the DENR, LGUs, and the Department of the Interior and Local Governments (USAID, 2011); the Southern Mindanao Integrated Coastal Zone Management Project (SMICZMP) to integrate upland and watershed management with coastal and marine habitats and resources (DENR-XI, undated); and multi-country partnerships such as the Coral Triangle Initiative (Coral Triangle Initiative on Coral Reefs, Fisheries and Food Security, undated). On the part of DA-BFAR, the Fisheries for Improved Sustainable Harvest Project (FISH) also devoted significant resources to local capacity-building, information dissemination, and networking (Fish Philippines, undated).

The provisions of the ICM policy are entirely consistent with the contents of ArcDev; it certainly influenced the thinking of the ICM policy planners as one of the key references for policy formulation (Adan, 2011). The ICM policy remains the main pillar of the DENR's oceans-related activities and sets the stage for a potentially significant shift in the Philippine's coastal and marine policies in the years to come.

The Commission on Maritime and Ocean Affairs and the National Coastwatch System

In 2007, concern over whether or not the Philippines would be able to meet the May 2009 deadline for the making of submissions concerning the continental shelf beyond 200 nautical miles before the Commission on the Limits of the Continental Shelf (CLCS) led to the issuance of Executive Order No. 612,[16] which created the Commission on Maritime and Ocean Affairs (CMOA) as an attached agency under the Office of the President. Among its priorities was to mobilize agencies and resources to ensure that the Philippines was able to prepare the corresponding submissions for the so-called 'extended' continental shelf areas. Under the chairship of the Executive Secretary, the CMOA was able to successfully shepherd to completion the Philippines' submission for a continental shelf beyond 200 nautical miles in the Benham Rise Region, on the eastern flank of Luzon facing the Pacific Ocean. The submission was filed with the CLCS a full month ahead of the original deadline, and quickly and finally validated by the CLCS in April 2012.

The flurry of submissions for continental shelf areas just prior to the original 2009 deadline, however, had an unexpected impact of exacerbating the delicate peace established in the South China Sea created by the 2002 Declaration of Conduct of the Parties in the South China Sea. Vietnam and Malaysia made submissions for shelf areas in the South China Sea, which were vehemently protested by China. In the course of making its protests, China officially adopted and publicized the latest version of its so-called 'Nine Dashed Lines' map, and claimed sovereignty and sovereign rights over the islands and adjacent waters, and relevant rights in other waters, of the South China Sea within the area enclosed by the lines.[17] This event marked a turning point in the South China Sea disputes, which saw the rapid deterioration of relations between the principal littoral states in the aftermath.

By 2011, a series of incidents raised anxieties over the security of the Philippines' maritime domains to the west, resulting in the issuance of Executive Order No. 57,[18] establishing the National Coast Watch System (NCWS). The NCWS was to be a 'central inter-agency mechanism for a coordinated and coherent approach on maritime issues and maritime security operations', with a National Coast Watch Council again chaired by the Executive Secretary, but with a National Coast Watch Center under the Philippine Coast Guard as its principal operating arm. The Council was principally charged with providing strategic direction and policy guidelines for maritime security operations, multinational and cross-border cooperation on maritime security, review of maritime security operations, policymaking for managing and security the country's maritime domain, capability planning and funding for maritime security missions, and coordination of maritime agencies.[19] The Order also suddenly abolished the CMOA and transferred all of the latter's policymaking functions to the Council.

The sudden change took most agencies by surprise, including even the Philippine Coast Guard, and the NCWS was still non-functional when a full-blown crisis arose in the summer of 2012. A months-long confrontation between Philippine and Chinese law enforcement vessels took place over fishing in Scarborough Shoal, eventually ending with the Chinese vessels taking control of the Shoal. The incident sundered Philippines–China relations and, the following year, led to the Philippines unilaterally instituting Annex VII arbitration proceedings against China. The case is still pending at time of writing.[20]

Bureaucratic wrangling delayed the implementation of the NCWS, which finally began moving faster in 2013 with the appointment of a full-time undersecretary charged with overseeing its operation and implementation. As of 2014, however, the NCWS is still preoccupied with pending operational concerns, primarily revolving around how to address the challenges posed by higher tensions over the South China Sea disputes. It has just begun the process of embarking on its policy-related functions, which definitely include reconsideration and updating of the NMP.

Meanwhile, continuing fragmentation with respect to aspects of marine policy other than environmental management and maritime security are appearing. In late 2013, Executive Order No. 154[21] created the Philippine Committee against IUU Fishing, and charged it with the implementation of the National Program of Action Against IUU Fishing (NPOA-IUU), submitted to the Food and Agriculture Organization of the United Nations (FAO). The Order clearly omits the existence of the NCWS, the broader composition and wider functions clearly of which include efforts to combat foreign fishing in Philippine waters, even though the NCWS is operationally oriented mainly toward maritime security issues. How these two bodies interact remains to be seen, because both are practically just starting their operations as of time of writing in 2014.

Current institutional arrangements

Rather than codifying marine policy, the NMP only introduced certain broad principles and highlights key objectives to guide policymakers and implementers. It did not alter the institutional arrangements that already existed over Philippine ocean sectors at the time of its issuance. Similarly, neither the proposed Archipelagic Development Framework nor the integrated coastal management strategy has introduced major changes to these institutional arrangements. At this point, the priority appears to be on testing the waters at the policy level and relying on inter-office coordination and cooperation before dealing with proposals for institutional reforms. Current institutional arrangements encompass the executive, legislature, judiciary, LGUs, and even civil society, which, broadly speaking, includes academic institutions, non-governmental organizations (NGOs), and people's organizations.

The president

The president, as chief executor of all state policies, is ultimately responsible for the formulation and administration of Philippine policies, including over the oceans. Much depends on the president's stated priorities and programme emphasis at any given time, and whether or not the leadership genuinely and personally embraces the cause of sustainable development of the oceans is an important and highly influential factor. Thus it was during President Ramos' term when major policy initiatives on the oceans took shape. These initiatives have lost momentum in succeeding administrations; unlike his successors, President Ramos had a special affinity for the marine environment. He was a recreational diver and a former military man acutely aware of the problems posed by the maritime borders of the country. His immediate successors have been concerned with other priorities and have not exhibited similar fondness for the sea.

The president makes decisions with the assistance of a presidential management staff that ensures that all documents necessary for a decision are available. All policy proposals are sent to the affected departments for technical review and policy assessment, the Department of Budget and Management for funding, if any, the Office of the Presidential Legal Counsel for legal review, and the Presidential Legislative Liaison Office for legislative review. The president may also call upon and consult presidential assistants and presidential advisers on issues within their expertise. This highlights the pivotal role of the middle levels of government in the decision-making process. The president must rely mainly upon the subordinate agencies' staff work and recommendations to determine the policy options open for presidential action (Aguirre, 1998: 41–8).

The departments

Almost all departments of the national government exercise primary mandates related to ocean governance, either directly (for example the Philippine Navy) or as a result of ocean-user sectors under their jurisdiction (such as fisheries under the Department of Agriculture). The major ocean governance functions of each relevant department are described briefly in Table 15.1. President Ramos established the CCMOA in 1994 precisely to coordinate the various maritime- and ocean-related functions of these departments.[22] The CCMOA provided all concerned departments with a high-level interagency mechanism to address and resolve overlapping issues and concerns in maritime matters at the Cabinet level. An interagency Technical Committee on Maritime and Ocean Affairs, comprising agency representatives with specialized orientation in maritime and ocean affairs, provided policy- and decision-making support. This mechanism was able to produce consensus among the departments on specific ocean-related issues and decisions. Executive Order No. 137,[23] issued in 1999, strengthened the system through the assignment of specifically defined functions, expanded the role of the Technical Committee, and created a Maritime and Ocean Affairs Center (MOAC) within the DFA as its secretariat. However, in 2001, President Joseph E. Estrada issued Executive Order No. 37,[24] which abolished the CCMOA and the Technical Committee, and transferred their functions to the DFA and MOAC, respectively. This transfer was a serious misstep, and resulted in a regression in the policymaking and decision-making structure for ocean management in the Philippines by creating redundant functions and potentially new jurisdictional turfs in favour of the DFA and MOAC. The resulting lack of progress in the preparation of continental shelf submissions eventually led to the transfer of the functions to the Commission on Maritime and Ocean Affairs (CMOA) under Executive Order No. 612 in 2007. This allowed the preparation of submissions to be spearheaded at the technical level, which eventually succeeded in securing recognition of the continental shelf beyond 200 nautical miles in the Benham Rise Region. The CMOA was, however, short-lived: it was abolished and replaced by the National Coast Watch Council in 2011 under Executive Order No. 57.

The legislature

The bicameral Congress of the Philippines is patterned after the US Congress and consists of a Senate and House of Representatives. The Senate comprises twenty-four senators, each with a six-year term, elected by popular vote. The House comprises 200 members elected from the legislative districts and fifty representatives from registered national, regional, and sectoral parties or organizations, who hold three-year terms.[25]

Various legislative committees are responsible for the enactment of national legislation relating to the coastal and marine areas, including committees on public works, defence and security, agriculture and fisheries, transportation, science and technology, and environment and natural resources. However, since the promulgation of the NMP, no comprehensive piece of

Table 15.1 Major ocean-related functions of key executive departments in the Philippine government

Department	Function
Environment and Natural Resources (DENR)	General environmental management functions
	Environmental protection by means of parks and protected areas
	Regulation of the use of foreshore areas
	Resource mapping and inventory
	Species protection
	Regulation of mining and other resource-extractive industries
	Coastal management
Agriculture (DA)	Fisheries management
	Coastal management
Transportation and Communication (DOTC)	Regulation of shipping
	Regulation of ports
	Regulation of seafarer sector
	Maritime security
National Defense (DND) and Armed Forces of the Philippines (AFP)	Maritime security and law enforcement
Foreign Affairs (DFA)	Foreign policy and relations
	Maritime security
Science and Technology (DOST)	Conduct/support for marine scientific research
	Capacity-building
Interior and Local Government (DILG) and Philippine National Police (PNP)	Supervision of coastal LGUs
	Maritime law enforcement
Energy (DOE)	Regulation of energy resource exploration and exploitation
	Energy development
National Economic Development Authority (NEDA)	Planning and development at national and regional levels
	Oversight over major foreign-assisted projects
National Security Council (NSC)	Implementation and monitoring of comprehensive security and national security policies
	Maritime security
Labor and Employment (DOLE)	Regulation of seafarer sector
Trade and Industry (DTI)	Regulation of businesses and trade
Tourism (DOT)	Regulation of national tourism activities
Justice (DOJ)	Prosecution of offenses
	Resolution of jurisdictional conflicts/issues between government agencies
	Regulation of customs
Public Works and Highways (DPWH)	Regulation of coastal infrastructure
Budget and Management (DBM)	Allocation of funding
Finance (DOF)	Sourcing of finances
Local government units (not a national department)	*Fisheries management*
	Environmental management

legislation on ocean governance has been enacted, although several significant sectoral laws were passed, including the Philippine Fisheries Code of 1998, two new Shipping Acts (the Domestic Shipping Act of 2004 and 2004 Amendments to the Overseas Shipping Act), and the establishment of several protected seascapes under the National Integrated Protected Areas System (NIPAS) Act of 1992.

The judiciary

The judiciary institutionally comprises the Supreme Court, the highest law of the land, and all lower courts. These lower courts comprise the Court of Appeals, a collegial and appellate body, the regional trial courts, which exercise general jurisdiction over most civil and criminal cases, and the metropolitan trial courts and municipal trial courts, which are courts of limited jurisdiction and assigned to the various cities and municipalities.[26]

The judiciary's role in ocean governance is mostly confined to the resolution of litigation (usually criminal cases) that arises as a result of violations of national laws and local ordinances. For example, practically all fisheries violations, such as the use of destructive gear or fishing without a permit, are considered as offences subject to criminal prosecution and punishable by either fine or imprisonment under the Fisheries Code of 1998. Occasionally, the courts may also be called upon to resolve jurisdictional issues between agencies and offices of government if this is an issue in litigation; but normally these questions are dealt with by the Department of Justice upon request of the agencies concerned.

The role of the courts is most relevant to issues of enforcement and implementation of ocean-related legislation. In recent years, however, the High Court has had a few opportunities to rule upon significant legal questions on ocean policy. In *Tano v. Socrates* (1997),[27] the Court upheld the power of cities and municipalities to enact local ordinances for the purposes of local marine environmental protection in accordance with the mandates of the Fisheries Code of 1998 and the Local Government Code of 1991. In the Concerned Citizens v. *MMDA* (2008),[28] the Supreme Court ordered not only the DENR to fully implement a plan for the rehabilitation, restoration, and conservation of Manila Bay, but also ten other national government agencies to carry out their respective legal mandates to prevent the pollution of the bay and to rehabilitate its fisheries.

The local government units

The LGUs are the frontline of coastal management, since they are the most basic units of government directly in contact with the problems and issues that the Philippines faces. Out of 1,541 LGUs, there are about 832 coastal municipalities and 25 coastal cities (DENR et al., 2000). In addition to their land territory, cities and municipalities exercise management jurisdiction over municipal waters extending 15 km from the shoreline for purposes of regulating municipal fisheries (mainly fishing using boats of 3 gross tons or less, and other forms of small-scale nearshore fisheries) under the Fisheries Code of 1998, as well as for residual environmental management purposes under the Local Government Code of 1991. The Local Government Code seeks to maximize local autonomy and provides LGUs with basic environmental management and service functions, in addition to being a focal point for the coordination and delivery of national government services. The combined jurisdictional scope of LGUs over both their land territories and adjacent waters, where the most intensive coastal and marine resource uses occur, places the greater burden of ocean governance upon them.

To compensate for the lack of capacity, many LGUs have opened their governance to cooperation and collaboration with NGOs, people's organizations, the private sector, and development agencies. From the 1980s to the 1990s, there was a plethora of coastal resource management projects across the country, encouraging sustainable coastal development and management of local coastlines. Many started out as pilot projects or small community-based projects seeking to address coastal poverty issues. Coastal LGUs now look to community-based coastal resource management projects as a tool for promoting sustainable livelihoods and local development.

Civil society

Since the late 1980s, civil society groups comprising academia, NGOs, and people's organizations have taken prominent roles in all aspects of governance, including coastal and marine areas. Academic institutions such as the University of the Philippines and Silliman University have become very active, through a range of local research projects with socio-economic and policy components, in influencing local-level management of coastal and marine resources, as well as national-level policy formulation and monitoring. A large number of NGOs and people's organizations also operate at the local level, implementing coastal and marine resource management projects in collaboration with the LGUs where the projects are located. The NGOs have exhibited their strengths in community organization, mobilization, and participatory planning and implementation processes. In many cases, they serve as vital links between the local government and the grass roots to ensure a degree of effectiveness and success. Civil society popularized community-based coastal resource management (CBCRM) as a framework of choice for local ocean governance.

The apparent success of CBCRM was the reason why it was adopted as the official strategy for the DENR-based Coastal Resource Management Program, and the DA-based Fisheries Resource Management Program in the early 2000s. These are the largest coastal resource management projects undertaken in the Philippines to date. They took different approaches to CBCRM: the former concentrated on capacity-building and technical support for local governments; the latter focused on community-organizing through NGOs and people's organizations, and the provision of technical support and livelihood assistance programmes. In both cases, the participation of civil society, in many varied forms, was firmly embedded in all efforts to improve local coastal management. Civil society complements the national and local governments best by providing needed capacity and assistance to LGUs and local communities in carrying out management initiatives.

Continuing challenges

Despite the issuance of the NMP in 1994, to date ocean management continues to be largely fragmented and uncoordinated at the national level. While there have been some improvements in the openness to new concepts and in governmental attitudes, which have allowed improvements in local-level governance and marine environmental protection, overall, integrated ocean management across the different maritime agencies continues to be problematic.

Inappropriate assignment of lead functions at the national levels

Generally, at the national level, three major departments have very clear and direct influence over the management of coastal and marine resources: the DA, through the Bureau of Fisheries and Aquatic Resources (DA-BFAR): the DENR, through component agencies such as the Environmental Management Bureau and the Protected Areas and Wildlife Bureau; and the DOTC, through the Maritime Industry Authority, the Philippine Ports Authority, and the

Philippine Coast Guard. To a lesser extent, other departments also exercise specific functions with respect to other marine resources or marine-related activities, such as the DOE, with respect to offshore energy, and the DOST, with regard to marine scientific research.

However, none of these key national institutions has taken a lead role. No single office or agency currently exercises direct management functions over any aspect of Philippine coastal and marine resources. Technically, at present, it should be the National Coast Watch Council under Executive Order No. 57, but the Council has yet to be fully operational in the broad task of ocean management. Further, the establishment of the Philippine Committee on IUU Fishing demonstrates that there has yet to be full appreciation of the need for an integrated approach.

Previous experiences also evidence the absence of institutional leadership at the national level. One example is the DFA, which was charged with implementing the NMP under Executive Order No. 37. Despite an impressive list of mandates under the same Order, the DFA suffers from an inherent problem: its purported ocean-related functions duplicate or overlap with the legislated primary functions of all other departments, even though it inherently does not have direct management functions and capabilities at the domestic level. Not only does every other department have its own ocean governance mandates, but each also has more dedicated resources and personnel to devote to the task. This stems from the fact that the original functions of CCMOA and the Technical Committee were actually functions meant to be exercised by a collegial body of government departments. As such, they were meant as functions of an integrated interdepartmental mechanism that could jointly exercise all of the powers of the various departments, particularly directed toward addressing issues and problems that straddled departmental mandates, and could benefit from the pooling of departmental powers and resources.

In this light, the wholesale transfer of CCMOA functions to the DFA under Executive Order No. 37 was clearly inappropriate and prone to institutional failure because of the inevitable jurisdictional conflicts and unsuitable functional and legal relationships that emerged. Further, the Philippines are an archipelago with extensive domestic maritime interests. Ocean governance is much more a domestic concern involving practical implementation of national laws and policies than an issue of observing international commitments. The DFA, with its principally international focus, is not the proper government department to manage the enormous task of ocean governance. It did not have a particularly prominent and direct domestic governance mandate or functions, unlike other offices, whether at the national level (such as the DENR and DILG), or at the local level through LGUs.

In 1994, leadership of the CCMOA was originally assigned by the president to the DFA because, at the time of its formation, the main concern was to ensure that international obligations under UNCLOS were anticipated and met. More than a decade later, most foreign policy dimensions remain unresolved; in some ways, they have become more difficult (for example the maritime disputes in the South China Sea) and constitute only one aspect of numerous challenges to Philippine ocean governance. The majority of the challenges are domestic in nature and, as the NMP and ArcDev rightly imply, are development-oriented: coastal management, maritime industrial sector management, maritime law enforcement, and the like. These are mainly domestic policy issues for which departments other than the DFA have the primary jurisdiction, direct mandates, appropriate personnel, and relevant resources.

Fragmentation of jurisdiction

There remains considerable sectoral fragmentation in the governing regimes of the various ocean-related resource sectors, resulting in jurisdictional competition among several agencies. Multisectoral bodies – such as the Fisheries and Aquatic Resource Management Councils for coastal fisheries,[29] and the protected areas management boards (PAMBs) for protected landscapes and seascapes[30] – have been favoured by law in recognition of the existence of overlaps and the need for coordination among different interested agencies and groups. Ad hoc interagency meetings are also often resorted to by government agencies as a means of exchanging information and, in some cases, to resolve programme implementation issues that involve their respective offices. These multisectoral bodies have provided good opportunities for greater intersectoral interaction and public participation. However, the underlying legal and jurisdictional framework remains unchanged; thus the impact that these multisectoral bodies have on actual decisions and policies depends greatly on the personalities involved. This makes the matter of harmonization and coordination through these bodies highly subject to the personal leadership skills and inclinations of their members at any given time.

Meanwhile, at the local level, cities and municipalities have broad management responsibilities that include their adjacent coastal waters and the coastal environment. While these LGUs could theoretically undertake ecosystem-based management by means of cooperative and coordinative arrangements that are allowed and encouraged by law, they are hobbled by the limited reach of their marine jurisdiction and the lack of a unifying and widely accepted legal framework for doing so in concert. There is a further problem with the fact that many municipal waters have still not been delineated and delimited even at present.[31] Joint management programmes need to be supported by contractual arrangements between the LGUs. These have been experimented with for purposes of bay-wide management in the fisheries sector,[32] but have not been applied to larger environmental issues such as siltation and watershed management.

Jurisdictional competition also occurs across geographic levels. For example, a seascape protected under the NIPAS Act of 1992 is under the DENR's supervision through the Protected Areas and Wildlife Bureau (PAWB). However, such areas may include municipal waters that are subject to intensive fishing activities and under LGU jurisdiction according to the Fisheries Code. While NIPAS allows the LGU to participate in the management of the protected area through membership in the area's multisectoral PAMB, the implied subordination of the LGU to the Board is not necessarily welcomed by locals. Further, the remaining jurisdiction of the LGU governs all other management issues outside the protected area (such as business activities) is not within the Board's official scope.

Very slow progress in legislative harmonization and integration

The continuing sectoral fragmentation of coastal and marine policy is the outcome of the government's inability to periodically and fairly review, evaluate, update, and modify previous policies to respond to the needs of the times. The lack of efforts is clearly seen in the number of antiquated laws that are still in force today. These include: the Public Land Act,[33] which has been the governing law over public lands, including coastal lands, since 1936; shipping laws that date back to portions of the Spanish Code of Commerce of 1888; the National Defense Act of 1935, promulgated at the birth of the Republic, which still governs national defence and security;[34] and various environmental laws that date back to the 1970s.[35]

One factor that exacerbates the policy confusion is the manner of amendment of laws undertaken by the legislature. Most often, laws that attempt to modify previous legislation are not worded specifically – that is, the legislature often resorts to implied or indirect amendments through the use of a standard repealing clause that simply states that all prior laws and issuances concerning the same subject matter are 'amended or modified accordingly', without specifying what they are. This makes the issue of whether or not a provision of law has been amended a matter of constant argument and debate. These arguments can be settled definitively only by litigation – and this only creates more confusion and conflict as contending parties seek to interpret the laws according to their own interests.

The need for an 'honest broker'

Generally speaking, Philippine ocean governance requires, but lacks, a coordinating mechanism in the form of an 'honest broker' – that is, an intermediary who can assist in settling jurisdictional issues without having its own self-serving interest in the outcome. Such a mechanism could impartially balance competing interests and help those interests to arrive at a common consensus on how particular management issues could be resolved, without having any stake in the outcome other than the desire to see the contending parties work together effectively. This honest broker is especially needed considering the intensive multiple uses of Philippine waters and the multiplicity of offices involved.

The original CCMOA, assisted by the Technical Committee on Maritime and Ocean Affairs, may be seen as the first real attempt to establish such an entity for ocean governance. As a collegial and interagency body, it was an appropriate high-level coordinating mechanism. All relevant agencies were represented equally. Therefore it could be the proper venue in which to conduct the necessary dialogue and to achieve the needed compromises among contending points of view between agencies. The CCMOA was able to exercise this function when it drafted and approved the NMP in 1994.

But it soon became apparent that the implementation of UNCLOS was not a simple task in ocean management. The full implications of effective implementation reached down to the domestic level. The CCMOA was initially hampered by the limitation of its mandate to UNCLOS. Executive Order No. 132 of 1999[36] strengthened the CCMOA and its Technical Committee and Secretariat. The Order provided them with new mandates designed explicitly to state the CCMOA's powers and focus as it related to maritime and ocean affairs at both national and international levels. In accordance with its renewed and expanded mandate, the CCMOA was able to approve an eleven-point Priority Work Program. The programme broadly defined an agenda for action by the interagency body and allowed the subdivision of the Technical Committee into various subcommittees that could focus on each item in the programme. The upgrading of the Secretariat into the MOAC would have provided additional support to CCMOA and the Technical Committee's work.

Unfortunately, in 2001, despite the already existing subcommittees and continuous activity of the Technical Committee, both the CCMOA and the Technical Committee were unceremoniously abolished. Their respective functions were transferred to the DFA and MOAC without prior consideration and consultation with either collegial body or their members. Only the DFA, through MOAC, provided any input on this executive action. The consolidation of powers into the DFA and MOAC effectively derailed all previous efforts at building up a high-level and impartial institutional mechanism for interagency cooperation. It created an even worse situation: not only did the mandates and functions now overlap and become redundant, but also the mandates and functions were granted to a department with no inherent and direct responsibility for ocean management at the implementation level. The original functions of the CCMOA and the Technical Committee under the 1999 Executive Order were designed for a collegiate body composed of the member departments. The CCMOA was based on the assumption that its collective mandate was founded on a 'pooling' of the individual mandates of the constituent departments and that implementation of collegial decisions would be undertaken by the member department directly concerned. This highlighted the mainly coordinative and facilitative function of both the CCMOA and its Technical Committee. The new Executive Order changed the status of the DFA from honest broker to competitor with duplicate ocean policy- and decision-making functions, but without actual capacity to implement them. Lacking the needed impartiality and detachment, and without either capacity or capability, it was difficult for the DFA to elicit enthusiastic commitment and substantial cooperation from other offices in exercising its ocean management functions. It was therefore not surprising that the eleven-point Priority Work Program was eventually not implemented. After all, ocean management is fundamentally domestic in nature and thus not central to the DFA's foreign relations functions. The task of ocean management remains firmly and properly within the realm of the other concerned departments.

With the promulgation of Executive Order No. 612, the abolition of MOAC and the transfer of functions to a Cabinet commission chaired by the Executive Secretary should have resolved most of these issues. Technically, the Executive Secretary is in a better position to be an honest broker capable of mediating between competing domestic and international departments and concerns. But whether or not this role can be realized ultimately depends on leadership and vision on the part of the Executive Secretary to ensure that maritime issues do not become flashpoints of jurisdictional competition among the different government agencies.

Lack of knowledge and capacity

Lack of basic knowledge and awareness about applicable policies for coastal and marine issues is the greatest obstacle to effective management. With so many laws and regulations possibly applicable to any particular ocean use, the possibility of conflicting uses and regulations is rather high. Wading through the applicable laws and regulations is a difficult and time-consuming process. For the most part, knowledge is compartmentalized among the different departments and offices of government, and extremely challenging for the ordinary citizen to comprehend. It is not unusual for government agencies to be fully knowledgeable about their respective mandates and functions, but not be overly concerned with other agencies that have overlapping or conflicting mandates and functions. While mechanisms for interagency cooperation do exist, and have been required by law in some cases, the task of properly interpreting the legal relationships and impacts of the different laws and rules is often a challenge.

The offices most urgently in need of knowledge and capacity are the LGUs. They are mandated to implement all national laws, including those on the environment, within their respective territories. At the same time, they must exercise local autonomy and local legislative and regulatory functions. How the LGUs are to undertake this task while remaining within their legal mandates and not transgressing into national jurisdiction has not been clarified by the law in general, especially for ocean management purposes. The devolution of powers and functions to LGUs in 1991 was not accompanied by clear instructions, especially in terms of the precise delineation of overlapping or similar functions and jurisdictions between national and local governments. The LGUs – especially their local executives and legislative councils – have not undergone extensive training on any aspect of coastal or marine management.

The general population, on the other hand, also requires extensive information on relevant laws, rules, and regulations. The ordinary citizen cannot be expected to know the massive amounts of legislation, executive issuances, rules, and regulations that prevail over any given ocean-related activity. Policies are expressed in legal instruments that are not easily accessible to the ordinary citizen. It is possible for entire communities to be unaware of the details of legislation that are published in only a few major newspapers, or local ordinances required by the Local Government Code that are posted only on bulletin boards.

Socio-economic conditions in the coastal areas linked to environmental conditions

Perhaps the greatest problem faced by Philippine ocean policy is the social and economic conditions along the coastal areas, which are inextricably linked with the condition of the coastal and marine environment. The commercial and municipal fisheries sectors compete within municipal waters for dwindling fisheries resources. Most fisheries in the Philippines are already fully exploited or overexploited, and this is caused and worsened by conditions in many of the poor coastal communities. Poorer segments of the population resort to destructive fishing methods, such as dynamite fishing, while commercial fishers turn to more efficient gear to provide more catch for less effort. Coral reefs are suffering from both destructive fishing and overfishing. They are also adversely affected by changes in the marine environment resulting from global warming. Mangrove areas have been degraded by conversion into fishponds and used for fuel-wood and other forest products (DENR et al., 2004: 16).

Detrimental changes in coastal habitats are also caused by coastal development, pollution, siltation, coastal mining, and quarrying activities. As coastal communities or their members seek to improve their lot by engaging in economic activities, the lack of a clear overarching framework for development allows the unsustainable uses and exploitation of available resources.

Power relations and vested interests

As borne out by years of experience and experimentation with coastal resource management, politics and personalities have a pervasive influence over management. Despite the similarity of situations, it is still possible for different sites to produce different results in terms of the seriousness of the problems and the effectiveness of solutions. A detailed social analysis encompassing the interactions between the major national government agencies, LGUs, political and economic interests, and personalities in key positions is necessary to properly understand and resolve management problems and issues at a particular site. Politics, vested interests, and personalities are 'wild cards' in coastal and marine management in the Philippines. Policy and institutional relationships and issues cannot be properly understood without an equally penetrating analysis into the political personalities, vested economic interests, and/or personal qualities of the government officials or employees involved. In the face of these factors, laws and bureaucracies lose much of their efficacy, since they undermine neither the substantive content nor the procedural efficiency of management rules and decisions, but rather the actual implementation of such rules and decisions.

Published case studies outline failures in coastal and marine management rooted in individuals and groups operating within coastal areas outside the bounds of the legal framework or ignoring management requirements (White, 1989; Arquiza, 1993; Gamalinda, 1993; Peñaranda, 1996). These are often driven by single-minded determination to maximize perceived gains and profits from the use of coastal and marine resources. In many cases, these individuals or groups rely on political patronage, intimidation, and political or economic positions to defend against local grievances, as well as directives of the management authority. Local elites who are politically connected can choose to act against or outside of the management framework, and are often able to do so with the acquiescence, or even assistance, of the responsible government officials.

In a developing country such as the Philippines, degradation of the coastal and marine environment may be linked directly to the uneven distribution of wealth and opportunities derived from coastal and marine resources. Any management initiative must include in its plan of action consideration of the social and political context within which the resource-use activities take place. This makes the task of coastal and marine management not only a matter of environmental engineering and legislative and administrative reforms, but also an issue of social reform. Not only does it entail reforms in government and administration, but it also requires a change in the individual attitudes and perceptions of the ordinary citizenry about the marine environment and its resources, and how they are used in society and by whom.

Key policy gaps

Several coastal and marine activities are not covered by adequate policy-based guidance. Some of these are vitally important, such as the management of the exclusive economic zone (EEZ). Only the rudiments of an EEZ management regime have been developed by either DA-BFAR, which certainly has an interest in doing so for offshore fisheries management purposes, or the DENR, which has plenary powers over environmental protection. The absence of an integrative and holistic EEZ management regime results in a lack of basic monitoring and surveillance systems for the EEZ.

Another major policy gap is with respect to tourism. In spite of the Philippines' reliance on tourism as an income-generating activity, there is a distinct absence of specific policy guidance on tourism in coastal and marine areas. These areas are among the most popular attractions of the country for the foreign market. Providing policy guidance for coastal and marine tourism is particularly problematic, since each city and municipality technically has the mandate and ability to engage its own tourism programmes without coordination and cooperation with neighbouring LGUs. Although there has long been a Department of Tourism (DOT), there was no defined national tourism law or policy until late 2009, when the Tourism Act was finally passed. But even under this new law, it is the provincial governments that are largely responsible for local tourism regulation, development, and coordination.

Coastal land use has also emerged as a major policy gap, particularly with respect to the narrow shore-land and foreshore areas that attract high-impact developments such as hotels and infrastructure, or residential housing and human settlements. Shorelines in the Philippines are changing drastically owing to the influx of either tourist or residential settlements into the narrow strips of public land along the coastline. Although the provinces, cities, and municipalities have the most important duties and responsibilities for legal regulation of this area, the concurrent jurisdiction of national agencies results in very loose control over coastal land use. This is manifested in a host of problems such as uncontrolled or haphazard shoreline development, unregulated coastal settlements, a loss of foreshore areas, non-compliance with shoreline setback requirements, and unregulated extraction of shore-based minerals (DENR et al., 2001: 25–33).

A key problem here is the absence of adequate monitoring criteria and specific mechanisms and modes of implementation as a result of the overly broad statement of the NMP's goals and objectives. This has not been helped by the current NMP implementation mechanism and the inherent weakness of the institution that is tasked with oversight functions. There has not been a systematic means of assessing policies and their impacts, much less a unified and coherent way of improving and reforming the coastal and marine management regime.

Other policy gaps include:

- sea lane management for both domestic and international vessels, with a view to protecting the marine environment and national security;
- near-shore seabed mining and quarrying, particularly of beaches and sandbars;
- thermal pollution in areas in which power plants and energy-intensive industries operate;

- regulation of recreational boating activities, especially in relation to their environmental impacts;
- land-based sources of pollution in coastal and marine waters;
- mitigation of coastal erosion, which is relevant to management of the foreshore and shore-lands;
- military uses of coastal and marine areas; and
- coastal population and infrastructure management.

The list will continue to lengthen in light of the extensive enumeration of management issues that have been identified in many international environmental instruments. Unfortunately, as happens in many other countries throughout history, a crisis is often required before national attention turns to addressing these policy gaps.

Good practices

Despite the daunting challenges, in the years since the promulgation of the NMP, several notable developments in the fields of coastal and marine management indicate that the Philippines still retains strong institutional building blocks for sustainable development of the oceans. The Archipelagic Development Framework (DENR et al., 2004: 26) identifies:

- a strong civil society;
- the emergence of socially and environmental conscious business groups;
- the increasing involvement of various sectors in environmental programmes;
- the existence of interagency mechanisms and multi-sectoral councils concerned with sustainable development;
- the mobilization of networks of NGOs and people's organizations;
- the use of innovative partnership models; and
- an enabling environment for community empowerment.

Most of the major encouraging developments are related to community-based coastal resource management (CBCRM), which is undertaken in many communities nationwide. This process engages communities in the stewardship of their ecosystems, together with capacity-building in local governments to develop good environmental governance. It is based on inclusive and participatory processes of research, planning, formulation, and implementation of interventions, often with the active involvement of a multi-sectoral group of citizens groups, organizations, and government offices. Current CBCRM programmes have turned to integrated, multi-sectoral, and participatory management approaches at the local level. The CBCRM approach itself has been adopted as a key coastal management strategy by government. Since the 1990s, both the DENR and DA-BFAR, through major projects funded by different development agencies, have respectively centred on building local technical resources and management capacities and providing assistance and training to LGUs and local community groups in pilot areas throughout the country.

Some socially and environmentally conscious business groups have also emerged to engage in monitoring environmental issues, as well as to advocate government reforms. For instance, one of the main mass media organizations in the country formed the *Bantay Kalikasan* (Environmental Guard), which makes use of television and radio to call the attention of government to environmental issues and impacts on citizens. Private financial institutions such as HSBC have begun to support environmental programmes and initiatives, including coastal environmental programmes.

The use of partnerships has been especially developed and refined by the IMO/UNDP/ GEF-funded Partnerships in Environmental Management for the Seas of East Asia (PEMSEA), based in the DENR. PEMSEA has successfully experimented with collaboration with local businesses in the management of marine pollution issues at its integrated coastal management (ICM) pilot site in Batangas Bay. Inclusive, multi-stakeholder partnerships, characterized by participation at all vertical and horizontal levels, bound by a clear and shared vision, and complemented by simultaneous bottom–up and top–down incentives, are a key ingredient to the success of the PEMSEA model of ICM (Kullenberg et al., 2006: 53–5). While most CBCRM projects initiated by NGOs often emphasize the participation of marginalized sectors in coastal communities, PEMSEA's ICM experience also highlighted the possible role and contributions of partnerships between the better-off private sector (such as locally based companies) and national and local government agencies. PEMSEA has also provided the Philippines with the groundwork and network to work towards a regional international management framework for ocean management, which may be seen as movement toward partnership in a larger context.

The inability of the legislature to pass harmonizing legislation has not deterred national and local agencies from seeking to cooperate. Joint memoranda and multi-sectoral consultative groups and meetings have been popular mechanisms for various offices seeking some acceptable level of coordination and harmonization. As an example, in 1995, law enforcement agencies such as the Philippine National Police, Coast Guard, Customs, National Bureau of Investigation, prosecution offices, DENR, and DA-BFAR entered into a memorandum of agreement on the terms of cooperation and coordination in fisheries law enforcement. In 2000, the DENR and DA issued a joint memorandum order that clarified some of their respective authorities and jurisdictions over the management of coastal and marine resources, particularly in light of the Fisheries Code of 1998.[37]

The penchant for multi-sectoral cooperation is not limited to ad hoc initiatives, but is actually embedded in some major legislation. The Local Government Code mandates the creation of local development councils, which are tasked with providing the overall guidance for socio-economic planning and development of LGUs. The Fisheries Code of 1998 mandates the creation of fisheries and aquatic resource management councils, which provide advice and recommendations to the local legislature and executive regarding fisheries. Protected areas management boards (PAMBs) are their counterpart in the management of protected areas. Some localities have advanced even further. For example, the Province of Bohol passed its own provincial Environmental Code and convened a provincial Coastal Law Enforcement Summit that resulted in the creation of coastal law enforcement councils. Cebu City and the Province of Negros created their own law enforcement alliances for the specific purpose of addressing problems in coastal management (DENR et al., 2004: 28).

Several NGOs and people's organizations have consistently taken an active part in CBCRM activities over the years. These organizations range from local counterparts of international NGOs to community cooperatives to academic institutions. These groups cooperate with government offices on a host of issues, such as law enforcement, information exchange, and local policy implementation. They are credited with increasing the public's awareness and interest in coastal and marine resource management issues. Acting through councils such as those mentioned, they are able to engage and mobilize local community members in multi-sectoral partnerships for coastal resource management (DENR et al., 2004: 28).

Conclusion

Philippine ocean policy has moved in fits and starts since the 1990s, relying on the approval of very general guidance documents, which inform policy- and decision-makers not necessarily etched in legislation. UNCLOS had an undeniable impact, having provided the opportunity for the creation of the NMP. The ArcDev that followed was similarly designed, although much longer and more detailed than the NMP. ArcDev, in turn, was motivated partly by the need to respond to a proposed regional sustainable management framework. Integrated ocean management appears to be much easier to express (although harder to achieve) through such guidance documents for the executive agencies, rather than the more contentious legislative action. Legislation has tended to be enacted either as a response to crisis or to be more sectorally oriented, with increased chances of not following the 'integrative' mode.

Integrated ocean management has found strongest justification in the economic sense – that is, rather than being based on political, organizational, or scientific considerations – which reinforces the link between sustainable development and integrated management. Its greatest challenge appears to be finding the appropriate balance and optimal formula for cooperation between national agencies and local government, because these two levels are the most vital, yet also the most prone to competition for authority to manage and decide coastal uses.

Left alone, executive agencies seem vulnerable to successive changes in modus operandi, and thus find it difficult to sustain an integrative framework and mode of policy- and decision-making over the long term. Long-term planning seems more effective when undertaken within a sectoral, rather than multi-sectoral or integrative, perspective.

As shown in the section on 'Continuing challenges', Philippine oceans policy still needs as lot of work, and will continue to evolve; given past experience, the next developments will probably focus on limited sectoral concerns again. Legislation may not be the proper tool with which to encourage such evolution, because it seems to take crises and controversy before the legislature can act. Thus the executive bears the burden of promoting improvements and refinements in order to address the continuing challenges of integrated ocean management.

Notes

* Mr Batongbacal acknowledges the assistance of the Pierre Elliot Trudeau Foundation, based in Montreal, Quebec, Canada, that enabled him to participate in The Ocean Policy Summit (TOPS) 2005 in Lisbon, Portugal. He acknowledges the comments and contributions of Mr Robert Jara of the Manila Bay Environmental Management Project and the Coastal Marine and Management Office of the Department of Environment and Natural Resources, Quezon City, Philippines.
 1 Executive Order No. 186, s. 1994, Expanding the Coverage of the Cabinet Committee on the Law of the Sea and Renaming it as the Cabinet Committee on Maritime and Ocean Affairs, 12 July 1994.
 2 Republic Act No. 3046 (1961), as amended by Republic Act No. 5446 (1968), established the existing baseline system of the Philippines prior to the conclusion of the negotiations for UNCLOS. The Acts used the straight baseline method in connecting the outermost points of the archipelago into a single unit, rather than the straight archipelagic baseline method contained in UNCLOS, Pt IV. Instead of using a standard 12 nautical mile limit extending from these baselines, the Philippines claimed a territorial sea of variable width extending from the baselines to the limits described in the Treaty of Peace between the United States of America and the Kingdom of Spain, signed in Paris on 10 December 1898 (the 1898 Treaty of Paris), which is shaped like an irregular rectangle, and the Convention between the United States of America and Great Britain Delimiting the Boundary between the Philippine Archipelago and the State of North Borneo, signed in Washington on 2 January 1930 (the 1930 US–UK Convention), which separated the islands of North Borneo from the southwestern Philippines. Subsequent legislation complicated this maritime zone configuration. Presidential Decree No. 1596 (1971) appended a polygonal zone of absolute sovereignty to the western side of the Treaty of Paris limits to encompass the Kalayaan Island Group, and Presidential Decree No. 1599 (1971) declared the country's EEZ to extend from the straight baselines and outward to 200 nautical miles. The EEZ declaration was apparently based on rules in the Convention, but is not consistent with either the Treaty of Paris limits or Presidential Decree No. 1596: see Lotilla (1995); Batongbacal (2001a).

Only recently, Republic Act No. 9522 (2009) amended the technical descriptions of the baselines to make them conform to the standards mentioned in UNCLOS, Pt IV.

3 The NMP uses the terms 'marine ecology' and 'marine environment' interchangeably. See CCMOA (1994: 7–8).

4 Although often referred to as the 'Treaty of Paris limits', the boundaries referred to are actually the conglomeration of various limits described in: the 1898 Treaty of Paris; the Treaty between the Kingdom of Spain and the United States of America for Cession of Outlying Islands of the Philippines, signed in Washington on 7 November 1900; and the 1930 US–UK Convention. Since 1978, official Philippine maps have also appended the boundaries of the Kalayaan Island Group described in Presidential Decree No. 1596.

5 See, e.g., *Philippine Daily Inquirer* (2008a; 2008b)

6 An Act to Amend Certain Provisions of Republic Act No. 3046, as amended by Republic Act No. 5446, to Define the Archipelagic Baseline of the Philippines and for Other Purposes, Republic Act No. 9522, 14th Congress, 2nd Session, 10 March 2009.

7 Local Government Code of 1991, Republic Act No. 7160, 5th Congress (Philippines), 5th Session, 10 October 1991.

8 Executive Order No. 533, s. 2006, Adopting Integrated Coastal Management (ICM) as a National Strategy to Ensure Sustainable Development of the Country's Coastal and Marine Environment.

9 Ibid., s. 1.

10 Ibid., Thirteenth Whereas Clause.

11 Ibid., s. 3.

12 Ibid., s. 6.

13 Ibid., s. 8.

14 Ibid., s. 11.

15 Philippine Fisheries Code of 1998, Republic Act No. 8550, 10th Congress (Philippines), 5th Session, 17 February 1998.

16 Executive Order No. 612, s. 2007, Reorganizing the Department of Foreign Affairs–Maritime and Ocean Affairs Center into the Commission on Maritime and Ocean Affairs under the Office of the President.

17 See the record of diplomatic communications on both the Vietnamese and Malaysian submissions online at http://www.un.org/depts/los/clcs_new/commission_submissions.htm [accessed 5 November 2014].

18 Executive Order No. 57, s. 2011, Establishing a National Coast Watch System, Providing for its Structure and Defining the Roles and Responsibilities of Member Agencies in Providing Coordinated Inter-Agency Maritime Security Operation and for Other Purposes.

19 Ibid., s. 3.

20 See the record of the proceedings of the Permanent Court of Arbitration (PCA) online at http://www.pca-cpa.org/showpage.asp?pag_id=1529 [accessed 5 November 2014].

21 Executive Order No. 154, s. 2013, Adopting a National Plan of Action to Prevent, Deter, and Eliminate Illegal, Unreported, and Unregulated Fishing, and for Other Purposes.

22 Executive Order No. 186, s. 1994.

23 Executive Order No. 137, s. 1999, Declaring the Month of July of Every Year as National Disaster Consciousness Month and Institutionalizing the Civil Defense Deputization Program.

24 Executive Order No. 37, s. 2001, Abolishing the Cabinet Committee on Maritime and Ocean Affairs, 24 September 2001.

25 1987 Constitution, Art. VI.

26 Ibid., Art. VIII.

27 *Tano v. Socrates*, Supreme Court of the Philippines, G.R. No. 110249, 21 August 1997.

28 *Metro Manila Development Authority (MMDA), et al. v. Concerned Citizens of Manila Bay, et al. (MMDA)*, Supreme Court of the Philippines, G.R. No. 171947–48, 18 December 2008.

29 Philippine Fisheries Code of 1998, ss. 68–79.

30 NIPAS Act of 1992, s. 11.

31 Although the DENR issued Department Administrative Order No. 17 in 2001, providing for the delineation and delimitation of municipal waters of all cities and municipalities, it was rescinded in 2004. A new Administrative Order No. 1 was issued by the Department of Agriculture in 2004, providing for the delineation and delimitation of municipal waters of only cities and municipalities without offshore islands. All municipalities that had already been delineated and delimited as of 2004, however,

were allowed to stand. This means that several cities and municipalities that have islands are now unfairly disadvantaged for being unable to define their jurisdictional boundaries. For a more detailed account, see Batongbacal (2006).

32 DA-BFAR's Fisheries Resource Management Project, for example, focuses on several key bays and gulfs that are subject to intensive fishing pressures. It has promoted the use of multi-sectoral and inter-governmental bay management councils by means of voluntary agreements between the littoral LGUs. See FRMP (2004).

33 Commonwealth Act No. 141 of 1936 and an Act to Amend Compile the Laws Relative to Lands of the Public Domain.

34 Commonwealth Act No. 1 of 1935, An Act to Provide for the National Defense of the Philippines, Penalizing Certain Violations Thereof, Appropriating Funds Therefor, and for Other Purposes.

35 For example, Presidential Decree No. 1586, the Environmental Impact Assessment System Law, was issued in 1978; Presidential Decree No. 1151, the Environment Code, was passed in 1977; Presidential Decree No. 1067, the Water Code, was issued in 1976; and Presidential Decree No. 979, the Marine Pollution Decree, was passed in 1976. A perusal of Philippine environmental laws will reveal that many of these laws are decades old and only a few of recent vintage.

36 Executive Order No. 132, s. 1999, Strengthening the Cabinet Committee on Maritime and Ocean Affairs and its Supporting Mechanisms, Establishing its Technical Committee, and for Other Purposes, 30 July 1999.

37 Joint Memorandum Order No. 1 of 2000, Delineating/Defining the Areas of Cooperation and Collaboration between the Department of Agriculture and the Department of Environment and Natural Resources in the Implementation of Republic Act No. 8550 (1998 Philippine Fisheries Code).

References

Adan, W. (2011) *The National Integrated Coastal Management Program: A Proposal for Compliance with the Mandate of Executive Order No. 533*, Special Report No. 00.21.07, Quezon City: Integrated Coastal Resources Management Project.

Aguirre, A. (1998) 'The Role of the Office of the President in Ocean Policy Making', *Ocean Law and Policy Series*, 2(1): 41–8.

Arquiza, Y. (1993) 'Trouble in Tubbataha', in E. Gamalind and S. Coronel (eds) *Saving the Earth: The Philippine Experience*, Makati City: Philippine Center for Investigative Journalism.

Batongbacal, J. (2001b) 'Archipelagic Studies: Charting New Waters', in E. Mann Borgese, A. Chircop, and M. McConnell (eds) *Ocean Yearbook 15*, Chicago, IL: University of Chicago Press.

Batongbacal, J. (2001a) 'The Maritime Territories and Jurisdictions of the Philippines and the United Nations Convention on the Law of the Sea', *Philippine Law Journal*, 76(2): 123–68.

Batongbacal, J. (2006) 'Delineation and Delimitation of Sub-National Maritime Boundaries: Insights from the Philippines', in A. Chircop, S. Coffen-Smout, and M. McConnell (eds) *Ocean Yearbook 20*, Chicago, IL: University of Chicago Press.

Bower, E. (2010) *The JMSU: A Tale of Bilateralism and Secrecy in the South China Sea,* online at http://csis.org/publication/jmsu-tale-bilateralism-and-secrecy-south-china-sea [accessed 27 April 2014].

Cabinet Committee on Maritime and Ocean Affairs (CCMOA) (1994) *The National Marine Policy*, Manila: Department of Foreign Affairs.

Coastal Environmental Program Management Office (DENR-XI) (undated) 'Southern Mindanao Integrated Coastal Zone Management Project', online at http://oneocean.org/denr-xi/smiczmp.html [accessed 28 April 2014].

Coastal Resource Management Project (CRMP) (undated) 'The Philippine Experience in Coastal Resource Management', online at http://www.oneocean.org/about_crmp/where_to.html [accessed 8 January 2005].

Coastal Resource Management Project (CRMP) (2004) *Completion Report: The Coastal Resource Management Project – Philippines, 1996–2004*, Cebu City: Department of Environment and Natural Resources.

Coral Triangle Initiative on Coral Reefs, Fisheries and Food Security (undated) 'About Us', online at http://www.cti.bmb.gov.ph/national-cti-coordinating-committee.html [accessed 28 April 2014].

DENR-XI *See* Coastal Environmental Program Management Office

Department of Environment and Natural Resources (DENR) and United Nations Development Programme (UNDP) (2007) *Sectoral Framework Plans: Coastal and Marine Resources Management, Vol. IV*, Quezon City: DENR.

Department of Environment and Natural Resources (DENR), Department of Agriculture (DA), Department of Interior and Local Government (DILG), Coastal Resource Management Project (CRMP), and United States Agency for International Development (USAID) (2000) *Philippine Coastal Management Guidebook Series, Vol. 1: Coastal Management Orientation and Overview*, Quezon City: DENR.

Department of Environment and Natural Resources (DENR), Department of Agriculture (DA), Department of Interior and Local Government (DILG), Coastal Resource Management Project (CRTMP), and United States Agency for International Development (USAID) (2001) *Management of Habitats and Protected Areas*, Philippine Coastal Management Guidebook Series No. 5, Cebu City: CRMP.

Department of Environment and Natural Resources (DENR), United Nations Development Programme (UNDP), and Marine Environment and Resources Foundation (MERF) (2004) *ArcDev: A Framework for Sustainable Philippine Archipelagic Development – Revaluing Our Maritime Heritage and Affirming the Unity of the Land and Sea*, Quezon City: DENR.

Fisheries Resource Management Project (FRMP) (2004) 'FRMP', online at http://www.frmp.org (no longer available)

Fish Philippines (undated) 'Fisheries Improved for Sustainable Harvest Project', online at http://www.oneocean.org/fish/the_project.html [accessed 28 April 2014].

Gamalinda, E. (1993) 'The Silent Revolution', in E. Gamalinda and S. Coronel (eds) *Saving the Earth: The Philippine Experience*, Makati City: Philippine Center for Investigative Journalism.

International Chamber of Commerce (ICC) (2004) 'Malampaya Deepwater Gas to Power Project', online at http://www.iccwbo.org/home/environment_and_energy/sdcharter/corp_init/icc-unep/main-pages/excellent/malampaya/long.asp [accessed 24 November 2004] (no longer available).

Kullenberg, G., Habito, C., and Lowry, K. (2006) *Performance Evaluation: Building Partnerships in Environmental Management for the Seas of East Asia (PEMSEA) – Terminal Evaluation Report*, Project No. RAS/98/G33/A/1G/19, Quezon City: PEMSEA.

Lotilla, R. P. M. (1995) *The Philippine National Territory: A Collection of Related Documents*, Quezon City: University of the Philippines Institute of International Legal Studies.

National Statistics Office (NSO) (2010) *Philippines in Figures*, Manila: National Statistics Office.

Pacific Rim Management Exponents, Inc. (PRIMEX), Bureau of Fisheries and Aquatic Resources (BFAR), Canadian International Development Agency (CIDA), and Oceans Institute of Canada (OIC) (1994) *Workshop Report on the Policy Analysis Workshop for Integrated Ocean Planning and Management Strategies and their Implementation for Philippine Fisheries (May 30 – June 2, 1994)*, Pasig City: PRIMEX.

Pangilinan, A. K. (1994) 'Introduction', in *National Marine Policy Research Project Working Papers*, Unpublished interagency collection, Foreign Service Institute, Manila.

Partnerships in Environmental Management for the Seas of East Asia (PEMSEA) (2003) *Sustainable Development Strategy for the Seas of East Asia: Regional Implementation of the World Summit on Sustainable Development Requirements for the Coasts and Oceans*, Quezon City: PEMSEA.

Peñaranda, V. (1996) 'The Betrayal of Maqueda Bay', in S. Coronel (ed.) *Patrimony: Six Case Studies on Local Politics and the Environment in the Philippines*, Pasig City: Philippine Center for Investigative Journalism.

Philippine Daily Inquirer (2008a) 'De Venecia Seeks Immediate OK of Bill Defining National Territory', 11 March.

Philippine Daily Inquirer (2008b) 'RP Baselines Bill Unacceptable to Int'l Community: Exec', 15 March.

Republic of the Philippines (2009) *National Plan of Action for the Coral Triangle Initiative on Coral Reefs, Fisheries and Food Security*, Quezon City: DENR Parks and Wildlife Bureau.

US Agency for International Development (USAID) (2011) *Environmental Governance Phase 2 Project Evaluation: Final Evaluation Report*, online at http://pdf.usaid.gov/pdf_docs/Pdacr988.pdf [accessed 28 April 2014].

White, A. (1989) 'Two Community-Based Marine Reserves: Lessons for Coastal Management', in C. Thia-Eng and D. Pauly (eds) *Coastal Area Management in Southeast Asia: Policies, Management Strategies, and Case Studies*, Kuala Lumpur: Ministry of Science, Technology, and Environment/International Center for Living Aquatic Resources Management.

16

NATIONAL MARINE POLICY

A Vietnamese case study

Nguyen Chu Hoi and Hoang Ngoc Giao[1]

Overview of Vietnam's national marine policy

Basic information

Vietnam has a total sea area approximately three times larger than its land area, which totals 331,700 km^2. Vietnam's marine area includes 2,773 nearshore islands, covering 1,636 km^2 of the seafloor, and two offshore archipelagos, the Spratly and Paracels. Among these islands, there are eighty-two islands larger than 1 km^2 and twenty-three larger than 10 km^2. Three have an area of 100 km^2 or more. The mainland is surrounded by the Bien Dong Sea (the South China Sea), which faces two gulfs: the Gulf of Tonkin to the north, and the Gulf of Thailand to the south. The average depth of Bien Dong Sea is 1,140 metres and its volume is approximately 3.928 × 10^6 km^3.

Vietnam's seas are rich in biodiversity, with some of 11,000 species of marine animals, plants, and seabirds recorded to date. Among them are 2,038 species of fish, with more than 110 species of commercial value. In Vietnam's seas, there are about 1,122 km^2 of coral reefs that are located in coastal waters bordering inshore and offshore islands, as well as along the coast of central Vietnam. Vietnam's key coastal marine ecosystems provide returns of some US$60–80 million per hectare per year (ADB, 1999), while income from fish and lumber from mangroves alone has been estimated at about $6,000 per hectare per year. Marine fish stocks are estimated at about 5.3 million tons, not including marine shrimp and squid, with an allowable fishing limit of some 2.3 million tons (Hoi, 2009). Coastal waters provide 80 per cent of the country's total fishery catch; in 2012, about 2.4 million tons of fish caught from coastal seas contributed over $6.1 billion to gross domestic product (GDP) exports (Hoi, 2013).

Oil and gas are important natural resources found on the continental shelf, with reserves totalling 10 billion tons of oil and 300 billion cubic metres of gas. In 2007, over 20 million tons of oil and more than 7 billion metric tons of gas were produced (Hoi, 2009); in 2010, about 15 million tons of oil and more than 9 billion metric tons of gas were produced (Hoi, 2013).

At present, 80 per cent of the country's tourists visit the coastal areas. There were about 3 million tourists in 2005 and some 10 million in 2010, and coastal tourism increases of some 20 per cent annually (Hoi, 2007; 2013). Vietnam also has great potential to develop sea ports along its long coastline. Vietnam plans to develop its seagoing fleets during the twenty-first century.

The political situation during the Indochina wars divided the country into two parts. Before 1975, the government of the Republic of Vietnam in the south of the country maintained a continuous presence on the Spratly and Paracel archipelagos. At that time, South Vietnam began talks with neighbouring countries on delimitation of offshore regions, including the continental shelf in the Gulf of Thailand and marine areas in the southeast. In the north of Vietnam, the Democratic Republic of Vietnam had confirmed its sovereignty over the sea. In a speech made to North Vietnam's Navy in 1961 (Vietnamese Navy and CCPE, 2007: 40), President Ho Chi Minh declared: 'In the past, we just had darkness and forests, now we have open sky and ocean. Our coasts, seas and islands are long, large and beautiful. We have to protect and preserve them.'

After the end of the war in 1975, Vietnam immediately joined the Third United Nations Conference on the Law of the Sea (UNCLOS III). In 1977, following the spirit and the letter of the UNCLOS III draft resolution, Vietnam declared sovereignty over its sea territories: a contiguous zone; a 200 nautical mile exclusive economic zone (EEZ); the continental shelf; and the Spratly and Paracel archipelagos in the Bien Dong Sea. Vietnam signed the United Nations Convention on the Law of the Sea (UNCLOS) in 1982. In that same year, the Vietnamese government also made a statement declaring its intention to use the system of straight baselines to measure the width of its sea territory, as provided for under the UNCLOS. Since 1982, Vietnam has had an EEZ of about 1,000,000 km², three times larger than its land area, and with a ratio of 100 km² of mainland per 1 km of coastline (Maritime Index = 0.01).

Vietnam is not the only country adjacent to the Bien Dong Sea. Neighbouring countries are China, the Philippines, Indonesia, Brunei Darussalam, Singapore, Malaysia, Thailand and Cambodia. As a result of geographical location and provisions in UNCLOS, these countries have overlapping maritime claims. These territories need to be delimited through dispute settlement measures provided for in UNCLOS. Together with its neighbours, Vietnam has made some progress in settling disputes of sovereignty over sea areas and in establishing rules of behaviour in the Bien Dong Sea, the Gulf of Tonkin, and the Gulf of Thailand.

Administratively, the government of Vietnam is divided into two levels: the central government and local governments. The local governments consist of authorities on three levels: province, district, and commune. In Vietnam's sixty-four provinces, twenty-eight coastal provinces or central cities and 145 coastal districts (including fourteen island districts) are situated along more than 3,260 km of coastline (not including island coastlines). The coastal area includes over half of Vietnam's major cities. As of 2013, more than 50 per cent of Vietnam's population of about 90,000,000 lives in coastal provinces.

An overview of national marine policy

The Vietnamese people have lived along the coasts and on the islands for centuries. They also reclaimed coastal areas for agriculture and aquaculture development. Fishing was conducted almost entirely in coastal and marine waters. Only after the end of the second Indochina war in 1975 was the government of Vietnam able to take all necessary political and legal measures to affirm its sovereignty over its sea territories and islands.

In 1986, Vietnam adopted an open-door policy and moved towards a market-oriented economy. During this time of change, the sea territories were recognized as one of the most important factors in Vietnam's economic development and national security. However, lack of management experience created many problems for Vietnam's seas and in its maritime activities. In recent times, the country has had to tackle the loss of biodiversity, coastal pollution, the degradation of marine ecosystems, the destruction of coastal habitats, overfishing, oil spills, and other coastal disasters, including climate change and sea-level rise, as well as conflicts regarding the multiple use of coastal resources.

Although Vietnam is a maritime nation, the formulation of its marine policy started only in the late 1990s. The awareness of all levels of society about the role of marine policy for sustainable marine development continues to be limited. The primary concern of Vietnam has been issues of national security, national sovereignty, and rights over maritime areas, rather than issues of environmental protection, resource conservation, and sustainable development in general.

The government's policy of moving towards a market economy has meant that it is paying more attention to Vietnam's sea areas to spur economic growth. There was a governmental socio-economic plan that supposedly included major coastal and island cities, port systems, and coastal and marine economic corridors, with an open-door mechanism to solicit investment from international donors (Thang, 2011). During the 1990s, there was much marine development, but without any coordination or integration. Major seaborne sectors, including fishery, marine transport, oil and gas exploration, and tourism, have been booming without any integrated coordination by the government. The notion of a truly national maritime policy was novel to Vietnam's government.

More recently, the government has become aware of the importance of coordinating and integrating all maritime and coastal activities into a national marine policy. At the same time, there has been more understanding of the importance of sustainable development, scientific research, environmental protection, biodiversity conservation, marine protected area (MPA) management, and the conservation and management of marine habitats and coastal wetlands. Other marine policy focus points are oil-spill emergency response plans, sustainable oil and gas development, sustainable fisheries, and safe and clean maritime transport. Currently, Vietnam is endeavouring to formulate an overall national marine policy. The integrated and inter-sector management approach is gaining increased attention from marine policymakers.

Although Vietnam has not yet developed an integrated national maritime policy, it does have a series of laws on sea-related activities in various sectors.

The primary laws related to sovereignty and protection of sovereign rights are the Statement on the Territorial Sea, the Contiguous Zone, the Exclusive Economic Zone and the Continental Shelf of 12 May 1977; the Statement on Spratly and Paracel Archipelagos in the Bien Dong Sea of 1977; and the Statement of 12 November 1982 of the Government of the Socialist Republic of Viet Nam on the Territorial Sea Baseline of Viet Nam. The Ordinance of 1997 on National Boundary Guards, including Coast Guards, the Law of 2003 on National Boundaries, and the Law of 2004 on National Security (in force 2005) include key principles defining marine boundaries and the responsibilities of public agencies in marine governance. These laws also deal with the definition of the contiguous zone, EEZ, and continental shelf, as well as passage rights in Vietnam's territorial sea, including foreign ships or boats.

In the marine fisheries and oil and gas sectors, Vietnam's government has enacted the Law of 2003 on Fisheries, which replaced the Ordinance of 1989 on Aquatic Living Resources Protection, and the Law of 1993 on Oil and Gas, which was amended in 2000. Decrees are issued to enact and enforce these laws. Decree No. 27/2005/ND-CP of 2005 contains detailed regulations and guidelines for enacting the Law of Fisheries. Decree No. 48/2000/ND-CP of 2000 contains guidelines on enacting the Law on Oil and Gas. These laws provide the legal framework for developing two key marine economic sectors for Vietnam. They also encourage governmental agencies, non-governmental organizations (NGOs), and internal and overseas donors to invest in capital and technology to promote fisheries and oil and gas activities in Vietnam, by providing protective ownership of capital, property, and other rights from the Vietnamese government. The Law on Fisheries has a wide scope, including provisions on fishing and aquaculture, post-harvest and transport, resource marketing, import–export of marine products, and fisheries investigation and surveys, as well as marine protected areas (MPAs) and integrated fisheries management and planning.

In order to regulate the navigation and tourism sectors, the government enacted the Code of Navigation of 1990, as amended in 2005, and the Ordinance of 1999 on Tourism. The Ordinance states that tourism is an important economic service, and focuses on coastal tourism and ecotourism. It links tourism with coastal and marine conservation. The Code of Navigation was prepared with reference to international conventions such as the 1973 International Convention for the Prevention of Pollution from Ships, as amended by the Protocol of 1978 (MARPOL), and provides the basis for fleet modernization. The Tourism Ordinance also supports marine economic sector development outside of tourism and refers to integration of tourism development with other marine sectors into the global economy in the near future.

In comparison with these economic sectors, there are few legal documents dealing with marine scientific research. In 1991, the government approved Decree No. 24/HDBT on Regulating Scientific Activities of Foreign Organizations and Ships in the Marine Waters of Vietnam. However, this decree does not mention domestic marine scientific research.

In 1994, Vietnam enacted the Law of 1994 on Environmental Protection, as amended in 2004. In 1998, the government issued Order No. 36-CT/TW on Strengthening Environmental Protection in the Period of Industrialisation and Modernisation. This order states that environmental protection is the responsibility of authorities at all levels, in all sectors, and of all citizens. The Criminal Code of 2003 also contains ten general crimes against the environment. The Law of 2009 on Biodiversity (in force since July 2009) includes regulations for biodiversity conservation planning and strategy development, biodiversity corridors, and establishing and managing a natural protected areas system with three subsystems for special use forests, inland waters (wetlands), and MPAs.

In 2009, the government approved a decree on integrated management of resources and environmental protection of seas and islands for the first time: Decree No. 25/2009/ND-CP of 2009 (in force since May 2009). The decree contains five chapters, with thirty articles focusing on matters including: principles and components; island/sea-use planning; plans and programmes of integrated management of resources and environment protection of seas and islands; basic surveys management of seas and islands; sea and island use management; prevention and control of marine pollution and on islands; response and improvement of coastal, island, and marine natural hazards and environmental incidents; and coastal environment protection and implementation measures.

Apart from the abovementioned laws, the Law of 2012 on the Seas of Vietnam, enacted in June 2012 by the National Assembly, is considered a basic law of the seas in Vietnam. It contains seven chapters and fifty-five articles affirming sovereignty, sovereign rights, and jurisdiction rights of Vietnam for the territorial sea, contiguous zone, EEZ, and continental shelf, as well as for the Spratly and Paracel archipelagos in the Bien Dong Sea. The law also regulates the activities in Vietnam's seas, including the development of marine economics and the governance, management, and protection of seas, islands, and coastal areas. For example, Chapter 1, article 7, of the law defines the basic principles of marine governance and management as being that:

1   the government unites governance and management for seas and islands;
2   the ministries, line sectors, coastal provinces, and central cities promote coastal and marine governance and management within the framework of its function, authority, and mission, as decided by the government; and
3   Vietnam maintains the importance of addressing the conflicting claims in Bien Dong Sea through peaceful solutions and negotiation.

This means that, in Vietnam's institutional system of coastal and marine governance, there is an existing integrated state management system for seas and islands, including coastal areas from the central level (such as the Vietnam Administration of Seas and Island, or VASI) and provincial agencies of seas and islands management (twenty-eight coastal provinces), as well as sectoral state management system for seas and islands such as the Ministries of Agriculture and Rural Development, Transportation, etc. Integrated state management will not take the place of sectoral management, but will link, coordinate, and regulate the behaviour of sectoral development relating to sea and island use and exploitation.

Currently, VASI, within the Ministry of Natural Resources and Environment, is preparing a Law of Marine Resources and Environment, which focuses on coastal, island, and sea-use management and governance in a cross-sectoral and coordinated manner, and applies coastal and marine spatial planning (CMSP).

The major institutions with responsibilities relevant to marine management are considered next.

Policymaking

The Central Committee of the Communist Party of Vietnam (CPV) drafts policies on ocean and coastal issues in the form of resolutions of the Plenum or instructions. The National Assembly of Vietnam annually considers and approves five-year socio-economic development plans. The Inter-ministerial Commission on Bien Dong Sea and Islands operates under the leadership of the deputy prime minister. The Provincial People's Committee is the local authority that is mandated to make policy for a province.

Law-making

The National Assembly of Vietnam issues laws, ordinances, and resolutions.

Planning and policy development

The primary ministries with responsibilities for marine planning and policy development are as follows:

- the Ministry of Planning and Investment (MPI), with a National Office of Sustainable Development;
- the Ministry of Agriculture and Rural Development (MARD), with a National Steering Committee on Marine Protected Areas (covering sustainable fisheries, living resources protection, and MPA management);
- the Ministry of Natural Resources and Environment (MONRE), with the Vietnam Administration of Seas and Islands (VASI) (covering coastal wetlands, marine resources, and environmental management, coastal and marine biodiversity, integrated coastal management, marine comprehensive surveys, and management of coastal, marine, and island use, national marine database development, and management)
- the Ministry of Foreign Affairs, with a Commission of National Frontier with Marine Department and Marine Policy Research Board;
- the Ministry of Sciences and Technology (MOST), with a five-year national programme of marine sciences and technology; and
- provincial people's committees, with line provincial departments, among which is a Sub-department of Seas and Islands under the Provincial Department of Natural Resources and Environment.

The CPV is the highest political institution, and is in charge of making strategic decisions and macro-policies in support of Vietnam's development. The National Assembly is not only a law-making body, but also approves socio-economic development strategies for Vietnam. The government, as an executive body, plays an important role in formulating and implementing Vietnam's marine policies. The Inter-ministerial Commission on Bien Dong Sea and Islands reports directly to the prime minister.

The policy development process

Before 1975

Pre-1874, prior to French colonization
The marine concerns of the Vietnamese people have appeared in stories and historical books on the Vietnamese dynasties controlling the state. The Van Don deep-sea port was situated on an offshore island of Quang Ninh province in Northern Vietnam, and trade and fishing vessels often came to offshore islands. Although government writings include references to the sea as early as 1874, the first official decree on ocean issues was the Law of 1926 on Fishing for Colony Nations. This law, which first appeared as a policy in 1888, was the first step towards the national and international laws that are in place today.

1874–1954, during French colonization
The Decree on Expanding Application of Fishing Laws for Colony Nations 1888 was issued in 1926. Other decrees issued during this time include the Decree of 1936 on the Ministry of Colonies on the Fisheries and Marine Waters of Indochina and the Decree of 1948 on Maritime Zones.

1954–76, during Vietnam's division into North and South
During this period, the country's major concern was national security and protection of sovereignty on islands and sea territories. The Geneva Agreement divided the country into two parts, with two separate marine territories delimited by a line perpendicular to the coastline. In 1957, Vietnam received Bach Long Vi Island in the Gulf of Tonkin from China after a period of cooperation and support. Before 1974, the government of the Republic of Vietnam in the south of Vietnam maintained a continuous presence on the Spratly and Paracel islands, on which Vietnam's feudal constituencies have been present since the seventeenth century. It also started talks with neighbouring countries on delimitation of marine regions and the continental shelf in the Gulf of Thailand. President Ho Chi Minh's 1961 speech to the Vietnamese people (see 'Basic information') expressed the will of Vietnamese people to have sovereignty over its seas.

1975–82

The government of Vietnam has asserted sovereignty and sovereign rights over its maritime areas, basing these assertions on the international treaties and practice, especially UNCLOS. After the union of Vietnam, the CPV assumed its position as the highest political institution, and as the only national policymaking body on social and economic aspects of life in Vietnam. The CPV's marine policy comprises the following.

- On 14 February 1976, a resolution agreed to at the Fourth Party Conference instructed the government 'to make production planning for all areas in prairies, highland and *marine areas; to develop fisheries as an important industrial sector; to develop a sea transportation fleet, to build and upgrade a sea port system*' (CPV, 1976, emphasis added).
- In its Statement on the Territorial Sea, the Contiguous Zone, the Exclusive Economic Zone and the Continental Shelf of 12 May 1977, Vietnam declared a 12 nautical mile territorial sea, a contiguous zone, a 200 nautical mile EEZ, and jurisdiction over the continental shelf. Article 1 of Vietnam's Constitution of 1980, amended 1992 and 2013, provides that 'the sovereignty and jurisdiction of Vietnam included: land, air space, *marine areas and islands*' (emphasis added).
- In its Statement of 12 November 1982 on the Territorial Sea Baseline of Viet Nam, the government of Vietnam declared the baseline system measuring its sea territories. Also in 1982, Vietnam, together with more than 100 other states, signed UNCLOS. Vietnam ratified the Convention on 25 July 1994.

1982–96

The Sixth Party Conference emphasized that 'Vietnam has land, marine areas which are suitable for developing fisheries, transportation . . . [W]e need to combine the development of economy and national defense, and marine economic development is defined as a leading sector of Vietnam' (CPV, 1986). The Sixth Plenum of the Party, dated 30 November 1987, 'reaffirmed the need for sovereignty protection on Spratly and Paracel Islands and to intensify the presence of Vietnam in the South China Sea and Spratly Islands' (CPV, 1987).

The Seventh Party Conference in 1991 approved the general direction of marine economic development, sovereignty, and national interest protection of Vietnam to 2000: 'Step by step to comprehensively exploit the potential of the sea, develop island economics, control territorial seas and the continental shelf, to exercise the national rights in the EEZ of Vietnam' (CPV, 1991).

On 6 May 1993, the Party issued Resolution No. 3–NQ/TW relating to marine economic development objectives and needs for the years to come. The Resolution emphasized that:

> Location and geographic characteristics of Vietnam in relation to the complicated situation of the region are good conditions for development on one hand, on the other hand, [they] require us to strengthen the marine economic development at the same time as enhancing the capability of sovereignty and marine interests, environmental protection, while striving for a marine economic state.

On 5 August 1993, the prime minister issued Order No. 330 to implement Resolution No. 3, with the following goals:

- comprehensive marine economic exploitation and island development;
- expansion of the sovereign rights of Vietnam on the EEZ and continental shelf; and
- development of the marine cultivation, fishing, and food-processing industries.

1996–present

The Eighth Party Conference defined the objectives of marine economic development as being:

To maximize the marine potentialities and *the advantages of developing marine and coastal zones in combination with national defence and security*; to build up *marine economic centres*, major cities, industrial areas, tourism, trading centres and port systems; to develop marine economic corridors along the coast.

(CPV, 1996, emphasis added)

However, Order No. 20-CT/TW of 22 September 1997 noted that, 'because of *the lack of comprehensive strategy for marine development* and the linkage of zones, sectors, national defence and security and socio-economics, science and technology, Vietnam is not in fact engaged in marine economic development' (CPV, 1997, emphasis added). Thus marine policy in Vietnam can be characterized as follows.

- The marine economic development policy is combined with strengthening national defence and security.
- The government has started to map marine economic areas – both the natural marine and coastal zones, and the economic centres, such as major cities, ports, and marine economic corridors.
- The CPV has acknowledged the lack of a comprehensive strategy, a lack of linkage among sectors and zones, and a lack of attention to social issues.

The government has approved the Strategy for Socio-Economic Development for 2010–2020 (CPV, 2011), a general policy paper on development in Vietnam. The National Strategy for Marine Economic Development towards 2020 (MPI, 2004a: 25) comprises the following elements: marine and island economic development; enhancing basic surveys for marine economic planning; intensifying farming and food processing; oil and gas exploration and exploitation; developing tourism and shipbuilding; marine protection; comprehensive and integrated economic development; maximizing the advantages of seaports and estuaries; promoting the development of sea areas; installing service facilities on islands for offshore development; and enhancing the marine economy, in combination with marine security protection. The Strategy defines a comprehensive survey as a scientific basis for zoning and marine economic planning. Marine protection is affirmed as a strategic requirement. The Strategy mainly concentrates on marine economic development; however, the government has not yet defined critical elements to ensure the integration of the multi-sector nature of the sea. Analysis of the Strategy also reveals that the government has not yet defined requirements to ensure the sustainable development of the marine resources and environment, and to deal with social issues on the mainland and islands.

In May 1997, the Development Strategy Institute of the Ministry of Planning and Investment submitted its *Report of the Comprehensive Plan for Marine and Island Economic Development to 2010* (MPI, 1997), with the following strategic targets:

- The development of marine economics, with a focus on sectors such as oil and gas, fisheries, salt farming, marine transportation, and tourism;
- the combination of marine economic development with national security and defence concerns; and
- the mapping of four economic sub-regions along the coasts (Mong Cai–Deo Ngang, Deo Ngang–Mui Ne, Mui Dinh–Ca Mau, Ca Mau–Ha Tien) and offshore areas outside of the territorial limit.

The Comprehensive Plan identified some policy initiatives for marine environmental protection, including:

- setting up and developing a tax policy for marine environmental protection;
- researching the environmental impact of development projects that might cause chemical or petrochemical pollution at sea ports;
- requiring an environmental protection action plan when a developer applies for a construction permit for industrial zones, export processing zones, and tourist areas;
- conducting an environmental impact assessment (EIA) for all socio-economic development projects and strategic environmental assessment (SEA) for all planning, strategy and policy projects in coastal areas;
- ensuring that exploitation of living resources does not exceed the allowable sustainable capacity of these resources;
- banning the development of catching facilities for nearshore fishing within 20 metres of the coast;
- setting up national programmes to protect mangrove forests and wetlands, and designing projects on sustainable use and restoration of degraded coastal and marine ecosystems; and
- conserving coastal and marine areas to protect and develop specific ecosystems and biological diversity (the focus areas including Ha Long Bay in Quang Ninh province, Cat Ba islands in Hai Phong city, Nha Trang Bay in Khanh Hoa province, Con Dao islands in Ba Ria-Vung Tau province, and Phu Quoc islands in Kien Giang province, as well as other MPA sites).

To some extent, this policy shows an integrated approach between marine economic development and protection of marine resources and the marine environment. The policy connects marine economic development with national security and defence. However, there are still major problems, such as the lack of sector integration, horizontal integration between central and local governments and communities, reference to stakeholders' participation in processes, and expression of principles for sustainable development.

In 2002, Vietnam prepared its own Agenda 21, focusing on strategic orientation for sustainable development in Vietnam. This document was approved by the prime minister in August 2004 (MPI, 2004b). After the approval, Agenda 21 for Vietnam was under trial implementation in six provinces and four priority sectors, including fisheries. To implement Agenda 21 for Vietnam, in 2004 the government established a National Office of Sustainable Development under the Ministry of Planning and Investment to help the prime minister in implementation. Recently, the National Office of Sustainable Development has planned a new implementation phase of Agenda 21 for Vietnam, which was presented at Rio+20 in 2012 (Government of Vietnam, 2012a).

In 2003, Vietnam agreed to implement the Sustainable Development Strategy for the Sea of East Asia (SDS-SEA) within the framework of the Partnership in Environmental Management for the Seas of East Asia (PEMSEA, 2003). The SDS-SEA provides a holistic approach for addressing both sectoral and cross-sectoral issues that have local, national, regional, and global implications through six major components of strategies (sustain, preserve, protect, develop, implement, communicate) and 227 action programmes, which are based on the prescriptions of global and regional instruments such as the World Summit on Sustainable Development (WSSD) Plan of Implementation, the United Nations Millennium Declaration, and Agenda 21. A number of activities in Vietnam were incorporated into the SDS-SEA in-country implementation plan (2003–11), supported by PEMSEA.

The review of the SDS–SEA implementation progress in Vietnam (2003–13) and the SDS–SEA implementation plan in Vietnam (2014–18) have been prepared by the VASI–MONRE and submitted to the PEMSEA.[2] Regarding the implementation of the SDS–SEA, the government approved the Strategy on Sustainable Use and Exploitation of Marine Resources and Environmental Protection until 2020, Vision towards 2030, on 6 September 2013.[3] The Strategy focuses on three key strategic breakthroughs:

- managing and increasing the sustainability of coastal, island, and marine use and exploitation by means of a science-based approach;
- promoting coastal zoning and marine spatial planning (MSP) to maximize the potential and advantages of Vietnam's seas and to mitigate the impacts of climate change, as well as to help to resolve multi-use conflicts regarding the coastal and marine environment and resources; and
- finalizing and implementing an integrated policy and institutional framework of coastal, island, and marine environmental and resources governance and management.

In 2007, the CPV approved the Strategy on Vietnam Seas toward 2020 (CPV, 2007). A multi-stakeholder approach was taken to ensure full participation at the formulation stage of the Strategy. The preparation process involved all line ministries, sectors, and national consultants. More than sixty thematic reports were prepared, detailing inputs from, and overviews of, the participants. The Strategy concluded that Vietnam has to become a strong seaborne nation and further enrich itself from the seas by 2020. This will require Vietnam to establish a powerful public authority, with the function of integrated and unified state management for seas and islands. The main objectives and directions for the Strategy of Vietnam Seas toward 2020 are to:

- increase the growth of the marine economy;
- address social concerns relating to marine affairs;
- promote inter-sectoral coordination in marine management;
- preserve key coastal and marine ecosystems;
- reform the current policy and institutional framework to include integrated and inter-sector approaches;
- identify and implement sustainable development strategies;
- reduce poverty among local communities; and
- link economic development with national defence, security, and safety for general national interests.

The Strategy should be regarded as a landmark for an integrated marine governance approach, focusing on the sustainability of marine economic development in Vietnam.

Evaluation of, and outlook on, the national marine policy

The following points summarize the characteristics of Vietnam's national marine policy:

1 In Vietnam, existing marine policies focus mainly on socio-economic development. National security and defence are also emphasized.
2 Existing marine policies should reflect more the interrelationship between economic development, environmental protection, and resource conservation.

3 Existing marine policies are mainly being recorded in the CPV's documents and issued as political declarations. The marine policies lack detail, are incomplete, and are not adequately based on scientific baseline research.

4 Vietnam has not managed to set up a comprehensive and integrated marine governance policy based on scientific knowledge, which meets with the principles of sustainable development.

5 In the absence of an integrated marine policy, procedures to enforce marine policies have also not been determined, particularly for marine development activities and for the protection of the marine environment and resources.

6 The lack of an integrated marine policy up to now impedes improvements to the legal framework for marine and coastal development in Vietnam. Laws on marine affairs are fragmentary and sector-based. They do not adequately reflect Vietnam's international commitment to sustainable marine development. There is no comprehensive legal framework for marine and coastal development.

7 The Ministry of Planning and Investment (MPI), as a coordinating agency, has the authority to draft countrywide policies and plans. Regarding marine affairs, the MPI is in charge of drafting the plan for marine economic development. Unfortunately, these documents cannot be considered to be an integrated marine policy based on the principles of sustainable development.

8 Practice shows that there is a lack of understanding of integrated marine policy at high political levels, as well as in society. The MPI and other sea-related governmental agencies place most of their attention on marine economic development rather than sustainable development. The public has weak awareness of marine sustainable development or of the need for an integrated marine policy. Vietnam's sustainable development is an issue of both public and government awareness.

9 The state power and machinery of Vietnam is built on the centralization principle: almost all processes are centralized. Thus centralization is a characteristic of planning and policy formulation, which are nearly always conducted in a top-down manner without effective participation of the stakeholders. National or provincial and community collaboration in national marine policies is therefore nearly non-existent.

10 The process of integrated marine policy development is an initiative being driven by the Decree of 2009 on Integrated Management of Resources and Environment Protection of Seas and Islands, the Law of 2012 on the Vietnam Seas, and the Strategy on Sustainable Use and Exploitation of Marine Resources and Environmental Protection until 2020 and Vision towards 2030, as well as the drafting of the Law of Marine Resources and Environment.

The following lessons need to be integrated into future marine policy development in Vietnam:

1 The sustainable development of the seas has to be based on maintaining the essentially natural functions and connectivity of the coastal and marine ecosystems. This means that ecosystem-based coastal and marine management should be promoted.

2 The sustainable development of the seas has to be linked with poverty reduction in coastal communities and livelihood improvements for fishery labourers and other people whose income depends on coastal and marine resources.

3 The sustainable development of the seas has to be set within a framework of integrated coastal, marine, and watershed management policy, because over 70 per cent of the environmental impacts on coastal areas are from land-based activities.

4 Sustainability and environmental considerations, including climate change and ocean change, have to be integrated into marine governance policy in each step of the policy development process.

5 Sustainable development of the seas requires both changing the awareness and behaviour of coastal communities and managers, and strong political support from the government at different levels.

6 Coastal and marine governance policies require new approaches and tools (such as integrated coastal management, coastal and marine spatial planning, marine function zoning, and co-management) to address the complicated cross-cutting problems, which can be internationally sensitive. International assistance is very necessary, especially in the form of technical support.

Overall, the outlook on developing the national marine policy of Vietnam will require:

- the development of a national marine action plan based on the objectives of the Strategy on Vietnam Seas toward 2020 (and new phase beyond) and the principles for sustainable development of Vietnam's seas;
- the Law of 2012 on the Vietnam Seas to be considered a basic comprehensive legal document on marine and island governance and management in Vietnam, taking into account the objectives of the Strategy on Vietnam Seas toward 2020, enforcement of which law will involve the development of, decrees that guide and provide more detail, such as decrees guiding sea-use planning and marine research ships or vessels in Vietnam marine waters;
- the review of existing legal documents relating to the seas and islands, to ensure effective law enforcement;
- the establishment of a specialized marine management institution, in the form of a ministry of marine affairs to coordinate and integrate sea-related activities and agencies;
- the prioritizing of investments to develop marine economic sectors with comparative advantages such as oil and gas, fisheries, tourism, and navigation, to determine what the driving force for other sectors will be in the future; and
- the development of a 'blue economy' in Vietnam to implement the objectives of the National Strategy on Green Growth, which was approved by the government in 2013.[4]

Major approaches to the development of national marine policy of Vietnam

The national marine policy of Vietnam should be developed using following approaches.

1 *Best scientific information* – The marine zones and boundaries of Vietnam must be based on the best available scientific surveys and legal documents.

2 *Sustainability* – Vietnam's seas are rich in biodiversity and are very sensitive to natural and man-made impacts, including climate change and sea-level rise. They are connected to coastal ecosystems and watersheds. Vietnam should apply a marine management system aimed at preventing coastal and marine pollution, especially from land-based activities. With the support of Japan and the Netherlands, Vietnam has started an integrated management plan in the Red River watershed in the north and the Dong Nai watershed in the south, but it is not linked to the integrated coastal management plan. A small grant fund on integrated watershed and coastal management for the Vu Gia-Thu Bon watershed in Central Vietnam (Quang Nam province and Da Nang city) is receiving the support of the

German Hans Seidel Foundation (HSF) and Mangroves for the Future (MFF). The policy recommendation on *Integrated Watershed and Coastal Management: A 'From Ridge to Reef', Approach in the Vu Gia-Thu Bon Watershed Case Study, Viet Nam*, has been prepared and is in press at time of writing.

3 *International integration* – Vietnam is still a developing country and needs to be proactively integrated into the Association of Southeast Asian Nations (ASEAN), East Asia, and Asia-Pacific Economic Cooperation (APEC) economies, as well as the international economy. Therefore international integration factors need to be taken into account while marine governance is developed.

4 *Comprehensiveness* – The diversity of coastal and marine resources has created a premise for multi-use development, which leads to differences of interest among coast and sea users. Therefore, besides the Law of 2012 on the Vietnam Seas, it is essential for Vietnam to promulgate a law of marine resources and environment that provides a comprehensive legal tool for regulating all marine activities relating to marine resource use and environmental protection. Such a law should cover conflicts mitigation in fishing, marine resource exploitation, and marine management.

5 *Inter-sector and integrated management* – Coastal and marine concerns are not isolated in a single sector, but are cross-cutting issues, including complicated interactions between natural, economic, social, and political systems. Marine governance for sustainability thus requires integrated and inter-sector management and includes a multi-stakeholder approach.

6 *Local community involvement* – In 1998, the prime minister enacted the Regulation of 1998 on Strengthening Democracy at the Grass Roots (Community) Level. This legal document protects the democratic rights of villagers and encourages poorer people to participate in public works at the commune or grass-roots level. According to the Regulation, villagers enjoy democratic rights, such as the right to know, the right to be consulted, the right to decide upon the will of villagers, and the right to monitor. These rights must be stressed during policy implementation in general, and in coastal and marine management in particular. At present, the Ministry of Agriculture and Rural Development is preparing to decentralize coastal spatial management and to use rights to the local community level to promote co-management of coastal marine environment and resources in the framework of the project on coastal resources for sustainable development (CRSD) by using the World Bank's loans and Global Environment Facility (GEF) support (2012–18).

Principles of marine sustainable development in practice

The main principles and approaches of marine sustainable development in Vietnam are:

- integrative and ecosystem-based approaches in coastal and marine governance and management;
- adaptive management in the new climatic regime and sea-level rise context;
- stakeholder involvement in the process of coastal and marine governance and management, especially in CMSP;
- capacity-building and human resources development for integrated coastal and marine management;
- governance focusing on the role of coastal and island local communities; and
- coastal and marine pollution prevention, with effective application of the EIA and SEA tools and the 'polluter pays' principle.

With international financial and technical assistance and the government's support, several integrated coastal management (ICM) projects have been launched. These have included:

- the National Project on Research and Development of ICM, which plans to maintain coastal sustainability and ecological safety (1996–2000);
- the Asian Development Bank (ADB) project on coastal management of the Bien Dong Sea (1999–2002), which is linked to poverty reduction in central Vietnam (2003–06);
- the PEMSEA project in Da Nang (2001–04) and the ongoing project on scaling up ICM for SDS-SEA implementation in Vietnam between VASI and PEMSEA (2014–18);
- the Vietnam–Netherlands project on integrated coastal zone management (ICZM), with three case studies (2001–05) and Phase II (2005–08);
- the former Ministry of Fisheries–WorldFish Centre project on facilitating ICM in Vietnam (2004–05);
- the National Oceanic and Atmospheric Administration (NOAA)–Vietnam project on building capacity for ICM in the Gulf of Tonkin, including the Phase I case study in Ha Long Bay (2003–05), the Phase II expanded studies for the coastal areas of Quang Ninh–Hai Phong, with a focus on a comprehensive development framework (2006–09), and the ongoing Phase III, focusing on coastal spatial use zoning and management planning (2011–14); and
- the European Union–Germany sub-regional project on coastal management, focusing on educational and training components (2003–06).

The former Ministry of Fisheries prepared a master plan for Vietnam's MPA system to 2010, with a vision towards 2020 (Ministry of Fisheries, 2006), with the first sixteen representative MPA sites listed, submitted to, and already approved by the prime minister on 26 May 2010.[5] Effectively managed MPA sites under international assistance include:

- Nhatrang Bay (2001–04), with support from the World Bank, the Global Environment Facility (GEF), the International Union for Conservation of Nature (IUCN), and the Danish International Development Agency (Danida);
- Cu Lao Cham (2003–09), with Danida support; and
- the Con Dao, Phu Quoc, and Cat Ba islands (1995–2007), with World Wildlife Fund (WWF) support.

The spatial planning approach initiative has been applied in the process of establishing a national system of MPAs in Vietnam (Hoi, 2014).

With IUCN support, the former Ministry of Fisheries has prepared a National Action Plan on Sea Turtle Conservation and Management to 2010 that has been approved and has begun to be effectively implemented (Ministry of Fisheries, 2004). It is for this reason that Vietnam is now recognized as a world leader in the protection and conservation of the sea turtle.

In the framework of regional cooperation, the joint United Nations Environment Programme (UNEP)–United Nations Development Programme (UNDP)–GEF project on prevention of environmental degradation in Vietnam's South China Sea has been finalized for Phase I and preparations for Phase II are under way. Many legal documents have been, and are being, prepared that focus on the following components of marine protection: mangrove forest preservation; seagrass bed, coral reef, and coastal wetland protection; land-based pollution prevention; and aquatic living resource conservation and management.

With UNEP support, some projects have been carried out, such as the project on marine pollution management from land-based sources (2009–11), with three components (an inventory of land-based pollution sources, development of a national plan of action, and a pilot site in Ha Long Bay), and the project on marine spatial planning (MSP) (2010–13), with two components (the development of technical guidelines on MSP and application of MSP in a pilot site).

The prime minister has approved sector and provincial socio-economic development strategies to 2010 and 2020 for marine transportation, fisheries, oil and gas, and tourism sectors, and in all twenty-eight coastal provinces and cities. Some sectors have implemented their respective strategies on sustainable development, for example the Strategy for Fishery Development toward 2020 (Government of Vietnam, 2012b), the National Strategy on Green Growth to 2020,[6] and the National Action Plan on Green Growth in period 2014–20.[7]

Five-year national programmes on marine investigation and research, with 100 different thematic projects annually, have been carried out since 1978 with government financing. Other projects have been undertaken by means of cooperation with China, Russia, the Philippines, the European Union, and ASEAN. These programmes provide a huge volume of marine scientific information that can be used in policy- and decision-making. The 2011–15 marine science research period is ongoing. A national network for coastal and marine environment monitoring has been established since 1995. Under this network, twenty-six parameters are measured in representative locations every three months. Data are used to prepare an annual report on the state of Vietnam's marine environment that is submitted to the National Assembly. The five-year programme of the Ministry of Natural Resources and Environment (MONRE) (2010–15) on marine sciences and technology in the service of integrated and unified state management for seas and islands has been implemented and is ongoing.

In order to implement the Strategy on Vietnam Seas toward 2020, the government of Vietnam formed the Vietnam Administration of Seas and Islands (VASI) under MONRE. The Administration is responsible for cooperatively conducting integrated and unified state management of coasts, seas, and islands, focusing on sea and island use management and ICM. Two ICM centres have been established under VASI: one in the north, in Ha Noi; and the other in the south, in Ho Chi Minh City. An ongoing national programme on comprehensive and basic surveys and management of marine resources and environment to 2010 and towards 2020 (named 'Project 47' and approved in 2006),[8] the national plan on marine-related international cooperation towards 2020 ('Project 80', approved in 2008),[9] and the national programme on ICM in fourteen central coastal provinces ('Programme 158', approved in 2007),[10] as well as the national target programme to respond to climate change (NTP, approved in 2008),[11] are also being undertaken.

The National Strategy on Biodiversity was prepared by MONRE in collaboration with line sectors and localities.[12] The Strategy focuses on five main components:

- the conservation of natural ecosystems;
- the conservation of wild species and precious, rare, and dangerous genes and species for agricultural development, including forestry and fishery;
- the reasonable utilization and benefit-sharing from ecosystems services and biodiversity;
- the control of activities that threaten biodiversity; and
- biodiversity conservation, in a context of climate change, towards green growth.

Regarding the coastal and marine component, the Strategy's objectives by 2020 are the effective preservation and protection of 0.24 per cent of Vietnam's EEZ area, the maintenance of existing areas of mangroves, seagrass beds, and coral reefs, and the restoration and replanting of 15 per cent of the degraded natural ecosystems.

Some conclusions and recommendations

1   Vietnam's seas are rich in natural resources and play a very important role in the country's economic development, as well as national and regional security.

2   Vietnam has a long-standing history of coastal and marine exploitation and use, but the modality of exploitation and use are still backward, other than in the oil and gas industry sector.

3   Vietnam's marine policy has been formulated in different periods of the country's development. After 1975, the government of Vietnam was able to take all necessary political and legal measures to affirm its sovereignty over its sea territories and islands. Notably, Vietnam has declared a 12 nautical mile territorial sea, a contiguous zone, a 200 nautical mile EEZ, and jurisdiction over the continental shelf (1977), the baseline system measuring its sea territories (1982), and signed (1982) and ratified (1994) UNCLOS.

4   Besides sectoral laws and policies (fisheries, navigation, oil and gas, tourism), the Law of 2003 on National Boundaries, the 2007 Strategy of Vietnam's Seas towards 2020, and the Law of 2012 on the Vietnam Seas are integral to marine governance and management laws and policies.

5   The institutional framework for integrated coastal and marine governance and management in Vietnam has been established, but is still weak in collaborative mechanisms and in the implementation process.

6   Generally, the enforcement of these laws and policies in Vietnam is limited and ineffective.

7   Vietnam should complete the Law of Marine Resources and Environment and apply the integrative approach, including tools such as ICM and MSP.

8   In the future, Vietnam can develop other laws on matters such as marine spatial use management, island use management, coastal spatial governance, and marine protected areas.

9   Capacity-building and development of the institutional and policy framework are needed in order to carry out integrated management and united governance for seas and islands in Vietnam.

10 Vietnam would benefit from improved regional and international cooperation, including sharing lessons and experiences between nations, along with concerns about integrated coastal, marine, and ocean governance and management.

Notes

1   The views expressed in this chapter are the views of the authors and do not necessarily reflect the views of the Government of Vietnam.
2   See PEMSEA (2012: 142–53).
3   Decision No. 1570/QD-TTg of 2013 on a Strategy on Sustainable Use and Exploitation of Marine Resources and Environmental Protection until 2020 and Vision towards 2030.
4   Decision No. 1393/QD-TTg of 2012 on a National Strategy on Green Growth.
5   Decision No. 742/QD-TTg of 2010 Approving the Plan for Establishment of a National Marine Protected Area Network.
6   Decision No. 1393/QD-TTg.
7   Decision No. 43/QD-TTg of 2014 on a National Action Plan of Green Growth in Vietnam in the Period 2014–2020.
8   Decision No. 47/2006/QD-TTg of 2006 on a National Programme of Comprehensive and Basic Surveys, and Management of Marine Resources and Environment towards 2010 and Directions to 2020.

9 Decision No. 80/2008/QD-TTg of 2008 on a Programme for Marine International Cooperation to 2020.
10 Decision No. 158/2007/QD-TTg of 2007 on the Integrated Coastal Zone Management Programme for North Central Region and Central Coastal Provinces until 2010 and Orientations until 2020.
11 Decision No. 80/2008/QD-TTg of 2008 on a National Target Programme to Respond to Climate Change.
12 Decision No. 1250/QD-TTg of 2012 on a National Strategy on Biodiversity.

References

Asian Development Bank (ADB) (1999) *Coastal and Marine Protected Areas Planning in Vietnam*, Hanoi: ADB.
Communist Party of Vietnam (CPV) (1976) *Document of the Fourth Party Conference*, Hanoi: National Political Publishing House (in Vietnamese).
Communist Party of Vietnam (CPV) (1986) *Document of the Sixth Party Conference*, Hanoi: National Political Publishing House (in Vietnamese).
Communist Party of Vietnam (CPV) (1987) *Plenum No. 6 (30 November)*, Hanoi: National Political Publishing House (in Vietnamese).
Communist Party of Vietnam (CPV) (1991) *Document of the Seventh Party Conference*, Hanoi: National Political Publishing House (in Vietnamese).
Communist Party of Vietnam (CPV) (1996) *Document of the Eighth Party Conference*, Hanoi: National Political Publishing House (in Vietnamese).
Communist Party of Vietnam (CPV) (2007) *Strategy on Vietnam Seas toward 2020*, Hanoi: National Political Publishing House (in Vietnamese).
Communist Party of Vietnam (CPV) (2011) *Strategy for Socio-Economic Development for 2010–2020*, Hanoi: National Political Publishing House.
Government of Vietnam (2012a) *Implementation of Sustainable Development: National Report at the United Nations Conference on Sustainable Development (RIO+20)*, online at http://sustainabledevelopment.un.org/content/documents/995vietnam.pdf [accessed 13 July 2014].
Government of Vietnam (2012b) *Strategy for Fishery Development toward 2020*, Hanoi: Agricultural Publishing House.
Hoi, N. C. (2007) 'Development of Sustainable Coastal and Marine Tourism in Vietnam', in *National Proceedings on Marine Economic Vision and Fisheries Development in Vietnam*, Hanoi: Agricultural Publishing House (in Vietnamese).
Hoi, N. C. (2009) 'The Basic Issues of State Management for Seas and Islands in Vietnam', *Journal on Natural Resources and Environment*, 65(3):12–16 (in Vietnamese).
Hoi, N. C. (2013) 'The Marine Economics in Vietnam from a Marine Resources and Environment Perspective', *Journal on Political Science*, 5: 30–41 (in Vietnamese).
Hoi, N. C. (2014) 'Application of Spatial Planning in Establishing a System of Marine Protected Areas for Sustainable Fisheries Management in Vietnam', *Journal of the Marine Biological Association of India*, 56(1): 28–33.
Ministry of Fisheries (2004) *The National Action Plan on Sea Turtle Conservation and Management to 2010*, Hanoi: Agricultural Publishing House (in Vietnamese).
Ministry of Fisheries (2006) *Planning for a Network of Marine Protected Areas in Vietnam to 2010 and towards 2020*, Hanoi: Vietnam Institute of Fisheries Economics and Planning (in Vietnamese).
Ministry of Planning and Investment (MPI) (1997) *Report of the Comprehensive Plan for Marine and Island Economic Development to 2010*, Hanoi: MPI Institute of Development Strategy.
Ministry of Planning and Investment (MPI) (2004a) *The National Strategy for Marine Economic Development towards 2020*, Hanoi: MPI (in Vietnamese).
Ministry of Planning and Investment (MPI) (2004b) *The Directions for Sustainable Development of the Viet Nam (Agenda 21 for Viet Nam)*, Hanoi: S&T Publishing House.
Partnerships in Environmental Management for the Seas of East Asia (PEMSEA) (2003) *Sustainable Development Strategy for the Seas of East Asia: Regional Implementation of the World Summit on Sustainable Development Requirements for the Coasts and Oceans*, Quezon City: PEMSEA.
Partnerships in Environmental Management for the Seas of East Asia (PEMSEA) (2012) *Regional Overview: Implementation of the Sustainable Development Strategy for the Seas of East Asia (SDS-SEA), 2003–2011*,

online at http://www.pemsea.org/sites/default/files/regional-sdssea-review_0.pdf [accessed 13 July 2014].

Thang, B. T. (2011) *Rapid and Sustainable Development of Vietnam Marine Economics*, Hanoi: Agricultural Publishing House (in Vietnamese).

Vietnam NAVY and Central Commission for Popularization and Education (CCPE) (2007) *Vietnam's Seas and Islands*, Hanoi: National Political Publishing House (in Vietnamese).

Legislation

Decree of 1926 on Expanding Application of Fishing Laws for Colony Nations 1888

Law of 1926 on Fishing for Colony Nations

Decree of 1936 of the Ministry of Colonies on the Fisheries and Marine Waters of Indochina

Decree of 1948 on Maritime Zones, State Records Management and Archives Departments of Vietnam

Statement on Spratly and Paracel Archipelagos in the Bien Dong Sea of 1977

Statement on the Territorial Sea, the Contiguous Zone, the Exclusive Economic Zone and the Continental Shelf of 12 May 1977

Constitution of 1980, amended 1992 and 2013

Statement of 12 November 1982 by the Government of the Socialist Republic of Viet Nam on the Territorial Sea Baseline of Viet Nam

Ordinance of 1989 on Aquatic Living Resources Protection

Code of Navigation of 1990, as amended in 2005

Decree No. 24/HDBT of 1991 on Regulating Scientific Activities of Foreign Organizations and Ships in the Marine Waters of Vietnam

Law of 1993 on Oil and Gas, as amended in 2000

Resolution No. 3-NQ/TW of 6 May 1993 on Strengthening Marine Economic Development

Law of 1994 on Environmental Protection, as amended in 2004

Ordinance of 1997 on National Boundary Guards

Order No. 20-CT-TW of 22 September 1997 on Increasing Marine Economic Development toward Industrialisation and Modernisation

Order No. 36-CT/TW of 1998 on Strengthening Environmental Protection in the Period of Industrialisation and Modernisation

Ordinance of 1998 on Vietnam's Marine Police

Regulation of 1998 on Strengthening Democracy at the Grass Roots (Community) Level

Ordinance of 1999 on Tourism

Decree No. 48/2000/ND-CP of 2000 Detailing the Implementation of the Petroleum Law

Criminal Code of 2003

Law of 2003 on Fisheries

Law of 2003 on National Boundaries

Law of 2004 on National Security

Decree No. 27/2005/ND-CP of 2005 Adopted to Specify and Guide the Implementation of a Number of Articles of the Fisheries Law

Decree No. 25/2009/ND-CP of 2009 on Integrated Management of Resources and Environment Protection of Seas and Islands

Law of 2009 on Biodiversity

Law of 2012 on the Vietnam Seas

17

A CASE STUDY OF INDIA'S POLICY AND LEGAL REGIMES ON OCEAN GOVERNANCE

Tony George Puthucherril

Introduction

India is the seventh largest country in the world and the second largest nation in Asia, with an area of 3,287,263 km². Bordered by Pakistan to the northwest, China, Nepal, and Bhutan to the north, and Bangladesh and Myanmar to the east, the country is home to more than 1 billion people, nearly 16 per cent of the world's population. Physically, the land mass can be divided into four relatively well-defined regions: the Himalayan mountains; the Gangetic river plains; the Deccan plateau; and the islands of Lakshadweep and the Andaman and Nicobar. This makes India one of the twelve mega-biodiversity countries, housing four of the global biodiversity hotspots.

India's bio-richness is reflected in its marine environment. Washed by the waters of the Indian Ocean to the south and two semi-enclosed seas, the Arabian Sea to the west and the Bay of Bengal in the east, India is endowed with vast marine resources, with a long coastline of about 7,500 km and a 2.02 million km² exclusive economic zone (EEZ) (Ministry of Earth Sciences, undated*a*). As the outer limit of the continental shelf extends beyond 200 nautical miles, the country stands to gain rights of exploration and utilization of resources in an additional area of about 1 million km² (DOD, 2002). Additionally, the network of major and minor rivers and their tributaries, which criss-cross the Indian land mass before discharging into the two major seas, have endowed the country with a wide variety of coastal ecosystems including estuaries, mangroves, beaches, backwaters, salt marshes, and lagoons (Swaminathan, 2005). The bay islands of the Andaman and Nicobar, and the atoll island group of Lakshadweep, add to the rich and unique marine and coastal biodiversity of the country. More importantly, they have also extended the country's sovereignty over ocean space to areas well beyond the mainland.

For administrative convenience, India is divided into twenty-nine states and seven union territories. Of these, nine maritime states and four union territories are situated on the coast, accommodating a burgeoning population. It is estimated that 50 per cent of the country's total population lives in fifty-three coastal districts. Eleven cities on the Indian coastline house more than 1 million people each, with the population of Mumbai alone being 20 million. The density of population in coastal areas varies from fewer than twenty-five persons per 1 km² to as many as 700 or more (Krishnamoorthy et al., 2001). To support this vast population, the coastal regions accommodate a huge infrastructure: thermal power plants, nuclear reactors, an

462

array of industries, mines, sewage treatment plants, etc. There are also twelve major ports and 187 minor or intermediate ports, which support over 90 per cent of India's foreign trade by sea (Department of Shipping, 2013). The location of ports and harbours makes the coastline a preferred destination for siting industries and power plants. Of late, India's coasts have been attracting an increasing number of tourists, with the result that several projects are being undertaken to set up adequate tourism-related infrastructure, often at the expense of the health of fragile ecosystems. Thus overpopulation and competing uses stress basic resources, leading to severe marine pollution and unsustainable coastal resource management.

As far as marine resources are concerned, India's marine ecosystem covers 2.1 million km^2 in area. Marine faunal biodiversity represents more than 15 per cent of the total fauna of the country (Government of India, 2001). The fishery resources in the EEZ stand assessed at 3.93 million metric tonnes, distributed as follows: inshore, 58 per cent; offshore, 34.9 per cent; and deep sea, 7 per cent. The major share of this resource is demersal (2.02 million tonnes), followed by 1.67 million tonnes of pelagic and 0.24 million tonnes of oceanic resources (Ministry of Agriculture, 2004). The ocean also has a wealth of organisms of pharmaceutical value.

The ocean also possesses tremendous energy potential that could be tapped. India has made strides in ocean thermal energy conversion technology with the launch of the Indian ocean thermal energy conversion (OTEC) programme in 1980 involving the installation of a 20 megawatt (MW) plant off the Tamil Nadu coast. The OTEC potential around India, from Mumbai in the west, down south, and up to Vishakhapatnam on the east coast, has been estimated at 50,000 MW. The wave energy potential is estimated to be around 6,000 MW, and an experimental energy power plant of 150 MW capacity has been installed at the Vizhinjam beach in Kerala. The identified economic tidal power potential is in the order of 8,000–9,000 MW, with about 7,000 MW in the Gulf of Cambay, about 1,200 MW in the Gulf of Kutch, and less than 100 MW in Sundarbans (EAI, undated). The country also has major hydrocarbon reserves in offshore regions of the Bombay High, the Gulf of Kutch, and the Godavari basin (Sharma and Sinha, 1994: 44). There are also huge mineral deposits in the country's EEZ.

India is thus endowed with a rich and diverse maritime estate over which different stakeholders lay claims, leading to intense competition amongst them to control these resources. The need for a principled approach to ocean governance was made clear by the tsunami of December 2004, which wreaked havoc in the coastal regions of South Asia as far as East Africa. Despite its high speed and magnitude, it is generally believed that the effects of the tsunami in many of the affected countries, particularly India, could have been considerably reduced had coastal zone regulations been followed in their true spirit (Kurien, 2005). The disaster, unprecedented in magnitude, exposed Indians to the fury of the sea for the first time in recent history. However, the greatest challenge that India faces from the oceans is yet to unfold, even though the signs are ominous: India is slated to be one of the countries that will be worst hit by climate change. As the menacing seas lay claim to more and lower-lying areas, islands are vanishing from India's map and the country is already burdened with its first set of climate migrants (Puthucherril, 2012). All of these issues pose new challenges for ocean governance. They mandate a sound policy and an efficient legal system for the management of ocean resources if sustainable ocean governance is to be ensured.

It is against this backdrop that this chapter analyses India's current policy and legal regimes on ocean governance. Throughout this contribution, the relevant policies and laws are discussed and their gaps are identified. The role of the Indian judiciary in shaping India's ocean policy and law is also highlighted, because the judicial contribution to the development of India's ocean policy and law is significant.

Ocean governance under the Indian Constitution

Oceans have always had a place of divinity in Indian mythology and religion. According to the ancient Indian text, the *Bhagavatha Purana*, the *devas* (gods) and the *asuras* (demons) jointly churned the ocean to bring forth *amruth* (the elixir of immortality) (Saigal et al., 1995). The discovery of the lost city of Dwaraka, India's Atlantis, is proof that great cities have flourished on the country's coasts since ancient times (Gaur et al., 2004). Indians maintained trade contacts with the Greeks and the Romans. As early as the third century BC, during the reign of Emperor Chandragupta Maurya, his minister, Kautilya, devoted an entire chapter in his treatise on statecraft, the *Arthasastra*, to shipping and other allied matters The *Arthasastra* mentions a well-organized board of admiralty and a naval department headed by the superintendent of ships, whose jurisdiction ran over the harbour and certain maritime zones beyond inland waters (Anand, 1983). However, the belief that ocean is God's resting place and that any human activity on it would disturb its tranquillity soon found its way into the Hindu psyche, and therefore anyone who crossed the seas was ostracized from his caste (Historic Alleys, undated). Nevertheless, some of the ancient Indian emperors did maintain strong navies with blue water capabilities and undertook naval expeditions to Southeast Asia (Chakravarti, 1930).

With colonization, and to facilitate trade, the British initiated a series of steps to develop India's maritime infrastructure. These included the establishment and development of major maritime cities along its eastern and western coasts. They also enacted important legislation, such as the Indian Fisheries Act 1897, the Indian Carriage of Goods by Sea Act 1925 and the Indian Ports Act 1908, which continue to operate to this day. These laws significantly impact on modern Indian Ocean governance. The idea that a coastal state could exercise jurisdiction in a small belt from the shore to a certain distance into the high seas was fortified during British rule, when the limit of the territorial sea was fixed at 3 nautical miles, even though this was found to be insufficient.[1]

Decolonization brought a radical transformation in the international stage: what was mainly a European setting evolved into a more inclusive world community. This profoundly influenced the future development of the law of the sea. However, the Constitution of India of 1950, being enacted prior to many of these developments, does not mirror the core principles espoused by the international law of the sea. Nonetheless, the Indian Constitution does wield considerable influence over the legal and regulatory framework of Indian ocean governance. Several articles and legal principles derived from the Constitution have a direct bearing on the development of India's ocean law and policy. This impact can be analysed from two broad angles. First, under the constitutional scheme on federal governance, the definition of 'India' becomes relevant regarding the jurisdiction of the country over the seas. It ordains whether, and to what extent, the centre and the states have competence to legislate on matters relating to fisheries, wildlife protection, and major and minor ports.

Second, the Constitution guarantees to citizens certain fundamental rights. The *primus inter pares* amongst these is article 21, which guarantees to every person 'the right to life'. Consequent to judicial interpretation, this now includes the right to a clean environment and access to natural resources.[2]

Clearly, the definition of 'India' under the Constitution focuses on the actual land mass. Article 1(3) provides that the 'territory of India' shall comprise the territories of states, the Union territories specified in the First Schedule, and such other territories as may be acquired. However, under article 297 of the Constitution (as it originally stood), all lands, minerals, and other things of value underlying the ocean within the territorial waters of India were vested in the Union and to be held for the purposes of the Union. Even though India ratified the 1982

United Nations Convention on the Law of the Sea (UNCLOS) in 1995, in 1963, influenced by the developments in the law of the sea, article 297 was amended to include the words 'or the continental shelf' after 'territorial waters'. In 1976, article 297 was further amended to enumerate which things of value within territorial waters or the continental shelf and resources in the EEZ were vested with the Union of India. This amendment also empowered the Parliament of India to specify from time to time, by law, the different maritime zones.

The 1950 Constitution assigns to the Union Parliament competence to make laws on matters including major ports, maritime shipping and navigation, lighthouses (including lightships, beacons, and other matters relating to the safety of shipping), piracy and crimes committed on the high seas, port quarantine, the regulation and development of oilfields and mineral oil resources, archaeological sites and remains, and carriage of passengers or goods by sea (Seventh Schedule, List I, entries 21, 25–28, 30, 53, and 67). Matters such as water and land are under the jurisdictional authority of the states (Seventh Schedule, List II, entries 17 and 18). Entry 57 of List I of the Seventh Schedule specifies that 'fishing and fisheries beyond territorial waters' is a Union subject, while entry 21 of List II speaks of fisheries as a state subject. Reading both entries together, it appears that control and regulation of fishing and fisheries within territorial waters falls within the exclusive province of the state, whereas those beyond territorial waters are the exclusive domain of the Union. Both the Union Parliament and the state legislatures could make law on matters enumerated in the 'Concurrent List' that are relevant to ocean governance (Seventh Schedule, List III, entries 17A, 17B, 31, 32, and 40). Residuary powers of legislation are vested in the Union Parliament (article 248). In certain cases, it can even legislate on matters that are reserved to the states (articles 249–252).

Although article 40 of the Constitution requires the state to take steps 'to organize *panchayats* [local self-government units] and endow them with the necessary powers and authority to enable them to function as units of self-government', this article has been inactive for a very long time. A revolutionary change came about with the Seventy-third Constitution Amendment Act of 1992, which recognized decentralization of powers at the grass-roots level. The amendment added a third tier to the existing structure of government and provides legal recognition to *panchayats* to function as institutions of self-governance comprising elected members and endowed with administrative and legislative powers. Under article 243G, the legislature of a state are to enact laws to empower the *panchayat* to prepare plans 'for economic development and social justice', apart from other matters specified in the Eleventh Schedule to the Constitution. Some of the topics in the Eleventh Schedule relevant to integrated coastal and ocean governance include fisheries, water management and watershed development, health and sanitation, and the maintenance of community assets. In the context of ocean governance, the decentralization scheme seems to be marginal or minimal; however, even this framework is yet to be operationalized effectively.

The Indian Constitution also exerts considerable influence in determining the domestic implementation of international agreements and treaties. Article 253 provides that:

> Parliament has power to make any law for the whole or any part of the territory of India for implementing any treaty, agreement or convention with any other country or countries or any decision made at any international conference, association or other body.

Furthermore, entry 14 of List I in the Seventh Schedule confers powers on the central government to enter into and implement treaties, agreements, and conventions with foreign countries. Thus treaty-making power falls within the exclusive prerogative of the Union government. Table 17.1 lists some of the major conventions related to ocean governance and their status with regards to India.

Table 17.1 Major multilateral environment agreements (MEAs) ratified and global programmes entered into by India

Agreement	Date of ratification by India
1971 Convention on Wetlands of International Importance (Ramsar Convention)	11 February 1982
1972 Convention for the Protection of World Cultural and Natural Heritage	4 November 1977
1973 Convention on International Trade in Endangered Species (CITES)	20 July 1976
1979 Convention on Migratory Species of Wild Animals (Bonn Convention)	1 November 1983
1985 Convention for Protection of the Ozone Layer (Vienna Convention)	18 March 1991
1987 Protocol on Substances that Deplete the Ozone Layer (Montreal Protocol)	19 June 1992
1989 Convention on Transboundary Movements of Hazardous Wastes and their Disposal	24 June 1992
1992 United Nations Framework Convention on Climate Change (UNFCCC)	1 November 1993
1997 Kyoto Protocol to the UNFCCC	26 August 2002
1992 Convention on Biological Diversity (CBD)	18 February 1994
2000 Cartagena Protocol on Biosafety to the CBD	11 September 2003
1994 United Nations Convention to Combat Desertification (UNCCD)	17 December 1996
1998 Rotterdam Convention on the Prior Informed Consent Procedure for Certain Hazardous Chemicals and Pesticides in International Trade	24 May 2005
2001 Stockholm Convention on Persistent Organic Pollutants	13 January 2006
1982 United Nations Convention on the Law of the Sea (UNCLOS)	June 1995
UNEP Regional Seas Programme South Asian Seas★	Members: India, Pakistan, Bangladesh, Maldives, Sri Lanka

★ *This is not yet a convention; rather, UNCLOS is considered to be the umbrella convention. However, there is a South Asian Seas Action Plan, which entered into force in 1997.*

Indian courts, in adhering to the principle of monism, have relied on international conventions and principles of customary international law not inconsistent with fundamental rights as guaranteed by the Constitution, to enlarge its meaning and content.[3] Indeed, it is mainly through the judicial process that international environmental law principles such as sustainable development,[4] the 'polluter pays' principle,[5] the precautionary principle,[6] public trust,[7] and intergenerational equity,[8] have become part of Indian environmental jurisprudence.

Demarcating the maritime estate and maritime boundary disputes

In pursuance of the mandate under the amended article 297, India enacted the Territorial Waters, Continental Shelf, Exclusive Economic Zone and Other Maritime Zones Act 1976. This legislation, the primary ocean law of India, delineates the maritime estate into distinct zones and sets out the rights and obligations that can be exercised in them. Under this Act, the central government is empowered to declare India's historic waters and claims a 12 nautical mile territorial sea (sections 3, 4, and 8), a 24 nautical mile contiguous zone (section 5), and a 200 nautical mile EEZ (sections 4 and 7). As far as India's continental shelf is concerned, it comprises the seabed and subsoil of the submarine areas that extends beyond the limit of the territorial waters throughout the natural prolongation of the land territory to the outer edge of the continental margin, or to a distance of 200 nautical mile from the baseline where the outer edge of the continental margin does not extend up to that distance (section 6). The central government exercises 'full and exclusive sovereign rights', consistent with the provisions of UNCLOS.

UNCLOS provides scientific and technical criteria regarding delineation of the outer limits of the continental shelf that lie beyond 200 nautical miles from the baselines of a coastal state (UNCLOS, Articles 76–85 and Annex II). Specifically, taking into account the inequity that could befall those states that seek to establish the outer edge of the continental margin in the southern part of the Bay of Bengal, consequent to the application of Article 76 of UNCLOS it was provided that the outer edge of the continental margin in this area could be established by straight lines not exceeding 60 nautical miles connecting fixed points at which the sediment thickness is not less than 1 km. The Statement also addresses the special geographic circumstances in determining the outer edge of the continental margin in the southern part of the Bay of Bengal. India has made a submission under Article 76(8) to the United Nations Commission on the Limits of the Continental Shelf (CLCS) regarding the establishment of the outer limits of its continental shelf that lie beyond the 200 nautical miles, in three regions: in the eastern offshore region in the Bay of Bengal (the outer limits are defined by straight lines not exceeding 60 nautical miles in length connecting 452 fixed points); in the western offshore region of Andaman islands; and in the western offshore region in the Arabian sea. As a coastal state in the southern part of the Bay of Bengal, India seeks to make a separate second partial submission of information and data to support its claims over the outer limits of continental shelf. The Union of Myanmar, Bangladesh, and the Sultanate of Oman have objected to India's continental shelf claims.

Determining India's land boundaries has always been a thorny issue. This is also true of oceanic boundaries. As a littoral state, India has five neighbouring maritime states. Under section 91(1) of the 1976 Maritime Zones Act, maritime boundaries between India and any state whose coast is either opposite or adjacent to it are to be determined by agreement. Pending such agreement, the maritime boundary is not to extend beyond the line every point of which is equidistant from the nearest point from which the breadth of the territorial waters of India and of such state is measured. Maritime boundary delimitation agreements have been entered into with almost all of the opposite maritime neighbours:

- *Indonesia* – the 1974 Agreement between the Government of the Republic of India and the Government of the Republic of Indonesia Relating to the Delimitation of the Continental Shelf Boundary between the Two Countries (US Department of State, 1975b), the 1977 Agreement between the Government of the Republic of India and the Government of the Republic of Indonesia on the Extension of the 1974 Continental Shelf Boundary between the Two Countries in the Andaman Sea and the Indian Ocean (DAOLOS, 1987a), and the 1978 Agreement between the Government of the Kingdom of Thailand, the Government of the Republic of India and the Government of the Republic of Indonesia Concerning the Determination of the Trijunction Point and the Delimitation of the Related Boundaries of the Three Countries in the Andaman Sea (DAOLOS, 1987b);
- *Maldives* – the 1976 Agreement between India and Maldives on Maritime Boundary in the Arabian Sea and Related Matters (US Department of State, 1978b), and the 1976 Agreement between Sri Lanka, India and Maldives Concerning the Determination of the Trijunction Point between the Three Countries in the Gulf of Mannar (DAOLOS, 1987c);

- *Myanmar (Burma)* – the 1986 Burma–India Agreement on the Delimitation of the Maritime Boundary in the Andaman Sea, the Coco Channel and the Bay of Bengal;
- *Sri Lanka* – the 1974 Agreement between Sri Lanka and India on the Boundary in Historic Waters between the Two Countries and Related Matters (US Department of State, 1975a), the 1976 Agreement between Sri Lanka and India on the Maritime Boundary between the two Countries in the Gulf of Mannar and the Bay of Bengal and Related Matters (US Department of State, 1978a), and the 1976 Agreement between Sri Lanka, India, and Maldives; and
- *Thailand* – the 1978 Agreement between the Government of the Kingdom of Thailand, the Government of the Republic of India, and the Government of the Republic of Indonesia.[9]

However, disputes continue to simmer with the country's adjacent neighbours, Bangladesh[10] and Pakistan.[11] The recent award of the Permanent Court of Arbitration (PCA) in the *Bay of Bengal Maritime Boundary Arbitration* (2014)[12] has finally resolved the controversies surrounding the maritime boundary between the two states. The PCA has identified the location of the land boundary terminus between Bangladesh and India, and has also determined the course of the maritime boundary in the territorial sea, EEZ, and on the continental shelf within and beyond 200 nautical miles.

An overview on the national policy on ocean governance

The adoption of UNCLOS in 1982 extended the economic jurisdiction of coastal states to 200 nautical miles from the coastline. For a developing country like India, this meant that nearly 2.02 million km^2, or nearly two-thirds of the Indian land mass, fell within the country's jurisdiction, wherein it could exercise exclusive utilization rights over living and non-living resources (DOD, 1982: para. 2). India was also recognized as a pioneer investor in an area of up to 150,000 km^2 in the deep seas for the recovery and processing of polymetallic nodules.

In 1981, India launched its Antarctic programme with a scientific expedition to this icy continent. At that point, oceanography was a relatively young science in a country that had far more pressing problems to deal with. However, policymakers realized that development and exploitation of sea resources could significantly improve the living standards of the people and fuel the growth of the economy. The dynamics of the cold war era – particularly the setting up of a military base by the United States and the United Kingdom in Diego Garcia, and the sending of an nuclear-armed aircraft carrier from the United States Seventh Fleet to the Bay of Bengal at the height of the liberation war of Bangladesh in 1971 – may also have strengthened India's resolve to extend its grasp over the Indian ocean. As a necessary prerequisite, it became imperative for India to master the science of ocean management. As a first step in this regard, the government of India, in 1981, created a Department of Ocean Development (DOD) directly under the Office of the Prime Minister by means of a presidential notification. One of the first tasks of the new department was to organize a series of workshops on themes ranging from marine instrumentation, ocean data management, ships and submersibles, ocean engineering, exploration and exploitation of seabed materials, and manpower requirements, to education, training and research facilitates – all cutting-edge areas of ocean science.

On the basis of these deliberations, the Department finalized the Ocean Policy Statement, which the Indian Parliament discussed and adopted in 1982 (DOD, 1982). The Statement basically charts the future course for the development of ocean science in India. It is a plain document of fifteen paragraphs, intertwined around the major theme of developing the science needed for ocean governance. Its underlying philosophy can be gauged from the statement that:

> The extension of national frontiers by an area of 2 million sq kms of ocean space and the consequent access to new sources of energy, minerals and food, requires great strides in ocean engineering . . . Marine science and technology has also to look beyond the current state-of-the-art to achieve major technological break-through in the future.
>
> *(DOD, 1982: para. 7)*

The Statement also notes that, '[f]or success in ocean development, the entire nation should be permeated by the spirit of enterprise and the desire to explore the frontiers of knowledge' (DOD, 1982: para. 1).

Saliently, the Ocean Policy Statement emphasizes the need for adequate knowledge of marine space, recognizing this as a fundamental prerequisite to the control, management, and utilization of sea resources. Thus it calls for creation of a database to coordinate the efforts of different agencies, preparation of an inventory of commercially exploitable fauna, surveying of the deeper portions of the ocean, and mapping and assessing the availability of minerals from the deep sea. It also emphasizes developing appropriate technologies to harness the both living and non-living resources, as well as infrastructure and systems to support this development. The Statement stresses the coordinating role of the DOD in implementing these policy objectives.

The 1982 Ocean Policy Statement remains the basic framework for ocean governance in India. In 2002, to further promote ocean science and technology, the DOD formulated the Vision Perspective Plan 2015 (DOD, 2002). The Plan aims to 'improve our understanding of the Ocean, specifically the Indian Ocean, for sustainable development of ocean resources, improving livelihood, and timely warnings of coastal hazards that will make India an exemplary steward of her people and ocean' (DOD, 2002). It brings to the fore the concept of 'one ocean' and seeks to realize the same through 'joint science programmes and observations at basin scales with countries in the region'(DOD, 2002). It calls for 'a well sampled ocean', which is to be achieved through a network that continuously observes physical, chemical, and biological variables in time and space. It focuses on improving ocean technology, implementing programmes in the Southern Ocean and the Antarctic, enhancing ocean awareness, conserving maritime heritage, managing living and non-living resources, protecting coastal environmental systems, conserving marine biodiversity and habitats, adopting measures to address climate change and rising oceans, and supporting island and remote coastal communities. Importantly, it calls for the creation of an Ocean Commission to act as a nodal agency to recommend legislation on all matters connected with the oceans, including the adoption of international conventions and recommendations.

In 2010, the Ministry of Earth Sciences set out the Vision and Prospective Plan for 10 Years in Ocean Sciences & Services (Ministry of Earth Sciences, 2010), which articulates a vision and prospective plan for ocean sciences and services under key themes: the role of the ocean in monsoon climate; routine forecasting of conditions in the Indian EEZ; natural hazards; environmental impact assessment (EIA); and bio-geochemistry. It also explains the infrastructure necessary to achieve these goals. Interestingly, it recognizes a major impediment to realization of these goals – namely, the lack of adequate manpower – and proposes the setting up of a National School of Oceanography.

Administratively, implementation of the Ocean Policy Statement goals initially fell to the DOD, which was converted into the Ministry of Ocean Development in February 2006. Then, on 12 July 2006, the Ministry of Ocean Development was reorganized as the Ministry of Earth Sciences, with the mandate to provide the nation with the best possible services in forecasting monsoons and other weather or climate parameters, and information on the state of the ocean, on earthquakes, tsunamis, and other phenomena related to earth systems, by means of well-integrated programmes. The Ministry also deals with science and technology for exploration and exploitation of ocean resources (living and non-living), and plays a nodal role in polar and Southern Ocean research. The Ministry of Earth Sciences is to look after atmospheric sciences, ocean science and technology, and seismology in an integrated manner. In addition, there is also the Earth Commission, which is responsible for formulating policies, overseeing its implementation in mission mode, and ensuring the necessary interdisciplinary integration.

In spite of the diverse nature of India's ocean areas, and its significant economic dependence on coastal and marine resources, there are several gaps in India's present ocean law and policy. From the preceding discussion, it is clear that the policy does not constitute a comprehensive ocean governance framework. Being of 1980s vintage, it is not of much use to a major maritime power like India, with its significant blue water capabilities and its ambition to utilize its marine resources for overall development, including energy security. The underlying objective of the 1982 Ocean Policy Statement is to ensure the development of the science and technology appropriate to ocean governance. It views ocean governance from the prism of science, but, in reality, ocean governance is a much more complex phenomenon.

India has been able to achieve considerable success in developing necessary science and technology, but the concept of ocean governance and the scenario in which such a policy is to be worked has, over the years, changed considerably. Science and technology are important concerns for sustainable ocean governance, but no longer the central concern. There are new issues that must be addressed if sustainable ocean governance is to be achieved, including environmental protection, allocation of fishing rights, protection of marine species and the conservation of marine biodiversity, marine pollution, and offshore oil and gas exploration regulation. Obviously, India needs to review its thirty-year old national ocean policy. When it was formulated, principles such as sustainable development, 'polluter pays', precaution, public trust, ecosystem-based approach, integration, community-based management, stewardship, and principled ocean governance had not yet entered the language of ocean governance, and they are absent from the Ocean Policy Statement. While the more recent Vision Statement refers to many of these principles, it provides no content for them in the Indian context. Further, the Vision and Prospective Plan for 10 Years in Ocean Sciences & Services is directed towards ocean sciences and services.

There is also an urgent need to revamp the present administrative structure on ocean governance. While the Ministry of Environment and Forests (MoEF) generally deals with environmental issues, the management of maritime territories for the purpose of resource use and development falls under the Ministry of Earth Sciences. Fisheries within territorial waters fall within the legislative competence of the states, while fisheries beyond territorial waters are under the jurisdiction of the Union Ministry of Agriculture. The management of marine non-living resources, including offshore oil and gas exploration and exploitation, falls under the purview of the Ministry of Petroleum and Chemicals. Clearly, the administrative apparatus on ocean governance in India is complicated by the overlapping responsibilities and jurisdictions assigned to numerous agencies. The confusion is compounded by the plethora of laws at both the federal and state levels, and, in certain cases, even at the level of the *panchayat*, which deal with different facets of ocean and coastal management. Such a disjointed arrangement does not augur well if a principled approach to ocean governance is what India aspires to. There is a need for an umbrella agency to coordinate and harmonize authorities at the federal, state, and *panchayat* levels, and to serve also as a forum for conflict resolution among the different stakeholders. Although the Vision Statement speaks about the need to constitute an Ocean Commission, there have been no steps to establish such a body. The Earth Commission, as presently constituted, is no substitute for an exclusive Ocean Commission with responsibility to manage India's ocean resources (Ministry of Earth Sciences, undated*a*).

Issues in ocean governance and management: Policy and legal responses

India has benefited immensely from globalization, with investment flowing into major industry-oriented infrastructural projects such as port construction, the setting up of heavy industries on coastal lands, and the creation of special economic zones (SEZs). Some of these projects have the potential to endanger marine biodiversity, as well as to affect traditional rights to fishing enjoyed by coastal communities. In permitting these projects, it is doubtful whether the state has dutifully essayed its role as a trustee of these resources. The politics involving the Coastal Regulation Zone (CRZ) Notification and dilution of its provisions on one pretext or another, opening up pristine ecosystems to development activities, introducing the coastal aquaculture law to negate the beneficence of the *Jagannath* dictum,[13] and encroachment into the legislative domain of the states by the union, are attributable to the lack of well-defined policies and laws on integrated ocean and coastal management.

Moving from coastal regulation to coastal management or mismanagement?

The need to protect the long coastline found concrete expression in the early 1980s during the tenure of then Prime Minister Indira Gandhi. She addressed letters to the chief ministers of all coastal states directing them to ensure that the coastal zone up to 500 metres from the high-tide line (HTL) was kept free from development activities of all kinds. However, owing to pressure from the tourism industry, the central government relaxed this ban in 1986. Later, in 1991, the government released a notification under the Environment (Protection) Act 1986 and its Rules, according special protection to the coastal areas of the country: the Coastal Regulation Zone (CRZ) Notification 1991.

Under the CRZ Notification, coastal stretches of sea, bays, estuaries, creeks, rivers, and backwaters influenced by tidal action up to 500 metres from the HTL were declared to fall under the CRZ (MoEF, 1991). Even the land between the low-tide line (LTL) and the HTL was included in the prohibited area. The notification classifies the CRZ area around the country into four zones: CRZ-I, which consists of ecologically sensitive areas; CRZ-II, which includes urban areas; CRZ-III rural areas; and CRZ-IV, which includes mainly the coastal stretches in the Andaman and Nicobar and the Lakshadweep islands. The Notification requires all coastal states to prepare coastal zone management plans that identify and classify the CRZ areas. Restrictions are imposed on the setting up or expansion of industries, operations, or processes in the CRZ. Thirteen activities – among others, establishing new industries and expanding existing industries; the manufacture, handling, storage, or disposal of hazardous substances; the setting up and expansion of fish-processing units; the setting up and expansion of mechanisms for disposal of waste and effluents; the harvesting or withdrawal of ground water and construction of mechanisms for the same – within 200 metres of the HTL are prohibited, subject to certain exceptions.

Although the CRZ Notification 1991 was issued after a lot of study, barely had the ink dried on its text when demands were made to amend the law. Introduction of this regulatory framework coincided with India's embrace of globalization and the initiation of a series of economic reforms. In this new scheme for economic development, coastal lands emerged as prized real estate. Accordingly, the Notification was diluted on one pretext or the other, and the mandatory coastal zone management plans and establishment of adequate institutional mechanisms to ensure effective implementation of the law were delayed, and in some cases were never put in place. Moreover, the repeated amendments – almost twenty-five to date – have rendered this

Notification nugatory. For instance, in 2002, an amendment was pushed through that permitted the setting up of non-polluting service industries in the CRZs of SEZs. This amendment also excluded notified port limits and notified SEZs from the no-development zone of 200 metres from the HTL in CRZ-III areas. Consequently, in the CRZ-III areas of notified SEZs, non-polluting industries, such as those relating to information technology, service sector industries, desalination plants, beach resorts, and related recreational facilities essential for the promotion of the SEZ, can be set up. As well, the exemption has triggered a frenzy of port approvals. It is estimated that there are proposals to set up nearly 300 ports, which translates to roughly a port every 20–25 km along the Indian coastline, virtually making a mockery of the prohibition (MoEF, 2013).

Despite all of the amendments to the 1991 Notification, very rarely was the public consulted. The protection of the CRZ proved to be a fertile ground for judicial intervention. The courts were called upon to determine a range of issues – from validity of amendments to the CRZ Notification,[14] to the setting up of a naval museum,[15] hospital,[16] sewage treatment plant,[17] and resorts.[18] However, the courts have not been able to resolve the development–conservation stalemate in the coastal law context, primarily because of the inconsistencies brought into the law by the amendments. In many cases, the courts favoured development, ignoring genuine environmental objections. For instance, in *Dahanu Taluka Environment Protection Group v. Bombay Suburban Electricity Supply Company Ltd* (1991),[19] the primary issue in dispute centred on the siting of a thermal power plant. The major grievance against the project was that the clearance was contrary to the Environmental Guidelines for Thermal Power Plants, which laid down two important criteria:

1   thermal power plants were not to be located within 25 km of the outer peripheries of metropolitan cities, national parks, wildlife sanctuaries, and ecologically sensitive areas; and

2   a distance of 500 metres from the HTL and a further buffer zone of 5 km from the seashore should be kept free of any thermal power station.

It was pointed out that the Dahanu site, being the only greenbelt left in that region, would receive emissions of pollutants and coal, and fly ash contaminants, which would adversely impact the local ecology. Brushing aside these objections, the court opined that the Guidelines were of general nature and that, in locating a thermal power plant in a particular region, the special features of that region had to be taken into account. In this case, electricity was to be supplied to Bombay suburban areas, and since the requirements of water supply dictated close access to the sea, it was only natural that the plant had to be located as near as possible to Bombay and the sea. Accordingly, the distances mentioned in the Guidelines were held not to be rigid and inflexible. Thus the court was reading the law so as to afford an exception when, in reality, the law did not provide for one.

In 2004, in a bid to bring coherence to the coastal law, the government appointed a committee to undertake a comprehensive review of the CRZ Notification. Based on its recommendations, the government introduced the Coastal Management Zone (CMZ) Notification 2008, with the objective of protecting and ensuring the 'sustainable development of the coastal stretches and marine environment through sustainable coastal zone management practices based on sound scientific principles' (MoEF, 2008: para. 2). Even though the CMZ Notification is indicative of the new thinking on coastal management, this law has been severely criticized by civil society for being the end of the road for coastal protection. For one, the strategy of the government to regulate or manage a vital ecosystem by means of notification rather than a fully fledged law, which would involve serious debate before the appropriate legislative body, has been severely decried. The striking feature of the 2008 Notification is that it introduces the concept of integrated coastal zone management (ICZM) into Indian coastal law. Owing to the widespread opposition to this law, the central government allowed this notification to lapse.

Recently, the central government, in a significant move to revamp the coastal law, introduced the CRZ Notification 2011, which supersedes the CRZ Notification 1991 (MoEF, 2011). The primary objective of the 2011 Notification is to provide for livelihood security to local coastal communities, to promote coastal environmental conservation and protection, and to promote sustainability based on scientific principles, taking into account the dangers posed by natural hazards and sea-level rise. The 2011 Notification is basically built on the framework provided by the 1991 Notification. Accordingly, it defines the CRZ as the land area from the HTL to 500 metres on the landward side along the sea front and the intertidal zone. Further, the CRZ also includes the water area between the LTL up to the territorial water limit of 12 nautical miles, thereby recognizing the land–sea interface of coastal regions.

The CRZ Notification 2011 retains the prohibitory scheme contemplated under the 1991 Notification, with a few more exemptions and clarifications. It reclassifies the coastal zone into five categories rather than four. The first three categorizations – namely, CRZ-I, II, and III – are almost identical to the categories under the 1991 Notification. Differences arise in respect of CRZ-IV, which formerly comprised Andaman and Nicobar, Lakshadweep and other islands; under the 2011 Notification, these island territories are excluded and subjected to a different legal framework. The CRZ-IV category now includes the water area from the HTL to 12 nautical miles on the seaward side. In addition, the 2011 Notification introduces a new category, CRZ-V, for areas that require special consideration and critically vulnerable areas. The Sunderbans biosphere reserve and certain other ecologically important areas are designated such. The 2011 Notification provides that these areas are to be managed on the basis of an integrated management plan. Thus ICZM seems to have been relegated to this narrow category, while for the other CRZ areas authorities are to prepare coastal zone management plans identifying and classifying the CRZ areas within their respective territories. The 2011 Notification also provides for regulation of land-based sources of marine pollution, adapting to shoreline changes, and introduces the concept of the hazard line as a tool to adapt to the problems posed by sea-level rise.

Despite these new additions, the fact is that these changes are only cosmetic; the law continues to be weak from the perspective of operationalizing ICZM goals and in preparing the Indian coastline to adapt to the challenges posed by climate change and sea-level rise. The MoEF, rather than enacting fully fledged legislation on the subject, has once again persisted with the strategy of managing a sensitive ecosystem by means of a notification, thereby retaining wide executive powers to amend the coastal law as and when it so desires. By doing so, it seems that the MoEF has once again sown the seeds of confusion. The CRZ Notification 2011 is not an ICZM-friendly law. The centrepiece of its provisions, where some form of integration is attempted, is the inclusion of 12 nautical mile water area along with the coastal land segment within its jurisdictional ambit. Apart from this spatial integration, there is very little in this law that furthers the concept of integration. Even though there are provisions in the 2011 Notification for the creation of a coastal zone management plan, such plans are more or less a coastal zone management map.

The CRZ Notification 2011 is not based on a principled approach to coastal and oceans management. There is no mention of any of the core principles that can help to advance the sustainable development of coastal areas and resources, such as the doctrine of public trust, 'polluter pays', and inter- and intra-generational equity. Even though elements of the precautionary principle can be discerned from some of its provisions, the pro-development inclination of the law defeats the underlying objectives of the precautionary rule. Thus it can be concluded that the battle to bring in coherence to the coastal law is far from over. The 2011 Notification is no improvement over the earlier coastal zone regulation notifications; rather, it dilutes the regulatory requirements of the previous law, opening pristine territory to rapacious development. It is only a question of time before we witness more rancorous court battles and controversies over this narrow strip of land and ocean space (Puthucherril, 2014).

Controlling sources of marine pollution

Globally, oceans are viewed as receptacles into which wastes can be dumped. Of late, several instances of algal blooms in the coastal waters on the east and west coasts of India have occurred, a pointer to its poor water quality, which is attributable mainly to the enormous quantities of wastes that are being dumped into them.

One of the most notorious cases on land-based sources of marine pollution (LBSMP) in India is that of Travancore Titanium Products, a public sector company owned by the State of Kerala. For decades, it has been dumping concentrated sulphuric acid and other pollutants into the Arabian Sea. One of the very few profit-making public sector units operating in industrially backward Kerala, it has been alleged that the state government has often looked the other way, enabling the company to secure almost an immunity from the operation of environmental laws. It was only when the Supreme Court Monitoring Committee on Hazardous Wastes started monitoring the company's operations that certain measures were taken to control the pollution.[20]

India operates one of the biggest shipbreaking industries in the world. The shipbreaking yard at Alang is the largest and, in its heyday, broke almost a ship a day. This industry brings in a huge amount of revenue to the government and to the operators of the shipbreaking units. However, the industry functions under extremely lethal and primitive conditions. Decrepit ships are generally driven at high tide at full speed on to the Alang beach and are broken down in the intertidal zone, with the tearing down and removal done almost entirely by human hand. Consequently, asbestos, tributyltin, polycholorinated biphenyls, and other harmful and carcinogenic substances get mixed with the coastal waters, severely contaminating the marine environment (Puthucherril, 2010). Interestingly, Pakistan and Bangladesh also have large shipbreaking yards. These three countries command almost 90 per cent of the business. In all three countries, there are no effective legal mechanisms to regulate the operation of the business. As a result, toxins from ships all over the world end up either in the Arabian Sea or in the Bay of Bengal. Studies have pointed out that elevated concentrations of butyltins in fish from Bangladesh could be associated with ship-scrapping activities. Similarly, in India, fish collected from the cities of Bombay and Calcutta show higher concentrations of butyltins than those from inland areas (Kannan et al., 1995). In spite of the growing awareness regarding the harmful consequences of persisting with the present harmful shipbreaking practices, there is a general reluctance on the part of these countries, all parties to UNCLOS, to adopt measures to control marine pollution and to introduce remedial measures.

Contrary to popular perception, it is municipal wastes, not industrial wastes, which form the single largest source of marine pollution in India. Approximately 7,663 million litres per day (MLD) of sewage is generated, while the industrial discharge stands at 1,719 MLD. Out of this huge quantity of municipal sewage, only 14 per cent (1,073 MLD) is treated, and a significant portion of the rest reaches the coastal waters untreated (TERI, undated). Although three-quarters of the volume of wastewater generated is from municipal wastes, industrial wastes contribute over half of the total pollutant load (MoEF, 1992). In addition, there are several non-point sources of marine pollution, for example run-off from agricultural inputs such as pesticides, insecticides, and fertilizers. India has no direct law to regulate LBSMP, but the Water (Prevention and Control of Pollution) Act 1974 has a bearing on the subject. This Act represents one of India's first attempts to deal comprehensively with an environmental issue. It seeks to prevent and control water pollution and to maintain or restore the wholesomeness of water through the establishment of water boards as nodal agencies for its effective implementation. The Act is comprehensive in its coverage and is applicable to streams, which, under section 2(j), includes rivers, water courses, inland waters, subterranean waters, and, more importantly, *sea or tidal waters*. The Act provides for a detailed regulatory system to prevent and control water pollution. It prohibits any person from knowingly causing or permitting poisonous, noxious, or polluting matter to enter into a stream, well, sewer, or on land. It also envisages a permit system or consent procedure to prevent and control water pollution. Several other laws, such as the CRZ Notification 2011 and the Insecticides Act 1968, also have a bearing on LBSMP.

In addition to this legislative framework, the role of the judiciary in controlling water pollution is also significant. Even though Indian courts have not had many opportunities to deal head-on with LBSMP, they have often sounded the wake-up call to the executive and the legislature on the necessity to act to maintain the health of rivers, thereby indirectly dealing with this issue. One of the earliest cases in this regard is the pollution of the River Ganges. In a trilogy of cases, the Supreme Court laid down a series of directions to cleanse the river of industrial and municipal sewage, which not only impaired the health of the river, but, more importantly, the ocean into which its waters ultimately flowed. Beginning with *M. C. Mehta v. Union of India* (1987),[21] which was initiated as a public-interest litigation to discipline the tanneries that were discharging effluents into the River Ganges, the Court adopted a pragmatic approach, directing the tanneries to set up in the minimum, primary treatment plants. In *Kanpur Municipalities* (1988),[22] the Court directed the Kanpur Nagar Mahapalika to expedite works initiated under the Ganga Action Plan to improve the sewerage system. Kanpur is one of the largest cities on the banks of the Ganges and was discharging nearly 274.50 MLD of sewage water into the river. Interestingly, although the directions issued to maintain the health of the river were mainly aimed at the Kanpur Municipality, the Court clarified that the directions, all things being equal, applied to all other *mahapalikas* and municipalities that have jurisdiction over areas through which the river Ganges flowed. Finally, in *M. C. Mehta (Calcutta Tanneries Matter) v. Union of India* (1997),[23] the Court concerned itself with the discharge of highly toxic chrome-based tanning effluents from the nearly 550 Calcutta tanneries. The Supreme Court directed the tanneries to relocate to a new leather complex set up by the state government. More importantly, the Court applied the 'polluter pays' principle and directed the tanneries to pay the costs to reverse the damage that they had caused to the environment.

Another significant case is *Vellore Tanneries* (1996),[24] which was directed against the pollution caused by the enormous discharge of untreated effluents by tanneries and other industries in Tamil Nadu into the river Palar, which in turn empties into the Bay of Bengal. The Supreme Court held that although the leather industry was a major foreign exchange earner and a major provider of employment, it had no right to destroy the ecology, degrade the environment, and pose health hazards. Frowning upon the traditional concept of development and ecology being opposed to each other, the Court imported the principle of sustainable development into Indian environmental jurisprudence. Outlining its salient features, the Court recognized the 'polluter pays' and precautionary principles as a part of the environmental law of the land. Voicing its concern over the environmental impacts of continued operation of the tanneries, the Court directed the central government to constitute an authority to implement the two principles.

The Court also imposed a pollution fine on all of the tanneries in the district, which sums, together with the compensation amounts recovered from the polluters, were to be transferred to an 'Environment Protection Fund' to be utilized for compensating the affected persons and also for restoring the damaged environment.

The courts have also addressed certain religious and social practices that contribute to marine pollution. In *V. Elangovan* (2004),[25] the Madras High Court held that there was no need for further orders, because the Pollution Control Board had already prevented immersion of idols made of plaster of Paris or those painted with toxic substances into the sea. However, it emphasized that even idols made of clay could be immersed only at designated disposal areas and under the supervision of a monitoring committee.

India operates several nuclear power stations in coastal states. The regular releases of reactor coolant water affect local ecology and, indirectly, the livelihood of fisher folk. Estimates show that while the fish yield from the Kalpakkam coast, which houses the Madras nuclear power station, stood at 1,171 tonnes before the plant opened, it has since declined to just 274 tonnes (Puthucherril, 2008a). Despite grave concerns, little action has been taken concerning nuclear power plants. In the wake of the 2004 tsunami, there was public panic regarding the safety of nuclear installations situated on the coasts. This led to public-interest litigation before the Bombay High Court.[26] One of the points raised in that case related to the several instances of radioactive nuclear waste leaks from Babha Atomic Research Centre into the Thane Creek, which, it was claimed, could adversely affect the marine environment and the coastal city of Mumbai. However, the Bombay High Court refused to look into the matter, because it was swayed by considerations of national security. Against the backdrop of the 2011 earthquake and tsunami in Japan, which caused serious damage and leakage of radiation from the Fukushima nuclear power plant, questions were again raised about the safety of the nuclear power plants in coastal regions. Of particular concern was the commissioning of the two Russian-built nuclear reactors at Koodankulam, near the Gulf of Mannar. Claims and counterclaims regarding environmental and health and safety aspects were agitated initially before the Madras High Court,[27] and subsequently on appeal to the Supreme Court.[28] In upholding the judgment of the Madras High Court, the Supreme Court virtually brushed aside the demand for a new environmental assessment, taking refuge behind the veil of policy. It was held that:

> . . . these issues are to be addressed to policy makers, not to courts because the destiny of a nation is shaped by the people's representatives and not by a handful of judges, unless there is an attempt to tamper with the fundamental Constitutional principles or basic structure of the Constitution.[29]

The Court rejected apprehensions expressed to hold that the '[n]uclear power plant is being established not to negate right to life but to protect the right to life guaranteed under Article 21 of the Constitution'.[30] Even though a utilitarian line of reasoning was adopted to justify the setting up of the nuclear plant, with emphasis on the public money spent on the project and the benefits that could ensue, overriding 'minor inconveniences', 'minor radiological detriments', and 'minor environmental detriments', the Court emphasized that the Atomic Energy Regulatory Board and the Ministry of Environment and Forests have a 'constitutional trust' to ensure that safety measures are taken before the plant commences its operation. Since this was a constitutional trust, the Court laid down fifteen directions to which the concerned authorities had to give effect prior to the commissioning of the plant.

India's geography may be redrawn in an effort to provide fresh water in arid regions of the country. With the escalating intensity of unavailable potable water looming large, interlinking of rivers – in particular, the River Ganges – has been mooted as a plausible solution to the water crisis. Even though the genesis of this project could be traced way back to the days of the British, in 2002 a passing observation by then President of India A. P. J. Abdul Kalam prompted public interest litigation before the Supreme Court in the matter of water-sharing between the states of Karnataka and Tamil Nadu.[31] This resulted in notices to the central and states governments, soliciting their views on the interlinking of rivers, to which only the central government and the state of Tamil Nadu responded. The Court subsequently passed an order for the central government set up a high-powered task force to build national consensus, to work out detailed plans, and to complete the entire interlinking network by 2016. The project is expected to extend irrigation to 150 million acres of land, to generate 60,000 MW of hydropower, to create 40 million jobs, and to reduce flooding. Even though the interlinking proposal was put on hold, it received a new lease of life when a new petition was instituted before the Supreme Court. In an elaborate judgment, the Court spoke on the benefits that could accrue if the project were implemented in a timely manner.[32] Despite the benefits, the river-linking project raises serious issues of environmental concerns that have transboundary ramifications. For instance, diverting water from the Ganges or the Brahmaputra could affect lower riparian nations such as Bangladesh and lead to increased salinization of deltas. From the perspective of LBSMP, the massive transfer of water from one river basin to another could affect the hydrological cycle, leading to diminished flow into the oceans, exacerbating marine pollution.

The legislative framework for controlling marine pollution in general is provided by the Maritime Zones Act 1976. Under sections 6(3)(d) and 7(4)(d), the central government has exclusive jurisdiction to preserve and protect the marine environment, and to prevent and control marine pollution within the continental shelf and EEZ. The Indian Coastal Guard patrols the maritime zones and is responsible for 'taking such measures as are necessary to preserve and protect that maritime environment and to prevent and control maritime pollution' (Coast Guard Act 1978, section 14(1)). The Offshore Areas Mineral (Development and Regulation) Act 2002 also empowers the central government to prescribe measures for preventing and controlling pollution, and for protecting the marine environment in offshore areas. Moreover, every holder of an operating right is liable for any pollution or damage to the marine environment consequent to activities relating to the operating right in offshore areas, and has to pay such compensation as may be determined by the administering authority, keeping in view the extent of pollution or damage.

India is flanked by two of the major oil transport chokepoints of the world – namely, the Strait of Hormuz and the Strait of Malacca, which lie off the west and east coasts of India respectively. Because of the narrowness of these lanes, they are accident-prone. India, a major importer of crude oil with a large number of oil tankers visiting its ports, is at a heightened risk of a major oil spill. Even though India has not yet ratified the 2001 International Convention on Civil Liability for Bunker Oil Pollution Damage, apart from certain provisions in the Merchant Shipping Act 1958, the Merchant Shipping (Prevention of Pollution of the Sea by Oil) Rules 1974 impose stringent rules to prevent oil pollution. India also promulgated the National Oil Spill Disaster Contingency Plan in 1996 to provide the basic framework and guidelines for a national response to a significant spill at sea. Under the Contingency Plan, the Coast Guard is designated as the central coordination authority for '[r]esponsibilities of co-ordination in the event of an oil spill at sea'. Port authorities retain responsibility for accidents within port limits. However, the inadequacy of these mechanisms was demonstrated in early August 2010 when two merchant vessels, *MSC Chitra* and *MV Khalijia-III*, collided off the Mumbai coast, resulting in an oil slick. Nearly 100 containers, some of them containing hazardous substances, also fell into the sea. Consequently, the Mumbai port was shut down for several days. The oil spill – one of the worst in India – contaminated marine life and destroyed nearly 300 hectares of mangroves.

India's coastlines are slowly falling prey to harmful predators from ballast water. The twelve major and 187 minor and intermediate ports that dot the Indian coastline are unwittingly becoming gateways for bio-invasion. For instance, nearly 5,000 ships call at the Mumbai Port annually, dumping approximately 2 million tonnes of ballast water (GloBallast, undated). It is estimated that eighteen non-native species of animals and plants have established themselves along India's coasts as a result. Another study that takes 1960 as the baseline (the Shipping Corporation of India was established the following year, heralding the birth of modern ship-ping activity in India) analysed the data collected from 111 papers published between 1960 and 2004 (Subba Rao, 2005). It notes that 205 taxa – representing thirty-two taxonomic groups, ranging from fungi to fish, which have never been reported previously in Indian waters – have been introduced from their native environs. This is not to suggest that ballast discharges are the sole medium for such introductions. Nevertheless, the study concludes that ballast has played a major role, particularly in cases in which the spread of biota is between geographically distinct or separate bodies of water, for example *Carcinus meanas* (Green crab) and *Mytlopsis saleii* (a kind of shellfish) between the Baltic Sea and the Atlantic Ocean, respectively, and Indian marine waters. Indirectly, these introduced species have affected the food chain. Moreover, there have been several recent harmful algal blooms in Indian oceanic waters. Even though the foremost cause of the spread of blooms is increased pollution, transport of exotic species via ballast water from ships is thought to be a trigger. Unfortunately, India has yet to enact a law to regulate these discharges or to accede to the 2004 International Convention for the Control and Management of Ships Ballast Water and Sediments (Puthucherril, 2008b).

India, therefore, does not have an effective regime to ensure ocean health. Apart from the general environmental laws applicable to the control of environmental pollution, there are no specific laws on LBSMP. Whatever measures are undertaken to control marine pollution are highly fragmented. Also, India has yet to ratify the 1972 Convention on the Prevention of Marine Pollution by Dumping of Wastes and Other Matter (the London Convention) and its 1996 Protocol. This is rather surprising, since India is a party to most of the major international environmental law instruments (see Table 17.1). There is also an urgent need to formulate legal measures to regulate ballast water discharge. India needs to rethink its strategy on marine pollu-tion and to formulate a comprehensive law to control it.

Protecting the rights of traditional fishermen and conserving marine species

Fishing, both inland and sea-based, provides a livelihood to millions in India. There are an estimated 3,302 marine fishing villages in India, with a total fishing population of 3.52 mil-lion, of whom 25.7 per cent are active fishers, 80.7 per cent of whom are full-time fishers (Central Marine Fisheries Research Institute, 2005). The potential of the fishing sector to gen-erate employment for coastal women through ancillary activities, and their empowerment, is also significant. In 2004–05, fisheries contributed 1.04 per cent to the national gross domestic product (GDP) (Rajagopalan, 2008). India also has huge fishing-related infrastructure. The Comprehensive Marine Fisheries Policy of 2004 notes that there are 1,896 traditional fish-land-ing centres, and thirty-three minor and six major fishing harbours, which serve nearly 208,000 traditional non-motorized craft, 55,000 small-scale beach-landing craft fitted with outboard motors, 51,250 mechanized craft (mainly bottom trawlers and purse seiners), and 180 deep-sea fishing vessels (Ministry of Agriculture, 2004).

With the introduction of large-scale mechanization in the fisheries sector heralding the 'Blue Revolution' in India, even though fish production increased significantly, modern harvesting techniques have had their share of pitfalls. Mechanized boats use nylon nets, which can sweep an area of its fish wealth, including the spawns, and destroy eggs, thus not allowing for regeneration, can increase turbidity in the sea bottom, thereby wiping out aquatic habitats and spawning grounds, and ultimately can lead to depletion of the resource. There have also been increasing encroachments into inshore waters by mechanized boats, leading to a sharp decline in the catch available to the traditional sector. Several incidents of gear entanglement and destruction of fishing gear used by artisanal fishermen have led to conflicting claims between the traditional and mechanized sectors over fishing grounds and fishing rights in several of the coastal states. Moreover, the introduction of highly destructive fishing devices, for example purse seining, led to a massive transfer of income from traditional fishermen to the few rich entrepreneurs for whom fishing is generally not the traditional source of livelihood. Thus traditional fishermen need the state protection afforded under article 46 of the Constitution of India.

One of the earliest national laws seeking to protect fisheries is the Indian Fisheries Act 1879, which is to be read as supplemental to other fisheries laws. The Act contains provisions that proscribe destruction of fish by explosives or through poisoning in inland and coastal waters. In addition, the state government is empowered to make rules to protect fish in selected waters for matters such as the dimension and kind of nets to be used, and the modes of using them, and prohibiting fishing in any specified water for a period not exceeding two years. Since fishing in territorial waters is primarily a state subject, several states have introduced 'spatial' remedial measures (restricting mechanized boats from harvesting within inshore waters), as well as 'temporal' measures (proscribing mechanized fishing during early monsoons), which have helped to safeguard not only the rights of the traditional fishermen, but also fisheries within territorial waters.

Although the states have jurisdiction over matters relating to fishing in territorial waters, in 2004 the Union Ministry of Agriculture issued the Comprehensive Marine Fishing Policy. The Policy raises several issues from the perspective of federal relations in ocean governance and fishing management. Prior to this, the Union government focused on development of the deep-sea sector, leaving coastal areas policy to be determined by the concerned state. However, non-integrated policy hampered realization of national objectives, and therefore the present Policy was issued, with the objective 'to bring the traditional and coastal fishermen also in to the focus together with stakeholders in the deep-sea sector so as to achieve harmonized development of marine fishery both in the territorial and extra territorial waters of our country' (Ministry of Agriculture, 2004: para. 1.1). The Policy seeks:

> . . . a focused endeavour from the coastal States and the Central Departments with full appreciation of the international conventions in force for conservation, management and sustainable utilization of our invaluable marine wealth, without losing its relevance to the food and livelihood security of the coastal communities, which totally depend on this.
>
> *(Ministry of Agriculture, 2004: Foreword)*

The Policy seeks to develop marine fisheries through an integrated approach. It calls for developing a master plan on infrastructural improvement, with emphasis on private sector initiatives. It calls for a review of the existing legal framework on fisheries operations, and for introduction of new legal mechanisms in fisheries harbour management and conservation of resources, as well as limits to access in fisheries, harmonization of existing national laws with the international law on fisheries, and development of 'mutually agreeable systems' with 'friendly neighbouring countries' to address the problem of small, mechanized boats straying into each other's territorial waters. Participation in regional fisheries management bodies, eco-labelling of marine products, use of information technology, and strengthening the marine fisheries database have also been identified as key areas to be attended to.

Since marine fishing is the sole livelihood of the majority of fishermen households, the Policy attaches top priority to their social security and economic well-being. It calls for strengthening the cooperative movement in the fishing sector, introducing uniformity in the different welfare schemes, encouraging greater participation of cooperatives, non-governmental organizations (NGOs), and local self-governments in the implementation of welfare schemes, providing identification cards to each household, introducing measures to eliminate middlemen, and developing programmes to improve safety at sea and housing schemes.

To curb overexploitation in inshore waters, the Policy advocates a 'stringent fishery management system'. It calls for a review of existing marine fishing regulation and suggests a new model Bill on coastal fisheries development and management. It also calls for measures such as closed seasons, a strict ban on destructive methods of fishing, the regulation of mesh size, the introduction of resource enhancement programmes, the registration of boat-building yards, the licensing of new fishing units, a prohibition on catching juvenile and non-targeted species, and the posting of observers on commercial fishing vessels. Importantly, the Policy considers the effect of environmental factors on the health of living resources and stresses the need for stringent implementation and enforcement of laws relating to environmental pollution. It calls for sensitizing fishermen to the harmful effects of land-based pollution on coastal waters and the introduction of mangrove-planting programmes, with community participation. Andaman and Nicobar and the Lakshadweep Islands have their own EEZs, and thus are vulnerable to illegal, unregulated, and unreported (IUU) fishing by foreign-flagged vessels. The Policy advocates several measures to protect fisheries and to augment fish production from the waters surrounding the islands.

One of the first coastal states to adopt legal measures to protect the rights of traditional fishermen and to conserve marine fish resources is Kerala. This state has successfully employed zoning as a management tool with which to achieve these twin objectives. Under section 4(1) of the Kerala Marine Fishing Regulation Act of 1980, the state government can regulate, restrict, or prohibit fishing in the territorial waters by notifying orders to this effect. These orders can specify areas in which fishing can be done by class or classes of fishing vessels, the number of vessels, fishing gear, the species that can be caught, and the fishing season. Section 4(2) specifies the objectives for which the powers under section 4(1) are to be exercised – namely, to protect the interests of those fishermen who use traditional fishing craft, to conserve fish and to regulate fishing on scientific basis, and to meet the need to maintain law and order in the sea. Several state policies to protect the rights of traditional fishermen and to augment fishery resources have been subjected to judicial scrutiny.[33] Following the Kerala model, several coastal states have introduced similar legislation, such as the Karnataka Marine Fishing (Regulation) Act 1986, the Orissa Marine Fishing Regulation Act 1982, and the Tamil Nadu Marine Fishing Regulation Act 1983, and these have also involved judicial intervention.[34] Despite these measures, degradation of fisheries and conflicts between subsistence fishers and mechanized commercial fishers continue.

Another important legislative measure to secure India's ocean space from poaching by foreign fishing vessels is the national Maritime Zones of India (Regulation of Fishing by Foreign Vessels) Act 1981. Under this Act, no foreign vessel can be used for fishing within any maritime zone of India except in accordance with a licence or permit. While the owner of a foreign vessel has to obtain a licence subject to conditions and restrictions, an Indian citizen who wishes to use such a vessel must obtain a permit. The Act empowers the Coast Guard or authorized officers to stop or board a foreign fishing vessel in any maritime zone of India and to search the vessel to ascertain whether or not the requirements of this law have been complied with. The officers have the power to seize and detain a vessel, including any fishing gear, fish, and equipment, and to require the master of the vessel to bring it to any specified port. They also have the power to arrest any person suspected to have committed an offence.

Exploitation of fishery resources in the territorial waters has reached breaking point, and consequently Indian fishermen who have benefited from mechanization are increasingly seeking out underexploited resources in the EEZ. In a move that can have far-reaching implications for fishing rights and federal relations in fisheries management, the Union Ministry of Agriculture introduced the draft Marine Fisheries (Regulation and Management) Act 2009 (draft Marine Fisheries Act). Basically, this draft seeks to repeal the Maritime Zones of India (Regulation of Fishing by Foreign Vessels) Act 1981 and to regulate all vessels engaged in either direct or indirect exploitation of fisheries resources in the maritime zones of India. Accordingly, it provides that no vessel shall engage in any fishing or fishing activity within any part of the maritime zones of India, except with the prior written permission of the central government. Indian fishing vessels are deemed to have permission for undertaking fishing within the limits of the territorial waters if there is compliance with all laws, rules, and regulations of the state under whose jurisdiction the relevant territorial waters fall. However, if the Indian fishing vessel undertakes fishing or any fishing activity in any maritime zone outside the territorial waters, then a permit is required under section 3. In cases in which an Indian fishing vessel violates section 3, the owner and the master of the fishing vessel can be punished with imprisonment or a fine. Certain exemptions from prosecution are provided for Indian vessels in cases in which the vessel has strayed innocently into the EEZ. For foreign fishing vessels, the period of imprisonment is the same, but the fines are higher.

Another major, but contentious, provision of the draft Marine Fisheries Act relates to the power of the central government to adopt fisheries management plans in any maritime zone for conservation and regeneration of fish stocks, and to ensure that fishing is done in an environmentally sustainable and peaceful manner. Accordingly, all permits granted under section 3 are subject to the fisheries management plan and, in case of any inconsistency, all permits are deemed amended by the plan so as to remove the inconsistency. Interestingly, the central government can make a fisheries management plan for territorial waters as well, provided that the concerned state government is consulted. The fishing community and political parties strongly objected to the introduction of this law, calling it 'an overt attempt to trample upon the rights of traditional fishermen while opening up India's rich exclusive economic zone to powerful foreign fishing trawlers and to big indigenous players' (PTI, 2009). Moreover, through its power to frame fisheries management plans, the central government can exercise greater control over the management of territorial waters, which may indirectly affect the power of states to manage such waters. Owing to the widespread opposition, this law has been mothballed for now.

Traditionally, artisanal fishermen have utilized beachfronts for their homes and as a base for their fishing activities – that is, for launching their catamarans, for boat storage, for fish drying, and for gear repair and storage. Access to the beachfront is critical for these fishers. The generally illiterate fishermen do not have rights over their land based on any documentary records, because most shore areas are held in the trusteeship of religious institutions or other community-based institutions. Consequently, the government has been able to acquire these coastal lands for developmental projects by exercising its power of eminent domain. Also, measures adopted by the government to stop the ingress of a receding sea, for example building sea walls, have adversely affected the access of local fishing communities to beach areas. In the aftermath of the tsunami and the large death toll attributed to the close proximity of fishing hamlets to the sea, the government of one of the worst affected Indian states, Tamil Nadu, issued a notification to move fishing communities away from the coast and to discourage settlements, so as to free beachfront land (Government of Tamil Nadu, 2005). The present policy of most coastal states is to discourage fishing communities from setting up hamlets or continuing habitation in close vicinity to the sea. Although justifiable on the ground of safety, NGOs have criticized these measures as an attempt to limit access claims by fishing communities and to make the land available for developmental projects, hotels, and special economic zones (SEZs).

Participatory decision-making has come to be recognized as a vital tool for effective governance. As noted earlier, the landmark step in India came about with the establishment of *panchayats* as the third tier of government through the Seventy-third Constitutional Amendment. Additionally, since 1994, participatory decision-making has been built into the environmental impact assessment (EIA) process by means of 'public hearing'. In 2006, a new notification replaced 'public hearing' with 'public consultation' (Government of India, 2006). Even though the new procedures were designed to improve participatory decision-making, the public consultation process has been turned into a forum for project proponents to prove the viability and eco-soundness of their proposals at any cost. The emphasis, it seems, is more on formal, rather than substantial, compliance with procedure. Often, practices such as bribing influential elders, politicians, and officials, providing wrong information about the project, and obstructing the presentation of opinions that may be critical to the project are resorted to. Of late, the state itself has resorted to violence to secure its interest in almost all public hearings, pointing to the hollowness of the whole exercise.

Although having a wider scope, the Scheduled Tribes and Other Traditional Forest Dwellers (Recognition of Forest Rights) Act 2006 is highly relevant to tribal communities that live in forests situated in coastal areas, such as those in the Sunbderbans Tiger Reserve, or in the Andaman and Nicobar Islands (the *Jarawas*). The law recognizes certain 'forest rights' in relation to forest-dwelling scheduled tribes and other traditional forest dwellers, including the right to fish. Forest rights holders also have a duty to protect wildlife, biodiversity, and ecologically sensitive areas.

It must also be pointed out that, when India decided to revamp its coastal law, the expert committee under the chairship of M. S. Swaminathan recommended, among other things, the enactment of a special legislation to protect the rights of traditional fisherfolk. The draft Traditional Coastal and Marine Fisherfolk (Protection of Rights) Act 2009 recognizes the right of ownership and access to areas by fisherfolk, traditional rights customarily enjoyed by traditional fisherfolk, the right of access to biodiversity, and the community right to intellectual property and traditional knowledge. An important step to protect the rights of traditional fisherfolk, this draft is yet to be enacted into a law.

Apart from fisheries, there are several other marine species that require protection. The Wildlife Protection Act 1972 imposes a total and complete prohibition on hunting of wild animals except in exceptional circumstances and is the pivotal legislation for protection of marine wildlife. Marine species specified in the Act's schedules include the snubfin dolphin, the green sea turtle, the whale shark, the sea cucumber, and various mollusca, sponges, and corals. The rights of traditional communities that have depended on these resources for sustenance remains protected. For example, if a fisherman residing within 10 km of a sanctuary or national park inadvertently enters its territorial waters on a boat that is not used for commercial fishing, the boat is not to be seized. The hunting rights conferred on the scheduled tribes of the Nicobar Islands are also protected.

Courts have also played an important role in conserving marine species by interpreting and illuminating various regulatory provisions of the Wildlife Protection Act 1972.[35] *Centre for Environmental Law World Wide Fund for Nature (WWF) India v. State of Orissa* (1999)[36] dealt with proposals for construction and developmental projects within and around a mangrove sanctuary that also includes a rookery for olive ridley sea turtles (*Lepidochelys olivacea*), an endangered species. In an elaborate judgment, the Orissa High Court upheld the concept of 'meaningful life' guaranteed by article 21 of the Constitution and held that the implementation of the projects could normally not be prohibited merely because of their possible impacts on the Bhitarkanika ecosystem, including the nesting area. The Court laid down guidelines for implementing the projects, including curbing migration of humans to the surrounding areas, the realignment of a road, measures to check illegal fishing, requiring vessels to use turtle extrusion devices and measures to prevent the poaching of sea turtles, and establishing a sea turtle research programme.

Coral reefs are diverse and vulnerable ecosystems. On the Indian subcontinent, reefs are distributed along the east and west coasts. Fringing reefs are found in the Gulf of Mannar, Palk Bay, and on the Andaman and Nicobar Islands. Platform reefs are seen along the Gulf of Kutch, and atoll reefs are found in the Lakshadweep archipelago. Increasing human population and anthropogenic pressures have severely affected coral distribution and biodiversity. Global warming has also led to adverse impacts on the survival of coral reefs; the bleaching phenomenon of 1998 in the Indian Ocean is reported to have caused considerable damage to coral reefs in the Indian coast. An interesting question placed before the Madras High Court in *State of Tamil Nadu v. Messers Kaypee Industrial Chemicals Private Ltd* (2005)[37] related to the use of corals for the manufacture of lime. Fisherfolk living in the vicinity of the Gulf of Mannar national marine park had long collected corals washed ashore to be supplied to a lime factory. However, it was alleged by the state that the corals were collected from the seabed by breaking the natural attachment, thereby causing extensive ecological damage, and therefore that legal action could be taken under the Wildlife Protection Act. Trawlers that moved inside the creeks for illegal fishing also damaged the corals. The respondents contended that the fisherfolk collected only dead corals, that death being attributable to the discharge of hot water from a nearby thermal power station and global warming. Striking a balance between these conflicting claims, the Madras High Court held that, as long as the respondents purchased dead corals that were washed ashore, the authorities could not interfere. However, if the seafarer were to kill the animal, the Wildlife Protection Act would be applicable.

Coastal aquaculture

Traditionally, aquaculture was practised by Indian fishermen by following the rice/shrimp rotating system, wherein rice was grown during a part of the year, while shrimp and other fish species were cultured during the remainder. As part of the Blue Revolution, this system was replaced by a more intensive method of shrimp culture, which enabled production of thousands of kilograms per hectare. Increasing numbers of coastal areas in the country were being brought under semi-intensive and intensive modes of shrimp farming, posing severe environmental risks.

Public-interest litigation brought before the Supreme Court sought to enforce the Coastal Zone Regulation (CZR) Notification 1991, and to stop intensive and semi-intensive types of prawn farming in ecologically fragile coastal areas.[38] Affirming the state's obligation to control marine pollution and to protect the coastal environment, the Supreme Court held that since the shrimp culture industry was neither directly related to the waterfront nor needing foreshore facility, the establishment of coastal shrimp farms up to 500 metres from the HTL and the line between the LTL and the HTL was prohibited under the CRZ Notification 1991. Even though shrimp aquaculture industry had the singular distinction of earning maximum foreign exchange for the country, the Court concluded that it caused considerable ruin of the local ecology. Accordingly, the Court ordered the demolition and removal of all aquaculture industries, except traditional and improved traditional types, operating in the CRZ. Relying on the precautionary and the 'polluter pays' principles, the Court also directed the central government to constitute an authority under the Environment (Protection) Act 1986, with powers to protect ecologically fragile areas threatened by the shrimp culture industry. Henceforth, any proposal to set up a shrimp industry or shrimp pond in ecologically fragile coastal areas has to pass a strict environmental test based on EIA and has to be scrutinized by this authority. The authority was also directed to assess the loss to the ecology in the affected areas and to seek compensation from the operators of intensive shrimp farms for affected persons and restoration of the environment.

The judgment threw nearly 300,000 aquaculture workers out of employment and resulted in a large number of review petitions to the Supreme Court. In *Gopi Aqua Farms v. Union of India* (1997),[39] the Supreme Court declined to entertain the petition. It was left to the executive and the legislature to resolve the issue. The government decided to amend the CRZ Notification 1991 to clarify that aquaculture was not intended to be a prohibited activity. Finally, the Coastal Aquaculture Authority Act 2005 was enacted, which provides for the establishment of a Coastal Aquaculture Authority and regulates coastal aquaculture activities. Section 3 of the Act prescribes guidelines grounded on the principle of responsible coastal aquaculture. The Act requires registration of coastal aquaculture farms, and empowers the Coastal Aquaculture Authority to make regulations for the construction and operation of aquaculture farms in coastal areas, to inspect these farms with a view to ascertaining their environmental impact, and to order removal or demolition of any coastal aquaculture farm that causes pollution, after a hearing with the occupier of the farm.

Conserving coastal wetlands and mangroves

India is a party to the 1971 Convention on Wetlands of International Importance (Ramsar Convention) and has twenty-five Ramsar sites, covering an area of 677,131 hectares (Khan, undated). Moreover, 103 wetlands have been identified as national wetlands, and programmes such as the National Lake Conservation Plan protect urban wetlands from developmental pressures. However, the country's wetland ecosystems are rapidly being lost or degraded, because the government has yet to formulate a clear and well-defined conservation policy. There have been no national laws on wetland conservation. One major reason for the lack of wetland conservation at the national level is that wetlands are often equated with land, which is a state subject under the Constitution. Consequently, any protective measures will have to emanate from the state level. Even though various state governments have laws that directly or indirectly deal with wetland conservation, such as the Kerala Land Utilization Order 1967, the approach is fragmented, with different government agencies and laws complicating the picture. Often, the response at the state level is inadequate and not timely, with the result that wetland degradation continues.

Among federal statutes, the Wildlife Protection Act 1972 defines land to include wetlands. The Act also provides for the establishment of sanctuaries and national parks, and these can include wetland areas. However, strategies for wetland conservation are limited. More importantly, the Environment (Protection) Act 1986 can be invoked to protect and conserve wetlands. 'Environment', as defined under section 2 of the Act, includes water, air, and land, and their interrelationship with human beings and other living creatures, plants and microorganisms, and property. The CRZ Notification and the Environment Impact Assessment (EAI) Notification passed under this Act are also relevant to wetland conservation. Recently, the central government offered the Wetlands (Conservation and Management) Rules 2010, as subordinate legislation under the Environment (Protection) Act 1986 (MoEF, 2010). Under these

Rules, 'wetlands' include coastal wetlands. Based on the significance of the functions performed, wetlands are classified into six categories. The Rules prohibit certain activities, including wetland reclamation, and subject other activities, such as withdrawal of water, to prior approval from the state government. There is also provision for the establishment of a Central Wetlands Regulatory Authority. Court cases relating to coastal wetland conservation generally have not served to protect these areas from coastal development initiatives.[40] Hopefully, these Rules may stem the tide of wetland degradation in the country.

The entire Indian coast was once rich in mangrove vegetation. Historically, the Kerala coast is described as a mangrove forest. However, most of the mangroves in this state have vanished (Swaminathan, 2005). Nonetheless, India still harbours extensive mangrove swamps in the alluvial deltas of the Ganges, Mahanadi, Godavari, Krishna, and Cauvery rivers, and on the Andaman and Nicobar group of islands. The mangroves of Sunderbans, home to the famous Royal Bengal tiger, are the largest single block of tidal holophytic mangrove in the world. It is estimated that the total area covered by mangroves in India is about 6,700 km^2, which amounts to about 7 per cent of the world's mangroves (Government of India, 2001). Apart from the threats that these unique ecosystems face from development projects, many are on the brink of extermination as a result of the effects of global warming. For instance, the Sunderbans run the risk of being lost to the rising sea. The primary forest management and conservation laws, the Indian Forest Act 1927 and the Forest Conservation Act 1980, are the legislative basis for mangrove protection. In addition, there are certain programmes on mangrove conservation (Kumar, undated). However, economic development projects continue to encroach on mangrove forests.[41]

The list of cases is not exhaustive, but it does illustrate certain trends in the law relating to conservation of coastal and marine ecosystems, the most striking being the absence of consistency in approach. The primary reason for this situation is the lack of well-defined policies and laws on conservation of wetlands and mangroves.

Marine protected areas

The coastal environments of India are exceptionally diverse and highly productive. Conservation of coastal and marine biodiversity is done through the designation of marine protected areas (MPAs), administered under the Wildlife Protection Act 1972. Additionally, section 37(1) of the Biological Diversity Act 2002 empowers the state government to designate, in consultation with local bodies, areas of biodiversity importance as 'biodiversity heritage sites'. This law identifies the following categories of protected area: national parks; sanctuaries; conservation reserves; community reserves; and tiger reserves. To date, no MPA has been declared under the last three categories.

Under section 18 of the 2002 Act, the state government can designate a sanctuary in areas within reserve forest and territorial waters if it considers the area to be of adequate ecological, faunal, floral, geo-morphological, natural, or zoological significance, warranting protection. Prior permission of the central government must be obtained by the state government, and the area of the territorial waters to be included within the sanctuary has to be determined by the Chief Naval Hydrographer of the central government and after taking measures to protect the interests of local fishermen. Additionally, the right to innocent passage of any vessel or boat through such territorial waters is not to be affected. Once a sanctuary has been decided upon and its area demarcated, no alteration of the boundaries can be made, except by resolution passed by the state legislature. Moreover, only certain categories of person are permitted to reside or enter into the sanctuary, in accordance with the conditions of a permit granted under section 28 – that is, for the investigation or study of wildlife, for photography, for scientific research, for tourism, and for transacting lawful business with any person residing in the sanctuary. The Act also empowers the state government to constitute national parks under section 35 under a similar process.

India has a network of 611 protected areas. These comprise ninety-six national parks, 510 wildlife sanctuaries, three conservation reserves, and two community reserves, all covering a total of 155,978.05 km² – or approximately 4.75 per cent of the geographical area of the country – including both terrestrial and marine ecosystems. Out of these, there are thirty-one MPAs, of which eighteen are fully under the marine environment, the remaining thirteen having both aquatic and terrestrial components. An additional 100 protected areas have terrestrial or freshwater ecosystems that border seawater or partly contain coastal and marine environments (Rajagopalan, 2008). All of the MPAs have been constituted either as national parks or wildlife sanctuaries under the Wildlife Protection Act. The Government of India has also constituted three marine biosphere reserves – namely, Sunderbans, the Gulf of Mannar, and the Great Nicobar – under the United Nations Educational, Scientific and Cultural Organization (UNESCO) Man and Biosphere Programme.

In spite of the special status and protections that these ecosystems enjoy, they are under great threat from developmental activities.[42] The Sethusamudram Link Canal is one of the most controversial projects that independent India has undertaken and highlights the need for a new strategy on marine environment protection. At present, there is no continuous navigation channel connecting the east and west coasts of India, which requires vessels to navigate around the Sri Lankan coast (the existing waterway is shallow owing to the presence of the Adam's Bridge/ Ram Sethu, a discontinuous chain of sandbars connecting the Indian subcontinent with Sri Lanka). Conceived during the days of the British, but never implemented, the Sethusamudram Ship Channel Project envisages the dredging of a channel through the Palk Straits between India and Sri Lanka, linking the Gulf of Mannar with the Palk Bay. Such a passage through India's territorial waters would save up to 424 nautical miles (780 km) and 30 hours of sailing time. The alignment of the Sethu Canal runs close to the Gulf of Mannar Marine Biosphere Reserve. Apart from the severe effects on the rich biodiversity of that area that would result from the required dredging, once completed traffic would increase significantly and this could seriously jeopardize the local marine ecosystems. In *O. Fernandes v. Tamil Nadu Pollution Control Board* (2004),[43] the petitioner moved for a direction to declare the public hearing, conducted in connection with the Sethu Project in the coastal districts, ineffective, since there was no compliance with the requirements of law. In that case, the public hearings were conducted on the basis of the Rapid Environment Impact Assessment Report and not on the basis of the Comprehensive Environment Impact Assessment Report. Moreover, the public hearings were characterized by pandemonium – shouting matches for political parties who were determined to uphold their political interests. Dismissing the petition, the Court held that the initiator of the project, the Tuticorin Port Trust, had submitted the requisite materials to the Tamil Nadu State Pollution Control Board, which, after being satisfied with them, had called for the public hearing. Characterizing the project as one in the national interest, the Court observed:

> [W]e should not obstruct the scientific and technical progress of the country in the name of environment protection. No doubt, the environment has to be protected, but at the same time. We must never overlook the basis aim of our country, which is to make India a powerful and modern industrial state. Today the real world is cruel and harsh. It respects power, not poverty or weakness. The truth is that Indians, despite being intelligent and industrious people, are not respected by Westerners, not because our skin is brown or black in colour, but because our country is poor. Nobody respects

the poor. When the Chinese and Japanese were poor people they were derisively called 'yellow' races by the Westerners, but today they are industrialized and powerful nations, and now nobody dares to call them that. Similarly, if we wish to get respect in the world community we must make our country highly industrialized and prosperous.[44]

The legal battle over the Sethu took an unprecedented turn when a public-interest litigation was filed again before the Madras High Court,[45] seeking a writ of mandamus to direct the government to implement the Sethusamudram Canal Project by following any other alternate alignment without affecting or destroying the historic Rama Sethu/Adams Bridge. According to legend, the Sethu was built by the *Vanarasena* (the monkey army), headed by Hanuman (the Monkey God), so that Lord Rama could cross over to Lanka to destroy the evil King Ravana, who had abducted Sita Devi, his wife. It was contended that Adams Bridge/Rama Sethu is a national monument and that the government had a constitutional duty to protect it. Accordingly, the Court directed the Union Ministry of Culture and Tourism to file a counter affidavit, explaining whether any study had been undertaken by the archaeological or any other department in respect of Adams Bridge/Rama Sethu and whether the bridge could be regarded as a national monument. The Union of India was also to explain whether the project could be implemented by resorting to some other route. The battle over the Sethu Project is now being waged before the Supreme Court of India, where an expert committee is examining the possibility of an alternative alignment (Legal Correspondent, 2008).

Concluding comments

The broad brush used in this chapter to describe India's tryst with ocean and coastal governance reveals that, despite being one of the earliest countries to have adopted an ocean policy and having taken a lead role in ratifying several international ocean and environmental law instruments, ocean governance in India still continues to be sectoral and highly fragmented. As seen, the Ocean Policy Statement of 1982 (DOD, 1982) and the recent Vision Perspective Plan 2015 (DOD, 2002) are documents that articulate the need for a sound science for effective ocean governance. At the same time, internationally the concept of ocean governance has evolved over the years to encompass the broader objective of a principled approach to ocean governance wherein sustainability is the key to be achieved by furthering integration, precaution, stewardship, the ecosystem-based approach, and community-based management. Ocean governance is no longer based solely on the development of scientific knowledge. India has definitely incorporated some of these concepts, in varying degrees, in several of its laws and policies on ocean and coastal management. However, most of these regimes have evolved in isolation, as a result of being the product of discursive thinking, and are responses to specific problems as and when they arise. This has led to jurisdictional fragmentation and an inability to translate the concept of principled ocean governance into actual practice, thwarting sustainable ocean and coastal governance.

India is at a critical juncture in its history. In the coming years, India's potential as a maritime entity may well be as great, if not greater, than its potential as a land-based actor. At the heart of the mismanagement of its oceanic and maritime estate are the pressures of development, aggravated by the inability of the present policy and legal regimes to resolve the conservation–development stalemate to further sustainable development. Dwarfing all of these concerns will be the dominating factor that will gravely impact on the oceanic and coastal environment – namely, climate change and the abnormal and substantial rise in sea level that is bound to result in the need to re-engineer India's ocean law and policy. Thus there is an urgent need for a shift from the present piecemeal approach to ocean governance toward a more holistic and comprehensive policy and legal framework for integrated ocean and coastal area management, geared to attain the stewardship and sustainable use and development of the country's marine territory and coastal areas.

Notes

1 *Annakumaru Pillai v. Muthupayal*, 27 ILR 1904 (Mad.) 551.
2 *Subhash Kumar v. State of Bihar* (1991)1 SCC 598.
3 *Visakha SPCA v. Union of India*, 2000 (6) ALD 539.
4 *Tehri Bandh Virodhi Sangarsh Samiti v. State of UP*, 1991 Supp. 1 SCC 44.
5 *Indian Council for Enviro-Legal Action v. Union of India* (1996) WP No. 664 of 1993.
6 *A. P. Pollution Control Board v. Prof. M. V. Nayudu* (1999) 2 SCC 718.
7 *M. C. Mehta v. Kamal Nath* (1997) 1 SCC 388.
8 *State of Tamil Nadu v. M. S. Hind Stone*, AIR 1981 SC 711, 750.
9 See also US Department of State (1981) for a list of maritime boundary agreements between India, Indonesia, and Thailand, 1971–79.
10 See Sharma and Sinha (1994: 113–15); ITLOS (2013); Jayewardene (1990: 193); Tanaka (2011).
11 See Shah (2009).
12 *In the Matter of the Bay of Bengal Maritime Boundary Arbitration between the People's Republic of Bangladesh and the Republic of India* (7 July 2014) PCA, Judgment online at http://www.pca-cpa.org/showpage. asp?pag_id=1376 [accessed 11 July 2014].
13 *S. Jagannath v. Union of India* (1997) 2 SCC 87.
14 *Indian Council for Enviro-Legal Action v. Union of India* (1996) WP No. 664 of 1993.
15 *Visakha SPCA v. Union of India*, 2000 (6) ALD 539.
16 *Citizens Interest Agency v. Lakeshore Hospital and Research Centre Pvt. Ltd*, 2003(3)KLT424.
17 *Visakhapatnam Municipal Corporation v. Government of India Ministry of Environment & Forests* (16 August 2001), Judgment online at http://www.indiankanoon.org/doc/1151324/ [accessed 2 February 2014].
18 *Goa Foundation, Goa v. Diksha Holdings Pvt. Ltd* (10 November 2000), Judgment online at http://indiankanoon.org/doc/1898591/ [accessed 2 February 2014].
19 (1991) 2 SCC 539.
20 *Research Foundation for Science, Technology and Natural Resource Policy v. Union of India*, WP No. 657 of 1995.
21 (1987) 4 SCC 463.
22 AIR 1988 SC 1115.
23 (1997) 2 SCC 411.
24 *Vellore Citizens Welfare Forum v. Union of India* (1996) 5 SCC 647.
25 *V. Elangovan v. The Home Secretary, State of Tamil Nadu* (2004) WP No. 25586 of 2004.
26 *Citizens for a Just Society, through its Vice President, K Pullaiah v. Union of India*, 2005 (5) Bom CR 316.
27 *G Sundarrajan v. India* (31 August 2012), WP No. 24770 of 2011 (Madras HC).
28 *G Sundarrajan v. India* (6 May 2013), CA No. 4440 of 2013 (SC India).
29 Ibid., [22].
30 Ibid., [184].
31 *In re Networking of Rivers*, 2002 (8) SCALE 195; 2003 (1) SCALE 2.
32 *In re Networking of Rivers* (27 February 2012), WP Civil No. 512 of 2002 (India SC).
33 *Cochin Trawl Net Boat Operators Association v. State of Kerala*, AIR 1992 Ker. 342; *Kerala Swathanthra Matsya Thozhilali Federation v. Kerala Trawlnet Boat Operators Association*, 1994 (5) SCC 28; *State of Kerala v. Joseph Antony*, 1994 AIR 721.
34 *Goa Environment Federation v. State of Goa*, WP No. 212 of 2000 (High Court of Bombay at Goa); *Abdul Hameed v. State of Maharashtra*, 2004 IND LAW Mum. 510.
35 *Sri Satyabrata Majhi v. State of Orissa*, 2000 (I) OLR 230.
36 AIR 1999 Ori. 14.
37 AIR 2005 Mad 304.
38 *S. Jagannath v. Union of India* (1997) 2 SCC 87.
39 AIR 1997 SC 3519.

40 *Goa Foundation v. Konkan Railway Corporation,* AIR 1992 Bom. 471; *People United for Better Living in Calcutta-Public v. State of West Bengal,* AIR 1993 Cal. 21; *Ramgopal Estates Private Ltd v. State of Tamil Nadu,* 2007 INDLAW MAD 964.

41 *Ajit D. Padiwal v. Union of India,* AIR 1998 Guj. 147.

42 *Essar Oil Ltd v. Halar Utkarsh Samiti,* AIR 2004 SC 1834; *T. N. Godavarman Thirumalpad v. Union of India,* 2006 INDLAW SC 123.

43 (2004) WP No. 33528 (High Court of Judicature at Madras).

44 Ibid., [17].

45 *Rama Gopalan v. Union of India* (2007) WP Nos 18076, 18223, & 18224 (High Court of Judicature at Madras).

References

Anand, R. P. (1983) *Origin and Development of the Law of the Sea: History of International Law Revisited,* The Hague: Martinus Nijhoff.

Central Marine Fisheries Research Institute (2005) 'Marine Fisheries Census for India', online at http://bobpigo.org/uploaded/bbn/sep_06/pages33-36.pdf [accessed 2 February 2014].

Chakravarti, P. C. (1930) 'Naval Warfare in Ancient India', *The Indian Historical Quarterly,* 4(4): 645–64.

DAOLOS   *See* United Nations Division for Ocean Affairs and the Law of the Sea

Department of Ocean Development (DOD) (1982) 'Ocean Policy Statement', online at http://www.dod.nic.in/dodpol.htm [accessed 23 January 2014].

Department of Ocean Development (DOD) (2002) *Vision Perspective Plan 2015,* New Delhi: DOD, on file with the author.

Department of Shipping (2013) 'Ports Wing', online at http://www.shipping.gov.in/index1.php?lang=1&level=0&linkid=16&lid=64 [accessed 20 January 2014].

Energy Alternatives India (EAI) (undated) 'Ocean Energy', online at http://www.eai.in/ref/ae/oce/oce.html [accessed 10 July 2014].

Energy and Resources Institute (TERI) (undated) *National Programme of Action for Coastal Pollution,* online at http://www.teriin.org/upfiles/projects/ES/2005EE25_20090104104444.pdf [accessed 2 February 2014].

Gaur, A. S., Sundaresh, and Tripati, S. (2004) 'An Ancient Harbour at Dwarka: Study Based on the Recent Underwater Explorations', *Current Science,* 86(9): 1256–60.

Global Ballast Water Management Programme (GloBallast) (undated) 'Ballast Water Hazard . . . India Awakens . . . ', online at http://www.globallastwaterindia.com/images/shell_brochure.pdf [accessed 2 February 2014].

Government of India (2001) *India's Second National Report to the Convention of Biological Diversity,* New Delhi: Ministry of Environment and Forests.

Government of India (2006) *Environmental Impact Assessment Notification 2006,* online at http://www.env-for.nic.in/legis/eia/so1533.pdf [accessed 1 February 2014].

Government of Tamil Nadu (2005) *Government Order (G.O.) Ms. No. 172, 30 March 2005,* online at http://www.tn.gov.in/tsunami/Projects/GOs/rev-e-172-2005.htm [accessed 1 February 2014].

Historic Alleys (undated) 'Hindus and the Ocean Taboo', online at http://www.historicalleys.blogspot.com/2009/01/hindus-and-ocean-taboo.html [accessed 20 January 2014].

International Tribunal on the Law of the Sea (ITLOS) (2013) 'Arbitral Proceedings between Bangladesh and India New Arbitrator Appointed', Press release, 19 July, online at http://www.itlos.org/fileadmin/itlos/documents/press_releases_english/PR_198_E.pdf [accessed 6 November 2014].

Jayewardene, H. W. (1990) *The Regime of Islands in International Law,* Leiden: Martinus Nijhoff.

Kannan, K., Tanabe, S., Iwata, H., and Tatsukawa, R. (1995) 'Butyltins in Muscle and Liver of Fish Collected from Certain Asian and Oceanian Countries', *Environmental Pollution,* 90(3): 284–7.

Khan, S. (undated) 'India Designates Six New Ramsar Sites at the 9th Conference of Contracting Parties', online at http://ramsar.rgis.ch/cda/en/ramsar-news-latest-india-s-latest-additions/main/ramsar/1-26-76%5E16874_4000_0__ [accessed 1 February 2014].

Krishnamoorthy, R., Devasenapathy, J., Thanikachalam, M., and Ramachandran, S. (2001) 'Environmental and Human Impacts on Coastal and Marine Protected Areas in India', in G. Visconti, Beniston, M., Iannorelli, E. D., NS Barba, D. (eds) *Global Change and Protected Areas,* Boston, MA: Kluwer Academic.

Kumar, R. (undated) 'Conservation and management of mangroves in India, with Special Reference to the State of Goa and the Middle Andaman Islands', online at http://www.fao.org/docrep/x8080e/x8080e07.htm [accessed 1 February 2014].

Kurien, J. (2005) 'Securing the Future against Tsunamis', *Economic & Political Weekly*, 8 January, p. 98.

Legal Correspondent (2008) 'Pachauri to Head Six-Member Experts Committee', *The Hindu*, 31 July, online at http://www.hinduonnet.com/2008/07/31/stories/2008073161591100.htm [accessed 9 February 2014].

Ministry of Agriculture (2004) *Comprehensive Marine Fishing Policy*, online at http://www.dahd.nic.in/fishpolicy.htm [accessed 20 January 2014].

Ministry of Earth Sciences (undated*a*) 'About Us', online at http://www.moes.gov.in/dodhead.htm [accessed 20 January 2014.

Ministry of Earth Sciences (undated*b*) 'Members of the Earth Commission', online at http://www.dod.nic.in/earthcom.htm [accessed 13 January 2014].

Ministry of Earth Sciences (2010) *Vision and Prospective Plan for 10 Years in Ocean Sciences & Services*, online at http://www.dod.nic.in/Vision-OceanSciences-updated.pdf [accessed 17 January 2014].

Ministry of Environment and Forests (MoEF) (1991) *Coastal Regulation Zone Notification*, online at http://www.envfor.nic.in/legis/crz/crznew.html [accessed 2 February 2014].

Ministry of Environment and Forests (MoEF) (1992) *Policy Statement for Abatement of Pollution*, New Delhi: Government of India.

Ministry of Environment and Forests (MoEF) (2008) *Coastal Management Zone Notification*, online at http://www.envfor.nic.in/legis/crz/so-1070(e).pdf [accessed 2 February 2014].

Ministry of Environment and Forests (MoEF) (2010) *Wetlands (Conservation and Management) Rules*, online at http://www.moef.nic.in/downloads/public-information/Wetlands-Rules-2010.pdf [accessed 1 February 2014].

Ministry of Environment and Forests (MoEF) (2011) *Coastal Regulation Zone Notification*, online at http://www.moef.nic.in/downloads/public-information/CRZ-Notification-2011.pdf [accessed 2 February 2014].

Ministry of Environment and Forests (MoEF) (2013) *Report of the Committee for Inspection of M/S Adani Port & SEZ Ltd. Mundra, Gujarat*, online at http://www.cseindia.org/userfiles/adani_final_report.pdf [accessed 2 February 2014].

PTI (2009) 'Marine Bill: AIADMK to Stage Demo at Delhi on December 18', *The Hindu*, 11 December, online at http://www.thehindu.com/news/national/tamil-nadu/marine-fisheries-bill-aiadmk-to-stage-demo-in-delhi-on-dec-18/article63536.ece [accessed 1 February 2014].

Puthucherril, T. G. (2008a) 'Harnessing the Atom: Strengthening the Regulatory Board for Nuclear Safety in India Based on the Canadian Experience', *Journal of Energy and Natural Resources Law*, 26(4): 553–8.

Puthucherril, T. G. (2008b) 'Ballast Waters and Aquatic Invasive Species: A Model for India', *Colorado Journal of International Environmental Law and Policy*, 19: 418.

Puthucherril, T. G. (2010) *From Ship Breaking to Sustainable Ship Recycling: Evolution of a Legal Regime*, Leiden: Martinus Nijhoff.

Puthucherril, T. G. (2012) 'Change, Sea Level Rise and Protecting Displaced Coastal Communities: Possible Solutions', *Global Journal of Comparative Law*, 1(2): 225–63.

Puthucherril, T. G. (2014) *Towards Sustainable Coastal Development: Institutionalizing Integrated Coastal Zone Management and Coastal Climate Change Adaptation in South Asia*, Leiden: Brill/Nijhoff.

Rajagopalan, R. (2008) *Marine Protected Areas in India*, Chennai: International Collective in Support of Fishworkers.

Saigal, K., Rajagopalan, R., and Ganesh, L. S. (eds) (1995) *Pacem in Maribus XXII*, Madras: International Ocean Institute Operational Centre.

Shah, S. A. (2009) 'River Boundary Delimitation and the Resolution of the Sir Creek Dispute between Pakistan and India', *Vermont Law Review*, 34: 357–413.

Sharma, R. C., and Sinha, P. C. (1994) *India's Ocean Policy*, New Delhi: Khanna Publishers.

Subba Rao, D. V. (2005) 'Comprehensive Review of the Records of the Biota of the Indian Seas and Introduction of Non-Indigenous Species', *Aquatic Conservation: Marine and Freshwater Ecosystems*, 15(2): 117–46.

Swaminathan, M. S. (2005) *Report of the Expert Committee on Coastal Regulation Zone Notification 1991*, New Delhi: Ministry of Environment and Forests.

Tanaka, K. (2011) *Indo-Bangladesh Maritime Border Dispute: Conflicts over a Disappeared Island*, ICE Case Studies No. 270, online at http://www1.american.edu/ted/ICE/taplatti.html [accessed 2 February 2014].

TERI   *See* Energy and Resources Institute

United Nations Division for Ocean Affairs and the Law of the Sea (DAOLOS) (1987a) 'Agreement between the Government of the Republic of India and the Government of the Republic of Indonesia on the Extension of the 1974 Continental Shelf Boundary between the Two Countries in the Andaman Sea and the Indian Ocean, 14 January 1977', in *The Law of the Sea: Maritime Boundary Agreements (1970–1984)*, New York: United Nations.

United Nations Division for Ocean Affairs and the Law of the Sea (DAOLOS) (1987b) 'Agreement between the Government of the Kingdom of Thailand, the Government of the Republic of India and the Government of the Republic of Indonesia Concerning the Determination of the Trijunction Point and the Delimitation of the Related Boundaries of the Three Countries in the Andaman Sea, 22 June 1978', in *The Law of the Sea: Maritime Boundary Agreements (1970–1984)*, New York: United Nations.

United Nations Division for Ocean Affairs and the Law of the Sea (DAOLOS) (1987c) 'Agreement between Sri Lanka, India and Maldives Concerning the Determination of the Trijunction Point between the Three Countries in the Gulf of Mannar, 23, 24, and 31 July 1976', in *The Law of the Sea: Maritime Boundary Agreements (1970–1984)*, New York: United Nations.

US Department of State (1975a) 'Agreement between Sri Lanka and India on the Boundary in Historic Waters between the Two Countries and Related Matters', in *Historic Water Boundary: India–Sri* Lanka, Limits in the Seas No. 66, online at http://www.state.gov/documents/organization/61460.pdf [accessed 15 May 2014].

US Department of State (1975b) 'Agreement between the Government of the Republic of India and the Government of the Republic of Indonesia Relating to the Delimitation of the Continental Shelf Boundary between the Two Countries (with Annexed Chart), 1974', *Continental Shelf Boundary: India–Indonesia*, Limits in the Seas No. 62, online at http://www.state.gov/documents/organization/61495.pdf [accessed 15 May 2014].

US Department of State (1978a) 'Agreement between Sri Lanka and India on the Maritime Boundary between the two Countries in the Gulf of Mannar and the Bay of Bengal and Related Matters', in *Maritime Boundaries: India–Sri Lanka*, Limits in the Seas No. 77, online at http://www.state.gov/documents/organization/58833.pdf [accessed 15 May 2014].

US Department of State (1978b) 'Agreement between India and Maldives on Maritime Boundary in the Arabian Sea and Related Matters', in *Maritime Boundary: India–Maldives and Maldives' Claimed 'Economic Zone'*, Limits in the Seas No. 78, online at http://www.state.gov/documents/organization/59587.pdf [accessed 15 May 2014].

US Department of State (1981) *Continental Shelf Boundaries: India–Indonesia–Thailand*, Limits in the Seas No. 93, online at http://www.state.gov/documents/organization/58818.pdf [accessed 15 May 2014].

PART III

Regional Ocean Policies

18

THE INTEGRATED MARITIME POLICY OF THE EUROPEAN UNION

Sylvain Gambert*

Introduction

The European Union, the world's most developed supranational cooperation of independent countries, was born essentially as a terrestrial and continental project. Its first step was the creation of the European Coal and Steel Community (ECSC) in 1951 to remove trade barriers for resources as coal, coke, steel, or scrap iron. The Treaty of Rome officially established the European Economic Community (EEC) on 1 January 1958, with the ambition to create a common market for the free movement of goods, workers, capital, and services on land. The most emblematic policy of the European Union, and still its largest budget, is none other than the Common Agricultural Policy (CAP) created in 1962.

The Union started to manage its maritime assets only later and in a piecemeal way, either as subdivisions of sectoral land-focused administrations or in reaction to external crises. In the 1970s, the first fisheries measures were developed under the umbrella of the CAP and aimed to create a free trade area in fish products. The Common Fisheries Policy (CFP) was then created in 1983. For shipping, it was the sinking of the *Erika* single-hull tanker off the coasts of Brittany in 1999, polluting 400 km of European coastlines and killing up to 300,000 birds, which prompted the European Union to propose the so-called 'Erika' legislative packages on maritime safety in 2003. Finally, it was the decline of European shipyards and the inefficiency of state aids in the 1990s that prompted the adoption of a new innovation-based shipbuilding policy in the early 2000s.

The marginality of maritime issues in EU governance reflected a more global 'maritime exceptionalism' (Suárez de Vivero et al., 2009: 633) that also pervaded the economic and political history of Member States. However, by the mid-2000s, that marginality appeared increasingly paradoxical – all the more so because Europe is geographically and economically a maritime continent. The surface of marine waters under the jurisdiction of the EU Member States is the world's most important, being far larger than their total land area. Maritime sectors are also key elements of the EU economy, with a gross added value of €500 billion per year generated by marine-based industries and services. Some 75 per cent of the European Union's external trade and around 40 per cent of its internal trade is transported by sea. Overall, maritime activities are responsible for 5.4 million direct jobs in Europe (European Commission, 2012a).

The creation of the Integrated Maritime Policy

The decision to give visibility and coherence to maritime affairs was not an isolated gesture, but part of the larger international trend towards applying a cross-sectoral and participatory approach to ocean governance. The annual reports of the UN Secretary-General on oceans and the law of the sea had repeatedly pointed out the problems affecting the oceans resulting from a lack of coordination and priority. In the wake of this global paradigm shift for maritime governance, the European Union decided to follow early starters, Canada and Australia, in developing a comprehensive maritime policy (Suárez de Vivero, 2007: 409). New EU President José Manuel Barroso, having developed an integrated maritime policy as prime minister of Portugal, made it a priority of the EU agenda (Wegge, 2011: 335). The strategic objectives of the European Commission for 2005–09 noted the particular need for an 'all-embracing maritime policy aimed at developing a thriving maritime economy and the full potential of sea-based activity in an environmentally sustainable manner' (European Commission, 2005a: 9). The intention was to reduce sectoral fragmentation, which had led to policy conflicts or incoherent decision-making.

Joe Borg, Commissioner for Fisheries, was asked 'to steer a new Maritime Policy Task Force with the aim of launching a wide consultation on a future Maritime Policy for the Union' (European Commission, 2005b: 1). The Task Force was endowed with a broad and extremely ambitious mandate (De Santo, 2010: 415; Wegge, 2011: 338), which resulted in the publication of a consultation document in June 2006, the Green Paper, highlighting the links between EU maritime-related activities and raising questions for public consultation (European Commission, 2006). The stakeholder consultation was an unexpected success, with almost 500 written contributions and more than 230 events. It formed the basis for the adoption of a Communication creating the Integrated Maritime Policy for the European Union (IMP) in October 2007 (European Commission, 2007). The Communication, called the 'Blue Book', was accompanied by a detailed action plan. European heads of states welcomed both in the European Council of 14 December 2007.

In the complex political, administrative, and institutional setting in which it was meant to function, the IMP had to be based on mechanisms that increase coordination and cooperation, rather than on the rearrangement of bureaucratic responsibilities. Instead of bringing together its directorates dealing with shipping, marine environment, fisheries, offshore energy, marine research, and others, into a new directorate-general for the seas and oceans, the Commission searched for the best way in which to bring together the expertise, ideas, and best practices from the different maritime units. Different governance mechanisms were set up to ensure political coherence, under the coordination of the renamed Directorate-General for Maritime Affairs and Fisheries (DG MARE). A steering group of Commissioners was established in 2005 to lead the early political development of the maritime policy. A permanent interservice group, representing the different directorates-general having an interest in maritime affairs, was established. Likewise, a Member States expert group was set up to gain insights into Member State activities and to allow the exchange of information concerning the development of their maritime policies, as well as that of the European Union. This group of experts was then also used as an examination committee to adopt implementing legislation on the IMP annual work programmes.

The strength of administrative sectoral division in European maritime affairs proved to be a concern for the other two main institutions, the Council of the European Union and the European Parliament. In the Council, working parties on, among other things, fisheries, environment, shipping, and industry usually prepare the agreements to be reached by ministers in a sectoral manner. In order to mirror the changes made in the European Commission, the Council created a horizontal 'Friends of Presidency' group on integrated maritime policy, the main function of which was to prepare Council conclusions to support the development of the IMP. By means of the essential role of the General Secretariat of the Council, which provided continuity to presidencies rotating every six months, the group has been institutionalized and now plays fully its function of co-legislator on legislative proposals. In order to escape the sectoral logic of the Council organization, the group was placed under the authority of the General Affairs Council, a configuration that usually holds responsibility over files that affect more than one EU policy.

In the European Parliament, support for, and coverage of, maritime policy has been very positive from the beginning. However, here, again, the sectoral division of committee work has hampered the development of a coherent approach to maritime affairs. The IMP has been dealt with mainly by the Committee for Transport and Tourism, as a persistence of maritime transport's ascendancy. In 2010, forty members of the European Parliament created a Seas and Coastal Areas Intergroup that tried to bring forward a coherent maritime agenda through informal exchanges of views.

An essential part of those developments was the entry into force of Directive 2008/56/EC of the European Parliament and the Council in 17 June 2008, establishing a framework for community action in the field of marine environmental policy. The Directive, commonly known as the Marine Strategy Framework Directive (MSFD), constitutes the environmental pillar of the IMP and was the first EU legislation to tackle environmental issues affecting marine ecosystems in Member States' economic exclusive zones (EEZ) – that is, marine pollution, depletion of marine biodiversity, destruction of wetlands, or the effects of extreme weather events caused by climate change. The MSFD aims to achieve good environmental status of the European Union's marine waters by 2020, and to protect the resource base upon which marine-related economic and social activities depend. The coherent package of the twin adoption of the IMP and MSFD in 2007 and 2008 sent the clear political signal that Europe's ability to achieve the full economic potential of its seas could only be backed by increasing protection and sustainable stewardship of the marine environment.

Blue growth and the institutionalization of the Integrated Maritime Policy

By 2010, when Commissioner Maria Damanaki took over the portfolio for Maritime Affairs and Fisheries, most of the Blue Book Action Plan had been implemented. The economic climate had also radically changed since the Green Paper had been drafted five years earlier. The European Union had launched its 'Europe 2020' strategy to unlock new sources of growth and to help the continent out of the economic crisis. The Commission thereafter launched a major study defining which would be the future scenarios and drivers for maritime growth in the long term. The Blue Growth Strategy was published on that basis in September 2012 (European Commission, 2012a). A month later, under the chairship of the Cyprus presidency, European ministers in charge of maritime affairs signed the 'Limassol Declaration', which endorsed blue growth as the maritime pillar of the Europe 2020 (Cyprus Presidency of the European Union, 2012). For the first time, an essential role was given to the seas in helping to revitalize Europe's economy.

The Strategy recognized that the blue economy forms a coherent whole, which is based on the interdependency of its individual sectors. In terms of transferable skills, shared infrastructure, or reliance on cross-fertilizing innovation, it translated into policymaking the acknowledgement that the blue economy is more than the sum of its parts. The Limassol Declaration marked a new step in what has been called the 'maritimization of the economy': the extension of maritime policies beyond their specialized, localized, and sectorally managed remit, where their geographical domain of application makes them exceptional, to become a normalized part of the national economy (Suárez de Vivero et al., 2009).

Within two years, the Blue Growth Strategy managed to place the blue economy on the agenda of Member States, regions, enterprise, and civil society. It delivered strategic guidelines for the sustainable development of EU aquaculture in 2013 (European Commission, 2013a), a legislative proposal for maritime spatial planning (MSP) in 2013 (European Commission, 2013b), and the adoption of initiatives for ocean energy, sustainable coastal tourism, and maritime innovation in 2014 (European Commission, 2014a; 2014b; 2014c). Support for research and technology development is also a key to unlocking the innovative potential of emerging maritime sectors. The European Union has firmly anchored blue growth as a focus area of its research and innovation programme for the years 2014 to 2020, focusing on observation systems, integrated response capacity to oil spills and marine pollution, climate change impacts on fisheries and aquaculture, and ocean literacy.

The Blue Growth Strategy launched a second phase for the IMP, during which its role shifted from an objective of integration toward a role of catalyst for maritime affairs – focusing on delivering concrete projects, launching new economic policy directions, and ensuring strong visibility. By moving from policy elaboration towards policy implementation, the IMP achieved its full institutionalization. The adoption of two financial instruments to provide funding for its policies strengthened that process. In particular, the European Maritime and Fisheries Fund adopted in 2014 embedded financial support to cross-border strategies around Europe's sea basins and integrated data management systems until 2020.

The regionalization of maritime policy

When the IMP was devised, it became clear that a regional structure to maritime policy would be required. The consultation process demonstrated that the success of maritime policy would depend on the support and sense of ownership of stakeholders, including regional actors already very active in developing integrated maritime actions. Furthermore, the diversity of maritime regions required differentiated policymaking.

The waters of the European Union and its Member States are divided into distinct eco-regions or maritime basins that have differing characteristics. The Baltic Sea is very shallow and has minimal tides. The Mediterranean is much deeper, but has minimal exchange with the Atlantic Ocean. The Black Sea is very deep, but lacks the oxygen levels required for a flourishing ecosystem, while the North Sea and the Atlantic have high tidal variation. Political realities also differ, such as the Mediterranean, shared with a large number of countries and in which very few EEZs are established or recognized, or the North Sea, where all but one surrounding countries are EU Member States and where political cooperation is historically established. In addition to obvious geographical reasons, existing networks and cooperation forums already existed at the regional level. This was especially the case for the existing four regional sea conventions: the 1992 Helsinki Convention on the Protection of the Marine Environment of the Baltic Sea Area; the 1992 Convention for the Protection of the Marine Environment of the North-East Atlantic (OSPAR Convention); the 1995 Barcelona Convention for the Protection of Marine Environment and the Coastal Region of the Mediterranean; and the 1992 Bucharest Convention for the Protection of the Black Sea against Pollution.

The regional approach to maritime affairs exemplified 'the scalar reorganisation of marine governance' affecting both the geographical and organization scale of EU marine governance (Thiel, 2013: 331). The shift was twofold. First, EU regional funding needed to take into account that maritime regions in different countries around the same sea may have more in common amongst themselves than with inland regions in their own countries. An efficient system to use regional funds and to share expertise was required across traditional national borders. Second, the maritime economy has its own needs in terms of financial support, research, employment, or innovation, which required a new approach rather than the mere extrapolation of regional and territorial policies.

To make better use of EU regional funds and to increase the cross-border sharing of best practices between coastal regions, the European Union launched regional and sea basin strategies. These strategies support transnational cooperation for the targeted use of European funds on commonly agreed maritime objectives. Since 2009, the Strategy for the Baltic Sea Region has implemented eighty projects under the leadership of its Member States. Achievements include the reduction of pollution from vessels, the development of sustainable short sea shipping, support for maritime clusters and maritime education, and the preparation of a regional climate change adaptation plan. The Commission has also launched successful strategies for the Atlantic Ocean in 2011 (European Commission, 2011) and for the Adriatic-Ionian Seas in 2012 (European Commission, 2012b). They aim to support maritime growth and job creation in those regions, with a strong focus on the establishment of a new, low-carbon, blue economy. The involvement of stakeholders – national, regional, and local authorities, the industry, civil society, and think tanks – is also central in contributing expertise and ideas to the drawing up of implementation action plans.

The role of maritime policy at the supranational level is to ensure appropriate regional articulation, rather than the development of a single plan applicable to all regions. This approach also applies to the implementation of environmental legislation – and especially the MSFD, which established European marine regions on the basis of geographical and environmental criteria. The Directive does not harmonize what 'good environmental status' means for marine waters throughout Europe with a single standard; instead, to take account of differing regions and coastal areas, Member States are required to develop specific strategies for their marine waters in cooperation with other Member States and non-EU countries.

Data management: A key role for policy success and cost-efficiency

Duplication of efforts in the collection of marine data and the absence of interoperability in their management often results from sectoral fragmentation. Since 2008, the Commission has launched two initiatives to rationalize and integrate marine and surveillance data between sectors and countries.

First, maritime surveillance encompasses several different objectives and realities that are managed by separate administrations: border control, safety, and security; fisheries control; customs; environment; or defence. National authorities use public money to collect separately some essential situational data, which are then not shared. The potential savings for the European Union and Member States are significant if one considers that the same data may currently be collected several times. More importantly, better access to maritime surveillance data would improve situational awareness at sea, toward more efficiency in preventing and responding to accidents, detecting illegal oil discharges, monitoring fishing activities, and safeguarding the environment. In 2010, the Commission presented and started implementing a 'roadmap' for the establishment of a common information-sharing environment for the EU maritime domain, which will integrate existing maritime surveillance systems and networks, and give all relevant authorities access to more information (European Commission, 2010). Member States expressed their strong support for the initiative at the General Affairs Council of May 2011.

Similarly, public marine data collection is mainly done by Member States for a specific purpose – for example to exploit marine resources, to monitor compliance with regulations, or as part of research projects. Data are collected through a multitude of formats and stored by a multitude of institutions, making publicly financed data very difficult to access and reuse. The aim of the European Marine Observation and Data Network (EMODnet) is to assemble marine data to make them interoperable and easily accessible. The added value of the European Union is clear as the Commission ensures coherence of format across borders and between different user communities. In 2012, the Commission announced its objective to prepare a multi-resolution digital seabed map of European waters by 2020 (European Commission, 2012c). Bringing together marine data from different sources reduces uncertainty in understanding the behaviour of the seas. It also helps industry, public authorities, and researchers to find data easily and to make more efficient use of them for the commercialization of new products and services. or for the protection of marine ecosystems.

The time for maritime spatial planning

Similar to other cases of integrated ocean policies in the United States, Canada, or Australia, maritime spatial planning (MSP) is the key tool that the Commission has developed to balance the development of maritime sectors, to minimize conflict over increasingly limited access to space, and to manage efficiently the cumulative impact of human activities on marine ecosystems. As most commonly defined, MSP is 'a public process of analysing and allocating the spatial and temporal distribution of human activities in marine areas to achieve ecological, economic, and social objectives that are usually specified through a political process' (Ehler and Douvere, 2009: 18). It provides a stable legal framework within which public authorities and stakeholders can coordinate their action, and investors can predict where future developments will be possible in the long term. Because the Blue Growth Strategy promotes sectors the development of which requires space, such as renewable energy, the European Union has worked on MSP as a tool with which to ensure that marine ecosystem and habitats, as well as the services that they perform, are taken into account in any development process from an early stage. Maritime spatial planning improves on existing ad hoc planning practices that give little consideration to the impact of economic activities on each other and, cumulatively, on the environment.

In 2013, after having conducted five years of studies and pilot projects, the Commission proposed legislation on MSP in the format of a framework directive. Directive 2014/89/EU of the European Parliament and of the Council of 23 July 2014 establishing a framework for maritime spatial planning entered into force in September 2014 (the MSP Framework Directive), and requires all Member States to develop MSP and to cooperate with their neighbours to coordinate their planning. An early study of existing MSP systems in Europe had shown a lack of transnational perspective, despite the clear interconnectedness of adjacent sea spaces (Douvere and Ehler, 2009: 87). Cross-border cooperation is essential, because marine ecosystems, fishing grounds, and marine protected areas (MPAs), as well as maritime infrastructures, such as cables, pipelines, shipping lanes, and oil, gas, and wind installations, among others, run across national borders. The planning of major investments such as offshore energy grids needs to be considered on a cross-border basis. Similarly, one of the main objectives of EU environmental legislation is the establishment of a coherent network of MPAs. Such an endeavour is impossible to complete without cross-border planning. This necessity is sharper in Europe, where the political landscape of national jurisdiction over marine waters remains extremely fragmented – for example the Baltic, a relatively small sea, is surrounded by nine countries, including eight EU Member States (Schaefer and Barale, 2011: 244).

The MSP Framework Directive will change maritime affairs in Europe profoundly. It will bring spatial coherence and support to the implementation of EU legislation, especially for environmental policy and the MSFD (De Santo, 2010: 418; Brennan et al., 2014: 362) and renewable energy policy (Jay, 2009: 498). Until then, only Germany, Belgium, and the Netherlands had planned all waters up to the limit of their EEZs. To comply with the Directive, the twenty-three coastal states of the European Union will need to complete such plans before 2021, improving the cross-sectoral coherence of their maritime activities, ensuring stakeholder participation, and setting up mechanisms for cooperation with their neighbours. The legal bases of the Directive (environment, transport, energy, and fisheries) make it one of the first cross-sectoral EU legislations and its requirements make it the world's first supranational legislation through which sovereign countries will have to coordinate their maritime planning. It cannot be overstated how important the future of its implementation will prove for marine policy worldwide.

Discussion and conclusion

In retrospect, the strong maritime heritage of Europe, on which the creation of the IMP was based, has paradoxically hampered its full development at the EU level. Key proposals made in 2006 on the creation of a European coastguard service and of a common maritime space were opposed by historical maritime nations and were not developed by the European Commission, in compliance with the principle of subsidiarity. In relation to MSP, some Member States have been vocal in stressing the limited competences of the European Union over planning on land. The 2013 legislative proposal on MSP faced similar subsidiarity controversies to those linked to the adoption of the European Spatial Development Perspective in 1999, the 'mother document' for terrestrial planning (Qiu and Jones, 2013: 189). However, the added value of the proposal, based on the transnational nature of marine ecosystems and maritime activities, prevailed.

The IMP has also been criticized by environmental NGOs for what has been perceived as a disproportionate emphasis on its economic agenda. The use of the word 'maritime' in MSP, rather than the more commonly used 'marine', has reinforced this interpretation (De Santo, 2010: 419; Brennan et al., 2014: 363). Although the ecosystem-based approach and the aim to reach good environmental status of EU marine waters both underpin any maritime objective developed by the European Union, the IMP is not solely an environmental policy. This runs contrary to the policy approach often developed for marine waters, based on the assumption that Member States should 'not need further encouragement from the European Commission in promoting growth in the maritime economy' (Qiu and Jones, 2013: 183). Technological innovation, especially in shipbuilding, robotics, or submersibles, makes economic operations possible further offshore and deeper than has ever been the case in history. Increased economic exploitation of the sea will inevitably ensue. Modern maritime policies need to engage with such evolutions, and to ensure that any economic development will be sustainable and will not lead to irreversible depletion of marine resources. The goal of an integrated policy is to channel new employment opportunities towards coastal communities that experienced the decline of traditional activities, such as fisheries or shipbuilding, and to ensure full compliance with the high environmental standards that the Union has set for itself.

There is no denying that building a truly horizontal policy that integrates decades-old administrative structures, silo organizational cultures, and neo-corporatist arrangements is a challenge that needs time and constant high-level political support to be achieved. The EU experiment has not escaped some of the pitfalls of integrated and collaborative management, including the relatively low interest of politicians for maritime issues, the incomparable scope of policy areas concerned, and the limits to stakeholder involvement in policymaking (Friedheim, 2000;

Koivurova, 2009; Gambert, 2010). While the 2006 Green Paper – as much a philosophical stance on governance as a concrete policy mandate (Suárez de Vivero, 2007: 410) – has not yet been fully delivered, the IMP did change deeply the management of maritime affairs in Europe and put an end to the 'maritime exceptionalism' referred to in the introduction. The IMP has built up common responsibilities and governance mechanisms for Europe's sea basins, has massively increased the political visibility of maritime issues, has developed initiatives to boost innovative sectors in which Europe has a competitive edge, has set up successful instruments to avoid duplication of efforts and incoherence in policymaking, and has adopted legal requirements to develop MSP in all Member States.

Some changes may be more important than they seem. The normalization of maritime affairs as a cross-sectoral policy not only requires difficult administrative transformation, but also a new legal approach. In accordance with the principle of conferral, the European Union may propose legislation only within the limits of the competences that are conferred by the Member States through the treaties. The overwhelming majority of EU legislative proposals are based on one single dominant treaty Article that reflects the sector to which the proposal's objectives are most linked. A cross-sectoral maritime policy contributes to multiple objectives that are inseparably linked, without one being incidental to the other. Therefore, and exceptionally, the three legislative acts that have been adopted under the EU IMP have been based on multiple legal bases, such as transport, industry, energy, fisheries, environment, research, or tourism, recognizing their simultaneous and equal contribution to these sectors. This, then, is a case of the foundations of a legal system being quietly altered under the necessity of policy evolution.

Notes

* Any views or opinions expressed in this chapter are solely those of the author and do not necessarily represent those of the European Commission.

References

Brennan, J., Fitzsimmons, C., Gray, T., and Raggatt, L. (2014) 'EU Marine Strategy Framework Directive and Marine Spatial Planning: Which is the More Dominant and Practicable Contributor to Maritime Policy in the UK?', *Marine Policy*, 43: 359–66.

Cyprus Presidency of the European Union (2012) *Declaration of the European Ministers Responsible for the Integrated Maritime Policy and the European Commission on a Marine and Maritime Agenda for Growth and Jobs (the Limassol Declaration)*, online at http://ec.europa.eu/maritimeaffairs/policy/documents/limassol_en.pdf [accessed 15 February 2014].

De Santo, E. M. (2010) 'Whose Science?' Precaution and Power-Play in European Marine Environmental Decision-Making', *Marine Policy*, 34(3): 414–20.

Douvere, F., and Ehler, C. (2009) 'New Perspectives on Sea Use Management: Initial Findings from European Experience with Marine Spatial Planning', *Journal of Environmental Management*, 90(1): 77–88.

Ehler, C., and Douvere, F. (2009) *Marine Spatial Planning: A Step-by-Step Approach toward Ecosystem-Based Management*, Intergovernmental Oceanographic Commission and Man and the Biosphere Programme, IOC Manual and Guides No. 53, ICAM Dossier No.6, Paris: UNESCO.

European Commission (2005a) *Strategic Objectives 2005–2009: Europe 2010 – A Partnership for European Renewal, Prosperity, Solidarity and Security*, COM (2005) 12 final, Brussels: European Commission.

European Commission (2005b) 'Commission to Consult on Future Maritime Policy for the Union', Press release, 2 March, online at http://europa.eu/rapid/press-release_IP-05-231_en.htm [accessed 18 November 2014].

European Commission (2006) *Green Paper on A Future Maritime Policy for the Union: A European Vision of the Oceans and Seas*, COM (2006) 275 final, Brussels: European Commission.

European Commission (2007) *An Integrated Maritime Policy for the European Union*, COM (2007) 575 final, Brussels: European Commission.

European Commission (2010) *Integrating Maritime Surveillance: Draft Roadmap towards Establishing the Common Information Sharing Environment for the Surveillance of the EU Maritime Domain*, COM (2010) 584 final, Brussels: European Commission.

European Commission (2011) *Developing a Maritime Strategy for the Atlantic Ocean Area*, COM (2011) 0782 final, Brussels: European Commission.

European Commission (2012a) *Blue Growth, Opportunities for Marine and Maritime Sustainable Growth*, COM (2012) 494 final, Brussels: European Commission.

European Commission (2012b) *A Maritime Strategy for the Adriatic and Ionian Seas*, COM (2012) 0713 final, Brussels: European Commission.

European Commission (2012c) *Marine Knowledge 2020: From Seabed Mapping to Ocean Forecasting*, COM (2012) 473 final, Brussels: European Commission.

European Commission (2013a) *Strategic Guidelines for the Sustainable Development of EU Aquaculture*, COM (2013) 229 final, Brussels: European Commission.

European Commission (2013b) *Proposal for a Directive Establishing a Framework for Maritime Spatial Planning and Integrated Coastal Management*, COM (2013) 133 final, Brussels: European Commission.

European Commission (2014a) *Blue Energy: Action Needed to Deliver on the Potential of Ocean Energy in European Seas and Oceans by 2020 and Beyond*, COM (2014) 8 final, Brussels: European Commission.

European Commission (2014b) *A European Strategy for More Growth and Jobs in Coastal and Maritime Tourism*, COM (2014) 86 final, Brussels: European Commission.

European Commission (2014c) *Innovation in the Blue Economy: Realising the Potential of Our Seas and Oceans for Jobs and Growth*, COM (2014) 254 final, Brussels: European Commission.

Friedheim, R. (2000) 'Designing the Ocean Policy Future: An Essay on How I Am Going to Do That', *Ocean Development and International Law*, 31(1–2): 187–9.

Gambert, S. (2010) 'Territorial Politics and the Success of Collaborative Environmental Governance: Local and Regional Partnerships Compared', *Local Environment*, 15(5): 467–80.

Jay, S. (2009) 'Planners to the Rescue: Spatial Planning Facilitating the Development of Offshore Wind Energy', *Marine Pollution Bulletin*, 60(4): 493–9.

Koivurova, T. (2009) 'A Note on the European Union's Integrated Maritime Policy', *Ocean Development & International Law*, 40(2): 171–83.

Qiu, W., and Jones, P. J. S. (2013) 'The Emerging Policy Landscape for Marine Spatial Planning in Europe', *Marine Policy*, 39: 182–90.

Schaefer, N., and Barale, V. (2011) 'Maritime Spatial Planning: Opportunities and Challenges in the Framework of the EU Integrated Maritime Policy', *Journal of Coastal Conservation*, 15(2): 237–45.

Suárez de Vivero, J. L. (2007) 'The European Vision for Oceans and Seas: Social and Political Dimensions of the Green Paper on Maritime Policy for the EU', *Marine Policy*, 31(4): 409–14.

Suárez de Vivero, J. L., Rodríguez Mateos, J. C., and Florido del Corral, D. (2009) 'Geopolitical Factors of Maritime Policies and Marine Spatial Planning: State, Regions, and Geographical Planning Scope', *Marine Policy*, 33(4): 624–34.

Thiel, A. (2013) 'Scalar Reorganisation of Marine Governance in Europe? The Implementation of the Marine Strategy Framework Directive in Spain, Portugal and Germany', *Marine Policy*, 39: 322–32.

Wegge, N. (2011) 'Small State, Maritime Great Power? Norway's Strategies for Influencing the Maritime Policy of the European Union', *Marine Policy*, 35(3): 335–42.

19

THE PACIFIC ISLANDS REGIONAL OCEAN POLICY AND THE FRAMEWORK FOR A PACIFIC OCEANSCAPE

'Many islands—one ocean'

Mary Power and Anama Solofa

Introduction

The Pacific Islands Regional Ocean Policy (PIROP) was approved by Pacific Island Forum leaders[1] in 2002. It underscored the importance of the ocean to Pacific Island nations and communities. PIROP was intended to serve as an overarching framework within which the various actions affecting oceans and coasts in the region could be viewed, to assess progress towards achievement of outcomes desired by the region. The associated Pacific Islands Regional Ocean Policy Framework for Integrated Strategic Action (PIROF-ISA) set forth both broad initiatives and specific actions that were needed to implement the Regional Ocean Policy. This chapter describes the evolution of the Pacific Islands Regional Ocean Policy since its endorsement by Pacific Island leaders, as well as the Framework for a Pacific Oceanscape as a catalyst for action for PIROP, which was endorsed by Pacific Island Forum leaders in 2010.

The geographic and demographic setting

The Pacific region comprises a vast ocean area greater than 30 million km² and is home to twenty-two Pacific Island countries and territories (PICTs) (see Figure 19.1), having a combined exclusive economic zone (EEZ) close to 20 million km². In contrast, the total land area is just over 500,000 km², of which Papua New Guinea accounts for 83 per cent, while Nauru, Pitcairn, Tokelau, and Tuvalu are each smaller than 30 km² (see Table 19.1). Nine of the PICTs have a sea area extending over 2 million km², while the combined maritime area under national jurisdiction exceeds the land area by more than fifty times. In addition, the Pacific Islands region includes areas of high seas that are fully enclosed by the EEZs of several island countries (Tuquiri, 2001). Although most of the countries and territories have declared a 200 nautical mile EEZ, most are still in the process of delimiting these. Currently, there are forty-eight shared boundaries (see Figure 19.1), of which twenty-seven overlapping boundaries remain to be negotiated. Indicative of the progress that PICTs are making in governing their maritime zones, seven bilateral agreements (among Cook Islands, Kiribati, the Marshall Islands, Nauru, Niue, Tokelau, and Tuvalu) and one trilateral agreement (among Kiribati, the Marshall Islands, and Nauru) were signed at the forty-third Pacific Islands Forum (PIF) leaders meeting in Rarotonga, Cook Islands in 2012 (PIFS, 2012). Eight PICTs – Cook Islands, Fiji, Federated States of Micronesia, Papua New Guinea, Solomon Islands, Palau, Tonga, and Vanuatu – have lodged submissions for claims for an extended continental shelf under Article 76 of the 1982 United Nations Convention on the Law of the Sea (UNCLOS), following an assessment of the strength of this potential.

Diversity within the region begins with its people: the three sub-regions of the Pacific comprise Melanesia to the west, Polynesia in the southeast, and Micronesia to the north. Geographic diversity within and between different PICTs can be seen in the atolls, low-lying coral islands, and high volcanic islands that make up the 7,500 islands of the region. Economic development status amongst PICTs also varies widely; metropolitan areas, symbolic of economic development, are often surrounded by communities that continue to engage in some level of subsistence living. Agriculture and fisheries, previously the main revenue earners for most countries in the region, are now competing with the expanding tourism industry. This region is characterized by small land masses, a high degree of ecosystem and species diversity, an extraordinary level of endemicity, a high degree of economic and cultural dependence on the natural environment, vulnerability to a wide range of natural and environmental disasters, and a diversity of cultures and languages, traditional practices, and customs focused on the marine and coastal environment.

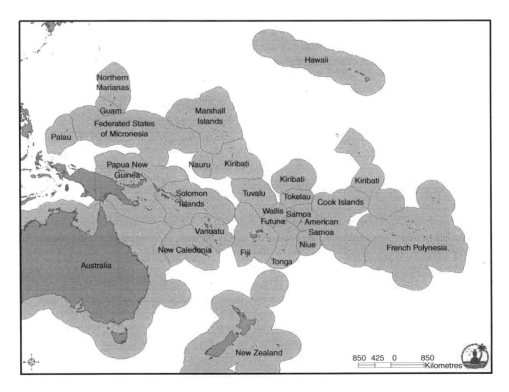

Figure 19.1 Pacific Island countries and territories (PICTs) and their EEZs

Source: Library and Information Resource Centre

Table 19.1 The twenty-two Pacific Island countries and territories

Country/territory	Land area (km²)	Area of 200 nautical mile zone (km²)	Estimated population (July 2007)
Independent Pacific Island countries			
Cook Islands	180	1,830,000	15,473
Federated States of Micronesia	702	2,978,000	109,999
Fiji	18,376	1,290,000	834,278
Kiribati	726	3,550,000	93,707
Marshall Islands	720	2,131,000	52,701
Nauru	21	320,000	9,930
Niue	258	390,000	1,587
Palau	500	629,000	20,162
Papua New Guinea	461,690	3,120,000	6,332,751
Samoa	2,934	120,000	179,478
Solomon Islands	29,785	1,340,000	503,918
Tonga	696	700,000	102,264
Tuvalu	26	900,000	9,701
Vanuatu	12,189	680,000	227,146
Pacific Island territories			
American Samoa	197	390,000	65,029
French Polynesia	3,521	5,030,000	260,072
Guam	549	218,000	173,995
New Caledonia	19,103	1,740,000	242,561
Northern Marianas	475	1,823,000	64,050
Pitcairn Islands	5	800,000	54
Tokelau	12	290,000	1,170
Wallis and Futuna	124	300,000	15,369

Sources: Gillett and Preston (1997); SPC (undated)

The importance of marine and maritime resources in the region

The monetary and subsistence economies of the PICTs are largely based on agriculture, fisheries, and tourism. Both oceanic fisheries and coastal tourism contribute significantly to foreign-exchange earnings for many countries. In addition, a majority of Pacific Islanders rely heavily on coastal fisheries for their subsistence needs. Because of the generally large EEZs, the oceanic fisheries sector and seabed non-living resources offer some of the few opportunities for significant economic returns.

The PICT fisheries are of two main types: export-oriented industrial oceanic fisheries, which are normally large-scale and high-technology-based and, for the most part, exploited by distant-water fishing nations (DWFN) rather than national fleets; and nearshore fisheries, which are usually small-scale, use low technology, and often for subsistence, but increasingly for local and export markets also. Oceanic fisheries (primarily for tuna) are undertaken by about 1,300 fishing vessels from twenty-one countries, a third of which are based in the Pacific Islands and employ 6–8 per cent of the labour force. The catch value from the region's oceanic fisheries is estimated at US$2 billion[2] in the western and central Pacific Ocean,[3] of which $800–900 million is taken in the waters of PICTs (Clark, 2005). These values are substantially greater after the catches are canned, loined, and processed in other ways. While most of the catch is still made by DWFN vessels, an increasing share is being taken by vessels based in PICTs (for example Papua New

Guinea catches exceed US catches). The value of catches by PICT vessels was estimated to have been $300 million in 2003 and increasing. Access fees generated approximately $78.5 million in 2007, an increase of roughly 25 per cent since 1999 (Gillett, 2009). Over 95 per cent of the access fees are earned by six countries (including approximately 30 per cent by Kiribati, which has one of the largest EEZs in the region), but earnings by individual countries fluctuate greatly (Clark, 2005). Additional economic benefits are derived from servicing foreign vessels.

Coastal fisheries remain critically important to food security and living standards. The coastal fishery is characterized by its subsistence nature, the involvement of women and the non-monetary sector, its importance for import substitution, and its substantial contribution to the health and welfare of the population. The intrinsic value of coastal fishery catches for food was estimated at $180 million in the late 1990s (Dalzell et al., 1996). The value of coastal fisheries for export products such as bêche-de-mer, aquarium fish, live reef food fish, and corals is approximately $50–80 million (Clark, 2005). Thanks largely to strong customary marine tenure, domestic food fisheries are in relatively good shape compared to reef fisheries in other parts of the world. However, these fisheries have little further commercial expansion potential and are facing a bleak future in the absence of effective intervention (SPC, 1999). Increasingly, the coastal habitats that support these fisheries are also under threat.

Fisheries as a whole may (the figures are very approximate) contribute about 10 per cent of the gross domestic product (GDP) of the Pacific Islands Forum (PIF) nations. However, this regional average is likely to be greatly underestimated owing to the varying national accounting processes used and the gross underestimation of the contribution of subsistence fisheries. According to official data in Pacific Island countries, the percentage contribution of fishing to GDP in 1999 (or the latest prior year available) ranges from 0.6 per cent in Papua New Guinea (PNG) to 12 per cent in Kiribati (ADB, 2002). However, an Asian Development Bank (ADB) study (2002) also found that contributions were generally grossly underestimated (ADB, 2002).

Aquaculture production is largely dominated by French Polynesia and New Caledonia, the source of 95.5 per cent of the value of aquaculture from all twenty-two PICTs in 2007 (Gillett, 2009). Contribution to GDP from this fisheries subsector, however, remains relatively small. Aquaculture exports in 2007 were dominated by pearl culture in French Polynesia ($123 million), shrimp culture in New Caledonia ($107 million), and pearl culture in the Cook Islands ($1.6 million), which were cumulatively responsible for about 23 per cent of all PICTs-sourced fishery exports (Gillett, 2009).

Tourism, much of it focused on marine- and coastal-based activities, contributed approximately $3.5 billion in 2005 to domestic earnings and is a major employer (SPTO, 2005). The future potential of tourism in the region is linked directly to the state of the coastal environment.

Maritime transport is also an integral consideration in developing a marine sector strategy. The ability to utilize and trade in the region's natural resources relies on available and effective maritime transport and marine infrastructure. The Pacific Ocean provides low-cost sea transportation routes between East and West, but maritime transport also poses a high degree of risk – both from local and passing traffic – as the catastrophic damage and costs of major oil pollution events elsewhere show.

While Pacific Island communities and governments have benefited significantly from a broad variety of resources and services that the Pacific Ocean has provided over generations, there remain significant areas of additional potential. Non-living marine resources currently contribute much less than the living resources to national economies; however, there is considerable potential in some areas. Sand and gravel (and, in some areas, coral) mining is important for local construction and often the only source of construction aggregate. There are substantial concentrations of deep-sea minerals within and adjacent to PICT EEZs. The major constraints to developing these resources at present include the economics of deep-sea mining, and the legal and boundary issues related to the ownership of these resources. Offshore hydrocarbon developments are currently important in Papua New Guinea, and there may be commercial potential in Solomon Islands, Fiji, Tonga, and Vanuatu.

Bio-prospecting for compounds of pharmaceutical or industrial value also offers significant potential. Many valuable compounds extracted from sea life are already utilized in medical and biotechnology fields, but the potential remains relatively untapped. Ocean energy offers a potential to replace the large import bills for fossil fuel energy. The potential of wave energy and ocean thermal energy conversion (OTEC) are likely to be considerable in the region. However, there are currently no pilot plants and there is no indication of the economics of such energy conversion plants.

Apart from the direct economic benefits that the Pacific Ocean provides the Pacific small island developing states (SIDS), there are regional and global benefits of a non-economic nature. While ecosystem services (that is, non-market-traded services) are important contributors to human welfare, there have been very few attempts to estimate their accrued value. Nevertheless, estimates broadly agree that the total global value of ecosystem services to be in the vicinity of $33 trillion annually (Costanza et al., 1997). Ecosystem services provided by the ocean include thermo-regulation, carbon sequestration, coastal buffering, and waste assimilation.

The Pacific Plan, the regional development agenda, clearly recognizes the significant role of the ocean and the goods and services that it provides in the development future of the region (PIFS, 2005). For example, the Plan establishes regional priorities, including maximizing sustainable returns from fisheries, enhancing shipping services for smaller island states, implementing regional tourism marketing and investment plans, developing strategies and legislation relating to maritime security and surveillance, and expanding technical and vocational education training programs in, among other things, seafaring and tourism (PIFS, 2005: para. 13).

A brief overview of the nature and evolution of regional ocean policy

The pre-eminent regional policy guidance on ocean and natural resource management comprises the Pacific Island Regional Ocean Policy and the Pacific Plan. The PICTs have also committed to a plethora of other policy and legal agreements and frameworks, at national, regional, and international levels, which relate to and have implications for the sustainable development and use of the islands, coasts, seas, and ocean within the region. Of key relevance to a framework for the Pacific oceanscape are the ratification of multilateral environmental and management agreements, and the endorsement of companion regional policy instruments for the sea, biological diversity, climate change, and pollution, as well as endorsement of more encompassing frameworks for sustainable development such as the 2002 Johannesburg Programme of Action on Sustainable Development and the Mauritius Strategy for the Further Implementation of the 1994 Barbados Programme of Action for the Sustainable Development of SIDS, which are complemented by regional instruments such as the Pacific Plan and PIROP.

The collaborative framework for marine and coastal resource and environmental governance in the Pacific Islands region is complex. Several conventions and agreements provide the international basis for regional and national action to address a diverse range of marine and coastal issues in the region. The extent to which these instruments are reflected in related national and regional policy varies considerably, and is determined by the political, economic, and social significance of the issue to be addressed and the capacity to participate in the negotiation of international instruments and subsequently design and manage appropriate national and regional responses (South and Veitayaki, 1999). For these international and regional instruments to achieve their objectives and their purpose, considered effort and support for the implementation of national policies and plans of action such as national sustainable development strategies or national development planning instruments, national biodiversity strategies and action plans (NBSAPs) and national adaptation programmes of action (NAPAs) is necessary.

For many Pacific Island countries, and the regional organizations that they mandate to manage coastal and marine activities, it is difficult to remain up to date with all international developments relating to coastal and marine affairs. The challenges arise principally because of the limited capacity in the region – in terms of both financial capacity and expertise. At the national and subnational levels, most environmental agencies and resource management institutions have a small staff. Although they possess considerable valuable local knowledge, they are generally under-resourced with respect to the technical capacity to support institutional arrangements for multi-sectoral management of resource use and the environment. The low level of financial support available to build capacity compounds the difficulties associated with establishing enduring institutions and collaborative arrangements for effective environmental governance. As such, the implementation of cross-sectoral engagement mechanisms, particularly at the national level, requires considerable strengthening.

Challenges are also created by the colonial basis for much of the national environment and natural resource law in the region. These legal frameworks often support a sectoral, rather than an integrated, approach to environment and natural resource management. Increased effort is required to create opportunities for harmonization, and to develop synergies between related national and regional environment and resource management and conservation law and policy.

There are encouraging signs that the sectoral approach to environmental governance, particularly at the national level, is changing. Many Pacific Island countries are establishing multi-sectoral working groups and task forces, which often include non-governmental organizations (NGOs) and the private sector, to address a wide range of national issues including waste management, sustainable development, biodiversity conservation, and tuna resources management. Hence the stage is set for an improved approach to ocean and coastal management. The PIROP was viewed as the regional platform to facilitate adoption of this approach at the time of its endorsement by Forum leaders, but that has not proven to be the reality. With lessons learned from the lack of action relating to PIROP and PIROF-ISA, the region now looks to the Framework for a Pacific Oceanscape with similar hopes for collaborative action for sustainable coastal and ocean management.

Policy development process

The Pacific Islands Regional Ocean Policy and the Framework for Integrated Strategic Action (PIROF-ISA) were developed by means of collaboration among many organizations and individuals, coordinated through the member organizations of the Marine Sector Working Group (MSWG) of the Council of Regional Organisations in the Pacific (CROP).[4] This occurred over a number of years, from the first direction for the Pacific leaders to the CROP agencies to commence this undertaking, through a series of consultations and iterations, including engagement of the international community, to final endorsement by the leaders of the final outcomes of the consultation and development process.

Building on the experience of developing and driving PIROP and PIROF-ISA, the Framework for a Pacific Oceanscape is seen as a catalyst for action for previous policies to protect, manage, maintain, and sustain the cultural and natural integrity of the ocean for current and, more importantly, future generations within the region and beyond. The 'Pacific Oceanscape' will drive the advancement of pride, leadership, learning, and cooperation across the ocean environment, with the overall intention of fostering ownership at various levels – local, national, regional, and international – to ensure the health and well-being of the Pacific Ocean and its peoples.

Initiation of the policy

At their 1999 annual meeting, the PIF leaders endorsed a list of recommendations emerging from the Pacific Regional Follow-up Workshop on the implementation of UNCLOS, convened in Tonga in 1999. One of the key recommendations – that a regional ocean policy be produced – was adopted as a regional-level initiative by the leaders (PIF, 1999).

Protection of the marine environment and management and conservation of marine resources are the focus of several sections of UNCLOS (Part V, Article 56, in relation to the protection and preservation of the marine environment, and Article 61, concerning the conservation of living resources; Part XI, Article 145, relating to the protection of the marine environment from harmful activities; and Part XII, Article 194, in relation to pollution from any source). The Convention suggests mechanisms for cooperation, including institutional arrangements that might support that cooperation. Article 197 states:

> States shall cooperate on a global basis, and as appropriate, on a regional basis, directly or through competent regional organisations, in formulating and elaborating international rules standards and recommended practices and procedures consistent with this Convention, for the protection and preservation of the marine environment, taking into account characteristic regional features.

All Pacific Island countries have ratified UNCLOS. The development and endorsement of PIROP was an example of Pacific Island states seeking to apply elements of the Convention at the regional level. The CROP MSWG was tasked with developing a regional ocean policy. The policy was subsequently endorsed by the thirty-third Pacific Islands Forum (PIF) meeting in 2002. The PIROP was then launched at the World Summit on Sustainable Development (WSSD) in 2002 as one of the key Pacific 'Type II' initiatives.[5]

The PIROP was not a legally binding document; rather, it articulated the principles and types of action that regional organizations should seek to programme and that countries in the region could consider in the development of national policies and national sustainable development plans. As a result, its development was not a particularly controversial process. As a regional-level framework, however, the key challenge was raising awareness of the policy at all levels of national government and assisting countries to implement the policy at the national level.

A key first step in the regional implementation process was the Pacific Regional Ocean Forum held in February 2004 to develop the Framework for Integrated Strategic Action. The Ocean Forum included not only government fisheries and foreign affairs officials, but also representatives of most sectors concerned with the ocean (tourism, shipping, fishing, environment, oceanography, etc.) and from many different societal groups (government, NGOs, private sector, academia, etc.), setting in motion a broad discussion on ocean governance. While it did not result in consensus, given the broad nature of the topic and range of stakeholders, it nonetheless provided an impetus for further progress. The final outcome of the Ocean Forum was a draft blueprint for further actions to implement the PIROP, the PIROF-ISA, which was circulated to member country and territory representatives and other Ocean Forum participants. Comments incorporated in a final draft were considered by the PIF leaders meeting in

Samoa in August 2004. The PIROF-ISA was presented to the international community at the Mauritius International Meeting on SIDS in late 2004. The value of the PIROP has been recognized internationally, with notes from the Mauritius meeting urging SIDS and their partners to develop regional initiatives collaboratively in a fashion similar to the process by which PIROP came into being (United Nations, 2005).

Five years after endorsement of the PIROF-ISA by Pacific Island leaders, the Pacific Oceanscape concept was introduced at the fortieth PIF meeting in Cairns, Australia, in August 2009 by the Republic of Kiribati, promoting its vision for a secure future for PICTs based on ocean conservation and management. Kiribati cited regional cooperation and strong Forum leadership as drivers for successful implementation of a Pacific Oceanscape, with filter-down effects to national development aspirations and priorities. Successful implementation of the Oceanscape concept would be further evidenced by improved management of critical issues, such as climate change effects and impacts on Pacific peoples, their islands, and their Ocean. In the Cairns communiqué:

> Leaders welcomed the Pacific Oceanscape concept and its companion Pacific Ocean Arc initiative tabled by Kiribati aimed at increasing marine protected area investment, learning and networking. Leaders tasked the Secretariat, together with relevant CROP agencies and key partners, to develop a framework for the Pacific Oceanscape, drawing on the Pacific Islands Regional Ocean Policy, as a priority area for attention under the Pacific Plan.
>
> *(PIFS, 2009: para. 69)*

The geographic scope of the Pacific Oceanscape and subsequent Framework mirrors that of PIROP in that the extent of the region includes not only the area within the 200 nautical mile EEZ boundaries circumscribing PICTs, but also the ocean and coastal areas that encompass the extent of the marine ecosystems that support the region. The 'ocean' is defined to include the waters of the ocean, the living and non-living elements within, the seabed beneath, and the ocean atmosphere and ocean–island interfaces.

Objectives and major principles

The PIROP presented a vision for a 'healthy ocean that sustains the livelihoods and aspirations of Pacific Island communities'. Accordingly, PIROF-ISA was premised on five key principles under the overarching umbrella of 'sound ocean governance' for the region:

- improving our understanding of the ocean;
- sustainably developing and managing the use of the ocean resources;
- maintaining the health of the ocean;
- promoting the peaceful use of the ocean; and
- creating partnerships and promoting cooperation.

In a similar fashion, the major principles and objectives of the Pacific Oceanscape Framework are based upon those of PIROP and the Pacific Plan. Using previous shortfalls in implementation as a cautionary tale, the Framework emphasizes the importance of utilizing pre-existing policies to facilitate collaboration, particularly in relation to integrated ocean management and biodiversity conservation. The Framework elaborates on the following principles:

- **Improving ocean governance** to engage leaders, decision-makers, resource custodians and other stakeholders to establish, strengthen, and implement appropriate and practical governance mechanisms that contribute to effective coordination and implementation for a healthy ocean that sustains the livelihoods of Pacific Island people
- **Sustainably developing and managing the use of ocean resources** to develop and embrace practices, approaches and processes that promote sustainable ocean resource use, development and management based on existing experiences and foreseeable levels of national funding and capacity to address challenges of isolation and infrastructure. In order to replenish, sustain and increase our knowledge base, it is necessary to generate new knowledge about the oceans upon which our way of life depends. Fundamental to the sustained generation of new knowledge and capacity is the continuing education of a cadre of scientists and policy makers. Educating and training people within the region is the best strategy for ensuring the continuity of marine understanding and replenishment of knowledge
- **Maintaining the health of the ocean** to reduce the negative impacts of human activities and implement measures that protect and conserve biodiversity by ensuring that the lack of full scientific certainty of the causes and effects of damage to the ocean should not be a reason for delaying action to prevent such damage and that polluters should bear the cost of pollution, wherein damage costs should be reflected in benefit cost assessments of actions affecting the ocean environment
- **Improving our understanding of the ocean** to improve the availability, management, use and dissemination of information targeted at better-informed decision-making and increased support for practical ocean management that embraces precautionary management approaches that are more robust where comprehensive scientific understanding and intensive monitoring are difficult.
- **Ocean security** has economic, environmental, political, and military dimensions which seek to discourage and reduce unacceptable, illicit, criminal or other activities that are contrary to regional and international agreements and threaten the oceans, the major source of livelihood for Pacific Island people.
- **Partnerships and cooperation** effective implementation will be founded on developing strong partnerships and, fostering cooperation and inclusiveness.

(Pratt and Govan, 2010: 55)

The objectives of the Framework for a Pacific Oceanscape cover three broad categories:

- **Integrated Ocean Management** to focus on integrated ocean management at all scales that results in the sustainable development, management and conservation of our island, coastal and ocean services that responds to Pacific Island countries development aspirations and, ensuring and maintaining environmental health and ecological function.
- **Adaptation to Environmental and Climate Change** to develop suitable baselines and monitoring strategies that will inform impact scenarios and specific understanding of environmental and climate change stressors. Only through empirical understanding can Pacific peoples develop and pursue effective, appropriate and sustained adaptation responses and solutions. Solutions need

to consider the full range of ocean and island environments and articulate the limits to adaptation and provide appropriate responses. Better information and understanding of these impacts will facilitate a confident and united engagement at regional and international levels.

- **Liaising, Listening, Learning and Leading** to articulate and use appropriate facilitative and collaborative processes, mechanisms and systems and research that results in the achievement of the objectives for Integrated Ocean Management and Adaptation to Environmental and Climate Change, while mindful of the interests, rights, responsibilities and differences between partners and stakeholders.

(Pratt and Govan, 2010: 57)

These objectives will, in turn, be used with strategic actions and priorities to address areas for immediate implementation under the Framework, including: jurisdictional rights and responsibilities; good ocean governance; sustainable development, management, and conservation; sustaining action; and adapting to a rapidly changing environment.

Institutional arrangements

The CROP MSWG maintains the lead role for implementing the policy at the regional level. At the national level, various CROP agencies[6] will assist governments with incorporating the principles and objectives of PIROP into their national development planning process, by means of mechanisms appropriate to the local institutional framework and policy environment.

The adoption of the Pacific Plan in 2005 by all sixteen member countries of the Pacific Island Forum brought renewed impetus towards cooperation in the Pacific region. This marks an important and ambitious step forward in regional cooperation in the Pacific region. The Pacific Plan focuses, in particular, on economic growth, sustainable development, good governance, and security for Pacific countries through regionalism. Implementation of PIROF-ISA will contribute toward delivery of the outcomes identified in the Pacific Plan.

The nature of the regional initiative established

The PIROF-ISA does not have legal status and is not intended to be a regulatory mechanism. The Policy and implementing Framework are intended to serve as a guide for regional coordination, integration, and collaboration on ocean issues, in keeping with the Regional Ocean Policy's goal of improving ocean governance and ensuring sustainable use of the ocean and its resources. It is envisaged that the Policy will be 'nationalized' through the development of national ocean policies embodying the principle and actions outlined and endorsed in the regional framework.

At national levels and through regional collaborative arrangements, Pacific Island countries have been attempting to formulate responses to increased threats to the marine environment for thirty years. At the regional level, twenty-five countries and territories collaborate on environment and resource management matters through six regional intergovernmental organizations.[7] This supports a highly developed collaborative framework for the management and conservation of ocean resources. However, unlike the high level of cooperation on tuna resources for which the region is globally renowned, associated ocean and coastal initiatives have not galvanized the same level of sustained action and harmonized support among Pacific Island countries.

The adoption of the PIROP by heads of government for Pacific Island countries therefore offers an opportunity to start a long-term process to address these constraints. The Policy, a global first, provides a broad regional framework within which national- and regional-level action will be designed to attempt to address some of these apparently intractable issues constraining the effective management of Pacific Islands' coastal ecosystems.

Problems, issues, or obstacles addressed by the regional ocean policy

Recognition of the need for a regional ocean policy stemmed from the growing awareness of leaders and decision-makers of the transboundary and dynamic nature of the ocean, the increasing number and severity of threats to its long-term integrity, and the reality that sustainable economic and social development of the region is dependent on wise use of the ocean and its resources. Lack of attention to policy, legal, and institutional reforms, as well as low priority for public investments and for enforcing regulations in private sector compliance, now place at risk not only coastal and marine ecosystems, but also the communities that depend on them for economic security and social stability. There is also strong recognition of the potential for fragmentation of programmes and for conflicting actions in different sectors as ocean-related activities increase, and of the associated need for increased regional collaborative arrangements among Pacific Island governments.

As identified in the regional process for input into the Mauritius International Meeting, Pacific SIDS generally are defined by their historic, cultural, and economic links to the oceans and seas. They continue to be heavily dependent on their marine resources, particularly for the sustainable livelihoods of coastal communities. The management of coastal and marine resources has slowly become integrated into broader ocean management strategies since the entry into force of UNCLOS. However, implementation continues to be impeded by financial constraints and a lack of capacity.

Studies have indicated that Pacific reefs are in peril from a range of threats (Vieux et al., 2004) and that this degradation has a cascading effect on communities and their economies (Hoegh-Guldberg et al., 2000). The World Bank has estimated that climatic fluctuations can devastate island economies, with damage to GDP varying from 4 per cent for high elevation islands to 38 per cent for low-lying islands of the Pacific (World Bank, 2000). As previously noted with regard to Pacific tuna fisheries, 90 per cent of the catch is taken by distant-water fishing nations (DWFN) fleets, with a subsequent dependence of the island states on Japanese, American, and European markets and their political considerations in terms of obtaining benefits from the exploitation of oceanic fisheries within their EEZs (Sydnes, 2002). This dependence is accentuated by the fact that, as noted by Tutangata and Power (2002), Pacific SIDS also face challenges from a globalizing world economy. Most notably, they may have little influence on global markets, and therefore need to work together to address externalities stemming from globalized trade and investment regimes.

The Pacific Plan lists as one of the region's key priorities the facilitation of international financing for sustainable development, biodiversity, and environmental protection and to counter the effects of climate change in the Pacific. The PIROP makes the case for integrated ocean management in the interest of future generations of Pacific Islanders, and provides an integrated framework for action of ocean initiatives. This has been recently recognized by one of the major donors to the region, the European Commission, in its engagement strategy for the Pacific–ACP (African Caribbean and Pacific) States (European Commission, 2006).[8]

Implementation, evaluation, and long-term outlook

Development of the PIROP has raised awareness at the highest political levels of the importance of the ocean and its resources for the long-term well-being of Pacific Island communities and also for the global good. Pacific Island countries have long had a strong coordinated voice and recognition in international forums as members of, and playing a lead role for, the collective global SIDS under the Alliance of Small Island States (AOSIS) banner. Through the SIDS Network, they have championed key environmental initiatives of global and regional significance, such as the 1997 Kyoto Protocol, the 1992 UN Framework Convention on Climate Change (UNFCCC), and the 1992 Convention on Biological Diversity (CBD). Now, however, they are starting to see themselves also as 'large ocean states', indicating the shifting perspective on the relative importance of the ocean as their natural resource base and an economic and cultural lifeline.

The PIROF-ISA has become a regional platform that underpins many ocean-related policy decisions. At their thirty-seventh meeting in Nadi in October 2006, Pacific Forum leaders issued a Declaration on Deep-Sea Bottom Trawling to Protect Biodiversity in the High Seas. The Declaration calls on states to move towards an appropriate legal framework to manage deep-sea bottom trawling to protect biodiversity in the high seas, reminding states of:

> . . . the Pacific Islands Regional Oceans Policy endorsed by Pacific Islands Forum leaders in 2002 which aims to ensure the future sustainable use of our oceans and its resources by Pacific Island communities and partners, and the need to establish high-level leadership on oceans issues.
>
> *(PIFS, 2006: para. 3)*

The PIROF-ISA has also become a regional platform with which to attract donors who are looking towards agreed policy frameworks within which to focus their development assistance efforts. The PIROP increasingly informs the corporate and annual work programmes of the key regional intergovernmental technical organizations in targeting their development assistance to member countries.

Implementation of PIROP at the national level was always foreseen to be, and has been, a slow process. It was envisaged that countries and states would react to PIROP as local conditions allowed. Further, a certain set of political and economic conditions would be necessary prerequisites for progress in operationalizing the PIROP principles in national development planning. Implementing PIROP will be an ongoing, dynamic, and long-term process. Several countries are now in the early phase of developing integrated national ocean policies or plans, including Fiji, the Solomon Islands, and the Cooks Islands.

Pacific Island leaders recognized the need for regional collaboration on ocean governance and endorsed the development of PIROP in 2002. The Policy was developed around the guiding principles of improving understanding of the ocean, developing and managing its resources in a sustainable manner, maintaining the health of the ocean, promoting its peaceful use, and creating partnerships and promoting cooperation.

Although not a legal document, PIROP was founded on international law. Following its endorsement, a forum of stakeholders was convened in 2004 to gather information relating to its implementation needs. This Pacific Island Regional Ocean Forum (PIROF) culminated in an implementation framework that outlined integrated strategic actions to be carried out in order to facilitate achievement of the guiding principles of the PIROP. The progress of PIROP and the PIROF-ISA has since faltered, with very little action and feedback recorded since almost immediately after the meeting of PIROF. It has been suggested that barriers to improving regional ocean governance stem from a number of areas – namely, lack of political will, the absence of integrated decision-making processes, limitations in available expertise and institutional capacity, shortages in accessible financial resources, and the influence of other external factors.

Another factor that may have contributed to the shift in priority across the region was the endorsement of an overarching regional development plan in 2005 by Pacific Islands Forum (PIF) leaders, called the Pacific Plan. The Pacific Plan is centred on four strategic objectives: economic growth, sustainable development, good governance, and security. The concept of 'regionalism' is promoted in the Plan as a facilitative tool for prosperity, prescribing the establishment of dialogues and processes between governments, and the pooling of national resources at the regional level, as well as regional integration facilitating market access between PICTs. Through regionalism, the Pacific Plan suggests, the four strategic objectives will be enhanced and stimulated to the benefit of the region and its people. The Pacific Plan does not directly reference the PIROP, but consideration of the implementation priorities emanating from the sustainable development and good governance goals of the Plan – developing national and regional management measures for sustainable development and management of fisheries, and enhancement of resource management governance mechanisms – reveals a connection to the PIROP initiatives. In theory, the Pacific Plan should not usurp the importance of the PIROP but, instead, underscore it.

Contributing research and education

There are several major regional initiatives under way that will contribute substantially to the regional knowledge base and also progressively to the initiatives articulated in the PIROF-ISA. A few key programmes are described here to indicate the nature and scale of activities, but this is not an inclusive list.

The Pacific Islands Global Ocean Observing System (PI-GOOS) and the Pacific Islands Global Climate Observing System (PI-GCOS) are both regionalized components of global programmes. The PI-GOOS is coordinated by the Pacific Islands Applied Geoscience Commission (SOPAC), and PI-GCOS, by the Secretariat of the Pacific Regional Environment Programme (SPREP). Both programmes collaborate with each other, and with researchers from within and outside the region, especially those with interests in science education, operational oceanography, marine scientific research, palaeo-environmental research, and remote sensing. Both programmes are compatible with the goals of PIROP, which encourages multilateral cooperation in order to improve understanding of the ocean and its ecosystems, whilst ensuring sustainable use of its resources by PICTs and external partners.

The South Pacific Sea Level and Climate Monitoring Project (SPSLCMP) aims to generate an accurate record of variance in long-term sea level for the South Pacific, and to establish methods to make these data readily available and usable by PICTs. Since 1991, the SPSLCMP has provided data, information products, and training to the international community and the PICTs through twelve high-resolution sea-level monitoring stations throughout the Pacific (Australian Government Bureau of Meteorology, undated). The Science Component of the Pacific-Australia Climate Change Science and Adaptation Planning program (2011-2014), which covers 14 island countries and East Timor, is a collaborative partnership between Australia, partner countries, and regional and non-government organizations in the western tropical Pacific which has helped fill the climate information and knowledge gap in the region (Australian Government, undated).

The Marine Resources Division of the Secretariat of the Pacific Community (SPC) undertakes a range of projects looking at the status and dynamics of coastal and ocean fisheries, which will contribute to further advancing the adoption of the ecosystem approach to fisheries in the region.

The SOPAC also has a major bathymetric mapping/habitat characterization and hydrodynamic modelling programme that provides information to sister CROP organizations and countries to support more informed decision-making in relation to the use and management of benthic and nearshore resources, especially non-living resources. This programme also supports a range of other applications such as tsunami modelling, pollution plume behaviour modelling, sedimentation assessment, and, of course, maritime boundary delimitation (including the extended continental shelf claims process.)

A key ocean-focused education project, SeaRead, focuses on primary, secondary, and tertiary institutions, and NGOs, for improved awareness of ocean issues across the region. This includes the introduction of relevant and practical ocean science and associated information for sustainable living in tropical Pacific Islands. Currently, the Fijian Ministry for Education is looking to introduce ocean science, tsunami warning, and other components of the SeaRead programme to social studies and science programmes in Fijian schools. Similar work has already been received positively in the Cook Islands and Samoa, with assistance from the Intergovernmental Oceanographic Commission (IOC) Perth Regional Program Office, the Pacific Office of the United Nations Educational, Scientific and Cultural Organization (UNESCO), and the New Zealand Institute of Water and Atmosphere. Through liaison with the International Tsunami Information Centre, tsunami warning education packages have been made available for immediate distribution to local Fijian schools.

Monitoring, evaluation, and adjustment

Monitoring is, of course, crucial to determine whether policies and plans are being implemented effectively and to establish whether the objectives of such activities are being met (in ecological, socio-economic, and other terms). At present, the CROP organizations are required to report annually to their governing councils on how their work programme has contributed to the implementation of PIROP in the preceding year. Additionally, the CROP MSWG is usually called upon to report collectively to the PIF leaders annual gathering on regional initiatives and on progress towards achieving the objectives of PIROP and PIROF-ISA. Overall, this remains something of an 'informal' process, in that it has not been institutionalized into a single existing CROP agency or a dedicated agency. Additionally, CROP agencies are expected to report regularly on how their respective work programmes contribute to the implementation of the Pacific Plan.

Outlook: Recommendations for improving the regional ocean policy/ programme

Like all broad-scale integrative processes, and especially one such as this that has no legal standing, it is difficult for any single organization or individual to maintain an overview of PIROP. Thus it has been difficult to maintain momentum and institutional enthusiasm for the whole process. Progress to date has relied predominantly on the efforts and commitment of a few key individuals.

It is likely that the most effective way forward, if implementation of PIROF-ISA is to continue to progress, would be for the region to devote specific resources to it – in particular, by setting up a PIROP office with a full-time staff member. A small, cost-effective office could be attached to one of the existing regional organizations. As well as providing dedicated motivation and capacity to organize whole-ocean-related activities, such an office could also function as the secretariat for the CROP MSWG. However, its initial function would be to assist specialized agencies in developing cross-cutting collaborations, and to assist participating PICTs in the process of developing, and harnessing support for, their own domestic ocean policies.

These national policies would be based on consultation with island stakeholders, and would incorporate local concerns, traditions, and specific national development policies. They would elaborate the general principles of the PIROP into a form that would take into account the very significant differences between each island group.

In the ideal scenario, countries could be supported to develop a national ocean management plan or policy that integrates all relevant institutions, takes a 'ridge-top to EEZ' focus, and has full political support and commitment. As impossible as this may seem, it has already been achieved to a degree in American Samoa. In August 2003, the American Samoan Governor signed Executive Order No. 004-2003, formally establishing the Ocean Resource Management Process and Plan. This document provides a structure for managing American Samoa's marine resources in a manner that balances ecological, economic, and cultural needs. It is a living document, in that it is designed to evolve and change with time (Hamnett et al., 2003).

Integrated ocean management must be seen as a long-term approach in Pacific SIDS. However, in the interim, we can take realistic and practical measures. What is needed immediately is improved institutional coordination both horizontally (across sectors) and vertically (local–provincial–national). This can be achieved at relatively low cost and with minimal institutional restructuring, provided that there is individual and political will.

At a regional scale, further action is now required across several areas, as follows:

1   Complete the forging of agreements between contiguous countries for delimitation and lodging of maritime boundaries as a matter of urgency.
2   Submit any claims to the United Nations Commission on the Limits of the Continental Shelf (CLCS) within ten years of a State's becoming a party to UNCLOS.
3   Further the work on the assessment of living and non-living seabed resources within national jurisdiction, including physical habitat characterization and mapping.
4   Establish effective monitoring, reporting, and enforcement, and control of fishing vessels, including by SIDS as flag states, and implement international plans of action to prevent, deter, and eliminate illegal, unreported, and unregulated (IUU) fishing and to manage fishing capacity.
5   Expand marine protected area (MPA) networks consistent with relevant international agreements and taking into account the programme of work on marine and coastal biological diversity adopted by the Conference of Parties (CoP) to the Convention on Biological Diversity (CBD) at its seventh session, and commitments made at the 2005 Mauritius International Meeting and the CBD COP 8.
6   Strengthen regional monitoring efforts of global ocean observing systems (Argos, PI-GOOS and PI-GCOS).

Recent developments in the Pacific Islands Regional Ocean Policy

Pacific Island leaders endorsed the Framework for a Pacific Oceanscape at the forty-first PIF meeting held in Port Vila, Vanuatu, on 4–5 August 2010 (PIFS, 2010). The Pacific Oceanscape Framework is the region's implementation tool for PIROP and the Pacific Plan (in relation to ocean issues). The Framework seeks to address six strategic priority areas:

- jurisdictional rights and responsibilities;
- good ocean governance;
- sustainable development, management, and conservation;
- listening, learning, liaising, and leading;

- sustaining action; and
- facilitating adaptation to a rapidly changing environment.

Since its endorsement of Pacific Oceanscapes, there have been a number of developments in ocean management in the Pacific that directly and indirectly advance the objectives and priority areas of the Framework.

Within the Framework, the establishment of a Regional Ocean Commissioner was an activity mandated under the strategic priority of good ocean governance, and this role was filled with the appointment of Tuiloma Neroni Slade, the Secretary General of the Pacific Islands Forum Secretariat (PIFS), in 2011. The Ocean Commissioner serves as an advocate for PIF members at international forums on ocean issues, while also coordinating partnerships and activities in the region that further key ocean priorities of the Pacific. Agencies of the Council of Regional Organisations of the Pacific (CROP), facilitated by the CROP Marine Sector Working Group (MSWG), provide technical and scientific support to the Ocean Commissioner.

Great strides have also been made in relating to the third strategic priority of the Pacific Oceanscape Framework – sustainable development, management, and conservation – with national commitments including, amongst others,

- the establishment of the Phoenix Ocean Arc between the governments of the United States and Kiribati (Pacific Islands Forum Secretariat, undated);
- a total ban on shark fishing in Tokelau (Techera and Klein, 2014);
- the creation of a shark sanctuary in Palau, followed by a more recent announcement to ban commercial fishing across its EEZ (Techera and Klein, 2014);
- the declaration of the Marae Moana Marine Park, covering an area of 1.1 million km^2 in the Cook Islands in 2012 (Cook Islands Marine Park, undated); and
- the establishment of Le Parc Naturel de La Mer de Corail marine sanctuary, covering 1.3 million km^2 of New Caledonia's EEZ in early 2014 (Conservation International, 2014).

Several regional projects are also being implemented that integrate traditional coastal resource management, build on marine spatial planning mechanisms, and use collaborative partnerships to manage resources in high seas pockets, including the Pacific-ACP Regional Legislative and Regulatory Framework for Deep Sea Minerals Exploration and Exploitation (SPC, 2012). In recent years, much attention has been given to exploration of deep-sea ecosystems, and PICTS are working with CROP agencies and international partners to develop policies and legislation that allow Pacific Island countries to benefit economically from the extraction of deep-sea minerals in ways that are sustainable and less harmful to the marine environment (SPC, 2012).

The objectives of the Pacific Oceanscape Framework, as a catalyst for activities that support PIROP and its associated Framework for Integrated Strategic Action (PIROF-ISA), continue to be integrated into ocean management activities at both the regional and national levels in the Pacific region. Its successful implementation, however, will require further strengthening of commitments by PICTs to implement the priority actions, with support not only from international partners, but also at all levels of government within each of the PICTS.

Notes

1 The 2002 Pacific Island Forum leaders comprised the prime ministers of the Cook Islands, the Federated States of Micronesia, Fiji, Kiribati, the Marshall Islands, Nauru, Niue, Palau, Papua New Guinea, Samoa, the Solomon Islands, Tonga, Tuvalu, Vanuatu, New Zealand, and Australia.
2 At prices in regional ports.

3 Includes Indonesia and the Philippines, and non–Forum Pacific Island territories.
4 The MSWG organizational members are: the Pacific Islands Forum Fisheries Agency (FFA); the Pacific Islands Forum Secretariat (PIFS); the Secretariat of the Pacific Community (SPC); the Secretariat of the Pacific Regional Environment Programme (SPREP); the South Pacific Applied Geoscience Commission (SOPAC); and the University of the South Pacific (USP).
5 The 'Type II' concept was established at the WSSD in Johannesburg in 2002. These partnerships provide for better collaboration between national, regional, and international stakeholders in the implementation of development activities. Such partnerships are developed and facilitated by Pacific regional agencies on behalf of their member countries. They aim to provide a single strategy as a mechanism for coordinating activities, so that lessons can be shared and gaps identified, and to align donor support in the interests of all members. Such partnerships are voluntary and non-binding.
6 There are ten CROP agencies: PIFS; the FFA; the Pacific Islands Development Programme (PIDP); the SPC; the SPREP; SOPAC; the South Pacific Tourism Organisation (SPTO); the USP; the Fiji School of Medicine (FSchMed); and the South Pacific Board of Education Assessment (SPBEA).
7 Namely, the SPREP, SOPAC, the FFA, the SPC, PIFS, and the USP.
8 The Pacific Island ACP states are the Cook Islands, the Fiji Islands, Kiribati, the Marshall Islands, the Federated States of Micronesia, Nauru, Niue, Palau, Papua New Guinea, Samoa, the Solomon Islands, Timor-Leste, Tonga, Tuvalu, and Vanuatu. (Timor-Leste has observer status at the Pacific Islands Forum.)

References

Asian Development Bank (ADB) (2002) *The Contribution of Fisheries to the Economies of Pacific Island Countries*, Manila: ADB.

Australian Government Bureau of Meteorology (undated) *South Pacific Sea Level and Climate Monitoring Project*, online at http://www.bom.gov.au/oceanography/projects/spslcmp/html [accessed 31 January 2015].

Australian Government (undated) *Pacific Climate Change Science*, online at http://www.pacificclimate-changescience.org/ [accessed 31 January 2015].

Clark, L. (2005) *Pacific 2020: The Potential of Fisheries to Contribute to Growth and Poverty,* Input paper to AusAid Roundtable Discussions, Brisbane, June.

Conservation International (2014) 'In New Caledonia, New Hope for a Thriving Ocean Economy with Historic Decree Creating the World's Largest Protected Area', Press release, 5 January, online at http://www.conservation.org/NewsRoom/pressreleases/Pages/NEW-CALEDONIA-WITH-HISTORIC-DECREE-CREATES-THE-WORLD%E2%80%99S-LARGEST-MARINE-PROTECTED-AREA.aspx [accessed 11 July 2014].

Cook Islands Marine Park (undated) 'What is Marae Moana?', online at http://www.maraemoana.gov.ck/index.php/about-marae-moana/what-is-marae-moana [accessed 11 July 2014].

Costanza, R., d'Arge, R., de Groot, R., Farberk, S., Grasso, M., Hannon, B., Limburg, K., Naeem, S., O'Neill, R. V., Paruelo, J., Raskin, R. G., Sutton, P., and van den Belt, M. (1997) 'The Value of the World's Ecosystem Services and Natural Capital', *Nature*, 387: 253–60.

Dalzell, P., Adams, T. J. H., and Polunin, N. V. C. (1996) 'Coastal Fisheries in the Pacific Islands', *Oceanography and Marine Biology: An Annual Review*, 34: 395–431.

European Commission (2006) *Communication from the Commission to the Council, the European Parliament and the European Economic and Social Committee: EU Relations with the Pacific Islands – A Strategy for a Strengthened Partnership*, COM (2006) 248 final, online at http://eur-lex.europa.eu/LexUriServ/LexUriServ.do?uri=COM:2006:0248:FIN:EN:PDF [accessed 7 November 2014].

Gillet, R. (2009) *The Contribution of Fisheries to the Economies of Pacific Island Countries and Territories*, Pacific Studies Series, online at https://www.ffa.int/system/files/Benefish%20Final%20as%20printed%20by%20ADB.pdf [accessed 7 November 2014].

Gillett, R., and Preston, G. (1997) *The Sustainable Contribution of Fisheries to Food Security in the Oceania Sub-Region of the Asia-Pacific Region: Review of Food Security Issues and Challenges in the Asia and Pacific Region*, online at http://www.fao.org/docrep/003/X6956E/x6956e09.htm [accessed 7 November 2014].

Hamnett, M., Anderson, C. L., and Cain, S. M. (2003) *American Samoa Ocean Resource Management Plan*, Honolulu, HI: Social Science Research Institute, University of Hawai'i at Mānoa.

Hoegh-Guldberg, O., Hoegh-Guldberg, H., Stout, D., Cesar, H., and Timmerman, A. (2000) *Pacific in Peril: Biological, Economic and Social Impacts of Climate Change on Pacific Coral Reefs*, Amsterdam: Greenpeace.

Pacific Islands Forum (1999) *Leaders Statement*, Thirtieth Pacific Island Leaders South Pacific Forum, 3–5 October, Palau.

Pacific Islands Forum Secretariat (PIFS) (2005) *The Pacific Plan for Strengthening Regional Cooperation and Integration*, Suva: PIFS.

Pacific Islands Forum Secretariat (PIFS) (2006) 'Annex B: Declaration on Deep-Sea Bottom Trawling to Protect Biodiversity in the High Seas', in *Forum Communiqué*, Thirty-Seventh Pacific Islands Forum, Nadi, Fiji, 24–25 October 2006, online at http://www.piango.org/PIANGO/RegMtgs/CSO/CSO2006_Final_Leaders_Forum_Communique_2006.pdf [accessed 24 November 2014].

Pacific Islands Forum Secretariat (PIFS) (2009) *Forum Communiqué*, Fortieth Pacific Islands Forum, Cairns, Australia, 5–6 August, online at http://www.forumsec.org/resources/uploads/attachments/documents/2009%20Forum%20Communique,%20Cairns,%20Australia%205-6%20Aug.pdf [accessed 24 November 2014].

Pacific Islands Forum Secretariat (PIFS) (2010) *Forum Communiqué*, Forty-First Pacific Islands Forum, Port Vila Vanuatu, 4–5 August, online at http://www.forumsec.org/resources/uploads/attachments/documents/2010_Forum_Communique.pdf [accessed 11 July 2014].

Pacific Islands Forum Secretariat (PIFS) (2012) *Forum Communiqué*, Forty-Third Pacific Islands Forum, Rarotonga, Cook Islands, 28–30 August, online at http://www.forumsec.org/resources/uploads/attachments/documents/2012%20Forum%20Communique,%20Rarotonga,%20Cook%20Islands%2028-30%20Aug1.pdf [accessed 24 November 2014].

Pacific Islands Forum Secretariat (undated) *Conservation Activities in our Region*, online at http://www.forumsec.org/pages.cfm/strategic-partnerships-coordination/pacific-oceanscape/conservation-activities-in-our-region.html?printerfriendly=true [accessed 31 January 2015].

Pratt, C., and Govan, H. (2010) *Our Sea of Islands, Our Livelihoods, Our Oceania Framework for a Pacific Oceanscape: A Catalyst for Implementation of Ocean Policy*, online at http://www.forumsec.org/resources/uploads/embeds/file/Oceanscape.pdf [accessed 7 November 2014].

Secretariat of the Pacific Community (SPC) (undated) 'PopGIS', *Statistics for Development*, online at http://www.spc.int/prism/data/popgis2 [accessed 11 July 2014].

Secretariat of the Pacific Community (SPC) (1999) *Project Proposal: South Pacific Region Comparative Assessment of Reef Fisheries Project*, Noumea, NCL: Marine Resources Division, Coastal Fisheries Programme, Secretariat of the Pacific Community.

Secretariat of the Pacific Community (SPC) (2012) *Pacific-ACP States Regional Legislative and Regulatory Framework for Deep Sea Minerals Exploration and Exploitation*, Prepared under the SPC-EU EDF10 Deep Sea Minerals Project, online at http://www.sopac.org/dsm/public/files/reports/SOPAC_RLRF_for_DSM_Final_12.07.12_.pdf [accessed 11 July 2014].

South, G. R., and Veitayaki, J. (1999) *Global Initiatives in the South Pacific: Regional Approaches to Workable Arrangement*, Asia Pacific School of Economics and Management Working Paper 99-1, online at https://digitalcollections.anu.edu.au/bitstream/1885/40961/4/so99-1.pdf [accessed 7 November 2014].

South Pacific Regional Tourism Organization (SPTO) (2005) *Economic Impact of Tourism in SPTO Member Countries*, Suva: SPTO.

Sydnes, A. K. (2002) 'Establishing a Regional Fisheries Management Organization for the Western and Central Pacific Tuna Fisheries', *Oceans and Coastal Management*, 44: 787–811.

Techera, E. J., and Klein, N. (eds) (2014) *Sharks: Conservation, Governance and Management*, New York: Routledge.

Tuqiri, S. (2001) 'Overview of an Ocean Policy for the Pacific Islands', Paper prepared for the Marine Sector Working Group, Council of Regional Organisations in the Pacific.

Tutangata, T., and Power, M. (2002) 'The Regional Scale of Ocean Governance: Regional Cooperation in the Pacific Islands', *Ocean and Coastal Management*, 45: 873–84.

United Nations (2005) *Report of the International Meeting to Review the Implementation of the Programme of Action for the Sustainable Development of Small Island Developing States Port Louis, Mauritius 10–14 January 2005*, New York: United Nations.

Vieux, C., Aubanel, A., Axford, J., Chancerelle, Y., Fisk, D., Holland, P., Juncker, M., Kirata, T., Kronen, M., Osenberg, C., Pasisi, M., Power, M., Salvat, B., Shima, J., and Vavia, V. (2004) 'A Century of Change in Coral Reef Status in Southeast and Central Pacific: Polynesia Mana Node, Cook Islands, French Polynesia, Kiribati, Niue, Tokelau, Tonga, Wallis and Futuna', in C. Wilkinson (ed.) *Status of Coral Reefs of the World: 2004, Vol. 2*, Townsville: Australian Institute of Marine Science.

World Bank (2000) *Cities, Sea, and Storms: Managing Change in Pacific Island Economies, Vol. IV – Adapting to Climate Change*, Washington, DC: Papua New Guinea and Pacific Island Country Unit, World Bank.

20

THE SUSTAINABLE DEVELOPMENT STRATEGY FOR THE SEAS OF EAST ASIA

Policy implications at local, national, and regional levels

Stella Regina Bernad and Chua Thia-Eng

Introduction

On 12 December 2003, ministers and senior officials of the twelve countries of East Asia[1] signed the Putrajaya Declaration and adopted the Sustainable Development Strategy for the Seas of East Asia (SDS–SEA) (PEMSEA, 2003a) as a regional marine strategy for achieving sustainable coastal and ocean development. The SDS–SEA was a product of more than three years of extensive national and regional consultations not only among the countries of the region, but also other stakeholders. The development of the marine strategy was largely based on relevant international conventions and other international and regional instruments, as well as the lessons learned from the coastal and ocean governance experiences in the region – particularly from a decade of efforts and activities undertaken by the Partnership in Environmental Management for the Seas of East Asia (PEMSEA).

The geographical scope of the marine strategy – that is, the Seas of East Asia – covers six large marine ecosystems: the South China Sea, the Gulf of Thailand, the East China Sea, the Yellow Sea, the Sulu-Celebes Sea, and the Indonesian Seas,[2] including their associated river and watershed systems (see Figure 20.1). The Seas of East Asia are bordered by twelve coastal nations, with China, Japan and the two Koreas on the north, and the Southeast Asian nations bordering the east, west, and south of these 'semi-enclosed seas'. Table 20.1 provides some vital statistics on the coastal resources of the twelve coastal countries bordering the Seas of East Asia.

A major task of PEMSEA before the completion of its second phase (1999–2007) was to provide the countries of the region with recommendations for:

- a policy framework for building partnerships in environmental protection and management of the Seas of East Asia;
- a regional arrangement for implementing international conventions in the East Asian Seas, including mode of operation and a sustainable mechanism;
- establishing a regional marine environment resource facility to implement regional initiatives in coastal and ocean governance;

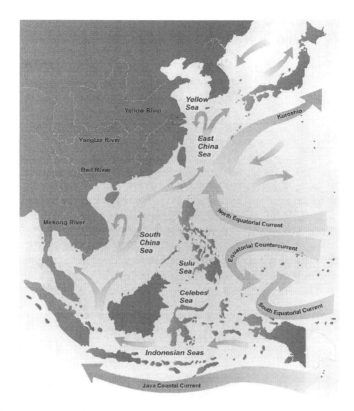

Figure 20.1 The Seas of East Asia

Source: Modified from Large Marine Ecosystems of the World (undated)

- conducting a policy conference on regional arrangements for implementing international conventions in the East Asian Seas; and
- developing a functional regional mechanism to sustain existing coastal and ocean management efforts (UNDP, 1999).

During the period 2000–03, at each of PEMSEA's intergovernmental meetings (meetings of the Programme Steering Committee), the countries progressively approved the development of the SDS-SEA from concept paper to the final draft for adoption (PEMSEA, 2000: 27–8; 2002a: 27–8; 2003b:22–3).The development of the SDS-SEA followed the process of consensus–building and consultation (PEMSEA, 2005a), including the formulation of the shared vision[3] of the concerned stakeholders, and their mission for achieving sustainable coastal and ocean development.[4] This was followed by the identification of the values that the people of East Asia attached to the seas and the threats to them. Figure 20.2 shows the type of values and threats, and how they are addressed through the Strategy to attain the shared vision. Relevant information arising from years of study and findings from different projects were used as the basis in this process.[5] The countries adopted the SDS-SEA by means of the Putrajaya Declaration at the Ministerial Forum in Malaysia held on 12 December 2003 at the East Asian Seas Congress (PEMSEA, 2003e).

Table 20.1 Statistics on coastal resources of PEMSEA countries

	Land Area km²	Water Area km²	Continental Shelf km²	Territorial Sea km²	Claimed EEZ	Coastline km	Biodiversity Mangrove Forests km²	Seagrass Species	No. of Coral genera	Wetlands of International Importance	Coastal Population
Brunei	5,270	500	7,074	3,157	5,614	161	197	4	4	X	100%
Cambodia	176,520	4,520	36,646	19,918	X	443	467	1	X	546	24%
China	9,326,410	270,550	810,387	348,090	X	14,500	0	5	36	5,884	24%
DPR Korea	120,410	130	26,251	12,654	72,755	2,495	0	X	X	X	93%
Indonesia	1,826,440	93,000	1,847,707	3,205,695	2,914,978	54,716	23,901	12	77	2,427	96%
Japan	374,744	3,091	304,246	373,381	3,648,393	28,751	0	8	75	837	96%
RO Korea	98,190	290	226,277	81,125	202,585	2,413	0	X	X	10	100%
Malaysia	328,550	1,200	335,914	152,367	198,173	4,675	1,659	9	72	383	98%
Philippines	298,170	1,830	244,493	679,774	293,808	36,289	23	19	74	684	100%
Singapore	682	10	714	744	X	268	X	11	66	X	100%
Thailand	511,770	2,230	185,351	75,876	176,540	7,066	5,092	14	68	5	39%
Vietnam	325,360	4,200	352,420	158,569	237,800	11,409	734	9	1	120	83%

Source: PEMSEA (2005c: 26)

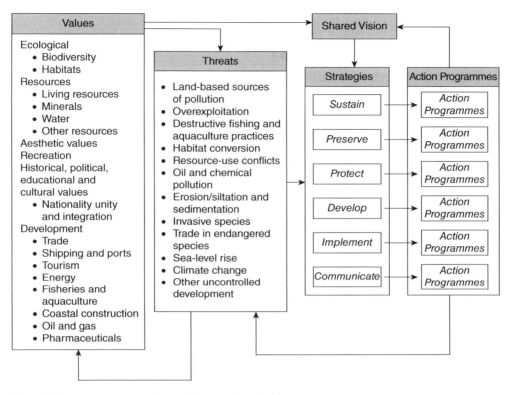

Figure 20.2 A strategic approach to achieving a shared vision

Source: PEMSEA (2003a: 49)

The SDS-SEA and international instruments

Incorporating relevant environment-related international instruments into the SDS-SEA

The SDS–SEA identifies the relevant international and regional instruments relating to the marine environment as among the pillars on which it is founded (PEMSEA, 2003a: 42). The ratification and implementation of international conventions on marine pollution was a component of the first phase of the PEMSEA Regional Programme (1994–99). During the second phase, the scope extended over international instruments relating not only to marine pollution, but also to the whole coastal and marine environment. By then, their importance in the management of the coastal and marine environment was well understood by the region's stakeholders.

During the first phase, the international instruments were treated individually, particularly with regard to national government obligations. During the second phase, the need for a more integrated approach was acknowledged, as was the role of local government in their implementation. When the development of the SDS–SEA began, the requirements of these instruments were the basis for the identified objectives and activities, in line with the shared vision.

The regional strategy began as the environmental strategy for the Seas of East Asia, with the general objective of ensuring the management of the marine and coastal environment of the region. However, as the first draft was completed, it became clear that it had integrated all sectors, as well as the economic and social aspects of development, and that it was really a sustainable development strategy, not only an environmental one.

The development of the SDS–SEA, not by design, occurred almost in parallel with the development of the World Summit on Sustainable Development (WSSD). This is because the SDS–SEA endeavoured to address the issues and problems affecting the Seas of East Asia, which in many ways reflect global concerns. Furthermore, the SDS–SEA used Agenda 21 (UNGA, 1992) as a framework for devising the necessary actions, hence its kinship with WSSD. Therefore the SDS–SEA became a regional strategy for implementing both the relevant provisions of Agenda 21 and those of the WSSD Plan of Implementation (WSSD, 2002). In addition, SDS–SEA indirectly implements the seventh Millennium Development Goal (MDG) on environmental sustainability and some provisions of the MDGs – namely, those on poverty, gender equality, health and global partnership for development.

The shared vision and mission for the Seas of East Asia region perform a very important function by setting a common direction for all stakeholders in the region. The SDS–SEA also identifies the desired changes (institutional, operational, and outcomes) expected to occur in the region (PEMSEA, 2003a: 38–9), thus providing the rationale for pursuing the actions identified in the Strategy.

Integrated management framework

The experience of PEMSEA in integrated management was put to good use in the development of the SDS–SEA. This marine strategy enshrined the integrated approach as its basic framework (PEMSEA, 2003a: 37). It represented the conviction that the provisions of Agenda 21 and WSSD related to coastal and ocean governance could be effectively implemented through a holistic and integrated planning and management approach.

The incorporation of the international and regional instruments into the appropriate places of the strategy was therefore an essential part of this approach. Their implementation was facilitated by contextualizing them vis-à-vis the main goals and objectives, and by avoiding the conventional approach in which the implementation of each convention is a stand-alone goal with coincidental connection to the others. In fact, the SDS–SEA provided an integrated management framework for incorporating various environmental concerns. It also facilitated common platforms for stakeholder cooperation and partnership at regional, national, and local levels.

The Strategic Action Statement of the SDS–SEA (see Box 20.1) established the six sub-strategies of the SDS–SEA – 'Sustain', 'Preserve', 'Protect', 'Develop', 'Implement', and 'Communicate' – and incorporated the various international and regional instruments, as appropriate. Each of them incorporated specific instruments, such as the 1992 Convention on Biological Diversity (CBD) and the Code of Conduct for Responsible Fisheries of the Food and Agriculture Organization of the United Nations (FAO, 1995). These instruments belong respectively with the 'Sustain' and 'Preserve' sub-strategies, while the International Maritime Organization (IMO) conventions on marine pollution fall under 'Protect'. The 'Develop' sub-strategy referred to the development of national policies on integrated coastal management (ICM) and the integration of environmental protection policies into the national development policies. This is where Agenda 21 and the WSSD Plan of Implementation are most clearly applied. It is clear in the overall context that 'Develop' is necessary for the effective implementation of all of the conventions.

Box 20.1   Strategic Action Statement of the SDS–SEA

The East Asian countries shall:

- Ensure **Sustain**able use of coastal and marine resources.
- **Preserve** species and areas of the coastal and marine environment that are pristine or are of ecological, social or cultural significance.
- **Protect** ecosystems, human health and society from risks occurring as a consequence of human activities.
- **Develop** economic activities in the coastal and marine environment that contribute to economic prosperity and social well-being while safeguarding ecological values.
- **Implement** international instruments relevant to the management of the coastal and marine environment.
- **Communicate** with stakeholders to raise public awareness, strengthen multisectoral participation and obtain scientific support for the sustainable development of the coastal and marine environment.

Source: PEMSEA (2003a: 46)

'Sustain', 'Preserve', and 'Protect' may be said to be more sectoral in substance, while 'Develop', 'Implement', and 'Communicate' are more inter-sectoral. However, both sets of sub-strategies have measures of inter-sectoralism and sectoralism for an overall integrative effect. By taking the integrated approach, it is possible to implement several conventions at the same time, their specific requirements integrated into one system. This approach will also help in the elimination of duplication of efforts. Figure 20.3 shows the relationship of the SDS–SEA sub-strategies with several international instruments.

Implementing the marine strategy

Regional implementation through a partnership arrangement

The SDS–SEA is being implemented through an evolutionary and incremental process. In considering the implementing mechanism, the participating countries went through a series of debates during the PEMSEA intergovernmental meetings of 2004–06 (PEMSEA, 2004: 7–11, 30–1; 2005b: 7–9, 25–6; 2006a: 9–10, 29–30), and finally agreed to enter into not a regional convention, but a non-binding agreement instead.

There were two reasons for this decision. First, country representatives felt that capacity disparity between the countries in the region, in terms of human and financial resource, was too wide. Such capacity gaps would not place all of the concerned countries on a level playing field in addressing environmental and economic development issues. Many of the transboundary issues could not be addressed collectively. Thus the conventional approach of equal responsibility might not be effective.

Second, PEMSEA's experience over the preceding sixteen years had shown that the responsibility of caring for the coastal and marine environment and using it in non-destructive ways lies not only with national and local governments, but also with other stakeholders – including members of civil society, the private sector, academia, and international stakeholders. These stakeholders needed to act as partners to address management challenges that cut across sectors and jurisdictional boundaries. A partnership arrangement for PEMSEA participating countries

	Rio Declaration	Agenda 21	UNCLOS	UNFCCC	CBD	Ramsar	CITES	Basel Convention	GPA on LBS	MARPOL	London Convention	CLC	FUND	OPRC	Ballast Water
Sustain															
1. Conservation and redress of biodiversity	✓							✓	✓						
1 1. Policy and strategic framework for biodiversity		✓	✓		✓	✓	✓								
2 2. Restore coastlines, habitats and resources		✓	✓		✓	✓	✓								
2. Quality of coastal waters	✓	✓	✓							✓					
3 1. Compatibility of freshwater and marine water uses						✓									
4 2. International water system management						✓									
3. Sustainable fisheries	✓									✓					
5 1. Fisheries management in subregional sea areas		✓	✓												
6 2. Responsible utilization of living resources		✓	✓		✓		✓								
7 3. Integration of fisheries into coastal management programmes		✓													
Preserve	✓													✓	✓
1. Marine protected areas															
8 1. Selection and prioritization of MPAs		✓	✓		✓	✓			✓	✓	✓				
9 2. Appropriate management regimes for MPAs		✓	✓		✓	✓			✓	✓					
2. Rare, threatened and endangered species		✓													✓
10 1. Regional accord for species protection			✓		✓	✓	✓		✓						
11 2. National recovery and management processes for species at risk			✓		✓	✓	✓		✓						
12 3. Regionwide safety nets for species at risk			✓		✓	✓	✓		✓						
3. Areas of social, cultural, historical and geological significance			✓												
13 1. Protect cultural and natural properties of regional value						✓									
14 2. Manage transborder cultural and natural heritage sites	✓					✓									
Protect															
1. LME/subregional sea management		✓									✓	✓	✓		✓
15 1. Intergovernmental cooperation on regional seas	✓		✓	✓	✓	✓		✓	✓	✓				✓	
16 2. National and local governments' roles in regional seas management	✓		✓	✓	✓	✓			✓	✓				✓	
2. Degradation from land-based activities	✓							✓	✓		✓				
17 1. Capability strengthening for protection from land-based activities		✓	✓						✓		✓				
18 2. Impact management programmes		✓	✓		✓				✓						
19 3. Historic approach to impact management		✓	✓		✓	✓	✓	✓	✓	✓	✓				
20 3. Adverse impacts of sea-based activities	✓				✓					✓	✓	✓	✓	✓	
21 1. Prevention of shipping pollution		✓	✓							✓				✓	✓
22 2. Control of ocean dumping		✓	✓								✓				
23 3. Countering accidental spills		✓	✓							✓				✓	
4. Land- and sea-based economic activities		✓	✓							✓	✓				✓
24 4. Pollution damage liability and compensation									✓						
25 1. Expediting recovery of costs and damages from oil spills	✓	✓	✓										✓		
26 2. Expediting cost recovery and damage compensation schemes	✓	✓			✓						✓	✓	✓	✓	✓
27 3. Innovative approaches to restoration of damages	✓	✓			✓						✓	✓	✓	✓	✓

Figure 20.3 (continued)

	Rio Declaration	Agenda 21	UNCLOS	UNFCCC	CBD	Ramsar	CITES	Basel Convention	GPA on LBS	MARPOL	London Convention	CLC	FUND	OPRC	Ballast Water
Develop															
1.Sustainable economic development							✓								
28 1. Promotion of national coastal and marine strategies and policies	✓	✓		✓	✓	✓			✓						
29 2. Mechanisms that promote public participation	✓	✓		✓	✓				✓	✓					
30 3. Integrated economic development and environmental management	✓	✓		✓	✓				✓	✓	✓				✓
2. ICM as a management framework						✓	✓				✓	✓	✓	✓	✓
31 1. Reducing non-sustainable usage of coastal and marine resources	✓	✓	✓	✓	✓	✓			✓		✓				
32 2. Harnessing knowledge and concern about the environment	✓	✓		✓	✓				✓						
33 3. Sustainable development programmes at local level	✓	✓		✓	✓				✓		✓			✓	
34 4. Ecological and social impacts of coastal urbanization	✓	✓		✓		✓			✓		✓				
3. Subregional growth areas				✓				✓	✓	✓					
35 1. Growth area implications on coastal and marine resources	✓	✓		✓											
36 2. Appropriate policies and guidelines	✓	✓		✓		?	?								
4. Sustainable financing and environmental investments		✓			✓				✓						✓
37 1. National policies and programmes for stable investment climate	✓			✓						✓					
38 2. Environmental investments at local level	✓			✓						✓					
39 3. Strengthening the role of the private sector										✓					
Implement															
1. National compliance with relevant conventions															
40 1. Desired management outcomes of international instruments	✓	✓	✓	✓	✓	✓	✓	✓	✓	✓	✓	✓	✓	✓	✓
41 2. Efficiency and effectiveness of implementation	✓	✓	✓	✓	✓	✓	✓	✓	✓	✓	✓	✓	✓	✓	✓
2. Regional cooperation and implementation of conventions	✓														
42 1. Enhancing synergies and linkages of regional level		✓	✓	✓	✓	✓	✓	✓	✓	✓	✓	✓	✓	✓	✓
43 2. Functional framework for regional cooperation		✓	✓	✓	✓	✓	✓	✓	✓	✓	✓	✓	✓	✓	✓
3. Local government implementation of conventions	✓														
44 1. Enabling local stakeholders' contribution to international instruments		✓	✓	✓	✓	✓	✓	✓	✓	✓	✓	✓	✓	✓	✓
Communicate															
1. Public awareness and understanding															
45 1. Good information exchanges between stakeholders	✓	✓	✓	✓	✓	✓	✓	✓	✓	✓	✓		✓	✓	✓
46 2. Strengthening the use of available information	✓	✓	✓	✓	✓	✓	✓	✓	✓	✓	✓		✓	✓	✓
2. Science and traditional knowledge															
47 1. Information technology as a vital tool	✓	✓		✓					✓	✓					
48 2. Utilizing science and traditional knowledge in policymaking	✓	✓	✓	✓	✓	✓			✓	✓	✓				
3. Mobilization of governments, civil society and private sector							✓								
49 1. Dissemination of reliable and relevant data	✓	✓	✓	✓	✓	✓	✓		✓	✓	✓	✓		✓	✓
50 2. Information sharing	✓	✓	✓	✓	✓	✓	✓		✓	✓	✓	✓		✓	✓
51 3. Stakeholder ownership	✓	✓		✓	✓				✓						

Figure 20.3 The SDS–SEA Strategy and international instruments

Notes: UNCLOS – 1982 United Nations Convention on the Law of the Sea; UNFCCC – 1992 United Nations Framework Convention on Climate Change; CBD – 1992 Convention on Biological Diversity;

Ramsar – 1971 Convention on Wetlands of International Importance Especially as a Waterfowl Habitat; CITES – 1973 Convention on the International Trade in Endangered Species of Wild Fauna and Flora; Basel Convention – 1989 Basel Convention on the Control of Transboundary Movements of Hazardous Wastes and their Disposal; GPA on LBS – Global Programme of Action for the Protection of the Marine Environment from Land-based Activities (UNEP, 1995); MARPOL – 1973 International Convention for the Prevention of Pollution from Ships, as amended by the 1978 Protocol; 1972 London Convention – Convention on the Prevention of Marine Pollution by Dumping of Wastes and Other Matter, as amended by the 1996 Protocol; CLC – 1969 International Convention on Civil Liability for Oil Pollution Damage; FUND – 1971 International Convention on the Establishment of an International Fund for Compensation for Oil Pollution Damage; OPRC – 1990 International Convention on Oil Pollution Preparedness, Response and Co-operation; Ballast Water – 2004 International Convention for the Control and Management of Ships' Ballast Water and Sediments

thus evolved as the regional implementing mechanism for the SDS–SEA, putting into practice partnerships as the second pillar of the SDS–SEA (PEMSEA, 2003a: 42).

A step-wise process, established during the development of the SDS–SEA and its adoption through a non-binding declaration, continued with the evolution of the implementing mechanism for SDS–SEA. This included the very important measure of identifying regional priorities for implementation. These priorities were embodied in a regional framework programme designed to strengthen national capacity, to foster environmental investments, to promote synergies, to build strategic partnerships, and to coordinate and catalyse national and international efforts for SDS–SEA implementation in a non-binding, but participatory, setting. Over the long term, a country- and stakeholder-owned, self-sustaining regional mechanism was expected to evolve.

In the meantime, the regional partnership mechanism continued to operate under the United Nations (UN) framework. While four participating countries (China, Japan, the Republic of Korea, and the Philippines) began to contribute to the Secretariat of the regional partnership mechanism and were later joined by other participating countries (East Timor and Singapore), the Partnership applied for the continuing support of the Global Environment Facility (GEF) for national and regional implementation of the SDS–SEA and to facilitate the operation of the implementing mechanism[6] under its third phase (2008–13). Other prioritized actions included national coastal and marine policy development, a scaling up of integrated coastal management practices, the networking of areas of excellence and marine affairs institutions, and the establishment of strategic partnership programmes (see Box 20.2).

Box 20.2 Components of the GEF/UNDP Regional Programme on the Implementation of the SDS–SEA

A   A functional regional mechanism for SDS–SEA implementation

B   National policies and reforms for sustainable coastal and ocean governance

C   Scaling up ICM programmes

D   Twinning arrangements for ecosystem-based management

E   Intellectual capacity and human resources

F   Investment and financing

G   Strategic partnership arrangements

Other PEMSEA partners and collaborators would implement other aspects of the SDS-SEA in accordance with their own mandates. For example, UN partners such as the IMO, the FAO, the United Nations Environment Programme (UNEP) Global Programme of Action for the Protection of the Marine Environment from Land-based Activities (GPA), the Intergovernmental Oceanographic Commission (IOC), and the International Atomic Energy Agency (IAEA) could contribute to the implementation of the SDS-SEA through implementation of their regional mandates. The synergies between them could enhance their own outputs, because they benefit from the sharing of information generated and capacity built through this process. Many, if not most, of the partners and collaborators were already implementing activities identified in the SDS-SEA. In this way, the partnership also interconnected with site- and issue-specific regional and sub-regional activities and mechanisms. Many potential partners and collaborators ran projects addressing specific issues that were sectoral or cross-sectoral in nature in specific geographical areas within the East Asian Seas region. For example, the sub-regional agreement on oil spill preparedness and response jointly developed by Cambodia, Thailand, and Vietnam in the Gulf of Thailand is considered to be an integral part of SDS-SEA implementation.

In July 2012, the country partners, during the Ministerial Forum held in the Republic of Korea, recognized that while progress has been made towards the Haikou targets (PEMSEA, 2006c), efforts needed to be redoubled. This would be done with a new SDS-SEA Implementation Plan (2012–16), adopted by means of the Changwon Declaration, *Toward an Ocean-Based Blue Economy: Moving Ahead with the Sustainable Development Strategy for the Seas of East Asia*, signed by ministers from ten countries of the East Asian Seas region during the 2012 East Asian Seas Congress on 9–13 July 2012 (PEMSEA, 2012b). This Plan addresses, among other things: the mainstreaming of SDS-SEA objectives, targets, and actions; a shifting from government–centred toward more inclusive coastal and ocean governance; converging sectoral initiatives and programmes in priority coastal, marine, and watershed areas within the framework of national ICM programmes (especially in relation to climate change adaptation and disaster risk reduction, biological diversity, water quality, food security, and investments); building technical and management capacity; and targeting research.

National implementation

At the national level, the participating countries agreed to develop and implement their ten–year national plan of actions for the implementation of the SDS-SEA. To a large extent, the type and level of national actions are in accordance with their national capacity. In other words, participating countries committed to implement the SDS-SEA within their own time frames and at their own pace. In most cases, where less national resources were available for implementation or political processes took a longer time to get through, implementation was limited to some priority areas of national concern.

Institutional arrangements for the implementation of the SDS-SEA

In 2006, the ministers of the eleven country partners[7] signed the Partnership Agreement in Haikou, China (PEMSEA, 2006c), and all of the partners agreed to the Partnership Operating Arrangements (PEMSEA, 2006d), adopting the following interrelated components of the regional institutional arrangements.

1   The *East Asian Seas (EAS) Partnership Council* is the governing body. The Council is conducted in two sessions. A Technical Session is attended by government representatives,

as well as concerned stakeholder partners, and focuses on technical matters relating to the implementation of the SDS-SEA. The Intergovernmental Session is limited to government representatives; this session is responsible for policy matters and adoption of the recommendations of the Technical Session. Each session elects its own chair. The Council chair is elected under a joint session.

2   The *PEMSEA Resource Facility (PRF)* is made up of two functional units: Secretariat Services and Technical Services. The PRF Secretariat Services acts as the Secretariat to the Partnership Council and the Executive Committee. It organizes the Partnership Council and Executive Committee meetings, coordinates SDS-SEA implementation at the national level, coordinates various networks set up by PEMSEA, facilitates information dissemination and capacity building, and prepares the triennial EAS Congress, Ministerial Forum, and other major workshops. The PRF Technical Services implements projects and programmes, conducts training courses, and provides technical assistance to interested countries and other technical supports.

3   The *Regional Partnership Fund (RPF)* is a trust fund built up from donor contributions and other income arising from the sale of goods (publications, software) and services from the PRF Technical Services. The Fund is used for specific activities toward attaining the goals and objectives of PEMSEA. By operationalizing the RPF, PEMSEA hopes to gradually shift from being fully dependent on the GEF to future reliance on multiple sources of financial income. The SDS-SEA and the Programme of Activities can provide a framework through which donor communities can identify the projects and activities that they want to support.

4   The *EAS Congress* takes place every three years, bringing together stakeholders, experts, regional partners, and other actors from around the world to evaluate progress in the implementation of the regional strategy, and to share their experience and exchange information or ideas in different areas of concern on the sustainable development of coasts and oceans. The event includes an international conference, a Ministerial Forum, exhibits, and other side events. A total of more than 4,700 participants have taken part in the last four congresses.

5   The *Ministerial Forum* is held to be an integral part of the EAS Congress and is attended by ocean-related ministers from the participating countries of PEMSEA. The Forum allows ministers to review the status of implementation of the SDS-SEA, renew commitments, and set new policy directions.

Figure 20.4 shows the key components of the implementing arrangements for the SDS-SEA and how they interrelate. The institutional arrangement facilitates PEMSEA's transformation from a fixed-term regional programme into a longer-term partnership arrangement, in order to be true to the third pillar of the SDS-SEA – that is, self-reliance and sustainability (PEMSEA, 2003a: 42).

While non-binding, the mechanism was undertaken with a long-term view and the expectation that it would develop and evolve as needed. Its future was carefully studied, particularly in consideration of the need to balance the stability of the conventional approach with the flexibility to accommodate the active participation of all stakeholders. This is to keep in view the fourth and final pillar of the SDS-SEA, 'synergy': attaining a multiplier and cumulative effect towards the achievement of the shared vision through the inclusion of all sectors, interests and issues. A case in point is the catalysing effects of PEMSEA's holistic approach to coastal and ocean governance. A United Nations Development Programme (UNDP)/GEF report entitled *Catalysing Ocean Finance* noted that PEMSEA ICM implementation and scaling-up initiatives

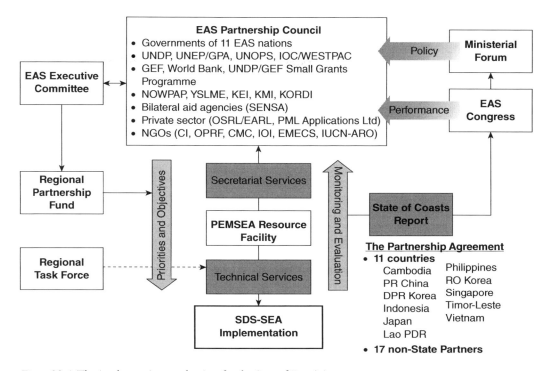

Figure 20.4 The implementing mechanism for the Seas of East Asia

have catalysed US$5–9 billion in public and private sector finance over a cumulative GEF investment of US$36.1 million during the last twenty years (UNDP/GEF, 2012).

These innovative approaches represented a paradigm shift in coastal and ocean governance. The regional mechanism facilitates establishment of linkages (synergies) with stakeholders at the international, regional, national, and local levels, with a view to enhancing regional coordination, promoting policy and functional integration, and establishing or linking up regional, sub-regional, and subnational agreements. The implementation of site- or issue-specific sub-regional agreements is strengthened through the cooperative framework of the SDS-SEA. This approach enables the region to streamline regional or sub-regional agreements and ensure a more effective, cohesive, and vision-focused regional cooperation. A variety of relationships in various collaborative modes or forms, such as the PEMSEA Network of Local Governments (PNLG), is encouraged as an integral part of the partnership. Such arrangements provide a stronger bond between partnering stakeholders, because they are based on common concerns.

The new regional mechanism is process- and results-oriented in establishing and consolidating its operational functions (see Figure 20.5). This allows the regional mechanism to mature through succeeding phases, ensuring effectiveness, trust, and commitment.

In 2009, eight of the country partners signed an agreement recognizing the international legal personality of PEMSEA with headquarters in the Philippines, making PEMSEA an international organization to focus on the implementation of the SDS-SEA (PEMSEA, 2009a).[8] In this manner, PEMSEA completed the process of finalizing a regional institutional arrangement for implementing the SDS-SEA.

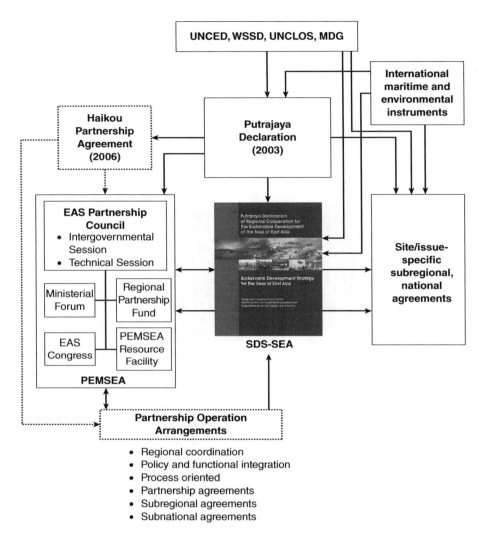

Figure 20.5 Process- and partnership-oriented regional mechanism for the Seas of East Asia

Source: Bernad et al. (2006)

Implications of the SDS-SEA on coastal and ocean policymaking

The SDS-SEA and the institutional arrangements for its implementation can significantly influence policy formulation on coastal and ocean governance at local, national, and regional levels. They influence regional policy in the following ways:

1 The broad management framework of the SDS-SEA requires interagency and multi-sectoral cooperation for implementing the various action programmes on coastal and ocean govern-ance. This creates a need and justification for reforming and streamlining sectoral policies, and facilitates stronger policy and functional integration on coastal and marine management

and cross-sector cooperation at the operational level. The ecosystem-based approach also encourages policymakers at local and national levels to consider cross-boundary management issues and to develop appropriate cooperative programmes to address them.

2   The regular Ministerial Forum provides an opportunity for policymakers to look at cross-boundary issues of the coasts and oceans, which often rank low in the national priority agenda. The state of the coasts report (PEMSEA, 2011) is an excellent tool with which policymakers can examine environmental sustainability issues from the standpoint of regional interest. The Ministerial Forum represents the governmental side of the partnership. Through this modality, the ministers periodically issue pronouncements on new policy directions and commitments of the partner states, thus building the necessary arguments for corresponding national coastal or marine policy development. A further advantage is that the ministers representing the countries come from different sectors (environment, development, fisheries, transportation, and marine affairs), and are each able to contribute different perspectives.

3   The 2003 Putrajaya Declaration is, in fact, a regional policy arising from the political commitments of the ministers to implement the regional strategy (PEMSEA, 2003e). Although it is not a legally binding agreement, the countries have the moral obligation to implement the Strategy to which they have committed.

4   The Haikou Partnership Agreement (PEMSEA, 2006c), which was signed by the concerned country ministers, and the Partnership Operating Arrangements (PEMSEA, 2006d), signed by the stakeholder partners, are certainly another form of regional policy committing to safeguard the functional integrity of the Seas of East Asia. The two documents provide the institutional framework for cooperation in the implementation of the SDS-SEA, ensuring that it will move forward. Furthermore, the Haikou Agreement identified the goals of the Partnership for 2015 as:

  i   ICM implemented in at least 20 per cent of the region's coasts by 2015;
  ii   national policies and action plans in at least 70 per cent of participating countries by 2015; and
  iii a rolling ten-year partnership programme and state of the coasts reports by 2009.

5   The Manila Declaration on Strengthening the Implementation of Integrated Coastal Management for Sustainable Development and Climate Change Adaptation in the Seas of East Asia Region (PEMSEA, 2009b) expressed the formal adoption of ICM, the approach used by PEMSEA, and the agreement to undertake specific actions to address climate change and to mitigate disasters.

6   The Changwon Declaration, *Toward an Ocean-Based Blue Economy: Moving Ahead with the Sustainable Development Strategy for the Seas of East Asia* (PEMSEA, 2012b), articulated that the new SDS-SEA Implementation Plan (2012–16) would be used to implement Rio+20 and other relevant commitments.

7   The EAS Partnership Council is designed to facilitate regional efforts in addressing common regional concerns. The Council is the modality for interface between regional and national policy, and, through this, the promotion of the development and/or reform of national policy.

Thus the regional commitment in the implementation of the regional strategy has a strong bearing on national coastal and ocean policy development.

At the national level, countries each develop their own national plans of action to implement the SDS-SEA. A major effort is to integrate existing national sector programmes and development plans related to coastal and ocean governance coherently into such plans of action, in line with the SDS-SEA. This encourages development of national coastal/marine/ocean policy as a means of ensuring stronger integration of policies and functions of line agencies.

The region has valuable national experiences in policy development in integrated coastal and ocean governance. These earlier experiences influenced the development of the SDS-SEA and later experiences are instructive in its implementation.

Box 20.3   Interface of the Rio+20 outcome document and the SDS-SEA and its Implementation Plan

Rio+20: The Future We Want

II   *RENEWING POLITICAL COMMITMENT*

    A   Reaffirming the Rio Principles and past action plans

    B   Advancing integration, implementation and coherence: assessing the progress to date and the remaining gaps in the implementation of the outcomes of the major summits on sustainable development and addressing new and emerging challenges

IV   *INSTITUTIONAL FRAMEWORK FOR SUSTAINABLE DEVELOPMENT*

    E   Regional, national, subnational and local levels

98. We encourage regional, national, subnational and local authorities as appropriate to develop and utilize sustainable development strategies as key instruments for guiding decision-making and implementation of sustainable development at all levels . . .

V.   *FRAMEWORK FOR ACTION AND FOLLOW-UP*

    A   Thematic areas and cross-sectoral issues
- Poverty eradication
- Food security and nutrition
- Water and sanitation
- Energy
- Sustainable tourism
- Sustainable cities and human settlements
- Oceans and seas
- Regional efforts
- Disaster risk reduction
- Climate change
- Biodiversity

SDS-SEA Regional Strategy

SDS-SEA AND MINISTERIAL FORUMS

SDS-SEA contains six strategies and 227 action programmes incorporating the principles, objectives, and actions that have been enshrined in global and regional agreements, including UNCLOS,

---

UNFCCC, Agenda 21, CBD, UNEP/GPA, WSSD Johannesburg Plan of Implementation (JPOI), UN MDGs, and various IMO Conventions

Ministerial Forums are convened triennially to review progress in achieving SDS-SEA objectives and to identify gaps, as well as changes and emerging trends in the Seas of East Asia. Each Ministerial Forum advances innovative concepts and targets for the region, which respond to national, regional, and global priorities and concerns.

SDS-SEA IMPLEMENTATION PLAN (2012–16)

Targets 1 (self-sustained regional implementation mechanism), 2 (national coastal and ocean policies in 70 per cent of participating countries), and 3 (ICM programmes covering 20 per cent of the region's coastline)

Action 1.1: Complete the transformation of PEMSEA into a self-sustained regional partnership mechanism for SDS–SEA implementation

Action 2.1: Achieve national coastal and ocean policies and supporting institutional arrangements . . . and integration of SDS–SEA objectives and targets into medium-term development and investment plans.

Action 3.1: Maximize local government capacity to effectively contribute to (SDS–SEA implementation) . . .

SDS-SEA IMPLEMENTATION PLAN (2012–16)

ICM scaling up (Target 3: ICM programmes covering 20 per cent of the region's coastlines)

Action 3.2: Realize climate change adaptation and disaster risk reduction measures in vulnerable coastal areas

Action 3.3: Integrate sustainable use of coastal and marine ecosystem services in biodiversity and fisheries hotspots

Action 3.4: Advance water supply conservation and management and pollution reduction and waste management in priority coastal and watershed area

Source: Lotilla (2012)

The interrelation of local and national integrated coastal management policy

Several local coastal management initiatives have a strong bearing on national coastal policy development. For example, while practising integrated coastal management (ICM), the city government of Xiamen in China adopted an integrated functional zoning scheme and enacted the Xiamen Marine Use Regulation. Together, these two regulatory instruments formed the backbone of integrated management that addressed multiple-use conflicts. The regulatory instruments addressed common property rights issues – that is, allocating use rights, while protecting the ownership right of the state. By means of the use of economic instruments, the city government was able to regulate the type and level of sea use, and thus resolved several outstanding use conflicts and effectively reversed the trend of environmental degradation (PEMSEA, 2006b: 52–3). This approach was adopted by national legislation in 2002. The legislation mandated the application of an integrated management approach to ensure 'rational use and sustainable development', and to enhance the benefits of use by adopting functional zoning schemes and a permit and users' fee system. The national legislation also addressed the issues related to open access of the coastal and marine areas of China (PEMSEA, 2003c: 53–8).

In the Philippines, learning and scaling up from the experiences of Batangas Bay, Manila Bay, and several other initiatives, the ICM approach has been adopted as a national policy by executive order.[9] This example fully demonstrates the 'bottom-up' approach in influencing national policy development and enactment. Local governments in the Philippines have considerable power over natural resources, although the national government agencies regulate many activities that take place in the coastal zone. Local governments are therefore very important to coastal management, because 822 of the 1,502 municipalities are coastal (World Bank Group, 2005: 1). The executive order obliges the national agencies to cooperate in the development and implementation of ICM programmes. In this way, it paves the way for capacity building among local governments to implement ICM, as well as to coordinate and integrate sectoral resources and efforts to achieve sustainable development objectives. Legislation on the same is currently pending in Congress.

Another developmental approach may be observed in the Republic of Korea. Here, the enactment of national legislation on ICM provides the legal basis and resources for the implementation of site-specific coastal programmes. The country has set up its legislative framework by means of the Coastal Management Act of 1998, the Wetlands Conservation Act of 1999, the Marine Ecosystems Conservation Act of 1997, and the Marine Environment Management Act of 2009, and has adopted a national ocean policy and an ocean agenda known as 'Ocean Korea 21'. Other countries, such as Indonesia, established coastal or ocean policy to promote coordination and integration of management measures in the coastal zone.

Institutional changes

National efforts towards sustainable use of the goods and services from the coasts and oceans have led to institutional reform in some countries in the region. Some examples are as follows.

In the Republic of Korea, the marine-related functions of the Ministries of Communication and Transportation, Agriculture, Forestry and Fishery, Environment, and Science and Technology were merged into the Ministry of Maritime Affairs and Fisheries in 1996, with the objective of enhancing global competitiveness (PEMSEA, 2003d: 24).

Three years later, Indonesia set up a new Ministry of Marine Affairs and Fisheries to oversee the development and management of coastal and marine resources, especially fisheries (PEMSEA, 2002b: 60–1). The country also passed legislation to decentralize resource management to local governments. Many local governments have yet to catch up with the required capacity for effective utilization of their newly delegated authority and mandates. Some of them, however, have already acquired the needed coastal management experience from their earlier involvement with ICM initiatives in the country. Therefore there is a critical mass of expertise and experience available in the country to embark on efforts to replicate ICM practices and their scaling up throughout the coastal areas.

In Vietnam, several national and regional efforts in coastal management have resulted in national institutional reform. The Ministry on Natural Resources and Environment was created, and the Vietnam Environmental Protection Agency (VEPA), which was formerly under the Ministry of Science, Technology and Environment, placed under it. An ICM division was established under the new ministry, with the objective of implementing a sustainable ICM programme (PEMSEA, 2003b: 8). A Law on ICM is identified in the government's list for legislation to be prepared after the approval of the Vietnamese Law of the Seas. The Viet Nam Administration of Seas and Islands (VASI) was established in March 2008 to coordinate the integrated and unified state management for seas and islands, including coastal areas (PEMSEA, 2012a).

In Japan, substantial efforts were devoted to the formulation of a national ocean policy and enactment of ICM-related legislation (PEMSEA, 2006e), until the Ocean Basic Law of Japan was finally approved by the *Diet* (Parliament) in 2007. Based on the policy, the Basic Plan on Oceans Policy was promulgated in 2008. Influenced by many developments, including the Fukushima nuclear disaster, the latter was revised in 2013 in the form of the Renewed Basic Plan on Ocean Policy, which will guide the country during the following five years (PEMSEA, 2013). Institutionally, the Ministry of Land, Infrastructure, Transportation and Tourism (MLIT) emerged from the integration of the Hokkaido Development Agency, the National Land Agency, the Tourism Agency, the Ministry of Transport, and the Ministry of Construction in the early part of the century. The mission of MLIT is to promote, in a comprehensive manner, national land policy, and to improve social capital, transport, and related policies. It has been mandated with the creation of a vision of the best design and utilization for Japan's national land, including coastal areas, and to realize that vision by combining and developing the different infrastructure improvement and transport policies consistent with sustainable development (PEMSEA, 2002b: 47). It should be noted that most environmental concerns are still within the mandate of the Ministry of Environment. It is foreseen that further integration with ocean management will result from the implementation of the ocean policy and the establishment of the Ocean Secretariat.

In China, most fishery and ocean agencies at provincial, municipal, and district levels were merged into a single fishery and ocean agency for each level. These agencies report to both sectoral line agencies at the central level. This means that the local fishery and ocean bureau, while reporting to the Department of Fisheries and the State Oceanic Administration separately at the central level, operates under the local government technically as one agency (Chen and Uitto, 2003: 67–80).

Conclusion

The new approach in regional cooperation for the Seas of East Asia:

- shifts coastal and ocean governance from being government-centred toward a more inclusive approach, involving both governments and stakeholder partners;
- consolidates regional efforts in achieving target-focused action programmes;
- mobilizes the human and financial resources of stakeholder partners through a common platform and framework for coastal and ocean governance;
- reorients existing fragmented projects and programmes related to coastal and ocean management through a common programme framework; and
- instils a dynamic process to enable advocacy, monitoring, and evaluation of progress and impacts.

The achievement of the SDS-SEA may be measured as the aggregate of all of the smaller successes and impacts derived from the implementation of each component part of the marine strategy at local and national levels by the governments and stakeholder partners. The impact of the SDS-SEA as a whole will be seen by knitting these achievements together. National and local policy reforms in coastal and ocean governance are increasing throughout the region. The outcome is expected to inspire and encourage national and local leaders to take more proactive roles in addressing cross-boundary environmental and resource use issues at the regional level. It will build momentum among the regional partners to continue the worthwhile efforts towards achieving the goals and objectives of the SDS-SEA.

At this point in time, the SDS-SEA and the ministerial declarations represent policy at the regional level. It is very soft policy, but the regional mechanism provides a stable base that can be built upon. Through this step-wise, incremental, non-binding, participatory approach, the lack of a regional convention has been turned into a positive and, so far, continues to grow its potential for success.

Notes

1 Brunei Darussalam, Cambodia, China, Democratic People's Republic (DPR) of Korea, Indonesia, Japan, Malaysia, Philippines, Republic of Korea, Singapore, Thailand, and Vietnam.
2 The Gulf of Thailand, originally considered to be part of the South China Sea in the SDS-SEA, was later reclassified as a separate large marine ecosystem.
3 The shared vision is stated as follows: 'The sustainable resource systems of the Seas of East Asia are a natural heritage for the people of the region, a medium of access to regional and global markets, and a safeguard for a healthy food supply, livelihood, economic prosperity and harmonious co-existence for present and future generations' (PEMSEA, 2003a: 35).
4 The mission is stated as follows: 'To build interagency, intersectoral and intergovernmental partnerships for achieving the sustainable development of the Seas of East Asia' (ibid.: 36).
5 The Joint Group of Experts on the Scientific Aspects of Marine Pollution (GESAMP), the Global International Waters Assessment (GIWA), and the South China Sea project, among others.
6 The GEF approved two modalities for support, the GEF/UNDP Regional Programme on the Implementation of the SDS-SEA and the GEF/World Bank Regional Revolving Fund Project.
7 Cambodia, China, DPR Korea, Indonesia, Japan, Lao People's Democratic Republic (PDR), Philippines, Republic of Korea, Singapore, Timor-Leste, and Vietnam.
8 Agreement Recognizing the International Legal Personality of the Partnerships in Environmental Management for the Seas of East Asia, signed on 26 November 2009 in Manila, by the representatives of Cambodia, China, DPR Korea, Indonesia, Lao PDR, Philippines, Republic of Korea, and Timor-Leste. A Headquarters Agreement between PEMSEA and the Philippines was signed in 2012 and is in the process of being ratified by the Philippine Senate.
9 Executive Order No. 533 of 6 June 2006 adopting integrated coastal management as a national strategy to ensure the sustainable development of the country's coastal and marine environment and resources and establishing supporting mechanisms for its implementation.

References

Bernad, S. R., Gallardo, K. R., and Thia-Eng, C. (2006) 'Regional Arrangement for the Implementation of SDS-SEA: A Partnership Approach', *Tropical Coasts*, 13(July): 1–64.

Chen, S., and Uitto, J. I. (2003) *Governing Marine and Coastal Environment in China: Building Local Government Capacity through International Cooperation*, China Environmental Series No. 6, Washington, DC: Environmental Change and Security Program (ECSP).

Food and Agriculture Organization of the United Nations (FAO) (1995) *Code of Conduct for Responsible Fisheries*, Rome: FAO.

Large Marine Ecosystems of the World (undated) 'Digital Maps', online at http://lme.edc.uri.edu/index.php?option=com_content&view=article&id=171&Itemid=60 [accessed 15 June 2006].

Lotilla, R. P. (2012) 'PEMSEA as a Regional Mechanism for the Implementation of the Ocean Agenda of Rio+20', Presentation at the East Asian Seas Congress, Changwon, Republic of Korea, July.

Partnerships in Environmental Management for the Seas of East Asia (PEMSEA) (2000) *Proceedings of the Seventh Programme Steering Committee Meeting: Dalian, P.R. China, 26–29 July 2000*, PEMSEA Meeting Report No. 1, Quezon City: PEMSEA.

Partnerships in Environmental Management for the Seas of East Asia (PEMSEA) (2002a) *Proceedings of the Eighth Programme Steering Committee Meeting: Busan, Republic of Korea, 19–22 March 2002*, PEMSEA Meeting Report No. 2, Quezon City: PEMSEA.

Partnerships in Environmental Management for the Seas of East Asia (PEMSEA) (2002b) 'Good Policy Practices in Environmental Management of the Coastal and Marine Area', In-house paper, Quezon City: PEMSEA.

Partnerships in Environmental Management for the Seas of East Asia (PEMSEA) (2003a) *Sustainable Development Strategy for the Seas of East Asia*, Quezon City: PEMSEA.

Partnerships in Environmental Management for the Seas of East Asia (PEMSEA) (2003b) *Proceedings of the Ninth Programme Steering Committee Meeting: Pattaya, Thailand, 6–9 August 2003*, PEMSEA Meeting Report No. 3, Quezon City: PEMSEA.

Partnerships in Environmental Management for the Seas of East Asia (PEMSEA) (2003c) *The Development of National Coastal and Marine Policies in the People's Republic of China: A Case Study*, Quezon City: PEMSEA.

Partnerships in Environmental Management for the Seas of East Asia (PEMSEA) (2003d) *Case Study on the Integrated Coastal Policy of the Republic of Korea*, Quezon City: PEMSEA.

Partnerships in Environmental Management for the Seas of East Asia (PEMSEA) (2003e) *Putrajaya Declaration of Regional Cooperation for the Sustainable Development of the Seas of East Asia*, online at http://www.pemsea.org/publications/putrajaya-declaration-regional-cooperation-sustainable-development-seas-east-asia-and-s [accessed 9 November 2014].

Partnerships in Environmental Management for the Seas of East Asia (PEMSEA) (2004) *Proceedings of the Tenth Programme Steering Committee Meeting: Xiamen, P.R. China, 25–29 October 2004*, PEMSEA Meeting Report No. 4, Quezon City: PEMSEA.

Partnerships in Environmental Management for the Seas of East Asia (PEMSEA) (2005a) *Consensus-Building for the Formulation of the Sustainable Development Strategy for the Seas of East Asia*, Quezon City: PEMSEA.

Partnerships in Environmental Management for the Seas of East Asia (PEMSEA) (2005b) *Proceedings of the Eleventh Programme Steering Committee Meeting: Siem Reap, Cambodia, 1–4 August 2005*, PEMSEA Meeting Report No. 5, Quezon City: PEMSEA.

Partnerships in Environmental Management for the Seas of East Asia (PEMSEA) (2005c) *Framework for National Coastal and Marine Policy Development*, Quezon City: PEMSEA.

Partnerships in Environmental Management for the Seas of East Asia (PEMSEA) (2006a) *Proceedings of the Twelfth Programme Steering Committee Meeting: Davao City, Philippines, 1–4 August 2006*, PEMSEA Meeting Report No. 6, Quezon City: PEMSEA.

Partnerships in Environmental Management for the Seas of East Asia (PEMSEA) (2006b) *A Perspective on the Environmental and Socioeconomic Benefits and Costs of Integrated Coastal Management: The Case of Xiamen, PR China*, Quezon City: PEMSEA.

Partnerships in Environmental Management for the Seas of East Asia (PEMSEA) (2006c) *Haikou Partnership Agreement on the Implementation of Sustainable Development Strategy for the Seas of East Asia*, online at http://www.pemsea.org/publications/haikou-partnership-agreement [accessed 9 November 2014].

Partnerships in Environmental Management for the Seas of East Asia (PEMSEA) (2006d) *Partnership Operating Arrangements for the Implementation of the Sustainable Development Strategy for the Seas of East Asia*, online at http://www.pemsea.org/sites/default/files/partnership-operating-arrangements.pdf [accessed 9 November 2014].

Partnerships in Environmental Management for the Seas of East Asia (PEMSEA) (2006e) 'OPRF Submits Proposal for a 21st Century Ocean Policy for Japan', *PEMSEA E-Updates*, 7(2): 3–4.

Partnerships in Environmental Management for the Seas of East Asia (PEMSEA) (2009a) *Agreement Recognizing the International Legal Personality of the Partnerships in Environmental Management for the Seas of East Asia*, online at http://www.pemsea.org/publications/agreement-recognizing-international-legal-personality-partnerships-environmental-manage [accessed 9 November 2014].

Partnerships in Environmental Management for the Seas of East Asia (PEMSEA) (2009b) *Manila Declaration on Strengthening the Implementation of Integrated Coastal Management for Sustainable Development and Climate Change Adaptation in the Seas of East Asia Region*, online at http://www.pemsea.org/eascongress/section-support-files/manila_declaration.pdf [accessed 9 November 2014].

Partnerships in Environmental Management for the Seas of East Asia (PEMSEA) (2011) *Guidebook on the State of the Coasts Reporting for Local Governments Implementing Integrated Coastal Management in the East Asian Region*, Quezon City: PEMSEA.

Partnerships in Environmental Management for the Seas of East Asia (PEMSEA) (2012a) *Regional Review: Implementation of the Sustainable Development Strategy for the Seas of East Asia (SDS-SEA) 2003–2011*, Quezon City: PEMSEA.

Partnerships in Environmental Management for the Seas of East Asia (PEMSEA) (2012b) *Toward an Ocean-Based Blue Economy: Moving Ahead with the Sustainable Development Strategy for the Seas of East Asia*, online at http://www.pemsea.org/publications/toward-ocean-based-blue-economy-moving-ahead-sustainable-development-strategy-seas-east [accessed 9 November 2014].

Partnerships in Environmental Management for the Seas of East Asia (PEMSEA) (2013) *Proceedings of the Fifth East Asian Seas Partnership Council Meeting: Manila, Philippines, 9–11 July 2013*, PEMSEA Meeting Report No. 11, Quezon City: PEMSEA.

United Nations Development Programme (UNDP) (1999) *UNDP–GEF Project Document: Building Partnerships for Environmental Protection and Management of the East Asian Seas*, New York: UNDP.

United Nations Development Programme (UNDP) and Global Environment Facility (GEF) (2012) *Catalysing Ocean Finance*, online at http://www.undp.org/content/undp/en/home/librarypage/environment-energy/water_governance/ocean_and_coastalareagovernance/catalysing-ocean-finance/ [accessed 9 November 2014].

United Nations Environment Programme (UNEP) (1995) *Global Programme of Action for the Protection of the Marine Environment from Land-based Activities*, UN Doc. UNEP(OCA)/LBA/IG.2/7, online at http://www.gpa.unep.org/index.php/about-gpa [accessed 9 November 2014].

United Nations General Assembly (UNGA) (1992) *Report of the United Nations Conference on Environment and Development, Vols I–III*, UN Doc. A/CONF.151/26/REV.1, Rio de Janeiro: UNGA ('Agenda 21').

World Bank Group (2005) *Philippine Environmental Monitor 2005: Coastal and Marine Resource Management*, Metro Manila: World Bank Group.

World Summit on Sustainable Development (WSSD) (2002) *Plan of Implementation of the World Summit on Sustainable Development*, United Nations Division for Sustainable Development, online at http://www.un.org/esa/sustdev/documents/WSSD_POI_PD/English/WSSD_PlanImpl.pdf [accessed 2 July 2014].

Conventions and treaties

1969 International Convention on Civil Liability for Oil Pollution Damage (1969) 973 UNTS 3, as amended by the protocols of 1976 and 1992

1971 Convention on Wetlands of International Importance Especially as a Waterfowl Habitat (1971) 996 UNTS 245 (Ramsar Convention)

1971 International Convention on the Establishment of an International Fund for Compensation for Oil Pollution Damage (1971) 1110 UNTS 57, as amended by the protocols of 1976, 1984, 1992 and 2003

1972 Convention on the Prevention of Marine Pollution by Dumping of Wastes and Other Matter (1972) 1046 UNTS 120 (London [Dumping] Convention), as amended by the 1996 London Protocol

1973 Convention on the International Trade in Endangered Species of Wild Fauna and Flora (1973) 993 UNTS 243, 244(E), 272 (F); 12 ILM 1085, 1088 (CITES)

1973 International Convention for the Prevention of Pollution from Ships (1973) 1340 UNTS 184 (MARPOL), as amended by the 1978 Protocol

1978 Protocol Relating to the International Convention for the Prevention of Pollution from Ships of 1973 (1978) 1340 UNTS 61

1982 United Nations Convention on the Law of the Sea (1982) 1833 UNTS 397 (UNCLOS)

1989 Basel Convention on the Control of Transboundary Movements of Hazardous Wastes and their Disposal (1989) 28 ILM 657

1990 International Convention on Oil Pollution Preparedness, Response and Co-operation, IMO Doc. OPPR/CONF/25, (1991) 30 ILM 735

1992 Convention on Biological Diversity (1992) 31 ILM 823 (CBD)

1992 United Nations Framework Convention on Climate Change (1992) ILM 849 (UNFCCC)

1996 Protocol to the Convention on the Prevention of Marine Pollution by Dumping of Wastes and Other Matter (1996) 36 ILM 1 (London Protocol)

2004 International Convention for the Control and Management of Ships' Ballast Water and Sediments, IMO Doc. IMO/BWM/CONF/36

21

OCEAN AND COASTAL POLICY PROCESSES IN SUB-SAHARAN AFRICA

Issues, opportunities, and strategic options

Indumathie Hewawasam and Bernice McLean, with contributions from Leopoldo Maraboli and Magnus Ngoile

Introduction

Africa is the world's second largest and second most populous continent, after Asia. With more than 1.1 billion people (as of 2013), Africa accounts for over 15 per cent of the world's population. According to the World Bank, the poverty headcount ratio at US$1.25 a day, as a percentage of the population, was as high as 46.8 per cent in 2011 (World Bank, undated). The late nineteenth century saw a 'scramble for Africa' by the European imperial powers and the creation of many colonial nation states. Post-colonial Africa has fifty-four fully recognized independent nations, many of whose borders were delineated during the colonial era. After a troubled series of decades through which many nations experienced military dictatorships, coups, apartheid, and periods of instability, several countries have achieved political stability and economic growth. These include Ghana, Mozambique, South Africa, Kenya, and Tanzania. The most populous nation in Africa, Nigeria hosts the world's largest proven oil reserves and is currently the largest economy in Africa. Overall, the economic growth in the continent has improved significantly, despite the global economic crisis in 2009 (World Bank, undated). This chapter contains a narrative of the economic and ecological importance of ocean and coastal areas in sub-Saharan Africa, the current and emerging threats to sustainability, some governance processes undertaken at the regional and national level to address these threats, and some options on how to move these processes forward as regional or sub-regional initiatives.

Why are oceans and coasts an important policy area for Sub-Saharan Africa?

Oceans and coasts are unique and complex environments that provide a number of essential ecosystem services and host a diverse array of commercially important industries. They also provide essential goods to coastal populations that depend directly on ocean and coastal resources for livelihoods. The coast is a distinctive system in which a range of considerations – biophysical, economic, social, and institutional – interconnect in a manner that requires a comprehensive and integrated management approach. Apart from the direct benefits, including commercial and subsistence food production, raw materials, transportation, tourism, recreational opportunities, and waste disposal, coastal ecosystems yield indirect benefits that are usually unrecognized or undervalued. These ecosystem services include erosion control, shoreline protection and buffering against storm surges, and nutrient recycling, filtering and processing by coastal wetlands.

Determining the value of these direct and indirect benefits can be useful to informing governance priorities and decisions. For instance, an effort was made by South Africa to assess the economic benefits of coastal areas. This was an exercise carried out during the development of its coastal management policy. Direct benefits obtained from coastal goods and services were estimated to contribute approximately 35 per cent (R168 billion) of the nation's annual gross domestic product (GDP). Indirect benefits, for example ecosystem services such as erosion control and waste treatment, were estimated at a further R134 billion annually (Republic of South Africa, 2000).

Major ocean and coastal sectors in Sub-Saharan Africa: Threats, opportunities, and issues

Threats to the ocean environment

The main threats to the ocean and coastal environment include: rapid population growth, combined with industrial and urban coastal development; increased deforestation and land clearing, leading to coastal erosion; uncontrolled and destructive fishing; pollution from domestic and industrial sources; and conversion of wetlands and fragile coastal ecosystems for development. Additional threats include: competition for natural resources and living space by the increased number of coastal communities; improperly planned and managed settlements; poorly planned and managed coastal tourism; lack of appropriate policy, regulatory, and institutional frameworks for urbanization and access to marine and coastal resources and areas; deficiencies in education on environment and hygiene; and inadequate incentives for conservation and environmental management (Hewawasam, 2002).

'Open access' conditions and greater fishing effort, with increasingly efficient technologies both for locating and processing fish stocks, are leading to overexploitation of the fisheries in both western and eastern Africa. In addition to increases in licensing – for example the United Republic of Tanzania issued some 200 licences in 2005 for tuna exploitation (URT, 2005) – fish stocks are exploited by illegal, unreported, and unregulated (IUU) vessels throughout the region. Other threats to ocean and coastal environments in eastern Africa include a significant increase in industrial trawling, especially on shallow continental shelf seas and seamounts, leading to decline in nearshore fisheries, degradation of coral reefs through overfishing and destructive fishing methods, climate-change effects, sedimentation, a decline in large marine fauna (manta rays, whale sharks, dugong, dolphins, sea turtles, sharks, and sawfish) resulting from unsustainable fishing, conflicts between resource users over access rights, and conflicts in marine protected areas (MPAs) and 'no take' zones. Insufficient capacity for sound governance of ocean and coastal areas and inadequate systems for monitoring exacerbate all issues. According to the World Wildlife Fund (WWF), the marine resources in Eastern Africa are under severe pressure from the rapid growth of human settlements and increasing global demand for natural resources such as oil and gas, tuna, and shrimps (WWF, undated*c*). The WWF draws the connection between overharvesting of marine life and the impact on the local poor, who depend on fish for protein and income. Other threats mentioned include the illegal harvesting of turtle eggs, pressure on coral reefs resulting from pollution and warming of sea surface temperatures, and the possible extinction of species such as the dugong owing to degradation of habitat.

Opportunities in the ocean and coastal environment

Fisheries

In most coastal nations in sub-Saharan Africa, fisheries are a major contributor to rural income and employment. The sector attracts considerable local and foreign investment, and contributes significantly to food security. In many countries, the sector is also a substantial source of foreign exchange and funding for public budgets. According to a study on the value of African fisheries, carried out in the framework of the NEPADFAO Fisheries Programme, the fisheries sector as a whole employs 12.3 million people on a full or part time basis, representing 2.1 percent of Africa's population (FAO, 2014). Similarly, in terms of food security, fish provides up to 70 per cent of the daily animal protein intake in some coastal countries in the region (such as Ghana and Sierra Leone). In 2011, the value added by the fisheries sector as a whole was estimated at more than US$24 billion or 1.26 percent of the GDP of all African countries. Although still developing, aquaculture produces an estimated value of almost US$3 billion per year (FAO 2014).

Exports from the fisheries sector in Mauritius amounted to R9.591 billion in 2011 (Republic of Mauritius, Ministry of Fisheries, 2011). Although local production is insufficient to cover market needs, it provides employment for some 11,000 people, in addition to returns for inhabitants of coastal regions. Revenue for the national budget of coastal nations in Africa is also generated through the issuance of licence fees for foreign vessels, import permit licence fees, and the sale of marine products. In 2010, 225 fishing licences were issued to foreign fishing vessels, generating fees totalling $1,684,000. A total of 2,592 fishing vessels called into Port Louis for trans-shipment, bunkering, repairs, maintenance, crew change, and dry-docking, generating significant revenue for the country. Mauritius is an attractive platform for trans-shipment and for value-added seafood activities. In Tanzania, the total revenue generated from fisheries in the exclusive economic zone (EEZ) amounted to $9.2 million in 2009, after the establishment of the Deep Sea Fishing Authority (World Bank, 2013). Prior to this, the two sides of the Union issued licences in an ad hoc and non-transparent manner.

In Mozambique, the industrial and semi-industrial fishing fleet exceeds 150 vessels, earning the economy over $100 million per year, mostly through the export of shrimp. According to the FAO, the value of fisheries exports in 2006 was $96,638,000 (FAO, 2007). The shrimp trawlers are also active in Tanzania and Kenya, although not to the same extent as in Mozambique, which hosts more expansive and productive shrimp habitats around river mouths.

Fisheries in the West African Marine Ecoregion generate some $400 million annually, making the sector the single most important source of foreign exchange in the region and a key source of revenue for economic and social development. Nearly 8 million people live along the West African coast. In Senegal alone, a country with a population of 12 million, the livelihoods of more than 600,000 men and women depend directly on fishing and fisheries-related industries. In Mauritania, landings by foreign industrial fleets top 600,000 tonnes annually, catching about 80 per cent of the fish. The catch of the artisanal sector is approximately 20 per cent. However, in Senegal, the artisanal fishers land about 80 per cent of the annual catch of approximately 400,000 tonnes (WWF, undated*a*).

Issues in the Sector

Throughout the region, the number of overexploited and depleted stocks is rising as a result of overcapacity. Total marine fish catch grew by more than 50 per cent during 1975–90, but there was a downturn in eastern Africa after 1990. In 2009, fish consumption in Africa was the lowest of all the regions of the world (FAO Fisheries and Aquaculture Department, 2012). The catch in southern Africa is far below the 3 million tonnes level of the 1950s (UNEP, 1999). In the Western Indian Ocean, total landings reached a peak of 4.5 million tonnes in 2006, but have declined slightly since, and 4.3 million tonnes were reported in 2010. A recent assessment has shown that narrow-barred Spanish mackerel (*Scomberomerus commerson*), a migratory species found in the Red Sea, Arabian Sea, Gulf of Oman, Persian Gulf, and off the coast along Pakistan and India, is overexploited. Catch data in this area are often not detailed enough for stock assessment purposes. However, the Southwest Indian Ocean Fisheries Commission (SWIOFC) conducted stock assessments for 140 species in its mandatory area in 2010 based on best-available data and information. Overall, 65 per cent of fish stocks were estimated to be fully exploited, 29 per cent overexploited, and 6 percent not fully exploited in 2009 (FAO Fisheries and Aquaculture Department, 2012). The FAO reports that, in the Canary Current large marine ecosystem (LME), a decline in catch from 2.3 million tonnes in 1990 to 1.8 million tonnes in 1999 was observed; a similar trend is shown in the Guinea Current LME, where the catch declined from 950,000 tonnes in 1990 to 900,000 tonnes in 1999 (FAO Fisheries and Aquaculture Department, 2012). Stocks within the Somali Coastal Current LME also reveal a ten-year decreasing trend, falling from 60,000 tonnes in 1990 to about 50,000 tonnes in 1999 (FAO Fisheries and Aquaculture Department, 2012).

Fisheries access agreements with distant-water fishing nations (DWFN) have the potential to bring much needed income to impoverished coastal nations in Africa. However, inequitable agreements have also put unsustainable pressures on limited fish stocks, leading to conflicts between local and foreign fleets. The situation is further exacerbated by improvements in fishing technology that have increased fishing efficiency. In the nearshore, with an ever-increasing number of boats searching for rapidly declining stocks, the industry has seen a dramatic increase in the use of destructive, habitat-destroying fishing techniques such as dynamite fishing, bottom trawling, and beach seining. The issue is exacerbated because of a lack of resources in many countries with which to monitor and enforce fisheries regulations. The increased fishing effort and destructive methods have also resulted in increased capture of endangered species such as marine turtles, juvenile fish, and a massive expansion of the trade in shark and ray fins (WWF, undated*b*). A deficit in the supply of fish is increasingly evident in the region and current regional demands remain unfulfilled. Per capita consumption of marine fish in the region has slumped by an annual average of 2.1 kg per person over the last twenty years, while imports have risen by 177 per cent over the same period. The supply deficit has resulted in an increase in fish prices and it is a key driving factor in the development of commercial aquaculture (Hecht, 2006: 41).

Aquaculture

There exists considerable scope for aquaculture development in sub-Saharan Africa despite current low levels of commercial operations in the region. According to the FAO, aquaculture contributed a record 42.2 per cent of fish produced by capture fisheries and aquaculture in 2012 (FAO, 2014). Despite the relative low levels of aquaculture production in sub-Saharan Africa compared to other regions, production increased from 4,243 tonnes in 1970 to 454, 691 tonnes in 2012. Nigeria leads aquaculture production in sub-Saharan Africa, with a total of 200,535 tonnes reported in 2010 (FAO Fisheries and Aquaculture Department, 2012). Mariculture is characteristically underdeveloped in the region as a whole, but is growing in selected countries. In Madagascar, black tiger shrimp (*Penaeus monodon*), and in the United Republic of Tanzania, Eucheuma seaweed, are thriving, while the production of niche species such as abalone (*Haliotis* spp.) is increasing in South Africa (FAO, 2010). Despite these isolated growth centres, several countries have identified the potential for the farming of prawns, fish, seaweed, or shellfish, and are preparing to initiate the development of a mariculture sector. Mariculture products include shrimps from Madagascar and Mozambique, seaweeds from Tanzania, Madagascar, and Mozambique, and abalone from South Africa. The value of these exported products comprises 95 per cent of the total mariculture revenue of the target countries and 33 per cent of the total value of aquaculture products in the region (Hecht, 2006: 24).

Issues in the sector

While non-commercial freshwater aquaculture is still considered essential to livelihood diversification strategies and provides for greater food security, a number of management-related factors challenge mariculture production. They include inadequate quality, a lack of appropriate type of extension services, the quality of fish seed and feed, disconnection between non-commercial producers and urban markets, excessive dependence on market forces, and private sector interest relative to freshwater aquaculture. Aquaculture therefore requires comprehensive business and environmental planning, in conjunction with effective means of extension and the strengthening of fish market chains, to include local small-scale entrepreneurs (Hecht, 2006). Where an adequate legal and regulatory framework and resources for monitoring and enforcement are lacking, negative environmental impacts of intensive aquaculture are more likely to manifest.

Coastal tourism

Coastal tourism contributes significantly to local economies of the region, particularly in eastern and southern Africa. In Kenya, coastal tourism accounts for a large and increasing proportion of foreign currency earnings. The potential for growth of this sector in Tanzania and Mozambique is significant. Some islands off the Tanzanian coast are demarcated for high-end tourism. For instance, the island of Mnemba, off the isles of Zanzibar, places a high value on ecological services offered, charging US$1,200 per night in peak season. Other privately run MPAs, such as Chumbe Island Coral Park, are also highly acclaimed for their aesthetic beauty, ecotourism, and environmental education values that are optimized. Issues do, however, exist with some fishing communities over benefits from these parks and restriction of access to fishing grounds. Such conflicts need to be resolved through consultation and agreement on trade-offs. Many lessons can be learned on conflict resolution from the process followed in Mafia Island Marine Park off the coast of mainland Tanzania.

In South Africa, tourism is positioned as one of the six core pillars of growth in the country's 'New Growth Path' framework. The government expects the sector to contribute to the development of rural areas and the culture industries by growing the economy and creating jobs. Tourism contributed approximately R189.4 billion (about US$24 billion) both directly and indirectly to the GDP in 2009 (Republic of South Africa, 2010). With 3,000 km of coastline and several world heritage sites, the coastal areas of South Africa contribute significantly to the growth in the sector.

Issues in the sector

Issues related to coastal tourism that require careful management attention include: the erosion of local cultural values; environmental degradation; conflicts between residents and the industry; conflicts among different sectors, such as fishing, mariculture, oil and gas development, and mining; a lack of sound policy and an institutional mandate to promote environmentally sustainable tourism; and inadequate regulations, zoning, and enforcement. The cruise ship industry is one aspect of coastal tourism that poses a serious threat to inadequately equipped small island developing states (SIDS) because of the pressures that it imposes in terms of pollution loads emitted into the marine environment, the insensitivity of visitors to local social and cultural values, and a rapid and considerable demand for essential and often scarce resources, such as fresh water. Other tourism-related issues include the inequitable level of revenue accrued to the coastal states from large-scale tourism that is largely organized and paid for in a visitor's country of origin, and increased restrictions and inadequate employment and benefits for local inhabitants, which creates unrest and social conflict.

Lack of spatial planning also creates conflicts between the tourism sector and other uses. An example is the expansion of seaweed farming in Tanzania, particularly in the Zanzibar islands. Tanzania is currently the biggest exporter of Spinosa and Cottonii, with Cottonii bringing higher prices. Cottonii needs to be grown in deeper, cooler waters. As the farms expand into the deeper marine areas, they obstruct tourism activities such as snorkelling and swimming. A number of donors and partners, such as the Tanzania Coastal Management Partnership (TCMP), the Government of Finland, the United States, the International Institute of Environment and Development (IIED), and the Marine and Coastal Environmental Management Project (MACEMP), supported by the World Bank and the Global Environment Facility (GEF), have supported studies, capacity building, and land-use planning in different parts of Tanzania, particularly along the coastal districts. The entity responsible for carrying out the planning activities is the National Land Use Planning Commission (NLUPC). The extent to which these plans are used and implemented is not clear. The NLUPC also has the mandate to ensure compliance with approved land use plans (URT, 2009).

In Mozambique, several donors, including the World Bank, the GEF, the United Nations Development Programme (UNDP), and others, promoted land-use planning in the context of poverty reduction, community empowerment, and decentralization. Since the early 1990s, the elite in the country and foreign investors have bought pristine lands along the coast, preventing coastal communities from having access to coastal and marine resources. As Tanner (2005) notes, implementation of the Land Law is only partially successful. The focus is on fast-tracking applications for private sector land use, which could lead to conflicts over resource access and use. According to Tanner:

> The progressive mechanism of the community consultation is being applied, but in a way that does not bring real local benefits – instead it gives a veneer of respectability to what is more like a European style enclosure movement, aimed to rationalize land use and place resources in the hands of a class that sees itself as more capable and better able to use national resources than the peasant farmers whose rights are legally recognized but still unprotected in practice.
>
> *(Tanner, 2005: 2)*

Oil and gas development

Mineral exploitation represents an enormous growth opportunity for sub-Saharan African coastal states, but also bears considerable risk. The Gulf of Guinea in West Africa is a significant producer of hydrocarbons. The Gulf of Guinea runs from Guinea, on Africa's northwestern tip, to Angola in the south, and includes oil-producing Nigeria, Ghana, Ivory Coast, Democratic Republic of the Congo, Congo (Brazzaville), Gabon, Cameroon, Equatorial Guinea, and Angola. Gulf of Guinea nations mostly supply European and American markets, although Angola supplies the Chinese as well. Nigeria is the biggest oil exporter, at 2.5 million barrels per day (bpd) in 2011, with Angola exporting an estimated 1.84 million bpd. Other producers include Equatorial Guinea (303,000 bpd), Congo (Brazzaville) (298,000 bpd), Gabon (244,000 bpd), Cameroon (62,000 bpd), and Ivory Coast (40,600 bpd). The data represent estimated average daily oil output for 2011 (KPMG, 2013). Reports indicate that many of these West African oil producers will increase production and exports over the coming years. Additionally, Ghana started producing in 2010, resulting in exponential economic growth in 2011.

While all of the major oil-producing countries in Africa report clear economic growth resulting from the boom in hydrocarbons, World Bank data show that many are at the bottom of socio-economic, governance, and equity indicators (World Bank, 2013; see Table 21.1). For example, Angola, one of the fastest-growing economies in Africa largely because of the major oil boom, also ranks in the bottom ten of socio-economic conditions in the world. With the exception of Ghana, indicators of poverty and access to improved water resources were significantly low.

Second only to Nigeria in Africa, the oil and gas sector accounts for more than 60 per cent of Angola's gross national product (GNP) and 90 per cent of government revenues. Oil production remains largely offshore and has few linkages with other sectors of the economy. Of all sub-Saharan African countries, Angola is the third largest trading partner of the United States, largely because of its petroleum exports. The United States imports about 4 per cent of its oil from Angola, a share that is projected to increase. By the same token, US companies account for more than half the investment in Angola, with Chevron-Texaco leading the way. Subsistence agriculture and dependence on humanitarian food assistance sustain the large majority of the Angolan population. Similarly in Nigeria, the primary oil producer in Africa for decades, 68 per cent of the population still lives on $1.25 a day.

In East Africa, several countries are expecting significant growth resulting from discoveries of significant reserves of oil and gas offshore. In late October 2013, Tanzania raised its estimate of recoverable natural gas reserves to 33 trillion cubic feet (tcf) from 28.74 tcf, following recent discoveries offshore. In fact, gas strikes off East Africa's seaboard have led to predictions that the region could become the world's third largest exporter of natural gas over the long term. According to the Tanzania Petroleum Development Corporation (TPDC), about eighty-one wells have been drilled offshore and onshore (TPDC, 2014). Tanzanian sources believe that the country has potentially 60 tcf of natural gas. While commercialization will still take time, the next decade is likely to see significant production of offshore gas. According to Ernst and Young, emerging exploration 'hot spots' in West Africa include Angola, Gabon, and Côte d'Ivoire (Ernst & Young, 2012). In Eastern Africa, emerging 'hot spots' include onshore preliminary discoveries in Uganda and Kenya, and offshore deposits in Tanzania. According to the International Monetary Fund (IMF), Tanzania has seen favourable natural gas exploration results (IMF, 2013). There appear to be good prospects that commercial quantities of natural gas will

Table 21.1 Growth relative to socio-economic indicators among a sample of oil-producing countries in the Gulf of Guinea

Country	Percentage of growth 2010–11	Population living below US$1.25/day (%)	Access to improved water resources % of total population 2010)
Nigeria	7.4	68	58
Angola	3.9	54.3	51
Ghana	14.4	28.6	86

be confirmed, 'resulting in multi-billion dollar foreign direct investments in Tanzania's natural gas sector over the next 5 years, and subsequent large export and budget revenue flows around the end of the decade' (IMF, 2013: 15). The IMF also reports that total proven and probable reserves are estimated at between 24 and 33 tcf; the challenge is to prepare the country for 'the gas economy and establish strong foundations to best take advantage of this potential resource wealth' (IMF, 2013: 15). The government has already developed a Natural Gas Policy and is finalizing a report outlining steps to be taken to prepare the country for a new gas economy. The framework will comprise design of a gas revenue management system integrated with the budget and possibly underpinned by a sovereign wealth fund (IMF, 2013).

Almost all of the countries in the Western Indian Ocean off the shores of eastern and southern Africa are experiencing serious efforts at oil and gas exploration. According to KPMG, East Africa could well be the 'next energy frontier' (KPMG, 2013). There are many challenges that may hamper this growth. Additionally, this growth may not necessarily reduce poverty or alleviate socio-economic pressures. There are risks that may actually increase these socio-economic pressures. Experience from Nigeria, Angola, and Equatorial Guinea clearly show that economic growth resulting from oil and gas production does not bring about a reduction of poverty in the absence of a sound environmental governance framework, social safeguards, and transparency in revenue management requirements. Experience from oil and gas exploration and production in the Gulf of Mexico may yield valuable lessons on associated environmental and coastal management. The offshore oil and gas extraction in the Gulf provides the United States with a large amount of domestic oil and natural gas, and represents a major focal area of intense industrial activity and investment. Exploration is intensifying at greater depths, with approximately 100 offshore rigs currently exploring reserves. These activities are conducted in accordance with comprehensive regulatory, oversight, and enforcement regimes, with technical, environmental, and coastal management dimensions.

Issues in the sector

Most of Africa's oil-producing countries lack refining capacity, which means that while they export crude oil, they still need to import refined oil at an additional cost. Other issues involve weak governance, poor maintenance and security issues, the displacement of people and their livelihoods, and a lack of compensation or alternative livelihoods. A key issue noted by KPMG (2013) is that the sector lacks linkages to other sectors and provides very little employment to the local people. The authors note that Angola employs less than 1 per cent of the workforce of the country in the oil industry. Experience from Nigeria demonstrates how displacement of people and their livelihoods led to rebellions, sabotage, and kidnappings.

According to the World Bank, the GDP in Equatorial Guinea is US$17.7 billion in 2012. However, the poverty headcount ratio at national poverty line (percentage of population in 2006) was as high as 76.8 per cent (World Bank, 2013). A BBC report in March 2014 cites Equatorial Guinea 'as a textbook case of the resource curse – or the paradox of plenty' (BBC, 2014). The report notes that while it has become one of sub-Saharan Africa's biggest oil producers and, in 2004, was said to have the world's fastest growing economy, the country ranks near the bottom of the UN Human Development Index and less than half the population has access to clean drinking water.

According to KPMG (2013), African oil-producing countries have has mixed success in saving excess revenues from oil and gas exploitation. In a few cases, it has helped to avert boom-and-bust cycles. Nigeria's Excess Crude Account (ECA) was backed by IMF fiscal reforms. Launched in 2003, the ECA helped Nigeria to strengthen the government budget.

However, over the years, it has failed to be a good mechanism of ring-fencing savings and provided no legal underpinning for the sharing of oil revenues among the different tiers of government. Corruption in the system had led to the disappearance of tens of billions of dollars. The report also notes that, in Angola, the $5 billion Fundo Soberano de Angola (FSDEA or sovereign wealth fund) has yet to demonstrate results (KPMG, 2013).

In Tanzania, the World Bank and GEF supported the MACEMP-proposed Marine Legacy Fund, which would protect and conserve marine areas and benefit from productive sectors such as fisheries, tourism, and oil and gas. The GEF and the government were expected to contribute the seed funds to establish the fund. While many of the key proponents in government favoured the idea, it is yet to be established. Given that many of the oil and gas deposits are reportedly offshore, establishing such a fund would have served as a depository for biodiversity offsets, as well as for ring-fencing excess hydrocarbon revenue.

The 2013 National Natural Gas Policy of Tanzania identifies several challenges in developing its gas industry (URT, 2013). They include the need for:

1   an effective institutional and legal framework to administer the industry;
2   human resources with the requisite skills and attitudes, discipline, and knowledge of the natural gas industry;
3   an enabling environment to attract local and foreign investment;
4   natural gas infrastructure and security;
5   a competitive and efficient domestic market for natural gas;
6   a transparent mechanism for the sound management of the natural gas revenues;
7   compliance with health, safety, and environmental standards; and
8   management of high public expectations and conflicting views of stakeholders (URT, 2013).

Coastal mining

Coastal mining involves offshore mining using dredges and related on-board and/or land-based upgrading facilities, and land-based mining using surface or underground technology and associated upgrading facilities. Offshore dredging is no longer commonly used owing to high operating and maintenance costs, and low technical performance resulting from the inherent dynamism of offshore conditions and outdated excavating technology. In light of the anticipated isolated offshore dredging projects, the environmental and coastal management issues associated with such activities are best addressed on a case-by-case basis, rather than through the establishment of frameworks for the entire mining sector. Land-based mining activities in coastal areas are, however, likely to continue in areas with promising mineral endowment and in areas with established mining activity. Such operations should be duly addressed and reflected in line with international best practice for applicable mining policy and respective mineral frameworks, including: legal and fiscal regimes; social and environmental considerations; and institutional structure, organization, and capacity. Emphasis should also be placed on the administration, management, monitoring, compliance, and enforcement, with particular attention to be paid to mitigation mechanisms for existing operations and mine closure.

In the absence of appropriate environmental and social safeguards, the degradation of the coastal environment will result in significant impacts on marine biodiversity and habitats. The absence of social safeguards will lead to the displacement of people and livelihoods. The lack of a sound governance framework also leads to lack of transparency and accountability to the larger society on how benefits from mineral exploration are shared or utilized equitably.

Issues in the sector

Many of the issues discussed in relation to the other sectors are equally applicable to this. While adequate and/or improved policy and mining sector regimes can be formulated on the basis of extensive international experience, considerable care must be taken to put in place adequate linkages and coordination with respective coastal governance frameworks, to ensure that the long-term social, economic, and environmental impacts of the mineral sector are considered and addressed.

Regional ocean and coastal policy processes in Africa

There is no single, all-encompassing regional ocean or coastal policy for the Africa region or for sub-Saharan Africa. There are, however, several ongoing and emerging policy and programmatic processes involving close collaboration among countries, especially among sub-regional coastal nations. Regional collaboration for governing the oceans and coasts in sub-Saharan Africa is recognized as an essential approach to overcome the conspicuous lack of human and financial resources in the region. Some level of regional and sub-regional collaboration already exists in the form of both agreements and cross-border programmes.

Regional conventions

There are two principal regional agreements related to coastal management in Africa. The Abidjan Convention for Cooperation in the Protection and Development of the Marine and Coastal Environment of the West and Central African Region and Related Protocols was adopted in 1984, and the Nairobi Convention for the Protection, Management and Development of the Marine and Coastal Environment of the Eastern African Region and Related Protocols were adopted in1996. In March 2010, the Nairobi Convention was amended under the Regional Seas Programme to become the Amended Convention for the Protection, Management and Development of the Marine and Coastal Environment of the Western Indian Ocean (UNEP, 2010; Dzidzornu, 2012).

The West and Central Africa Region (WACAF) Action Plan was adopted in 1981 and came into force in 1984. The Abidjan Convention was adopted in 1981 and came into force in 1984. The Nairobi Protocol Concerning Cooperation in Combating Marine Pollution in Cases of Emergency in the Eastern African Region was adopted 1981 and came into force in 1984. The objectives of the Abidjan Convention are to protect the marine environment, coastal zones, and related internal waters falling within the jurisdiction of the states of the West and Central African region. The Convention also contains a Protocol concerning cooperation in combating pollution in cases of emergency (UNEP, undateda). The work programme for signatories to the Abidjan Convention includes assessments of coastal erosion and activities for improving the management of coastal ecosystems, with a special focus on mangroves and oil pollution.

The objectives of the Nairobi Convention are very similar to those of the Abidjan Convention, and propose to protect and manage the marine environment and coastal areas of the eastern African region. Adopted at Nairobi on 21 June 1985 and entering into force in May 1996, the Nairobi Convention recognizes the economic and social value of the marine and coastal environment of the eastern African region. It aims to address the threats to the marine and coastal environment, its ecological equilibrium, resources, and legitimate uses posed by pollution and by the insufficient integration of an environmental dimension into the development process. The Nairobi Convention also has two Protocols and an action plan (UNEP, undatedb). The Protocols relate to protected areas and wild fauna and flora in the eastern African region, and cooperation in combating marine pollution in cases of emergency in the eastern African region. The action plan details objectives for the protection, management, and development of the marine and coastal environment of the eastern African region.

Development and implementation of these two regional agreements has, however, proven to be slow. Formulation of the agreements began in the early 1980s, under the auspices of the United Nations Environmental Programme's Regional Seas Programme (UNEP RSP). The Eastern African Regional Seas Action Plan was signed in 1985. This signalled the beginning of regional cooperation for strengthening management of marine and coastal resources. Ministerial commitment was made by countries in the region to sustainable coastal development and management in Arusha, Tanzania, in 1993, and again in the Seychelles in 1996. In addition to these two meetings, national workshops and regional integrated coastal management workshops were held to facilitate discussion among national experts, decision-makers, and stakeholders on options for addressing priority coastal issues. Recommendations emerged during these workshops for initiating national processes for developing policy for coastal management.

A GEF-supported initiative entitled 'The African Process for the Development and Protection of the Coastal and Marine Environment in Sub-Saharan Africa' (the African Process) emerged out of priorities highlighted at two conferences held in 1998: the Pan African Conference on Sustainable Integrated Coastal Management (PACSICOM), in Mozambique; and the Cape Town Conference on Cooperation for Development and Protection of the Coastal and Marine Environment in Sub-Saharan Africa. These two events stimulated a unified political awareness amongst African governments for the need to develop an integrated approach towards sustainable development of coasts and oceans in Africa (McLean, 2010).

The underlying motivation for the African Process was the recognition of the need for regional cooperation to maximize capacity to address the many social, economic, and environmental problems that are either transboundary in nature or common to most countries. The initiative drew from scientific expertise within individual countries to identify priority areas for action, and led to the incorporation of a coastal and marine sub-component of the environment component of the New Partnership for Africa's Development (NEPAD) Action Plan. It was institutionalized through the establishment of the NEPAD Coastal and Marine Secretariat (CosMar), based in Nairobi (Dzidzornu, 2011a).

New Partnership for Africa's Development (NEPAD)

The NEPAD is a vision and programme of action for the development of the African continent formulated by African leaders through the African Union. NEPAD was finalized on 23 October 2001. It is a comprehensive integrated development plan that addresses key social, economic, and political priorities for the sustainable growth of Africa. Its goals are to promote accelerated growth and sustainable development, to eradicate widespread and severe poverty, and to halt the marginalization of Africa in the globalization process. While providing admirable ideals, NEPAD has been criticized for neglecting the needs of the people, being designed without involving civil society, and failing to mainstream gender issues (McLean, 2010).

NEPAD has developed an environmental action plan to address priority environmental issues. The framework proposes four strategic directions: capacity building for environmental management; securing political will to address environmental issues; mobilizing and harmonizing international, regional, and national resources, conventions, and protocols; and supporting best practice or pilot programmes. The action plan is organized in clusters of programmatic and project activities to be implemented over an initial period of ten years. The priority sectors and cross-cutting issues in this action plan include combating land degradation, drought, and desertification, protecting wetlands, combating invasive species, protecting marine and coastal resources, the cross-border conservation of natural resources, and addressing climate change. The plan of action builds upon the related problems of pollution, forests and plant genetic resources, fresh water, capacity-building, and technology transfer (NEPAD, 2003; Dzidzornu, 2011b). NEPAD currently functions as an African Union strategic framework for pan-African socio-economic development and manages a number of programmes in six theme areas. Within the Agriculture and Food Security theme, NEPAD is implementing the Partnership for African Fisheries (PAF) and is working through the Comprehensive African Fisheries Reform Strategy, to improve the sustainability of Africa's fisheries.

NEPAD and the FAO are implementing a joint project aimed at promoting the enhanced contribution of fisheries and aquaculture to poverty alleviation, food security, and economic growth through improved and sustainable management of the fishery and aquaculture sectors. This partnership aims to focus on environmental sustainability of ecosystems in both marine and inland fisheries.

Large marine ecosystems programmes

The GEF and its implementing partners, the World Bank, UNDP, and UNEP, support several large marine ecosystem (LME) programmes in Africa to address the restoration and sustainability of these ocean areas. These include: the Canary Current LME off the coast of northwest Africa, bordering seven countries and the disputed territory of the Western Sahara; the Guinea Current LME, bordering sixteen countries in West Africa; the Benguela Current LME (BCLME), bordering three countries in southwest Africa; the Aghulas Current LME, bordering seven countries in the southwestern Indian Ocean; and the Somali Coastal Current LME, bordering three countries in eastern Africa. These programmes are committed to ecosystem-based assessment and management practices, in line with Chapter 17 of Agenda 21 (UNGA, 1992). They aim to improve global coastal health, to reduce pollution, and to restore depleted biomass yields. Projects are based on the five-module assessment and management methodology targeting productivity, fisheries, pollution, ecosystem health, socio-economics, and governance (NOAA, undated).

Under the BCLME, the GEF supported an ecosystem-based programme in partnership with the governments of Angola, Namibia, and South Africa to promote integrated management, sustainable development, and environmental protection. The BCLME is based on the understanding of the fragmented nature of coastal and marine resource management as a legacy of the colonial past and of subsequent political turmoil in Namibia, Angola,and South Africa (Shannon and O'Toole, 2003). Key objectives emerging from a transboundary diagnostic analysis (TDA) ranged from sustainable management and utilization of resources, to assessment of seabed mining and drilling impacts, policy harmonization, protection of vulnerable species and habitats, and capacity strengthening. The Strategic Action Plan (SAP), signed by ministers representing fisheries, environment, and mines and energy sectors of Angola, Namibia, and South Africa, laid out guidelines and a policy framework for the integrated and sustainable management of the BCLME. The principal cooperative actions agreed to by the three governments were to:

1   apply the precautionary principle;
2   develop and promote anticipatory actions (such as oil spills and harmful algal blooms contingency plans);
3   stimulate the use of clean water technologies;
4   promote the use of economic and policy instruments that foster sustainable development (such as the 'polluter pays' principle);
5   include environmental and health considerations within all relevant policies and sectoral plans;
6   promote cooperation among states bordering the BCLME;

7   encourage interests by other states in the southern African region; and

8   promote transparency, and public and private participation (UNDP, 2007).

The Benguela Current Commission (BCC) was established to strengthen regional cooperation and to address the gaps in current knowledge. On 13 March 2013, the Benguela Current
Convention was signed by the three states, marking the establishment of the first multilateral
commission to be based on the LME approach to ocean governance.

Taking efforts towards financial sustainability, a Strategic Partnership for a Sustainable
Fisheries Investment Fund focused on the LMEs of sub-Saharan Africa has been initiated to
assist the coastal countries in the region to meet the targets for sustainable fisheries set by the
World Summit on Sustainable Development (WSSD) (World Bank, 2006a). This initiative
aims to complement the existing regional LME projects funded by the GEF and to ensure the
health of the fisheries resource base upon which many lives and livelihoods depend. The lead
implementing agency is the World Bank Group and the main executing agencies are the African
Union, FAO, and World Wildlife Fund (WWF) (UNDP, 2007).

While the LME programmes generate significant amounts of data, the perception still exists
in political circles that they are science-oriented and have little to do with policies and political
priorities such as growth and poverty reduction. Intergovernmental structures such as the BCC
and stronger linkages with the African Union may help to eliminate this perception and ensure
the integration of science into policy processes.

The Western Indian Ocean Marine Science Association

The Western Indian Ocean Marine Science Association (WIOMSA) is a non-governmental
and non-profit regional organization for promoting the educational, scientific, and technological development of all aspects of marine sciences throughout the Western Indian Ocean
(WIO) region. The Association gathers and disseminates marine science information. It
coordinates meetings to foster marine science development and information exchange, and
to enhance communication among marine scientists and other professionals involved in the
advancement of marine science research and development in the WIO region (WIOMSA,
undated). The Association supports several research and capacity development programmes,
including: the Marine Science for Management (MASMA) programme, which supports
research on current themes; the Marine Research Grant (MARG) programme, which provides upcoming scientists with a reliable and flexible mechanism to turn their ideas into
research projects and an opportunity for presentation of research findings; and Western
Indian Ocean Certification of Marine Protected Area Professionals (WIO-COMPASS),
a certification programme to assess and certify marine protected area (MPA) professionals
in the WIO region. The Association has, in a move towards financial sustainability, established a trust to provide funding to promote cooperation in the WIO region in all aspects
of coastal and marine sciences, as well as management and sustainable development in the
region (Humphrey and Francis, 1997).

Land-based sources of marine pollution (LBMPs)

According to the WWF, more than 80 per cent of marine pollution is from land-based sources
(WWF, undatedb). The sources range from oil spills to run-off from cities and industry,
coastal mining, fertilizers, plastics, untreated sewage, and toxic chemicals. While there are a
large number of international treaties addressing the topic of marine pollution, including the
Global Programme of Action for the Protection of the Marine Environment from Land Based

Activities (GPA) (UNEP, 1995), the 1973 International Convention for the Prevention of Pollution from Ships, amended by Protocol of 1978 (MARPOL), and the 1972 Convention on the Prevention of Marine Pollution by Dumping of Wastes and Other Matter, as amended by the Protocol of 1996 (the London [Dumping] Convention), their implementation is hampered by many challenges.

The GPA provides practical guidance for national and/or regional authorities to devise and implement sustained actions to prevent, reduce, control, and/or eliminate marine degradation from land-based activities (UNEP, 1995). Adopted in 1995, the GPA aims to prevent the degradation of the marine environment from land-based activities, and to assist states in taking actions individually or jointly within their respective policy priorities and resources to control and/or eliminate degradation of the marine environment. Outcomes will contribute to ensuring the protection of human health and promoting the conservation and sustainable use of marine living resources. The recommendations and principles contained in the GPA are used to focus policy decisions, regional initiatives, and international cooperation to protect human health and marine environmental resources (EPA, undated).

Several protocols have been adopted in sub-Saharan Africa signifying commitment to cooperate on preventing pollution of the marine environment. The first Protocol associated with the Abidjan Convention was adopted in 1981. The Protocol Concerning Cooperation in Combating Pollution in Cases of Emergency came into force in 1984. The second, more recent, Protocol Concerning Cooperation in the Protection and Development of the Marine and Coastal Environment from Land-Based Sources and Activities (LBSA) in the Western, Central, and Southern Africa Region was adopted in June 2012. The first Protocol associated with the Nairobi Convention was adopted in 1985, the Protocol Concerning Co-operation in Combating Marine Pollution in Cases of Emergency in the Eastern African Region. In March, 2010, the Protocol for the Protection of the Marine and Coastal Environment of the Western Indian Ocean from Land-Based Sources and Activities was adopted.

Additionally, a project entitled 'Addressing Land-based Activities in the Western Indian Ocean' (WIO-LaB) has been undertaken under the Nairobi Convention to address issues related to LBMPs in the WIO region. Eight countries in the region participated in the project – namely, Comoros, Kenya, Madagascar, Mauritius, Mozambique, Seychelles, South Africa. and Tanzania. The project was supported by UNEP, the Government of Norway, and the GEF. A TDA was carried out on land-based activities in the WIO region. According to the report, LBMPs in the WIO region are primarily associated with urban areas and discharges from larger catchments into the sea (UNEP et al., 2009). The report mentions that rapid urbanization in the region will lead to smaller coastal towns and suburban areas joining up with the main cities, creating larger continuous urban zones and larger 'hotspots' (UNEP et al., 2009).

The project also studied five priority transboundary problems linked to marine pollution in the WIO region: microbial contamination, suspended solids, chemical pollution, solid waste, and eutrophication. Recommendations include:

1   develop specific management tools and regional best practice guidelines for industrial wastewater and solid waste management;
2   apply the 'polluter pays' and 'cradle to grave' principles and other economic incentives for waste management;
3   develop guidelines for ports and harbours for on- and off-loading, and disposal of waste and used oil and oil-related products;
4   develop guidelines for agricultural activities;
5   institute monitoring and assessment programmes;

6   undertake regional training programmes; and

7   develop a web-based regional information management system that includes appropriate technology, institutional, and policy frameworks, and sustainable financial mechanisms (UNEP et al., 2009).

Tanaka (2006: 574) describes the regulation of LBMPs as an 'acute tension between economic development and environmental protection under international law'. He further notes that, on the one hand, regional conventions develop approaches and legal techniques limiting the margin of discretion of states, while, on the other, the application of these approaches and legal techniques must be reconciled with economic, political, and social factors specific to each state. Thus the validity and effectiveness of legal frameworks in this field rely essentially on the sound balance between the requirement of the environmental protection and each state's need for economic, social, and political development (Tanaka, 2006: 574). This statement holds true for many of the poorer coastal nations and is articulated not as a 'tension', but as a priority in several of the coastal and ocean management documents, including the recently approved National Ocean Policy in South Africa.

Apart from the 1982 United Nations Convention on the Law of the Sea (UNCLOS), which contains specific provisions on preventing pollution of the marine environment, other rules, regulations, standards, recommended practices, and procedures are contained in non-binding frameworks such as the GPA. They seek to influence the behaviour of sovereign governments, national entities, companies, and individuals. The legal framework for combating marine pollution from land-based sources has expanded significantly, and has become more complex and sophisticated. Identifying the interlinkages among the regional protocols with global policy documents and national development plans is important in determining their comparative advantages and weaknesses. Incorporation of LBMPs within a regional ocean policy would enable greater specificity in terms of both principles and recommendations. Examples are the protocols in the wider Caribbean, such as the 1999 UN Protocol Concerning Pollution from Land-Based Sources and Activities to the Convention for the Protection and Development of the Marine Environment of the Wider Caribbean Region, which refers to 'inequalities in economic and social development among countries of the Wider Caribbean Region'. In addition to the inequities between the countries in a region, additional challenges include:

- a lack of guidelines and capacity for enforcement and compliance;
- a lack of awareness and prioritization at the decision-making level;
- a lack of awareness and participation of the public;
- the absence of resources and power at the local level despite decentralization;
- the absence of cooperation among the various government agencies to tackle the problem;
- the reluctance of authorities to regulate economic sectors during their growth phase;
- the link to large populations of poor communities who do not have access to basic sanitation facilities; and
- the need to establish wastewater and solid waste management facilities in urban centres.

Regional fisheries programmes

There are several regional programmes that collaborate to promote sound governance of fisheries in Africa's ocean areas. In western Africa, regional programmes include the International Commission for the Conservation of the Atlantic Tuna (ICCAT), the Fishery Committee for the Eastern Central Atlantic (CECAF), the Sub-Regional Commission on Fisheries (SRCF), the Regional Fisheries Committee for the Gulf of Guinea (COREP), the South East Atlantic Fisheries Organization (SEAFO), and the Commission for the Conservation of Southern Bluefin Tuna (CCSBT). In eastern Africa, regional fisheries programmes include the Indian Ocean Tuna Commission (IOTC), the Western Indian Ocean Tuna Organization (WIOTO), and the South West Indian Ocean Fisheries Commission (SWIOFC). To promote sound monitoring and governance of the fisheries in the southwest Indian Ocean, the European Commission supported a programme for the South African Development Community (SADC) countries. Continuation of this work is anticipated by the countries themselves, as in the case of Tanzania, with support from the World Bank and the GEF, and the regional fisheries programme supported by the World Bank and currently under development. The following sections briefly outline the mandates of a few of these organizations.

The South West Indian Ocean Fisheries Project

While global trends in fish landings for most of the world's main fishing regions are negative, the Western Indian Ocean has maintained a steady rate of increase in total landings (Dzidzornu, 2011a). This has largely been a result of the increased harvest of tuna and mackerel-like species, with recent additions of 'new' deep-water species such as toothfish and orange roughy. While the total catch is relatively modest, this region is seen by distant-water fleets as an opportunity to offset their decreased landings from other regions. The Indian Ocean is surrounded by developing nations, with close to half the world's population residing in countries that border it. These coastal nations are anxious to promote economic growth and to meet the expectations and demands of their societies, which escalate during times of drought, climate change, and unsettled socio-economic conditions. The southwest Indian Ocean region enjoyed a spate of academic and scientific research interest in the past. Sadly, the majority of the research was donor-driven, uncoordinated, and not internalized by the relevant government agencies.

In response to a request for assistance by several countries bordering on the Western Indian Ocean, the World Bank (with support from the GEF) initiated a multinational fisheries management and development programme. The South West Indian Ocean Fisheries Project (SWIOFP), utilizing an ecosystem and transboundary approach, was adopted to assist littoral countries with the assessment and management of offshore resources in the Agulhas and Somali Current LMEs. The SWIOFP aimed to build human and institutional capacity in the nine participating countries and to forge a regional approach to resource management. It encompassed the 200 nautical mile exclusive economic zones (EEZs) of Madagascar, Kenya, Tanzania, Mozambique, Mauritius, Seychelles, Comoros, and South Africa (SWIOFP undated). The Project was particularly successful in building regional capacity for fisheries management, including through establishing a network of fisheries researchers and managers, and developing a regional management framework. The SWIOFC member countries agreed to reform the Commission (see below), promoting it from an advisory body to a regional fisheries management organization (RFMO) of the coastal states, which enables it to take binding decisions on fisheries management. Given the important achievements of the Project, SWIOFC member countries requested a follow-on programme, the South West Indian Ocean Fisheries Governance and Shared Growth Programme (SWIOFish).

SWIOFish aims to continue to support regional integration around fisheries management, while expanding the approach beyond research to strengthen sector governance and harness the value of coastal and marine fisheries to national economies. The proposed programme is expected to encompass a fifteen-year period, using World Bank and GEF resources, as well as support from other donors and trust funds. Initial investment in the SWIOFish1 will support Tanzania, Mozambique, and Comoros, and regional organizations in charge of project coordination. Other SWIOFC members, such as Seychelles, South Africa, Kenya, Mauritius, Madagascar, and Somalia, are expected to join when they are ready. SWIOFish1 is expected to be presented to the World Bank's Board later this year.[1]

The South West Indian Ocean Fisheries Commission

The role of the South West Indian Ocean Fisheries Commission (SWIOFC) is to promote responsible and sustainable fishing in the southwestern Indian Ocean. The Commission will function as an advisory body to promote the sustainable development and utilization of coastal fishery resources of the region, as well as responsible management and regional cooperation on fisheries policy.

According to the FAO, the current membership of SWIOFC includes Comoros, France, Kenya, Madagascar, Maldives, Mauritius, Mozambique, Seychelles, Somalia, South Africa, United Republic of Tanzania, and Yemen, whose territories are situated wholly or partially within the SWIOF region (FAO Fisheries and Aquaculture Department, undated). FAO studies show that, in the entire West Indian Ocean, the larger region encompassing the zone in which SWIOFC will operate, 75 per cent of fishery resources are currently being fished at their maximum biological productivity, while the remaining 25 per cent are overexploited and require better management. It is known that catches have increased by over 10 per cent during the last decade, with landings in 2001 (319,000 tonnes) representing an all-time high. The majority of fishing boats operating in the southwestern Indian Ocean are distant-water fleets from Spain, the Taiwan Province of China, Japan, France, and Uruguay (FAO Fisheries and Aquaculture Department, undated).

The Indian Ocean Tuna Commission

The Indian Ocean Tuna Commission (IOTC) has thirty-two contracting parties, including the Comoros, Eritrea, European Union, France, Guinea, Kenya, Madagascar, Maldives, Mauritius, Mozambique, Seychelles, Sierra Leone, Sudan, Tanzania, and the United Kingdom. Member countries are required to describe the actions that they have taken under national legislation to implement conservation and management measures adopted by the Commission. These reports of implementation need to be submitted sixty days prior to the annual meeting of the IOTC. There is a Compliance Committee constituted of policymakers (commissioners) and fisheries monitoring, control, and surveillance (MCS) practitioners. The Committee's responsibilities include: reviewing all aspects of compliance with IOTC conservation and management measures; reviewing information relevant to compliance from IOTC subsidiary bodies and from reports of implementation; and identifying problems related to the effective implementation of, and compliance with, IOTC conservation and management measures, and making recommendations to the Commission on how to address these problems. The IOTC manages a range of tuna and related species, including yellow fin tuna, skipjack tuna, big eye tuna, albacore tuna, long tail tuna, Kawakawa, frigate tuna, bullet tuna, narrow barred Spanish mackerel, Indo-Pacific king mackerel, Indo-Pacific blue marlin, black marlin, striped marlin, Indo-Pacific sailfish, and swordfish (IOTC, undated).

The International Commission for the Conservation of the Atlantic Tuna

The International Commission for the Conservation of the Atlantic Tuna (ICCAT) is responsible for the conservation of tuna and tuna-like species in the Atlantic Ocean and adjacent seas.

It undertakes scientific research on biometry, ecology and oceanography, conditions and trends of stocks, studies on 'by-catch' and related research. It was established in 1969 and currently has forty-nine participants. Cooperating non-contracting parties include Chinese Taipei, Guyana, Netherlands Antilles, and Colombia (ICCAT, undated).

National ocean and coastal policy processes: Regional linkages

Despite the large number of regional policy initiatives, very few countries have developed comprehensive coastal or ocean policy processes and fewer have managed to fully institutionalize such integrated programmes into broader governance frameworks. Increasingly, however, coastal nations are realizing the need for a comprehensive approach to ocean and coastal governance and are opting to craft national policies. These nations are approaching marine management from a perspective that tries to take into consideration the multiple issues affecting oceans and coasts. This section discusses some key initiatives.

South Africa

Development and management of South Africa's marine and coastal resources have undergone a major shift in emphasis at the national level in the last decade. The ad hoc, sector-specific management approaches of the 1970s were replaced by top-down regulatory control in the 1980s, which were based principally in the natural sciences and justified in terms of 'conservation' (Glavovic, 2000). The political transition in South Africa, beginning in the early 1990s, provided the impetus for consideration of the political and socio-economic dimensions in all of the country's coastal areas. In line with government priorities, emphasis was placed on addressing past inequities, and the new coastal policy outlines the importance of recognizing the value of the coast (Republic of South Africa, 2000).

The White Paper for Sustainable Coastal Development in South Africa was officially launched in June 2000, with support from Britain's Department for International Development (DfID), and aims to achieve sustainable coastal development through integrated coastal management (Republic of South Africa, 2000). Many recognize the White Paper as being a thorough document encompassing a holistic vision for sustainable coastal development. It remains to be seen, however, whether it will remain simply a 'good policy document', or will effectively assist decision-makers in achieving the challenge of environmentally and socially sustainable development in coastal areas. Several major issues impinge upon the successful implementation of the policy. These include:

- continued social and economic inequities faced by most coastal inhabitants;
- the need for more capacity (including human and financial resources) at all levels of government for strengthened coastal governance;
- continuation of conflicting and uncoordinated coastal activities despite a reorientation towards sustainable development;
- an inadequate legal framework for implementation of the policy, with an emphasis on some existing regulations being somewhat contradictory to the directives outlined in the coastal management policy; and
- rejection by some sectors of government, business, and industry of the importance of the sustainable development agenda (Glavovic, 2000).

A series of key factors enabled the transformation towards sustainable coastal management in South Africa:

1 transition from apartheid to democracy, which provided a window of political opportunity;
2 greater partnership between the government and civil society;
3 financial support provided by the British government;
4 alignment of the policy with the government's poverty eradication agenda, thus moving the coastal agenda from the 'political periphery towards centre stage'; and
5 vision, commitment and activism of a core group of individuals representing diverse sectors and organizations (Glavovic, 2008).

Despite the policy rhetoric promoting decentralization and a co-management/partnership approach to the management and development of coastal resources, a significant challenge was encountered in terms of the willingness and capacity to devolve decision-making powers and actual responsibilities to local actors and institutions (Hauk and Sowman, 2001). This resulted in a delay in implementing enabling legislation. The policy principles and objectives were finally enacted in 2009 in the form of the National Environmental Management: Integrated Coastal Management Act (ICMA), No. 24 of 2008 (McLean, 2010). The ICMA seeks:

> To establish a system of integrated coastal and estuarine management in the Republic, including norms, standards and policies, in order to promote the conservation of the coastal environment, and maintain the natural attributes of coastal landscapes and sea-scapes, and to ensure that development and the use of natural resources within the coastal zone is socially and economically justifiable and ecologically sustainable; to define rights and duties in relation to coastal areas; to determine the responsibilities of organs of state in relation to coastal areas; to prohibit incineration at sea; to control dumping at sea, pollution in the coastal zone, inappropriate development of the coastal environment and other adverse effects on the coastal environment; to give effect to South Africa's international obligations in relation to coastal matters . . .
>
> *(ICMA, Full title)*

The Marine Living Resources Act (MLRA), No. 18 of 1998, as amended by the Marine Living Resources Amendment Act, No. 68 of 2000, governs the exploitation of marine resources below the high-water mark in South Africa. The MLRA was implemented in September 1998 by the Republic of South Africa to allow previously excluded communities full access to the fishing industry and to prepare for the free trade and deregulated markets that dominate the international marketplace. The government aimed to make the industry more internationally competitive and equitable. The guiding principles of the new Act are taken directly from the White Paper on Marine Resources, which stressed that all of the natural marine living resources of South Africa, as well as the environment in which they exist, are a national asset and the heritage of all South Africa's people (Republic of South Africa, 1997). Prior to the promulgation of the MLRA, the wealth of marine resources was exploited for the benefit of a very small minority. By way of example, more than 90 per cent of South Africa's most lucrative fishery, the hake deep-sea trawl fishery, was shared between five companies (Kleinschmidt et al., 2005). The MLRA embraces principles of ecosystem management, equity, employment generation, and the rule of international law. Specific goals according to the MLRA relate to the needs to:

- achieve optimum utilization and ecologically sustainable development of marine living resources;
- conserve marine living resources for both present and future generations;
- apply precautionary approaches;

- utilize marine living resources to achieve economic growth, human resource development, capacity building, employment creation, and a sound ecological balance and development;
- protect the ecosystem as a whole, including species that are not targeted for exploitation;
- preserve marine biodiversity;
- minimize marine pollution;
- achieve broad participation in the decision-making processes;
- meet any relevant obligation under international agreement or applicable rule of international law; and
- restructure the fishing industry to address historical imbalances and to achieve equity within all branches of the fishing industry.

The White Paper on the National Environmental Management of the Ocean (NEMO) Policy was approved by Cabinet in December 2013. The policy aims to protect and conserve South Africa's ocean environment, as well as to promote sustainable development for present and future generations. The broad policy objectives are interdependent and include: enhancing ocean environmental information; improving ocean environmental knowledge; improving ocean environmental management; and enhancing ocean environmental integrity by means of cooperating at the national, regional, and global level. Key concepts within the broad objectives include: improving the competitiveness and effectiveness of existing activities taking place within the marine jurisdiction, while researching and developing innovative and responsible future uses; maintaining and improving marine ecosystems' resilience; conserving biodiversity and restoring degraded habitat; participating and strengthening involvement in global and regional developments; and spatial planning as key to managing large ocean spaces. The overarching national objective of job creation and social benefits are integrated within NEMO, in which it is suggested that South Africa's rich marine resources, both living and non-living, should be utilized optimally and sustainably in order to promote job creation and increase social benefits, and in which the importance of the ocean ecosystem in impacting on human livelihoods, food security, agriculture, trade and industry is acknowledged (Republic of South Africa, Department of Environmental Affairs and Tourism, 2013).

Once adopted, NEMO will enable South Africa to move from sectoral ocean management planning toward a coordinated environment management and, over time, toward integrated ocean management. The policy processes in South Africa provide leadership in the subregion and are also closely linked to regional initiatives such as in the Benguela Current LME (BCLME) and South West Indian Ocean Fisheries Project (SWIOFP), and are expected to enhance regional management of transboundary stocks and improve marine scientific research.

Tanzania

Tanzania has been a key participant in regional and international processes to promote an integrated approach to coastal management. The nation played a leadership role in the formulation of the Nairobi Conventions and in the subsequent Pan African Conference on Sustainable Integrated Coastal Management (PACSICOM) and Cape Town Conferences on sustainable governance of ocean and coastal resources management. Tanzania also played a key role in the GEF-supported African Process. Tanzania is one of only a few countries in sub-Saharan Africa to have instituted a national governance process for coastal and ocean areas. This was achieved through a cooperative process to develop an integrated coastal management policy known as the Tanzania Coastal Management Partnership (TCMP). This partnership was formed between Tanzania's National Environmental Management Council (NEMC) and the University of

Rhode Island's Coastal Resources Center, and was supported by the United States Agency for International Development (USAID). The process resulted in the adoption by Cabinet of a national integrated coastal environmental management strategy for the Tanzania mainland.

The impetus for developing an integrated approach for coastal and ocean resource management at the national level in Tanzania emerged both from regional momentum towards developing national coastal policies and from experience from a number of local coastal management efforts before the mid-1990s (McLean, 2010). Agreement was reached in 1995 to embark upon a national integrated coastal management programme to effectively address coastal and marine problems (Ngoile, 2002). The policy development process was highly participatory. The TCMP developed a number of issue profiles, ranging from mariculture, to tourism and coastal livelihoods. The document *Initial Elements of a Coastal Policy: Proposed Vision, Principles, Goals and Strategies* was prepared for discussion by TCMP (TCMP, 1999a). Key policy elements were then captured in a Green Paper, *Options for a National Integrated Coastal Management Policy*, and submitted to Cabinet in November 1999 (TCMP, 1999b). After much deliberation, the government decided that there was no need for a separate policy or statute for integrated coastal management and that it would instead be a specific strategy under the National Environmental Policy (URT, 1997). The National Environmental Policy serves as an umbrella framework, providing guidance for managing the broader environment including the coastal and marine areas. It is also the framework for making changes that are needed to bring environmental considerations into mainstream decision-making in the country.

The National Integrated Coastal Environmental Management Strategy (NICEMS) aims to:

- improve the decision-making process for sustainable coastal development by providing guidance on resource use and allocation at both the national and local levels;
- harmonize decision-making among different sectors towards environmentally sound management of coastal resources;
- support decentralization to conserve, protect, and develop coastal resources; and
- support human and institutional capacity development at various levels.

The NICEMS is applicable to mainland Tanzania. Zanzibar has its own Integrated Coastal Management Strategy. While the NICEMS has inter-sectoral committees in place to harmonize decision-making in the different sectors towards sustainable coastal development, it is still only a framework. While it does not have the needed political clout to enforce decision-making in difficult scenarios involving conflict between resource users, the combination of the strategy with the national Environmental Management Act (EMA), 2004, provides the needed clout for taking action against environmental, as well as coastal, management violations. The EMA provides the:

> . . . legal and institutional framework for sustainable management of environment; to outline principles for management, impact and risk assessments, prevention and control of pollution, waste management, environmental quality standards, public participation, compliance and enforcement; to provide a basis for implementation of international instruments of [*sic*] environment; and to provide for implementation of the National Environment Policy . . .
>
> *(EMA, Full title)*

The EMA clarifies several non-transparent provisions under the old National Environmental Management Act (NEMA)Act, 1983, and the National Environmental Management Council (NEMC) is provided greater powers for compliance and enforcement.

During preparation of the World Bank and GEF supported Tanzania Marine and Coastal Environmental Management Project (MACEMP), district-level action planning was seen as the most effective mechanism with which to implement the strategy options at the subnational level. This approach was initiated with support from USAID, and implemented in three of the mainland districts to prioritize coastal issues and to identify specific goals and objectives. The MACEMP replicated and scaled up the district-level action planning approach in the remaining coastal districts (World Bank, 2005).

With respect to its offshore marine resources, Tanzania has enacted two key instruments to govern its exclusive economic zone (EEZ): the Territorial Sea and Exclusive Economic Zone Act (TSEEZA), 1989; and the Deep Sea Fishing Authority Act (DSFAA), 1998. These instruments apply to both the Tanzania mainland and the Zanzibar islands. The TSEEZA provides for implementation of UNCLOS and in the exercise of the sovereign rights, to make provisions for conservation and management of marine resources. The Act vests Tanzania with sovereign rights over the EEZ, including the rights to explore, exploit, conserve, and manage the living and non-living natural resources, and any economic activity within the EEZ. Jurisdiction is also given to Tanzania to ensure 'the protection and preservation of the marine environment' (section 9(2)(iii), in accordance with Article 56(1)(b) of UNCLOS). The TSEEZA allows the Minister of Foreign Affairs, who is responsible for the EEZ, to make regulations for conservation measures to protect the marine living resources. This provision creates an opportunity for establishment of marine protected areas (MPAs) within the EEZ (Ruitenbeek et al., 2005).

The DSFAA is complementary to the TSEEZA and applies to both mainland Tanzania and Zanzibar. It was enacted in 1998, but was not operationalized until 2007, when the Tanzanian Parliament passed key amendments to the Act. Subsequently, regulations were formulated in 2009 to include a benefit-sharing formula between mainland Tanzania and Zanzibar, which was endorsed by both parties (World Bank, 2013). The Deep Sea Fishing Authority (DFSA), established under section 4(1) of the 1998 Act, is intended to oversee implementation of its provisions and will operate through executive, advisory, and management committees. The functions of the authority are: to promote, regulate, and control fishing in Tanzania's EEZ; to regulate licensing of persons and vessels intending to fish in the EEZ; to initiate, implement, and ascertain the enforcement of policies on deep-sea fishing vessels; to formulate and coordinate programmes for scientific research in respect of fishing; to formulate fisheries policies; and to negotiate and enter into any fishing or other contract, agreement, or any kind of fishing cooperation with any government, international organization, or other institution in pursuance of the provisions of the Act. A high degree of collaboration is required between the two sides of the Union to operationalize the Act in an equitable and acceptable manner. The MACEMP supported the passage of needed legislative amendments, regulations, and institutional reforms for improved governance of fisheries in Tanzania's EEZ (World Bank, 2013).

The ocean and coastal policy processes in Tanzania provide leadership in the sub-region for greater interaction and participation with regard to transboundary resources management. The lessons learned from MACEMP have been incorporated into the Kenya Coastal Development Project and within the South West Indian Ocean Fisheries Governance and Shared Growth Programme (SWIOFish). These interventions, combined with regional ecosystem-based efforts such as BCLME and Aghulas and Somali Current LMEs (ASCLME), are expected to promote stronger governance processes in managing the valuable marine and coastal resources of the sub-region.

Namibia

Several other coastal nations have initiated developing integrated coastal policies or plans. Namibia initiated the Namibian Coast Conservation and Management (NACOMA) Project to promote an integrated coastal zone management system, with the support of the GEF (Government of Namibia, 2006). This programme supported the strengthening of policy, legal, and institutional frameworks for sustainable ecosystem management of the Namibian coast. The marine environment of Namibia falls within the Benguela Current system. This ocean system is rich in pelagic and demersal fish populations, and is supported by plankton production, driven by intense coastal upwelling. The Namibian coast is approximately 1,500 km long and is hyper-arid desert along its entire length. The coastal zone is sparsely populated and unsuitable for agriculture. The marine environment is thus free from the high levels of pollution commonly associated with large urban communities and is considered relatively pristine except for the deposition of sediment in the water column from diamond mining along the southern coast. Fishing is the third largest sector of the Namibian economy, after agriculture and mining.

The NACOMA Project supported a highly participatory national integrated coastal zone management (ICZM) policy development process to guide national, regional, and local planning and management processes. This also involved an effort to mainstream ICZM and biodiversity conservation principles and processes within national, provincial, local, and sectoral frameworks, such as the government's Vision 2030 (World Bank, 2006b). On 27 March 2013, the National Policy on Coastal Management for Namibia was launched in Swakopmund by the Minister of Environment and Tourism. According to the Ministry of Environment and Tourism (MET), the National Policy provides a framework that aims:

- to achieve the specific targets of the national development plans for sustainable economic growth, employment creation and reduced inequalities in income;
- to strengthen governance of Namibia's coastal areas to realize long-term national goals defined in Vision 2030;
- to strike a balance – to improve the quality of life of coastal communities, while maintaining the biological diversity and productivity of the country's coastal ecosystems; and
- to guide management actions in coastal resource use and allocation (Government of Namibia, 2013).

Both South Africa and Namibia prioritize the need for economic growth, employment creation, and reduced inequalities in income in their ocean policies, involving the need to balance maintenance of coastal and ocean ecosystems and biodiversity with the needs of the poor populations.

Seychelles

The Seychelles provides important lessons for the region in governing ocean space and the valuable resources therein. The Seychelles Fishing Authority (SFA) is the government executive agency in charge of management and conservation of living marine resources. The SFA was set up in August 1984 at a time of intense development, especially in foreign industrial tuna fishing. The vision of the SFA is to develop the fishing industry to its fullest potential and to safeguard the resource base for sustainable development. Policy objectives include conservation and management of marine resources, employment generation, maximization of revenue, promotion of an integrated economy, enhancement of food security, and safety at sea.

Violations of provisions of the consolidated Fisheries Act 1986 incur heavy fines up to US$175,020. The capacity of the SFA encompasses academic and operational fisheries research, fisheries oceanography, aquaculture, coral reef research, coastal zone sciences and management, marine pollution, natural resource assessment and economics, marine and coastal geographical information systems, institutional and policy analysis, and development of natural resources of the coastal and marine areas (SFA, undated).

Seychelles signed a fishery access agreement with the European Community (EC) in 2004 that provides for an increase in revenues from fishing activities in Seychelles waters from €4.6 million for 46,000 tonnes of fish to €5.5 million for 55,000 tonnes. This sum is paid annually in advance by the EC. Part of the revenue collected will be reinvested in the development of the fishing port and the fisheries industry. The number of licences issued to long liners has been reduced from twenty-seven to twelve, and licence fees to be paid in advance have increased by 50 per cent. Members have agreed that issuing fewer licences granted to long liners would help to conserve species such as albacore (European Commission Fisheries, 2010). The experience of fisheries licensing in the Seychelles is very pertinent to the other coastal nations in the region aiming to promote sustainable commercial fisheries in their EEZs. The government of the Seychelles supports technical assistance, including study tours by neighbouring nations, as a means by which to gain exposure to the policy and institutional mechanisms for governance of the ocean space.

According to Robinson and Shroff (2004), the Seychelles' vast EEZ of 1.37 million km² of productive tropical waters provides numerous and diverse prospects for the future of a marine-resource-based economy. Compared to many tropical and temperate marine fisheries, in which stock collapses, habitat destruction, and other factors threaten national food security and the social and economic fabric of coastal communities, the challenges for the Seychelles seem to be less onerous. Given limitations in human, technological, and financial capital, there is a critical need for greater collaboration between government, the private sector, and regional and international partnerships (Robinson and Shroff, 2004).

Challenges for regional ocean and coastal policy

Most coastal nations in Africa have adopted the key international agreements and principles relating to ocean and coastal management, and many are forging ahead with developing regional and sub-regional policy processes. However, most of these nations lack the financial, technical, or human resource capabilities to pursue the common vision needed for formulating and adopting such a common policy. In addition, the ocean and coastal policy agenda remains but one among numerous national priorities. On the other hand, the ecological, social, economic, and political diversity in the region presents challenges for a single ocean and coastal policy to be comprehensively applicable and acceptable. The following discussion enumerates some of the challenges to the formulation of a common ocean and coastal policy for the Africa region.

Ecological diversity and differing approaches in ecosystem management

The African continent is one of the most diverse in terms of ecological diversity and complexity. Ocean diversity is characterized by several large marine ecosystems (LMEs): the Canary Current LME, the Guinea Current LME, the Benguela Current LME, the Agulhas Current LME, and the Somali Coastal Current LME. Environmental variability contributes significantly to the diversity, abundance, and distribution of marine and coastal populations. This natural diversity, together with the variation in coastlines, EEZ extent, and densities of populations along the coastline have resulted in the adoption of different national approaches towards the utilization, management, and protection of marine and coastal resources. Furthermore, the degree to which marine and coastal sectors contribute to the economies of different coastal nations varies, which results in variations in the degree of importance (and therefore public budgets, human resources, and research attention) extended to these sectors.

Socio-economic conditions

Socio-economic conditions, the level of economic development, and political stability also vary greatly among Africa's coastal nations. Nigeria and Ghana have recorded high economic growth rates over the period 2010–11. However, in Nigeria, 68 per cent of the population is still living on less than US$1.25 a day. In comparison, in Ghana, which discovered oil reserves as late as 2010, has only 28.6 per cent of its population living on less than $1.25 (World Bank, 2013). Some nations are considered to be 'failed states', for example Somalia, in which ocean and coastal resource management is conducted in the absence of a policy, legislative and institutional framework. Overexploitation by illegal, unreported, and unregulated (IUU) fishing vessels, piracy, and illegal dumping of toxic wastes in the Somalian marine environment are the result of lack of governance in a nation ruled by warlords. Angola's twenty-seven-year-long civil war has ravaged the country's political and social institutions, and displaced an estimated 1.8 million people. Angola is setting out on a trajectory to economic recovery through the exploration and exploitation of its oil reserves.

In comparison, South Africa is a middle-income country with an abundant supply of resources, well-developed financial, legal, communications, energy, and transport sectors, and a stock exchange that ranks among the ten largest in the world. The country also has an extensive infrastructure, supporting the efficient distribution of goods to major urban centres throughout the region. South Africa's per capita GDP, corrected for purchasing power parity, positions the country as one of the fifty wealthiest in the world. While, on one level, South Africa is designated a developed nation, enormous disparities in income persist. The dual economy and ongoing social problems characterize South Africa as a developing nation. When compared to neighbouring countries, however, South Africa's technical and human resource skills in some sectors of government and the private sector, the proactive free market, and pro-equity policies and robust growth during the past decade, place South Africa on a different level with regard to development of marine and coastal resources.

The variation in socio-economic conditions also contributes to the degree of importance assigned to coastal and marine management by decision-makers. In countries in which meeting basic needs is a priority, integrated coastal management – an extended, complex and resource-intensive process – may not be assigned high priority when compared to the immediate pressures of health, education, and enterprise development for poverty reduction.

The 2012 Integrated Coastal Zone Management (ICZM) Protocol to the amended Nairobi Convention seeks to strengthen the legal framework of the Nairobi Convention for more effective management of marine and coastal ecosystems across sectors and national boundaries to achieve sustainable development. The Conference of Parties of the Nairobi Convention, in partnership with the Indian Ocean Commission, organized seven intergovernmental meetings of the Ad hoc Legal and Technical Working Group on ICZM to develop the Protocol to provide a framework for addressing a number of threats to the marine and coastal environment. The threats include:

> . . . anthropogenic pressures such as growing intensity of human settlements and unsustainable socioeconomic activities; natural disasters and climate change; and lack

of adequate coordination of various sectors that have contributed to the haphazard coastal development, habitat degradation and a decline in ecosystem services in the WIO region.

(UNEP, 2012: Summary)

Divergences in governance systems

Governance systems in sub-Saharan Africa vary from dictatorships and authoritarian regimes to successful democracies. Some systems retain remnants from colonial governance processes, particularly with respect to legal systems, institutional frameworks, and language. While much of eastern Africa is English-speaking, the official language of Mozambique is Portuguese and several of the island nations in the southwest Indian Ocean are French-speaking. In West Africa, Anglophone coastal nations exist side by side with Francophone and Lusophone nations. Legal and institutional frameworks and trade patterns are often structured around ties to the former colonial powers. As suggested earlier, a broad spectrum of governance approaches can be observed among the coastal nations.

In coastal mining regions that are endowed with rich mineral deposits, governance can be complicated by the formal or informal structure of players and interests, the role of rural local authorities, and corruption. In some countries, these become the main ingredients for significant informal small-scale mining, trading, and exporting, which in turn also contributes to the absence of social responsiveness and environmental performance. In this context, it is worth noting that, in such situations, the establishment of adequate policy, sectoral regimes, and enforcement instruments are usually insufficient incentives to prevent informal mining, illegal trading, and irresponsible social and environmental performance. Increased efforts and a dedicated emphasis on enhancing transparency are required. These can usually be facilitated through initiatives based on tribal structures and/or organizations based on community or cooperative frameworks, which can foster both commitment and the creation of shared interests.

Challenges to good governance

Governance is a challenge in many nations in Africa. The level of corruption and lack of transparency, however, varies greatly across the sub-region. As mentioned earlier, the discovery of large mineral deposits in the mid-1990s and the exploitation thereof made Equatorial Guinea the world's fastest-growing economy by 2004. However, the corruption watchdog Transparency International has ranked the Corruption Perceptions Index (CPI) of Equatorial Guinea at 169 (out of 177) and the Corruption Perceptions Score at 19 (scores range from 0 for 'highly corrupt' to 100 for 'very clean') (Transparency International, 2013). Despite calls for more transparency, oil revenues are maintained as a state secret.

Several other countries struggle continuously with political upheaval and civil unrest, including Somalia, Sudan, and Eritrea. In comparison, Tanzania, Mozambique, South Africa, and Namibia have remained relatively stable over the last decade. West Africa is largely stable of late, except for ongoing conflicts in countries such as Côte d'Ivoire.

Financial and capacity problems

Worldwide, experience shows that transformation of policy processes are gradual and multi-faceted, involving large investments of financial and human capital. Many African nations have the political will to formulate ocean and coastal policy, as evidenced by the number of policy processes in the region and statements by high-level delegates at international forums on ocean and coastal resource management. There is, however, a lack of financial resources and technical capacity to formulate and implement these policies. Even in the more well-to-do countries in sub-Saharan Africa, it is a challenge to secure government budgetary resources for ocean and coastal planning. For ministries of economic and planning, the priorities are just too many. There is also insufficient understanding of the importance and economic values to be gained through such planning. The intersectoral nature of such planning involving a multitude of sectors is often seen as an expensive, time-consuming, and difficult task. In many countries, these processes have been donor-initiated and -supported. An example is Tanzania, where the coastal management process was supported by USAID through the University of Rhode Island. While every effort was made to involve different levels of government and civil society, including the private sector and non-governmental organizations (NGOs), endorsement by government took an inordinately long time, indicating the lack of political will in some parts of government. This work, however, paved the way for the broader marine and coastal initiative supported by the World Bank and the GEF: the MACEMP. In summary, it is clear that raising awareness, prioritizing issues, and obtaining consensus require more significant investments in capital, time, and human resources than are often available.

Human resources in many of the African states are characteristically overextended. Representation in the international forums on ocean and coastal processes is often limited to two or less individuals who become known as 'policy champions'. Their time becomes increasingly scarce, and unless their exposure and training is shared among colleagues and juniors, such involvement may become unsustainable. A critical mass of people needs to be developed and trained to work on these issues. While the region benefited over the past decades from capacity-building efforts by international donors in the marine sciences, limited attention was given in this respect to policy development, legislative reform, economics, fiscal reform, and overall governance issues. The efforts of the European Commission through the South African Development Community (SADC) to support monitoring, control, and surveillance (MCS) resulted in improved revenues and strengthened management of the fishery resources of the sub-region.

Number and range of priorities

The African continent faces serious challenges in dealing with poverty, income disparity, stagnant growth, lack of access to markets, and limited financial and human resources. Other challenges that exacerbate key priorities include HIV/AIDS, a lack of access to health services, limited access to water and arable land, land tenure conflicts, unsustainable population growth, migration, improperly planned urbanization, desertification, deforestation, and general environmental degradation. Erosion pollution and other issues facing the marine and coastal areas have already been discussed. Inequities, distributional issues, lack of services, and governance problems are some key macro issues. The ocean and coastal agenda is therefore only one among a long list of priorities. This agenda item is also often not a traditional priority, and the contribution of the ocean and coasts to the GDP may not be adequately recognized unless the nation is a small island developing state (SIDS). For instance, in Tanzania, revenue from freshwater fisheries from the Lake Victoria is more than twenty times that of the revenue from offshore fisheries. Similarly, revenue from coastal tourism is only a fraction of the revenue from the world famous national parks. Attention of authorities and allocation of resources generally matches the revenue. Education, health, infrastructure, energy, HIV/AIDS, establishment of governance systems following war situations, and investment in agriculture and industry pose competing priorities.

In most priority-setting initiatives involving poverty alleviation, environmental management is often at the bottom of national priorities even though environmental sustainability is one of the Millennium Development Goals (MDGs). In many cases, commercial interests and growth at any cost overrides the need for environmental oversight, conservation, and equity in benefit sharing.

Most of the policy processes discussed in this chapter have been promoted and largely or fully financially supported by external agencies and donors. While political will does exist in specific government entities mandated with responsibility for specific ocean or coastal sectors, the issue area may not be recognized as a sufficiently high priority for decision-makers in finance and economic planning, which are the entities that allocate resources for policy initiatives.

As noted by Hewawasam et al. (2008), Tanzania has established a comprehensive legislative and regulatory framework for managing marine and coastal resources. However, a number of sectoral policies have overlapping or complementary governance regimes. The authors note that complexities of managing the marine and coastal environment require coordination and feedback mechanisms among implementing agencies, decision-makers, and beneficiaries, greater transparency in decision-making processes, and greater capacity and empowerment of local government representatives and communities (Hewawasam et al., 2008).

Small island developing states

The entire land area of SIDS is the coastal zone, which means that the whole island is influenced by and influences the ocean. The issues faced by SIDS regarding EEZ management, fisheries, tourism, and climate change have different, more significant, implications than those of their mainland neighbours. For instance, while the inhabitants of the coastal areas of mainland Tanzania and the two islands of Zanzibar (Pemba and Unguja) rely heavily on fisheries for subsistence, those on Zanzibar face additional livelihood challenges in terms of access to resources such as fuel, potable water, and goods and services. They are also more vulnerable to storm events, natural disasters, and dynamic natural ocean processes. Given the marginal status of many SIDS, they can also be a lot less resistant to economic and social impacts from transient industries such as tourism and political conflicts. Comprehensive ocean policies therefore need to recognize the additional contextual challenges faced by SIDS and the different management choices needed when compared to mainland coastal states.

Criteria for formulating or strengthening regional ocean policy

Many coastal nations in sub-Saharan Africa have embarked on a path of regional integration, economic reform, technological modernization, and democratic participation. As a result, some economies in Africa have maintained sustained growth and their populations have seen improved living standards. However, alongside – and sometimes as a result of – this growth, many nations in Africa have seen violence, starvation, and displacement. Along with the opportunities, these nations have seen significant risks, particularly the potential for social unrest and environmental damage. Some countries have addressed these issues, but most have either ignored them or addressed them only in a piecemeal fashion. According to the World Development Report 2014:

> [T]he solution is not to reject change in order to avoid risk but to prepare for the opportunities and risks that change entails. Managing risks responsibly and effectively has the potential to bring about security and a means of progress for people in developing countries and beyond.
>
> *(World Bank, 2014: 3)*

As noted, while there is no single regional ocean and coastal policy in the Africa region, several sub-regional policy processes are under way or emerging. A range of national efforts to strengthen governance processes for ocean and coastal resources management are also under preparation or in the early stages of implementation. Admittedly, many of these processes focus on critical issues affecting particular sectors, and perhaps not all of the issues and opportunities are recognized. Given these ecological, economic, developmental, social, and political divergences, the existing processes do, however, reflect the level at which such initiatives are feasible within the available financial and technical resource constraints. The following section outlines some of the essential elements in a regional ocean policy.

Strengthened policy and legal framework

The overexploitation of fisheries, sedimentation, and pollution of marine and coastal ecosystems resulting from increased urbanization, industrial development, and overall population pressure emphasize the urgent need to strengthen ongoing ocean policy processes. Significant levels of financial and technical resources are needed to address these complex and emerging issues and to scale up existing initiatives. Furthermore, according to the International Monetary Fund (IMF), the Asian Development Bank (ADB), and KPMG, East Africa is about to experience a hydrocarbon-induced economic boom. Much of these deposits will be in pristine marine environments. A serious issue is that this natural gas exploration is currently being pursued without adequate scientific studies to look at the vulnerability of marine environments, threatened species, and newly discovered deep-water marine life living in hydrothermal vents in the Indian Ocean. According to Mutch (2012), the estimated volume of the gas reserves in the offshore areas of Tanzania are estimated to be 100 trillion cubic feet (tcf) and petroleum reserves are estimated at 600,000 barrels per day (bpd).

Unless the necessary governance systems are established to maximize and ensure equitable sharing of the benefits from these economic opportunities, and sufficient capacity is built within government systems to enter into equitable negotiations, the opportunities to support the growth and development agenda will be missed. Moreover, unless appropriate environmental and social safeguards are put in place, further degradation of ocean and coastal ecosystems is likely, resulting in the possible collapse of entire species or ecosystems. Given the existing dependence of populations on these resources, the repercussions could mean displacement of the poor, loss of livelihoods, and severe negative impacts on other productive ocean and coastal sectors.

Political will and recognition of issues and opportunities from ocean sectors

There is a need for a broad cross-section of decision-makers to recognize and highlight ocean and coastal areas as a priority for growth and poverty reduction. Lead agencies need actively to engage the ministries of finance and economic planning in understanding the issues, challenges, and opportunities that the ocean offers for growth and stability in the region. Increased efforts are essential for mainstreaming the ocean and coastal policy agenda within macro policies and programmes and elevating it within the poverty reduction and growth agendas, since resources from multilateral, bilateral, or private sources respond to these policy processes. Political will is also necessary to maintain broad-based, sectoral buy-in for the long-term processes required for ensuring strong governance of ocean and coastal areas.

Capacity enhancement for improved governance

Coupled with political will, a concerted and lasting investment in capacity enhancement is needed for sustainable management of ocean and coastal resources and areas. Capacity needs to be viewed in broad terms to include proficiency in an array of topics and themes, including scientific research, environmental and biodiversity management, resource economics, marine spatial planning, and the use of market-based instruments. Capacity needs to be enhanced for promoting certification and branding, financial management (including risk pooling), micro-finance, entrepreneurship development, communication, and protected area management. At a macro level, capacity needs to be enhanced for greater transparency in legislation and regulatory mechanisms, monitoring, control, and surveillance, and performance monitoring and evaluation. Capacity needs to be strengthened at all levels: regional, national and, in particular, local and community, where interventions have the most impact. Most nations in Africa have embraced decentralization of decision-making as a priority. However, there is often a divergence between policy pronouncements and real devolution of power, with an associated transfer of resources to the subnational levels. Enhancing capacity should also include the development of clear 'rules of the game' if the public sector is to mobilize the private sector in a more meaningful manner. Given that public funding is likely never to be sufficient to address the range of issues affecting ocean and coastal areas, skills are also needed to mobilize international private capital towards financial sustainability.

Public awareness about the threatened state of marine resources is also critical. Training programmes for the media and educational institutions can be powerful tools in generating awareness at all levels of the importance of ocean and coastal policy and the opportunities that can arise from sound governance. Such enhanced awareness will, in turn, empower the public to take an active role in managing and protecting the oceans and coasts for their own benefit and that of future generations of users.

Partnerships for more efficient and effective management of resources

Management of ocean and coastal resources is a complex task involving the science relating to marine life, governance mechanisms, socio-economic data, market information and economic instruments, and international best practice on licensing, concessions, and commercial arrangements. Managers need to engage a broad range of stakeholders, including civil society, NGOs, the private sector, national and subnational entities, and regional governance bodies. Management measures need to have sufficient flexibility to respond adaptively to changing social realities, markets, and natural phenomena. Partnerships and co-management arrangements are essential to legitimizing and regulating 'informal' or traditional exploitation arrangements and to ensuring equity in benefit-sharing. Links to emerging global issues, such as maritime security, research into alternative energy sources such as wind and tidal energy, and bio-technology, adding value to marine products and the 'Blue Economy', need to be investigated.

Sound governance

While the aforementioned are all critical elements of sound governance of oceans and coasts, another key element is to ensure that coastal nations and their inhabitants benefit from the exploitation of the ocean and coastal resources on a sustainable basis. Economists might argue that all exports improve the balance of trade and raise the country's level of production, as well as its GNP. Theoretically, this means increased growth and employment, and a reduction in poverty, increased revenues with which to improve service delivery, and overall improvements in quality of life. However, experience with the exploitation of diamonds in Sierra Leone, oil in Equatorial Guinea, and oil and gas in Nigeria reveals that exploitation of natural resources does not always benefit the population in this way; exploitation may rather result in loss of livelihood through restricted access, poorer quality of life through industrial pollution and environmental degradation, and reduction of social services owing to weak governance and corruption. Experience also shows that investments in mining and oil and gas production often fail to create large-scale employment opportunities because of international competition and the efficiency of the industry, which necessitate advanced technology and a specialized workforce. Wealth from these industries may increase, but is often concentrated among an elite few, while the growth agenda circumvents the poor.

The responsibility to manage ocean and coastal resources soundly, sustainably, and for the benefit of the nation and civil society is a responsibility of governments and those involved in ocean-related industries, such as fisheries, tourism, aquaculture, mining, and oil and gas extraction. Ensuring transparency of access agreements, and licence and concession arrangements, will improve accountability and ensure that benefits are available to the broader society. Formal arrangements are necessary to ensure equitable benefit-sharing and improvement of the performance of the relevant sectors and protection of livelihoods of those whose access to the resource may be constrained. Elimination of corruption within the various ocean sectors should be a national priority. Regional entities already engaged in these sectors and the international community could support national governments in this effort. Improved governance can also assist weak states in confronting 'neo-colonialism' by interest groups involved in the unsustainable exploitation of fisheries, oil and gas, and mineral reserves.

Preparedness for the impacts of climate change

Climate change has become a major threat in sub-Saharan Africa, especially as it impacts on fisheries, biodiversity, and tourism, as well as livelihoods. However, few countries are prepared for the impacts of climate change and even natural disasters. While most policymakers are exposed to information from international conferences on climate change, few have put in place policy mechanisms for addressing the impacts. The SIDS are generally ahead of the mainland coastal countries in this respect, given the highly visible impacts of climate change on these vulnerable countries. Regional ocean policies can assist in integrating climate change into national ocean policy.

Towards a regional ocean policy for sub-Saharan Africa

If a single, shared ocean and coastal policy for the continent is impossible or impractical because of the numerous challenges outlined earlier, would it be possible for sub-regional policies to evolve and collaborate? This may, in fact, be the soundest option to be given serious consideration, given the large number of national and sub-regional ocean processes under way, as well as the conventions and protocols under the United Nations Environmental Programme's Regional Seas Programme (UNEP RSP). A sub-regional ocean policy also makes much sense, given existing similarities and linkages between ecosystems, levels of economic development, social and cultural traditions, and related realities. Given the high values associated with migratory fisheries, and the impacts of transboundary pollution, trans-frontier trade, and exponential growth expected in the hydrocarbon sector, there is every reason for countries in this sub-region to work together.

It should be possible to build upon the political will expressed within the different regional conventions, the ongoing scientific research supported by many donors, the activities under way within the LMEs, and national-level ocean and coastal governance efforts. There are several options and possible differences of opinion as to who should adopt the lead role on the policy processes. The LME programmes might be appropriate vehicles for developing or implementing ocean and coastal policy agendas. Both the BCLME and the ASCLME programmes tackle some policy elements relating to fisheries. In order for the LME programmes to lead the development of regional ocean policy, they must be perceived as more than 'environmental', or 'scientific', or 'fisheries' programmes. Further, the traditional clientele may need to be more broad-based and existing governance regimes must be strengthened, with particular links to key decision-making entities including ministries of finance.

An alternative option could include integration of the ocean and coastal policy agendas within the mandates of the regional economic organizations. For example, the objectives of the SADC include: economic growth, poverty alleviation, enhancing the standard and quality of life, and supporting the socially disadvantaged through regional integration; promoting and defending peace and security; promoting self-sustaining development on the basis of collective self-reliance and the interdependence of member states; and promoting and maximizing productive employment and utilization of resources in the region. The SADC also incorporates environmental, as well as social, goals within the specific objectives. The ocean and coastal agenda may fit in quite well within these goals. However, the SADC has no mandate to operate in West Africa. The same constraint applies to regional fisheries organizations, in addition to their mandate being limited to fisheries management. Similarly, in West Africa, there are many regional economic organizations. Among them, the Economic Community of West African States (ECOWAS) adopted an Environmental Policy in 2008 that includes references to sustainable management of coastal, island, and marine ecosystems (ECOWAS, 2008). The argument for one regional agency to spearhead an ocean policy will still not be possible with ECOWAS, since the organization's mandate is limited to West Africa.

Clearly, a regional or sub-regional ocean policy agenda should be driven by Africans themselves if it is to be acceptable to the various constituencies within the continent. The African Union is well placed to advance sub-regional ocean and coastal policies, building on the policy agendas initiated with the Abidjan and Nairobi Conventions. It is fully consistent with the mission of the Union to be 'driving the African integration and development process in close collaboration with African Union Member States, the Regional Economic Communities and African citizens' (African Union, undated). It is also consistent with the Union's mandate under the 'key areas of concern', one of which is the 'management of coastal and marine resources'. The African Union also recently adopted the concept of 'Blue Economy'.[2] As noted by the Vice President of Seychelles Danny Faure, who led the Seychelles delegation at the Union's Assembly of Heads of State in 2014, '[t]he Blue Economy, encapsulating all of the potential of our oceanic resources, offers us a platform for Africa's transformation both in terms of Agenda 2063 and in terms of the post 2015 Development Agenda and the sustainable development goals' (Faure, 2014). A significant opportunity exists for the African Union to take leadership to collaborate with the different national and regional ocean agencies to develop a policy agenda that is acceptable to all of the different constituencies in the sub-regions. Greater collaboration among coastal nations in the region in maintaining regional governance frameworks would help to reduce the number and magnitude of challenges involved in decisions regarding the rational exploitation and use of ocean and coastal resources.

Notes

1 Author's personal communication with a World Bank team leader, 2014.
2 The 'Blue Economy' concept is a 'cross-cutting initiative aim[ing] to provide global, regional and national impact to increase food security, improve nutrition, reduce poverty of coastal and riparian communities and support sustainable management of aquatic resources' (FAO, 2013).

References

African Union (undated) 'Vision and Mission', online at http://www.au.int/en/about/vision [accessed 9 November 2014].

British Broadcasting Corporation (BBC) (2014) 'Equatorial Guinea Profile', online at http://www.bbc.co.uk/news/world-africa-13317174 [accessed 19 February 2014].

Dzidzornu, D. M. (2011a) 'Ocean Policy in Africa and Treaty Aspects of Marine Fisheries Cooperation, Management, and Environmental Protection', in A. Chircop, S. Coffen-Smout, and M. McConnell (eds) *Ocean Yearbook 25*, Leiden/Boston, MA: Martinus Nijhoff.

Dzidzornu, D. M. (2011b) 'Economic Development and Environmental Protection in Africa: The Constitutional, Conventional and Institutional Contexts', in F. N. Botchway (ed.) *Natural Resource Investment and Africa's Development*, Cheltenham: Edward Elgar.

Dzidzornu, D.M. (2012) 'Marine Environmental Protection under the Nairobi and Abidjan Regimes: Working towards Functional Revitalization', in A. Chircop, S. Coffen-Smout, and M. McConnell (eds) *Ocean Yearbook 26*, Leiden/Boston, MA: Martinus Nijhoff.

Economic Community of West African States (ECOWAS) (2008) *ECOWAS Environmental Policy*, online at http://events.ecowas.int/wp-content/uploads/2013/03/Environ-PUBLICATION-ENG-1.pdf [accessed 24 June 2014].

EPA   *See* United States Environmental Protection Agency

Ernst and Young (2012) 'African Oil and Gas: Driving Sustainable Growth', online at http://www.ey.com/Publication/vwLUAssets/African_oil_and_gas:_driving_sustainable_growth/$FILE/EY-African_oil_and_gas-driving_sustainable_growth.pdf [accessed 19 May 2013].

European Commission Fisheries (2010) 'Bilateral Talks, New Protocol with the Republic of Seychelles', Press release, 4 June, online at http://ec.europa.eu/fisheries/news_and_events/press_releases/040610/index_en.htm [accessed 9 November 2014].

Faure, D. (2014) 'The Blue Economy: Key to Africa's Vision for the Future', online at http://www.mfa.gov.sc/static.php?content_id=36&news_id=695 [accessed 19 May 2014].

Food and Agriculture Organization of the United Nations (FAO) (2007) *The State of World Fisheries and Aquaculture 2006*, online at http://www.fao.org/docrep/009/A0699e/A0699e00.htm [accessed 8 October 2010].

Food and Agriculture Organization of the United Nations (FAO) (2010) The State of World Fisheries and Aquaculture (SOFIA), 2008–2010, online at http://www.fao.org/fishery/sofia/en [accessed 8 October 2010].

Food and Agriculture Organization of the United Nations (FAO) (2013) 'Blue Growth Initiatives/Global Blue Economy', online at http://www.fao.org/bodies/council/cl148/side-events/blue-economy/en/ [accessed 19 May 2014].

Food and Agriculture Organization of the United Nations (FAO) Fisheries and Aquaculture Department (undated) 'South West Indian Ocean Fisheries Commission', online at http://www.fao.org/fishery/rfb/swiofc/en#Org-OrgsInvolved [accessed 19 May 2014].

Food and Agriculture Organization of the United Nations (FAO) Fisheries and Aquaculture Department (2012) *The State of World Fisheries and Aquaculture 2012*, online at http://www.fao.org/docrep/016/i2727e/i2727e.pdf [accessed 19 February 2014].

Food and Agriculture Organization of the United Nations (FAO) Fisheries and Aquaculture Department (2014) 'The State of World Fisheries and Aquaculture 2014', online at www.fao.org/3/a-i3720e.pdf [accessed 26 January 2015].

Glavovic, B. C. (2000) *Building Partnerships for Sustainable Coastal Development: The South African Coastal Policy Experience – The Process, Perceptions and Lessons Learned*, Cape Town: Department of Environmental Affairs and Tourism, Republic of South Africa.

Glavovic, B. C. (2008) 'Sustainable Coastal Development in South Africa: Chasm between Rhetoric and Reality', in R. R. Krishnamurthy, B. Glavovic, A. Kannen, D. R. Green, R. Alagappan, H. Zengcui, S. Tinti, and T. Agardy (eds) *Integrated Coastal Management: The Global Challenge*, Singapore: Research Publishing Services.

Government of Namibia (2006) 'Namibian Coast Conservation and Management (NACOMA) Project Description', online at http://www.met.gov.na/programmes/nacoma/nacoma.htm [accessed 20 September 2006] (no longer available).

Government of Namibia (2013) *National Policy on Coastal Management for Namibia*, Swakopmund: Government of Namibia.

Hauck, M., and Sowman, M. (2001) 'Coastal and Fisheries Co-Management in South Africa: An Overview and Analysis', *Marine Policy*, 25(3): 173–85.

Hecht, T. (2006) *Regional Review on Aquaculture Development, 4. Sub-Saharan Africa – 2005*, FAO Fisheries Circular No. 1017/4, Rome: FAO.

Hewawasam, I. (2002) *Managing the Marine and Coastal Environment of Sub-Saharan Africa: Strategic Directions for Sustainable Development*, Washington, DC: World Bank.

Hewawasam, I., McLean, B., and Ngoile, M. (2008) 'Benefitting People and Ecosystems through Ocean and Coastal Governance: Lessons from the Experience in Tanzania', in R. R. Krishnamurthy, B. Glavovic, A. Kannen, D. R. Green, R. Alagappan, H. Zengcui, S. Tinti, and T. Agardy (eds) *Integrated Coastal Zone Management: The Global Challenge*, Singapore: Research Publishing Services.

Humphrey, S., and Francis, J. (eds) (1997) *Sharing Coastal Management Experience in the Western Indian Ocean: Proceedings of the Experts and Practitioners Workshop on Integrated Coastal Area Management for Eastern Africa and the Island States*, Zanzibar: WIOMSA.

Indian Ocean Tuna Commission (IOTC) (undated) 'The Commission', online at http://www.iotc.org/about-iotc [accessed 20 May 2014].

International Commission for the Conservation of Atlantic Tunas (ICCAT) (undated) 'Contracting Parties', online at http://www.iccat.es/en/contracting.htm [accessed 9 May 2014].

International Monetary Fund (IMF) (2013) *United Republic of Tanzania: Fifth Review under the Policy Support Instrument, First Review under the Standby Credit Facility, and Requests for a Waiver of Nonobservance of a Performance and Assessment Criterion, Rephasing of the Standby Credit Facility Arrangement and Modification of Performance and Assessment Criteria*, IMF Country Report No. 13/12, online at http://www.imf.org/external/pubs/ft/scr/2013/cr1312.pdf [accessed 20 May 2014].

Kleinschmidt, H., Moolla, S., and Diemont, M. (2005) 'A New Chapter in South African Fisheries Management', Press release, 25 April, online at http://www.mcm deat.gov.za/press/commercial_fishing_rights_applications_2005.pdf [accessed 20 September 2006] (no longer available).

KPMG (2013) *Oil and Gas in Africa: Africa's Reserves, Potential and Prospects*, online at https://www.kpmg.com/Africa/en/IssuesAndInsights/Articles-Publications/Documents/Oil%20and%20Gas%20in%20Africa.pdf [accessed 19 July 2013].

McLean, B. (2010) 'Understanding the Dynamics of National Coastal Policy Change: Policy Narratives for South Africa and Tanzania', Unpublished PhD Dissertation, University of Delaware, Newark, DE.

Mutch, T. (2012) 'East African Oil and Gas: Proper Environmental Planning Needed to Avoid Disaster', *African Arguments*, 28 September, online at http://africanarguments.org/2012/09/28/east-african-oil-and-gas-proper-environmental-planning-needed-to-avoid-disaster-%E2%80%93-by-thembi-mutch/ [accessed 9 November 2014].

National Oceanographic and Atmospheric Association (NOAA) (undated) 'Large Marine Ecosystems of the World', online at http://lme.edc.uri.edu/ [accessed 19 May 2014].

New Partnership for Africa's Development (NEPAD) (2003) *Action Plan for the Environment Initiative of NEPAD*, online at http://www.nepad.org/climatechangeandsustainabledevelopment/knowledge/doc/1463/action-plan-environment-initiative [accessed 16 January 2007].

New Partnership for African Development (NEPAD) Partnership for Fisheries (PAF) (undated) 'What is PAF?', online at http://www.nepad.org/foodsecurity/fisheries/about [accessed 8 October 2010].

Ngoile, M. N. (2002) *The Challenges of Integrating Marine Science with Coastal Management in the Western Indian Ocean Region*, online at http://hdl.handle.net/1834/831 [accessed 3 December 2008].

Republic of Mauritius, Ministry of Fisheries (2011) 'Seafood Hub, Economic Outlook for the Fisheries Sector', online at http://fisheries.gov.mu/English/DOCUMENTS/SEAFOOD.PDF [accessed 9 June 2014].

Republic of South Africa (1997) *Marine Fisheries Policy White Paper*, online at https://www.environment.gov.za/sites/default/files/legislations/marine_fisheries.docx [accessed 15 January 2007].

Republic of South Africa (2000) *White Paper for Sustainable Coastal Development in South Africa*, online at http://www.polity.org.za/html/govdocs/white_papers/coastal/index.html [accessed 15 January 2007].

Republic of South Africa (2010) 'Minister Launches the National Tourism Sector Strategy', Press release, online at http://www.tourism.gov.za/AboutNDT/Publications/Tourism%20Minister%20launches%20National%20Tourism%20Sector%20Strategy.pdf [accessed 8 October 2010].

Republic of South Africa, Department of Environmental Affairs and Tourism (2013) *South Africa's Policy on National Environmental Management of the Oceans*, Cape Town: Department of Environmental Affairs and Tourism.

Robinson, J., and Shroff, J. (2004) 'The Fishing Sector in Seychelles: An Overview, with an Emphasis on Artisanal Fisheries', *Seychelles Medical and Dental Journal*, 7(1): 52–6.

Ruitenbeek, J., Hewawasam, I., and Ngoile, M. (eds) (2005) *Blueprint 2050: Sustaining the Marine Environment in Mainland Tanzania and Zanzibar*, Washington, DC: World Bank.

Seychelles Fishing Authority (SFA) (undated) 'About Us', online at http://www.sfa.sc/aboutus.jsp [accessed 20 May 2014].

Shannon, L. V., and O'Toole, M. J. (2003) 'Sustainability of the Benguela: Ex Africa Semper Aliquid Novi', in K. Sherman, L. M. Alexander, and B. D. Gold (eds) *Large Marine Ecosystems: Stress, Mitigation and Sustainability*, Washington, DC: American Association for the Advancement of Science.

South West Indian Ocean Fisheries Project (SWIOFP) (undated) 'South West Indian Ocean Fisheries Project', online at http://www.swiofp.net [accessed 19 May 2014].

Tanaka, Y. (2006) 'Regulation of Land-Based Marine Pollution in International Law: A Comparative Analysis between Global and Regional Legal Frameworks', *Zeitschrift fuer Auslaendisches Oeffentliches Recht und Voelkerrecht* [*Heidelberg Journal of International Law*], 66(3): 535–74.

Tanner, C. (2005) *The Changing Politics of Land in Africa: Domestic Policies Crisis Management and Regional Norms*, Pretoria: University of Pretoria.

Tanzania Coastal Management Partnership (TCMP) (1999a) *Initial Elements of a Coastal Policy: Proposed Vision Principles, Goals and Strategies*, Working Document No. 5018, Dar es Salaam: TCMP Support Unit.

Tanzania Coastal Management Partnership (TCMP) (1999b) *Options for a National Integrated Coastal Management Policy*, Working Document NO. 5026, Dar es Salaam: TCMP Support Unit.

Tanzania Petroleum Development Corporation (TPDC) (2014) *Status of Deep Exploration Wells up to July 2014*, online at http://www.tpdc-tz.com/Deep_Exploration_Wells_Status_2012.pdf [accessed 18 November 2014].

Transparency International (2013) 'The 2013 Corruption Perceptions Index', online at http://www.transparency.org/cpi2013 [accessed 19 May 2014].

United Nations Development Programme (UNDP) (2007) *Implementation of the Benguela Current LME Action Programme for Restoring Depleted Fisheries and Reducing Coastal Resources Degradation: The Initiation Plan*, online at http://iwlearn.net/iw-projects/3305 [accessed 11 March 2015].

United Nations Environment Programme (UNEP) (undated*a*) 'Abidjan Convention Protocols', online at http://abidjanconvention.org/index.php?option=com_content&view=article&id=99&Itemid=199 [accessed 19 May 2014].

United Nations Environment Programme (UNEP) (undated*b*) 'Nairobi Convention Protocols', online at http://www.unep.org/NairobiConvention/The_Convention/Protocols/index.asp [accessed 19 May 2014].

United Nations Environment Programme (UNEP) (1995) *Global Programme of Action for the Protection of the Marine Environment from Land-based Activities*, UN Doc. UNEP(OCA)/LBA/IG.2/7, online at http://www.gpa.unep.org/index.php/about-gpa [accessed 9 November 2014].

United Nations Environment Programme (UNEP) (1999) 'Africa: Marine and Coastal Areas', in *GEO-2000 Global Environment Outlook*, online at http://www.unep.org/geo/GEO2000/english/0057.htm [accessed 4 April 2014].

United Nations Environment Programme (UNEP) (2010) *Final Text of the Amended Nairobi Convention for the Protection, Management and Development of the Marine and Coastal Environment of the Western Indian Ocean*, online at http://www.unep.org/NairobiConvention/docs/Final_Act_Nairobi_Amended_Convention&Text_Amended_Nairobi_Convention.pdf [accessed 9 May 2014].

United Nations Environment Programme (UNEP), Nairobi Convention Secretariat, the Council for Scientific and Industrial Research (CSIR), and the Western Indian Ocean Marine Science Association (WIOMSA) (2009) *Regional Synthesis Report on the Status of Pollution in the Western Indian Ocean Region*, Nairobi: UNEP.

United Nations General Assembly (UNGA) (1992) *Report of the United Nations Conference on Environment and Development, Vols I–III*, UN Doc. A/CONF.151/26/REV.1, Rio de Janeiro: UNGA ('Agenda 21').

United Republic of Tanzania (URT) (1997) *National Environmental Policy*, Dar es Salaam: URT.

United Republic of Tanzania (URT) (2005) *National Strategy for Growth and Reduction of Poverty II*, Dar es Salaam: URT.

United Republic of Tanzania (URT) (2009) *National Land Use Framework Plan 2009–2029: Workshop Report*, online at http://www.nlupc.org/index.php/highlights/more/national_land_use_framework_plan_2009_-_2029_workshop_report/ [accessed 10 May 2013] (no longer available).

United Republic of Tanzania (URT) (2013) *National Natural Gas Policy*, Dar es Salaam: URT.

United States Environmental Protection Agency (EPA) (undated) 'Protecting the Marine Environment', online at http://www2.epa.gov/international-cooperation/protecting-marine-environment [accessed 19 May 2014].

Western Indian Ocean Marine Science Association (WIOMSA) (undated) 'About Us', online at http://www.wiomsa.org/index.php?option=com_content&view=article&id=48&Itemid=63 [accessed 2 May 2010].

World Bank (undated) 'Poverty and Equity Data, Regional Dash Board, Sub-Saharan Africa', online at http://povertydata.worldbank.org/poverty/region/SSA [accessed 8 June 2014].

World Bank (2005) *Tanzania Marine and Coastal Environment Management Project: Project Appraisal Document*, Report No. 32500-TZ, Washington, DC: World Bank.

World Bank (2006a) 'Strategic Partnership for a Sustainable Fisheries Investment Fund in the Large Marine Ecosystems of Sub-Saharan Africa', online at http://siteresources.worldbank.org/INTARD/Resources/Factsheet_3_SSA_FINAL.pdf?resourceurlname=Factsheet_3_SSA_FINAL.pdf [accessed 28 August 2006].

World Bank (2006b) *Namibian Coast Conservation and Management Project: Project Appraisal Document*, Report No. 31307–NA, Washington, DC: World Bank.

World Bank (2013) *Implementation Completion Report for Tanzania Marine and Coastal Environment Management Project*, Report No. ICR2754, Washington, DC: World Bank.

World Bank (2014) *World Development Report*, Washington, DC: World Bank.

World Wildlife Fund (WWF) (undated*a*) 'Depletion of Fisheries Could Affect Millions in West Africa', online at http://wwf.panda.org/what_we_do/where_we_work/west_africa_marine/area/fisheries/ [accessed 19 May 2014].

World Wildlife Fund (WWF) (undated*b*) 'Marine Problems: Pollution', online at http://wwf.panda.org/about_our_earth/blue_planet/problems/pollution/ [accessed 18 November 2014].

World Wildlife Fund (WWF) (undated*c*) 'Safeguarding the Natural World: West Indian Ocean', online at http://www.wwf.org.uk/what_we_do/safeguarding_the_natural_world/oceans_and_coasts/west_indian_ocean/ [accessed 19 May 2014].

Appendix A
A GUIDE TO COMPARATIVE CASE STUDIES ON NATIONAL OCEAN POLICIES

1. INTRODUCTION: BASIC INFORMATION AND OVERVIEW OF NATIONAL OCEAN POLICY

A. Basic information

- Geographic and demographic information, e.g. total land area, length of coastline, maritime zones, and human population
- Maritime boundaries, including a map
- Status and utilization of marine and maritime resources, including economic data and conflicts among uses
  - Why the oceans and coasts are an important policy area
  - Extent to which the ocean policy is part of the nation's sustainable development strategy

B. Brief overview of nature and evolution of national ocean policy

- Statement of the nature of existing ocean policy or the status of development of new policy
- Factors that gave rise to a national ocean policy initiative, e.g. multiple-use conflicts; environmental problems; increasing public demand for nature conservation and sustainable exploitation of natural resources; the need to expand coastal management efforts further offshore; opportunities for economic development; results from large-scale research programmes, such as large marine ecosystems (LMEs), Global International Waters Assessment (GIWA), and the Global Ocean Observing System (GOOS)
- Stakeholders/key players involved, i.e. which groups were in favour and which against
- Conflicts in the development of a national ocean policy, i.e. what obstacles were faced in national ocean policy formulation and implementation

2. POLICY DEVELOPMENT PROCESS

Description of how the policy development started, i.e. of the approach followed in establishing a national ocean policy, including information on objectives, the principles that guided the process, and the steps followed during the preparatory/formulation phase

A. *Initiation of the policy*

How was the policy initiated and by whom?

- Was it legislatively mandated?
- Was it led by a prime minister?
- Was it led by a Commission?
- Was there another mode of initiation?

B. *Objectives*

What objectives are addressed in the policy?

- To foster sustainable development of ocean areas?
- To protect biodiversity and vulnerable resources and ecosystems?
- To harmonize existing uses and laws?
- To coordinate the actions of the many government agencies that are typically involved in ocean affairs?

C. *Major principles*

What major principles were adopted/followed in the national ocean policy?

- Accountability?
- Adaptive management?
- Best available science?
- Ecosystem-based management?
- Indigenous rights?
- Integrated approach?
- Intergenerational and intra-generational equity?
- Multiple uses?
- Participatory governance?
- Precautionary approach?
- Protection of biodiversity?
- Stewardship?
- Subsidiarity?
- Sustainability?
- Timeliness?
- Transparency?

D. *Institutional arrangements*

What institutional arrangements and processes were followed to formulate the national ocean policy? How well did they work?

- Who was in charge of formulating the policy?
- What processes were followed and how is the political process being managed?
- How were stakeholders involved?
- What approaches were adopted to harmonize sectorial issues and to resolve conflicts?

What specific strategies/interventions were followed in the decision-making process?

- Consultation at ministerial and inter-ministerial levels?
- Organization of formal/informal advisory groups?
- Definition by the leading agency of a draft vision for the nation's ocean, guiding values and principles, main goals and objectives of the policy, and its scope and geographical application?
- Establishment of thematic working groups?
- Assessment of existing ocean-related policies and gap analysis?
- Formulation of policy options, including legislative and non-legislative?
- Various iterations of consultation with relevant constituencies?
- Issuance/adoption of a formal ocean policy?

3. NATURE OF THE POLICY AND LEGISLATION ESTABLISHED

A. *Nature of the resulting policy*

Is the resulting policy administratively based or legislatively based? Or both?

Has a framework ocean law been adopted or is it planned?

Does it incorporate constitutional or normative principles for ocean governance?

Describe the legal framework of the national ocean policy. If there is a law, does it integrate sectorial legislation?

B. *Implementation of principles: Detailed assessment*

What have been the major results of the programme?

Using a set of evaluative criteria such as the following, how well did the programme achieve national ocean policy objectives?

- Integration
    - Integration among sectors
    - Environment–development integration
    - Integration among different levels of government
    - Spatial integration (land–sea integration)
    - Science–management integration
    - How has it been implemented?

- Precautionary principle/approach
    - Fisheries
    - Aquaculture
    - Land-based and marine pollution
    - Marine conservation
    - How has it been implemented?

- Ecosystem-based management
    - Transboundary agreements and arrangements for managing shared marine species and ecosystems

- ○ Nature and adequacy of national ocean science capabilities
- ○ Consideration of ecosystem impacts/relationships in fisheries management and environmental impact assessment (EIA)/strategic environmental assessment (SEA) approvals
- ○ How has it been implemented?

- Public participation and community-based management

  - ○ Public involvement in marine resource management decision-making processes, e.g. EIA/SEA of coastal/marine projects, programmes, and plans; procedures for licensing of economic activities; provisions for alternative dispute resolution; policy development; and planning of coastal and marine uses
  - ○ Empowerment of the public through co-management or community-based management initiatives

C. Authority at national level

Who is in charge of implementing the policy? Is there a lead agency, interim body, or a unit in the prime minister's office that serves as a focal point for implementation?

How is the implementing structure organized? Is there an advisory group, or a secretariat or staff, in charge of the policy/programme?

D. National/subnational division of authority and interaction

What is the division of authority over ocean issues among national and subnational levels of government?

How is the authority over land and over water jurisdictions divided?

Are there general conflicts and competition among levels of government? Are there existing policies that address intergovernmental conflicts? What processes are present in addressing these conflicts? Are these processes working well?

E. Domestic implementation of international agreements

1   1982 UN Convention on the Law of the Sea (UNCLOS)

- ○ Has the country established zones of offshore jurisdiction in accord with the Convention, i.e. a 12 nautical mile territorial sea, a 24 nautical mile contiguous zone, a 200 nautical mile exclusive economic zone (EEZ), and an extended continental margin?
- ○ Are there any major divergences from UNCLOS provisions, e.g. requiring notice and authorization for foreign vessels carrying hazardous cargoes before entering the territorial sea or EEZ?

2   1993 Convention on Biological Diversity (CBD)

- ○ Has the country established a network of marine protected areas (MPAs)?
- ○ Has the country enacted legislation to protect endangered marine species?

3   1973 International Convention for the Prevention of Pollution from Ships, as amended by the 1978 Protocol (MARPOL)

    ○   Has the country implemented the mandatory annexes for oil pollution from ships (Annex 1) and noxious liquid substances carried in bulk (Annex 2)?

    ○   How is the country addressing pollution in the optional annexes, particularly sewage and garbage from ships?

4   1972 Convention on the Prevention of Marine Pollution by Dumping Wastes and Other Matter, as amended by the 1996 Protocol

    ○   Does the country regulate ocean dumping through a permitting system?

    ○   Has the country adopted the 1996 Protocol, which establishes a 'safe list' of wastes that may be dumped at sea?

5   Global Programme of Action for the Protection of the Marine Environment from Land-based Activities (GPA)

    ○   Has the country developed a national action plan to address land-based pollution and activities?

    ○   Has the country identified priorities for action in controlling land-based activities?

F. Enforcement

What organizational structures, methods, and resources does the country have to enforce provisions of national ocean policy?

Does the country give out violations and fines?

Does the country compile a report on compliance?

G. Research and education

Has the country conducted scientific research and established technological development programmes?

Does the country have an ocean chapter in the national state of the environment report?

Does the country have educational, training, and awareness programmes (building an 'ocean ethic')?

H. Financing

How is the national ocean programme financed/what are its sources of funds?

Does the country have adequate funds to administer the national ocean programme?

Does the country have funds available to develop related operational projects?

4. IMPLEMENTATION, EVALUATION, AND LONG-TERM OUTLOOK

A. *Review of problems, issues, or obstacles encountered by the policy/programme*

What were the major problems/obstacles encountered in the implementation of the national ocean policy? Is the policy able to address these adequately? If not, do these problems continue to affect the management and conservation of oceans, coasts, and islands, and how?

B. *Monitoring, evaluation, and adjustment*

What systems of monitoring and assessment/evaluation are used?

What feedback process, learning, and adjustments are being carried out by the programme? How are lessons identified and applied in programme implementation and how are factors leading to successful/unsuccessful interventions recognized? How well are the goals of the national ocean policy attained? Specifically, how well are economic security, maritime security, and social stability attained through implementation of the policy?

What indicators are used to measure success of the national ocean policy initiative? How were they developed?

C. *Outlook*

What are the short- and long-term outlooks for the national ocean programme?

Is this programme sufficiently rooted that it will be sustained for the next ten years?

5. RECOMMENDATIONS FOR IMPROVING THE NATIONAL OCEAN POLICY/PROGRAMME

What are some ideas for improving the national ocean programme in the country, e.g. promoting national-level action and promoting subnational-level action (provincial, local)?

Appendix B

A GUIDE TO COMPARATIVE CASE STUDIES ON REGIONAL OCEAN POLICIES

1. INTRODUCTION: BASIC INFORMATION AND OVERVIEW OF REGIONAL OCEAN POLICY

A. Basic information

- Geographic and demographic information, e.g. total land area, length of coastline, maritime zones, and human population
- Maritime boundaries, including a map
- Status of and utilization of marine and maritime resources, including economic data and conflicts among uses
  - ○ Why the oceans and coasts are an important policy area
  - ○ Extent to which the ocean policy is part of the region's sustainable development strategy

B. Brief overview of nature of and evolution of regional ocean policy

- Statement of the nature of an existing ocean policy or the status of the development of a new policy
- Factors that gave rise to a regional ocean policy initiative, e.g. multiple-use conflicts; environmental problems; increasing public demand for nature conservation and sustainable exploitation of natural resources; the need to expand coastal management efforts further offshore; opportunities for economic development; results from large-scale research programmes, such as large marine ecosystems (LMEs), Global International Waters Assessment (GIWA), and the Global Ocean Observing System (GOOS)
- Stakeholders/key players involved, i.e. which groups were in favour and which against
- Conflicts in the development of a regional ocean policy, i.e. what obstacles were faced in regional ocean policy formulation and implementation

2. POLICY DEVELOPMENT PROCESS

Description of how the policy development started, i.e. of the approach followed in establishing a regional ocean policy, including information on objectives, the principles that guided the process, and the steps followed during the preparatory/formulation phase

A. Initiation of the policy

How was the policy initiated and by whom, e.g. by a regional organization, by a group of countries?

B. Objectives

What objectives are addressed in the policy?

C. Major principles

What major principles were adopted/followed in the regional ocean policy?

D. Institutional arrangements

What institutional arrangements and processes were followed to formulate the regional ocean policy? How well did they work? (There might not be any lead agency; rather, there may be a networked approach.)

3. NATURE OF THE REGIONAL INITIATIVE ESTABLISHED

A. Nature of the resulting initiative (policy, agreement, or programme)

Measures to be implemented by the policy, agreement, or programme

B. Implementation of the initiative (the regional initiative may still be in the formulation phase)

What have been the major results of the programme?

Using a set of evaluative criteria, how well did the programme achieve regional ocean policy objectives?

C. Enforcement

Organizational structure, methods, and resources to enforce provisions of regional ocean policy

Summary of violations and fines

Report on compliance

D. Research and education

Does the regional area conduct scientific research and technological development programmes?

Does the region have an 'ocean' chapter in the regional state of the environment report?

Does the region have educational, training, and awareness programmes (building an 'ocean ethic')?

E. Financing

How is the regional ocean programme financed/what are its sources of funds?

Does the region have adequate funds to administer regional ocean programme?

Are there funds available to develop related operational projects?

4. IMPLEMENTATION, EVALUATION, AND LONG-TERM OUTLOOK

A. Review of problems, issues, or obstacles addressed by the policy/programme

What were the major problems/obstacles addressed by the regional ocean policy? Is the policy able to address these adequately? If not, do these problems continue to affect the management and conservation of oceans, coasts, and islands, and how?

B. Monitoring, evaluation, and adjustment (the regional initiative may still be in the formulation phase)

What systems of monitoring and assessment/evaluation are used?

What feedback process, learning, and adjustments are being carried out by the programme? How are lessons identified and applied in programme implementation and how are factors leading to successful/unsuccessful interventions recognized?

How are goals of the regional ocean policy attained? Specifically, how are economic security, maritime security, and social stability attained through implementation of the policy?

What indicators are used to measure success of the regional ocean policy initiative? How were they developed?

C. Outlook

What is the short- and long-term outlooks for the regional ocean programme?

Is this programme sufficiently rooted that it will be sustained for the next ten years?

5. RECOMMENDATIONS FOR IMPROVING THE REGIONAL OCEAN POLICY/PROGRAMME

Offer some ideas for improving the regional ocean programme

Appendix C
SYNOPSES OF NATIONAL AND REGIONAL OCEAN POLICY CASE STUDIES

This appendix briefly summarizes each of the fifteen national ocean policies and four regional ocean policies covered in the study (in alphabetical order), providing a snapshot of the nature of the ocean policy, as well as its current status, recent developments, and highlights.

National ocean policies

Australia

Australia's Oceans Policy was developed by the Department of the Environment and Heritage and launched in 1998 during the International Year of the Ocean, with the goal of coordinating marine activities in Australia to create an effective and efficient oceans management regime. Australia's Oceans Policy provided an umbrella framework for the integrated and ecosystem-based planning and management of Australia's marine jurisdiction. It is an administratively based policy with no supporting legislation. The Oceans Policy lists nine broad goals, which were developed to meet the general expectations of the Australian community and to provide the scope to address the range of issues facing the marine environment. Australia's Oceans Policy incorporates eight major planning principles: ecosystem-based management and integration; adaptive management; stewardship and transparency; indigenous rights; the precautionary approach; best available science; ecological sustainable development; and multiple-use management.

In addition to the federal Australian government, the management of Australia's marine jurisdiction is shared between seven state and territory governments, which were not signatories to Australia's Oceans Policy. Therefore any internal cross jurisdictional arrangements had to be negotiated in the political arena. The policy has to take into consideration all of the existing legislative instruments that govern Australia's oceans. The piece of legislation that is most relevant to the Oceans Policy is the Environment Protection and Biodiversity Conservation Act 1999 (EPBC Act). Marine bioregional planning – the new planning model designed in 2004 for the ongoing implementation of the Oceans Policy – sits within the framework of section 176 of the EPBC Act.

A number of key administrative bodies have been involved in the development and implementation of the Oceans Policy, including: the Marine and Biodiversity Division of the

Department of the Environment and Water Resources, which was responsible for coordinating the implementation of Australia's Oceans Policy; the Oceans Board of Management, which comprises the heads of those federal departments with an interest in marine matters; and the National Oceans Advisory Group, which is made up of representatives from key marine sectors and interest groups. With the 2004 government reorganization, some of these bodies were abolished, for example the National Oceans Ministerial Board.

The primary tool for implementing the Oceans Policy was originally intended to be regional marine planning, but this has evolved into bioregional plans in an attempt to focus on key sustainability objectives. The bioregional plans set objectives, strategies, and actions for biodiversity conservation in the region. Australia completed the bioregional planning process in 2012, leading to a bioregional plan for each of the five bioregions, and implementation is under way. A national network of marine protected areas (MPAs) has also been established, overseen by the Director of National Parks, and with the associated overarching marine policy assigned to a newly created Wildlife, Heritage and Marine Division.

Although the Oceans Policy has not been rescinded or superseded, its future appears uncertain. In terms of the institutional arrangements put in place in its implementation, only the Oceans Policy Science Advisory Group (OPSAG) still exists, which continues to coordinate the sharing of scientific information between government bodies and the broader marine science community in Australia. Under the new government, there is no standing intergovernmental body in charge of natural resource management issues generally or oceans management issues specifically. Currently, federal, state, and territorial ministers meet as needed, in recognition of the fact that the states are sovereign and capable of delivering their responsibilities, with appropriate accountability.

Brazil

Brazil's National Policy for Sea Resources (PNRM) was instituted in 1980 by means of a presidential directive. It aims to connect the various marine and coastal sectorial policies and plans, and provides initiatives and programmes for implementation under the National Coastal Management Plan (PNGC). Even though it was created during the Brazilian military regime, the PNRM has been adopted by subsequent regimes, providing a strong governance instrument for the coastal area. The PNRM is reviewed biannually by the Interministerial Commission for Sea Resources (CIRM)

In 2005, a Presidential Decree revised parts of the PNRM, which resulted in the consideration of several new issues, including a focus on marine biodiversity and exploration of non-living marine resources. The PNRM is executed in a decentralized manner, and involves the participation of states, cities, and civil society organizations. The implementation of PNRM's national ocean and coastal management programmes is coordinated, in most cases, by the CIRM and the Environment Ministry. The programmes of implementation are regulated and adopted by means of successive Sectorial Plans for Sea Resources (PSRMs), which are based on the precautionary principle and the integrated management of coastal and marine environments. The PNRM also incorporates the concepts of sustainability and recognizes the need for non-centralized and participative planning processes to facilitate integration and to ensure the achievement of its proposals. Examples of PSRMs include the National Plan of Coastal Management and the Evaluation Programme of the Sustainable Potential of Living Resources in the Exclusive Economic Zone.

Sectorial plans developed under the PNRM involve a variety of institutions predominantly focused on environmental research and management, including the federal and state agencies for environmental control, universities and research institutes, and other civil society groups. Recommendations to improve the oceans policy of Brazil include organizational modifications, effective implementation processes, avenues for community involvement, and the establishment of government priorities.

In 2013, a federal meeting entitled 'Oceans and Society' was held to evaluate and celebrate twenty-five years of implementation of the PNGC in Brazil; a new National Policy for the Conservation and Sustainable Use of the Brazilian Marine Biome was also proposed in Congress, and the CIRM established a working group to address shared use of the ocean and the application of marine spatial planning.

Canada

Canada was the first nation to develop a national oceans policy by initiating the Oceans Act, which entered into force in 1997 and laid the foundation for a comprehensive oceans policy in the country. The Oceans Act was developed and renewed through formal federal, provincial, territorial, aboriginal, and public consultation processes, which led to Canada's 2002 Oceans Strategy, the country's overarching oceans policy framework for modern oceans management, and serves as guidance for the development and updating of sector-based policies and processes, developed in collaboration with federal agencies, provincial and territorial governments, affected aboriginal organizations, coastal communities, and other persons and bodies, including those bodies established under land claims agreements.

The Act assigns to the Department of Fisheries and Oceans (DFO) the responsibility for coordinating existing policy and statutory instruments, and for formulating a national vision and guiding principles for oceans management, within which existing and emerging policies and laws would be interpreted and implemented. The Act defines the guiding principles of integrated management, sustainable development, and the precautionary approach, provides the mandate to develop programmes to implement these principles, and situates the DFO's existing regulatory and management authorities within the context of oceans management. The Act also recognizes other mandated authorities and provides guidance on how their mandates should be delivered within the marine environment.

During the five years immediately following the proclamation of the Oceans Act, the ocean management programmes called for in the statute were piloted in the field to better define the policy guidance required; these informed the development of the federal Oceans Action Plan (2005–07). The Plan outlined and provided for the funding of priority areas for action under four major themes: international leadership, sovereignty, and security; integrated oceans management for sustainable development; health of the oceans; and science and technology. In 2007, the government further committed five years of funding to specific elements of the broad oceans agenda. In 2014, a new National Conservation Plan was announced, which commits CAN$252 million over five years to conservation initiatives, including CAN$37 million to marine and coastal conservation, to support progress toward the 2020 target of protecting 10 per cent of marine and coastal areas under the Convention on Biological Diversity, among other related goals.

The Oceans Act requires the Minister of Fisheries and Oceans to lead the development and implementation of integrated management plans for Canada's coastal and marine waters, and that mandate was initially met largely through six large ocean management area (LOMA) planning efforts. The Canadian integrated management process has focused on the identification of social, cultural, and economic objectives or desirable targets that subnational and local governments, stakeholders, and the public wish to achieve in the planning area. There has been widespread community involvement in the ocean and coastal management process in Canada.

Identified parties include the federal government, provincial/territorial/local authorities, aboriginal organizations and communities, industry, non-governmental organizations (NGOs), community groups, and the academic and research community. The five LOMA plans that have been developed stand out for their general and aspirational character and not as examples of effective marine spatial planning.

The DFO is now moving toward a bioregional planning approach for twelve ocean bioregions and the Great Lakes, but it remains unclear how planning will proceed. No legislated planning procedures exist, nor has clear administrative guidance for bioregional planning been issued.

The Oceans Act also calls for the development and implementation of a national system of marine protected areas (MPAs). To achieve this mandate, a Federal Marine Protected Areas Strategy was developed, along with efforts to broaden the national network to include provincial and territorial MPAs, resulting in the National Framework for Canada's Network of Marine Protected Areas. Canada, however, is far from completing the establishment of an MPA network, mainly as a result of the complexity of the process, which involves multiple steps requiring, to a greater or lesser extent, the involvement of multiple government authorities and meaningful consultation with affected parties.

India

India's early initiatives in ocean governance were focused on exercising jurisdiction over its exclusive economic zone (EEZ) – nearly 2.02 million km^2 – as a result of the adoption of the United Nations Convention on the Law of the Sea (UNCLOS) in 1982. Created in 1981 under the Office of the Prime Minister, India's Department of Ocean Development (DOD) organized a series of workshops on various scientific themes. Deliberations on these themes underpinned the development of the Ocean Policy Statement, which was discussed and adopted by the Indian Parliament in 1982. The Statement basically charted the future course for the development of ocean science, which the government considered essential to ocean governance in India.

India's oceans policy does not yet constitute a comprehensive ocean governance framework. India has been able to achieve considerable success in developing the necessary science and technology, but the concept of ocean governance and the scenarios in which such a policy may be effectively promoted have evolved considerably over time. While the need to address the interlinked issues of environmental protection, allocation of fishing rights, protection of marine species and the conservation of marine biodiversity, marine pollution, and offshore oil and gas exploration regulation has been recognized, it is difficult to accommodate cross-sectoral approaches to marine resource management under the present administrative framework for ocean governance in India.

The need to protect India's long coastline found concrete expression in the early 1980s during the tenure of Prime Minister Indira Gandhi, when she directed all coastal states to ensure that the coastal zone up to 500 m from the high tide line (HTL) be kept free from development activities of all kinds. However, owing to pressure from the tourism industry, the central government relaxed this ban in 1986. Later, in 1991, the government issued a notification under the Environment (Protection) Act, 1986, and its Rules to accord special protection to the coastal areas of the country. Under the 1991 Coastal Regulation Zone (CRZ) Notification, coastal stretches of sea, bays, estuaries, creeks, rivers, and backwaters influenced by tidal action up to 500 m from the HTL were declared to fall under the CRZ.

In 2004, in an effort to bring coherence to the coastal law, the government appointed a committee to undertake a comprehensive review of the CRZ Notification. In 2008, based on the committee's recommendations, the government introduced the Coastal Management Zone Notification, with the objective of protecting and ensuring the sustainable development of the coastal stretches and marine environment through sustainable coastal zone management practices based on sound scientific principles. Recently, in a significant move to revamp the coastal law, the central government has released the CRZ Notification of 2011, which replaces and overhauls the 1991 Notification. The primary objectives of the 2011 Notification are to provide livelihood security to local coastal communities, to promote conservation and protection of coastal stretches, their unique environments, and marine areas, and to promote sustainability based on scientific principles, taking into account the dangers posed by natural hazards and sea-level rise. The 2011 Notification classifies the coastal zone into five categories, which are to be managed within the framework of coastal zone management plans. It also provides for critically vulnerable areas to be managed on the basis of an integrated management plan. Apart from this spatial integration, there is limited application of the concept of integration in this law. And even though elements of the precautionary principle can be discerned from some of its provisions, it can be (and is, in this volume) argued that the pro-development inclination of the law defeats the underlying objectives of the precautionary principle.

Jamaica

Jamaica recognized the need for the effective management of coastal resources quite early in the twentieth century. Coastal management legislation, in the form of the Beach Control Act of 1956, addressed the need for marine protected areas (MPAs), as well as the preservation of coral reefs and mangroves. In 1976, following the first United Nations Conference on the Human Environment (Stockholm, 1972), Jamaica attempted to address the need for an umbrella environmental management agency. The Natural Resources Conservation Department (NRCD) was formed to protect general environmental quality, and was organized and given new and additional technical capabilities in the areas of aquatic resource and wildlife management. However, prior to the ratification of UNCLOS in 1983, there was no policy and integrated approach to the management of coastal and ocean resources. The NRCD graduated to become a statutory body, the Natural Resources Conservation Authority (NRCA), which was tasked with the management of ocean and coastal areas, and the implementation of an integrated coastal zone management (ICZM) programme. In 1997, the NRCA, with the assistance of the Swedish International Development Cooperation Agency (Sida), began a series of national consultations for the development of a policy on ICZM. Guidelines for activities in the coastal area and an emphasis on policy and legislation characterized the efforts of the NRCA.

In 1996, the Jamaica Cabinet endorsed and established an Interagency Advisory Committee, which later evolved into the National Council for Ocean and Coastal Zone Management in 1998, to address the competing and complementary issues and players involved in Jamaica's coastal and oceans resources. That same year, the Council, with the support of the NRCA, led the policy formulation process that resulted in the endorsement of the 2002 Green Paper on Coastal and Oceans Management. Jamaica's Ocean and Coastal Zone Management Policy provides for enhancement of institutional capacities, integrated planning and management, prevention of environmental degradation, and community-based participatory approaches, among other things. The policy is guided by the precautionary principle, the 'polluter pays' principle, intergenerational equity, and the global and regional division of environmental impacts. In order to implement the policy, specific recommendations for action for each identified sustainable development issue were determined and institutional responsibility was assigned to the relevant agencies.

The Council has no legal basis, hindering its ability to influence decision-making at critical crossroads in the implementation of ocean and coastal policy. To facilitate and promote strict adherence to the Policy, the Council formed the Cays Management Committee in 2008, charged with making recommendations for the development of a comprehensive policy and administrative arrangement for the management of Jamaica's islands, cays, and associated eco-systems. A Cays Management Policy was scheduled to be completed in 2013, which aims to protect their ecological integrity and to promote their sustainable use. The Ocean and Coastal Zone Management Policy is currently under review, to evaluate the state of implementation and to make recommendations for more effective means of decision-making.

Japan

In the follow-up to the 1992 United Nations Conference on Environment and Development (UNCED), Japan began the slow process of designing a national ocean policy with a multi-purpose and comprehensive approach. Following parallel initiatives by government and private bodies, the *Diet*, Japan's parliament, passed the Basic Act on Ocean Policy on 20 April 2007, which entered into force on 20 July 2007. This ocean policy is based on preservation, utilization, and understanding of the ocean, along with the need to implement the policy strategically with an international perspective. The Act provides for the establishment of a Headquarters for Ocean Policy, with the Prime Minister as its head, and the Chief Cabinet Secretary and the Minister in Charge of Ocean Policy as deputies, as well as an Advisory Council, all of which were charged with the drafting and implementation of a Basic Ocean Plan and the coordination of activities covering a number of areas, including the exploitation and utilization of marine resources and the exclusive economic zone (EEZ), among others. The Cabinet approved the first Basic Plan on Ocean Policy in March 2008, after public comment, setting out in a comprehensive and systematic manner the basic directions of Japan's ocean policy and providing the policy measures that the government should pursue.

During the initial half-decade of implementation that followed, the Basic Act and Plan on Ocean Policy produced:

- more detailed plans for the implementation of the Basic Plan in several areas, including the development of marine energy and mineral resources, and basic guidelines for the conservation and management of remote islands, exploration for mineral resources, conduct of marine scientific research in the EEZ and the continental shelf, establishment of marine protected areas, and promotion of marine renewable energy development;
- several new laws under the Basic Act and Plan, such as an Anti-Piracy Act, a revised Mining Act, new laws for the protection of coastal areas, amendment to the Maritime Transportation Act and the Act on Seafarers, as well as an Act on the passage of foreign-flagged ships through the territorial sea and internal waters; and
- several other measures that require actions by several ministries and agencies, as well as close coordination among them.

In April 2013, the Cabinet approved a new Basic Plan on Ocean Policy, which covers another period of roughly five years. Part 2 of the new Basic Plan sets out the measures necessary for achieving the objectives under each of the twelve basic policy measures defined by the Basic Act on Ocean Policy in a specific manner. Part 3 further stipulates those measures that are required for pursuing policy measures in a comprehensive and systematic manner, such as strengthening of the functions of the Advisory Council and of the Secretariat of Ocean Policy Headquarters in implementing policy plans in a more comprehensive and effective way, as well as clarifying the roles of local governments, communities, and civil society, and promoting active information-sharing with the public.

Mexico

Mexico developed an integrated ocean policy in 2006 in the form of the National Environmental Policy for the Sustainable Development of Oceans and Coasts (NEPSDOC) addressing the need for coordination among the federal agencies that had previously worked sectorally and in isolation. It contained the main strategies and guidelines that define the environmental policy with regard to the oceans and coasts of Mexico. In 2006, this public policy was implemented as a joint working programme within SEMARNAT (*Secretaría de Medio Ambiente y Recursos Naturales*). The NEPSDOC serves as the forum for discussing all federal policies related to the integrated management of oceans and coasts through the Agenda del Mar project under the President's Office. Its main objectives are: the development of a strategy for the integrated management of oceans and coasts; strengthening of coordinated actions among coastal and marine-related institutions; promoting social and economic welfare through environmental and biodiversity conservation, and the sustainable use of coastal and marine resources; strengthening the legal, normative, and administrative framework for the management of oceans and coasts; and the development of an information system dedicated to oceans and coastal issues. The NEPSDOC has been implemented following the principles of ecosystem-based management, multiple use, sustainability, participatory governance, the precautionary approach, adaptive management, and biodiversity protection.

Under the leadership of SEMARNAT's Environmental Policy, Sectoral and Regional Integration Division (DGPAIRS), the ocean and land use planning environmental policy instrument (Land and Sea Use Planning, or LSUP) has become the key to sustainable management that promotes green development in a cross-cutting fashion and the adaptation of policies to climate change. Within the territorial sea and the Mexican exclusive economic zone (EEZ), the marine LSUP has become a soft law agreement for the numerous actors and users of the ocean space, involving the three government levels, academia, NGOs, organized social groups, stakeholders, and the private sector. To date, two marine LSUPs have been decreed (Gulf of California, and Gulf of Mexico and Caribbean), a third is under public consultation (Northern Pacific), and the Southern Pacific Marine LSUP is under development.

The Temporal Environmental Employment Program (TEEP) was created to alleviate the economic suffering caused by natural disasters mainly for the fisheries sector. The TEEP has since evolved into a powerful incentive to promote numerous actions in favour of the coastal zone environment, including actions that lead to the development of coastal resilience to climate change. New environmental public policies that foster a better understanding of the ocean dynamics and complexity through intra-governmental coordination have also been adopted.

Some lingering issues remain, including differences between sectorial objectives and integrating criteria, the power relationships given the current legal framework of Mexico, and the resistance of some stakeholders because of a lack of coordination, among others.

New Zealand

The increase in human pressures on the marine environment of New Zealand, along with increasing difficulty in utilizing outdated frameworks, initiated the development of a National Ocean Policy (NOP) in 2000, when the New Zealand Cabinet agreed to its development as the basis for an overhaul of the existing system. This was to be a 'cross-government', or 'whole of government', exercise involving officials and ministers from all key government departments, working cooperatively. The Cabinet delegated authority to manage development of the policy to an ad hoc group of six cabinet ministers, which established a Ministerial Advisory Committee on Oceans Policy (MACOP), to assist ministers in their role of overseeing the development of the oceans policy and, in particular, to undertake a public consultation process. The policy to be developed by this Committee was intended to focus on the integrated management of oceans under the country's jurisdiction. New Zealand has a number of sectorial policies and laws, but more capacity, inter-agency cooperation, and national guidance are necessary for further progress toward a national ocean policy. Currently, more than fourteen government agencies are involved with various aspects of marine affairs. There is no consistent goal for the management of the marine environment beyond 12 nautical miles. There are also inconsistent statutory approaches to the reconciliation of competing interests, environmental protection, and public participation in decision-making processes. The existing piecemeal system has resulted in a lack of integration among legislation, policy, decision-making, and activities in the marine environment.

The NOP was (until 2005) in the process of being developed in three stages:

1   the definition of the goals, values, and principles to guide the process;
2   defining the strategic and operational policy framework for the policy, and
3   the implementation of the policy.

Stage 1 was completed in 2001; in 2003, Stage 2 was in the process of being implemented when the government decided to suspend the NOP process, pending resolution of some politically sensitive, overlapping issues, including ownership of the public foreshore and seabed, and public access to the coast. The NOP process was reactivated in November 2005, but no significant developments have occurred since the government announced the restart. The NOP Secretariat did not produce a 'draft' NOP document for public comment and the momentum that had built up during the initial consultation phase was lost. In 2008, a new national government was elected to office. The priority of this government is to improve the regulatory regime for the environmental impact of human activities within the exclusive economic zone (EEZ). It has not expressly indicated an interest in the NOP process.

Norway

The concept of integrated ocean management has been debated among Norwegian academia for more than three decades. The last decade was characterized by a drive towards greater coordination among the various marine policy sectors. The evolution of scientific knowledge, first-hand experience with problems related to pollution and over-fishing, the requirements of international agreements, and increasing conflicts among different uses of the oceans are important forces in this regard. A critical juncture was a report to the *Storting* (the Norwegian parliament) in 2002 that outlined principles for a more integrated and ecosystem-oriented marine policy. As a result, the process to develop an integrated management plan for the Norwegian part of the Barents Sea, and the sea areas to the south to the Lofoten Islands, by 2006 was initiated. In the same period, the government set in motion work to develop a more modern legislative framework for the oceans.

The new and comprehensive Oceans Resources Act (ORA) was adopted by the Storting in 2008 and implemented in 2009. The integrated management plans, applied to the Barents Sea region since 2006, the Norwegian Sea in 2009, and the North Sea in 2013, represent the first steps in the development of integrated oceans management in Norway. The ORA consolidates all relevant provisions for the management of living marine resources into a single piece of legislation, and lists a series of principles and concerns that are to be taken into consideration in the management of resources and genetic material. Among these are the precautionary approach, the ecosystem-based approach, the optimum utilization and allocation of resources, the effective control of harvesting, the implementation of international law, transparency in decision-making, and regard for the Saami culture.

In addition to the ORA, a new Nature Diversity Act was adopted in 2009, the general objective of which is to conserve nature, including the territorial waters of the oceans under Norwegian jurisdiction. The principle of sustainable use is central to the law, and it is to be carried out by means of the establishment of conservation objectives for nature types and species, and the enactment of certain environmental principles (the precautionary principle, the 'polluter pays' principle, and the principle of responsible technologies and practices).

Philippines

The primary motivation for establishing an oceans policy in the Philippines was the potential of the oceans to contribute to national economic development. The Philippines was one of the first countries in the Asia-Pacific to adopt a national marine policy (NMP) document in the form of the Sustainable Archipelagic Development Framework in 1994, which was further developed into the Archipelagic Development Policy, and expressed the guiding principles for management and development of the country's ocean space. More than a decade later, a new and broader Archipelagic Development Policy was proposed and circulated. However, the Philippines effort has been hampered by an array of administrative and political obstacles that have prevented the NMP from fully realizing its potential.

In June 2006, the Philippine government endorsed an integrated coastal management (ICM) policy with the establishment of a national ICM programme. The ICM policy identifies a number of elements and best practices as basic guidelines for both national- and local-level management bodies. Objectives include sustainable development, integrated management, raising national marine consciousness, a marine research programme, the 'polluter pays' principle, and the establishment of quality schools and training in maritime issues, as well as maritime security, a viable fisheries programme, and technological development, among others. The Department of Environment and Natural Resources is charged with developing the national ICM programme in consultation with other government agencies and relevant stakeholders. Upon completion, the programme will provide principles, strategies, development priorities, action plans, national targets, and a national ICM coordinating mechanism to assist local governments and stakeholders as they develop their ICM plans.

In March 2007, a Cabinet-level commission led by the Office of the President was organized to address maritime concerns. However, this commission was replaced in 2011 by an inter-agency council tasked less with addressing policy and more with operational concerns. Ocean management in the Philippines continues to be largely fragmented and uncoordinated at the national level. While there have been some improvements in the openness to new concepts and in governmental attitudes that have allowed for improvements in local-level governance and marine environmental protection, in general integrated ocean management across the different maritime agencies continues to be problematic. Despite the challenges, in the years since the promulgation of the NMP, several notable developments in the fields of coastal and marine management indicate that the Philippines still retains strong institutional building blocks for the sustainable development of the oceans.

Portugal

In 1998, the Portuguese government initiated a process for the formulation of a national strategy for the ocean, triggered by several major projects and ventures such as the hosting of the 1998 Lisbon World Exposition (EXPO'98), themed 'The Oceans: A Heritage for the Future', the adoption by the United Nations General Assembly (UNGA) of the proposal submitted by Portugal through the Intergovernmental Oceanographic Commission (IOC) of the United Nations Educational, Scientific and Cultural Organization (UNESCO) to declare 'the 1998 International Year of the Ocean', and the presentation by the Independent World Commission on the Oceans of the report *The Ocean, Our Future*. The aim of this process was to define effective mechanisms and institutional arrangements for the sustainable development and conservation of maritime areas under sovereignty/national jurisdiction, addressing multiple uses of the sea in a scientific and balanced way, and to stimulate investment in new activities to enhance competitiveness, innovation, and economic growth.

In response to changing ocean national governance requirements, institutional adjustments, new domestic mechanisms, and new approaches were required to reinforce national capacities. As a result, the Secretary of State for Ocean Affairs was established in 2004 under the authority of the Ministry of State, National Defence and Ocean Affairs, which was subsequently revamped and is presently functioning as Secretary of State of the Sea under the Ministry of Agriculture and Sea.

In this context, the Portuguese government adopted a National Ocean Strategy (2006–16) in 2006. Based on the need for a more coordinated action in ocean affairs, a major step was taken in 2007 when the Council of Ministers approved the creation of an Interministerial Commission for Ocean Affairs (CIAM). Among other functions, CIAM coordinates, follows up, and evaluates the implementation of the National Ocean Strategy (2006–16). With a view of facilitating effective inter-ministerial and cross-sectoral coordination in ocean affairs, the Prime Minister chairs CIAM. The Commission is now supported by the Directorate General for Maritime Policy (DGPM), operating under the Secretary of State of the Sea, which incorporates previous functions assigned to other supporting bodies.

Portugal also influenced the development of regional maritime policy in the European Union through a Joint Contribution to the Green Paper on EU Maritime Policy submitted by Portugal, Spain, and France to the European Commission in 2005. Under the Portuguese presidency of the European Union, the EU Integrated Maritime Policy was adopted in October 2007 by the European Commission and endorsed by the heads of state and government in the European Council in December 2007.

The National Ocean Strategy 2006–16 was updated in the form of the National Ocean Strategy (2013–20), taking into account, inter alia, the European Union's strategies, policies, and financial cycles. In 2014, the Portuguese government adopted the Framework Law for National Maritime Spatial Planning and Management to provide an adequate framework for the use of the national maritime areas, in order to ensure the objectives of sustainable development. Overall, the progressive evolution of the Portuguese institutional system to a more effective ocean governance framework reflects, to a certain degree, the 'experimental nature' of the policymaking/institution-making process, which put into place new domestic mechanisms and approaches with which to address current oceans issues.

Russian Federation

In 2001, the Russian Federation developed and authorized its Marine Doctrine for the period to 2020 with the purposes of protecting state sovereignty, sovereign rights, and high seas freedoms. In the Doctrine, Russia declares its national interests in the world ocean, and recognizes priorities and principles for a functional national marine policy. Previously, in 1998, Russia approved the Federal Target Programme (FTP) 'World Ocean', which aimed to change the existing narrow, sectorial approach to a more integrated, effective nationwide system of regulation and management of marine activities. Financing of the FTP came from federal budgetary, and non-budgetary allocations. The 'World Ocean' programme addressed a number of issues, including the potential use of marine resources, the development of transport and communications, the maintenance of sovereign rights and jurisdiction in Russia's coastal waters and international marine waters, control over the marine environment, and the means to address natural and human-induced emergencies, as well as other specific problems related to the provision of safe and sustainable livelihoods both in coastal regions and throughout Russia as a whole.

The coordinating functions of the 'World Ocean' programme were carried out by the Interagency Commission on the FTP 'World Ocean' and its Scientific-Expert Council, and subsequently (since 2005) by the Marine Board of the Government of the Russian Federation. The programme was implemented in three stages: 1998–2002, 2003–07, and 2008–13. As each stage was implemented, changes to the programme were made accordingly. Subsequently, the Strategy for the Development of Marine Activities of the Russian Federation by 2030, adopted in 2010, identified major challenges and provided perspectives for the development of marine activities in the country, outlined strategic goals and targets, and established milestones in the development of marine activities.

United Kingdom

In the United Kingdom, coordination of government agencies involved in ocean affairs has been a long-standing aim in some areas of interest since the 1970s. Agency coordination has been well established in a limited number of these areas, such as search and rescue (SAR) at regional and national levels, defence in general, and fisheries protection. However, inter-agency coordination has never been well developed among central government departments. On the other hand, significant marine sectorial policy development has taken place in the UK since the early 1990s, significantly influenced by EU law, especially in fisheries, waste disposal and pollution, conservation, and integration measures.

Consideration of integrated, comprehensive coastal and marine policy in the UK began in 2002 with the publication of the Marine Stewardship Report, a Marine Bill in the 2005–06 parliamentary sessions, and the White Paper in March 2007. Stewardship, sustainability, an ecosystem approach, best available science, integration, stakeholder involvement, the precautionary approach, and protection of biological diversity are some of the principles endorsed in the Marine Stewardship Report for inclusion in a national ocean policy.

The Marine and Coastal Access Act 2009, the Marine (Scotland) Act 2010, and the Marine (Northern Ireland) Act of 2013 comprise the basis for legislative policy and new tools for marine and coastal conservation and management in the UK, providing a means for improved management and protection of marine species and habitats. Funding comes from the Consolidated Fund of HM Treasury. Marine conservation is one of the principal issues that have been addressed under the Acts, with the aim of expanding the current network of marine protected areas (MPAs) in UK waters. The Acts include provisions for the management of fisheries within the territorial sea, marine spatial planning, licensing of marine activities, and improving marine conservation by means of measures such as the development of marine conservation zones (MCZs) in England and MPAs in Scotland, as well as the establishment of the Marine Management Organization (MMO) for English waters and Marine Scotland for Scottish waters.

An Inter-departmental Group on Coastal Policy, chaired by the Department for Environment, Food and Rural Affairs (Defra), and including representatives from each central government department and the devolved administrations, exists to exchange information on government policy related to coastal matters. The practical implementation of marine policy is primarily administratively based. A large number of government departments, agencies, and other bodies are involved in an array of terrestrial and marine sectors, and are responsible for the management and regulation of many different activities related to marine policy. In 2011, the UK Parliament, the Scottish Government, the Welsh Assembly Government, and the Northern Ireland Executive adopted the UK Marine Policy Statement, which aimed to provide a framework that is more cohesive than the sectorial system and which has led to the creation of regional marine plans.

United States

Over the past twenty years, NGOs and academic groups had articulated the need to go beyond the sector-by-sector approach to ocean policy in the United States to address multiple-use conflicts, to preserve ecosystems, and to take advantage of new economic opportunities in the ocean. The call for comprehensive reform was given serious attention in the year 2000, when two ocean commissions were created to examine all aspects of US ocean policy. The first, established in 2000, was a privately funded commission, the Pew Oceans Commission, which examined, in particular, environmental issues related to ocean degradation and resource decline, and their underlying causes. Around the same time – and with a heightened awareness of the need to reform US ocean policy – the US Congress enacted the Oceans Act of 2000, creating the US Commission on Ocean Policy, tasked with conducting research and giving recommendations to Congress and the President for a coordinated and comprehensive national ocean policy. The US Commission on Ocean Policy released its report, *An Ocean Blueprint for the 21st Century*, on 20 September 2004. A Presidential Executive Order No. 13366 of 17 December 2004 subsequently established a Cabinet-level Committee on Ocean Policy under the Council on Environmental Quality, to provide advice on the establishment or implementation of policies concerning ocean-related matters.

In 2009, under the first Obama Administration, a presidential memorandum established an Interagency Ocean Policy Task Force. Based on the recommendations from the Task Force, President Obama signed Executive Order No. 13547 in 2010 to establish the National Policy for the Stewardship of the Ocean, Our Coasts, and the Great Lakes (the National Ocean Policy), aimed at ensuring the protection, maintenance, and restoration of the health of ocean, coastal, and Great Lakes ecosystems and resources. The Executive Order required the formation of a Cabinet-level National Ocean Council (NOC), representing an interagency assembly led by the Council on Environmental Quality (CEQ) and the White House Office of Science and Technology Policy (OSTP) to oversee implementation of the policy. The NOC will develop strategic action plans for each of the priority objectives of the National Ocean Policy to help to coordinate actions; existing legal authorities will carry out the implementation.

The nine priorities of the National Ocean Policy promote a healthy and productive ocean zone and cover a wide array of issues. Ecosystem-based management and coastal and marine spatial planning (CMSP) are both foundational tools included among the priorities. Environmental priorities include the reduction of harmful land-based impacts on the ocean, addressing the changing conditions of the Arctic, and climate change and ocean acidification adaptation. Strengthening the observation, mapping, and infrastructure of the oceans, coasts, and Great Lakes for domestic and international observation efforts is also a priority, because the implementation of the policy is based on best available science. Finally, in order to implement the policy effectively, support and improvement of inter-jurisdictional coordination are prioritized, along with the utilization of regional plans in nine regions of the United States, to achieve ecosystem protection and restoration, sustainable ocean use, and public education.

Marine planning is a high priority of the National Ocean Policy, and is undergoing a phased implementation. The NOC is currently organizing itself and delegating responsibilities regarding this tool, while strategic action plans are under development to further guide spatial planning efforts. Creation of regional planning bodies in nine US regions to create marine spatial plans is under way, albeit at a slower pace than anticipated.

Vietnam

Although Vietnam considers itself a maritime nation, the formulation of its marine policy started only in the late 1990s. The primary concerns of the government of Vietnam have been those of national security, national sovereignty, and rights over maritime areas, rather than issues of environmental protection, resource conservation, and sustainable development in general. During the 1990s, there was significant marine development, but without any coordination or integration. It was only recently that the government became aware of the importance of coordinating and integrating all maritime and coastal activities into a national marine policy. At the same time, there had been increased understanding of the importance of sustainable development, scientific research, environmental protection, biodiversity conservation, marine protected area (MPA) management, and the conservation and management of marine habitats and coastal wetlands.

Although Vietnam has not yet developed an integrated national maritime policy, it does have a series of laws on sea-related activities in various sectors that focus primarily on economic development and national security. In an effort to develop a comprehensive policy based on scientific knowledge, an overall integrated maritime policy, the Strategy on Vietnam Seas towards 2020 was formulated and approved by the Central Committee of the Communist Party of Vietnam (CPV) in 2007. The preparatory process involved a number of government ministries, marine sectors, and national consultants, and has already generated more than sixty reports on various themes related to the marine environment. The objectives of the 2020 Strategy include economic growth, intersectorial coordination, ecosystem preservation, integrated approaches, sustainable development, addressing social concerns and poverty reduction in local communities, and the linking of economic development with security and national defence.

Regional ocean policies

East Asia

The Seas of East Asia are bordered by twelve coastal nations, with China, Japan, South Korea and North Korea on the north, and the Southeast Asian nations (Cambodia, Philippines, Singapore, Indonesia, Timor-Leste, Japan, Vietnam, Laos) bordering the east, west, and south of these 'semi-enclosed seas'. The Seas of East Asia cover six large marine ecosystems – the South China Sea, the Gulf of Thailand, the East China Sea, the Yellow Sea, the Sulu-Celebes Sea, and the Indonesian Seas, including their associated river and watershed systems.

In 2003, the ministers of the countries of East Asia adopted the Sustainable Development Strategy for the Seas of East Asia (SDS-SEA), a regional marine strategy for the sustainable development of oceans and coasts. The development of the Strategy was largely based on relevant international conventions and other international and regional instruments, as well as the lessons learned from the coastal and ocean governance experiences in the region – particularly from a decade of efforts and activities undertaken by the Partnership in Environmental Management for the Seas of East Asia (PEMSEA), with the funding support of the Global Environment Facility, national governments in the region, and other national and international partners. Included in the Strategy were the development of a policy framework for building partnerships, a regional marine environment resource facility, and a regional mechanism to support existing efforts. Key components of the Strategy were the identification and incorporation of international instruments and conventions that pertain to the region, and identification of the need for a more integrated approach toward these instruments, with sub-strategies to sustain, preserve, protect, develop, implement, and communicate shared goals.

The development of the SDS-SEA involved consensus-building and consultation, including the formulation of the Shared Vision of the concerned stakeholders and their mission for achieving sustainable coastal and ocean development. This was followed by the identification of the values that the people of East Asia attach to the seas and the threats to them. The SDS-SEA was a product of more than three years of extensive national and regional consultations, not only among the countries of the region, but also with other stakeholders. From 2000 to 2003, at each of PEMSEA's intergovernmental meetings, the countries progressively approved the development of the SDS-SEA from concept paper to the final draft for adoption.

The SDS-SEA is being implemented by means of an evolutionary and incremental process. In considering the implementing mechanism, the participating countries went through a series of debates during the PEMSEA intergovernmental meetings in 2004 and ultimately agreed to enter not into a regional convention, but into a non-binding agreement instead. In 2006, during a ministerial meeting in Haikou, China, the countries involved in the SDS-SEA entered into a non-binding agreement that allows for widespread stakeholder participation and the step-wise development of the Strategy's goals. Each country has its own time period for implementation of the SDS-SEA goals based on national capacity and areas of concern; most participating countries agreed to develop and implement their ten-year national plans of action for the implementation of the SDS-SEA. PEMSEA has completed the process of finalizing a regional institutional arrangement for implementing the SDS-SEA. In 2009, eight countries had signed an agreement recognizing the international legal personality of PEMSEA with headquarters in the Philippines, making PEMSEA the international organization focusing on the implementation of the SDS-SEA. Owing to variation of capacity among the participating countries, the Global Environment Facility (GEF) continues to assist countries in the implementation of SDS-SEA. Additional assistance comes from the Food and Agriculture Organization of the United Nations (FAO), the Intergovernmental Oceanographic Commission (IOC), the International Atomic Energy Agency (IAEA), the International Maritime Organization (IMO), and the UN Global Programme of Action for the Protection of the Marine Environment from Land-based Activities (UNEP/GPA).

In July 2012, the country partners, during the Ministerial Forum held in the Republic of Korea, recognized that while progress has been made towards the Haikou targets, efforts needed to be redoubled. This would be achieved with a new SDS-SEA Implementation Plan (2012–16), adopted under the Changwon Declaration, which advocates for an ocean-based blue economy through increased support for implementation of SDS-SEA.

European Union

The European Union was born essentially as a terrestrial and continental project. It started to manage its maritime assets only in the 1980s and in a piecemeal way, either as subdivisions of sectorial land-focused administrations or in reaction to external crises. By the mid-2000s, the marginality of maritime affairs in EU politics became increasingly paradoxical – all the more so because Europe is geographically and economically a maritime continent. The surface of marine waters under the jurisdiction of the EU Member States is the world's most important, being far larger than the total land area of the Union. Maritime sectors are also key elements of the EU economy, with a gross added value of €500 billion per year generated by marine-based industries and services.

In June 2006, a Green Paper on a Future Maritime Policy for the Union highlighted the links between EU maritime-related activities and raised questions for public consultation. The stakeholder consultation was an unexpected success, with almost 500 written contributions and more than 230 events, and formed the basis for the creation of the Integrated Maritime Policy for the European Union (IMP) in October 2007. In March 2008, the Marine Strategy Framework Directive (MSFD) entered into force. It is aimed at achieving good environmental status of EU waters and protecting the resource base upon which marine-related economic and social activities depend. The adoption of the IMP and MSFD in 2007 and 2008 sent the political signal that Europe's ability to achieve the full economic potential of its seas should be backed by increasing protection of the marine environment.

In 2012, the Blue Growth Strategy was launched to give the ocean an essential role in the European economy. The Strategy managed to place the blue economy on the agenda of EU Member States, regions, enterprise, and civil society. Within two years, it delivered strategic guidelines for the sustainable development of EU aquaculture, legislation for maritime spatial planning, and the adoption of initiatives for ocean energy, sustainable coastal tourism, and maritime innovation. The Blue Growth Strategy launched a second phase, during which the focus shifted towards delivering concrete projects, launching new economic policy directions and ensuring strong visibility. By moving from policy elaboration towards policy implementation, the IMP achieved its full institutionalization.

The IMP is underpinned by three main cross-cutting tools: the building of marine knowledge, integrated surveillance, and maritime spatial planning (MSP). Since 2008, the Commission has launched initiatives to rationalize and integrate marine and surveillance data among sectors and countries. The potential savings for the European Union and its Member States are significant if one considers that the same data may currently be collected several times. More importantly, better access to maritime surveillance data improves situational awareness at sea. In 2014, the European Union adopted legislation on MSP. It requires each Member State to develop MSP in its exclusive economic zone (EEZ) and to cooperate with its neighbours to coordinate that planning. The MSP Directive will bring spatial coherence to maritime activities and support to the implementation of EU legislation, especially for environmental policy and renewable energy policy.

Although building a EU maritime policy that integrates decades-old administrative structures has proved to be a challenge that needs time and constant high-level political support, the IMP has already managed to build up common responsibilities and governance mechanisms for Europe's sea basins, and has strongly increased the political visibility of maritime issues at the European regional level.

Pacific Islands

The sea dominates the lives of Pacific Island peoples. The monetary and subsistence economies of the Pacific Island countries and territories (PICTs) are largely based on agriculture, fisheries, and tourism. Both oceanic fisheries and coastal tourism contribute significantly to the foreign-exchange earnings for many countries. A majority of Pacific Islanders rely solely, or largely, on coastal fisheries for their subsistence needs. At their 1999 annual meeting, the leaders of the Pacific Islands Forum endorsed a list of recommendations emerging from the Pacific Regional Follow-up Workshop on the implementation of UNCLOS. One of the key recommendations – that a regional ocean policy be produced – was adopted as a regional-level initiative by the leaders.

The Marine Sector Working Group of the Council of Regional Organizations in the Pacific (CROP MSWG) was tasked with developing a regional ocean policy. The policy was subsequently endorsed by the Thirty-third Pacific Islands Forum in 2002. The Pacific Islands Regional Ocean Policy (PIROP) was then launched at the World Summit on Sustainable Development (WSSD) in 2002 to serve as a framework for the region, along with an integrated strategic action (ISA) plan that provides initiatives to implement the policy. Although PIROP is not legally binding, it details the principles and actions that regional organizations should adopt to achieve sound ocean governance and that nations should consider in the development of national policies and sustainable development plans. The five key principles of PIROP are: improving understanding of the ocean; encouraging sustainable development; maintaining ocean health; using the ocean for peaceful uses; and promoting cooperation and partnerships. The adoption of the Pacific Plan in 2005 by all sixteen member countries of the Pacific Islands Forum brought renewed impetus towards cooperation in the Pacific region.

The traditional approach of many Pacific Island countries to marine and coastal issues has been sectorial, but this is changing. Many countries have developed multi-sectorial interest groups and forums, including participants from the private sector and NGOs, to assist them with integrated policy development. PIROP is viewed as a way in which to implement these policies on a regional basis. At the regional level, the policy is being implemented by the CROP MSWG, whereas at the national level CROP agencies assist governments with incorporating the PIROP framework into their national laws and regulations.

Pacific Island leaders endorsed the Framework for a Pacific Oceanscape at the Forty-first Pacific Islands Forum Meeting in Port Vila, Vanuatu, held on 4–5 August 2010.[1] The Pacific Oceanscape Framework is the region's implementation tool for PIROP and the Pacific Plan (in relation to ocean issues). The Framework seeks to address six strategic priority areas: jurisdictional rights and responsibilities; good ocean governance; sustainable development, management, and conservation; listening, learning, liaising, and leading; sustaining action; and facilitating adaptation to a rapidly changing environment. Since its endorsement, there have been a number of developments in ocean management in the Pacific that directly and indirectly advance the objectives and priority areas of the Pacific Oceanscape Framework. The position of Regional Ocean Commissioner was established in 2011 and the Secretary General of the Pacific Islands Forum Secretariat appointed to the role. Progress has been made in sustainable development, management, and conservation initiatives in Tokelau, Palau, Cook Islands, and New Caledonia. Several regional projects are being implemented that integrate traditional coastal resource management, building on marine spatial planning (MSP) mechanisms and using collaborative partnerships to manage resources in the high seas, including the Pacific–African, Caribbean and Pacific (ACP) Regional Legislative and Regulatory Framework for Deep Sea Minerals Exploration and Exploitation. The successful implementation of the Pacific Oceanscape Framework, however, will require further strengthening of commitments by PICTs to implement the priority actions, with support not only from international partners, but also at all levels of government within each of the PICTs.

Sub-Saharan Africa

Aquaculture, fisheries, coastal tourism, oil and gas development, and coastal mining are all important marine sectorial issues throughout Africa. However, no single, all-encompassing regional ocean or coastal policy exists, although several coastal nations have initiated the development of integrated coastal and ocean policies or plans. For example, South Africa has adopted the White Paper on National Environmental Management of the Ocean (2014), and the White Paper for Sustainable Coastal Development in South Africa (2000); Tanzania has adopted the National Integrated Coastal Environment Management Strategy (2002); Namibia has adopted the National Policy on Coastal Management for Namibia (2013).

Sub-regional agreements, include the Abidjan Convention for Cooperation in the Protection and Development of the Marine and Coastal Environment of the West and Central African Region and Related Protocols (1984) and the Convention for the Protection, Management and Development of the Marine and Coastal Environment of the Eastern African Region (1996), revised as 'The Amended Convention for the Protection, Management and Development of the Marine and Coastal Environment of the Western Indian Ocean' in 2010. The objectives of the Abidjan Convention are to protect the marine environment, coastal zones, and related internal waters that fall within the jurisdiction of the states of the West and Central African region. The Abidjan Convention also contains a Protocol concerning 'Cooperation in Combating Pollution in Cases of Emergency'. The objectives of the Nairobi Convention are very similar to those of the Abidjan Convention, and propose to protect and manage the marine environment and coastal areas of the Eastern African region. Formulation of these agreements began in the early 1980s under the auspices of the Regional Seas Programme of the United Nations Environment Programme (UNEP). Development and implementation of the two regional agreements has, however, proceeded at a slow pace. An integrated coastal zone management (ICZM) Protocol to the amended Nairobi Convention is currently being developed by the Conference of Parties of the Nairobi Convention, in partnership with the Indian Ocean Commission, to strengthen the legal framework for more effective management of marine and coastal ecosystems across sectors and national boundaries.

In 2001, the New Partnership for Africa's Development (NEPAD) was established as an integrated development plan to provide a vision and programme of action for the sustainable development of the African continent and the eradication of poverty. NEPAD has developed an environmental action plan to address priority environmental issues, with four strategic directives: capacity-building for environmental management; securing the political will to address environmental issues; mobilizing and harmonizing international, regional, and national resources, conventions, and protocols; and supporting best practice/pilot programmes.

A number of Large Marine Ecosystem (LME) projects are ongoing in Africa, aimed at addressing restoration and sustainability of the ecosystem, including the Agulhas and Somali Currents LME, the Benguela Current LME, and the Gulf of Guinea LME. The Benguela Current LME established the Benguela Current Commission; the Benguela Current Convention was signed by Angola, Namibia, and South Africa in 2013.

The African continent faces serious challenges in dealing with poverty, income disparity, stagnant growth, poor health, lack of access to markets, and limited financial and human resources. The ocean and coastal agenda is therefore only one among a long list of priorities. This agenda item is also often not a traditional priority, and the contribution of the ocean and coasts to the GDP may not be adequately recognized. While political will does exist in specific government entities mandated with responsibility for specific ocean or coastal sectors, the issue area may not be recognized as a sufficiently high priority for decision-makers in finance and economic planning. Furthermore, the complexities of managing the marine and coastal environment require coordination and feedback mechanisms among implementing agencies, decision-making, and beneficiaries, greater transparency in decision-making processes, and greater capacity and empowerment of local government representatives and communities.

Regional collaboration on the issues of aquaculture, fisheries, coastal tourism, oil and gas development, and coastal mining is viewed as a vehicle with which to address the lack of capacity, including in the areas of funding and manpower. The ecological, social, economic, and political diversity in the region presents further challenges for a single regional ocean and coastal policy to be acceptable and applied comprehensively in the region. In addition, ocean and coastal policy agenda remains but one among numerous national priorities. Regional and cross-border agreements currently assist countries in tackling pressing marine issues, as well as provide support for incorporating concepts such as sustainable development, government transparency, and the precautionary principle into plans and practices. Many of these agreements receive financial support from external aid organizations, such as the Global Environment Facility (GEF) and United Nations Development Programme (UNDP), as well as support for capacity-building.

Note

1 Pacific Islands Forum Secretariat (2010) *Forum Communiqué*, Forty-first Pacific Islands Forum, 4–5 August, Port Vila, Vanuatu, online at http://www.forumsec.org/resources/uploads/attachments/documents/2010_Forum_Communique.pdf [accessed 11 July 2014].

Appendix D

RECOMMENDATIONS ON ENHANCING INTEGRATED, ECOSYSTEM-BASED NATIONAL AND REGIONAL OCEAN POLICIES MADE BY THE GLOBAL OCEAN FORUM IN THE RIO+20 PROCESS*

Enhance Integrated, Ecosystem-based Ocean Governance at National and Regional Levels

- Scale up the practice of integrated oceans governance to all countries and regions around the world. Given the nature of the added challenges that will need to be faced in ocean and coastal areas and in Small Island States as a result of climate change, it is imperative that EBM/ICM efforts be scaled up and collective investments significantly increased.

National Level

- Scale up national programmes to include larger portions of the coastal zone and ocean under national jurisdiction.
- Further develop and implement (with funding) integrated coastal and ocean laws, e.g., through Ocean Parliamentarians.
- Further strengthen integrated institutions and decision processes for the coast and ocean
- Incorporate and apply Marine Spatial Planning, aiming to achieve, in national waters and regional areas, the Convention on Biological Diversity's Aichi target of protecting at least 10 per cent of marine and coastal areas by 2020.
- Address persistent poverty and inequality in large parts of the coastal areas of the developing world.
- Bring mitigation and adaptation to climate change in coastal areas under the framework of existing ICM/EBM institutions. Extensive capacity development of national and local/regional officials will need to take place to develop and apply climate mitigation and adaptation strategies.

- Mitigate climate change and sustain coastal resources through protection and restoration of coastal carbon sinks ("Blue Carbon").
- Facilitate the development of renewable sources of energy (e.g. offshore wind, wave, and tidal energy).
- Promote sustainable ocean and coastal livelihoods, "blue" green job creation, public private partnerships, and local level and community-based management.
- Address the issues (legal, humanitarian, economic, ecological) of possible displacement of millions of coastal and island peoples.

Regional Level

- Encourage and further develop the regional efforts described in this volume, involving regional and national political entities in the European Union, the East Asia region, the Pacific Islands, and the Africa region
- Encourage and assist the key role played by the Large Marine Ecosystem Programs (LMEs) and the Regional Seas Programmes in harmonizing actions of governments in transboundary contexts.
- Encourage the development and implementation of ICM/EBM protocols in regional seas programmes and their implementation at the national level, following the Mediterranean example.
- Encourage application of EBM/ICM approaches by the full range of bodies responsible for management of resources at the regional level, such as Regional Fishery Management Organizations, and other regional resource management arrangements.

Financing

- Provide sufficient financing for developing countries and Small Island Developing States to cope with the effects of climate change. Current financing estimates for coastal adaptation are woefully inadequate and need to be revised.
- Provide adequate financing to support the capacity development and public education that is vitally needed for integrated oceans governance and associated climate change and biodiversity issues.

Capacity Development

- Build capacity for ocean and coastal management in a transformative era, toward the Blue Economy and Blue Society
- Provide long-term capacity development in ICM/EBM including climate change issues and biodiversity issues, incorporating leadership training:

  ○ Enhance capacity for exercising leadership for high-level national decision makers and Ocean Parliamentarians
  ○ Strengthen or create university programmes to educate the next generation of leaders
  ○ Enhance the capacity of local decision makers

- Share best practices and experience on ICM/EBM, through networking and other measures. A network of National Ocean Officials should be promoted.
- Certify good practice in ICM/EBM, following the PEMSEA model.

Improve the International Regime for Integrated Ocean Governance

Extend EBM/ICM Principles and Approaches to Marine Areas Beyond National Jurisdiction

• Established EBM/ICM principles and approaches must be applied to the 64% of the ocean that lies beyond national jurisdiction (ABNJ) to address multiple use conflicts, manage new uses, and protect vulnerable ecosystems and marine biodiversity. While there has been growing consensus on the use of useful approaches such as Environmental Impact Assessments (EIAs) and establishment of networks of marine protected areas, more attention needs to be focused on institutional aspects – who will implement EIAs, manage marine protected areas, address conflicts, etc.? As in EBM/ICM decision processes under national jurisdiction, authority needs to be vested in existing or new institutions and a process for multiple-use decision making needs to be established.

Integrated Oceans Governance at the UN

• Elevate oceans to the highest levels of the UN system to enable a cross-cutting approach and appropriate and timely response to major threats and opportunities. For oceans, focused attention at the highest political levels – the UN Secretary-General – is needed.

Note

★ Extracted from B. Cicin-Sain (2014) 'Rio+20 Implementation and Oceans: A Perspective', *Environmental Policy and Law*, 44(1–2): 142–51.

INDEX

Note: The following abbreviations have been used – *f* = figure; *n* = note; *t* = table